The chemistry of
amidines and imidates

THE CHEMISTRY OF FUNCTIONAL GROUPS

A series of advanced treatises under the general editorship of
Professor Saul Patai

The chemistry of alkenes (published in 2 volumes)
The chemistry of the carbonyl group (published in 2 volumes)
The chemistry of the ether linkage (published)
The chemistry of the amino group (published)
The chemistry of the nitro and nitroso group (published in 2 parts)
The chemistry of carboxylic acids and esters (published)
The chemistry of the carbon–nitrogen double bond (published)
The chemistry of amides (published)
The chemistry of the cyano group (published)
The chemistry of the hydroxyl group (published in 2 parts)
The chemistry of the azido group (published)
The chemistry of acyl halides (published)
The chemistry of the carbon–halogen bond (published in 2 parts)
The chemistry of the quinonoid compounds (published in 2 parts)
The chemistry of the thiol group (published in 2 parts)
The chemistry of the hydrazo, azo and azoxy groups (published in two parts)
The chemistry of amidines and imidates

The chemistry of
amidines and imidates

Edited by

SAUL PATAI

The Hebrew University, Jerusalem

1975

JOHN WILEY & SONS

LONDON — NEW YORK — SYDNEY — TORONTO

An Interscience ® Publication

Library of Congress Cataloging in Publication Data:
Main entry under title:

The Chemistry of amidines and imidates.

    (The Chemistry of functional groups)
    'An Interscience publication.'
    Includes indexes.
    1. Amidines—Addresses, essays, lectures. 2. Imides
—Addresses, essays, lectures. I. Patai, Saul.
II. Series.
QD341.A7C63 1975 547'.04 75–6913
ISBN 0 471 66923 7

Set by William Clowes & Sons, Limited, Great Yarmouth
and printed in Great Britain by J. W. Arrowsmith, Ltd.,
Bristol.

Contributing authors

R. H. De Wolfe   University of California at Santa Barbara, U.S.A.

C. C. Farnoux   The University, Paris, France

G. Fodor   West Virginia University, Morgantown, West Virginia, U.S.A.

L. A. Garner   West Virginia University, Morgantown, West Virginia, U.S.A.

J.-A. Gautier   The University, Paris, France

F. Grambal   Palacký University, Olomouc, ČSSR

R. J. Grout   The University, Nottingham, England

G. Häfelinger   University of Tübingen, Germany

H. Lund   University of Aarhus, Denmark

C. G. McCarty   West Virginia University, Morgantown, West Virginia, U.S.A.

M. Miocque   The University, Paris, France

D. G. Neilson   The University, Dundee, Scotland

B. A. Phillips   West Virginia University, Morgantown, West Virginia, U.S.A.

W. H. Prichard   The City University, London, England

J. Ševčík   Palacký University, Olomouc, ČSSR

R. Shaw   Stanford Research Institute, Menlo Park, California, U.S.A.

K. M. Watson   The University, Dundee, Scotland

Foreword

The present volume deals with the chemistry of amidines and imidates and other imidic acid derivatives. Its presentation and general organization is on the same lines as those of the other volumes of the series, as described in the 'Preface to the Series' printed on the following pages.

The plan of this volume included several more chapters than are actually appearing. Two of these, on 'Cyclic Amidines' and on 'Amidoximes' did not materialize, while a third one on 'Imidoyl Halides' has been cancelled since it was found that no significant development occurred in the field since the publication of Dr. R. Bonnett's chapter on the same subject in the volume 'The Chemistry of the Carbon–Nitrogen Double Bond' in 1970. An additional chapter on 'Imidines and Diamidides' arrived too late for inclusion in the present volume and will go to press soon in a supplementary volume to the series on 'The Chemistry of Double Bonded Groups'.

Jerusalem, February 1975   SAUL PATAI

The Chemistry of Functional Groups
Preface to the series

The series 'The Chemistry of Functional Groups' is planned to cover in each volume all aspects of the chemistry of one of the important functional groups in organic chemistry. The emphasis is laid on the functional group treated and on the effects which it exerts on the chemical and physical properties, primarily in the immediate vicinity of the group in question, and secondarily on the behaviour of the whole molecule. For instance, the volume *The Chemistry of the Ether Linkage* deals with reactions in which the C—O—C group is involved, as well as with the effects of the C—O—C group on the reactions of alkyl or aryl groups connected to the ether oxygen. It is the purpose of the volume to give a complete coverage of all properties and reactions of ethers in as far as these depend on the presence of the ether group, but the primary subject matter is not the whole molecule, but the C—O—C functional group.

A further restriction in the treatment of the various functional groups in these volumes is that material included in easily and generally available secondary or tertiary sources, such as Chemical Reviews, Quarterly Reviews, Organic Reactions, various 'Advances' and 'Progress' series as well as textbooks (i.e. in books which are usually found in the chemical libraries of universities and research institutes) should not, as a rule, be repeated in detail, unless it is necessary for the balanced treatment of the subject. Therefore each of the authors is asked *not* to give an encyclopaedic coverage of his subject, but to concentrate on the most important recent developments and mainly on material that has not been adequately covered by reviews or other secondary sources by the time of writing of the chapter, and to address himself to a reader who is assumed to be at a fairly advanced post-graduate level.

With these restrictions, it is realized that no plan can be devised for a volume that would give a *complete* coverage of the subject with *no* overlap between chapters, while at the same time preserving the readability of the text. The Editor set himself the goal of attaining *reasonable* coverage with *moderate* overlap, with a minimum of cross-references between the chapters of each volume. In this manner, sufficient freedom is given to each author to produce readable quasi-monographic chapters.

The general plan of each volume includes the following main sections:

(a) An introductory chapter dealing with the general and theoretical aspects of the group.

(b) One or more chapters dealing with the formation of the functional group in question, either from groups present in the molecule, or by introducing the new group directly or indirectly.

(c) Chapters describing the characterization and characteristics of the functional groups, i.e. a chapter dealing with qualitative and quantitative methods of determination including chemical and physical methods, ultraviolet, infrared, nuclear magnetic resonance and mass spectra; a chapter dealing with activating and directive effects exerted by the group and/or a chapter on the basicity, acidity or complex-forming ability of the group (if applicable).

(d) Chapters on the reactions, transformations and rearrangements which the functional group can undergo, either alone or in conjunction with other reagents.

(e) Special topics which do not fit any of the above sections, such as photochemistry, radiation chemistry, biochemical formations and reactions. Depending on the nature of each functional group treated, these special topics may include short monographs on related functional groups on which no separate volume is planned (e.g. a chapter on 'Thioketones' is included in the volume *The Chemistry of the Carbonyl Group*, and a chapter on 'Ketenes' is included in the volume *The Chemistry of Alkenes*). In other cases, certain compounds, though containing only the functional group of the title, may have special features so as to be best treated in a separate chapter, as e.g. 'Polyethers' in *The Chemistry of the Ether Linkage*, or 'Tetraaminoethylenes' in *The Chemistry of the Amino Group*.

This plan entails that the breadth, depth and thought-provoking nature of each chapter will differ with the views and inclinations of the author and the presentation will necessarily be somewhat uneven. Moreover, a serious problem is caused by authors who deliver their manuscript late or not at all. In order to overcome this problem at least to some extent, it was decided to publish certain volumes in several parts, without giving consideration to the originally planned logical order of the chapters. If after the appearance of the originally planned parts of a volume it is found that either owing to non-delivery of chapters, or to new developments in the subject, sufficient material has accumulated for publication of an additional part, this will be done as soon as possible.

The overall plan of the volumes in the series 'The Chemistry of Functional Groups' includes the titles listed below:

The Chemistry of Alkenes (*published in two volumes*)
The Chemistry of the Carbonyl Group (*published in two volumes*)
The Chemistry of the Ether Linkage (*published*)
The Chemistry of the Amino Group (*published*)
The Chemistry of the Nitro and the Nitroso Group (*published in two parts*)
The Chemistry of Carboxylic Acids and Esters (*published*)
The Chemistry of the Carbon–Nitrogen Double Bond (*published*)
The Chemistry of the Cyano Group (*published*)
The Chemistry of Amides (*published*)
The Chemistry of the Hydroxyl Group (*published in two parts*)
The Chemistry of the Azido Group (*published*)
The Chemistry of Acyl Halides (*published*)
The Chemistry of the Carbon–Halogen Bond (*published in two parts*)
The Chemistry of the Quinonoid Compounds (*published in two parts*)
The Chemistry of the Thiol Group (*published in two parts*)
The Chemistry of the Carbon–Carbon Triple Bond
The Chemistry of Amidines and Imidates (*published*)
The Chemistry of the Hydrazo, Azo and Azoxy Groups (*published in two parts*)
The Chemistry of the Cyanates and their Thio-derivatives (*in preparation*)
The Chemistry of the Diazonium and Diazo Groups (*in preparation*)
Supplementary Volume on the Chemistry of Double-bonded Groups (*in press*)

Advice or criticism regarding the plan and execution of this series will be welcomed by the Editor.

The publication of this series would never have started, let alone continued, without the support of many persons. First and foremost among these is Dr. Arnold Weissberger, whose reassurance and trust encouraged me to tackle this task, and who continues to help and advise me. The efficient and patient cooperation of several staff-members of the Publisher also rendered me invaluable aid (but unfortunately their code of ethics does not allow me to thank them by name). Many of my friends and colleagues in Israel and overseas helped me in the solution of various major and minor matters, and my thanks are due to all of them, especially to Professor Z. Rappoport. Carrying out such a long-range project would be quite impossible without the non-professional but none the less essential participation and partnership of my wife.

The Hebrew University,   SAUL PATAI
Jerusalem, ISRAEL

Contents

CHAPTER **1**

General and theoretical aspects of amidines and imidic acid derivatives

G. HÄFELINGER

Universität Tübingen, Germany

I. INTRODUCTION

The amidine group (**1**) is the nitrogen analogue of carboxylic acids and esters (**2**) which are reviewed in a previous volume of this series[1].

(1a)   **(1b)**   **(2)**

It combines the properties of an azomethine-like C=N double bond[2] with an amide-like C—N single bond[3] with partial double bond character as indicated by the mesomeric form (**1b**).

Amidines are strong bases. The protonation occurs on the imino nitrogen[4,5] leading to the symmetrical amidinium ion (**3**) which is stabilized by resonance as is the isoelectronic carboxylate ion (**4**).

(3)   **(4)**

In strong acidic media a second cation (**5**) is formed[6,7,8] which has a localized carbon nitrogen double bond whereas in strong alkaline solutions

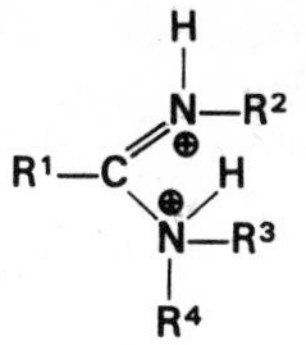

(5) 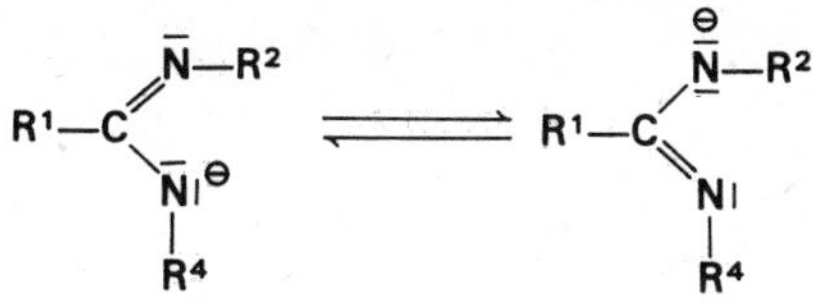   **(6)**

an anion (**6**) may be obtained[7].

The amidines may be classified into five general types depending on the number and distribution of the substituents on the nitrogen atoms:

(a). Unsubstituted

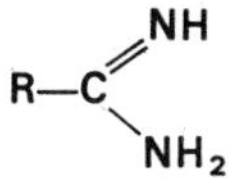

(b). Monosubstituted

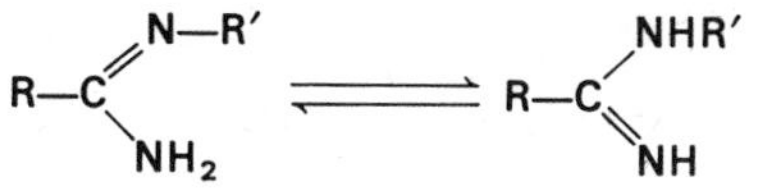

(c). *N,N'*-Disubstituted

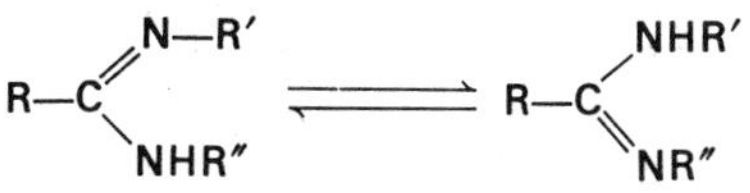

(d). *N,N*-Disubstituted

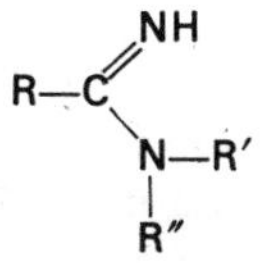

(e). Trisubstituted

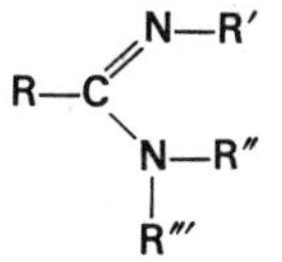

Of these types, monosubstituted and disubstituted amidines (with different substituents on the nitrogen atoms) may exhibit tautomerism. Numerous attempts have been made to isolate the two tautomeric forms but apparently they have all failed[9,10].

Experimental results favouring the possibility of tautomerism are:

(1) A single amidine results from a reaction designed to prepare two tautomeric forms;

(2) The alkylation of a monoalkylated amidine yields only two products (the *N,N′*-dialkylated and the *N,N*-dialkylated amidine);

(3) The hydrolysis of *N,N′*-dialkylated amidines produces a mixture of amides and amines;

(4) Spectroscopic evidence (see Section V.C).

Besides tautomerism *cis–trans* isomerism with respect to the carbon–nitrogen double bond as well as rotational isomerism around the C—N single bond may occur in all types of the amidines listed.

The preparation and the chemistry of amidines are reviewed by Shriner and Neumann[10]. Some amidines are very useful drugs and their pharmaceutical use has been summarized elsewhere[11-15]. From the theoretical point of view the amidine group has received very little attention.

Derivatives of imidic acid (7) are imidates (8) (also termed imino ethers, imido esters or imidic acid esters), thioimidates (9), imidoyl chlorides (10), amidrazones (11), and imidines (12).

Imidic acid (7) is the tautomeric form of amides which is not observed in the free form[16]. However, the derivatives (8) to (12) in which the iminole form is fixed by substitution (R″ = alkyl or aryl, R′ = H, alkyl or aryl) are well known. Imidates are monoacid bases whose preparation and chemistry has been reviewed by Roger and Neilson[17].

II. PHYSICO-CHEMICAL PROPERTIES

A. Molecular Structure

No structural determination has been performed on compounds which contain the unsubstituted amidine or amidinium group. In all cases investigated at least one substituent is present which may take part in the π-system of the amidine or amidinium group thus altering the bond lengths by conjugation.

I. Amidines

The best structural approach to an unsubstituted amidine in the crystalline state is formamidoxime (13)[18,19]. The oxygen substituent on nitrogen does not affect greatly the π-system of the amidine group since the N—O π-bond order is negligibly small[20].

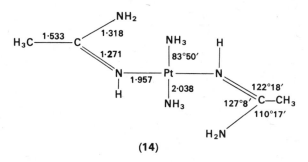

(13)*

The heavy atom skeleton of the formamidoxime molecule is completely planar showing a short C=N double bond (1·29 Å) which is only slightly longer than a pure unconjugated C=N double bond (1·27 Å)[21]. The C—N single bond distance (1·33 Å) is appreciably shorter than a pure single bond (1·47 Å)[21,22], but it corresponds to the C—N distance in amides (average 1·322 Å)[22].

The planarity, the elongation of the double bond and the shortening of the single bond reflect the effect of the amidine resonance in the π-system indicated by mesomeric structures (1a) and (1b).

The angles around the central carbon atom in (13) show an appreciable deviation from the expected value of 120° for a sp^2-hybridized carbon atom which may be due to the asymmetric substitution. In compounds (14) and (15) the angles are also unequal with large NCN angles of 127° and 131° respectively but the sizes of the other two angles are reversed in the two compounds.

(14)

* In this section bond distances are in Ångström units and standard deviations with respect to the last digit are given in brackets.

(15)

In diamminebis(acetamidine) platinum(II) chloride monohydrate (**14**)[23], which contains the planar amidine group asymmetrically complexed to a transition metal, the bond lengths are slightly shorter than in (**13**). The planar azobis(N-chloroformamidine) (**15**)[24] shows less accurately determined bond lengths in agreement to those in (**13**) and (**14**). The averages of these determinations yield 1·280 Å and 1·326 Å for the C=N double and C—N single bonds, respectively, in the amidine group.

2. Amidinium Salts

The structures of tetramethylformamidinium phosphonate (**16**)[25] and tetramethylformamidiniumphosphonic anhydride (**17**)[26] both containing the amidinium group as an inner salt, were very accurately determined by X-ray diffraction.

(16)   **(17)**

The amidinium group is planar, showing two equivalent CN bonds of 1·330 Å length but the $N(CH_3)_2$ groups are twisted out of the amidinium plane by about 25°. The angles around the central carbon atom are very close to 120°.

N,N′-Bis-(4-ethoxyphenyl)acetamidinium bis-p-nitrophenyl phosphate[27] contains in the cation (**18**) also two equivalent C—N bonds of 1·318 Å which are shorter than those in (**16**) and (**17**).

Both phenyl rings are twisted out of the plane of the amidine group. The one being *trans* to the methyl group is twisted by 57° whereas the other is rotated by 78° so that their resonance interaction with the amidinium group is of minor importance. The C—CH$_3$ bond distance is shorter than in (14).

(18)

In Table 1, C—N bond distance of some compounds are collected which contain the amidinium group bearing a substituent X at the central carbon atom. This substituent may take part in the π-system of the amidinium group by means of a free electron pair as indicated by the resonance form (19c).

(19a)   (19b)   (19c)

The two C—N bond lengths of each of the compounds in Table 1 are not significantly different in their limit of error, therefore only the average value is given in Table 1. These distances average to 1·314 Å in agreement with the CN distance of 1·318 Å in (18). However, they are shorter than the 1·330 Å in (16) and (17) which may be elongated by the steric repulsion and twisting of the N(CH$_3$)$_2$ groups. The cross-conjugation introduced by the resonance form (19c) leads to no measurable elongation of the C—N bond distance. It seems that 1·316 Å is a good estimate of the C—N bond lengths in the unconjugated amidinium group. In agreement with theoretical predictions (see Section III) this value is shorter than the amide-like C—N bond (1·326 Å) in amidines. The NCN angle of the compounds in Table 1 is always close to the theoretical value of 120° but the other two NCX angles are usually different.

B. Dipole Moments

I. Amidines

The dipole moments of amidines collected in Table 2 lie in the range from 2·2 to 3·4 D. The mesomeric moment of the amidine group

TABLE 1. Structural parameters of compounds containing the amidinium structure as determined by X-ray diffraction

$$\begin{array}{c} H_2N \\ \overset{\oplus}{C}\!-\!X \\ H_2N \end{array}$$

Compound	X	Average CN—bond length (Å)	Standard deviation $\sigma \cdot 10^3$(Å)	NCN-Angle (degrees)	Reference
Uronium phosphate	OH	1·331	9	120·2	28
Uronium nitrate	OH[a]	1·313	1	121·7	29
	OH[b]	1·306	3	121·7	30
	OH[c]	1·308	3	121·5	30
O-Methyluronium chloride	OCH₃	1·309	11	123·8	31
Thiuronium nitrate	SH	1·307	5	121·6	32
S-Methylthiuronium sulphate	SCH₃	1·311	9	122·5	33
S-Methylthiuronium p-chlorobenzoate	SCH₃	1·324	12	119·8	34
Formamidinium disulphide dichloride	SSC(NH₂)₂$^{\oplus}$	1·308	8	122·5	35
Azidoformamidinium chloride	N₃	1·308	4	123·1	36
Guanidinium chloride	NH₂	1·323	6	119·7	37
Guanidinium aluminiumsulphate	NH₂	1·316	7	—	38
Guanidinium chromosulphate	NH₂	1·320	8	—	38
Mean		1·314			

[a] Neutron diffraction. [b] X-ray diffraction using Cu-radiation [c] X-ray diffraction using Mo-radiation

TABLE 2. Dipole moments of some amidines

Compound	Formula	Solvent	Dipole moment (debyes)	Reference
2-Ethyl-Δ^2-imidazoline	$R = C_2H_5$	Benzene	3·42	39
2-Phenyl-Δ^2-imidazoline	$R = C_6H_5$	Benzene	3·08	39
Diaza-1,5-bicyclo[4.3.0]nonene-5		Benzene	3·29	39
Diaza-1,5-bicyclo[5.4.0]undecene-5		Benzene	3·41	39
N,N-Dimethylbenzamidine		Benzene	2·83	40, 41
		Dioxane	3·00	40, 41
N,N'-Diphenylformamidine		Dioxane	2·20	42

(Me_2N—C=N) was calculated[39] to be 1.9 ± 0.2 D. This value is between that of the amide group (Me_2N—C=O) with 1.09 D and the thioamide group (Me_2N—C=S) with 2.45 D[43], indicating an intermediate degree of conjugation in the amidines. However, this sequence is not in agreement with the results of the measurements of the height of rotational barriers around the CN single bond (see Section VI,C).

The dipole moment of N,N-dimethylbenzamidine (20) is assigned to the predominance of the E-configuration (20b)[44] in solution[41].

(20a) Z-isomer   (20b) E-isomer

The van der Waals radii show that in the E-form (20b) the amidine group may be planar but the phenyl group is then twisted with respect to this plane. This conclusion is reasonable since the energy of activation for rotation of the dimethylamino group is 18.2 kcal/mol[45] whereas the rotational barrier of the phenyl–carbon bond is assumed to be less than 5 kcal/mol[41]. In the Z-isomer (20a) the steric overcrowding of the N—H and N—CH_3 groups forces the dimethylamino group to rotate out of the plane of the amidine group leading to an energetically unfavourable reduction of the amidine π-conjugation.

2. Imidates

The dipole moments of imidates shown in Table 3 are lower than those of amidines. For the MeO—C=N group a mesomeric moment of 1.4 D. was derived[48] which is also lower than that of the amidines. This shows that the conjugation in the imidate group, as indicated by the resonance forms (21a) and (21b), is not so important as in the case of amidines,

(21a)   (21b)

although the conjugation shown favours a planar arrangement of the imidate group.

For noncyclic imidates four planar configurations (22a)–(22d) are possible which result from cis–trans isomerism with respect to the C=N double bond (E and Z) and restricted rotation around the C—O bond with partial double bond character.

The vectorial analysis of the dipole moments of phenyl substituted

(22a) E, trans   (22b) Z, trans   (22c) E, cis   (22d) Z, cis

imidates in comparison with their *p*-chloro derivatives showed, in agreement with the interpretation of dipole moments by Lumbroso[46,48], that generally the *E,trans* configuration (22a) is the dominant form of noncyclic imidates in solution[49]. This result contradicts the interpretation[51] of the nuclear magnetic resonance spectra of methyl acetimidates [(22): $R^1 = R^3 = CH_3$, R^2 = alkyl or phenyl] which have been explained by the exclusive predominance of the *Z*-form without clarifying the conformation with respect to the C—O bond.*

The n.m.r. spectrum of phenyl-*N*-methylacetimidate [(22): $R^1 = R^2 = CH_3$, $R^3 = C_6H_5$] showed[52] the presence of the *E*- and *Z*-form in the ratio 2:1, supporting again the predominance of the *E*-isomer (22a).

C. Basicity

The basicity of amidines is measured by their pK_a value (equation 1) where round brackets denote activities.

$$pK_a = -\log \frac{(H^+) \cdot (\text{base})}{(\text{base}^{\oplus} - H)} \qquad (1)$$

In Table 4 some characteristic pK_a values of nitrogen bases are collected. It shows that unsubstituted amidines are stronger bases than aliphatic amines while imidates are weaker. Since protonation occurs at the lone pair of the sp^2-hybridized imino nitrogen[4,5] which due to its higher degree of *s*-character, is less basic than the lone pair of the sp^3-hybridized nitrogen of aliphatic amines, one might expect a decrease in basicity. The observed increase in basicity results from the complete delocalization of charge in the amidinium cation (23) and hence its stabilization.

(23)

The effect of phenyl-substitution at the imino nitrogen or amino nitrogen on basicity (see Table 5) shows that protonation involves the imino nitrogen lone pair.

* See 'Note Added in Proof' on p. 84.

TABLE 3. Dipole moments of some imidates

Compound	Solvent	Dipole moment (debyes)	Reference
$H_3C-C(=NH)-OC_2H_5$	Benzene	1·42	46
	Dioxane	1·44	46
	Dioxane	1·33	47
	Triethylamine	1·51	48
$H_3C-C(=NCH_3)-O-C_6H_4-R$, R = H	Benzene	2·03 ± 0·3	49
R = Cl	Benzene	2·86 ± 0·2	49
H_3C-oxazoline	Benzene	1·30 ± 0·2	49
phenyl-oxazoline, R = H	Benzene	1·27 ± 0·4	49
R = Cl	Benzene	2·21 ± 0·3	49

TABLE 3 (continued)

R = H	Benzene	1·54	46
R = H	Benzene	1·70	50
R = H	Dioxane	1·52	46
R = H	Triethylamine	1·48	48
R = Pr—O	Benzene	2·43	46
R = Pr—O	Dioxane	2·51	48
R = NO_2	Benzene	3·80	50
R = H	Benzene	1·64 ± 0·3	49
R = Cl	Benzene	1·74 ± 0·3	49
	Benzene	1·02 ± 0.3	49
	Benzene	1·22 ± 0·4	49

TABLE 4. pK_a-Values of some nitrogen bases

Compound	Formula	pK_a	Solvent[a]	Reference
Ammonia	NH_3	9·245		53
Methylamine	CH_3NH_2	10·624		53
Aniline	$C_6H_5NH_2$	4·65		54
Acetamidine	$CH_3-C{\begin{smallmatrix}NH\\NH_2\end{smallmatrix}}$	12·40		55
N,N'-Diphenylacetamidine	$CH_3-C{\begin{smallmatrix}NC_6H_5\\NC_6H_5\\H\end{smallmatrix}}$	8·30		55
Benzamidine	$C_6H_5-C{\begin{smallmatrix}NH\\NH_2\end{smallmatrix}}$	11·6		56
		11·23	50% aqueous ethanol	57
		11·1	75% aqueous ethanol	58
N,N-Di-n-butylbenzamidine	$C_6H_5-C{\begin{smallmatrix}NH\\NBu_2\end{smallmatrix}}$	11·27	50% aqueous methanol	59

TABLE 4 (continued)

Phenyl N-methylacetimidate	$CH_3-C(=NCH_3)-OC_6H_5$	6·2	52
Methyl benzimidate	$C_6H_5-C(=NH)-OCH_3$	5·60	60
Methyl benzthioimidate	$C_6H_5-C(=NH)-SCH_3$	5·84	60

[a] Water at 25°C unless otherwise stated.

G. Häfelinger

TABLE 5. Effect of phenyl-substitution on basicity in benzamidines

Compound:

$$C_6H_5—C \begin{matrix} \overline{N}—R^1 \\ \\ \overline{N}—R^2 \\ | \\ R^3 \end{matrix}$$

	pK_a	Solvent	Reference
(24) N,N-Di-n-butylbenzamidine: $R^1 = H$; $R^2 = R^3 = n$-Bu	11·27	50% aqueous methanol	59
(25) N-n-Butyl-N-phenylbenzamidine: $R^1 = H$; $R^2 = $ n-Bu; $R^3 = C_6H_5$	10·40	50% aqueous methanol	59
(26) N,N-Dimethyl-N'-phenylbenzamidine: $R^1 = C_6H_5$; $R^2 = R^3 = CH_3$	7·8	50% aqueous ethanol	61

The introduction of a phenyl group on the amino nitrogen in compound (25) causes a reduction in basicity by a factor of about 10 relative to (24). But the introduction of a phenyl group on the imino nitrogen in compound (26) reduces the basicity by a factor of about 1000.

The reason for this drastic effect on basicity is that the imino phenyl group is not taking part in the π-system of the amidine since it is twisted with respect to this plane. Consequently, the phenyl-π-system is oriented so that it may overlap with the lone pair of the sp^2-hybridized imino nitrogen which is therefore in (26) less available for protonation. This is comparable to the situation in aniline which is 10^6 times less basic than methylamine (see Table 4).

The same effect is shown in pK_a-values of acetamidine and N,N'-diphenyl acetamidine (see Table 4) where the pK_a-difference of 4·1 units is nearly the same as the sum (4·4 units) of amino and imino phenyl substitution in (25) and (26).

N,N-Dialkyl substitution in N,N-di-n-butyl benzamidine (Table 4) shows only a slight effect on basicity as the protonation occurs on the remote N-nitrogen. N-Alkyl-monosubstitution in (28) and (29) of Table 6 shows only a slight reduction in basicity relative to (27), by 0·3 units, whereas the N-phenyl-monosubstitution in (30) shows reduction in basicity by 3·1 units. This result may be well explained by the conclusion of Prevorsek[63] who found by inspection of infrared spectra that N-alkyl monosubstituted amidines exist mainly as tautomers (33) whereas N-phenyl substituted amidines exist as tautomers (34).

$$R-C \overset{\bar{N}H}{\underset{NH\text{-alkyl}}{\diagup}}$$

(33)

$$R-C \overset{\bar{N}-\text{phenyl}}{\underset{\bar{N}H_2}{\diagup}}$$

(34)

In the tautomeric form (33) of compounds (28) and (29) the alkyl group shows only a slight effect as the protonation occurs on the remote imino nitrogen. But in the tautomeric form (34) the twisted phenyl group affects the imino nitrogen lone pair directly by conjugation leading to the reduction in basicity in (30). The electron donating p-ethoxy group reduces this conjugation and increases therefore the basicity in (31) whereas the electron attracting p-chloro substituent shows the reverse effect in (32) (see Table 6).

TABLE 6. pK_a-Values of some substituted p-phenylbenzamidines (in 50% aqueous ethanol at 20°C)[62].

Compound	pK_a
(27) p-Phenylbenzamidine	11·09
(28) N-n-Butyl-p-phenylbenzamidine	10·73
(29) N-Cyclohexyl-p-phenylbenzamidine	10·76
(30) N-Phenyl-p-phenylbenzamidine	7·95
(31) N-p-Ethoxyphenyl-p-phenylbenzamidine	8·12
(32) N-p-Chlorophenyl-p-phenylbenzamidine	7·74

III. THEORETICAL CONSIDERATIONS

A. Hückel Method

The Hückel (HMO) method[64] of semiempirical calculations for π-electron systems is a crude quantum mechanical approximation[65,66]. Due to the long list of neglects (neglect of electron spin, neglect of electron repulsion and electron correlation, neglect of σ-electrons) and the empirical choice of integral parameters (neglect of overlap integrals, all Coulombic integrals for carbon equal and all resonance integrals β for carbon–carbon bonds equal) the HMO calculation adopts the character of a well defined model in which the 'theoretical' considerations of the π-electron properties refer not to real molecules but to models for these. This allows the calculation of model properties in a consistent manner, and the comparison of these properties with experimental results may help to interpret trends in real molecules.

For all of the following calculations we apply the σ–π separation[66], i.e. we assume a planar skeleton of localized σ-bonds, constructed from overlap of sp^2-hybridized atomic orbitals, which are considered as a rigid nonpolarizable core building a field for a delocalized π-system which is obtained from overlapping p-orbitals that are perpendicular to the plane of the σ-skeleton. The planarity of the amidine and amidinium group is confirmed by the experimental structural determinations (see Section II, A) but the experimentally determined angles may deviate from the assumed theoretical value of 120° for sp^2-hybrid orbitals.

I. Amidinium cations

In amidinium cations the planar σ-skeleton is formed by overlap of two sp^2-hybridized nitrogen atoms with a sp^2-hybridized carbon atom. The π-system consists of three overlapping p-orbitals which contain four π-

electrons. According to mesomeric forms (35) the positive charge is equally distributed on both nitrogens. The π-electron system of (35) corresponds

(35)   (36)

to the allylic anion type π-system (36) for which the result of a HMO-calculation [65] is given in Figure 1.

There is obtained a bonding molecular orbital (MO), a nonbonding MO, and an antibonding MO containing, besides the nodal plane of the p-orbitals, no nodal plane, one nodal plane, and two nodal planes. Two electrons with antiparallel spin occupy the bonding MO and two the non-bonding MO leading to a π-bond energy of $2 \cdot 828\,\beta$.

The π-electron density q_μ (equation 2) is $1 \cdot 5$ on the end atoms

$$q_\mu = \sum_{j=1}^{n} b_j c_{j\mu}^2 \qquad (2)$$

n = number of MO's = number of AO's
b_j = occupation number of MO j
$c_{j\mu}$ = coefficient of MO j at the centre μ

and $1 \cdot 0$ on the middle atom leading to a charge density ζ_μ (equation 3) of $-0 \cdot 5$ on both end atoms and zero on the middle atom.

$$\zeta_\mu = Z_\mu - q_\mu \qquad (3)$$

Z_μ = nuclear charge of the atom μ ($=1$ for atoms contributing one electron to the π-system, and 2 for atoms contributing two electrons to the π-system)

This result indicates that the negative charge is only and equally distributed on the two end atoms as indicated by the mesomeric forms (36).

The π-bond order $p_{\mu v}$ (equation 4) is $0 \cdot 707$ for both bonds

$$p_{\mu v} = \sum_{j=1}^{n} b_j c_{j\mu} c_{j v} \qquad (4)$$

indicating that both π-bonds are equivalent.

The change from the allyl anion π-system to the amidinium cation system is performed by the replacement of the two end carbon atoms by two equivalent nitrogen atoms. In HMO theory the introduction of a heteroatom X is represented by a change of the Coulombic integral α_x (equation 5) and the bond integral β_{cx} (equation 6).

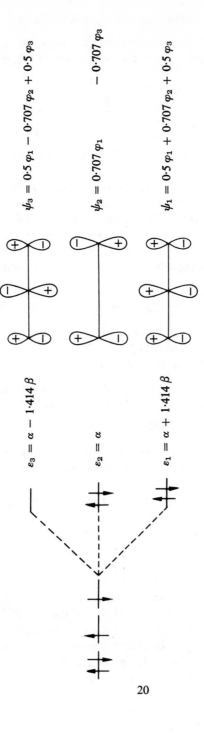

$\psi_3 = 0.5\,\varphi_1 - 0.707\,\varphi_2 + 0.5\,\varphi_3$

$\psi_2 = 0.707\,\varphi_1 \qquad\qquad - 0.707\,\varphi_3$

$\psi_1 = 0.5\,\varphi_1 + 0.707\,\varphi_2 + 0.5\,\varphi_3$

$\varepsilon_3 = \alpha - 1.414\,\beta$

$\varepsilon_2 = \alpha$

$\varepsilon_1 = \alpha + 1.414\,\beta$

$E_\pi^{\text{tot}} = 4\,\alpha + 2.828\,\beta;\ E_\pi^{\text{bond}} = 2.828\,\beta$
$q_1 = q_3 = 1.500;\ \zeta_1 = \zeta_3 = -0.5$
$q_2 = 1.000;\ \zeta_2 = 0.0$
$p_{12} = p_{23} = 0.707$

FIGURE 1. Results of the HMO-calculation for the allyl anion.

20

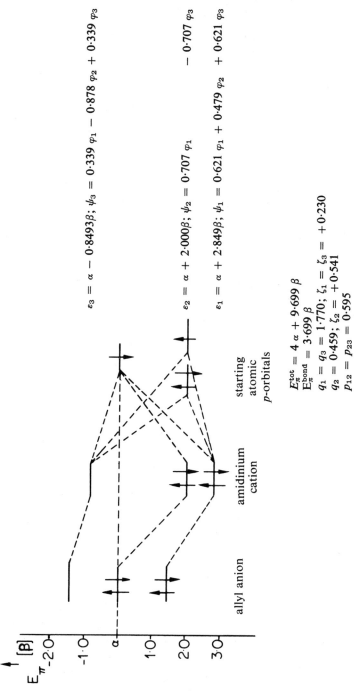

21

$$\varepsilon_3 = \alpha - 0{\cdot}8493\beta; \quad \psi_3 = 0{\cdot}339\ \varphi_1 - 0{\cdot}878\ \varphi_2 + 0{\cdot}339\ \varphi_3$$

$$\varepsilon_2 = \alpha + 2{\cdot}000\beta; \quad \psi_2 = 0{\cdot}707\ \varphi_1 \qquad\qquad\qquad - 0{\cdot}707\ \varphi_3$$

$$\varepsilon_1 = \alpha + 2{\cdot}849\beta; \quad \psi_1 = 0{\cdot}621\ \varphi_1 + 0{\cdot}479\ \varphi_2 + 0{\cdot}621\ \varphi_3$$

$E_\pi^{tot} = 4\ \alpha + 9{\cdot}699\ \beta$
$E_\pi^{bond} = 3{\cdot}699\ \beta$
$q_1 = q_3 = 1{\cdot}770; \quad \zeta_1 = \zeta_3 = +0{\cdot}230$
$q_2 = 0{\cdot}459; \quad \zeta_2 = +0{\cdot}541$
$p_{12} = p_{23} = 0{\cdot}595$

FIGURE 2. Result of the HMO-calculation for amidinium cations with the heteroatom parameters from equation 7.

$$\alpha_X = \alpha_C + h_X\beta_{CC} \tag{5}$$

$$\beta_{CX} = k_{CX}\beta_{CC} \tag{6}$$

A reasonable set of nitrogen parameters[21, 65] for the amidinium π-system is given in equation (7).

$$h_{=\overset{\oplus}{\underset{|}{N}}-} = 2\cdot0; \qquad k_{C=N^\oplus} = 1\cdot1 \tag{7}$$

The results of the HMO-calculation with these parameters are given in Figure 2.

The introduction of the two equivalent electronegative nitrogen atoms lowers the π-electron energy of all three MO's. But the non-bonding MO remains still non-bonding as it has the energy value of the nitrogen atomic orbitals. The π-bond energy is now greater than that of the allyl anion indicating a greater degree of thermodynamic stability. The charge density shows that the positive charge is distributed over all three atoms with the highest degree of positive charge on the central carbon atom. This result is not indicated by the two mesomeric formulas (35), The π-bond order with 0·595 is lower than that of the allyl anion.

2. Amidines

In the amidine group the two nitrogens are no longer equivalent. One is contributing two π-electrons to the π-system, the other only one as is seen from mesomeric form (37a). This is reflected by different heteroatom-

parameters (equations 8) for the HMO calculation for the π-system.

$$\begin{aligned} h_{=\overset{}{\underset{}{N}}-} &= 0\cdot8 & k_{C=N} &= 1\cdot1 \\ h_{-\overset{}{\underset{|}{N}}-} &= 1\cdot5 & k_{C-N} &= 1\cdot0 \end{aligned} \tag{8}$$

The results of the calculation are given in Figure 3.

The π-bond energy with 2·895 β is very close to that of the allyl anion (2·828 β). The formerly non-bonding MO ε_2 is now intermediate in energy between the two different nitrogen atomic p-orbitals. The π-bond order indicates the non-equivalence of the two C—N bonds corresponding to the mesomeric form (37a). There is obtained one bond with a high double bond character ($p_{12} = 0\cdot789$) and a single bond with some double bond character ($p_{23} = 0\cdot520$) which is lower than the π-bond order in the

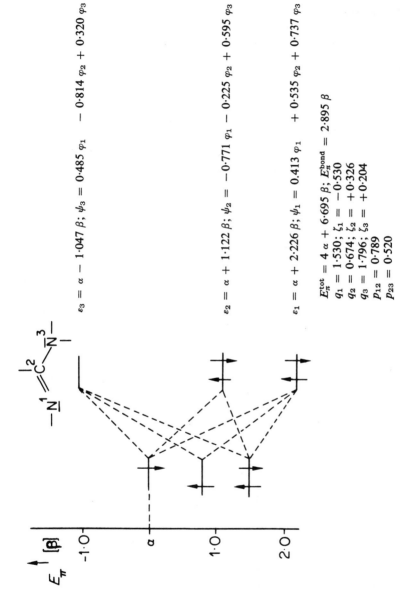

$\varepsilon_3 = \alpha - 1\cdot047\,\beta;\; \psi_3 = 0\cdot485\,\varphi_1 \qquad - 0\cdot814\,\varphi_2 + 0\cdot320\,\varphi_3$

$\varepsilon_2 = \alpha + 1\cdot122\,\beta;\; \psi_2 = -0\cdot771\,\varphi_1 - 0\cdot225\,\varphi_2 + 0\cdot595\,\varphi_3$

$\varepsilon_1 = \alpha + 2\cdot226\,\beta;\; \psi_1 = 0\cdot413\,\varphi_1 \qquad + 0\cdot535\,\varphi_2 + 0\cdot737\,\varphi_3$

$E_\pi^{tot} = 4\,\alpha + 6\cdot695\,\beta;\; E_\pi^{bond} = 2\cdot895\,\beta$
$q_1 = 1\cdot530;\; \zeta_1 = -0\cdot530$
$q_2 = 0\cdot674;\; \zeta_2 = +0\cdot326$
$q_3 = 1\cdot796;\; \zeta_3 = +0\cdot204$
$p_{12} = 0\cdot789$
$p_{23} = 0\cdot520$

FIGURE 3. Results of the HMO-calculation for the amidine group with the heteroatom parameters from equation 7.

amidinium group ($p = 0.595$). The π-electron charge distribution shows a negative partial charge for N^1 in (37b), but the positive charge is distributed over the other two atoms with a higher value for C^2.

3. Imidic acid derivatives

In imidic acid derivatives (38) the bonding situation is comparable to that of amidines as indicated by (38a) and (38b). They have the same asymmetric π-systems as amidines.

(38a)   (38b)

The results of the HMO-π-electron calculation with the heteroatom parameters [65, 68] (equation 9) are given in Table 7.

$$
\begin{array}{ll}
h_{=\bar{N}-} = 0.8; & k_{C=N} = 1.1 \\
h_{-\bar{N}-} = 1.5; & k_{C-N} = 1.0 \\
h_{-\bar{\underline{O}}-} = 2.1; & k_{C-O} = 1.0 \\
h_{-\bar{\underline{S}}-} = 0.5; & k_{C-S} = 0.43 \\
h_{-\overline{\underline{Cl}}|} = 2.0; & k_{C-Cl} = 0.4 \\
 & k_{-\bar{N}-\bar{N}-} = 0.8 \\
 & k_{=\bar{N}-\bar{N}-} = 1.4
\end{array}
\tag{9}
$$

In all cases we have a comparable distribution of MO's, two bonding MO's and one antibonding MO. The π-bond energy is decreasing in the series amidines > imidates > thioimidates > imidoyl chlorides, suggesting a lowering of thermodynamic stability in this order. In the same order the negative charge on the double bond nitrogen and the positive charge on the heteroatom X are decreasing indicating an increasingly better representation of the ground state by the mesomeric form (38a).

For amidrazones two tautomeric forms (39) and (40) are possible.

(39)   (40)

Table 8 shows the results of the HMO-calculation for both forms with the parameters of equation 9. For both isomers three bonding MO's and one antibonding MO filled with six π-electrons are obtained. The π-bond energy is higher for the tautomer (40) indicating a higher degree of stability In this compound the charge distribution is more smoothed out.

TABLE 7. Results of the HMO-calculation for imidic acid derivatives with the heteroatom parameters from equation 9.

Compound		j	$\varepsilon_j - \alpha$ (β)	c_{j1}	c_{j2}	c_{j3}	$E_\pi^{\text{tot}\,a}$ (β)	E_π^{bond} (β)	q_1	q_2	q_3	ζ_1	ζ_2	ζ_3
Imidates	X = O	1	2·614	0·267	0·440	0·857	7·793	2·793	1·493	0·648	1·859	−0·493	0·352	0·141
		2	1·283	−0·821	−0·361	0·441								
		3	−0·996	0·503	0·822	0·266								
Thioimidates	X = S	1	1·627	0·779	0·586	0·223	4·326	2·526	1·431	0·699	1·871	−0·431	0·301	0·121
		2	0·536	−0·329	0·079	0·941								
		3	−0·863	0·533	−0·807	0·254								
Imidoyl chlorides	X = Cl	1	2·131	0·249	0·301	0·920	7·218	2·418	1·372	0·656	1·973	−0·372	0·344	0·027
		2	1·478	−0·790	−0·487	0·373								
		3	−0·809	0·561	−0·820	0·117								

Structure:

$$\overset{2}{C}\!\!\left(\!\!\begin{array}{c}\\ \end{array}\!\!\right)\!\!-\!\!\overset{3}{\underline{X}}\!\!- \qquad -\overset{1}{\underline{N}}$$

^{a} 3α are subtracted.

TABLE 8. Results of the HMO-calculations for amidrazones.

Amidrazones	j	$\varepsilon_j - \alpha$ (β)	c_{j1}	c_{j2}	c_{j3}	c_{j4}	$E_\pi^{tot\,a}$ (β)	E_π^{bond} (β)
(39)	1	2·600	0·230	0·376	0·726	0·528	9·750	2·950
	2	1·547	−0·676	−0·459	0·034	0·576		
	3	0·728	0·517	−0·034	−0·594	0·615		
	4	−1·075	−0·472	0·804	−0·346	0·107		
(40)	1	2·840	0·247	0·331	0·630	0·658	10·119	3·319
	2	1·936	−0·839	−0·366	0·120	0·384		
	3	0·283	0·410	−0·499	−0·501	0·577		
	4	−1·259	−0·258	0·713	−0·581	0·295		

Amidrazones	j	q_1	q_2	q_3	q_4	ζ_1	ζ_2	ζ_3	ζ_4
(39)	1 2 3 4	1·555	0·707	1·761	1·977	−0·555	0·293	0·239	0·023
(40)	1 2 3 4	1·866	0·983	1·324	1·826	−0·134	0·017	−0·324	0·174

[Structure (39): a carbon C (positions 1,2) double-bonded, with =N (1), N—N (3,4) chain]

[Structure (40): a carbon C (positions 1,2), N, N—N (3,4) chain]

^{a} 4α are subtracted.

4. Calculation of bond lengths

The π-bond orders are related to bond lengths. Comparing HMO π-bond orders, calculated with the heteroatom parameters (equations 9), with experimental bond lengths, the following empirical linear equations (10) have been determined by linear least squares methods:

$$\begin{aligned}
\text{CN}^{68}&: d_{\mu\nu} = 1·478 - 0·236\,p_{\mu\nu}; & \sigma = 0·030 \\
\text{CO}^{69}&: d_{\mu\nu} = 1·431 - 0·256\,p_{\mu\nu}; & \sigma = 0·022 \\
\text{CS}^{68}&: d_{\mu\nu} = 1·804 - 0·234\,p_{\mu\nu}; & \sigma = 0·021 \\
\text{NN}^{68}&: d_{\mu\nu} = 1·437 - 0·179\,p_{\mu\nu}; & \sigma = 0·025
\end{aligned} \tag{10}$$

where σ = Standard deviation in Å.

In Table 9 the calculated π-bond orders and bond lengths predicted by means of equations (10) are collected. The agreement with experimental

TABLE 9. HMO π-Bond orders and bond lengths

Compound	Formula	μ	ν	π-Bond order $p_{\mu\nu}$	Bond length		References
					$d_{\mu\nu}$(Å) calc.	$d_{\mu\nu}$(Å) exp.	
Imines		1	2	0·940	1·256	1·270 ± 0·015	21
Amidinium cations		1	2	0·595	1·337	1·316	Section II, A, 2
Amidines		1	2	0·789	1·292	1·288	Section II, A, 1
		2	3	0·520	1·355	1·334	
Amides		1	2	0·827	1·219	1·235 ± 0·005	22
		2	3	0·463	1·369	1·333 ± 0·005	22
Imidates		1	2	0·828	1·282		
		2	3	0·437	1·304		
Thioimidates		1	2	0·861	1·275		
		2	3	0·410	1·708		
Imidoyl-chlorides		1	2	0·919	1·262		
		2	3	0·191	—		
Amidrazones		1	2	0·759	1·299		
		2	3	0·556	1·347		
		3	4	0·074	1·424		
		1	2	0·368	1·391		
		2	3	0·829	1·283		
		3	4	0·343	1·377		

data, as far as available, is satisfactory. The deviations are generally less than the standard deviations given for the equations (10).

The C=N double bond length is decreasing in the order: amidinium cations > amidines > amidrazones > imidates > thioimidates > imidoyl chlorides > imines. The C—N single bond length decreases in the order amides > amidines > amidinium cations.

5. Effect of phenyl substitution on amidines

The π-bond energies of phenyl-substituted planar amidines listed in Table 10 may only be compared directly for systems of equal size. The prediction is that the isomer (**41**) of phenylformamidine is more stable than benzamidine and the isomer (**42**). This agrees with the experimental result[63] that N-phenyl amidines occur as the tautomeric form (**41**). However, in the real molecules the phenyl group is likely to be twisted out of the plane of the amidine group leading to additional overlap with the sp^2-hybridized nitrogen lone pair (see Section II, C). In this case the σ–π-separation may not be applied, but the prediction of π-bond energy for the hypothetical planar molecules (**41**) and (**42**) agrees with experimental findings. For N-phenylbenzamidine the tautomeric form (**43**) is predicted to be more stable than (**44**). Again the real molecule is probably not planar.

The comparison of π-systems of different size may be possible by means of the properties conjugation energy per phenyl substituent (C/k) or specific π-bond energy[70] (E_π^{bond}/N). (C/k) predicts a decrease in stability from N-phenylbenzamidine (**43**) through phenylformamidine (**41**) and benzamidine to N,N'-diphenylbenzamidine whereas the specific π-bond energy predicts an increase in stability with increasing size of the π-system.

B. Pariser–Parr–Pople Method

The Pariser–Parr–Pople method[71–73] (PPP method) is a semi-empirical self-consistent field calculation for π-electrons which considers the inter-electronic repulsion explicitly. The principles of the PPP method are given in several text books[74–76] therefore here only the most important equations are given.

The PPP method as a π-electron method also makes use of the σ–π-separation (see Section III, A). The wave function Ψ for the ground state of a closed shell molecule with N π-electrons built from N p-orbitals and leading to N MO's ψ_j is written as a normalized Slater determinant (equation 11).

$$\Psi = [(N)!]^{-1/2} \det \{\psi_1(1)\alpha(1)\psi_1(2)\beta(2)\psi_2(3)\alpha(3) \ldots$$
$$\psi_{N/2}(N-1)\alpha(N-1)\psi_{N/2}(N)\beta(N)\}$$
$$= |\psi_1\bar{\psi}_1\psi_2\bar{\psi}_2 \ldots \psi_{N/2}\bar{\psi}_{N/2}| \tag{11}$$

The MO's ψ_j are constructed from a linear combination of N atomic p-orbitals φ_μ (equation 12).

$$\psi_j = \sum_{\mu=1}^{N} c_{j\mu}\varphi_\mu \tag{12}$$

The LCAO–MO coefficients $c_{j\mu}$ are determined by solution of a set of secular equations (13).

$$\sum_{\mu=1}^{N} c_{j\mu}F_{\mu\nu} = \varepsilon_j \sum_{\mu=1}^{N} c_{j\mu}S_{\mu\nu} \qquad (\nu = 1, 2 \ldots N) \tag{13}$$

By use of the zero-differential-overlap approximation the overlap integrals $S_{\mu\nu}$ are neglected unless $\mu = \nu$ in which case they are equal to unity. All two-electron integrals (equation 14) which depend on the overlapping of charge distributions of different atomic orbitals are neglected.

$$\iint \varphi_\mu^*(1)\varphi_\lambda(1) \frac{e^2}{\tau_{12}} \varphi_\mu^*(2)\varphi_\sigma(2) \, d\tau_1 \, d\tau_2 = \delta_{\mu\lambda} \, \delta_{\nu\sigma} \, \gamma_{\mu\nu} \tag{14}$$

In equation (14) $\delta_{\mu\lambda}$ is the Kronecker delta (equals 1 for $\mu = \lambda$ and 0 for $\mu \neq \lambda$) and $\gamma_{\mu\nu}$ represents the Coulomb electronic repulsion of an electron in the AOφ_μ and an electron in φ_ν (equation 15).

$$\gamma_{\mu\nu} = \iint \varphi_\mu^*(1)\varphi_\mu(1) \frac{e^2}{\nu_{12}} \varphi_\nu(2)\varphi_\nu^*(2) \, d\tau_1 \, d\tau_2 \tag{15}$$

The matrix elements $F_{\mu\nu}$ of the secular equations (13) in the zero-differential-overlap approximation are given by equations (16) and (17).

$$F_{\mu\mu} = U_\mu + \tfrac{1}{2} q_\mu\gamma_{\mu\mu} + \sum_{\nu(\neq\mu)}^{N} (q_\nu - Z_\nu)\gamma_{\mu\nu} \tag{16}$$

$$F_{\mu\nu} = \left\{ \begin{matrix} \beta_{\mu\nu}^{\text{core}} \ (\mu,\nu \ \text{bonded}) \\ 0 \qquad (\mu,\nu \ \text{nonbonded}) \end{matrix} \right\} - \tfrac{1}{2} P_{\mu\nu}\gamma_{\mu\nu} \tag{17}$$

These equations contain the empirical parameters:

U_μ = valence state ionization potential
$\gamma_{\mu\mu}$ = one-centre Coulomb repulsion integrals
$\gamma_{\mu\nu}$ = two-centre Coulomb repulsion integrals
$\beta_{\mu\nu}^{\text{core}} = \int \varphi_\mu^* H^{\text{core}}\varphi_\nu \, d\tau$ = Coulomb integral

which may be determined from experimental properties. The π-electron densities q_μ and the π-bond orders $p_{\mu\nu}$ have the same definition as in the HMO-theory, equations (2) and (4), respectively. As the matrix elements

TABLE 10. π-Bond energies of phenyl substituted amidines

Amidine	Formula	$E_\pi^{bond}(\beta)$	Conjugation energy^a $C(\beta)$	$\dfrac{C}{k}(\beta)$	E_π^{bond}/N^b (β)
Formamidine		2·895			0·965
Benzamidine		11·275	0·380	0·380	1·253
Phenylformamidine		11·285	0·390	0·390	1·254
		11·241	0·346	0·346	1·249

TABLE 10 (continued)

N-Phenylbenzamidine **(43)**	19·680	0·785	0·392	1·312
(44)	19·625	0·730	0·365	1·308
N,N′-Diphenylbenzamidine	28·033	1·138	0·379	1·335

[a] $C = E_\pi^{bond} - k \cdot E_\pi^{bond}$ (benzene) $- E_\pi^{bond}$ (amidine) $= E_\pi^{bond} - k \cdot 8\beta - 2\cdot895\ \beta$ $(k =$ number of phenyl substituents)

[b] Specific π-bond energy $(N =$ number of p-orbitals)

(16) and (17) in equation (13) depend on an initial choice of coefficients $c_{j\mu}$ the secular equations (13) must be solved by an iterative procedure. As starting coefficients one takes those of a HMO calculation.

The one-centre Coulomb repulsion integrals $\gamma_{\mu\mu}$ are taken as the difference between the valence state ionization energy and the electron affinity[71]. The two-centre Coulomb repulsion integrals $\gamma_{\mu\nu}$ are approximated by Mataga's formula (equation 18)[77]

$$\gamma_{\mu\nu} = \frac{14\cdot397}{r_{\mu\nu} + \dfrac{28\cdot794}{\gamma_{\mu\mu} + \gamma_{\nu\nu}}} \qquad (18)$$

in dependence of the interatomic distance $r_{\mu\nu}$. The empirical parameters[78] used for the calculations are collected in Table 11.

In Table 12 the results of the PPP calculations with the parameters of Table 11 are given. By Koopmans' theorem[81] the negative value of the SCF-molecular orbital energies ε_j is equal to the ionization energy for removal of one electron out of the occupied MO. The calculated ionization energy of the highest occupied MO is increasing in the series:

allyl anion < benzamidine < thioimidate < amidine < imidate <
(1·855 eV)   (9·609 eV)   (10·787 eV)   (10·849 eV) (11·486 eV)
                      amide   <   imine
                 (11·791 eV)   (12·200 eV)

The total π-electron energy is not obtained simply as a sum of the doubly occupied SCF molecular orbital energies ε_j, since this procedure would count the electronic repulsion twice, but by means of expressions[73] (19).

$$E = \frac{1}{2} \sum_{\mu} \sum_{\nu} p_{\mu\nu}(H_{\mu\nu} + \mathbf{F}_{\mu\nu}) \qquad (19)$$

Symbols: $H_{\mu\mu} = U_\mu$; $H_{\mu\nu} = \beta_{\mu\nu}$; $\mathbf{F}_{\mu\nu}$ see equations (16) and (17)

The values of the LCAO-coefficients $c_{j\mu}$ are altered by the PPP calculation relative to those of the HMO-calculation, but the symmetry properties of the MO's stay the same. The π-electron densities q_μ are changed in such a way that the charge distribution is smoothed out. In all imidic acid derivatives the middle carbon atom is bearing a positive charge which is smaller than that given by the HMO calculations. The π-bond orders $p_{\mu\nu}$ are altered as usual in SCF-calculations so that the bonds with high double bond character increase their bond orders and those with single bond character decrease their values with respect to that of the HMO calculation.

TABLE 11. Empirical parameters for PPP calculations[78,79]

Atom	Valence state	U_μ (eV)	$\gamma_{\mu\mu}$ (eV)	Bond	$\beta_{\mu\nu}^{core}$ (eV)	Bond distances (Å)
C	tr tr tr π	11·16	11·13	C—C	−2·32	1·397
N	tr²tr tr π ($=\bar{N}-$)	14·12	12·34	C=N	−2·55	1·290
N⁺	tr tr tr π ($-\bar{N}-$ H)	28·59	16·63	C—$\bar{N}$	−2·32	1·334
O	tr²tr²tr π ($=\bar{O}$)	17·70	15·23	C=O	−2·60	1·235
O⁺	tr²tr tr π ($-\bar{O}-$)	33·90	18·60	C—$\bar{O}$—	−2·32	1·304
S⁺	tr²tr tr π ($-\bar{S}-$)	22·88	11·90	C—$\bar{S}$	−1·50	1·708

TABLE 12. Results of PPP calculations for amidines and related compounds[a]

Compound	Formula	j	ε_j (eV)	$c_{j\mu}$			Total π-electron energy (eV)	q_μ			$p_{\mu\nu}$	
				c_{j1}	c_{j2}	c_{j3}		q_1	q_2	q_3	p_{12}	p_{23}
Imine		1	− 12·200	0·778	0·628		− 27·645	1·211	0·789		0·978	
		2	− 1·217	− 0·628	0·778							
Allyl anion		1	− 6·005	0·492	0·718	0·492	− 56·624	1·484	1·032	1·484	0·707	0·707
		2	− 1·855	− 0·707	0·0	0·707						
		3	5·906	− 0·508	0·696	− 0·508						
Amidine		1	− 14·595	0·278	0·483	0·831	− 96·232	1·318	0·810	1·872	0·901	0·390
		2	− 10·849	0·763	0·415	− 0·496						
		3	− 0·582	− 0·584	0·772	− 0·253						

TABLE 12 (continued)

	5	0·605	0·223	−0·187	−146·833	1·332	0·826	1·881	0·852	0·369
Benzamidine	−9·609									
Imidate 1	−17·060	0·131	0·324	0·937	−77·769	1·281	0·934	1·931	0·934	0·289
2	−11·486	0·789	0·538	−0·296						
3	−0·811	−0·600	0·778	−0·186						
Thioimidate 1	−12·875	0·506	0·556	0·659	−61·988	1·268	0·936	1·929	0·936	0·291
2	−10·787	−0·614	−0·303	0·728						
3	−0·876	−0·605	0·774	−0·188						
Amide 1	−14·961	0·361	0·496	0·789	−73·193	1·454	0·692	1·854	0·845	0·438
2	−11·791	0·772	0·314	−0·552						
3	−0·780	−0·522	0·809	−0·271						

^{a} Calculations were performed with the program QCPE 71·2 by J. E. Bloor and B. R. Gilson[80] on a CDC 3300 computer at the 'Zentrum für Datenverarbeitung' of the University of Tübingen.

TABLE 13. Calculated and experimental dipole moments (PPP method)

Class of compounds	Calculated dipole moments (debyes)	Compound	Experimental dipole moments (debyes)	Reference
Imine	1·28	CH_3—CH=N—n-Bu	1·62	82
Allyl anion	3·25			
Amidine	2·49	(5-membered ring with N, N–H, C_2H_5)	3·42	39
Benzamidine	2·60	C_6H_5—C(=NH)—$N(CH_3)_2$	2·83	40
Imidate	1·99	CH_3—C(=NH)—OC_2H_5	1·42	46
Thioimidate	2·01			
Amide	3·26	H—C(=O)—NH_2	3·71	83

In Table 13 the calculated dipole moments are compared with experimental values (see Section II, B) as far as available. The agreement is not too good, because in the PPP π-electron theory the effect of σ-bond moments is completely neglected, but the order of magnitude of experimental dipole monents is predicted correctly.

IV. ELECTRONIC SPECTRA

A. Amidines and Amidinium Salts

The unconjugated amidine group should give rise to a $n \rightarrow \pi^*$ and a $\pi \rightarrow \pi^*$ transition in the ultraviolet or vacuum ultraviolet region of the absorption spectrum. However, free acetamidine dissolved in water or methanol shows no absorption maximum above 200 nm[84] (see Figure 4). The reported maxima of acetamidine obtained by dissolving an acetamidine hydrochloride in aqueous sodium hydroxide solution at 224 nm ($\varepsilon =$ 4000)[85] or 219 nm ($\varepsilon = 1100$)[86] could not be reproduced[84]. So the weak $n \rightarrow \pi^*$ transition is not observable and falls probably like the $\pi \rightarrow \pi^*$ transition into the vacuum ultraviolet range below 200 nm. The PPP calculation with the parameters of Table 11 for the singlet transition energies of amidine with inclusion of configuration interaction (CI) predicts the lowest $\pi \rightarrow \pi^*$ transition to occur at 179 nm with an oscillator strength of 0·527 (see Table 14).

N-p-Chlorophenylacetamidine shows an absorption band at 236 nm with $\varepsilon = 8100$[85] (see Table 15). This band corresponds to the high intensity, short wave-length band of p-chloroaniline[85] (290 nm, $\varepsilon = 1700$ and 239 nm, $\varepsilon = 8500$) and the probable occurrence of a low intensity band around 290 nm in the substituted acetamidine may have been overlooked. The similarity of the spectrum of the amidine derivative to that of p-chloroaniline suggests that the N-phenyl substituent is not taking part in the conjugation of the amidine π-system since it is twisted out of the plane of the amidine group leading to aniline-like overlapping of the phenyl-π-system with the sp^2-hybridized nitrogen lone pair of electrons.

On protonation the band is shifted hypsochromically to 228 nm and lowered in intensity ($\varepsilon = 7000$), an effect which is also observed with N-phenyl-substituted formamidines[91]. Benzamidine shows two transitions in the ultraviolet region (see Table 15 and Figure 4), a weak band at 268 nm ($\varepsilon = 810$) and a stronger band at 228 nm ($\varepsilon = 13,800$). The spectrum closely resembles that of benzoic acid[92] (273 nm, $\varepsilon = 970$ and 230 nm $\varepsilon = 11,600$). The weak band corresponds to the weak α-band ($^1A_{1g} \rightarrow {}^1B_{2u}$) and the strong band to the p-band ($^1A_{1g} \rightarrow 1B_{1u}$) of benzene[93]. The PPP calculation with inclusion of CI predicts also two $\pi \rightarrow \pi^*$

G. Häfelinger

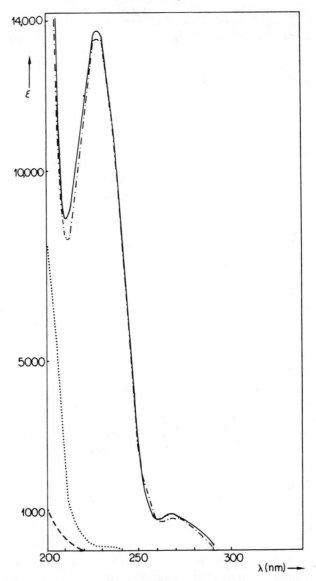

FIGURE 4. Ultraviolet spectra of amidines and amidinium hydrochlorides in methanol[84]: (———) benzamidine; (—·—·—) benzamidinium hydrochloride (containing 1·8 mol water of crystallization); (— — — —) acetamidine; (· · · ·) acetamidinium hydrochloride.

TABLE 14. $\pi \rightarrow \pi^*$ Transition energies calculated by use of PPP theory

Compound	Triplet transition energies[a]		Singlet transition energies[a]			f^b	Experimental absorption maxima[c] (nm)	Reference
	(eV)	(nm)	(eV)	(cm^{-1})	(nm)			
Imine	2·365	524	8·069	65080	153·5	0·544		
Amidine	2·428	511	6·926	55860	179	0·527	170(8000)	87
	(2·611)	(475)	(7·037)	(56760)	(176)			
Benzamidine 4 → 6	1·915		4·501	36300	275	0·007	268 (970)	84
	(3·000)		(5·300)	(42750)	(234)			
5 → 6			5·093	41080	245	0·571	228(13800)	84
			(5·167)	(41680)	(240)			
Imidate	2·441	508	7·580	61140	164	0·554		
	(2·509)	(494)	(7·630)	(61540)	(162·5)			
Thioimidate			6·322	51000	196	0·343		
			(6·476)	(52230)	(191·5)			
Amide	2·730	464	7·316	59000	169·5	0·472	181	88
			(7·419)	(59840)	(167)			

[a] Calculated by use of configuration interaction (CI) between all single excited structures. Values without CI in brackets.
[b] Oscillator strength.
[c] Molar extinction coefficient in brackets.

TABLE 15. Ultraviolet absorption maxima of amidines and amidinium salts

Compound	Solvent	Amidine λ_{max} (nm)	Amidine ε_{max}	Solvent	Amidinium hydrochloride λ_{max} (nm)	Amidinium hydrochloride ε_{max}	Reference
N-p-Chloro-phenylacetamidine	Water	236	8100	Water	228	7000	85
Benzamidine	Water	268	810	Water	—	—	86
	Methanol	229	9100	Methanol	228	8500	84
		268	960		269	820	
		228	13800		228	13600	
N-Aryl-substituted benzamidines:							
Phenyl	50% aqueous ethanol	236	14300				89
p-Tolyl		270	11800				89
p-Chlorophenyl		260	16800				89
p-Bromophenyl		234	16400				89
p-Iodophenyl		236	18300				89
		220sh	—				
p-Nitrophenyl		324	8200				89
		230sh	—				
m-Nitrophenyl		260	16800				89
		224sh	—				
N,N'-Diaryl-substituted formamidines:							
Diphenyl	Benzene	284	19500				90
Di(*p-n*-butoxyphenyl)	Benzene	296	24500				90
	Dioxane	294	27400				90
	Cyclohexane	291	24800				90

transitions (see Table 14), a weak one at 275 nm, $f = 0\cdot007$ and a strong band at 245 nm, $f = 0\cdot571$. The first band corresponds to the transition from the occupied fourth MO to the lowest empty MO 6 which is shifted bathochromically by consideration of CI, whereas the second transition from the highest occupied MO 5 to the lowest empty MO 6 is shifted hypsochromically by CI. The $n \to \pi^*$ transition is not observed as it is probably obscured by the $\pi \to \pi^*$ transitions.

Surprisingly benzamidinium hydrochloride shows practically the same absorption spectrum as the free base (see Table 15 and Figure 4). This is contrary to the finding in the case of N-phenyl-substituted amidines indicating the importance of conjugation of the sp^2-nitrogen lone pair, on which protonation occurs, with the twisted phenyl group. This is not possible in the benzamidine case although the phenyl ring may also be twisted out of the plane of the amidine group.

N-Aryl-substituted benzamidines (see Table 15) show one or two absorption bands between 300 and 220 nm. Contrary to the statement of Ševčik[89] these spectra are not analogous to the corresponding N-benzylidene anilines which show generally three strong bands (i.e. benzylidine–aniline[94,95] λ_{max}[nm] (ε). 310 (8200); 256 (16,000); 212 (18,000)).

N,N'-Diarylformamidines (Table 15) show a strong band about 284 nm, $\varepsilon = 19,500$ which may be due to the conjugation of both phenyl rings through the amidine π-system. Substituents in the benzene rings cause a bathochromic shift of this band while twisting of the benzene rings out of the amidine plane by introduction of *ortho*-substituents results in a hypsochromic shift[96]. The spectra are analogous to those of the corresponding triazenes[97].

B. Imidates

The PPP calculation predicts for the unconjugated imidate group a $\pi \to \pi^*$ transition at 164 nm (see Table 14). In agreement with this prediction a series of imidates derived from aliphatic acetylene carbonic acids show no absorption maximum above 205 nm (see Table 16). Extension of the conjugation by a phenyl substituent leads to the two absorption bands of monosubstituted benzenes, a weak band at 290 nm, $\varepsilon = 1300$ and a strong band at 260 nm, $\varepsilon = 26,300$. Methylbenzimidate shows also the two absorption bands of monosubstituted benzene derivatives, now at 270 nm, $\varepsilon = 900$ and 230 nm, $\varepsilon = 12,600$. This spectrum closely resembles that of benzamidine (see Table 15) and that of benzamide[16] (270 nm, $\varepsilon = 900$ and 228 nm, $\varepsilon = 9100$) but is different from that of N,N-dimethylbenzamide[16,98,99]. The conclusion[98,100] that benzamide

TABLE 16. Ultraviolet absorption bands of imidates

Compound	R	Solvent	λ_{max} (nm)	ε	Reference
$R-C \equiv C-C(=NH)-OCH_3$	t-Bu	Ethanol	<205		16
	n-Pentyl	Ethanol	<207		16
	n-Hexyl	Ethanol	<207		16
	Phenyl	Ethanol	290 / 260	1300 / 26,300	16
$C_6H_5-C(=NH)-OR$	CH$_3$	Methanol	270 / 230	900 / 12,600	16
	C$_2$H$_5$		268	500	98
$H-C(=N-R)-OC_2H_5$	Phenyl		273 (sh) / 240	560 / 10,000	98
	o-Tolyl		290 (sh) / 255	560 / 5000	98
	m-Tolyl		290 (sh) / 250	450 / 5600	98
	p-Tolyl		290 (sh) / 252	630 / 7100	98
	o-Anisidyl		285 / 250	3600 / 4000	98
$CH_3-C(=N-CH_3)-OC_6H_5$		Acetonitrile	267·5		99

therefore exists in solution mainly in the iminol form (46) was shown by Grob and Fischer[16] not to be conclusive, since N,N-disubstituted benzamides show abnormal light absorption because of steric interference of *ortho*-hydrogens with the N-alkyl substituents leading to non-planarity

$$C_6H_5-C{\overset{O}{\underset{NH_2}{}}} \rightleftharpoons C_6H_5-C{\overset{OH}{\underset{NH}{}}}$$

(45)   (46)

of the π-system. In ethyl N-arylformimidates the long wavelength absorption band is only recognized as a shoulder; the shorter wavelength band lies around 250 nm.

Summarizing these results one may state that neither the unconjugated amidine nor the unconjugated imidate group leads to an observable absorption band in the ultraviolet region of the absorption spectrum. Phenyl substitution leads to two bands, a weak long wavelength band and a strong shorter wavelength band characteristic of monosubstituted benzene derivatives, but showing no specific absorption due to the amidine or imidate group.

V. INFRARED AND RAMAN SPECTRA

A. Spectral Data for Simple Amidines

I. Acetamidine

Free acetamidine was prepared as a colourless oil by Davies and Parsons[101] who determined its infrared spectrum[102] and assigned 16 of the 24 normal vibrations which are listed in Table 17.

The planar molecule has C_s-symmetry, i.e. all vibrations are both infrared and Raman active. The spectrum shows the vibrations characteristic for an unsubstituted amidine. Its NH vibrations are broadened by intermolecular association. The imine NH vibration is assigned at 3429 cm^{-1}, while the asymmetric and symmetric NH$_2$ group vibrations occur at 3330 cm^{-1} and 3226 cm^{-1}, respectively. The NH$_2$ deformation vibration at 1608 cm^{-1} is higher for the amine group than that of the imine group at 1460 cm^{-1}. The C=N double bond vibration occurs at 1650 cm^{-1} whereas the CN single bond vibration is located at 1192 cm^{-1}. There is no strong coupling between these vibrations; but the latter seems to be coupled with the NH deformation vibration of the imine group.

TABLE 17. Infrared frequency assignments of liquid acetamidine[102] (Frequencies in cm^{-1}, relative intensities in brackets; v = valence vibration; δ = in plane deformation vibration; r = in plane rocking vibration; t = twisting vibration; ω = out of plane wagging vibration; a = antisymmetric vibration; s = symmetric vibration.)

Frequency	Assignment
3429 (9)	v (NH)
3330 (9)	v_a (NH$_2$)
3226 (9)	v_s (NH$_2$)
1650 (7)	v_a (NCN)
1608 (7)	δ (NH$_2$)
1460 (3)	δ (NH) + v_s (NCN)
1429 (3)	δ_a (CH$_3$)
1368 (2)	δ_s (CH$_3$)
1192 (5)	v_s (NCN) + δ(NH)
1124 (1)	r (NH$_2$)
1044 (1)	r (CH$_3$)
1002 (1)	v (CC)
862 (3)	t (C—NH$_2$) ?
800 ⎫	⎧ δ (NCH)
│	ω (NCN)
450 ⎭	⎩ t (C=NH)

2. Acetamidinium cation

The infrared spectra of the acetamidinium cation both with the chloride and in salts with complex anions of the type $[MCl_6]^{2-}$ as well as the spectrum of the deuterated cation were assigned in terms of normal vibrations assuming C_{2v}-symmetry by Mecke and Kutzelnigg[102] who improved the assignments given by Davies and Parsons[102].

The 27 normal vibrations listed in Table 18 show clearly that the protonation occurs on the imino nitrogen leading to a symmetrical structure for the amidinium cation with C_{2v}-symmetry if one assumes free rotation of the methyl group. The NH$_2$ valence vibrations occur at 3417 cm^{-1} and 3368 cm^{-1} in the hexachloroplatinate. They are lowered in the chloride to 3220 cm^{-1} and 3080 cm^{-1} due to hydrogen bonding to the anion. On deuteration the vibrations are split into the four theoretically-expected vibrations and shifted to 2560 cm^{-1}, 2531 cm^{-1} and 2425 cm^{-1}, 2397 cm^{-1}, respectively.

The planar NH$_2$ deformation vibrations are found in the hexachloroplatinate at 1667 cm^{-1} and 1555 cm^{-1} and are shifted on deuteration to 1176 cm^{-1}. The C—N valence vibrations are now strongly coupled leading to an asymmetrical vibration at 1690 cm^{-1} and a symmetrical one at 1520 cm^{-1}. But the shift on deuteration shows that they are also coupled with the NH_2 deformation vibrations. The C—C valence vibration occurs at 880 cm^{-1}.

3. Force constants calculation for the acetamidinium cation

The calculation of force constants of a valence force field including interactions of molecular frequencies is a problem which in general has no unique solution. Therefore a number of different potential functions, all of which are in agreement with the observed molecular frequencies have been calculated by means of an analog computer by Mecke and coworkers[105].

The geometry assumed for the amidinium cation is given in Figure 5. The C—N bond distance used was longer than the 1·316 Å derived in Section II, A, 2, which may affect the calculation.

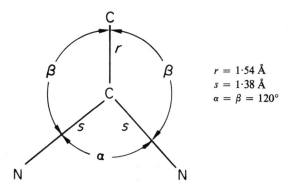

$$r = 1\cdot54 \text{ Å}$$
$$s = 1\cdot38 \text{ Å}$$
$$\alpha = \beta = 120°$$

FIGURE 5. Coordinates of the heavy atoms of the planar acetamidinium cation. (From Beckmann, Gutjahr and Mecke, *Spectrochim. Acta*, **20**, 1295 (1964), with permission.)

The six skeletal normal vibrations for a planar molecule of the type XYZ_2 with C_{2v}-symmetry are shown in Figure 6 together with the corresponding wave numbers for the acetamidinium cation[103]. In Table 19 four sets out of 26 calculated sets are given which seem to reflect best the bonding situation. The C—C single bond force constant is about 4·5 mdyn/Å, whereas the CN force constant with 9·1 mdyn/Å is rather close to the value of 10·6 mdyn/Å for a pure C=N double bond[106]. It is interesting to note that the resistance to angle deformation is greater for the NCN angle (f_α) than for the CCN angle (f_β).

G. Häfelinger

TABLE 18. Frequency assignment of the acetamidinium cation in different salts in Hostaflon oil and Nujol[103]. (Band positions in cm^{-1}, relative intensities in brackets, sh = shoulder.)

ν_i	Irreducible representation[a]	Assignment[b]	Cl⁻ Raman[104] (aq. soln)	Anion Cl⁻	Anion [PtCl₆]²⁻	Anion [PtCl₆]²⁻ (with deuterated cation)
1	A_1	νNH_2	3370 (1)	3220 (10)	3417 (9)	2560 (9)
2		νNH_2	3256 (1)	3080 (10)	3368 (10)	2425 (6)
3		νCH_3	2880 (4)	- ?	?	?
4		δNH_2	?	?	1667 (sh)	1176 (4)
5		ωCN	1519 (5)	1511 (6)	1520 (6)	1499 (7)
6		δCH_3	1378 (3)	1378 (8)	1379 (7)	1381 (6)
7		δNH_2	1155 (6)	1145 (2)	1200 (4)	943 (3)
8		ωCC	880 (10)	880 (0)	876 (1)	829 (3)
9		Δ	534 (4)	535 (5)	540 (4)	480 (5)
10	A_2	γNH_2	?	inactive	inactive	inactive
11		γNH_2	?			
12		τCH_3	?			

TABLE 18 (continued)

B_1	13	νNH_2	3370 (1)	3220 (10)	3417 (9)	2531 (10)
	14	νNH_2	3256 (1)	3080 (10)	3368 (10)	2397 (10)
	15	νCH_3	2998 (3)	!	2950 (0)	2950 (0)
	16	ωCN		1687 (10)	1690 (10)	1642 (10)
	17	δNH_2	1568 (1·5)	1587 (1)	1555 (1)	1176 (4)
	18	δCH_3	1444 (2)	1425 (5)	1411 (6)	1412 (5)
	19	ρNH_2	1094 (4)	1105 (3)	1081 (3)	943 (3)?
	20	ρCH_3	1028 (3)	1026 (1)	1020 (0)	1020 (0)
	21	Δ	445 (2)	445 (3)	450 (2)	?
					510 (2)?	
B_2	22	νCH_3	2998 (3)	?	2950 (0)	2950 (0)
	23	δCH_3	1444 (2)	1425 (5)	1411 (6)	1412 (5)
	24	γCH_3		1057 (2)	1053 (1)	1050 (3)
	25	Γ	706 (2)?	?	720 (0)?	621 (5)
	26	γNH_2	?	720 (8)	610 (8)	450 (8)
	27	γNH_2	?	?	?	?

[a] Point group C_{2v}; A_1, A_2, B_1, B_2 = irreducible representations; A = symmetric and B = antisymmetric with respect to C_2-axis, A_1, B_1 and B_2 are i.r. and Raman active, A_2 is only Raman active.

[b] ν = valence vibration to hydrogen; ω = valence vibration between heavy atoms; δ = in plane deformation vibration; ρ = rocking vibration; Γ = out of plane deformation vibration; Δ = planar skeleton deformation vibration.

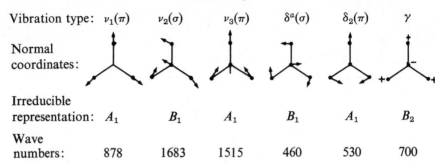

Vibration type:	$\nu_1(\pi)$	$\nu_2(\sigma)$	$\nu_3(\pi)$	$\delta^a(\sigma)$	$\delta_2(\pi)$	γ
Irreducible representation:	A_1	B_1	A_1	B_1	A_1	B_2
Wave numbers:	878	1683	1515	460	530	700

FIGURE 6. The 6 skeletal normal vibrations of a molecule of the type XYZ_2 with C_{2v}-symmetry and the corresponding wave numbers for acetamidinium cation. (From Beckmann, Gutjahr and Mecke, *Spectrochim. Acta*, **20**, 1295 (1964), with permission.)

B. C=N Double Bond Vibrations

I. Amidines

The C=N valence vibrations of amidines of different structural types are collected in Table 20. The double bond vibrations fall all in the range from 1658 cm^{-1} to 1582 cm^{-1}. They are lower than the C=N vibration of unconjugated imines (i.e. n-propylidene propylimine: 1679 cm^{-1} in CCl_4[108]).

The C=N vibrations in C-phenyl-substituted amidines are generally lower than those of the corresponding C-alkyl-substituted compounds, which is in agreement with the calculated lowering of HMO π-bond orders, for example: $p_{C=N}$ (amidine) = 0·789 and $p_{C=N}$ (benzamidine) = 0·727.

The C=N double-bond vibration is strongly affected by molecular association, especially in the case of N-monosubstituted and N,N'-disubstituted amidines. Therefore it is necessary to state the experimental conditions and to compare only the frequencies observed in the non-associated state[109]. N,N-Disubstituted amidines always absorb at lower frequencies than the corresponding N-monosubstituted derivatives[63].

2. Imidic Acid Derivatives

The C=N double bond vibration in nonconjugated imidates (see Table 21) lies in the range from 1670 to 1646 cm^{-1} which is close to the value of unconjugated imines. This indicates, in agreement with the high value of the HMO π-bond order of 0·828, that resonance between the structures (**47**) and (**48**) is not so important as in the case of amidines. As usual,

TABLE 19. Force constants of a valence force field for the acetamidinium cation[105]

f_r	f_s	f_α	t_β	f_{rs}	f_{ss}	$\frac{1}{r} f_{r\alpha}$	$\frac{1}{s} f_{s\alpha}$	f_γ
4·520	9·10	1·158	0·81	0·358	0·9	−0·389	−0·217	0·618
4·277	9·10	1·306	0·81	0·723	0·9	−0·389	−0·435	0·618
4·801	9·10	1·157	0·81	0·233	0·9	−0·584	−0·217	0·618
4·587	9·10	1·284	0·81	0·569	0·9	−0·584	−0·435	0·618

f_r = CN bond force constant; f_s = CC bond force constant; f_α, f_β and f_γ = angle deformation force constants; f_{rs} and f_{ss} = bond–bond interaction force constants; $f_{r\alpha}$ and $f_{s\alpha}$ = bond angle interaction force constants (compare Figure 5). All constants are in mdyn/Å except f_α and f_β which are in mdyn·Å.

TABLE 20. C=N Valence vibrations (cm^{-1}) of amidines R^1—C

$$R^2\!-\!N\!=\!C\!-\!N\!\begin{matrix}R^3\\R^4\end{matrix}$$

in the infrared spectrum. (Values in brackets are Raman spectral data)

Type of compounds	R^1 = CH$_3$	Reference	R^1 = Phenyl	Reference
(a) Unsubstituted				
R^2 = R^3 = R^4 = H	1650 (liq.)	102	1630 (KBr)	84
(b) Monosubstituted				
R^2 = R^4 = H; R^3 = CH$_3$	1615 (CHCl$_3$)^a	63	1612 (CHCl$_3$)	63
R^3 = R^4 = H; R^2 = C$_6$H$_5$	1640 (CHCl$_3$)^b	63	—	
(c) N,N'-Disubstituted				
R^3 = H; R^2 = R^4 = CH$_3$	1658 (CHCl$_3$)^c	107	$\left.\begin{matrix}1651\\1640\text{sh}\end{matrix}\right\}$(CCl$_4$) $\left.\begin{matrix}1655\text{sh}\\1635\end{matrix}\right\}$(CHCl$_3$)	108 108
R^3 = H; R^2 = C$_6$H$_5$; R^4 = CH$_3$	$\left.\begin{matrix}(1652)\\(1636)\end{matrix}\right\}$(CHCl$_3$)	109	$\begin{matrix}1620\\(1621)\end{matrix}$ (CHCl$_3$)	110 111
R^3 = H; R^2 = R^4 = C$_6$H$_5$	$\left.\begin{matrix}1655\\1633\end{matrix}\right\}$(CHCl$_3$)	63	1630 (KBr)	112
	$\left.\begin{matrix}1652\\1639\end{matrix}\right\}$	113	$\left.\begin{matrix}(1635)\\(1624)\end{matrix}\right\}$	113

TABLE 20 (continued)

(d) N,N-Disubstituted				
$R^2 = H; R^3 = R^4 = $ —CH$_2$—CH$_3$	1588 (KBr)^d	63	—	
$R^2 = H; R^3 = $ alkyl; $R^4 = $ aryl			—	
$R^2 = H; R^3 = R^4 = $ aryl			—	
(e) Trisubstituted				
$R^2 = R^3 = R^4 = CH_3$	—		1621 (CCl$_4$)	108
			1614 (CHCl$_3$)	108
$R^2 = C_6H_5; R^3 = R^4 = CH_3$	(1623–1619) (CHCl$_3$)	109	1594 (CHCl$_3$)	110
			(1582)	
$R^2 = R^3 = C_6H_5; R^4 = C_2H_5$	(1616)	111	—	
$R^2 = R^3 = CH_3; R^4 = $ —⟨C$_6$H$_4$⟩— OCH$_3$	—		1624 (CCl$_4$)	108
			1618 (CHCl$_3$)	108
$R^2 = R^3 = R^4 = C_6H_5$	(1633)	113	—	

^{a} $R^1 = $ Benzyl; $R^3 = C_2H_5$
^{b} $R^1 = $ Benzyl
^{c} $R^1 = H; R^2 = R^4 = $ Cyclohexyl
^{d} $R^1 = $ —CH—C$_6$H$_5$
　　　　　|
　　　　CH$_3$

TABLE 21. C=N Valence vibrations (cm^{-1}) of imidic acid derivatives

Functional group	Unconjugated	Monoconjugated	Diconjugated

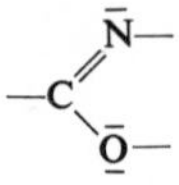

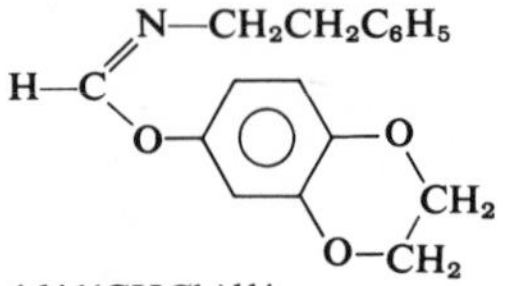

Row 1 (functional group $\overset{-}{N}$— / —C—$\overset{=}{O}$—):

Unconjugated: structures with
N—CH$_2$CH$_2$C$_6$H$_5$; H—C (benzodioxane type) 1646(CHCl$_3$)[114]

Monoconjugated: C$_6$H$_{13}$—C≡C—C(=NH)OC$_2$H$_5$ 1621[16]

Diconjugated: C$_6$H$_5$—C(=N—C$_6$H$_5$)OCH$_3$ 1665(CCl$_4$)[108] ; 1663(CHCl$_3$)[108]

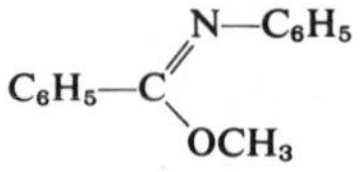

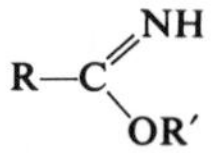

Unconjugated: R—C(=NH)OR′   1655–1652(Raman)[115]
Monoconjugated: C$_6$H$_5$—C(=NH)OR   1653–1648(Raman)[115]

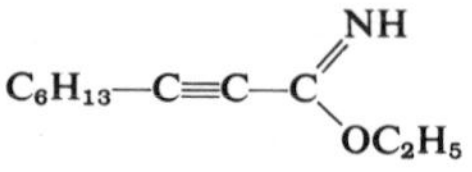

Unconjugated: CH$_3$—C(=N—CH$_2$)(O—CH$_2$)   1670[116]
Monoconjugated: C$_6$H$_5$—C(=N—CH$_2$)(O—CH$_2$)   1650[116]

Row 2 (functional group $\overset{\oplus|}{N}$— / —C—$\overset{=}{O}$—):

Unconjugated: CH$_3$, $\overset{\oplus}{N}$—CH$_2$CH$_2$C$_6$H$_5$; H—C (benzodioxane type) 1659(CHCl$_3$)[114]

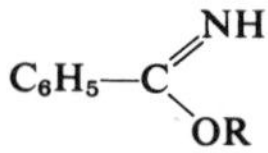

Row 3 (functional group $\overset{-}{N}$— / —C—$\overset{=}{S}$—):

Monoconjugated: C$_6$H$_5$—C(=N—CH$_3$)S—CH$_3$ 1622(CHCl$_3$)[114]
Diconjugated: C$_6$H$_5$—C(=N—C$_6$H$_5$)S—CH$_3$ 1611(CHCl$_3$)[114]

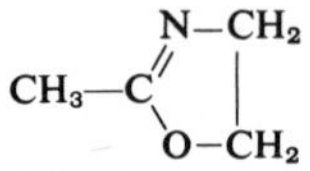

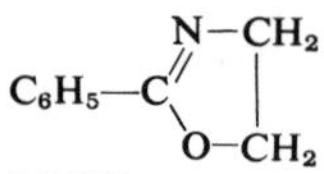

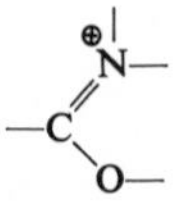

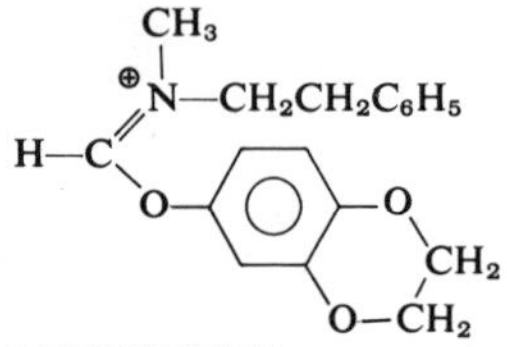

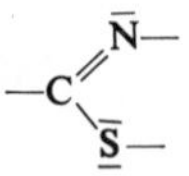

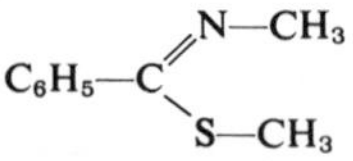

Unconjugated: H$_3$C—C(=N—CH$_2$)(S—CH$_2$)   1640[117]
Monoconjugated: C$_6$H$_5$—C(=N—CH$_2$)(S—CH$_2$)   1613[117]

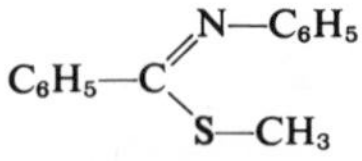

Row 4 (functional group $\overset{\oplus|}{N}$— / —C—$\overset{=}{S}$—):

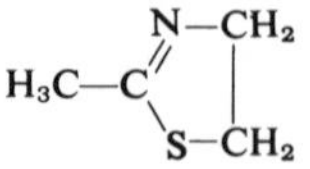

Unconjugated: H$_3$C—C(=$\overset{\oplus}{N}$(CH$_3$)CH$_3$)S—CH$_3$ 1607(CHCl$_3$)[114]
Monoconjugated: C$_6$H$_5$—C(=$\overset{\oplus}{N}$ morpholine)S—CH$_3$ 1580(CHCl$_3$)[114]
Diconjugated: C$_6$H$_5$—C(=$\overset{\oplus}{N}$(CH$_3$)C$_6$H$_5$)S—CH$_3$ 1562(CHCl$_3$)[114]

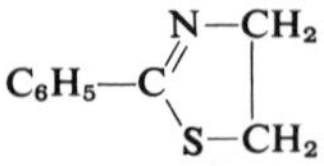

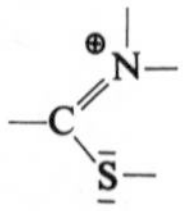

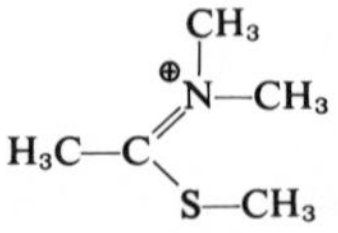

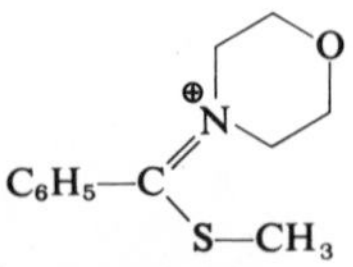

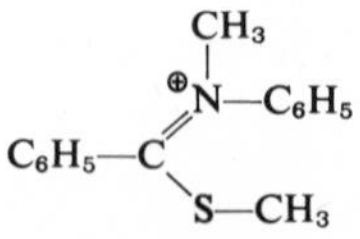

conjugation lowers the vibration frequency except in the case of the di-conjugated methyl N-phenylbenzimidate[108]. Quarternization raises the frequency of the C=N double bond.

$$\text{(47)} \qquad\qquad \text{(48)}$$

For thioimidates the values of the C=N vibrations are lower than in the corresponding imidates contrary to the prediction by HMO π-bond order which is higher (0·862) for the thioimidate. PPP calculations give practically the same π-bond orders for both compounds (imidate: 0·934 and thioimidate: 0·936) so that the difference in the C=N vibration frequency does not necessarily reflect a difference in π-electronic structure. It may also be caused by the effect of the greater mass of the sulphur atom on the skeletal vibration.

Quarternization of the nitrogen leads now to a reduction of the C=N vibration frequency indicating the important contribution of the meso-meric structure (50) to the ground state.

$$\text{(49)} \qquad\qquad\qquad \text{(50)}$$

C. Tautomerism

N-Monosubstituted and N,N'-disubstituted amidines may show tautomerism as indicated in Section I. The infrared spectra of both types of compounds have been thoroughly inspected and interpreted by Prevoršek[63,118].

I. N-Monosubstituted amidines

N-Phenylamidines show two NH bands[63] appearing near 3500 and 3400 cm^{-1}. The observed frequencies are in close agreement with those found for the asymmetrical and symmetrical valence vibrations in form-amide[119] at 3533 and 3411 cm^{-1}, respectively, which indicates that N-phenylamidines contain a terminal amino group, i.e. almost exclusive dominance of the tautomeric structure (52).

Additional evidence for the presence of a terminal amino group is the appearance of a band near 1600 cm^{-1}. This band moves to lower frequencies in solution and is reduced greatly in intensity or disappears fully in deuterated species, indicating that this mode involves deformation of

(51)   (52)

hydrogen atoms. The bending vibration of terminal NH_2 groups in a large number of compounds occurs in this region of the spectrum[119] whereas the deformation vibrations of =NH and —NH—R groups generally are found at lower frequencies.

In N-mono-alkyl substituted amidines three bands are observed in the region of the NH bond-stretching vibration[63]. One band at 3510 cm^{-1} is very weak, while the other two bands near 3450 and 3310 cm^{-1} are much stronger. The band at 3450 cm^{-1} is assigned to the NH stretching vibration of a secondary amino group, whereas that at 3310 cm^{-1} corresponds to the NH-valence vibration of an imino group. The weak band at 3510 cm^{-1} is assigned to the asymmetric stretching vibration of a primary amino group.

The conclusion derived from infrared spectra is that N-alkylamidines exist in chloroform solution as a tautomeric mixture of forms (53) and (54) with predominance of the tautomer (53). The occurrence of a weak

(53)   (54)

band at 1640 cm^{-1} and a strong band at 1615 cm^{-1}, both corresponding to C=N double bond vibrations is in agreement with this interpretation.

Grivas and Taurins[120, 121] recorded the infrared spectra of trichloro- and trifluoroacetamidines. They concluded that that these exist in nonpolar solvents only in the imino form (53) irrespective of the presence of a N-alkyl or N-aryl substituent. But a reassignment of these data by Moritz[122] showed that in all cases these compounds exist as a mixture of the imino (53) and the amino tautomer (54), with N-alkyl substituted compounds preferring the imino form (53) and N-aryl substituted derivatives occurring predominantly in the amino form (54). This is in agreement with the results of Prevoršek[63, 118] stated above.

2. N,N'-Disubstituted amidines

The infrared spectra of N,N'-disubstituted amidines show, in dilute chloroform solutions, two bands in the NH-stretching vibration region[63, 118] near 3450 and 3380 cm^{-1}. The high frequency band is always stronger and its position changes little whereas the low frequency band varies in intensity and frequency depending on the substituents on the nitrogen atoms. It is strongest with amidines having identical substituents on the nitrogen atoms and diminishes in intensity if one of the substituents is replaced by a more or less electronegative substituent. The region of the C=N double bond vibration shows also two bands at about 1655 and 1633 cm^{-1}, but sometimes only one asymmetric band. The relative intensities of these double bands do not depend on the concentration, so that the possibility is eliminated that the lower frequency bands are due to an associated form. Therefore there are either two forms of the monomer or a single form responsible for the doubling of vibration bands in the NH and C=N region. Consequently the following possibilities of two forms of the monomer may be discussed:

(a) *Cis–trans* isomerism with respect to the C=N double bond leading to the pairs of compounds (55) and (56) or (57) and (58).

However, the activation energy of isomerization around the C=N double bond is estimated to be too high for the existence of an equilibrium between the *cis* and *trans* forms at room temperature.

(b) Rotational isomerism with respect to the C—N single bond which has considerable double bond character, leading to the pairs of isomers (55) and (57) or (56) and (58). The occurrence of form (56) seems to be quite unlikely because of steric strain involved with the bulky aromatic substituents (R' = R" = phenyl or naphthyl). The isomerism between (55) and (57) was suggested by Shigorin and Syrkin[113] to be responsible for the observed doubling of vibration bands. Such rotational isomerism has been observed also with secondary amides[123] but in this case the splitting of the NH bands is about 30 cm^{-1}, i.e. smaller than that of the amidines (about 70 cm^{-1}) and no splitting of the C=O double bond vibration is observed.

(c) Tautomerism between forms (59) and (60) cannot explain the doubling of bands, since identical configurations result when the substituents on nitrogen are equivalent. Prevoršek[63] suggested a tautomerism

(59)   (60)

between form (61) and (62) which involves formal rotational isomerism of both the single and the double bonds, but the proton transfer would occur without change of the spatial positions of the substituents. At the moment no decision is possible whether explanation b or c is to be preferred.

(61)   (62)

D. Molecular Association

Liquid acetamidine shows very broad NH vibration bands in its infrared spectrum in the range between 3500 and 3200 cm^{-1} which are broadened by molecular association through hydrogen bonds[102]. However, no model is suggested for the network of hydrogen bonds. The Raman spectrum of N,N'-diethylacetamidine shows a strong dependence of its C=N—vibration on the solvent[109,111]. In dioxane solution, a band is observed at 1675 cm^{-1} which is shifted in hexane to 1592 cm^{-1}. The band observed in dioxane is assigned to the free amidine molecule (probably hydrogen bonded to a dioxane molecule). In hexane, a solvent of low dielectric constant, one assumes the formation of dimeric associates (63).

(63)   (64)

In the liquid state three bands are observed. The strongest band at 1635 cm^{-1} is assigned to molecular associates of type (64). The other two weak bands are attributed to the free form and the dimeric form (63). In the infrared spectrum the wave numbers are slightly higher at 1685, 1640 and 1595 cm^{-1}. For N,N'-diphenylformamidine in benzene solution the formation of cyclic dimers of type (63) was also suggested [124]. The molecular weight determination in dependence of the concentration shows in benzene at 6°C a degree of association up to 1·5. Association is lower, but still appreciable in naphthalene solution at 80°C [125]. The sterically hindered N,N'-di-o-chlorophenyl- and N,N'-di-o-tolyl-formamidines are not associated under these conditions [125, 126]. A series of N,N'-diaryl substituted acetamidines and benzamidines exhibit weak molecular associations in naphthalene at 80°C as indicated by their molecular weight vs. concentration curves [127]. These curves show clearly no association for trisubstituted amidines demonstrating that the molecular association is due to hydrogen bonding.

Solid N,N'-diphenylacetamidine in KBr pellets shows a broad NH vibration band from 3350 down to 2500 cm^{-1} with maxima around 3250 and 3050 cm^{-1}. This was taken to indicate the formation of a cyclic dimeric structure of type (63)[112]. No association was observed in the case of N,N'-diphenylbenzamidines which was explained by the steric overcrowding of the bulky phenyl groups[112] allowing no hydrogen bonding.

VI. NUCLEAR MAGNETIC RESONANCE SPECTRA

A. Proton Magnetic Resonance Spectra

Chemical shift data of the ^{1}H-n.m.r. spectra of formamidines are collected in Table 22. The formyl hydrogen signal occurs in the range

TABLE 22. ^{1}H-Nuclear magnetic resonance spectra of some formamidines and related compounds

Compound	R =	τ_{CH} (p.p.m.)		Hydro-chloride in D$_2$O	τ_{NCH_3} (p.p.m.)		Hydro-chloride in D$_2$O	Reference
		CDCl$_3$	C$_6$D$_6$		CDCl$_3$	C$_6$D$_6$		
H—C(=O)—OCH$_3$		1·92			6·23			128a
H—C(=O)—N(CH$_3$)$_2$		1·98			7·12 7·03			128b

TABLE 22 (continued)

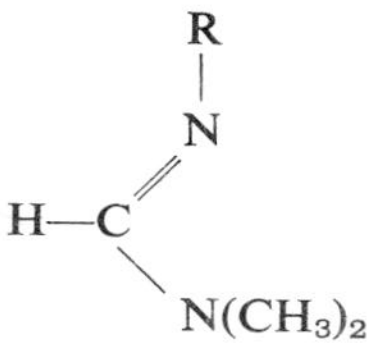

R—N=CH—N(CH₃)₂

R							Ref.
t-Bu	2·71			7·21			129
β-Phenylethyl	2·87	3·16	2·57	7·25	7·52	{6·99 / 7·11}	130
Phenyl	2·61	—ᵃ	—	7·17	7·50	—	130
p-Tolyl	2·50	2·80	1·88	7·10	7·46	{6·88 / 6·69}	130
p-Fluorophenyl	2·69	2·96	—	7·07	7·52ᵇ	—	130
p-Chlorophenyl	2·58	2·93	1·78	7·04	7·53ᵇ	{6·62 / 6·52}	130
p-Methoxyphenyl	2·78	2·95	1·88	7·19	7·50	{6·84 / 6·66}	130
p-Cyanophenyl	2·63	3·10	1·57	7·05	{7·68ᵇ / 7·40ᵇ}	{6·69 / 6·52}	130
p-Nitrophenyl	2·39	3·00	—	6·09	{7·70 / 7·37}	—	130

ᵃ Obscured by aromatic protons
ᵇ Broadened

between $\tau = 2\cdot39$ and $2\cdot87$ p.p.m., at higher field than in dimethyl-formamide ($1\cdot98$ p.p.m.) or in methyl formate ($1\cdot92$ p.p.m.). This value is shifted to lower field ($\tau = 1\cdot57 - 2\cdot57$) on protonation to formamidinium salts.

The amide-like dimethylamino group gives at room temperature rise to one signal around $7\cdot2$ p.p.m. at slightly higher field than in dimethyl-formamide. In deuterated benzene at room temperature the signal is broadened and occurs sometimes, in dependence on ring substituents, as a doublet[130]. In chloroform solution splitting is observed on cooling[131] indicating the magnetic nonequivalence of the two methyl groups due to restricted rotation around the amide-like CN bond.

In the corresponding amidinium salts in D_2O the signals of the dimethyl-amino group are shifted to lower field and occur always as a doublet due to restricted rotation already at room temperature.

The n.m.r. data of acetamidines and their salts are given in Table 23. The C-methyl group leads to a signal in the range from $\tau = 7\cdot90$ to $8\cdot16$ p.p.m. which is shifted on protonation to lower field ($7\cdot69$ p.p.m.). The protons of the N-methyl group absorb around $7\cdot2$ p.p.m. For N,N-dimethylacet-amidine in chloroform solution this signal was not split into a doublet on cooling to $-40°C$ indicating still rapid rotation around the C—N bond[132]. On protonation the signal is shifted to lower field and split into a doublet at room temperature. Due to the higher HMO π-bond order of amidinium salts ($p = 0\cdot595$) relative to amidines ($0\cdot520$), in the salts the rotation is already restricted at room temperature. N,N-Dimethyl-N'-aryl-substituted acetamidines[135] show the splitting of N-methyl signals in deuteroacetone solution in the temperature range from -30 to $-60°C$.

B. Geometrical Isomerism of Amidines and Amidinium Cations

I. Cis–trans isomerism of amidines

All structural types of amidines may show *cis–trans* isomerism with respect to the C=N double bond. But in no case was experimental evidence obtainable for the simultaneous occurrence of both forms[135].

(65) *Z*-form   **(66)** *E*-form

As in the case of aldimines[136] in the ^{1}H-n.m.r. spectra of N,N-dimethyl-N'-arylacetamidines (**67**) (from **65** or **66** with R = R″ = R‴ = CH₃; R' = aryl) no splitting of the C-methyl group signal in dependence on temperature was observed[135], indicating the absence of a temperature-dependent isomerization between (**65**) and (**66**). Only the freezing of rotation around the C—N bond may be seen on cooling. Whether (**65**) or (**66**) is the predominant and more stable form depends on the nature and of the steric requirement of the various substituents. The structural determination of formamidoxime and azo-bis(N-chloroformamidine) (see Section II, A) proves that these compounds exist in the Z-form (**63**). The same structure was suggested for N-aryl trichloroacetamidines[122]. However, dipole moment measurements of N,N-dimethylbenzamidine have been interpreted to show the occurrence of the E-form (**66**) with the phenyl ring twisted out of the plane of the amidine group[41]. The same configuration was also assigned[135] to the compounds (**67**).

2. Rotational isomerism with respect to the C—N single bond

The C—N single bond in amidines and amidinium salts has appreciable double bond character so that rotation around this bond is restricted. But since the energy of activation for this rotation (i.e. in N,N-dimethylbenzamidine 18·2 kcal/mol[45]) is below 23 kcal/mol no isolation of the corresponding isomers is possible[137, 139].

a. *N-Alkyl-substituted amidines and corresponding salts.* N-Ethyl- and N-benzyl-trichloroacetamidine show in the infrared spectrum the absorption bands of the primary amino group due to tautomer (**68**) and also a double absorption for the NH group of tautomer (**69**) which was attributed to the occurrence of the rotational isomeric forms (**69a**) and (**69b**)[122].

$$Cl_3C-C\overset{\displaystyle NH_2}{\underset{\displaystyle N-R}{\Big\langle}}$$

(68)

$$Cl_3C-C\overset{\displaystyle NH}{\underset{\displaystyle \underset{\displaystyle R}{\overset{\displaystyle |}{N-H}}}{\Big\langle}} \qquad Cl_3C-C\overset{\displaystyle NH}{\underset{\displaystyle \underset{\displaystyle H}{\overset{\displaystyle |}{N-R}}}{\Big\langle}}$$

(69a)   **(69b)**

Models show that isomer (**69b**) is less sterically hindered.

TABLE 23. ^{1}H—Nuclear magnetic resonance spectra of some acetamidines and their salts (in $CDCl_3$ or as stated).

Compound	R =	τ_{C-CH_3} (p.p.m.)	τ_{N-CH_3} (p.p.m.)	τ_{N-H} (p.p.m.)	Reference
$CH_3-C(=O)-N(CH_3)_2$		7·92	$\left.\begin{array}{c}7·06\\6·98\end{array}\right\}$		128c
$CH_3-C(=NH)-N(CH_3)_2$		7·90 (HCONH$_2$)	7·05 (HCONH$_2$)		132
$CH_3-C(NH_2)=\overset{\oplus}{N}(CH_3)_2$ Cl$^{\ominus}$		7·69 (H$_2$O)	$\left.\begin{array}{c}6·88\\6·75\end{array}\right\}$(H$_2$O)		132
$CH_3-C(NH_2)=\overset{\oplus}{N}(CH_3)(H)$ Cl$^{\ominus}$		7·61 (D$_2$O)	7·09 (D$_2$O)		133

TABLE 23 (continued)

Structure			Ref.
$\overset{CH_3}{\underset{CH_3}{}}N-H,\ CH_3-C^{\oplus},\ \overset{CH_3}{\underset{H}{}}N-CH_3\ \ Cl^{\ominus}$	$\left\{\begin{array}{l}7\cdot04\\6\cdot88\end{array}\right\}(D_2O);$ $\left\{\begin{array}{l}7\cdot14\\7\cdot06\end{array}\right\}(DMSO)$		133
		$\left\{\begin{array}{l}0\cdot76\\-0\cdot12\end{array}\right\}(DMSO)$	134
$CH_3-C^{\oplus}\overset{NH_2}{\underset{NH_2}{}}\ \ Cl^{\ominus}$	7·68 (D₂O)		84, 133
		$\left\{\begin{array}{l}1\cdot15\\0\cdot66\end{array}\right\}(DMSO)$	134
R—⬡—N=C(CH₃)N(CH₃)₂			
OCH₃	8·15	6·98	135
CH₃	8·15	6·97	135
H	8·17	7·01	135
Cl	8·16	6·98	135
COCH₃	8·15	6·97	135
NO₂	8·08	6·93	135

TABLE 24. Distribution of *cis–trans* isomers of *N*-methylacetamidinium salts in solution[138].

Compound	Solvent	(70a)	(70b)
(70), X = Cl	D_2O	96%	4%
(70), X = NO_3	D_2O	96%	4%
(70), X = NO_3	DMSO-d_6	97%	3%

For *N*-methylacetamidinium chloride and nitrate the isomer distribution between the *cis*- and *trans*-forms (70a) and (70b) as determined by n.m.r. spectroscopy[138] is given in Table 24. The Z-configuration (70a) is favoured very much over the E-isomer (70b), probably as the result of steric repulsion between the *cis*-methyl groups in (70b).

(70a) *Z*-isomer   (70b) *E*-isomer

b. *N,N'-Disubstituted acetamidinium cations.* In amidinium salts both CN bonds have equal HMO π-bond order, the value (0·595) being intermediate between that of the C=N double bond (0·789) and that of the amide-like single bond (0·520) of amidines. Consequently, on salt formation, the barrier of rotation around the CN bond is raised in amidinium cations whereas the activation energy of *cis–trans* isomerization of amidines, assuming a rotation mechanism, is lowered making both isomerization processes undistinguishable in the cation. In proton n.m.r. spectra of certain *N,N'*-disubstituted formamidinium trifluoroacetates in trifluoroacetic acid signals of both the *E,E*-isomer (71) and *E,Z*-isomer (72) in different ratios are observed[139].

The signals of the protons H^a to H^e can be assigned on the basis of line shape and coupling constants. (e.g. $J_{H^aH^b} \cong J_{H^cH^e} \cong 14$ c.p.s., $J_{H^dH^e} \cong 6$ c.p.s.; H^a, H^c and H^d are doublets broadened by quadrupole coupling; H^b is a triplet and H^e is a quartet.) Steric interaction between the Z-aryl and NH^c groups in (72) will force the aryl ring to orient so that H^c lies above the aryl-π system. Accordingly, H^c shows an upfield shift of about 1 p.p.m. relative to the H^a resonance because of the ring anisotropy.

Integration of appropriate n.m.r. signals provides a ready measure of

(71) E,E-isomer   (72) E,Z-isomer

$$R = \text{—} \bigcirc \text{—X} \quad \text{or} \quad t\text{-Bu}$$

the equilibrium concentrations of (71) and (72). In the case of R = t-Bu the large steric requirements of the t-butyl group force it to take up the E-position in all cases. The N,N'-di-t-butylformamidinium cation exists exclusively in the E,E-form of type (71).

The equilibrium constants collected in Table 25 show a striking aryl substituent effect. In the case of the symmetrically substituted salts (K), the E,Z-form (72) is favoured by an entropy factor which is not present in the N-aryl-N'-butyl formamidinium salts (K'). Nevertheless the substituent effect is similar in both series. Electron donating groups as well as ortho substituents favouring non-planarity stabilize the non-planar E,Z-structure (72). The coplanar E,E-structure (71) is stabilized by electron-attracting substituents. These results are contrary to predictions of resonance theory which lead to the conclusion that the coplanar form (71) should be stabilized by electron-donating substituents as for example the para-methoxy group. The possibility of an attractive N–H^c–π-interaction in the non-planar form (72) is discarded as in the N-2,6-dimethylaryl-N'-t-butylformamidinium salts the equilibrium is essentially independent of substituent effects in the 4-position. It appears likely that the E,Z-configuration (72) is stabilized by favourable dipolar interactions[139]. The equilibrium constants for meta- and para-substituted N-aryl formamidinium cations are correlated by a Hammett plot which yields ρ-values of -0.75 ($r = 0.917$) and -0.83 ($r = 0.989$) for the N,N'-diarylformamidinium (K) and the N-t-butyl-N'-arylformamidinium cations (K'), respectively.

C. Rotational Barriers

The substituents on the singly bound nitrogen in amidines and related compounds are magnetically nonequivalent. Therefore the measurement of temperature dependence of n.m.r. spectra allows the determination

TABLE 25. Equilibrium constants^a for *cis–trans* isomerization for *N,N'*-diarylformamidinium $\left(K = \dfrac{(72)}{(71)},\ R = \text{aryl}\right)$ and *N*-aryl-*N'*-*t*-butylformamidinium trifluoroacetates[139] $\left(K' = \dfrac{(72)}{(71)},\ R = \text{t-Bu}\right)$

Substituent (X)	K	K'	Substituent (X)	K	K'
H	1·00	0·48	4-Methoxy	1·90	1·00
3-Acetyl	0·65	0·28	2-Methyl	2·50	2·20
4-Acetyl	0·25	0·09	3-Methyl	1·10	0·62
3-Trifluoromethyl	0·65	0·25	4-Methyl	1·40	0·68
4-Trifluoromethyl	0·40	0·20	2,4-Dimethyl	3·0	2·1
3-Chloro	0·65	0·38	2,6-Dimethyl	large^b	—
4-Chloro	0·85	0·32	2,4,5-Trimethyl	large^b	2·4

^{a} N.m.r. spectra of 10 mol-% solute in trifluoroacetic acid were determined with a Varian A-60 instrument at $38 \pm 2°C$ for the diaryl series and with a HA-60 instrument for the *N*-aryl-*N'*-*t*-butyl series. Equilibrium constants are estimated to be accurate to ± 0.02 units for the aryl, *t*-butyl series and ± 0.10 units for the diaryl series.
^{b} No *E,E*-isomer (71) could be detected ($K \geq 20$).

of coalescence temperature and activation energy parameters for rotation around the C—N single bond. For unsubstituted acetamidinium chloride in DMSO the activation energy lies in the range from 9 to 25 kcal/mol[133].

I. *N,N*-Dimethyl-substituted amidines

In compounds of type (73) the activation energy for rotation of the dimethylamino group should be related either to the π-bond order as a ground state property or to the loss of π-electron energy ΔE_π (equation 20)

$$R-C \overset{\displaystyle X}{\underset{\displaystyle N(CH_3)_2}{<}} \qquad X = S, O, \overset{\oplus}{N}H_2, NR' \quad (73)$$

which is a measure of the energy of the transition state in which the

$$\Delta E_\pi = E_{\pi,\text{RCXNH}_2} - (E_{\pi,\text{RCX}} + 2\alpha_N) \tag{20}$$

dimethylamino group is rotated by 90 degrees out of the molecular plane allowing no π-electron interaction between the two parts of the functional group. The HMO properties calculated by Sandström[142] are collected in Table 26.

TABLE 26. HMO-Properties of compounds of type (73)[142].

X	p_{CN}	ΔE_π (in β units—see equation 20)
S	0·455	0·636
$\overset{\oplus}{N}H_2$	0·484	0·632
O	0·422	0·532
NR	0·363	0·449

The difference in π-electron energy predicts for the height of the rotational barrier the sequence $S > \overset{\oplus}{N}H_2 > 0 > NR$ for compounds of type (73). The experimental results summarized in Table 27 confirm the predicted sequence. The calculated HMO π-bond order is lower for the thioamide than for the amidinium group in contrast to experimental findings, so that this ground state property is not so well suited for comparison with experimental activation energies. But both properties as well as the experimental data show that the rotational barrier is higher in amidinium cations than in amidines.

G. Häfelinger

TABLE 27. Activation parameters[a] for rotation around CN—bonds with partial double bond character

Compound	R =	Solvent	T_c (°C)	E_a (kcal/mol)	log A	$\Delta G^{\ddagger}$ (kcal/mol)	$\Delta H^{\ddagger}$ (kcal/mol)	$\Delta S^{\ddagger}$ (e.u.)	Ref.
S‖ R—C—N(CH₃)₂	H	neat		27·9	13·6		27·3		140
	CD₃	DMSO-d_6		25·9 ± 0·9	14·6 ± 0·5	23·4	25·3 ± 0·9	+6·3 ± 2·1	138
O‖ R—C—N(CH₃)₂	H	neat	119	20·5 ± 0·2	12·7	21·0	19·9 ± 0·2	−1·7	141
	CD₃	DMSO-d_6		20·3 ± 0·3	14·1 ± 0·2	18·5	19·7 ± 0·3	+4·1 ± 0·8	138
NH‖ R—C—N(CH₃)₂	C₆H₅	CDCl₃	−53	18·2			17·6		45
NH₂ ⊕‖ R—C X⊖ —N(CH₃)₂	CH₃, X = Cl	Formamide		19·6 ± 1·0	12·7 ± 0·5		19·0		132
	CD₃, X = Cl	DMSO-d_6		22·8 ± 0·7	13·5 ± 0·4	21·8	22·2 ± 0·7	+1·4 ± 1·9	138
	CD₃, X = NO₃	DMSO-d_6		21·3 ± 0·3	12·7 ± 0·2	21·5	20·7 ± 0·3	−2·6 ± 0·7	138

[a] $\log k_r = -\dfrac{E_a}{2\cdot303RT} + \log A$; $\Delta G^{\ddagger} = -2\cdot303\,RT\log\left[\dfrac{hk_r}{kT}\right]$; $\Delta H^{\ddagger} = E_a - RT$; $\Delta S^{\ddagger} = \dfrac{\Delta H^{\ddagger} - \Delta G^{\ddagger}}{T}$; T_c = coalescence temperature.

2. Trisubstituted amidines

The free energies of activation for rotation around the C—N single bond in trisubstituted amidines collected in Table 28 lie in the range from 11 to 16 kcal/mol. N'-t-Butyl-N,N-dimethyl formamidine has a lower value than the corresponding N'-aryl-derivatives in which the free energy of activation is raised by electron attracting substituents[131]. The same substituent effect is observed with N'-aryl-substituted benzamidines[144] showing the dependence on the increase in π-bond order as indicated by the mesomeric forms (74a) to (74c).

(74a)   (74b)   (74c)

The activation parameters for N'-p-nitrophenyl-N,N-dimethyl formamidine are quite different in chloroform and benzene solution. The relatively large enthalpy as well as the large positive entropy of activation in benzene solution indicate that the solvent may stabilize the ground state by some specific interaction (possibly with the nitro group) which is relaxed in the transition state[130, 131].

In trisubstituted benzamidinium cations[143] the free energy of activation for rotation is raised to 20·4 kcal/mol, as against 12–13 kcal/mol in the corresponding amidines. A possible example of the existence of stable rotational isomers was reported by Raison[145], who found that N,N,N'-trimethyl-N'-phenylbenzamidinium iodide could be obtained in two states with distinct melting points. But as the n.m.r. spectra of both forms are identical, this must be a case of crystalline modifications[143]. The barrier of rotation about the C—$N(CH_3)_2$ bond is 14·2 kcal/mol which is raised with carboxylic acids as solvent, probably as a result of differences in association or solvation.

D. Heteronuclear Magnetic Resonance

1. Carbon-13 nuclear magnetic resonance spectra

The ^{13}C-n.m.r. spectra collected in Table 29 show that the central carbon atom in amidinium cations resonates at higher field than in the corresponding amides or carboxylic acids. Connected carbon atoms are also shifted in the same direction. In benzamidine the central carbon atom

TABLE 28. Activation parameters for rotation around CN bonds in trisubstituted amidines

Compound	R	Solvent	T_c (°C)	E_a (kcal/mol)	log A	$\Delta G^{\ddagger}$ (kcal/mol)	$\Delta H^{\ddagger}$ (kcal/mol)	$\Delta S^{\ddagger}$ (e.u.)	Ref.
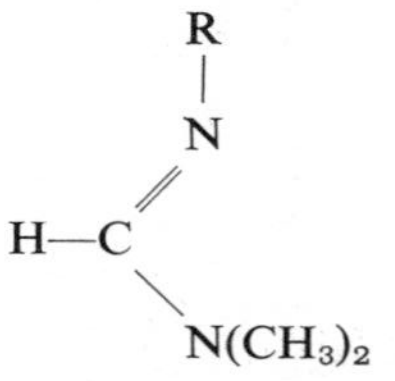	t-Bu	Toluene-d_8		13.0 ± 0.7	13.6	11.9 ± 0.6	12.4 ± 0.7	1.6 ± 2.8	129
	t-Bu	Pyridine-d_5	−48	11.0 ± 0.7	12.5	12.4 ± 0.8	10.8 ± 0.7	-5.4 ± 2.7	129
	4-Nitro-phenyl	CHCl$_3$		10.9		15.9 ± 0.3	10.3 ± 0.7	-20 ± 3	131
	4-Nitro-phenyl	Benzene		22.8		15.2 ± 0.6	22.2 ± 1.0	$+21 \pm 3$	131
	4-Tolyl	CHCl$_3$		12.7		14.1 ± 0.2	12.1 ± 0.3	-8 ± 1	131
	4-Tolyl	Benzene		13.2		13.9 ± 0.2	12.6 ± 1.3	-4 ± 3	131
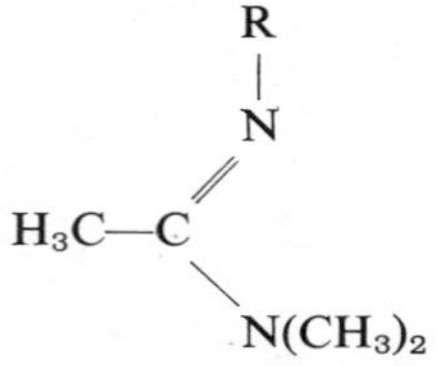	4-Methoxy-phenyl	Acetone-d_6	−60			11.3			135
	4-Tolyl	Acetone-d_6	−55			11.6			135
	4-Phenyl	Acetone-d_6	−53			11.65			135
	4-Chloro-phenyl	Acetone-d_6	−45			12.1			135
	4-Acetyl-phenyl	Acetone-d_6	−41			12.4			135
	4-Nitro-phenyl	Acetone-d_6	−33			12.8			135

R—N=CH—N(CH$_3$)$_2$

R—N=C(CH$_3$)—N(CH$_3$)$_2$

TABLE 28 (continued)

C_6H_5—C(=N—R)—N(CH₃)₂

R	Solvent					Ref
Benzyl	CH₂Cl₂	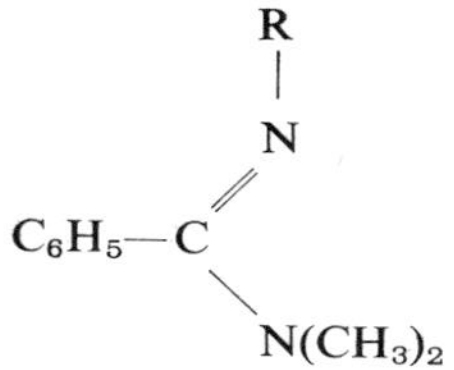 −40		12·0		143
Benzyl, Hydrochloride	PhNO₂	130		20·4		143
Phenyl	CH₂Cl₂	−16		13·0		143
Phenyl, Hydrochloride	PhNO₂	135	20·8	20·4	20·0	143
Benzoyl	PhCl	8	15·2	13		45
Benzenesulphonyl	PhCl	70	16·4			45
Diphenoxyphosphoryl	PhCl	83	17·6			45

C_6H_5—C(=N—R)—N(i-Pr)₂

R	Solvent					Ref
Phenyl	CDCl₃	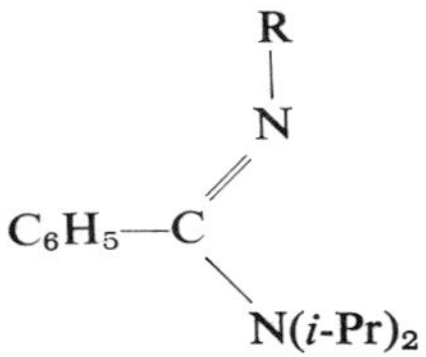 −13·9	12·6	12·6 ± 0·8	c.0	144
4-Nitrophenyl	CDCl₃	10·1	13·9	13·3 ± 1·0	c.0	144

TABLE 28 (continued)

Compound	R	Solvent	T_c (°C)	E_a (kcal/mol)	log A	$\Delta G^{\ddagger}$ (kcal/mol)	$\Delta H^{\ddagger}$ (kcal/mol)	$\Delta S^{\ddagger}$ (e.u.)	Ref.
	Phenyl, X = H	CDCl₃	−21·9			12·1	12·3 ± 1·2	c.0	144
	Phenyl, X = OCH₃	CDCl₃	−30·4			11·7	8·9 ± 0·6	11 ± 2	144
	4-Nitro-phenyl, X = H	CDCl₃	−4·8			13·1	17·7 ± 1·7	17 ± 6	144
	4-Methoxy-phenyl, X = H	CDCl₃	−17·0			12·3	16·2 ± 1·2	15 ± 5	144
	X = J	CH₂Cl₂	1			14·2			143
	X = J	CF₃COOH	63			17·8			143
	X = J	HCOOH	54			17·5			143
	X = BF₄	CH₂Cl₂	31			15·6			143

TABLE 29. ¹³C-Nuclear magnetic resonance spectra of amidines and related compounds

Compound	Solvent	Chemical shift-δ relative to TMS (p.p.m.)					Reference
		C^1	C^2	C^{ortho}	C^{meta}	C^{para}	
$\overset{1}{H}COOH$	H_2O	166·3					146
$\overset{1}{H}COO^{\ominus}NH_4^{\oplus}$	H_2O	171·4					146
$\overset{1}{H}CONH_2$	H_2O	167·6					147
$\overset{1}{H}C(NH_2)_2^{\oplus}Cl^{\ominus}$	D_2O	157·3					84ᵃ
$\overset{2}{H_3}C\overset{1}{-}COOH$	H_2O	177·2	21·1				146
$\overset{2}{H_3}C\overset{1}{-}COO^{\ominus}NH_4^{\oplus}$	H_2O	181·7	24·0				146
$\overset{2}{H_3}C\overset{1}{-}CONH_2$	H_2O	178·1	22·3				147
$\overset{2}{H_3}C\overset{1}{-}C(NH_2)_2^{\oplus}Cl^{\ominus}$	D_2O	168·5	18·6				84ᵃ
(phenyl)$\overset{2}{}\overset{1}{-}COOH$	CCl_4	174·9	130·6	130·0	128·5	133·6	148
(phenyl)$\overset{2}{}\overset{1}{-}C(NH_2)_2^{\oplus}Cl^{\ominus}$	D_2O	166·2	127·1	128·4	127·5	134·4	84ᵃ
(phenyl)$\overset{2}{}\overset{1}{-}C(=NH)NH_2$	$CDCl_3$	164·6	126·9	128·5	124·3	140·4	84ᵃ

ᵃ The ¹³C-n.m.r. spectra have been obtained by a Bruker HFX 90 spectrometer operating at 22·62 MHz. The accumulation of spectral was performed with the time-average instrument of the Fabri-Tek Company and the Fourier-transformation with a PDP 8-I computer of the Digital Company. As internal standard and lock signal TMS was used in $CDCl_3$ solution and 1,4-dioxane in D_2O solution. The temperature of the measurements was 27°C.

G. Häfelinger

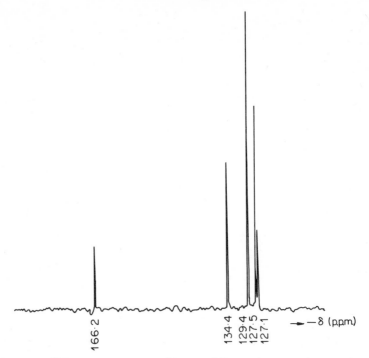

FIGURE 7. ^{13}C-n.m.r. spectrum of benzamidine hydrochloride in D_2O[84].

absorbs at higher field than in the benzamidinium cation, the spectrum of which is shown in Figure 7. The assignments of the peaks are given in Table 29. There is an uncertainty regarding the assignment of the absorption of the single C-atoms C^2 and C^{para} which might be reversed. The assignment chosen gives agreement with that of benzoic acid[148].

2. ^{15}N—H Coupling constants.

The isotope ^{15}N has no nuclear quadrupole moment and a nuclear spin of 1/2 which leads to spin–spin coupling with adjacent hydrogen atoms. The ^{15}N—C—H coupling through a sp^3-hybridized carbon atom is small (only 0·6 c.p.s. in [^{15}N]benzalmethylamine[149]) but the ^{15}N=C—H coupling through a sp^2-hybridized carbon atom is larger. In N,N-dimethyl-$^{15}N'$-phenylformamidine[150] (75) the coupling constant is 2·4 c.p.s., whereas in [^{15}N]benzalaniline[150] the corresponding coupling constant is 3·8 ± 0·1 c.p.s. Large ^{15}N—C—H coupling constants around 8·4 c.p.s. have been observed in $^{15}N, ^{15}N$-dimethyl-N-arylformamidines (76).

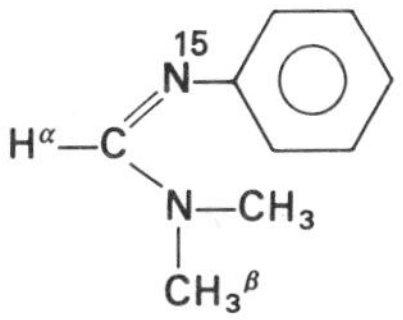

$$J_{15N=C-H^\alpha} = 2\cdot4 \pm 0\cdot1 \text{ cps}$$
$$J_{15N-CH_3{}^\beta} = 0$$

(75)

Substitution by the electron-donating p-methoxy group (**76b**) does not affect the magnitude of the ^{15}N—C—H coupling constant, but the electron attracting p-nitro group (**76c**) leads to a decrease of the coupling constant. The comparison of $J_{15_{NCH}}$ in formamides[151–153] (15–19 c.p.s. with that of formamidines (7·5–8·4 c.p.s.) indicates that the magnitude of ^{15}N spin coupling with a neighbouring proton on a sp^2 hybridized α-carbon atom varies directly with the electronegativity of the atom linked to carbon through a double bond[150].

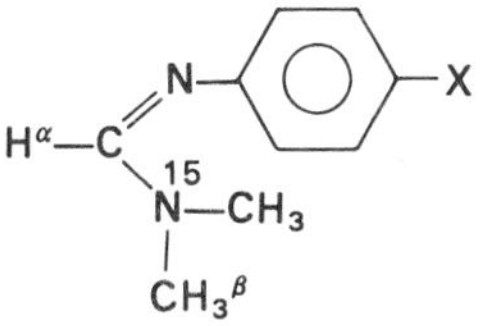

(74)

(76a),   X = H: $J_{15NCH^\alpha} = 8\cdot4 \pm 0\cdot1$ c.p.s.
$$J_{15NCH^\beta} = 0$$

(76b), X = OCH$_3$: $J_{15NCH^\alpha} = 8\cdot3 \pm 0\cdot1$ c.p.s.
$$J_{15NCH^\beta} = 0$$

(76c),   X = NO$_2$: $J_{15NCH^\alpha} = 7\cdot5 \pm 0\cdot1$ c.p.s.
$$J_{15NCH^\beta} = 0$$

VII. MASS SPECTRA

The mass spectrum of N,N-dimethyl-N'-phenyl formamidine[154, 155] presented in Figure 8 shows an intense molecular ion M$^+$ peak and also a significant [M—1]$^+$ peak. Deuterium labelling demonstrated that one of the *ortho* hydrogen atoms of the phenyl group is lost. The fragmentation scheme[158] (Scheme 1) explains this fact by the formation of benzimidazolium ions (**77**).

Both the [M—H]$^+$ and the [M—CH$_3$]$^+$ ion may lose an HCN molecule forming ions of mass 120 and 106, respectively. ^{15}N-labelling showed[154]

G. Häfelinger

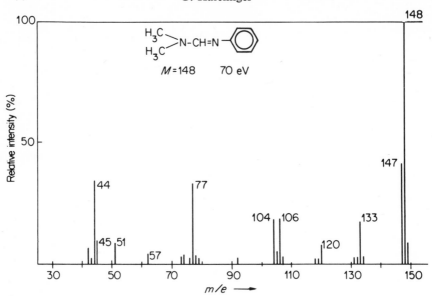

FIGURE 8. Mass spectrum of *N,N*-dimethyl-*N'*-phenyl formamidine[155]. (From Grützmacher and Kuschel, *Org. Mass Specrometry*, **3**, 605 (1970), with permission.)

that exclusively the N-atom of the amino group is removed from the $[M—H]^+$ ion. This is understandable since in this ion an HCN molecule is already preformed. In the $[M—CH_3]^+$ ion, whose structure is not known precisely, either the nitrogen of the imino group or the nitrogen of the amino group may be eliminated. This was explained by a migration of the methyl group in the $[M—CH_3]^+$ ion before the loss of HCN[154]. However deuterium labelling shows that a part of the ions with $m/e = 106$ are formed not only by elimination of HCN from the $[M—CH_3]^+$ ion but also by elimination of CH_3CN from the $[M—H]^+$ ion (**78**) which is formed by methyl group migration[155].

The mechanism of the cyclization reaction leading to the $[M—H]^+$ ion was further investigated by a study of the effect of substituents at the phenyl group on the appearance potential and the intensity of the $[M—H]^+$ ion as well as on the ionization potential of the molecular ions.

The intensity of the $[M—H]^+$ ions is reduced by *p*-hydroxy and *p*-methoxy substituents and to a lesser extent by *p*-methyl and *p*-chloro groups. An increase in intensity is only observed with *m*-carbomethoxy, *m*-acetyl and *m*-chloro substituents whereas other substituents show only a slight change in intensity. The appearance potential is only slightly affected

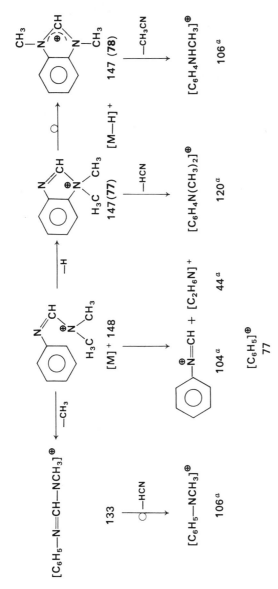

[a]Formula proved by peak matching

SCHEME 1. Fragmentation scheme of the N,N-dimethyl-N'-phenylformamidine radical cation (From Grützmacher and Kuschel, *Org. Mass Spectrom.*, 3, 605 (1970), with permission.)

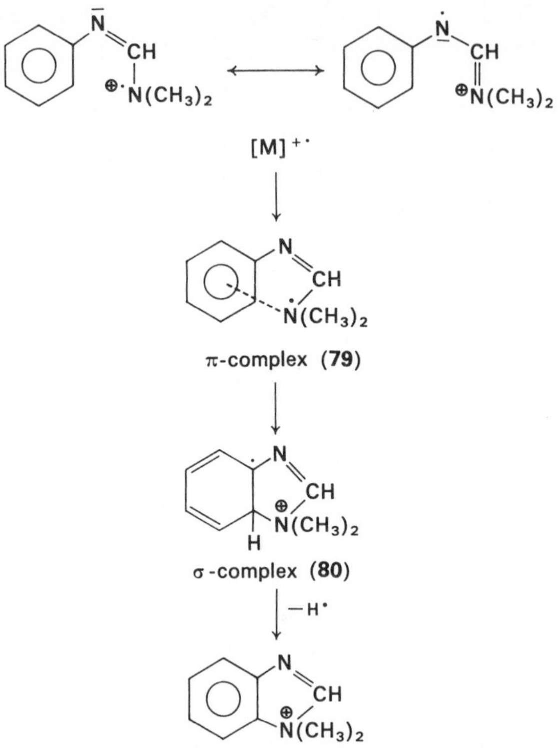

[M]$^{+\cdot}$

π-complex (**79**)

σ-complex (**80**)

$-H^{\cdot}$

[M—H]$^{+}$ (**77**)

SCHEME 2. Cyclization to produce [M—H]$^{+}$ ions in the mass spectrum of
N,N-dimethyl-N′-phenylformamidine. (From Grützmacher and Kuschel,
Org. Mass Spectrometry, **3**, 605 (1970), with permission.)

by different substituents. This substituent effect is explained by the proposed cyclization mechanism as depicted in Scheme 2. The cyclization
starts with the formation of a π-complex (**79**) which presumes the presence
of a positive charge centred at the amino nitrogen. However, electron
donating substituents in the *para* position favour electron distributions
with the positive charge centred in the aromatic part of the molecule as
indicated by the resonance forms (**81b**) and (**81c**). Therefore these substituents restrict the formation of the π-complex and lead to a reduction of
the intensity of the [M—H]$^{+}$ ion.

   In the second step a σ-complex (**80**) is formed which is similar to the
σ-complex of electrophilic aromatic substitution but in this case the positive
charge remains outside the aromatic ring, so that the appearance potentials are only slightly changed by different substituents. The hydrogen is

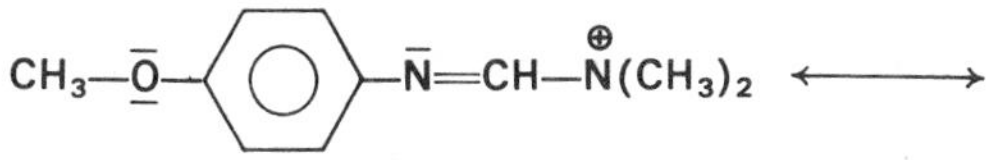

(81a)

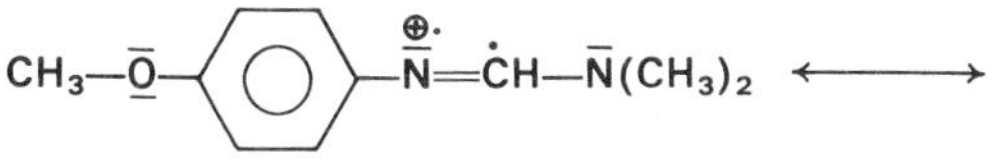

(81b)

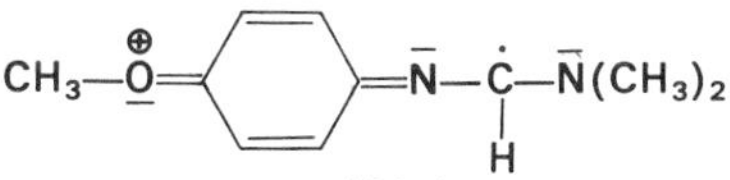

(81c)

lost from (78) by homolytic bond scission which is also not very much influenced by polar substituent effects.

The increase in intensity of the [M—H]$^+$ ion with m-carbomethoxy, m-acetyl and m-chloro substituents is due to the ability of these groups to accept the radical electron in the mesomeric form (80b) which may lead to easy abstraction of the hydrogen atom to yield (82).

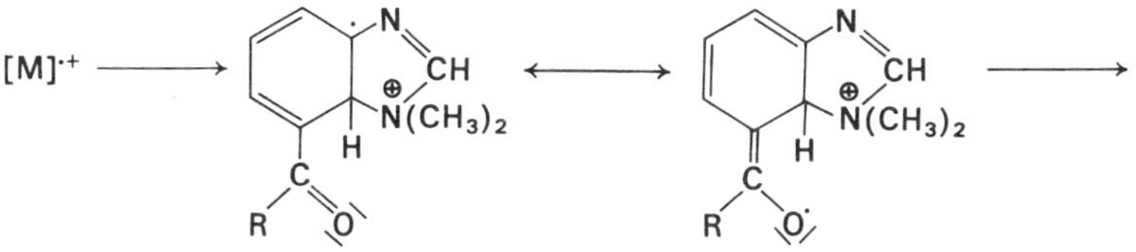

(80a)   (80b)

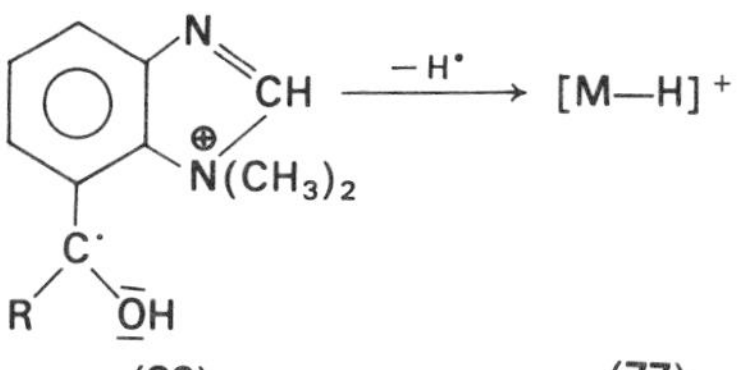

(82)   (77)

VIII. REFERENCES

1. *The Chemistry of Carboxylic Acids and Esters*, (Ed. S. Patai), Interscience Publ., London, 1969.
2. *The Chemistry of the Carbon–Nitrogen Double Bond*, (Ed. S. Patai), Interscience Publ., London, 1970.
3. *The Chemistry of Amides*, (Ed. J.Zabicky), Interscience Publ., London, 1972.

4. A. Kotera, T. Morita, S. Aoyagi, Y. Kakiuchi, S. Nagakura, and K. Kume, *Nippon Kagaku Zassi*, **82**, 302 (1961); *Chem. Abstr.*, **55**, 15128e (1961).
5. Z. B. Kiro, Yu. A. Teterin, L. N. Nikolenko, and B. I. Stepanov, *Zh. Org. Khim.*, **8**, 2573 (1972); *Chem. Abstr.*, **78**, 15210z (1973).
6. A. Hantzsch, *Ber. Deutsch. Chem. Ges.*, **63**, 1782 (1930).
7. M. E. Runner, M. L. Kilpatrick, and E. C. Wagner, *J. Am. Chem. Soc.*, **69**, 1406 (1947).
8. R. C. Neuman, Jr. and G. S. Hammond, *J. Phys. Chem.*, **67**, 1659 (1963).
9. R. Delaby, J. V. Horisse and S. H. Renard, *Bull. Soc. Chim. Fr.*, **11**, 227 (1944).
10. R. L. Shriner and F. W. Neumann, *Chem. Rev.*, **35**, 351 (1944).
11. E. F. Rosenberg, *Ann. Intern. Med.*, **25**, 832 (1946).
12. A. G. Jerez, *Farmacother. Actual* (*Madrid*), **4**, 702 (1947); *Chem. Abstr.* **42**, 3076 (1948).
13. F. N. Fastier, *Pharmacol. Revs.*, **14**, 37 (1962).
14. E. Hieke, *Pharmazie*, **18**, 653 (1963).
15. A. Kreuzberger, *Progr. Drug. Res.*, **11**, 356 (1968).
16. C. A. Grob and B. Fischer, *Helv. Chim. Acta.*, **38**, 1794 (1955).
17. R. Roger and D. G. Neilson, *Chem. Rev.*, **61**, 1794 (1961).
18. D. Hall and F. J. Llewellyn, *Acta Cryst.*, **9**, 108 (1956).
19. D. Hall, *Acta Cryst.*, **18**, 955 (1965).
20. G. Häfelinger, *Chem. Ber.*, **103**, 3370 (1970).
21. G. Häfelinger, *Chem. Ber.*, **103**, 2902 (1970).
22. L. E. Sutton, *Tables of Interatomic Distances and Configuration in Molecules and Ions*, Special Publications Nos. 11 (1958) and 18 (1965) of the Chemical Society, Burlington House, London.
23. N. C. Stephenson, *J. Inorg. Nucl. Chem.*, **24**, 801 (1963).
24. J. H. Bryden, *Acta Cryst.*, **11**, 158 (1958).
25. J. J. Daly, *J. Chem. Soc.*, *Dalton Trans.*, 1048 (1972).
26. F. Sonz and J. J. Daly, *J. Chem. Soc.*, *Dalton Trans.*, 2267 (1972).
27. M. Sax, J. Pletcher, C. S. Yoo, and J. M. Stewart, *Acta Cryst.*, **B27**, 1635 (1971).
28. R. V. G. Sundra-Rao, J. W. Turley, and R. Pepinsky, *Acta Cryst.*, **10**, 435 (1957).
29. J. E. Worsham, Jr. and W. R. Busing, *Acta Cryst.*, **B25**, 572 (1969).
30. S. Harkema and D. Feil, *Acta Cryst.*, **B 25**, 589 (1969).
31. J. N. Brown and E. A. Meyers, *Acta Cryst.*, **B 26**, 1178 (1970).
32. D. Feil and W. Song Loong, *Acta Cryst.*, **B 24**, 1334 (1968).
33. C. H. Stam, *Acta Cryst.*, **15**, 317 (1962).
34. O. Kennard and J. Walker, *J. Chem. Soc.*, 5513 (1963).
35. C. Villa, G. Manfredotti, M. Nardelli, and U. Tani, *Acta Cryst.*, **B 28**, 356 (1972).
36. H. Henke and H. Baeringhausen, *Acta Cryst.*, **B 28**, 1100 (1972).
37. D. J. Haas, D. R. Harris, and H. H. Mills, *Acta Cryst.*, **19**, 676 (1965).
38. B. J. B. Schein, E. C. Lingafelter, and J. M. Stewart, *J. Chem. Phys.*, **47**, 5183 (1967).

39. C. Pigenet and H. Lumbroso, *Bull. Soc. Chim. Fr.*, 3743 (1972).
40. H. Lumbroso, C. Pigenet, H. Rosswag, and G. Schwenker, *C. R. Acad. Sci. Paris, Ser. C*, **266**, 1479 (1968).
41. H. Lumbroso and G. Pifferi, *Bull. Soc. Chim. Fr.*, 3401 (1969).
42. O. A. Osipov, A. M. Simonov, V. I. Minkin, and A. D. Garnovskii, *Dokl. Akad. Nauk. SSR*, **137**, 1374 (1961); *Chem. Abstr.*, **55**, 24173f (1961).
43. H. Lumbroso, C. Pigenet, and P. Reynaud, *C. R. Acad. Sci. Paris, Ser. C*, **264**, 732 (1967).
44. J. E. Blackwood, C. L. Gladys, K. L. Loening, A. E. Petrarca, and J. E. Rush, *J. Am. Chem. Soc.*, **90**, 509 (1968).
45. G. Schwenker and H. Rosswag, *Tetrahedron Lett.*, 2691 (1968).
46. H. Lumbroso, D. M. Bertin, and P. Reynand, *C. R. Acad. Sci. Paris, Ser. C*, **261**, 399 (1965).
47. W. D. Kumler and C. W. Porter, *J. Am. Chem. Soc.*, **56**, 2549 (1934).
48. H. Lumbroso and D. M. Bertin, *Bull. Soc. Chim. Fr.*, 1728 (1970).
49. O. Exner and O. Schindler, *Helv. Chim. Acta*, **55**, 1921 (1972).
50. O. Exner, V. Jehlicka, and A. Reiser, *Coll. Czech. Chem. Commun.*, **24**, 3207 (1959).
51. R. M. Moriarty, C.-L. Yeh, K. C. Ramey, and P. W. Whitehurst, *J. Am. Chem. Soc.*, **92**, 6360 (1970).
52. M. Kandel and E. H. Cordes, *J. Org. Chem.*, **32**, 3061 (1967)
53. R. P. Bell, *The Proton in Chemistry*, Cornell University Press, Ithaca, New York, 1959, p. 65.
54. N. F. Hall and M. R. Sprinkle, *J. Am. Chem. Soc.*, **54**, 3469 (1932).
55. G. Schwarzenbach and K. Lutz, *Helv. Chim. Acta.*, **23**, 1162 (1940).
56. A. Albert, G. Goldacre, and J. Phillips, *J. Chem. Soc.*, 2240 (1948).
57. J. N. Baxter and J. Cymerman-Craig, *J. Chem. Soc.*, 1490 (1953).
58. J. Goerdeler, D. Krause-Loevenich, and B. Wedekind, *Ber. Chem. Ges.*, **90**, 1638 (1957).
59. E. Lorz and R. Baltzly, *J. Am. Chem. Soc.*, **71**, 3992 (1949).
60. R. H. Hartigan and J. B. Cloke, *J. Am. Chem. Soc.*, **67**, 709 (1945).
61. J. A. Smith and H. Taylor, *J. Chem. Soc. B*, 66 (1969).
62. J. Cymerman-Craig, M. J. Parker, and P. Woodhouse, *J. Chem. Soc.*, 3050 (1953).
63. D. C. Prevorsek, *J. Phys. Chem.*, **66**, 769 (1962).
64. E. Hückel, *Z. Phys.*, **70**, 204 (1931).
65. A. Streitwieser, Jr. *Molecular Orbital Theory for Organic Chemists*, J. Wiley and Sons, New York, 1961.
66. E. Heilbronner and H. Bock, *Das HMO-Modell und seine Anwendung*, Verlag Chemie, Weinheim/Bergstr, 1968.
67. C. A. Coulson and E. T. Stewart, *The Chemistry of Alkenes*, (Ed. S. Patai), Intersience Publishers, London, 1954, p. 106.
68. G. Häfelinger, *Tetrahedron*, **27**, 1635 (1971).
69. G. Häfelinger, *Chem. Ber.*, **103**, 2941 (1970).
70. G. Häfelinger, *Tetrahedron*, **27**, 4609 (1971).
71. R. Pariser, *J. Chem. Phys.*, **21**, 568 (1953).
72. R. Pariser and R. G. Parr, *J. Chem. Phys.*, **21**, 466, 767 (1953).

73. J. A. Pople, *Trans. Faraday Soc.*, **49**, 1375 (1953).
74. R. G. Parr, *The Quantum Theory of Molecular Electronic Structure*, W. A. Benjamin, New York, 1966.
75. L. Salem, *The Molecular Orbital Theory of Conjugated Systems*, W. A. Benjamin, New York, 1966.
76. M. J. S. Dewar, *The Molecular Orbital Theory of Organic Chemistry*, McGraw-Hill, New York, 1969.
77. K. Nishimoto and N. Mataga, *Z. Phys. Chem. NF.*, **13**, 140 (1957).
78. M. J. S. Dewar and T. Morita, *J. Am. Chem. Soc.*, **91**, 796 (1969).
79. M. J. S. Dewar and N. Trinajstić, *J. Am. Chem. Soc.*, **92**, 1453 (1970).
80. J. E. Bloor and B. R. Gilson, *Can. J. Chem.*, **43**, 1761 (1965).
81. T. Koopmans, *Physica*, **1**, 104 (1933).
82. K. B. Everard and E. L. Sutton, *J. Chem. Soc.*, 2318 (1949).
83. R. J. Karland and E. B. Wilson, *J. Chem. Phys.*, **27**, 585 (1957).
84. G. Häfelinger and G. Westermayer, unpublished results.
85. J. C. Gage, *J. Chem. Soc.*, 221 (1949).
86. S. F. Mason, *J. Chem. Soc.*, 2071 (1954).
87. C. Sandorfy in reference 2, p. 47.
88. R. Bramston-Cook and J. O. Erickson, *Varian Instrument Applications*, **7**, 6 (1973).
89. J. Ševčik, *Monatsh. Chem.*, **100**, 1307 (1969).
90. W. Bradley and I. Wright, *J. Chem. Soc.*, 640 (1956).
91. Z. B. Kiro, Y. A. Teterin, L. N. Nikolenko, and B. I. Stepanov, *Zh. Org. Khim.*, **8**, 2573 (1972); *Chem. Abstr.*, **78**, 135210z (1973).
92. L. Doub and J. M. Vandenbelt, *J. Am. Chem. Soc.*, **69**, 2714 (1947).
93. J. Petruska, *J. Chem. Phys.*, **34**, 1120 (1961).
94. P. Brocklehurst, *Tetrahedron*, **18**, 299 (1962).
95. G. Häfelinger, *Dissertation*, Universität Tübingen, 1965.
96. Z. B. Kiro and B. I. Stepanov, *Zh. Org. Khim.*, **8**, 2402 (1972); *Chem. Abstr.*, **78**, 110018e (1973).
97. P. Grammaticakis, *C.R. Acad. Sci. Paris*, **245**, 2307 (1957).
98. Mme Ramart-Lucas and M. M. Grunfeld, *Bull. Soc. Chim. Fr.*, **4**, 478 (1937).
99. M. Kandel and E. H. Cordes, *J. Org. Chem.*, **32**, 3061 (1967).
100. A. Hantzsch, *Ber. Deutsch. Chem. Ges.*, **64**, 661 (1931).
101. M. Davies and A. E. Parsons, *Chem. Ind. (Lond.)*, 628 (1958).
102. M. Davies and A. E. Parsons, *Z. Phys. Chem. NF*, **20**, 34 (1959).
103. R. Mecke and W. Kutzelnigg, *Spectrochim. Acta*, **16**, 1216 (1960).
104. D. N. Shigorin, *Zhur. Fiz. Khim.*, **25**, 789 (1951); *Chem. Abstr.*, **74**, 5254d (1953).
105. L.Beckmann, L. Gutjahr, and R. Mecke, *Spectrochim. Acta*, **20**, 1295 (1964).
106. C. Sandorfy in reference 2, p. 37.
107. M. T. Leplawy, D. S. Jones, G. W. Kenner, and R. C. Sheppard, *Tetrahedron*, **11**, 39 (1960).
108. J. Fabian and M. Legrand, *Bull. Soc. Chim. Fr.*, 1461 (1956).
109. J. Fabian, M. Legrand, and P. Poirier, *Bull. Soc. Chim. Fr.*, 1499 (1956).
110. J. Fabian, V. Delaroff, and M. Legrand, *Bull. Soc. Chim. Fr.*, 287 (1956).

111. D. N. Shigorin and Y. K. Syrkin, *Zh. Fiz. Khim.*, **23**, 241 (1949); *Chem. Abstr.*, **43**, 6081h (1949).
112. P. Sohár, *Acta Chim. Sci. Hung.*, **54**, 91 (1967).
113. D. N. Shigorin and Y. K. Syrkin, *Izvest. Akad. Nauk S.S.S.R. Ser. Fiz.*, **9**, 225 (1945); *Chem. Abstr.*, **40**, 1831 (1946).
114. J. D. S. Goulden, *J. Chem. Soc.*, 997 (1953).
115. K. W. F. Kohlrausch and R. Seka, *Z. Phys. Chem.*, **B 38**, 72 (1937).
116. W. Seeliger, E. Aufderhaar, W. Diepers, R. Feinauer, R. Nehring, W. Thier, and H. Hellmann, *Angew. Chem.*, **78**, 913 (1966).
117. W. Otting and F. Drawert, *Chem. Ber.* **88**, 1469 (1955).
118. D. C. Prevoršek, *Bull. Soc. Chim. Fr.*, 788 (1958).
119. W. J. Orville-Thomas and A. E. Parson, *Trans. Faraday Soc.*, **54**, 460 (1958).
120. J. C. Grivas and A. Taurins, *Can. J. Chem.*, **37**, 795 (1959).
121. J. C. Grivas and A. Taurins, *Can. J. Chem.*, **39**, 414 (1961).
122. A. G. Moritz, *Spectrochim. Acta.*, **20**, 1555 (1964).
123. R. A. Russell and H. W. Thompson, *Spectrochim. Acta*, **8**, 138 (1956).
124. N. E. White and M. Kilpatrick, *J. Phys. Chem.*, **59**, 1044 (1955).
125. R. M. Roberts, *J. Am. Chem. Soc.*, **72**, 3608 (1950).
126. R. M. Roberts, *J. Am. Chem. Soc.*, **78**, 2606 (1956).
127. L. Hunter and J. A. Marriott, *J. Chem. Soc.*, 777 (1941).
128. *NMR-Spectra Catlog*, Varian Associates, Palo Alto, California, 1962 and 1963: (a) Vol. I, no 9; (b) Vol. I., no. 39; (c) Vol. II, no. 421.
129. D. L. Harris and K. M. Wellman, *Tetrahedron Lett.*, 5225 (1968).
130. J. P. Marsh and L. Goodman, *Tetrahedron Lett.*, 683 (1967).
131. D. J. Bertelli and J. T. Gerig, *Tetrahedron Lett.*, 2481 (1967).
132. R. C. Neuman, Jr. and L. B. Young, *J. Phys. Chem.*, **69**, 2570 (1965).
133. G. S. Hammond and R. C. Neuman, Jr., *J. Phys. Chem.*, **67**, 1655 (1963).
134. R. C. Neuman, Jr., G. S. Hammond, and T. J. Dougherty, *J. Am. Chem. Soc.*, **84**, 1506 (1962).
135. D. Leibfritz, *Dissertation*, Universität Tübingen, 1970.
136. G. J. Karabatsos and S. S. Lande, *Tetrahedron*, **24**, 3907 (1968).
137. H. Kessler, *Angew. Chem.*, **82**, 237 (1970).
138. R. C. Neuman, Jr. and V. Jonas, *J. Phys. Chem.*, **75**, 3532 (1971).
139. K. M. Wellman and D. L. Harris, *Chem. Commun.*, 256 (1967).
140. A. Loewenstein, A. Melera, P. Rigny, and W. Walter, *J. Phys. Chem.*, **68**, 1597 (1964).
141. M. Rabinowitz and A. Pines, *J. Am. Chem. Soc.*, **91**, 1585 (1969).
142. J. Sandström, *J. Phys. Chem.*, **71**, 2318 (1967).
143. J. S. McKennis and P. A. S. Smith, *J. Org. Chem.*, **37**, 4173 (1973).
144. Z. Rappoport and R. Ta-Shma, *Tetrahedron Lett.*, 5281 (1972).
145. C. G. Raison, *J. Chem. Soc.*, 3319 (1949).
146. R. Hagen and J. D. Roberts, *J. Am. Chem. Soc.*, **91**, 4504 (1959).
147. D. E. Dorman and F. A. Bovey, *J. Org. Chem.*, **38**, 1719 (1973).
148. G. L. Nelson, G. C. Levy, and J. D. Cargioli, *J. Am. Chem. Soc.*, **94**, 3089 (1972); G. A. Olak and A. M. White, *J. Am. Chem. Soc.*, **89**, 7072 (1967).
149. G. Binsch, J. B. Lambert, B. W. Roberts, and J. D. Roberts, *J. Am. Chem. Soc.*, **86**, 5564 (1964).

150. A. K. Bose and I. Kugajevsky, *Tetrahedron*, **23**, 1489 (1967).
151. B. Sunners, L. H. Piette, and W. G. Schneider, *Can. J. Chem.*, **38**, 681 (1960).
152. A. J. R. Bown and E. W. Randall, *J. Mol. Spectrosc.*, **13**, 29 (1964).
153. M. T. Rogers and L. A. La Planche, *J. Phys. Chem.*, **69**, 3648 (1965).
154. A. K. Bose, I. Kugajevsky, P. T. Funke, and K. G. Das, *Tetrahedron Lett.*, 3065 (1965).
155. H.-F. Grützmacher and H. Kuschel, *Org. Mass Spectrom.*, **3**, 605 (1970).
156. C. O. Meese, W. Walter, and M. Berger, *J. Am. Chem. Soc.*, **96**, 2259 (1974).

Note added in proof (See page 11)

In a recent publication Walter and coworkers[156] determined the $E:Z$ ratio of four noncyclic imidates (**21c–21f**) in carbontetrachloride and deuteromethanol solution using the spin–spin coupling constants between C- and N-methylprotons and lanthanide shift reagents in ^{1}H-n.m.r. spectra.

The results, summarized in Table 30, show clearly the predominance of E-forms (**22a** or **22c**) in both solvents indicating that the conclusions of Moriarty and coworkers[51] are incorrect. Additional measurement of the dipole moment of methyl-N-methylacetimidate (**21c**) in CCl_4 at 20°C yielded $1 \cdot 14 \pm 0 \cdot 08$ D which is in good agreement with a calculated value of $1 \cdot 04$ D for the dipole moment of the $E, trans$-form (**22a**) for compound (**21c**) and with the general rule stated by Exner and Schindler[49].

TABLE 30. N.m.r. spectroscopic determination of the percentage of E-diastereomers of imidates (**22a** or **22c**)

Imidates	E (%)	
	in CCl_4	in CD_3OD
(**21c**): $R^1 = R^2 = R^3 = CH_3$	100	95
(**21d**): $R^1 = R^3 = CH_3$; $R^2 = C_2H_5$	100	95
(**21e**): $R^1 = R^2 = CH_3$; $R^3 = C_6H_5$	69	56
(**21f**): $R^1 = t$-Bu; $R^2 = R^3 = CH_3$	87	71

CHAPTER **2**

Constitution, configurational and conformational aspects, and chiroptical properties of imidic acid derivatives

G. Fodor and B. A. Phillips*

Department of Chemistry, West Virginia University, Morgantown, West Virginia 26506, U.S.A.

* Present Address: Division of Science and Mathematics, Fort Valley State College, Fort Valley, Georgia 31030, U.S.A.

I. INTRODUCTION

Imidic acids (**1a**) are the tautomeric forms of carboxamides (**1b**). Although the free acids are not known (discussed later), their esters, i.e. the alkyl and aryl imidates (**2**); their amides, the amidines (**3a** and **3b**); their halides, better known as imidoyl halides (**4**); their hydrazides, the amidrazones (**5a** and **5b**); and their hydroxyamides, the amidoximes (**6a** and **6b**) and the *N,N*-bis-imidoacylamines, the imidines (**7**) are well known and shall be treated in this book according to their relative importance.

We prefer using Barton's[1] definition for structure as a summary of our knowledge on constitution, configuration and conformation. The constitution, i.e. the sequence of how atoms are linked together, is rather unambiguous in the field of imidates. For example, methyl acetimidate formed from acetonitrile upon the acid-catalysed addition of methanol, could hardly be other than 2 ($R^1 = R^2 = CH_3$), the way leading to it being indicative. A more delicate question was the kind and nature of bonds, C—N single vs C=N double bond in some imidic acid derivatives being the most debated. That is why we are limiting main heading II to the discussion of tautomeric equilibria in some of the classes where such a phenomenon is possible: (1a) ⇌ (1b) and (1c) ⇌ (1d), (3a) ⇌ (3b), (5a) ⇌ (5b), (6a) ⇌ (6b), (7a) ⇌ (7b) ⇌ (7c), etc., while a somewhat different sort of tautomerization appears in imidoyl halides and the imidates.

II. CONSTITUTION: TAUTOMERISM OF IMIDIC ACID DERIVATIVES

A. Imidic Acids

The question as to whether amides exist as amides (1b) or as imidols (1a), lactims (1c) or lactams (1d) was on the wrong track for decades, mostly due to three factors: (a) the way in which u.v. spectroscopy was applied in a formal way in analogies, (b) conclusions drawn from relative yields of products derived from chemical reactions of individual tautomers, and (c) lack of modern methods. This problem has been adequately dealt with in this series[2,3] and in Wheland's monograph[4]. Without questioning Hantzsch's[5] and Ramart-Lucas'[6] basic contributions to amide–imidol and lactam–lactim tautomerisms by u.v. studies, Grob and Fischer[7] made it clear that the amide (and lactam) forms prevail over the imidic forms 1c and 1d. Steric inhibition of resonance in a number of simple aroyl amides (8) causes spectral changes that were previously interpreted[5,6] as evidence in favour of the imidic acid form.

$$
\begin{array}{c}
Ph \diagdown \qquad \diagup CH_3 \\
C\!-\!N \\
O \diagup \quad \diagdown CH_3
\end{array}
$$

(8)

B. Alkyl and Aryl Imidates

Alkyl and aryl imidates, known since Pinner[8], do not exhibit a similar tautomerism. Their O → N rearrangement, discovered by Chapman, shall be dealt with in McCarty's chapter[9] in this volume. However, there are a few cases[10,11] where with an α-cyano group, enamine-like ketene-aminohemiacetal tautomers (9 ⇌ 10) have been detected by i.r. and p.m.r. spectroscopy, in solvents like DMSO. The amount of the enamine tautomer varies with solvent and temperature in systems 11 ⇌ 12.

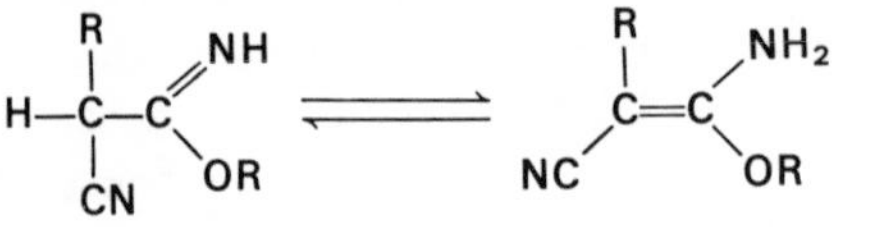

(9)(85%)   (10)(15%)

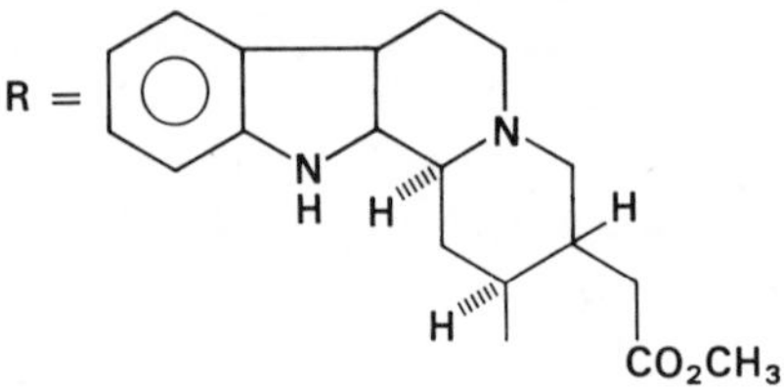

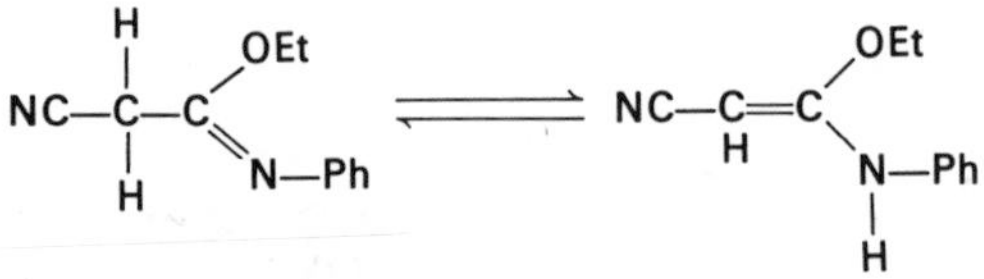

(11)   (12)

C. Amidines

The tautomerism of amidines has been most extensively studied, therefore we shall discuss it in detail. Amidines of general structure 13 can be divided into five classes—unsubstituted, monosubstituted, symmetrical, or unsymmetrical disubstituted, and trisubstituted amidines—depending on which or all of the R^1 through R^4 groups are H, alkyl or aryl, or

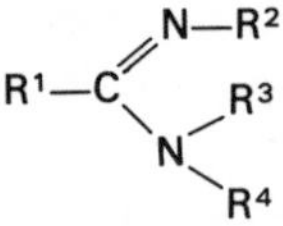

(13)

cycloalkyl. The *constitutions* of all types of amidines are unequivocally determined by one of the major synthetic routes: (1) ammonolysis or aminolysis of imidic esters; (2) addition of ammonia or of (primary or secondary) amines onto nitriles, and (3) conversion of amides into α-chloroimidates and subsequent ammonolysis or aminolysis. All these methods are dealt with in chapters of this volume, see also reference books Houben-Weyl[12], Rodd, and a review by Rogers and Neilson[13]. No appreciable amount of work has been done by X-ray crystallography, the only case reported[14] being the N,N'-bis(4-ethoxyphenyl)acetamidinium ion which showed protonation at the imino nitrogen.

The reaching of a tautomeric equilibrium is either acid- or base-catalysed.

As it has been known since Pyman's early work[15,16] two tautomeric amidines are interconverted either directly or by protonation and subsequent deprotonation. For example, **14** and **15** are in a tautomeric equilibrium[17,18]. The conjugate acid **16** and base **17** are stabilized by resonance.

$$HNR^2{\cdots}C{-}R^1$$
$$\overset{+}{H}N{-}R^3$$

(16)

$$R^2N{=}C{-}R^1 \quad\quad \xrightarrow{\;\;K_T\;\;} \quad\quad R^2N{-}\overset{H}{\underset{}{C}}{-}R^1$$
$$HNR^3 \quad\quad\quad\quad\quad\quad N{-}R^3$$

$+H^+$ K_1 $-H^+$ $\quad$ K_2 $+H^+$ $-H^+$

(14) $\quad\quad\quad\quad\quad\quad\quad\quad$ **(15)**

K_3 $-H^+$ $+H^+$ $\quad\quad$ $-H^+$ K_4 $+H^+$

$$R^2N{\cdots}C{-}R^1$$
$$\overset{-}{N}{-}R^3$$

(17)

The rate constant for tautomerism K_T results from the equilibrium constants $K_1 - K_4$ for protonation and deprotonation, i.e. from the acidity constants.

$$K_T = \frac{[14]}{[15]} = \frac{K_1}{K_2} = \frac{K_4}{K_3}; \quad K_1 = \frac{[14][H^+]}{[16]}; \quad K_2 = \frac{[15][H^+]}{[16]};$$

$$K_3 = \frac{[17][H^+]}{[14]}; \quad K_4 = \frac{[17][H^+]}{[15]}$$

Consequently, K_T depends on the relative acidities and basicities of tautomers **14** and **15**. Since (free) enthalpy of protonation and deprotonation,

$$\Delta G = RT \ln K_a$$

the tautomer which has the lower energy will prevail in the equilibrium and if the acidity constants K_a were known, the concentration of the same could be calculated. However, K_T is strongly solvent dependent since equations for K_1 through K_4 contain activities and *not* concentrations.

Relative acidities and basicities[19] of tautomers **14** and **15** (also of **18** and **19**) depend on relative electronegativities of substituents R^2 and R^3 since they give upon protonation the same resonance-stabilized cations and upon deprotonation (except for N,N'-trisubstituted amidines) the same anions are formed.

(18)   **(19)**

Trisubstituted amidines do not have any possibility for tautomerization, and only contributions by dipolar resonance hybrids can be detected by n.m.r. measurements which would reveal restricted rotation around the C—N bond. This is particularly so in the protonated N,N-disubstituted amidinium ion where one can predict that **20** rather than **21** would be the more stable cation. N.m.r. data would indicate lower field chemical shifts

(20)   **(21)**

in **20** for protons in R^2 and R^3 than in **21**. I.r. and n.m.r. spectra would clarify whether an N^+H_2 or a basic NH_2 group is present. Other amidines present different situations, in consequence of the character of the R groups.

No general statements are warranted as to the K_T unless all constitutional factors were considered. For example, 2-aminopyridines by analogy with aminoarenes, were often regarded as being prevalently in the 'aromatic' form as against the 'semiquinonoid' tautomer, α-pyridoneimide. Both can be regarded as N-substituted amidines. Inductive and mesomeric effects of the N-substituents shall determine the position of the equilibrium, *via* the relative basicity these factors lend to the two nitrogens. Substituent R^1 on carbon seems to have little if any effect[20]. For example, in benzamidines **22** and **23** ($R^1 = C_6H_5$) the phenyl ring is forced out of the amidine plane, therefore, it has no mesomeric effect[21]. Thus groups R^2 and R^3 determine relative basicities of the nitrogen, and hence also k_T.

The effect of groups R^2 and R^3 can be parallel or competitive. Taking hydrogen as R^3 one may more easily determine the effect of the character of R^2 upon the equilibrium between 'amino' **22** and 'imino' **23** forms.

$(22a, 23a: R^3 = H)$

$(22b, 23b: R^3 = OH)$

(22)   **(23)**

The acidic center is always the single-bonded amino-N, that bears in all (but trisubstituted) amidines one hydrogen atom and has no δ^+ charge. The basic center is therefore the double-bonded imine nitrogen with a partial negative charge. Increase in electronic density by substitution of the amino-nitrogen shall hence diminish the acidity of one tautomeric form while decrease leads to stronger acidity. On the other hand, basicity of one tautomeric form shall be increased upon substitution causing an increase in electron density at the imine nitrogen, and shall decrease upon decrease of electron density. A R^2-substituent having a $-I$ effect which diminishes the electron density on nitrogen should give preponderance to the 'amino-tautomer' **22**. Accordingly, amidoximes[22], $(R^3 = OH)$ exist in the hydroxylamine tautomeric from **22b**. A group R^2 with a $+M$ effect that should have an opposite effect will not influence the equilibrium since the amidine system is unable to take up more electrons. The $+I$ effect of the R^2 group should then give preference to the imide form **23** and this is indeed the case in most N-alkylamidines[22], while recent dipole moment studies have also shown that N-phenylamidines prevail in the amino form **22**. In the latter case there is, however, the resonance between N-phenyl and the N—C bond as an additional factor.

D. N-Trisubstituted Amidinium Salts

N-Trisubstituted amidines do not display tautomerism. However, once protonated the location and shift of that proton may be significant and can be proven. The hydrobromide of compound **25**, prepared from N-benzoylpiperidine *via* the bromoiminium[23] bromide **24** and 3-chloroaniline could exist in either form **25a**, **25b**, or **25c**. It actually proved to be a uniform compound[24]. The equilibrium between **25b** and **25c** was expected to be shifted towards **25b** in view of the much weaker basicity of the aryl-substituted amine nitrogen. For the same reason one would not expect the

—N̈H—lone electrons in **25a** to be involved in resonance with the $\diagdown$ N$^+$=C $\diagup$

double bond as indicated by dotted arrows in **25a**. Rather, based on previous studies one might have expected **25b** to be more resonance-stabilized than **25a**. I.r. studies revealed neither =N$^+$H (nor —N$^+$H) absorption

(24)   (25a)   (25b)

(25c)

bands in the 1350–2750 cm^{-1} region while significant —ṄH absorption was found[24] at 3350 cm^{-1} which is only compatible with **25a**.

The NH resonance appeared at δ 12·27 (HN$^+$= usually appears around 14·0) while in the N-phenyl derivative it is at δ 10·20. The strong deshielding of H$_{(2)}$ and H$_{(6)}$ methylene protons (δ 4·35 and 3·40) is due to an adjacent quaternary nitrogen, in addition to being deshielded by a *cis*-placed amino group at H$_{(2)}$ and deshielded on H$_{(6)}$ by a *cis*-phenyl group. All this fits with **25a** as the preponderant tautomer.

A series of iminium-type amidinium bromides with methyl group(s) at C$_{(2)}$ and/or C$_{(6)}$, different N′-aryl and N′-alkyl groups, showed very similar spectral characteristics consistent with structure **25a**.

E. Arylsulfonylamidines

Several attempts have been made to isolate the tautomeric forms of arylsulfonylamidines. The claim[25] that the phenylsulfonylamidines **26a,b,c** (α-forms) and their imide tautomers **27a,b,c** (β-forms) were isolated and the α-form converted into β- was later questioned[26, 27]. First, u.v. spectra were taken and the constitution assignments suggested to be reversed[26]. Lately, it was shown[27] that the products described earlier were not pure and the separation led to **26c** (or **27c**) and 4-acetamidobenzenesulfonamide, a by-product in the synthesis of the sulfonylamidine **26**.

Furthermore, N-(3-chlorophenylsulfonyl)-acetamidine (**28**) and N-(3-chlorophenylsulfonyl)-[N'-^{15}N]acetamidine were prepared to ascertain the N-atoms involved in the C—N bond.

(**26**)   (**27**)

X = (a) NO_2, (b) NH_2, (c) NHAc

Compound **26** gave an i.r.-spectrum with a ν_{max} 1650 cm^{-1} which disappeared on deuteration. This band was assigned to the amino-internal deformation mode since other bands (C=C, C=N) were not affected by deuteration and the deformation mode of imides is very weak. This supported the 'amino form' **26**, (analogous to **22**). The ^{15}N-derivative was prepared with ^{15}N at the terminal position. Thus the quadrupole associated with ^{14}N was eliminated, also ^{14}N—H coupling was produced which enabled distinction of the amino and imino forms. The ^{1}H spectrum of the imino form should show only one of the broad resonances split into a doublet while both signals should show coupling in the amino form. Indeed, both protons are coupled in the ^{15}N-derivative to the ^{15}N atom ($-I = 1/2$), giving rise to doublets $J_{N-H} \simeq 93$ Hz; each doublet being further split by geminal coupling $J_{NH} \simeq 2\cdot4$ Hz. This suggests that **26** is an equilibrium mixture of the two geometrical isomers of the amino form with some imine, but the predominant tautomer is best represented[27] as a resonance hybrid of **28a** and **28b**.

(**28a**)   (**28b**)

F. Amidrazones

Amidrazones are expected to show an effect opposite to amidines, i.e. preference of tautomer **30** since the amino nitrogen (which stand here for R^2 in our general formulae **22** ⇌ **23**) has a degree of hybridization with a smaller s-contribution than an sp^3 carbon. Meanwhile, both nitrogen atoms in the amidine system have sp^2-hybrid orbitals and are therefore more electronegative than R^2.

For *N*-thioacyl, *N*-sulfonyl, and for heterocyclic derivatives of amidines we shall refer to Schwenker's review article[17].

$$H_2N—N{=}C—R^1$$
$$\underset{H \quad\quad R^3}{\overset{|}{N}}$$

$$H_2N—NH—C—R^1$$
$$\underset{R^3}{\overset{|}{N}}$$

(29)   **(30)**

G. N-Halogenoamidines

N-Halogenoamidines (**31a** ⇌ **31b**) (and guanidines) were spectroscopically (n.m.r.) studied[28] in view of the difficulty of analysing NH signals

$$CH_3—\underset{H}{\overset{|}{N}}—\underset{N—X}{\overset{||}{C}}—Ph$$

$$CH_3—N{=}C—Ph$$
$$\underset{NH—X}{}$$

(31a)   **(31b)**

X = Cl, Br

because of quadrupole coupling association and exchange effects in the amidines proper. Usually there were no NH—CH couplings observed while in the *N*-halogeno derivatives this could be done. Since **31** is a weak base the exchange rate is low.

The *N*-methyl signal appeared at δ 2·82 (in $CDCl_3$) and 2·68 (DMSO-d_6) as a doublet (J 5·0 and 5·5, respectively) in the *N*-chloroamidine and at δ 2·85 in the bromoamidine which proves that **31a** was the predominant tautomer, in contrast with amidoximes[17].

H. Imidoyl Halides

N-[4-Nitrobenzyl]benzimidoyl chloride (**32**) and 4-nitro-*N*-benzyl-benzimidoyl chloride (**38**) give[29] a 92:8 equilibrium mixture in the presence of triethylamine in benzene. I.r. spectroscopy was used for analysis upon hydrolysis of **32** and **38** to the amides.

$$Ph—C\overset{N—CH_2—C_6H_4NO_2\text{-}p}{\underset{Cl}{}}$$

$$Ph—CH_2—N\overset{C^{C_6H_4NO_2\text{-}p}}{\underset{Cl}{}}$$

(32)   **(38)**

Isomeric α-chloroazomethines **34** and **35** were assumed as intermediates for this base-catalysed tautomerization and the following complete scheme was suggested.

In view of the easy dissociation[23] of imidoyl halides into nitrilium salts it seems likely that loss of halogen precedes the loss of proton leading to the

$$
\mathbf{32} \underset{+H^+}{\overset{-H^+}{\rightleftarrows}}
\begin{array}{c} Ph \\ C == N - C \\ Cl \quad\quad Ar \end{array}
\overset{(33)}{}
\underset{-H^+}{\overset{+H^+}{\rightleftarrows}}
\begin{array}{c} PhCH-N=C \\ Cl \quad\quad Ar \end{array}
\overset{(34)}{}
$$

$$
\mathbf{38} \underset{-H^+}{\overset{-H^+}{\rightleftarrows}}
\begin{array}{c} Ph \quad\quad Cl \\ C == N == C \\ H \quad\quad Ar \end{array}
\overset{(37)}{}
\underset{+H^+}{\overset{-H^+}{\rightleftarrows}}
\begin{array}{c} Ph \quad H \\ C = N - C - Ar \\ H \quad\; Cl \end{array}
\overset{(36)}{}
\rightleftarrows
\begin{array}{c} Ph-C=\overset{+}{N}=C-Ar \\ H \quad\quad H \\ Cl^- \end{array}
\overset{(35)}{}
$$

Ar = 4-nitrophenyl

diarylidene ammonium salt **35**. Therefore, **34** and **36** are not intermediates but rather the nitrilium ions **39** and **40**. Intermediate **35** could be detected by 2,4-DNPH which gave equal amounts of the benzaldehyde and 4-nitrobenzaldehyde dinitrophenylhydrazones. There was no tautomerization observed in the absence of triethylamine, which is an indication that somewhere, either at **32** and **38** or at **39** and **40** deprotonation had to occur. Therefore, **33, 34, 36,** and **37** are not necessarily involved in the process (see scheme below).

$$
\begin{array}{c} Ph \\ C=N-CH_2Ar \\ Cl \end{array}
\overset{(32)}{}
\underset{+Cl^-}{\overset{-Cl^-}{\rightleftarrows}}
Ph-C\overset{+}{\equiv}N-CH_2-Ar
\overset{(39)}{}
\rightleftarrows
\begin{array}{c} Ph-\overset{+}{C}=N=C \\ H \quad\quad Ar \end{array}
\overset{(35)}{} H
$$

$$
PhCH_2-N=C-Ar
\overset{(38)}{} Cl
\underset{-Cl^-}{\overset{+Cl^-}{\rightleftarrows}}
Ph-CH_2-\overset{+}{N}\equiv C-Ar
\overset{(40)}{}
$$

In the absence of triethylamine, amidrazones were obtained with 2,4-DNPH, no appreciable degree of tautomerization occurred. The only argument in favour of the α-chloroazomethines was that the equilibrium mixture from **32** gave, upon neutralization of the base and treatment with 2,4-DNPH, 9·5% and 7·6% of the two arylhydrazones, similarly to imidoyl chloride **38** (9·7 and 7·7%, respectively). This, however, points as the authors also admit, *either* to the iminium ion **35** *or* to the α-chloro-azomethines—of which we prefer the first alternative.

III. CONFIGURATION OF IMIDIC ACID DERIVATIVES

A. Amidines

The first valid proof for geometrical isomerism in amidines was produced quite recently. The condensation of methyl benzothiazole-2-iminocarboxylate[30] with α-aminoacetic esters gave two amidines **41** and **42** in which the 'amino tautomer' was stabilized by hydrogen bonding to

(41) (*anti*)   **(42a)** (*syn*)

R = H, CH$_2$CH(CH$_3$)$_2$, CH$_2$Ph
R^1 = H, C$_2$H$_5$, (CH$_3$)$_3$C

a heterocyclic nitrogen. Therefore, no significant amount of the corresponding 'imino' tautomers were present to complicate the picture.

However, there was another enol–keto tautomerism that overlapped in the enolate–ammonium zwitterion **42b** in neutral medium. At room

(42b)   **(43)**

temperature the *anti*-isomer **41** prevailed while in boiling methanol the *syn*-enols **42b** were exclusively formed. The latter are easily cyclized to imidazolones **43**, while the free carboxylate-enols (**42b**, R^1 = H) are not. Another important feature is that **41** (R^1 = alkyl) is optically active while **42b** is inactive. I.r. spectra of the amino tautomers **41** and **42** were expected

to show ν_{as} (C=N—) close to 1640 cm^{-1}; δ (NH$_2$) 1510–1520 cm^{-1};

ν_{as} (H_2N) and ν_s (NH_2) at 2400 and 3300 cm^{-1}, respectively. In the enolized *syn* NH_2 enols there are two bands, around 2740 and 2580 cm^{-1}, characteristic of $\nu_{N^+H_3}$. In the *anti*-NH_2 isomers, on the contrary, the $\nu_{C=O}$ is around 1730 cm^{-1} and ν_s (NH_2) and ν_{as} (NH_2) at 3260 and 3390 cm^{-1}.

The n.m.r. spectra of the *anti* (NH_2) amidines (R = alkyl or benzyl) show the methine proton α- to the carboxyl group, as expected around δ 4·00–5·00, while in the *syn* NH_2 enols (**42b**) this was absent and five exchangeable protons were indicated (for N^+H_3 and two OH's) in the carboxyl and four such protons in the δ 5·00 region for the ester.

All these facts together with an extensive u.v. spectral study proved that *syn–anti* geometrical isomers of an amidine were indeed isolated[30].

A more recent study[24] refers to geometrical isomerism of a trisubstituted amidinium salt **45**. A similar case has already been discussed from the point of view of tautomerism: N,N-pentamethylene-N'-chlorophenyl-benzamidinium bromide (**25**). Compound **45** is the α-methyl derivative of **25** which was made from 1-benzoyl-2-methylpiperidine *via* the benziminium[23] bromide **44**. The tribromides of **44**, formed as the mixture of geometrical isomers around the $\overset{\diagdown}{\underset{\diagup}{C}}=\overset{+}{\underset{\diagdown}{N}}\diagup$ bond, were separated by fractional crystallization. The major product was then reduced with ethylene (or cyclohexene) to the monobromide. Configurational assignment was made based upon strong deshielding by bromine of the 2-methyl in one isomer and shielding by a neighbouring phenyl group in the other. The major product proved to be the *syn* (Br/CH$_3$) isomer of **44**. Upon the action of 3-chloroaniline a stereoisomeric mixture of **45**-hydrobromides has formed, probably *via* an addition–elimination process, the transition state of which allowed free rotation around the $\overset{\diagdown}{\underset{\diagup}{C}}=N-$ bond. Although three tautomeric forms, analogous to **25a**, **25b**, and **25c** were possible, only *anti*-NH/CH$_3$ **45** and its *syn*-isomer in a 68:32 ration were detected by 250 MHz n.m.r. spectroscopy. The approach to the prevalent tautomeric form was similar to that of **25** except that in **45** two very similar n.m.r. curves were obtained with slight shift of the major signals, all of which have been identified by double irradiation. Thus upon irradiation of the 2-methyl doublet at δ 1·32 a multiplet at 3·96 partially collapsed, indicating $H_{(1)}$. Similarly, irradiation of the minor doublet and δ 1·65 resulted in simplification of the pattern of the $C_{(2)}$ methine proton at δ 5·60. Inversely, the higher intensity $H_{(6)}$ methylene resonated at δ 4·88, the lower one at δ 3·40. These differences of chemical shifts are so significant that they allowed correct configurational assignments to be made. Upon deprotonation only N,N'-trisubstituted amidine **46** was formed which, upon reprotonation, gave a different composition of the geometrical isomers than the one observed during their formation from the stereochemically pure α-bromoiminium salt **44**.

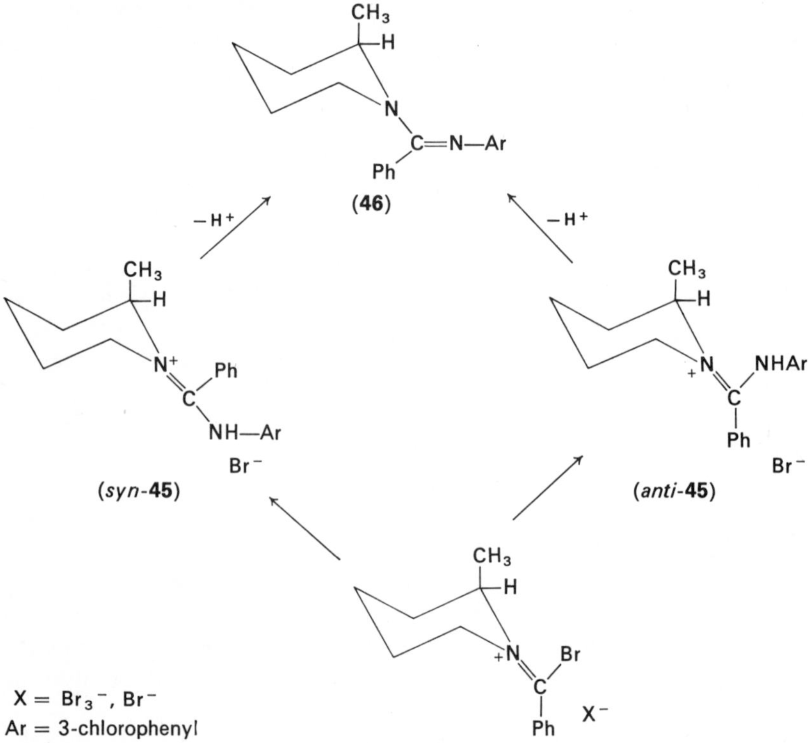

(**46**)

(*syn*-**45**)

(*anti*-**45**)

X = Br$_3^-$, Br$^-$
Ar = 3-chlorophenyl

(*anti*-**44**)

The axial position of the 2-methyl group was not rigorously proven; it was based on analogies with related allylic compounds and on the relatively low-field resonance of the H$_{(2)}$ methine. Complete separation of *syn* and *anti*-**45**-hydrobromides has not yet been achieved *.

B. Geometrical Isomerism in Imidates

A review by McCarty[32] has covered the literature concerning *syn–anti* isomerism of imidate derivatives up to 1969. However, since that time several papers investigating geometrical isomerism of *O*-methyl imidates and *S*-methyl thioimidates have appeared.

Moriarty and co-workers[33] have observed the n.m.r. spectra of a series

* *Syn* and *anti* refer here to CH—CH$_3$ to C—Ph.

TABLE 1. Chemical shifts for proton resonance in cyclic *O*-methyl imidates[33]. Reprinted with permission from R. M. Moriarty and co-workers, *J. Amer. Chem. Soc.*, **92**, 6360 (1970). Copyright by the American Chemical Society.

Ring size	N.m.r. signals, δ		
	OCH_3	$C{=}N{-}CH_2$	$N{=}C{-}CH_2$
5	3·72	3·57	
6	3·50	3·40	2·17
7	3·48	3·41	2·36
9	3·52	3·41	2·32
11	3·54	3·40	2·27
12	3·51	3·28	2·22
13	3·52	3·25	2·20
16	3·51	3·20	2·21

of cyclic *O*-methyl imidates (Table 1) from ring size five to 16, as well as of four open chain derivatives, i.e. *N*-methylacetimidate (**47a**), *N*-phenyl-acetimidate (**47b**), *N*-ethylacetimidate (**47c**) and *N*-*n*-butylacetimidate

(a, R = CH₃)
(b, R = C₆H₅)
(c, R = C₂H₅)
(d, R = *n*-C₄H₉)

(**47d**) (Table 2). Models indicate that the five-, six-, and seven-membered rings are restricted to the *syn* configuration (**48**), while for larger rings there is a possibility of incorporating the *anti*-imino group (**49**).

(48)   (49)

In comparing the n.m.r. spectra of imidates **47a** and **47b** (Table 2) one observes a downfield shift of 0·23 p.p.m. for the OCH_3 signal in (b) relative to (a), while the position of the OCH_3 resonance is shifted by only 0·02 p.p.m. The authors suggest that the shift results from the deshielding effect of the phenyl group in the *anti** configuration. The n.m.r. spectra of imidates **47a** and **47b** are invariant over a temperature range of -100

* In all cases *syn* and *anti* refers to the relative position of the CCH_3 and N—R groups.

TABLE 2. Chemical shifts for proton resonance in open chain imidates (47)[33]. Reprinted with permission from R. M. Moriarty and co-workers, *J. Amer. Chem. Soc.*, **92**, 6360 (1970). Copyright by the American Chemical Society.

Compound	N.m.r. signals, δ			
	OCH_3	$C{=}N{-}CH_3$	$C{=}N{-}CH_2$	$N{=}C{-}CH_3$
(a)	3·50	2·92	—	1·78
(b)	3·73	—	—	1·76
(c)	3·50	—	3·17	1·77
(d)	3·50	—	3·14	1·76

to +120°C. This could indicate either a high barrier to inversion (>23 kcal/mol) and configurationally stable *anti* diastereomers for compounds **47a,b,c** and **d**, or a low barrier with a large thermodynamic preference for the *anti*-form.

Evidence in favour of a high barrier in these *O*-methyl imidates came from a study[33] of the temperature-dependent n.m.r. spectra of their conjugate acids. At room temperature in 100% sulphuric acid *N*-methylacetimidate (**47a**) is converted completely to the protonated form with J_{NHCH_3} equal to 5 Hz. Upon heating at 80°C a gradual doubling of the OCH_3, CCH_3, and each coupled $NHCH_3$ resonance occurred. Equilibrium was established after heating at 80°C for 90 h. These results indicate that in the protonated form the stabilities of the *anti* and *syn* form had reversed. While the *anti* is the exclusive form for imidate **47a**, the ratio of the *anti* (**50**) and *syn* (**51**) forms of the conjugate acid is 1:2. Similarly for imidate **47b** the ratio of the conjugate acids is *anti*:*syn*, 1:3.

The authors interpreted the apparent high barrier to inversion at nitrogen in these compounds as resulting from interorbital electron repulsion between the nonbonding electrons on oxygen and the electrons localized in a *p* orbital on nitrogen in the transition state for the $sp^2 \rightleftharpoons$

$sp \rightleftarrows sp^2$ inversion process. Protonation relieves this interaction and rotational equilibrium of the conjugate acid yields isomers in which relative stabilities are chiefly dependent upon steric interaction.

Table 1 indicates that in the cyclic series the CH_2—N=C resonances undergo a transition between ring size 11 and 12. The signal goes from a constant value of around δ 3·40 in 11- and smaller membered rings to δ 3·28 in the 12-, δ 3·25 in the 13- and δ 3·20 in the 16-membered rings. Significantly, the CH_2—N=C resonance in the open chain N-ethyl-acetimidate (47c) and N-n-butylacetimidate (47d) appears at δ 3·17 and 3·14, respectively. Since the open chain imidates possess the *anti* configuration the transition of the CH_2—N=C signal in the cyclic series indicates a change from the *syn* (51) to the *anti* (50) form in the 12-, 13-, and 16-membered rings. Similarly to the open chain derivatives the cyclic imidates did not show any temperature dependence in the n.m.r. from -100 to $+120°C$.

When the cyclic compounds of 5–9 membered rings were heated at 80°C for 90 h no equilibration was observed. This agrees with a *syn* from for both the imidate and conjugate acid. However, the 11-, 13-, and 16-membered systems did undergo equilibration and the *anti:syn* ratios were found to be 1:1, 1:2, and 1:1·5, respectively. These results indicate an *anti*-configuration for the free imidate for the 13- and 16-membered rings, which upon protonation goes to a mixture of the predominant *syn* conjugate acids. The 11-membered ring is of *syn* configuration in the imidate form while when protonated the *syn* and *anti* forms are of about equal energy.

The assumption that interorbital repulsion is the governing factor in destabilizing the *syn* form of the O-methyl imidates was supported by the observation of the existence of interconverting diastereomers in the S-methyl thioimidate series[34]. The coulombic repulsion in the S-methyl thioimidates is less than in the O-methyl imidates partly because of the greater length of the sulfur–carbon bond relative to the oxygen–carbon bond which increases the distance over which the interaction takes place.

The n.m.r. data for the cyclic thioimidates used in this study and the open-chain thioimidates 52 and 53 are given in Table 3. The *syn* configuration must exist purely for steric reasons in the 5- to 9-membered rings and this is revealed by a sharp singlet in their n.m.r. spectra for the S—CH_3 proton (Table 3). In the 10-membered ring a new peak appears

(52)   (53)

TABLE 3. Chemical shifts for cyclic and open-chain S-methyl imidates[34].

Ring size	N.m.r. signal, δ			
	SCH$_3$		C=N—CH$_2$	N=C—CH$_2$
	syn	*anti*		
7	2·16	—	3·58	2·40
8	2·18	—	3·63	2·37
10	2.15	2.35	3·60	2·48
11	2·21	2·38	3·45	2·45
12	2·17	2·38	3·44	2·48
13	2·20	2·38	3·43	2·47
16	2·20	2·40	3·36	2·50

Compound	SCH$_3$		C=N—CH$_2$		N=C—CH$_3$ or N=C—CH$_2$CH$_3$	
	syn	*anti*				
52	2·22	2·40	3·30	3·42	2·27	2·49
53	2·28	2·45	3·30		2·20	2·30

downfield due to 10% of the presumably *anti* isomer. Isomer ratios for the cyclic and linear S-methyl thioimidates are given in Table 5. The activation parameters for the exchange between the cyclic *syn* and *anti* diastereomers were determined using the $W_{1/2}$ method[35] and are found in Table 4.

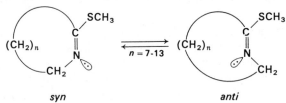

syn   *anti*

TABLE 4. Activation parameters for cyclic and linear S-methyl imidates (at 352·3 K)[34]

Ring size	$\Delta G^{\ddagger}$(kcal/mol)	$\Delta H^{\ddagger}$(kcal/mol)	$\Delta S^{\ddagger}$(e.u.)
11	18·8	17·9	−2·6
12	21·2	21·2	+3·8
13	19·0	19·0	+0·3
16	21·7	21·7	+3·6
Compound			
53	19·4	19·4	−2·2

TABLE 5. Isomer ratio for cyclic and linear S-methyl imidates[34]

Ring size	Ratio of conjugate acids		Ratio of free imidates	
	syn	anti	syn	anti
7	100	—	100	0
9	100	—	100	0
10	90	10	90	10
11	80	20	20	80
12	30	70	25	75
13	35	65	25	75
16	50	50	50	50
Compound				
52	45	55	—	—
53	55	45	50	50

The behavior of the conjugate acids of the S-methyl thioimidates was also investigated. The similarity in the syn:anti ratio of the free thioimidate and conjugate acid (Table 5) may indicate that steric effects are the dominant factor in determining the diastereomeric ratios in the free thioimidates in contrast to the behavior of the O-methyl imidates.

C. Imidoyl Halides and α-Halogenoiminium Salts

Geometrical isomerism about the carbon–nitrogen double bond in imidoyl halides has not yet been extensively investigated. This may be attributed to the fact that imidoyl halides are difficult to handle, for they

Z (anti)   E (syn)   X = F, Cl, Br

are readily hydrolysed to the corresponding amides upon exposure to atmospheric moisture. There are two different conventions for assigning configurations to imidoyl halides: Terms Z and E refer to proximity or remoteness of the N-substituted R^2 considering the sequence rule and the halogen atom while anti and syn indicate the steric relationship of the nitrogen substituent versus the carbon substituent.

It was reported that the addition of fluorine to benzylidene-t-butylamine (54) at −78°C afforded[36] a crude mixture of fluoramine (55) and the anti imidoyl fluoride (56a). Although anti imidoyl fluoride (56a) could not be isolated, its structure was inferred from spectra of the crude mixture of it and fluoramine (55) (Table 6). On standing, exposed to glass or silicagel,

TABLE 6. Spectral data of imidoyl fluorides **56a** and **56b**[36]. Reprinted with permission from R. F. Merritt and F. A. Johnson, *J. Org. Chem.*, **32**, 416 (1967). Copyright by the American Chemical Society.

	(56b)	(56a)
$\nu FC = N$, cm^{-1}	1686	1718
Φ^a	$+67\cdot0$ (singlet)	$+39\cdot0$ (singlet)
δ^b	$1\cdot50$ (doublet, $J = 1$ c.p.s.)	$1\cdot40$ (singlet)

[a] Φ = p.p.m. from internal CCl$_3$F.
[b] δ = p.p.m. from internal standard TMS (10% solutions in CDCl$_3$).

the fluoramine (**55**) loses fluorine to form the *syn*-imidoyl fluoride (**56b**, spectral properties in Table 6).

This is the only report of the characterization of stable, separated *syn–anti* isomers of a noncyclic imidoyl halide. Reportedly[37], the reaction of 1-pentene with benzonitrile in the presence of hydrogen fluoride gave a mixture of *syn* and *anti* N-2-pentylbenzimidoyl fluoride (**57a** and **57b**). The fluorine n.m.r. spectra showed two singlet peaks of equal intensity which indicates a 1:1 mixture of the two forms. No imidoyl chloride has been separated into stable *syn–anti* forms and no geometrical isomerism has been detected thus far for imidoyl bromides. The configuration of the carbon–nitrogen double bond has been determined for many classes of compounds even though the two stereoisomers have not been isolated[32]. This has been the case for imidoyl chlorides. Based on dipole moment data[38] it has been reported that N-(p-nitrophenyl)-p-nitrobenzimidoyl chloride (**58**) exists only in the *anti* form. The dipole moments calculated for both forms of imidoyl chloride (**58**) are shown below. The dipole moment measured in benzene was found to be $1 \cdot 20 \pm 0 \cdot 5$ D, a value which correlated very well with the *anti* form.

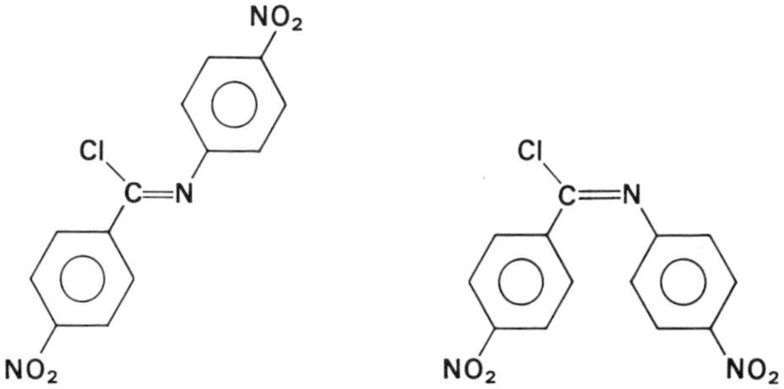

$\mu = 1 \cdot 40$ $\mu = 5 \cdot 90$

(*anti*-**58**)   (*syn*-**58**)

A more recent study[39] of the dipole moments of aromatic imidoyl chlorides also indicate that they exist in the *anti* (or Z) configuration. The dipole moments for the Z- and E-configurations of six aromatic imidoyl chlorides (Table 7) were calculated by vector addition of bond moments and compared with the measured values. The calculated moments are consistent with the Z-configuration, a conclusion that was also supported

TABLE 7. Dipole moments of some imidoyl chlorides in benzene at $25°C$[39]. Reprinted with permission from A. Dondoni and O. Exner, *J. Chem. Soc. Perkin II*, 1908 (1972).

Compound	Measured μ/D	Calculated μ/D	
		Z	E
$PhC(Cl)=NPh$	$1 \cdot 16$	$1 \cdot 19$	$1 \cdot 95$
$p\text{-}ClC_6H_4C(Cl)=NPh$	$0 \cdot 25$	$0 \cdot 41$	$0 \cdot 62$
$p\text{-}O_2NC_6H_4C(Cl)=NPh$	$3 \cdot 34$	$3 \cdot 31$	$2 \cdot 68$
$Ph(Cl)=N—(CH_2)_3CH_3$	$0 \cdot 54$	$0 \cdot 89$	$2 \cdot 18$
$p\text{-}ClC_6H_4C(Cl)=N(CH_2)_3CH_3$	$0 \cdot 96$	$0 \cdot 70$	$0 \cdot 92$
$p\text{-}O_2NC_6H_4C(Cl)=N(CH_2)_3CH_3$	$3 \cdot 78$	$3 \cdot 61$	$2 \cdot 60$

by a graphical method of analysis. The dipole moments of those compounds which are below 1 debye are less reliable because of the uncertainties in correction for atomic polarization and in the value of molar refraction.

Hence the *anti* configuration has been proven for aromatic imidoyl chlorides, which seems to be of general validity since aliphatic and aromatic aldimines[40] are preferentially also present in the same configuration. The stability of the Z-form seems to be controlled by strong steric repulsion which disfavors the E-form.

Although the *anti*-configuration predominates there is some evidence that in imidoyl chlorides an equilibrium exists between the *syn* and *anti* forms. It was reported that the n.m.r. spectrum of N-methylbenzimidoyl chloride (59) in SO_2 solution gave a broad singlet at δ 2·95 at −20°C that split into two singlets at 2·92 and 3·02 at −50°C. This was attributed[41] to the 'freezing out' of the *syn* and *anti* isomers 59a and 59b. This observation also indicates that the energy barrier for the interconversion of

(59a)   (59b)

isomers 59a and 59b is lower than would be expected based on an analogy with similar systems. One explanation suggested recently[23] is a rapid dissociation of one of the geometrically isomeric imidoyl halides into the nitrilium halide (60) and recombination of the ions into either the *syn* or *anti* configuration. No indication for the existence of geometrical isomers

(60)

at the C=N bond was found in the n.m.r. spectrum of N-methylbenzimidoyl bromide (61) at −85°C, in toluene-d_8. This may be due either to

(61)   (62)

preferential formation of one of the geometrical isomers, or to a rapid *syn–anti* interconversion, or to coincidental overlap of the chemical shifts of both isomers. However, the n.m.r. spectrum of imidoyl bromide 61 was identical to that of N-methylbenzonitrilium fluorosulfate (**62**, δ_{Me} 4·1) in SO_2 at $-29°C$. This indicates that in SO_2, a highly polar solvent, imidoyl bromide **61** tends to be completely dissociated. In acetonitrile-d_3 nitrilium salt **62** and imidoyl bromide **61** show very different N–Me signals (δ 3·95 and 3·50, respectively).

Further evidence for the equilibrium between nitrilium halides and imidoyl halides was provided[23] by the reaction of N-methylbenzimidoyl chloride (**59**) and bromide (**61**) with methyl fluorosulfate. The reaction gave rise to three products; N,N-dimethyl-α-halobenziminium fluorosulfate (**63**) and N-methylbenzonitrilium fluorosulfate (**62**) in addition to the methyl halide. N-Methylbenzimidoyl chloride (**59**) gave 33% N-alkylation and 67% nitrilium salt formation while N-methylbenzimidoyl bromide gave 20 and 80%, respectively. The methylation of halide in

$$2 \underset{X}{\overset{Ph}{>}}C{=}N{-}CH_3 + 2\,CH_3OSO_2F \longrightarrow \underset{X}{\overset{Ph}{>}}C{=}\overset{+}{N}\underset{CH_3}{\overset{CH_3}{<}}\; + \; Ph C{\equiv}\overset{+}{N}{-}CH_3 + CH_3X$$

(59, X = Cl)   **(63)** $^-OSO_2F$ **(62)** $^-OSO_2F$
(61, X = Br)

preference to alkylation was explained by an imidoyl halide–nitrilium halide equilibrium in which the halide ion is rapidly methylated and the equilibrium is shifted towards the nitrilium salt.

$$\underset{X}{\overset{Ph}{>}}C{=}N{-}CH_3 \rightleftharpoons Ph{-}C{\equiv}\overset{+}{N}{-}CH_3 \quad X^-$$

CH_3OSO_2F ↙   ↘ CH_3OSO_2F

$$\underset{X}{\overset{Ph}{>}}C{=}\overset{+}{N}\underset{CH_3}{\overset{CH_3}{<}}\; ^-OSO_2F \qquad Ph{-}C{\equiv}\overset{+}{N}{-}CH_3 + CH_3X$$

(63)   **(62)** $^-OSO_2F$

Alternately, attack of Magic Methyl upon the halogen in **59** or **61** cannot be ruled out either. The resulting intermediate dialkyl halo-iminium ion (**64**) could break down to give the nitrilium salt **62** and methyl halide.

$$H_3C{-}N{=}C\underset{Ph}{\overset{\overset{\delta+}{X}\cdots CH_3{----}\overset{\delta-}{OSO_2F}}{}} \longrightarrow Ph{-}C{\equiv}\overset{+}{N}{-}CH_3 + CH_3X$$

(64)   **(62)** $^-OSO_2F$

The expected *N*-methylation of the weakly basic imidoyl halide **61** to the fluorosulfate **63** varies with the electronegativity of the *N*-substituent of the imidoyl halide.

In summary, *syn–anti* isomers of imidoyl fluorides have been detected at room temperature. One example of an imidoyl chloride, *N*-methyl-benzimidoyl chloride, has been reported to separate into *syn–anti* forms at −50°C in sulfur dioxide, while for imidoyl bromides no geometrical isomers have thus far been detected. This trend follows the tendency of halides to ionize and lends some support to the mechanism[23] of inversion at nitrogen *via* an imidoyl halide–nitrilium halide equilibrium.

α-Haloiminium salts can be regarded as quaternary salts derived from α-haloimidates. The first cases of the occurrence of geometrical isomers were recently reported[42, 44] and are dealt with in a different context in this chapter, since α-bromoiminium bromides have been intermediates for geometrical isomers of certain amidinium salts, e.g. **45**. It may be added that *syn* and *anti* isomers of these monobromides as well as of tribromides tend to equilibrate rapidly at a measurable rate in solution, possibly *via* the geminal α,α-dibromoalkyldialkylamines **65**. This change has been monitored by n.m.r., although no appreciable life-time of **65** could be substantiated[42]. Kinetic data have not yet been reported.

(*syn*-**45**)   (**65**)   (*anti*-**45**)

IV. CONFORMATION OF IMIDIC ACID DERIVATIVES

A. Restricted Rotation in Amidines

The existence of conformational isomers due to restricted rotation around the C—N bond in amides has been extensively investigated[43]. However, similar restricted rotation in amidine systems has attracted considerably less attention. As in amides, hindered rotation has been attributed to the partial double bond character of the C—NR^{1}R^2 bond, for which structure **68** is responsible. Recently n.m.r. spectroscopy has been used to determine the barrier to rotation between **66** and **67**. Although the period of rotation at room temperature is relatively short (in most cases) on the n.m.r. scale at lower temperatures it is long, resulting in separate signals for the otherwise equivalent N-substituents R^1 and R^2. It is not within the scope of this review to discuss in detail the application of n.m.r. to chemical rate processes; several extensive reviews have appeared on this subject[44, 45].

I. Restricted rotation in benzamidines

The barriers to rotation about the C—N bond that have been reported for benzamidine derivatives are listed in Tables 8 and 9.

In 1968 rotational barriers were reported[46] for a series of N,N-dimethylbenzamidines (69) in which the N'-substituent was H(a), COPh(b), SO$_2$Ph(c), and PO(OPh)$_2$(d). The values were 18·2, 15·2, 16·4, and 17·6 kcal/mol, respectively (Table 8). The authors observed a loose correlation between the height (magnitude) of the rotational barrier and the electro-

R	Type 69
H	a
COPh	b
SO$_2$Ph	c
PO(OPh)$_2$	d

negativity of the N'-substituent. Electron attracting groups should increase the barrier by giving more importance to the zwitterionic structure 70 that increases the double bond character of the C—N(CH$_3$)$_2$ bond. The value obtained for N,N-dimethylbenzamidine (69a) was corrected by

3 kcal/mol (making it 15·2 kcal/mol) because of the effects of hydrogen bonding. Based on the relatively small differences in the rotational barriers of benzamidines 69a–d it was concluded that the electronegativity of the N'-substituent had little effect on the rotational barrier.

Other authors[47] did not agree that the electronegative influence of a SO$_2$

TABLE 8. Rotational barriers for benzamidines of the Ph—C=N—R type (69)[46, 51]

$$Ph-C=N-R$$
$$|$$
$$N(CH_3)_2$$

Compound 69	R	ΔG (kcal/mol)	$\Delta\nu$ (Hz)	T_c (°C)	Solvent	Reference
a	H	18·2	—	−53	$CDCl_3$	46
b	COPh	15·2	—	8	C_6H_5Cl	46
c	SO_2Ph	16·4	—	70	C_6H_5Cl	46
d	$PO(OPh)_2$	17·6	—	83	C_6H_5Cl	46
e	Ph	13·0	24	−16	CH_2Cl_2	51
f	CH_2Ph	12·0	33	−40	CH_2Cl_2	51

TABLE 9. Rotational barriers for benzamidines of the p-XC_6H_4C=N—C_6H_4Y-p type, in deuteriochloroform[52]

$$\underset{\underset{NR_2}{|}}{}$$

Compound	R	X	Y	$\Delta\nu$ (Hz)	T_c (°C)	$\Delta G^{\ddagger}$ (kcal/mol)	$\Delta H^{\ddagger}$ (kcal/mol)	$\Delta S^{\ddagger}$ (e.u.)
74	CH(CH₃)₂	H	H	54	− 13·9	12·6	12·6 ± 0·8	c. 0
75	CH(CH₃)₂	H	NO₂	48	+ 10·1	13·9	13·3 ± 1·0	c. 0
76	(CH₂)₅	H	H	65	− 21·9	12·1	12·3 ± 1·2	c. 0
77	(CH₂)₅	OCH₃	H	62	− 30·4	11·7	8·9 ± 0·6	11 ± 2
78	(CH₂)₅	H	NO₂	55·5	− 4·8	13·1	17·7 ± 1·7	17 ± 6
79	(CH₂)₅	H	OCH₃	66	−17·0	12·3	16·2 ± 1·2	15 ± 5

group (i.e., in benzamidine **69c**) was negligible. They suggested that the phenyl group of *N,N*-dimethyl-*N'*-benzenesulfonylbenzamidine (**71**) caused considerable steric crowding in the planar state (**71a**), which would raise its energy, and therefore diminish the barrier to rotation. To support

steric destabilization

$$C{-}\overset{\ominus}{N}{-}SO_2Ph$$
$$\overset{\oplus}{N}$$
$$CH_3 \quad CH_3$$

(**71a**)

their theory the authors prepared as a cyclic analogue 5-(*N,N*-dimethyl-amino)-1,3,4-oxathiazole-3,3-dioxide (**72**) and *N,N*-dimethyl-*N'*-chloro-methanesulfonyl formamidine (**73**) and reported their rotational barriers

$$O_2S{-}N$$
$$H_2C \quad C{-}N \begin{matrix} CH_3 \\ CH_3 \end{matrix}$$
$$O$$

(**72**)

$$\begin{matrix} H \\ N(CH_3)_2 \end{matrix} C{=}N{-}SO_2CH_2Cl$$

(**73**)

to be 17·9 and 23·3 kcal/mol, respectively. Steric influences on the rotational barrier are reduced to a minimum in amidine **72**; thus, the observed barrier indicates the presence of a partial double bond between the ring and the dimethylamino group, presumably due to the electronegativity of the ring SO_2 group. It was argued further that since no barrier could be determined for *N,N*-dimethylacetamidine[48], while *N,N*-dimethyl-*N'*-*p*-tolyl- and *p*-nitrophenylformamidine gave barriers of 14·1 and 15·4 kcal/mol, respectively[49], the barrier to rotation is dependent on the electronegativity of the *N'*-substituent, at least when no severe steric interactions are possible. For *N,N*-dimethyl-*N'*-chloromethanesulfonyl-formamidine (**73**), in which steric interactions of the dimethylamino group are insignificant[50], Jakobsen and Senning[47] obtained the highest rotational barrier that has been measured so far in any amidine system (23·3 kcal/mol).

McKennis and Smith[51] reported that the n.m.r. spectra of *N,N*-dimethylbenzamidines (**69**) bearing an *N'*-phenyl- (**e**), *N'*-benzyl- (**f**), *N'*-ethyl- (**g**) and *N'*-*t*-butyl- (**h**) showed singlets for the *N*-methyl protons at ambient temperatures. At lower temperatures, however, the spectra of benzamidines **69e**, **69f**, and **69g** showed two singlets of equal intensity for these protons. The rotational barriers about the C—N(CH₃)₂ bond in amidines (**69e**) and (**69f**) were determined by an n.m.r. approximate method to be 13 and 12 kcal/mol, respectively (see Table 8).

$$Ph-C \Big\langle \begin{matrix} N-R \\ N(CH_3)_2 \end{matrix}$$

(**69**)

R = Ph	**69e**
CH$_2$Ph	**69f**
C$_2$H$_5$	**69g**
C(CH$_3$)$_3$	**69h**

The barriers to rotation in amidines **69e** and **69f** are of special interest with respect to the fact that the barrier in N,N-dimethylbenzamidine (**69a**) (18·2 kcal/mol) is not much smaller than that of N,N-dimethyl-N'-benzenesulfonylbenzamidine (**69c**), even after a correction for the estimated contribution of H-bonding. As already mentioned Jakobsen and Senning[47] have suggested that this may not be due to insensitivity to electronic effects, as originally reported by Schwenker and Rosswag[46], but to steric effects resulting from the vastly different space requirements of the H and PhSO$_2$ groups. McKennis and Smith[51] compared the rotational barrier for N,N-dimethyl-N'-benzylbenzamidine (**69f**) (12 kcal/mol) to that reported for N,N-dimethyl-N'-benzenesulfonylbenzamidine (**69c**) (16·4 kcal/mol). The spatial requirement of a benzyl group is much closer to that of the benzenesulfonyl or benzoyl group than to hydrogen, but the electronic effects are much different. The suggestion was made that the small difference in the rotational barrier between amidines (**69f**) and (**69c**) confirms that the influence of electronic factors is not large. The lack of n.m.r. evidence for restricted rotation in benzamidine (**69f**) may be due to the rather small value for the magnetic nonequivalence of the methyl groups.

Rappoport and Ta-Shma[52] observed the temperature-dependent n.m.r. spectra of benzamidines **74–80** in deuteriochloroform (Table 9). At room temperature the n.m.r. spectrum showed only one doublet and one septet for the two isopropyl groups of benzamidines **74** and **75**. On cooling, the doublet first broadened, then separated into two singlets below the coalescence temperature (Table 9), and then gave two sharp doublets. The broad singlet of the α-methylene protons of the piperidine ring of amidines **76–79** broadened at room temperature and separated into two broad singlets on cooling.

Therefore, it seemed as if the transformation of the two magnetically nonequivalent alkyl groups at low temperature to two equivalent groups at higher temperature may be due either to (a) *syn–anti* isomerization (equation 1) or (b) to restricted rotation around the carbon–nitrogen single bond (equation 2). The conclusion that the only barrier observed was

$$p\text{-}XC_6H_4C{=}NC_6H_4Y\text{-}p$$
$$|$$
$$NRR'$$

(74) X = Y = H, R = R′ = CH(CH₃)₂

(75) X = H, Y = NO₂, R = R′ = CH(CH₃)₂

(76) X = Y = H, RR′ = (CH₂)₅

(77) X = CH₃O, Y = H, RR′ = (CH₂)₅

(78) X = H, Y = NO₂, RR′ = (CH₂)₅

(79) X = H, Y = CH₃O, RR′ = (CH₂)₅

(80) X = Y = H, R = Ph, R′ = CH₃

$$\tag{1}$$

syn   anti

$$\tag{2}$$

that indicated by equation 2 was based on the following:

(a) In all systems the two magnetically different species observed at low temperatures were in a 1:1 ratio. This would be expected for equation 2 since the R groups in all the amidines used are identical (except for amidine **80** for which no barrier was observed). However, it is highly improbable that for all the compounds the *syn* and *anti* isomers would have equal energies.

(b) No change in the methoxyl signal was observed in cooling for benzamidines **77** and **79**, as expected from equation 2. Such change was observed, however, for the *syn–anti* isomerization of *N*-arylketimines[53].

(c) The values obtained for the rotational barrier, 12–14 kcal/mol, are in the range obtained for similar amidine systems[49, 54].

(d) Since structure **81a** is responsible for the barrier to rotation, the barrier height should be raised by electron-attracting Ar′ groups, while it

(81a)   (81b)

TABLE 10. Rotational barriers[56] for $C_6H_5{-}C({=}X){-}N(CH_3^{A})(CH_3^{B})$

Compound	N-CH_3 n.m.r. signal		T_c(°C)	Solvent	E_{act} (kcal/mol)
	$\delta_A{}^a$	$\delta_B{}^a$			
82 (X = S)	2·73	3·30	83	C_6H_5Cl	15·4
83 (X = Se)	2·64	3·29	107·5	C_6H_5Cl	21·1

[a] The n.m.r. signals are reported at a temperature of 35 °C.

should be lowered by electron-donating aryl groups due to the contribution of structure **81b**. However, both the lateral shift and rotation mechanisms[32] for *syn–anti* isomerization show a positive ρ value (1·5–2·2)[55] for substituents on the nitrogen and a positive, but lower ρ value (0·1–0·35)[32] for substituents on the carbon. The increase in ΔG in changing the N'-phenyl-(**74**) to N'-p-nitrophenyl-(**75**) ($\rho = -1{\cdot}5$ for the pair (**74**)(**75**) at -30°C), and the decrease in ΔG on changing the C-phenyl-(**76**) to C-p-methoxyphenyl (**77**) ($\rho = -1{\cdot}35$ for the pair (**76**)(**77**) at -30°C) fits only equation 2*.

In trifluoroacetic acid at 30°C amidine **74** shows a pair of methyl doublets ($\Delta\nu$ = 51 Hz) and two methine septets ($\Delta\nu$ = 31 Hz). No spectral change was observed upon heating to 65°C. The barrier for internal rotation in acidic solvents is apparently much higher than in $CDCl_3$. This behavior is anticipated by the mechanism proposed by Rappoport and Ta-Shma[52] if the amidine is protonated on the negatively charged nitrogen atom of (**81a**).

Only benzamidine **80** showed no temperature-dependent spectra down to -70°C. This can be attributed to a greater free energy difference than 2 kcal/mol between the two conformers enabling only the thermodynamically more stable conformer to be observed, or to a rotational barrier either too high or too low to be measured by the n.m.r. technique.

Schwenker and Rosswag[56, 17b] reported barriers to rotation about the $\mathrm{C{-}N}\langle$ bond for the sulfur and selenium analogues of dimethylbenzamidines, **82** and **83**, respectively. These barriers and other pertinent data are found in Table 10.

* Data were not divided by 2·3 in the original publication. The corrected data should be as in this text. The authors are indebted to Dr. Z. Rappoport for this information.

2. Restricted rotation in formamidines

The barriers to rotation that have been reported for formamidine derivatives are listed in Table 13.

Marsh and Goodman[57] examined the n.m.r. spectra of a number of N'-aryl-N,N-dimethylformamidines (**84**) in deuteriochloroform and deuterated benzene. These authors recognized a striking difference in the

(**a**, Ar = C_6H_5)

(**b**, Ar = p-$CH_3OC_6H_4$)

(**c**, Ar = p-NCC_6H_4)

(**d**, Ar = p-ClC_6H_4)

(**e**, Ar = p-$CH_3C_6H_4$)

(**f**, Ar = p-$O_2NC_6H_4$)

appearance of the N-methyl resonances in these two solvents. Table 11 lists some of the compounds studied along with pertinent n.m.r. data.

In deuteriochloroform the N-methyl signals of all the formamidines (**84**) appeared as a six-proton sharp singlet. However, in deuterated benzene the N-methyl protons were usually represented by a broad singlet. In addition, the appearance of the N-methyl peak was strongly affected by substituents in the N'-phenyl group. When electron withdrawing groups were present, as in formamidines (**84c**) and (**84f**) the N-methyl groups showed two distinct singlets in deuterated benzene (Table 11), while with electron donating groups, e.g. formamidines (**84b** and **84e**) the deuterated

TABLE 11. Chemical shifts for the N-methyl protons for some formamidines of the Ar—N=CHN(CH₃)₂ type[a] (**84**)[57]

Type 84	Ar	Solvent	Chemical shift (τ)
(**a**)	C_6H_5	C_6D_6	7·50[b]
		$CDCl_3$	7·17
(**b**)	p-$CH_3OC_6H_4$	C_6D_6	7·50
		$CDCl_3$	7·19
(**c**)	p-NCC_6H_4	C_6D_6	7·68, 7·40[b]
		$CDCl_3$	7·05
(**d**)	p-ClC_6H_4	C_6D_6	7·53[b]
		$CDCl_3$	7·04
(**e**)	p-$CH_3C_6H_4$	C_6D_6	7·46
		$CDCl_3$	7·10
(**f**)	p-$O_2NC_6H_4$	C_6D_6	7·70, 7·37
		$CDCl_3$	6·09

[a] All chemical shifts relative to TMS.
[b] Broadened.

benzene spectra differed very little from that obtained in deuteriochloroform. The resolution of the N-methyl peaks in deuterated benzene was
also strongly temperature dependent. For example, at 59°C N'-(p-chlorophenyl)-N,N-dimethylformamidine (84d) showed a sharp N-methyl singlet
at 7·64 τ. Upon cooling +0–10°C this peak broadened and split into two
singlets at 7·50 and 8·01 τ.

The authors attributed this combination of solvent and substituent
effect to the formation of a complex between benzene and the formamidine
derivatives. The observed temperature dependence of the spectra in benzene was considered as reflecting the stability of the complex.

Shortly after that paper[57] appeared Bertelli and Gerig[49] reported that
the chemical shift phenomenon observed by the former authors was not
due to solvent complexation but rather to restricted rotation about the
C—N(CH$_3$)$_2$ bond. The n.m.r. spectra of N'-(p-nitrophenyl)-N,N-di-
methylformamidine (84f) at various temperatures (Figure 1) exhibited a
typical coalescence sequence in both benzene and chloroform, a fact that
was attributed to the equilibrium 85 ⇌ 86. The barrier to rotation arises

from additional π-bonding due to dipolar contributors such as 87 in the
ground state. The n.m.r. spectra of N'-(p-tolyl)-N,N-dimethylformamidine
(84e) at various temperatures exhibited curves essentially identical to
those shown in Figure 1, except that the coalescence temperatures are
lower. The barriers to rotation for formamidines 84e and 84f along with
pertinent n.m.r. data are given in Table 13.

Although Marsh and Goodman[57] were partially correct in assigning
some unusual benzene-substrate complexes in some cases, they seem to
have overlooked the fact that different solvents affect the relative chemical

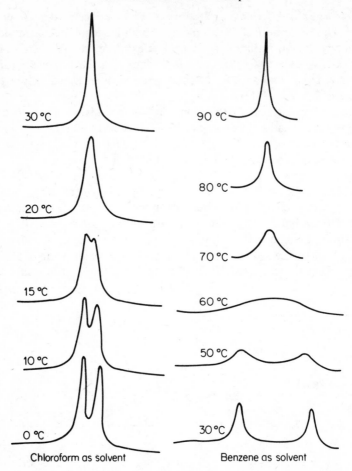

FIGURE 1. N.m.r. coalescence spectra for the N-methyl signals of N'-(p-nitrophenyl)-N,N-dimethylformamidine (**84f**).

shifts which change the $\Delta\nu$ term in the rate expression and thus the coalescence temperature. Therefore, in the n.m.r. spectrum of formamidines (**84f**) in chloroform solution, $\Delta\nu$ is small enough that the coalescence temperature is below ambient. However, in benzene solution the $\Delta\nu$ term is large, the coalescence temperature is above ambient and the methyl signals are resolved.

Harris and Wellman[54] were the first to measure successfully the activation parameters for hindered rotation in a trialkylated formamidine, namely N'-t-butyl-N,N'-dimethylformamidine (**84g**). The nuclear magnetic

TABLE 12. Kinetic data for the exchange process, $88 \rightleftarrows 89$[54]

Toluene-d_8		Pyridine-d_5	
T(K)	k	T(K)	k
248·7	125·0	247·5	111·0
241·5	78·9	244·0	64·6
238·5	42·8	241·5	52·7
233·0	23·4	236·0	44·9
227·5	13·9	235·5	31·2
225·5	7·3	233·0	37·5
220·7	5·2	230·5	15·9
		229·5	15·5
		225·0	9·0
		223·5	7·2
		221·0	6·4
		216·0	3·7
		212·0	2·4

resonance method was used to determine the barrier to rotation, $88 \rightleftarrows 89$ in toluene-d_8 and pyridine-d_5. In both solvents the n.m.r. signal of the

(88)   **(89)**   **(90)**

N-methyl protons was found to be temperature dependent. Kinetic data for the exchange process $88 \rightleftarrows 89$ for both solvents are found in Table 12. The low free energy of activation for this equilibrium (Table 13) implies that there is less double bond character of the C—N(CH₃)₂ bond, arising from electron delocalization (see structure **90**) than in other similar systems. The small entropies of activation that were observed (1·6 e.u. ± 2·8 in toluene-d_8 and −5·4 ± 2·7 in pyridine-d_5) do not indicate any unusual solvent–solute interaction as has been previously suggested.

3. Restricted rotation in amidinium salts

a. *Benzamidinium salts.* Recently McKennis and Smith[51] reported that the ambient temperature n.m.r. spectra of benzamidinium salts (**91a–91e**) showed two equally intense signals for the N-methyl protons (Table 14).

TABLE 13. Rotational barriers for formamidines of the H—C=N—R type (84)[47,49,54]

$$\underset{\underset{N(CH_3)_2}{|}}{}$$

Compound	R	ΔG (kcal/mol)	$\Delta \nu$ (Hz)	T_c (°C)	Solvent	Reference
84f	p-$O_2NC_6H_4$	15·9	4·7	—	$CHCl_3$	49
84f	p-$O_2NC_6H_4$	15·2	42·5	—	C_6H_6	49
84e	p-$CH_3C_6H_4$	14·1	4·7	—	$CHCl_3$	49
84e	p-$CH_3C_6H_4$	13·9	51·5	—	C_6H_6	49
84g	$C(CH_3)_3$	11·9	—	—	Toluene-d_8	54
84g	$C(CH_3)_3$	12·4	—	—	Pyridine-d_5	54
73	SO_2CH_2Cl	23·3	2·6	145	o-Dichlorobenzene	47

TABLE 14. N.m.r.[a] and rotational barrier data for some N,N-Dimethylbenzamidinium hydrohalides, type **91**[51]. Reprinted with permission from J. S. McKennis and P. Smith, *J. Org. Chem.*, **37**, 4173 (1972). Copyright by the American Chemical Society.

Compound	δ_{N-CH_3}	δ_{N-CH_3}	$\Delta\nu$ (Hz)	T_c (°C)	ΔG (kcal/mol)	Solvent
a, PhC=N—Ph $\qquad$N(CH$_3$)$_2$·HCl	6·4 (6·06)[b]	7·0 (6·83)[b]	47	135	20·4	PhNO$_2$
b, PhC=N—CH$_2$Ph $\qquad$N(CH$_3$)$_2$·HCl	6·47 (6·13)[b]	7·20 (6·90)[b]	47	130	20·1	PhNO$_2$
c, PhC=N—C$_2$H$_5$ $\qquad$N(CH$_3$)$_2$·HCl	6·5	7·1	36	—	—	PhNO$_2$
d, PhC=N—C(CH$_3$)$_3$ $\qquad$N(CH$_3$)$_2$·HCl	6·27	7·05	—	—	—	—
e, PhC=N—CPh$_3$ $\qquad$N(CH$_3$)$_2$·HI	6·15	7·0	44	120	19·7	C$_2$H$_2$Cl$_4$

[a] All values refer to chemical shifts in CHCl$_3$.
[b] Chemical shifts in PhNO$_2$.

(a, R = Ph, X = Cl)

(b, R = CH$_2$Ph, X = Cl)

$$\text{Ph—C} \underset{\text{N(CH}_3)_2}{\overset{\text{N—R}}{\diagdown}} \cdot HX$$

(c, R = C$_2$H$_5$, X = Cl)

(d, R = C(CH$_3$)$_3$, X = I)

(e, R = CPh$_3$, X = I)

(91)

These signals were not due to NH coupling resulting from protonation on the sp^3 nitrogen since the two signals presented in deuterium oxide. A similar behavior was observed for the hydrochloride salts of some N,N-dimethyl-N'-arylformamidines[57]. Assuming free rotation about the C—N(CH$_3$)$_2$ bond these absorptions could arise from an equal mixture of the two geometric isomers 92 and 93. However, an increase in the size of R should increase the amount of isomer 92 relative to 93. Furthermore

$$\underset{\text{N(CH}_3)_2 \quad \text{H}}{\overset{\text{Ph} \diagdown \quad \diagup \text{R}}{\text{C}=\overset{+}{\text{N}}}} \text{X}^-$$

$$\underset{\text{N(CH}_3)_2 \quad \text{R}}{\overset{\text{Ph} \diagdown \quad \diagup \text{H}}{\text{C}=\overset{+}{\text{N}}}} \text{X}^-$$

(92)   (93)

differences in the chemical shifts of the R groups that might be expected based on the existence of isomers 92 and 93 were not observed.

The preceding observations were accounted for by restricted rotation about the C—N(CH$_3$)$_2$ bond. The barriers for benzamidinium salts 91a, 91b, and 91d were determined by the variable-temperature n.m.r. technique to be 20·4, 20·1, and 19·7 kcal/mol, respectively (Table 14). The authors attributed the small effect of the N'-substituent to the predominance of steric effects.

A tentative assignment of the position of the methyl groups (endo 94a, exo 94b) in the protonated amidines was made on the basis of the n.m.r. spectrum of N-benzyl-N-methyl-N'-phenylbenzamidinium chloride (94).

$$\text{Ph—C} \underset{\underset{\underset{\text{CH}_2\text{Ph}}{|}}{\text{N—CH}_3}}{\overset{\overset{\overset{\text{Ph}}{|}}{\text{N—H}}}{\diagdown}} +$$

$$\text{Ph—C} \underset{\underset{\underset{\text{CH}_3}{|}}{\text{N—CH}_2\text{Ph}}}{\overset{\overset{\overset{\text{Ph}}{|}}{\text{N—H}}}{\diagdown}}$$

(94a)   (94b)

The ambient temperature spectrum of the salt in deuteriochloroform showed the existence of two rotational isomers in the ratio of 3·8:1. The rotamer in lower concentration possesses the higher field methyl signal and the lower field benzyl signal. Assuming that this rotamer has the bulkier group (CH_2Ph) in the endo position (**94b**) then the high and low field geminal N-methyl signals for the protonated amidines can be assigned to the exo- and endomethyl groups, respectively (Table 14).

When an amidine is protonated, both nitrogens can share the positive charge and would be expected to do so if a planar configuration was possible. If the protonated nitrogen possesses a substantial charge the rotational barrier for the C—NHR bond should increase proportionally to the magnitude of the share. McKennis and Smith reported[51] that in none of the ambient temperature spectra of the protonated benzamidines was there evidence for the existence of the two possible rotational isomers **95a** and **95b**. Even when the N-substituents were larger than the N_1-substituents, as in N,N-dibenzyl-N'-methylbenzamidinium chloride, two peaks were observed for the N-substituents and only one for the N'-substituent.

(95a)   (95b)

The importance of steric effects upon the conformation of amidinium derivatives was demonstrated by examining the low temperature n.m.r. spectra of N,N,N',N'-tetrasubstituted amidinium salts **96–98**. Restricted

(96)   (97)   (98)

rotation about the C—$N(CH_3)_2$ bond was observed in each compound (benzamidinium salt **96** gave a rotational barrier of 15·5 kcal mol, Table 15) by the nonequivalence of the N-methyl signals. However, hindered rotation about the C—NR_2 bond was not observed even down to $-50°C$. This could be due to either a substantially lower barrier to rotation about the C—NR_2 bond rather than about the C—$N(CH_3)_2$ bond, corresponding to localization of the charge mainly on the nitrogen of the C—$N(CH_3)_2$ bond, or a higher barrier that fixes the molecule in conformation **95a** or **95b**, exclusively. Space filling models indicate that the Z isomer (**95b**) is sterically more crowded than the E isomer (**95a**). Therefore, if charge delocalization is important, isomer **95a** would be expected to predominate.

Raison[58] reported that N,N,N'-trimethyl-N'-phenylbenzamidinium

TABLE 15. Rotational barriers for some quaternary amidinium salts[51]. Reprinted with permission from J. S. McKennis and P. Smith, *J. Org. Chem.*, **37**, 4173 (1972). Copyright by the American Chemical Society.

Compound	$\Delta\nu$ (Hz)	T_c (°C)	ΔG(kcal/mol)	Solvent
99a	15	1	14·2	CH_2Cl_2
99a	9	?	17·8	CF_3CI_2H
99a	6	54	17·5	HCO_2H
99b	17	31	15·6	CH_2Cl_2
96	23	31	15·5	CH_2Cl_2
97	—	>40	—	CH_2Cl_2

iodide (**99a**) could be obtained in two forms, with distinct melting points. McKennis and Smith studied[51] the n.m.r. spectra of the benzamidinium iodide (**99a**) and tetrafluoroborate (**99b**). At ambient temperatures two N—CH_3 absorptions were observed in the ratio 2:1, while at lower

$$Ph—C \genfrac{}{}{0pt}{}{N(CH_3)Ph}{N(CH_3)_2} {}^{+} \quad X^{-} \qquad \begin{array}{l}(\textbf{99a, X = I})\\(\textbf{99b, X = BF}_4)\end{array}$$

temperatures the absorption due to the geminal *N*-methyl protons split into two broad peaks (ratio ca. 1:1) but the *N'*-methyl peak remained a singlet (slightly broadened) down to −50°C. The rotational barrers determined for salt **99a**, in several solvents, and **99b** are given in Table 15. These results indicate that the two form of *N,N,N'*-trimethyl-*N'*-phenyl-benzamidinium iodide (**99**) obtained by Raison must have been different crystalline modifications, rather than different rotamers.

  b. *Formamidinium salts.* Relatively little has been done in the area of restricted rotation in formamidinium salts. As previously mentioned McKennis and Smith[51] observed restricted rotation about the C—N(CH_3)$_2$ bond in *N,N,N'*-trimethyl-*N'*-benzylformamidinium iodide (**98**). In 1964 Ranft and Dähne[59] reported the temperature-dependent n.m.r. spectra of formamidinium salts **100a–100d** in deuteriochloroform and concluded that rotation about the C=N double bond in these compounds is hindered. For example, at ambient temperatures the n.m.r. spectrum of *N,N,N'*-

(**100a**)   (**100b**)   (**100c**)   (**100d**)

tetramethylformamidinium perchlorate (**100a**) showed one singlet for the
N-methyl protons. Upon cooling to $-20°C$ this singlet gradually
broadened and split into two singlets of equal intensity. Using these data
the authors calculated the barrier to rotation in formamidinium salt **100a**
to be $17·5 \pm 1·5$ kcal/mol.

Marsh and Goodman[57] reported that the room temperature n.m.r.
spectra of formamidinium hydrochlorides of type **101** showed the N-
methyl protons as two sharp singlets of approximately equal intensity.

$$(101) \qquad (102)$$

The authors attributed this to the existence of the salt in solution with the
positive charge concentrated on the dimethylamino nitrogen (**102**),
making the two methyl groups nonequivalent because of the expected
cis–trans isomerism.

It should, however, be pointed out that the observations made by Marsh
and Goodman[57] are also consistent with the exclusive existence of conformation **103a** or **103b**. Due to steric interactions between the endo methyl
and aryl groups of isomer **103b** it would be expected that the isomer **103a**

$$(103a) \qquad (103b)$$

which is less sterically crowded will predominate. Therefore, if charge
delocalization is important the two N-methyl absorptions in the n.m.r.
may be due to the exclusive existence of conformational isomer **103a**.

It was reported[54] that protonation of N,N-disubstituted formamidines
(Table 16) in trifluoroacetic acid yields a conjugate acid which can exist
in either a *trans*-(**104a**) or *cis*-(**104b**) configuration. The n.m.r. spectra of
these formamidines gave distinguishable signals for protons H_a through
H_e. Although no chemical shifts were reported those protons were assigned
on the basis of line shape and coupling constants (e.g., $J_{H_aH_b} \cong J_{H_cH_e} \cong$
14, $H_{H_dH_e} \cong 6$ Hz; H_a, H_c, and H_d are doublets broadened by quadrupole
coupling, H_b is a triplet and H_e is a quartet). Steric interaction between

TABLE 16. *Cis–trans* equilibrium constants[a] $\left(K = \dfrac{[cis]}{[trans]} \right)$ for *N,N'*-diaryl-formamidinium and *N*-aryl-*N'*-*t*-butylformamidinium trifluoroacetates (type **105**)[54]

Substituent (X)	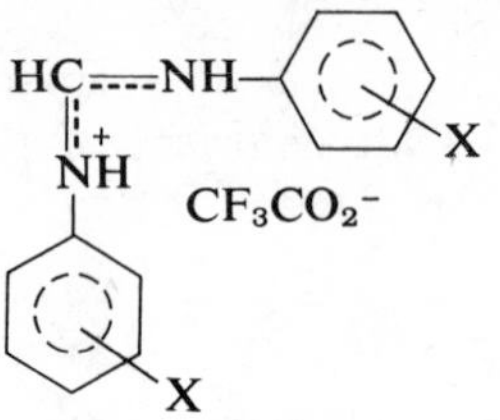 HC⋯NH⟨⟩X / NH⟨⟩X $CF_3CO_2^-$	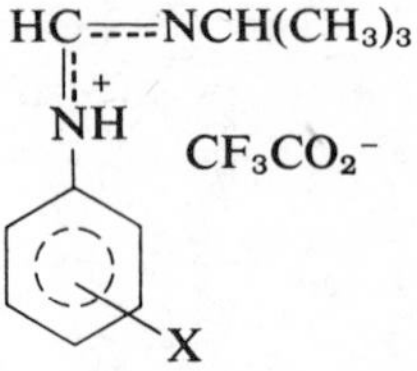 HC⋯NCH(CH₃)₃ / NH⟨⟩X $CF_3CO_2^-$
H	1·00	0·48
3-Acetyl	0·65	0·28
4-Acetyl	0·25	0·09
3-Trifluoromethyl	0·65	0·25
4-Trifluoromethyl	0·40	0·20
3-Chloro	0·65	0·38
4-Chloro	0·85	0·32
4-Methoxy	1·90	1·00
2-Methyl	2·50	2·20
3-Methyl	1·10	0·62
4-Methyl	1·40	0·68
2,4-Dimethyl	3·0	2·1
2,6-Dimethyl	≥20[b]	—
2,4,5-Trimethyl	≥20[b]	2·4

[a] N.m.r. spectra of 10% solute in TFA at 38 ± 2°C.
[b] In these cases no *trans*-isomer could be detected.

H_c and the endo aryl group in isomer **104b** causes the aryl moiety to orient in a way that H_c lies directly over the aryl π-system. Consequently H_c is shielded and the signal shows an upfield shift of about 1 p.p.m.

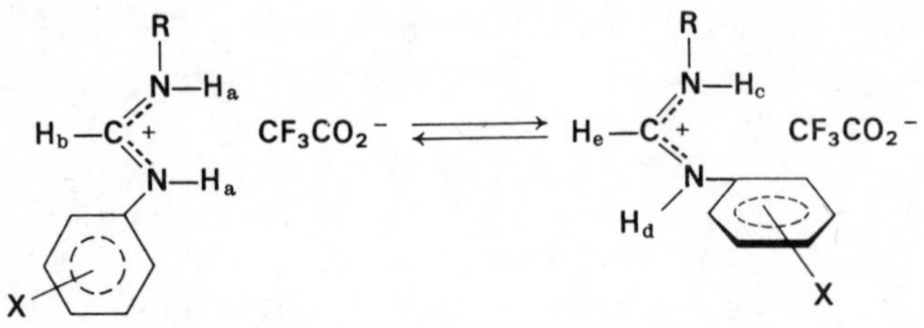

(104a)   (104b)

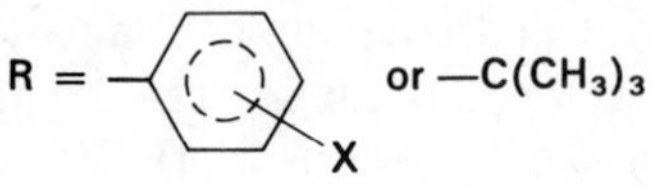

$R = -$⟨⟩X   or $-C(CH_3)_3$

Integration provided a measure of the concentration of the rotational isomers **104a** and **104b**. The results, which are summarized in Table 16, show a significant aryl-substituent effect upon the equilibrium. The equilibrium constants for *meta-* and *para-*substituted compounds can be correlated by a Hammett plot which gives ρ-values of −0·75 and −0·83 for the aryl and *t*-butyl series, respectively. Electron donating groups were found to stabilize the *cis*-configuration. This is illustrated in the methyl substituted arylformamidinium salts which show an increase in the *cis*-isomer as the number of methyl groups increase (Table 16). The authors state that this substituent dependence eliminates a simple steric effect, a conclusion which is supported by the fact that the conjugate acid of *N,N'*-di-t-butylformamidine exists wholly in the *trans*-configuration in TFA.

Harris and Wellman[54] concluded that because of the polar nature of the substituents about the partial carbon–nitrogen double bond it appears likely that the *cis*-configuration is stabilized by favorable dipolar interactions, however, the exact cause of the preferred *cis*-orientation was not revealed.

c. *Acetamidinium salts.* In 1963 Hammond and Neuman[60] observed hindered rotation in amidinium salts **106**. The n.m.r. spectra of salts

(106)

(107)

(108)

(a, R = H, X = Cl)
(b, R = H, X = NO₃)
(c, R = CH₃, X = Cl)

106a, **106b** and **107** in anhydrous DMSO (Table 17) showed two nitrogen proton signals in the ratio of 1:1. The authors attributed this to restricted rotation due to partial double bond character resulting in magnetically nonequivalent endo (H_a) and exo (H_b) protons. The single N—H proton signal observed for amidinium salt **108** (Table 17) is consistent with the above interpretation since the nitrogen protons are restricted to the exo

TABLE 17. N.m.r. results for some amidinium ions (**106–108**) at 30°C[60]. Reprinted with permission from G. S. Hammond and R. C. Neuman, Jr., *J. Phys. Chem.*, **67**, 1655 (1963). Copyright by the American Chemical Society.

Compound	Solvent	δ_{N-H_a}	δ_{N-H_3}	δ_{N-CH_3}
106a	DMSO	8·84	9·35	—
106b	DMSO	8·84	9·35	—
106c	DMSO	9·23	10·13	2·86; 2·94[a]
106c	H_2O	—	—	2·96; 3·12[a]
106c	14% H_2SO_4	7·1	7·6	2·99; 3·13[a]
106c	60% D_2SO_4	—	—	3·07; 3·23[a]
107	DMSO	8·87	9·13	—
108	DMSO	—	10·12	—

[a] Doublet.

(H_b) positions. Based on the similarity of the chemical shifts of the exo protons of amidinium salts **106c** and **108** (Table 17) the authors tentatively assigned the lower field N—H signal to the exo protons and the higher field N—H signal to the endo N—H protons.

The N—H signals of acetamidinium chloride (**106a**) in DMSO were found to be temperature dependent. The two signals observed at 30°C (δ 8·84 and 9·35) slowly coalesced to one broad singlet upon heating to above 100°C.

The initial signals (37°C) give a transverse relaxation time, $T_2 \cong 0·03$ sec while the final signal (115°C) gives $T_2 \cong 0·01$ sec. Using these T_2 data $E_a = 9 \pm 2$ kcal and 25 ± 8 kcal mol^{-1} were found. Since T_2 probably decreases in some regular manner with increase in temperature, the true activation parameters lie somewhere between these extreme values. Although the barrier of rotation is not known with adequate precision the value is within the same range as those of amides (7–18 kcal/mol).

There are three possible conformations for *N,N'*-dimethylacetamidinium chloride (**109a**, **109b**, and **109c**). Hammond and Neuman[60] suggest that

(109a)   (109b)   (109c)

the n.m.r. of this compound in several solvents (Table 17) indicates that conformation **109a** is the only detectable form present in solution. In each solvent two separate N—CH$_3$ peaks of equal area are observed. In

$$\overset{\text{X}}{\underset{\|}{}}$$

TABLE 18. Rotational barriers for $CH_3C\!-\!N(CH_3)_2$ in formamide[48]. Reprinted with permission from R. C. Neuman, Jr. and V. Jonas, *J. Phys. Chem.*, **75**, 3532 (1971). Copyright by the American Chemical Society.

Compound	X	E_a (kcal/mol)
Dimethylthioacetamide	S	43.7 ± 5.6
Dimethylacetamide	O	24.7 ± 0.8
Dimethylacetamidinium chloride	NH_2^+	19.6 ± 1.0
Dimethylacetamidine	NH	*a*

a Not measurable.

DMSO and 14% H_2SO_4, in which two N—H peaks of equal area are observed, each N—CH$_3$ signal is split into a doublet ($J = 5$ Hz). The magnitude of this coupling constant implies that splitting of the N—CH$_3$ signal is due to a proton on the same nitrogen atom. The spectra could also be interpreted as arising from equal mixtures of conformations **109b** and **109c**, however, models indicate considerable steric interaction between the endo methyl groups of **109b**.

The barrier to rotation about the central C—N bond in *N,N*-dimethylacetamidinium chloride has been determined[48] and compared to the analogous thioamide and amide (Table 18). It was not possible to determine a rotational barrier for *N,N*-dimethylacetamidine because the N—CH$_3$ protons gave rise to only one signal while cooling to $-40°C$ in chloroform. The authors suggested that the lack of nonequivalent methyl groups in *N,N*-dimethylacetamidine is due to rapid rotation about the C—N(CH$_3$)$_2$ bond. These results indicate that the contribution of the dipolar canonical form **110** to the ground state of these compounds appears to increase in the order N < O < S.

$$\begin{array}{cc} X^- & CH_3 \\ \diagdown & \diagup \\ C\!=\!N^+ \\ \diagup & \diagdown \\ CH_3 & CH_3 \end{array}$$

(110)

Neuman and Young[48] also suggested that values of $J(^{13}CH_3)$ might reflect the relative rotational barrier for these systems, however, in a late publication[61] Neuman and Jonas demonstrated that such a general correlation is not valid.

A comparison of the rotational barriers for compounds **111–113a** in DMSO was made by Neuman and Jonas[61]. These deuterated compounds were synthesized in order to obtain relatively symmetric N(CH$_3$)$_2$ n.m.r.

TABLE 19. Activation parameters for C—N rotation in $CD_3C(X)N(CH_3)_2$ in DMSO-d_6[61]. Reprinted with permission from R. C. Neuman, Jr. and V. Jonas, *J. Phys. Chem.*, **75**, 3532 (1971). Copyright by the American Chemical Society.

Compound	E_a (kcal/mol)	ΔF^* (25°C) (kcal/mol)	ΔH^* (kcal/mol)	ΔS^* (e.u.)
111 (X = O)	$20 \cdot 3 \pm 0 \cdot 3$	$18 \cdot 5$	$19 \cdot 7 \pm 0 \cdot 3$	$+4 \cdot 1 \pm 0 \cdot 8$
112 (X = S)	$25 \cdot 9 \pm 0.9$	$23 \cdot 4$	$25 \cdot 3 \pm 0.9$	$+6 \cdot 3 \pm 2 \cdot 1$
113a (X = $NH_2^+ NO_3^-$)	$21 \cdot 3 \pm 0 \cdot 3$	$21 \cdot 5$	$20 \cdot 7 \pm 0 \cdot 3$	$-2 \cdot 6 \pm 0 \cdot 7$
113b (X = $NH_2^+ Cl^-$)	$22 \cdot 8 \pm 0 \cdot 7$	$21 \cdot 8$	$22 \cdot 2 \pm 0 \cdot 7$	$+1 \cdot 4 \pm 1 \cdot 9$

signals amenable to analyses using the total line shape equation of Gutowsky and Holm[62]. Table 19 summarizes the results. The values obtained were compared with the results of molecular orbital calculations[63] and solvation effects were also discussed.

(**111**, X = O)
(**112**, X = S)
(**113a**, X = $NH_2^+NO_3^-$)
(**113b**, X = $NH_2^+Cl^-$)

These authors also synthesized *N*-methylacetamidinium chloride (**114**, X = $NH_2^+Cl^-$) and nitrate (**114**, X = $NH_2^+NO_3^-$) and determined the isomer distribution between **114a** and **114b** in different solvents. These data are compared with those obtained for the analogous amide (**114**, X = O) and thioamide (**114**, X = S). In all cases the N—CH_3 signals

(**114a**)　　　　　　　(**114b**)

corresponding to **114a** and **114b** can be identified with the isomer by the substantially greater C—CH_3, N—CH_3 spin-coupling observed in the *trans* configuration **114a**. The similarity in isomer distribution (Table 20) for the various X groups suggests that the potential steric interaction between the endo H and CH_3 groups of the amidinium salts (**114**, X = $NH_2^+Cl^-$, $NH_2^+NO_3^-$) is relatively unimportant.

4. Restricted rotation in amidoximes

Cramer and DeRyke[65a] have recently succeeded for the first time in determining the barrier to rotation in an amidoxime derivative. Table 21 gives the rotational barriers for 2,2'-iminobis(acetamidoxime) **115a**, abbreviated IBO, and its Ni(II) **115b** and Zn(II) **115c** complexes. The

$$\overset{\text{X}}{\overset{\|}{}}$$

TABLE 20. Isomer distribution for the compounds $CH_3\overset{\text{X}}{\overset{\|}{C}}NHCH_3$ (**114a** and **114b**) in solution[61]. Reprinted with permission from R. C. Neuman, Jr. and V. Jonas, *J. Phys. Chem.*, **75**, 3532 (1971). Copyright by the American Chemical Society.

X	Solvent	**114a** (%)	**114b** (%)	Reference
O	H_2O	97	3	64
S	CCl_4	97	3	64
	C_6H_6	97	3	64
$NH_2^+ Cl^-$	D_2O	96	4	61
$NH_2^+ NO_3^-$	D_2O	96	4	61
	DMSO-d_6	97	3	61

crystal structures of the Ni(II) and Cu(II) complexes **115b** and **115c** of acet-amidoxime have shown that the $C-NH_2$ bond length is comparable to many amides. The barrier to rotation about the $C-NH_2$ bond of IBO (10·5 kcal/mol) is among the lowest ever reported for an amide or an amidine. This low barrier is consistent with the failure of Neuman and Young[48] to observe magnetic nonequivalence of the methyl groups in *N,N*-dimethylacetamidine as low as $-40°C$.

$$HN\underset{CH_2C=N-OH}{\overset{CH_2C=N-OH}{<}}$$

with NH_2 groups on top ($CH_2C=N-OH$) and bottom ($CH_2C=N-OH$, NH_2)

(115a)

Coordination of either Zn^{2+} or Ni^{2+} causes the rotational barrier to increase by about 1 kcal/mol (Table 21) or about 10%. This increase in the barrier of IBO is comparable to that observed in other systems. For example, Gore and co-workers[65b] reported that coordination of BF_3 with

TABLE 21. Rotational barriers for amidoximes[65a]. Reproduced by permission of the National Research Council of Canada from the *Canadian Journal of Chemistry*, **51**, 892–895 (1973).

Compound	Solvent	$\Delta\nu$ (Hz)	T_c (°C)	ΔG^* (kcal/mol)
115a, IBO	DMF	25 ± 1	-60 ± 2	$10·5 \pm 0·3$
115b, $Zn(IBO)_2(NO_3)_2$	MeOH	15 ± 1	-47 ± 2	$11·5 \pm 0·3$
115c, $Ni(IBO)_2(NO_3)_2$	DMSO	3000 ± 300	$+17 \pm 2$	$11·8 \pm 0·4$

N,N-dimethylformamide increased the barrier to 21·9 kcal/mol from 20·9. The authors concluded that coordination of a Lewis acid at site X favors structure **116**, causing an increase in the double bond character of the C—NH_2 bond, and consequently increasing the rotational barrier.

(116)

B. Conformation of Imidic Esters

Alkyl carboxylates and their sulfur analogs show generally a rather rigid antiperiplanar conformation. Planarity is due to resonance within the ester function. Imidate esters being closely related to esters, similar conformational preference was expected. However, an additional geometrical isomerism can overlap in imidates depending on the optimum conformation of the substituent on the nitrogen. As a consequence, four possible distinctly different conformations are to be considered in imidic esters, i.e. an anti-*periplanar* (*ap*) and a syn-*periplanar* (*sp*) conformation within the E and Z forms[66]. Other authors prefer the *anti–cis, anti–trans, syn–cis, syn–trans* convention previously used. Z refers to proximity of OR^3 and R^2 on nitrogen while anti-*periplanar* and syn-*periplanar* indicate the OR^3/R^1 conformation; E means sterically remote OR^3 and R^2 groups. The *anti–syn, cis–trans* combination seems, however, less confusing, since both terms, *syn* and *cis*, refer to the same substituent. *Syn* and *anti*

E ap	*Z* ap	*E* sp	*Z* sp
(*ac*)	(*sc*)	(*at*)	(*st*)

means geometrical configuration of the OR^3 group and the substituent on nitrogen (R^2) while *cis* and *trans* indicate the conformation of R^3 on the oxygen with reference to R^2, the nitrogen substituent.

One recent study by Lumbroso and Bertin[67] dealt with N-unsubstituted imidates ($R^2 = H$). Dipole moments have been measured in benzene and calculated for different conformers, see Table 22. Ethyl acetimidate seems to be preferentially in the *syn–cis* (*sc*) and/or in the *anti–cis* (*ac*) conformation. The latter is preferable to (*sc*) since there is an attraction between the O-ethyl group and the nitrogen atom, while there is no such force operating between the methyl and N—H(—N). The Me---H distance (of the ethyl

TABLE 22. Reproduced with permission from H. Lumbroso and D. M. Bertin, *Bull. Soc. Chim. France*, 1728 (1970).

No.	Compound	Calculated				Found μ (D)	Suggested preferred conformer
		M (sc)	M (st)	M (ac)	M (at)		
117b	$MeC(=NH)OEt$	1·5	2·8	1·6	3·6	1·42	(ac)
118	$PhC(=NH)OEt$	1·7	2·6	1·2	3·2	1·54	(ac)
119	$4\text{-}O_2NC_6H_4C(=NH)OEt$	5·1	4·0	(3·0)	1·7	3·80[a]	(st)
120	$4\text{-}PrOC_6H_4C(=NH)OEt$	1·8	3·3	2·5	4·39	2·43	(ac)

[a] In dioxane.

TABLE 23. Reprinted with permission from O. Exner and O. Schindler, *Helv. Chim. Acta*, **55**, 1921 (1972).

No.	Compound	R_D^{20} (cm³)	μ (5%)[a] (D)	μ (15%)[b] (D)	Suggested preferred conformer
117a	Ph—C(=NH)OMe	40·01	1·66	1·61	E ap (ac); some Z ap (sc)
121	4-ClC₆H₄C(=NH)OMe	44·63	1·77	1·70	E ap (ac); some Z ap (sc)
122	PhC(=NMe)OMe	44·85	1·29	1·13	E ap (ac)
123	4-ClC₆H₄C(=NMe)OMe	49·86	0·96	0·82	E ap (ac)
124	MeC(=NH)OPh	44·41	2·05	2·00	E ap (ac)
125	MeC(=NH)OC₆H₄Cl-4	49·38	2·88	2·84	E ap (ac)
126		21·94	1·32	1·28	Z ap (sc) (0·84)[b]
127		42·74	1·31	1·23	Z ap (sc) (0·84)[b]
128		47·5	2·23	2·18	Z ap (sc) (1·85)[b]
129		25·47	1·05	0·99	E ap (ac) (0·97)[b]
130		31·49	1·26	1·18	E ap (ac) (0·97)[b]

[a] Corrections applied to the calculations of μ from ∞P_2 and R_D.
[b] μ, calculated for the conformation indicated.

group) for the (*sc*) conformer being 0·9 Å, i.e. less than the van der Waals distance. Similar considerations hold true for ethyl benzimidate and for 4-propoxybenzimidate (**120**) while ethyl-4-nitrobenzimidate (**119**) seems to be closest to the *syn–trans* (*sr*) conformation.

A somewhat more complete dipole moment and molecular refraction study including a number of *N*-substituted imidates has recently been published by Exner and Schindler[68] (Table 23). Measurements were made in benzene.

The configuration and the conformation of the cyclic oxazolines (**126–128**) being rigid *Z ap* (*sc*), may serve as reference. However, the calculated and the observed dipole moments are widely different (by 0·4 D). The semirigid lactim ethers **129** and **130** have definite *E* (*anti*) configuration at the C=N bond and their favorable conformation seems to be one with R³ *anti* to the *N*-substituent, based on close μ values.

Comparison of Tables 22 and 23 gives a similar picture of the conformations of ethyl and the methyl imidates **117a** and **117b**. It also indicates that, at least in benzene solution, *E ap* (*anti–cis*) conformation is generally valid for simple *N*-substituted alkyl and aryl imidates, irrespective of the alkyl or aryl substituent R¹ on carbon.

I. Conformation and reactivity of the imidates

There is still some controversy concerning the preferred conformation of imidates. Their *N,N*-disubstituted derivatives have at least unambiguous and rigid geometry around the —C=N⁺— bond, hence, their conformation, i.e. the orientation of the *p*-orbital of oxygen is more easily ascertained as the only variable. A very recent work[69] has pointed out that conformers (**133a** and (**133b**) of imidate salts react stereospecifically in hydrolysis.

In those compounds two identical R groups are attached to the iminium nitrogen; therefore, no configurational problems arise. As depicted in the Scheme I, 133a gives conformer 134a of the tetrahedral ortho acid amide-intermediate, in which the two new lone-pair orbitals on the oxygen of the OR^1 group and the nitrogen are both oriented anti-periplanar to the C—OH bond. There, the oxygens of the hydroxyl and of the OR^1 group each possess an orbital oriented *anti*-periplanar to the C—N bond so the cleavage of the C—N bond is possible by an orbital-assisted mechanism leading to an ester and a secondary amine. The nitrogen does not have an orbital *anti*-periplanar to the C—OR^1 bond thus breaking of the C—OR^1 bond as a higher energy process will not occur. In consequence, the *trans* (R'/R'') conformer should give specifically 134a that, in turn, should be decomposed in one direction, as indicated in the Scheme 1, to the ester.

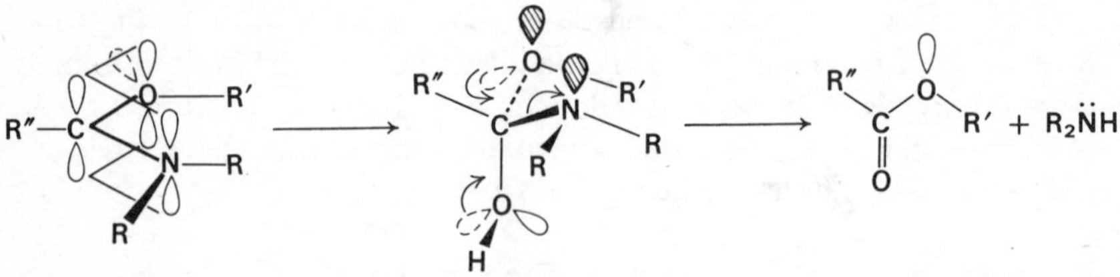

(133a)   (134a)

SCHEME 1

On the other hand, attack of hydroxide ion upon the *cis* (R'/R'') conformer (133b) of the imidate is expected to produce conformer 134b of the intermediate. Here there is no nitrogen orbital periplanar to the C—OR^1 bond and no OR^1 oxygen orbital *anti*-periplanar to the C—N bond. As a consequence this conformer is not supposed to react unless rotating into conformer 134a, which as stated previously, shall break down to an ester and a secondary amine. Alternatively, rotamer 134c would have both at the nitrogen and at the OH an orbital *anti*-periplanar to the C—OR^1 bond, hence it should immediately collapse into an amide and an alcohol (Scheme 2).

This means that the *cis*-imidate salt 133b can be hydrolysed by base either to the ester or into the amide depending on the ratio of 'new' conformers 134a and 134c. The assumption was checked with the conformationally rigid cyclic imidate salt 135 which upon hydrolysis with base (in MeCN, 20°C, 2 min) gave 95% aminoester (136) and 5% of the hydroxyamide 137. After 30 min O → N acyl migration of 135 took place quantitatively to 136. Similarly, the oxazolinium salt 138 gave, upon

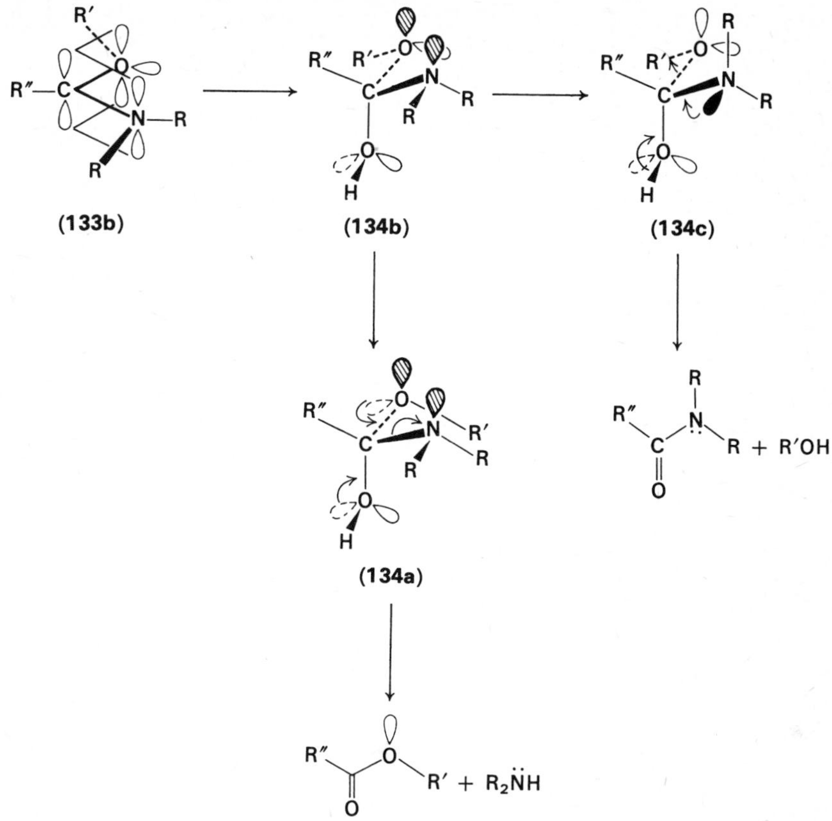

SCHEME 2

hydrolysis with sodium carbonate, exclusively the aminoester (139) which at higher pH rearranged to the hydroxyamide (140). The intermediate (141) has two factors that determine selective cleavage of the C—N bond: the orbital orientation that the 1,3-diaxial interaction between the pseudo-axial R^1 and the OH groups. Turning to the non rigid systems, e.g. 142, variation of bulkiness of R" caused significant shift in the ester:amide ratio (Table 24).

One expects a steric repulsion between methyl and the O-ethyl group in the cis-conformer (142b) and between the ethyl and R-groups in the anti-conformer (142a). Increase in bulkiness of R should hence result in a relatively lower population of 142b and this is reflected by product analysis (Table 24).

(135)   (136)   (137)

(138)   (139)   (140)

(141)

(142a)   (142b)

TABLE 24. Basic hydrolysis of imidate salts, type 142[69]. Reproduced by permission of the National Research Council of Canada from the *Canadian Journal of Chemistry*, **51**, 1665–1669 (1973).

R	% Ester[a]	% Amide[a]
H	50	50
CH$_3$	81	19
C$_6$H$_{11}$	50[b]	50[b]
(CH$_3$)$_3$C	> 98	—
C$_6$H$_5$	> 98	—

[a] Yields were estimated by p.m.r. spectroscopy analysis.
[b] Yields were estimated by v.p.c. analysis.

V. CHIROPTICAL PROPERTIES OF AMIDINES

The application of ORD to configurational studies of amidines has been limited to the hydrochloride salts of some α-hydroxy derivatives (**143** and **144**), their transition metal complexes, and some cyclic derivatives. The open-chain amidines were prepared by the Pinner synthesis[8, 13, 70–74]

(143)   (144)

and resolved *via* the mandelic acid salts[73–77]. It was not possible to isolate a noncyclic optically active amidine base from its hydrochloride, although in several instances the (±)-bases were stable when pure[73].

In order to determine the absolute configuration of any of the amidinium chlorides, reference compounds of known absolute configuration were needed. This was achieved in the case of (−)-mandelamidinium chloride (**145**) by synthesis from amygdalin, which also yields D-(−)-mandelic

(145)   (146)

(147)

acid on hydrolysis indicating that the (−)-amidinium chloride also has the D-configuration[78]. (+)-Lactamidinium chloride (**146**) has the D-configuration as it can be converted into the D-(+)-benzimidazole (**147**)[79]. In addition, hydrolysis of the (−)-amidinium chlorides (**148**, R=CH₃ or CH₃CH₂—) yields the corresponding (−)-acids (**149**) which belong to the D-series[74,80] and has an *R* configuration according to the Cahn–Ingold–Prelog convention. Thus the (−)-amidinium chlorides belong to the D-series also.

(148)   (149)

A. Optical Rotatory Dispersion of Open-Chain α-Hydroxyamidines

With the reference configurations established, Emerson and co-workers[81] observed the optical rotatory dispersion curves of a series of α-hydroxy-acids related to mandelic acid and the corresponding amidinium chlorides. The D-α-hydroxy-acids gave a positive Cotton effect related to the carboxyl $n-\pi^*$ absorption band at c. 205 nm. Full Cotton-effect curves could be obtained by conversion of the acids into their morpholine–thiourea derivatives[82].

The major extrema observed in the ORD measurements of amidinium chlorides derived from mandelic and related acids (**150–164**) are given in Table 25 while the u.v. data are given in Table 26. The ORD curves of the various amidinium chlorides were examined in methanol and water but the results were somewhat irregular and made further correlations difficult. For example, in water D-(−)-mandelamidinium chloride (**148**) gave a trough at 233 nm ($[\varphi]^\circ = -2860$) but the full Cotton effect curve could not be measured. Similar results were obtained for (−)-o-chloro- (**160**)

(**145**)   (**160**)   (**161**)

and (−)-o-bromomandelamidinium chloride (**161**) (−2490 tr, 227 nm; −2230 tr, 233 nm in water, respectively), see Table 25. The first extremum of the Cotton effect, when it can be reached, is at about 220 nm, corresponding to an absorption band for the $H_2NC{=}N^+H_2$ group at 190–200 nm; however, this region is complicated by the presence of phenyl absorptions. Curves measured in methanol showed similar tendencies but water is preferred as it is more transparent in the 200 nm region.

Since some ORD curves for the amidinium chlorides are not as definitely positive or negative as those of the corresponding acids, assignment of configuration by chemical evidence (hydrolysis) is in some cases preferable to the ORD evidence. By comparison with curves of compounds of known configuration the majority of amidinium chlorides described in Table 25 which have negative curves were assigned the D- or R- configuration[81]. When the aromatic ring in these amidinium salts bears an alkoxy-substituent the ORD curves increase in complexity and in addition to the low wavelength extrema an extremum in the aromatic absorption region (250–285 nm) is observed (Figure 2). Since the u.v. spectrum of this region for the series of aliphatic α-hydroxyamidinium chlorides (**165**, R = Me,

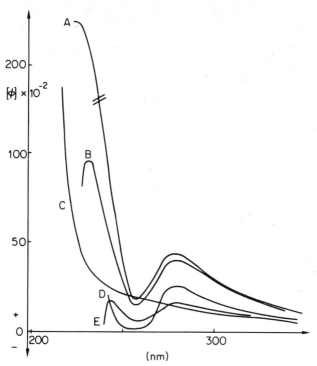

FIGURE 2. The ORD curves of A, (+)-*o*-methoxy-(**155**); B, (+)-*o*-ethoxy-(**156**); D, (+)-*p*-methoxy-(**159**), and E, (+)-*p*-ethoxymandelamidinium chloride (**158**), along with the parent (+)-mandelamidinium chloride (**145**), C, run in methanol. Reproduced with permission from D. G. Neilson, Optical Rotatory Dispersion of Alpha-hydroxy Amidines and their Transition Metal Complexes in Some Newer Physical Methods of Structural Chemistry Symposia Proceedings, 1967, p. 186. United Trade Press Ltd.

Et, *n*-Pr, *iso*-Pr) is featureless[84], these extrema must be attributed to optically active aromatic absorption bands.

$$\begin{array}{c} H \\ | \\ R-\underset{\underset{OH}{|}}{C}-\underset{\underset{NH_2}{|}}{\overset{+}{C}}=NH_2Cl^- \end{array}$$

(165)

The ORD data[81] for lactamidinium chloride (**146**)[76] as well as β-aryl-substituted derivatives **166** and **167** (Table 27), are not directly comparable with amidinium chlorides in which the aryl group is in the α-position (Figure 3). The introduction of a phenyl group into the lactamidinium systems, either in the α or β-position, causes a reversal of the sign of

TABLE 25ᵃ. ORD data of amidinium chlorides of the general formula

$$R\text{—}\overset{\displaystyle R'}{\underset{\displaystyle OH}{C}}\text{—}\overset{}{\underset{\displaystyle NH_2}{C}}=N^+H_2\ Cl^-$$

and having the D-(or R)- configurationᵇ,⁷⁹. Reproduced with permission from D. G. Neilson, Optical Rotatory Dispersion of Alpha-hydroxy Amidines and their Transition Metal Complexes in Some Newer Physical Methods of Structural Chemistry Symposia Proceedings, 1967, p. 186. United Trade Press Ltd.

Compound	R	R'	Solvent	Extrema						Lowest wavelength reached [φ]°	λ(nm)	Reference for absolute configuration
				[φ]°	λ(nm)	[φ]°	λ(nm)	[φ]°	λ(nm)			
145	H	C₆H₅	MeOH	—	—	—	—	—	—	-12,500!	217	83
			H₂O	—	—	—	—	-2860 tr	233	-2680!	227	83
150	Me	C₆H₅	MeOH	—	—	-1060 tr	250	—	—	+540!	227ᶜ	73
			H₂O	—	—	-320 tr	250	+3650 pk	222	+2830!	220ᶜ	73
151	Et	C₆H₅	MeOH	—	—	-1420 tr	243	—	—	-1170!	234	74
			H₂O	—	—	-840 tr	246	+1170 pk	222	0!	218	74
152	Me	m-MeC₆H₄	MeOH	—	—	—	—	—	—	-1290!	238	d
			H₂O	—	—	-860 tr	259–244	+1390 pk	223	0!	220	d
153	Me	p-MeC₆H₄	MeOH	—	—	-1130 tr	256	—	—	-995!	244	e
			H₂O	—	—	-890 tr	256	+1330 pk	231	+900!	299	e
154	H	p-MeC₆H₄	MeOH	—	—	—	—	—	—	-2340!	247ᶜ	f
			H₂O	—	—	—	—	—	—	-2030!	236ᶜ	f

TABLE 25. (cont.)

No.		R	Solvent									Ref
155	H	o-MeOC$_6$H$_4$	MeOH	-4360 tr	280	-1690 pk	261	$-17{,}000$ tr	229	$-15{,}100!$	227	[g]
156	H	o-EtOC$_6$H$_4$	MeOH	-3910 tr	283	-1450 pk	258	-9450 tr	234	$-7900!$	231[c]	[h]
157	H	m-EtOC$_6$H$_4$	MeOH	Fine structure	292–263	-4860 tr	240	—	—	$-3580!$	237	[h]
158	H	p-EtOC$_6$H$_4$	MeOH	-1370 tr	281	-320 pk	260	-1770 tr	246	0!	243[c]	[i]
159	H	p-MeOC$_6$H$_4$	MeOH	-2440 tr	278	0 pk	262	—	—	$-1900!$	245	[h]
160	H	o-ClC$_6$H$_4$	MeOH	—	—	—	—	$-18{,}950$ tr	225	$-9250!$	220[c]	[h]
			H$_2$O	—	—	—	—	-2490 tr	227	0!	224[c]	[h]
161	H	o-BrC$_6$H$_4$	MeOH	—	—	—	—	-5600 tr	231	$-4830!$	227[c]	[h]
			H$_2$O	—	—	—	—	-2230 tr	233	$+6250!$	222[c]	[h]
162	H	3,4-(MeO)$_2$C$_6$H$_3$	MeOH	-2260 tr	280–270	-4000 tr	260	0 pk	249	$-2500!$	240	[h]
163	H	2,3-(MeO)$_2$C$_6$H$_3$	MeOH	—	—	-1140 to 1480 sh	265–244	—	—	$-590!$	235[c]	[h]
164	H	2,4-(Cl$_2$)C$_6$H$_3$	MeOH	—	—	—	—	—	—	$-5300!$	233	[h]
			H$_2$O	—	—	—	—	-2910 tr	238	$-2760!$	236	[h]

[a] Taken from reference 81.
[b] All compounds have negative rotation at 546 nm.
[c] Measured as enantiomer.
[d] By comparison to curves of compounds 150 and 151 in water.
[e] By comparison to curves of compounds 151 and 152 in water.
[f] Tentative assignment on basis of negative curve.
[g] By comparison to curves of compound 156.
[h] From curves of copper complexes (Table 28).
[i] By comparison to curves of compound 159.

TABLE 26. Ultraviolet absorption spectra[79] of amidinium chlorides of general formula

$$R-\underset{\underset{OH}{|}}{C}-\underset{\underset{NH_2}{|}}{C}=N^+\,H_2Cl^-$$

Reproduced with permission from D. G. Neilson, Optical Rotatory Dispersion of Alpha-hydroxy Amidines and their Transition Metal Complexes in Some Newer Physical Methods of Structural Chemistry Symposia Proceedings,1967, p. 186. United Trade Press Ltd.

Compound	R	R'	Solvent	Absorption maxima λ (nm) log ε in parentheses			
145	H	C_6H_5	MeOH	252(2·18)	258(2·33)	264(2·27)	267(2·07)
			H_2O	191·5(4·94)			
150	Me	C_6H_5	MeOH	253(2·18)	259(2·31)	265(2·25)	268·5(2·04)
			H_2O	196(5·68)			
146	H	CH_3	H_2O	194(4·42) Featureless in the region: 250–280 nm			
152	Me	$m\text{-}MeC_6H_4$	MeOH	261 sh	267(2·65)	274(2·56)	
153	Me	$p\text{-}MeC_6H_4$	MeOH	222 sh(3·93)	257(2·33)	263(2·37)	269(2·25)
154	H	$p\text{-}MeC_6H_4$	MeOH	220 sh	277(3·29)		
156	H	$o\text{-}EtOC_6H_4$	MeOH	220 sh	277(3·78)		
			H_2O	191–194(5·11)			
157	H	$m\text{-}EtOC_6H_4$	MeOH	230 sh(3·74)	278–285(3·31)		
			H_2O	189(4·93)			
158	H	$p\text{-}EtOC_6H_4$	MeOH	234·5(3·99)	276·5(3·09)	282·5(3·00)	
			H_2O	196(5·43)			
159	H	$p\text{-}MeOC_6H_4$	MeOH	234·5(3·95)	275·5(3·11)	281·5(3·04)	
			H_2O	192(4·80)			
160	H	$o\text{-}ClC_6H_4$	MeOH	263·5(2·90)	268·5(3·00)	276(2·89)	
161	H	$o\text{-}BrC_6H_4$	MeOH	259·5(2·45)	265·5(2·53)	272·5(2·39)	
164	H	$2,4\text{-}(Cl_2)C_6H_3$	MeOH	220 sh	263(2·31)	273(2·40)	281·5(2·27)
166	Me	$CH_2C_6H_5$	MeOH	253(2·12)	259(2·24)	265(2·14)	

TABLE 27[a]. ORD data of amidinium chlorides[b, 81]. Reprinted with permission from T. R. Emerson and co-workers, *J. Chem. Soc.*, 4007 (1965).

Compound	Solvent	Extremum (trough)		Lowest wavelength reached		Absolute configuration	Reference
		$[\alpha]°$	λ	$[\alpha]°$	λ (nm)		
146[b]	MeOH	—	—	−1330!	217		76
	H₂O	—	—	−1290!	216	L	
166	MeOH	−17,100	215	−14,700!	212		77[c]
	H₂O	−10,400	220	−6620!	213	D	
167	MeOH	−22,800	214	−12,800!	212		—
	H₂O	−10,800	217	−8800!	213	d	

[a] Taken from reference 81.
[b] All have negative rotations at 546 nm.
[c] Enantiomer.
[d] Configuration unknown but ORD curve compares well with that for amidinium salt 166; compound 165 is therefore probably D at the α-hydroxy center.

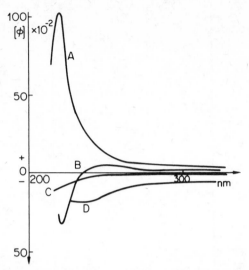

FIGURE 3. The ORD curves of A, (+)-α-benzyllactamidinium chloride (166); B, (+)-atrolactamidinium chloride (150); C, (−)-lactamidinium chloride (146) and D, (−)-mandelamidinium chloride (145) run in water. Reproduced with permission from D. G. Neilson, Optical Rotatory Dispersion of Alpha-hydroxy Amidines and their Transition Metal Complexes in Some Newer Physical Methods of Structural Chemistry Symposia Proceedings, 1967, p. 186. United Trade Press Ltd.

rotation for compounds of related configuration. No Cotton effect was observed for (−)-lactamidinium chloride above 217 nm although the authors postulate that it is likely, from the plain curve obtained, to be positive for the D-isomer.

(146)   (166)   (167)

B. Optical Rotatory Dispersion of Metal Complexes

It has been known that the copper complexes of α-amino acids[85] and α-hydroxy acids[86] give Cotton effect curves in the visible region that could be used for the correlation of configuration of the parent acids. It is also known that α-hydroxy amidines from complexes with transition metals, and in particular that copper complexes have a broad absorption band of low extension coefficient ($\varepsilon < 50$) in the visible range[87,88]. Neilson and Peters[71] suggested that complexes derived from optically active α-hydroxy amidines might prove useful in assigning configurations to those amidines which could not be correlated by chemical means.

The ORD curves of the copper complexes of α-hydroxy amidines[81, 89] of type **168** were found to show significant features in the 590 and 220 nm region, and a few have extrema at 270 nm (Table 28). The sign of the Cotton effect at *c*. 590 nm is positive for complexes derived from the

(168)

D-series of amidines and negative for the L-series (see Figure 4). This affords a useful method for the determination of the absolute configuration of amidines particularly since measurements of the ORD curves of the parent compounds do not always give a clear answer as to their configuration. The ORD curves of the copper complexes permitted that D-configuration be assigned to (−)-2-methoxy-, 2-ethoxy-, 2-chloro-, 2-bromo-, 4-methoxy-, 4-ethoxy-, 3-ethoxy, and 2,4-dichloromandel-amidinium chlorides which could not be correlated chemically *via* the parent acids owing to their susceptibility to base catalysed racemization. The copper complex of D-(+)-lactamidinium chloride prepared *in situ* gave a positive Cotton effect (pk ≅ 620 nm; tr ≅ 470 nm) and also illustrated the greater value of the ORD curves of the copper complexes over that of the parent amidines for the correlation of configuration[81].

The ORD curves of the nickel complex of D-(−)-*o*-bromomandel-amidinium chloride (+98 pk, 568 nm; −270 tr, 435 nm) was also studied[79]. The D-configuration was again associated with a positive Cotton effect in the visible region; however, the nickel complexes were in some cases difficult to synthesize and therefore are less useful than the copper analogues.

The behavior of these complexes in the 270 nm region (a feature that can probably be ascribed to a d → d transition of copper[88]) appears at first sight to be more complex than that at the longer wavelengths. The 2-chloro-, 2-bromo-, and 2,4-dichloromandelamidine copper complexes (Table 28) all show distinct extrema in the 270 nm region, whereas in the case of the 2- or 3-ethoxy- or 4-methoxyamidine copper complexes no extrema are observed. However, in these latter compounds, the amidine ligands themselves exhibit Cotton effect curves in the 270 nm region (Table 25) which are in the opposite sense to those derived from the copper transition. Emerson and co-workers[81] suggested that in the alkoxyamidine copper complexes the opposing Cotton effects cancel each other resulting

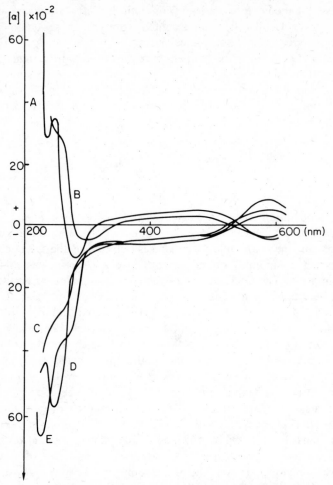

FIGURE 4. The ORD curves run in methanol of the copper complexes derived from A, L-(+)-*o*-chloro-; (B), L-(+)-*o*-bromo-; (C), D-(−)-*o*-methoxy-; (E), D-(−)-*o*-ethoxy-; and (D), D-(−)-mandelamidinium chloride. Reproduced with permission from D. G. Neilson, Optical Rotatory Dispersion of Alpha-hydroxy Amidines and their Transition Metal Complexes in Some Newer Physical Methods of Structural Chemistry Symposia Proceedings, 1967, p. 186. United Trade Press Ltd.

in no observable extrema in the 270 nm region. Support for this may be drawn from the CD curves (Figure 5) of *o*-ethoxymandelamidinium chloride and its copper complex (Figure 6). Since the halogen substituted ligands are featureless in the 270 nm region, as expected no cancelling effect is observed.

TABLE 28[a]. ORD data for copper complexes formed from amidinium chlorides,

$$\left[R-\underset{Ph}{\overset{O^-}{C}}-\underset{NH_2}{C}=N^+H_2\ Cl^- \right]_2 Cu\cdot 2H_2O,$$

and having the D- (or R)-configuration[b,81]. Reprinted with permission from T. R. Emerson and co-workers, J. Chem. Soc., 4007 (1965).

R	Sub-stituent	Extremum (peak) $[\phi]°$	λ (nm)	Extremum (broad shoulder or trough) $[\phi]°$	λ (nm)	Extremum (peak) $[\phi]°$	λ (nm)	Extremum (trough) $[\phi]°$	λ (nm)	Extremum $[\phi]°$	λ (nm)	Lowest wavelength reached $[\phi]°$	λ (nm)	Reference for absolute configuration of the amidinium chloride
H[c]	—	+2320	588	-2420 to -3680	483-300	—	—	-24,600	245	-16,900 pk	234	-20,200!	227	83
Me	—	+620	588	-1450 to -1650	450-313	—	—	—	—	—	—	-15,750!	242	73
Et	—	+830	595	-1780 to -1960	460-312	—	—	—	—	—	—	-23,200!	235	74
H	o-OMe	+780	589	-1950 to -2200	472-400	—	—	—	—	—	—	-35,200!	227	d
H	o-OEt	+1500	591	-2700 to -3860	442-300	—	—	—	—	—	—	-26,800!	223	d
H	o-Cl	+1600	598	-1640	467	+51,500	278	-15,900	240	-35,200 tr	226	-37,500!	225	d
H	o-Br	+2000	585	-1340	455	+2540	282	-16,800 sh	252-244	-11,000 pk	233	-25,200!	220[e]	d
H	o-OMe	+1080	585	-2420 to -2480	457-330	—	—	-14,500	251	-29,000 tr	224	-8750!	240[e]	d
H	p-OEt	+1280	589	-2080 to -1610	459-311	—	—	-15,200	253	—	—	-10,700!	242[e]	d
H	n-OEt	+1860	585	-3250 to -3440	420-313	—	—	—	—	—	—	-29,700!	236	d
H	2,4-Cl₂	+2120	580	-1290	461	+8600	283	—	—	—	—	-32,400!	240[e]	d

[a] Taken from reference 81.
[b] The solvent used was methanol and all compounds had a negative rotation at 546 nm.
[c] This complex contains O-2 H₂O; hence molecular rotations quoted could be up to 10% too high.
[d] Deduced from the sign of the C.E. at 590 nm.
[e] Measured as the enantiomer.

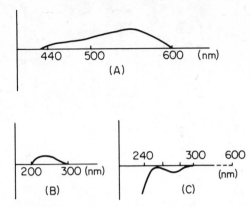

FIGURE 5. CD curves of the copper(ii) complex of D-(−)-*o*-ethoxymandelami-dinium chloride (A and B) and of the parent D-(−)-amidinium chloride (C). Reproduced with permission from D. G. Neilson, Optical Rotatory Dispersion of Alpha-hydroxy Amidines and their Transition Metal Complexes in Some Newer Physical Methods of Structural Chemistry Symposia Proceedings, 1967, p. 186. United Trade Press Ltd.

Difference curves were plotted over the 200–400 nm region by sub-tracting twice the ORD curve of the amidinium chloride from the ORD curve of the copper complex (Figure 6). This led to a positive Cotton effect at 270 nm for complexes derived from the D-series of amidinium chlorides. The authors pointed out, however, that this is an approximation, since there is no guarantee that the contribution to the ORD due to the ligand amidine is exactly matched by that of the complex.

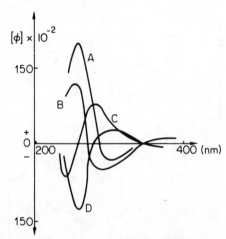

FIGURE 6. ORD difference curves (in methanol) calculated for the copper(ii) complexes of (A), L-(+)-*o*-ethoxy-; (B), L-(+)-*o*-bromo-; (C), D-(−)-*o*-chloro-; and (D), D-(−)-*o*-methoxymandelamidinium chloride.

C. Optical Rotatory Dispersion of Cyclic α-Hydroxyamidines

The amidine functional group also appears in heterocyclic systems such as imidazolines and pyrimidines. Neilson and coworkers[77, 78] prepared a series of optically active 2-α-hydroxyalkyimidazolines (**169–173**)

R =	MeCH(OH)	**(169)**
	PhCH(OH)	**(170)**
	PhCMe(OH	**(171)**
	PhCEt(OH)	**(172)**
	PhCH$_2$CMe(OH)	**(173)**

and plotted their ORD curves. In the range studied (400–250 nm) all gave plain curves with a molecular rotation less than 1000° at 300 nm. It was noted, however, that a marked shift in rotation occurred on going from neutral to acid media (Table 29). There is a positive shift in rotation on protonation of the imidazole nucleus when the configuration at the α-hydroxy- centre is known to be L. This has been used by Ewing and Neilson[77] to assign the L-configuration to (+)-2-α-benzyl-α-hydroxy-ethyl)imidazoline (**173**), its (+)-amidinium chloride, and its hydrolysis product (+)-α-benzyl-lactic acid. Dirkx and de Boer[90] have reported similar results for the closely related ephedrines which show a positive shift of the ORD curves on going from acid to neutral conditions when the absolute configuration at the α-centre is *R*. In addition L-α-amino acids show a positive shift on protonation[91, 92].

TABLE 29. Shift in rotation of some 2-α-hydroxyalkylimidazolines at 5461 Å on changing from neutral to acid media[77,78]. Reprinted with permission from D. F. Ewing and D. G. Neilson, *J. Chem. Soc.*, 770 (1965); D. G. Neilson, D. A. V. Peters, and L. H. Roach, *J. Chem. Soc.*, 2272 (1962).

Compound	$[\alpha]°$ (Solvent)	$[\alpha]°$ (Solvent) + HCl
D-(**171**)	+119·2 (EtOH)	−24·6 (EtOH)
D-(**172**)	+105·6 (EtOH)	−27·1 (EtOH)
D-(**173**)	−44·6 (MeOH)	−56·4 (MeOH)

Kadin and coworkers[93] have prepared an optically active compound in which the amidine group is part of an aromatic system. (−)-Benzimidazole **174**, which from the configuration of its precursor belongs to the D-series, exhibits a series of extrema in its ORD curve (−3430 tr, 286 nm; +13,200 pk, 256 nm; −43,600 tr, 227 nm). Further work will have to be done in this area, however, before any comment on these extrema can be made.

$$PhCH(OH)-C \underset{\underset{H}{\overset{N}{\vert}}}{\overset{N}{\diagup}} \bigcirc$$

(174)

The preparation and resolution of four 2-(α-hydroxybenzyl)-1,4,5,6-tetrahydropyrimidines (**175–178**) has been reported by Neilson and co-

	R	R'
175	Br	H
176	Cl	H
177	OMe	H
178	H	Me

workers[94]. (−)-Tetrahydropyrimidine **178** exhibited a positive shift of the ORD curve on going from neutral to acid media. The L-configuration was therefore assigned to the (+)-hydrochloride of tetrahydropyrimidine **178**. This configuration was confirmed by the ORD curve of the copper complex derived *in situ* from the (+)-hydrochloride of **178** which gave a negative Cotton effect in the visible region (−690 tr, 654 nm; +1180 pk, 483 nm).

The ORD curves of the hydrochloride salts of tetrahydropyrimidines **175–178** were plain positive over the region examined (down to 235 nm). The ORD curves of the hydrochloride of tetrahydropyrimidines **175** and **176**, which paralleled those of the S-(+) amidinium chlorides **160** and **161** absorbed too strongly to be examined below *c.* 235 nm, in which region a Cotton effect might well be expected. Although the *o*-methyoxytetrahydropyrimidine (**177**) hydrochloride absorbed too strongly below *c.* 280 nm, the (+)-form of **177** as its (−)-mandelate was transparent, and gave an intense Cotton effect ($a = 52\cdot3$) in the 270 nm region (+2230 pk, 285 nm; −3000 tr, 256 nm).

A similar result was observed for the amidrazone hydrochlorides **179** and **180** which gave plain curves down to 285 nm at which point light

(**179**, R = H)
(**180**, R = OMe)

absorption was too great to permit further examination[79]. However, the (+)-*o*-methoxyamidrazone-(−)-mandelate derivative gave a Cotton effect ($a = +116$) in the aromatic region (+4260 pk, 283 nm; −7390 tr, 263 nm).

VI. REFERENCES

1. D. H. R. Barton, in *Perspectives in Organic Chemistry*, A. R. Todd, Ed., Interscience, London, 1956, pp. 68–96.
2. M. B. Robin, F. A. Bovey and H. Basch, in *The Chemistry of the Amides*, J. Zabicky, Ed., John Wiley and Sons, New York, N.Y., 1970, pp. 37–38; 40–53.
3. S. Forsén and M. Nilsson, in *The Chemistry of the Carbonyl Group*, Vol. 2, J. Zabicky, Ed., John Wiley and Sons, New York, N.Y., 1970, pp. 159, 188.
4. G. W. Wheland, *Advanced Organic Chemistry*, 3rd ed., Wiley, New York, N.Y., 1960, Chapter 14.
5. A. Hantzsch, *Chem. Ber.*, **64**, 661 (1931).
6. P. Ramart-Lucas, *Bull. Soc.-Chim. France*, **3** (5), 723 (1946).
7. B. Fischer and C. A. Grob, *Helv. Chim. Acta*, **39**, 417 (1956).
8. A. Pinner, *Die Iminoäther und ihre Derivate*, R. Oppenheim, Berlin, 1892,
9. C. G. McCarty in this volume, Chapter 4.
10. L. Töke, G. Blaskó, L. Szabó and Cs. Szántay, *Tetrahedron Letters* (**24**), 2459 (1972).
11. H. Ahlbrecht and C. Vonderheid, *Chem. Ber.*, **106**, 2009 (1973).
12. H. Henecka and P. Kurty in *Houben-Weyl, Methoden der Organischen Chemie*, E. Müller, Ed., Sauerstoffverbindungen III, Thieme, Stuttgart, 1953, pp. 647–713.
13. R. Roger and D. Neilson, *Chem. Revs.*, **61**, 179 (1961).
14. C. S. Yoo, Univ. Microfilms, 71–22, 685; *Diss. Abstr. Int. B*, 1971, 32 (3) 1420.
15. F. L. Pyman, *J. Chem. Soc.*, 3359 (1923).
16. R. Burtles and F. Pyman, *J. Chem. Soc.*, 361 (1923).
17. G. Schwenker and K. Bösl, *Die Pharmazie*, **24**, 653 (1969); *cf.* G. Schwenker and H. Rosswag, *Tetrahedron Lett.*, 4237 (1967); 2691 (1968); G. Schwenker and R. Kolb, *Tetrahedron*, **25**, 5437 (1969).
18. R. A. Katritzky and J. M. Lagowski, in *Advances in Heterocyclic Chemistry* Vol. 1, Academic Press, New York, N.Y., 1963, p. 317.
19. C. R. Ganellin, *J. Pharm. Pharmacol.*, **25** (10), 787 (1973).
20. D. Prevorsek, *J. Phys. Chem.* **66**, 769 (1962).
21. H. Lumbroso, C. Pigenet, H. Rosswag and C. Schwenker, *C.R. Acad. Sci. Paris*, **266**, 1479 (1968).
22. D. Prevorsek, *J. Chim. physique Physico-Chim. biol.*, **55**, 840 (1958).
23. B. A. Phillips, G. Fodor, J. Gál, F. Létourneau and J. J. Ryan, *Tetrahedron*, **29**, 3309 (1973).
24. J. Saporetti, *M.S. Thesis*, West Virginia University, Morgantown, West Virginia, 1974.
25. H. J. Barber, *J. Chem. Soc.*, 101 (1943).
26. R. B. Tinkler, *J. Chem. Soc.* (*B*), 1052 (1970).
27. J. C. Danilewicz, M. J. Sewell and J. C. Thurman, *J. Chem. Soc.* (*C*), 1704 (1971).
28. A. Heesing and G. Maleck, *Tetrahedron Lett.*, 3851 (1967).
29. K. Bunge, R. Huisgen and R. Raab, *Chem. Ber.*, **105**, 1296 (1972).

30. P. Baudet and D. Rao, *Helv. Chim. Acta*, **53**, 1011 (1970).
31. R. A. Johnson, *J. Org. Chem.*, **33**, 3627 (1968) and papers cited therein.
32. C. G. McCarty, in *The Chemistry of the Carbon–Nitrogen Double Bond*, S. Patai, Ed., Interscience, London, 1969, p. 363.
33. R. M. Moriarty, Chin-lung Yeh, K. C. Ramey, and P. W. Whitehurst, *J. Amer. Chem. Soc.*, **92**, 6360 (1970).
34. Ei-lan Yeh, R. M. Moriarty, Chin-lung Yeh, and K. C. Ramey, *Tetrahedron Lett.*, 2655 (1972).
35. K. C. Ramey, D. J. Louick, P. W. Whitehurst, W. B. Wise, R. Mukherjee, and R. M. Moriarty, *Org. Magn. Res.*, **3**, 201 (1971).
36. R. F. Merritt and F. A. Johnson, *J. Org. Chem.*, **32**, 416 (1967).
37. J. R. Norell, *J. Org. Chem.*, **33**, 1619 (1970).
38. B. Greenberg and J. G. Aston, *J. Org. Chem.*, **25**, 1894 (1960).
39. A. Dondoni and O. Exner, *J.C.S. Perkin II*, 1908 (1972).
40. D. Pitea, D. Grasso, and G. Favini, *J. Chem. Soc. (B)*, 2290 (1971); G. J. Karabatsos and S. S. Lande, *Tetrahedron*, **24**, 3907 (1968).
41. G. A. Olah and T. E. Kiovsky, *J. Amer. Chem. Soc.*, **90**, 4666 (1968).
42. B. A. Phillips, *Ph.D. Dissertation*, West Virginia University, Morgantown, West Virginia, 1972.
43. W. E. Stewart and T. H. Siddall, III, *Chem. Rev.*, **70**, 517 (1970).
44. F. A. Bovey, *High Resolution Nuclear Magnetic Resonance Spectroscopy*, Academic Press, New York, N.Y., 1968, Chapter VII.
45. C. S. Johnson, Jr., in *Advances in Magnetic Resonance*, Vol. I, J. S. Waugh, Ed., Academic Press, New York, N.Y., 1965.
46. G. Schwenker and H. Rosswag, *Tetrahedron Lett.*, 2691 (1968).
47. H. J. Jakobsen and A. Senning, *Chem. Comm.*, 1245 (1968).
48. R. C. Neuman, Jr. and L. B. Young, *J. Phys. Chem.*, **69**, 2570 (1965).
49. D. J. Bertelli and J. T. Gerig, *Tetrahedron Lett.*, 2481 (1967).
50. J. Sandström, *J. Phys. Chem.*, **71**, 2318 (9167).
51. J. S. McKennis and P. A. Smith, *J. Org. Chem.*, **37**, 4173 (1972).
52. Z. Rappoport and R. Ta-Shma, *Tetrahedron Lett.*, 5281 (1972).
53. D. Y. Curtin, E. J. Grubbs, and C. G. McCarty, *J. Amer. Chem. Soc.*, **88**, 2775 (1966).
54. D. L. Harris and K. M. Wellman, *Tetrahedron Lett.*, 5225 (1968).
55. A. Lidén and J. Sandström, *Tetrahedron*, **27**, 2893 (1971).
56. G. Schwenker and H. Rosswag, *Tetrahedron Lett.*, 4237 (1967).
57. J. P. Marsh, Jr. and L. Goodman, *Tetrahedron Lett.*, 683 (1967).
58. C. G. Raison, *J. Chem. Soc.*, 3319 (1949).
59. J. Ranft and S. Dähne, *Helv. Chim. Acta*, **47**, 1160 (1964).
60. G. S. Hammond and R. C. Neuman, Jr., *J. Phys. Chem.*, **67**, 1655 (1963).
61. R. C. Neuman, Jr., and V. Jonas, *J. Phys. Chem.*, **75**, 3532 (1971).
62. H. S. Gutowsky and C. H. Holm, *J. Phys. Chem.*, **25**, 1228 (1956).
62. (a) J. Sandström, *J. Phys. Chem.*, **71**, 2318 (1967); (b) C. H. Bushweller, P. E. Stevenson, J. Golini, and J. W. O'Neil, *J. Phys. Chem.*, **74**, 1155 (1970).
64. J. Sandström and B. Uppström, *Acta Chem. Scand.*, **21**, 2254 (1967).
65a. R. E. Cramer and R. DeRyke, *Can. J. Chem.*, **51**, 892 (1973).
65b. E. S. Gore, D. J. Blears and S. S. Danylak, *Can. J. Chem.*, **43**, 2135 (1965).

66. W. Klyne and V. Prelog, *Experientia*, **16**, 521 (1960).
67. H. Lumbroso and D. M. Bertin, *Bull. Soc. Chim. France*, 1728 (1970).
68. O. Exner and O. Schindler, *Helv. Chim. Acta.*, **55**, 1921 (1972).
69. P. Deslongchamps, Cl. Lebreux and R. Raillefer, *Can. J. Chem.*, **51**, 1665 (1973).
70. R. Roger and D. G. Neilson, *J. Chem. Soc.*, 627 (1960).
71. D. G. Neilson and D. A. V. Peters, *J. Chem. Soc.*, 4455 (1963).
72. D. G. Neilson and L. H. Roach, *J. Chem. Soc.*, 1658 (1965).
73. D. G. Neilson and R. Roger, *J. Chem. Soc.*, 688 (1959).
74. D. G. Neilson and R. Roger, *J. Chem. Soc.*, 3181 (1961).
75. R. Roger, *J. Chem. Soc.*, 1544 (1935).
76. D. G. Neilson and D. A. V. Peters, *J. Chem. Soc.*, 1309 (1962).
77. D. F. Ewing and D. G. Neilson, *J. Chem. Soc.*, 770 (1965).
78. D. G. Neilson, D. A. V. Peters, and L. H. Roach, *J. Chem. Soc.*, 2272 (1962).
79. D. G. Neilson, *Optical Rotatory Dispersion of Alpha-hydroxy Amidines and Their Transition Metal Complexes*, in Some Newer Phys. Methods Struct. Chem. Symp. Proc., page 186, 1967.
80. S. Mitsui, S. Imaizumi, Y. Senda and K. Konnu, *Chem. and Ind.*, 233 (1964).
81. T. R. Emerson, D. F. Ewing, W. Klyne, D. G. Neilson, D. A. V. Peters, L. H. Roach and R. J. Swan, *J. Chem. Soc.*, 4007 (1965).
82. C. Djerassi, K. Undheim, and A. Weidler, *Acta Chem. Scand.*, **16**, 1147 (1962).
83. S. Reid, *Ph.D. Thesis*, University of St. Andrews, Scotland (1949).
84. D. A. V. Peters, *Ph.D. Thesis*, University of St. Andrews, Scotland (1963).
85. P. Pfeiffer and W. Christeleit, *Z. physiol. Chem.*, **245**, 197; **247**, 262 (1937).
86. P. Karrer and J. Heyer, *Helv. Chim. Acta.*, **20**, 407 (1937).
87. R. O. Gould, R. F. Jameson, and D. G. Neilson, *Proc. Chem. Soc.*, 314 (1960).
88. R. O. Gould and R. F. Jameson, *J. Chem. Soc.*, 15 (1963).
89. L. H. Roach and D. G. Neilson, *J. Chem. Soc.*, 4004 (1965).
90. I. P. Dirkx and Th. J. de Boer, *Rec. Trav. Chim.*, **83**, 535 (1964).
91. O. Lutz and B. Jirgensons, *Chem. Ber.*, **63**, 448 (1930).
92. O. Lutz and B. Jirgensons, *Chem. Ber.*, **64**, 1221 (1931).
93. S. B. Kadin, H. J. Eggers, and I. Tamm, *Nature*, **201**, 639 (1964).
94. D. G. Neilson, I. A. Khan, and R. S. Whitehead, *J. Chem. Soc.* (*C*), 1853 (1968).

CHAPTER **3**

Detection and determination of imidic acid derivatives

W. H. PRICHARD

The City University, London, England

INTRODUCTION

Despite the great biological importance of the C=N group as a structural unit, the physico-chemical data available are considerably less than for the analogous C=O and C=C groups. Data pertaining to the chemical determination of amidine and imidic acid derivatives are sparse. This may be a result of the relatively unimportant role that these compounds play in industrial processes. At the present time they have little application in the pharmaceutical, polymer or plastics industries although they have not remained exempt from study.

Interest has been stimulated in imidates because of their relatively easy conversion into substituted triazines[1, 2] which are potentially useful in that they afford preparative routes to dyestuffs, whitening agents, resins, pesticides, surface active agents and medicines[3].

Amidines may be quantitatively estimated by chemical means[4]. The compound is treated with 25 cm^3 0·1M-iodine to precipitate iodamidines.

$$
R-C \begin{array}{c} \diagup\!\!\!\!= NH \\ \diagdown NH_2 \end{array} + I_2 \longrightarrow R-C \begin{array}{c} \diagup\!\!\!\!= NI \\ \diagdown NH_2 \end{array} + HI
$$

The solution is made alkaline and the solid separated by filtration. The amount of iodine in the filtrate is determined by acidifying an aliquot of filtrate and titrating the liberated iodine with 0·5M-sodium thiosulphate.

In an alternative method[4] the amidine is precipitated as mercuriamidine by means of 25 cm^3 0·1M buffered mercury(II) acetate. After filtration

$$
R-C \begin{array}{c} \diagup\!\!\!\!= NH \\ \diagdown NH_2 \end{array} + Hg(OOCCH_3)_2 \longrightarrow R-C \begin{array}{c} \diagup\!\!\!\!= N-HgOOCCH_3 \\ \diagdown NH_2 \end{array} + CH_3COOH
$$

the excess mercury ions are determined by titrating an aliquot of filtrate with 0·02M-ammonium thiocyanate.

A new qualitative reaction for the identification of amidoximes[5] consists in precipitation of the complex ions formed between iron(III) and the amidoxime with potassium thiocyanate or ammonium thiocyanate as a dark brown, sparingly soluble compound in water. The reaction is stated to be specific for amidoximes and may be applied to mixtures containing amidoximes, phenols, nitrites and cyanhydrins.

The above groups of compounds apart, the other compounds of this chapter are poorly served by quantitative methods. The Kjeldahl estimation of nitrogen may be employed although this method is not without the disadvantage that total nitrogen may be extremely difficult to liberate and estimate. The estimation of chloride ion by precipitation as silver chloride has been employed for imidate hydrochlorides[6]. However, two major difficulties were the complete elimination of occluded hydrochloric acid from the crystalline material and the inherent instability of the hydrochloride.

The general chemistry of amidoximes[7], amidrazones[8], formazans[9], imidates[10], and imidoyl halides[11] is very well reviewed, but the amount of physico-chemical data contained in these reviews is relatively small.

It is extremely important in studies of amidrazones that the literature is very carefully scrutinized[3]. In several instances the term hydrazidine has been employed in the literature to describe compounds which are amidrazones[12, 13]. The nomenclature employed here follows reference 8.

It may also be relevant to the relative unimportance attached to the compounds considered in this chapter that the literature does not contain references to reviews of specific physico-chemical methods, e.g. infrared or n.m.r. spectra. The relevant data have to be gleaned from the original literature and in this context this chapter does not give a totally comprehensive coverage. It should be obvious to the reader that the literature cited is mainly from the last 10–12 years. Because of the lack of review articles of specific methods the reader will find references to earlier work, where they exist, in the references cited.

It should also be noted that certain physico-chemical methods, in particular all aspects of chromatography, have not been applied with any enthusiasm to the subject matter of this chapter.

II. AMIDINES

A. Infrared and Raman Spectra

The infrared spectra of amidines have been a fertile source of study for many years. Before the advent of the widespread use of nuclear magnetic resonance (n.m.r.) techniques infrared spectroscopy was the predominant instrumental method for characterization purposes.

The first thorough study of the structure of simple amidines by infrared methods was performed upon acetamidine and its hydrochloride salt[12].

The study attempted to resolve unambiguously the structure of amidines by a complete assignment of the spectrum for pure liquid acetamidine and for its hydrochloride as a solid. The Raman effect for acetamidine hydrochloride which had been recorded several years previously[13] was also used in the assignment. As a result of the study it was suggested that the nitrogen valencies in acetamidine were approximately planar and that a similar situation existed in the acetamidinium ion.

Whilst study of the simplest member of any series of compounds is of the greatest value, for characterization purposes it is the systematic study of many molecules of similar structure which allows tables of infrared group wavenumbers to be formulated. To this end several studies of amidines have been made. N,N'-Disubstituted amidines have been studied by several workers. One study[14,15] was devoted to a study of amidines of the type $CH_3C(:NR)NHR'$ where R and R' were phenyl (substituted and unsubstituted) and naphthyl. The results are summarized in Table 1.

It was also noted[14,15] that replacement by deuterium of the hydrogen atoms attached to nitrogen affected four bands in the 1500–1200 cm^{-1} region. This was interpreted as indicating coupling between nitrogen hydrogen vibrational modes and the heavy atom skeleton.

A series of N-alkyl-N'-phenylbenzamidines have also been studied[16]. The findings were very much in line with those observed in Table 1. The solution data were recorded for approximately 0·02M solutions in carbon tetrachloride for all the molecules studied and some were also examined in chloroform solution (0·06M). Very little change in the wavenumbers of the ν(N—H) modes were observed upon changing solvent. The values fall in the ranges 3463–3438 and 3408–3370 cm^{-1}. The ν(C=N) mode value was recorded as a 0·06M solution in chloroform. The mode fell in the range 1627–1620 cm^{-1} for the seven molecules for which values were recorded.

The infrared spectra of 13 N-substituted trichloroacetamidines as pure liquids or solids have been recorded[17]. The spectra of the majority of these compounds were also recorded as solutions in tetrachloroethylene. In solution, the ν(N—H) mode was recorded in the range 3520–3455 cm^{-1} and the ν(=N—H) mode in the range 3415–3350 cm^{-1}. The ν(C=N) mode lay between 1672 and 1593 cm^{-1} showing little sensitivity to changes in phase. The nitrogen–hydrogen bending mode of the amino group was observed as a very strong absorption in the range 1627–1510 cm^{-1} which was displaced by 1–5 cm^{-1} upon dilution. This was taken as indicative that the mode was the NH deformation (or amidine II mode). Other assignments were made but their usage for characterization purposes is not particularly definitive.

N,N'-Diphenylbenzamidine and N,N'-diphenylacetamidine have been studied by infrared spectroscopic methods and an assignment made of the spectra[18]. The former compound exhibited a weak N—H stretching mode band lying between 3400 and 3200 cm^{-1} with a maximum at approximately 3310 cm^{-1}. The latter compound possessed a totally different N—H stretching band in that it lay between 3350 and 2500 cm^{-1} with two maxima at 3250 and 3050 cm^{-1}. By comparison with the infrared spectra of

TABLE 1. The infrared wavenumbers for a series of N,N'-disubstituted amidines of the form $CH_3C(:NR)NHR'$

Compound		$\nu(N—H)$ Solid	Deuterated solid	Chloroform solution	
R	R'				
R^1	R^1	3245		3448	3390
		3205	2314		
		3105			
R^2	R^1	3236		3448	3370
		3184	2300		
		3012			
R^3	R^1	3278		3448	3390
		3205	2304		
		3105			
R^4	R^3	3290		3446	—
		3246	2358		
		3236			
		3115			
R^5	R^3	3400	2518	3448	3378
R^5	R^5	3412	2518	3448	3378
		$\nu(C\!\!=\!\!N)$			
R^1	R^1	1628	1618	1655	
				1633 (sh)	
R^2	R^1	1628	1618	1652	
			1610		
R^3	R^1	1636	1615	1647	
		1628	1605		
R^4	R^3	1636	1615	1647	
		1631	1600		
R^5	R^3	1644	1639	*	
		1628	1623		
R^5	R^5	1647	1639	1655	
		1620	1615	1620	

* Not studied; sh = shoulder; $R^1 = C_6H_5$, $R^2 = p\text{-}C_6H_4Cl$; $R^3 = p\text{-}C_6H_4OCH(CH_3)_2$; $R^4 = o\text{-}C_6H_4OCH(CH_3)_2$; $R^5 = \beta\text{-}C_{10}H_7$

carboxylic acid dimers it was concluded that N,N'-diphenylacetamidine existed as a mixture of predominantly strongly hydrogen-bonded cyclic dimer and a lesser amount of weakly hydrogen-bonded polymer association, whereas N,N'-diphenylbenzamidine existed solely as the weakly hydrogen-bonded polymeric association and no dimer formation.

The $\nu(C\!\!=\!\!N)$ and $\delta(NH_2^+)$ modes have been studied in a series of amidines and guanidines and their hydrochlorides[91]. The shift of the $\nu(C\!\!=\!\!N)$ mode in the salts from its position near 1660 cm^{-1} was attributed to the interaction of the mode with an in-plane deformation, $\delta(NH_2^+)$, and was studied as a function of the nature of the group attached to the carbon atom bearing the imino group.

In contrast to some of the observations recorded above it has been stated that the effects of molecular association are particularly marked for the $\nu(C{=}N)$ mode in the spectra of N,N'-disubstituted amidines[20]. In a Raman effect study of diethyl acetamidine it was noted[21] that the position of the $\nu(C{=}N)$ mode absorptions depended not only on the physical state of the molecules but also on the solvent if studied in solution. In dioxan a Raman shift was observed at 1675 cm^{-1} whereas in n-hexane the shift was observed at 1592 cm^{-1}. In the liquid state three shifts were observed, the strongest being at 1635 cm^{-1}.

It was further noted in N,N'-diarylsubstituted amidines that when the carbon nitrogen double bond is conjugated to a benzene ring the shifts observed for chloroform solutions occur as doublets (separation 7–13 cm^{-1} lying between 1660 and 1635 cm^{-1}. The presence of *syn* and *anti* isomers has been suggested to account for their existence.

In the case of N,N,N'-trisubstituted amidines[21] the $\nu(C{=}N)$ is observed as a Raman shift between 1633 and 1613 cm^{-1} when the carbon–nitrogen double bond is conjugated to a benzene ring.

It should be noted that care should be exercised when making direct comparisons between Raman and infrared vibrational data especially if elements of symmetry exist within the molecular systems being considered.

B. Ultraviolet and Visible Spectra

Ultraviolet and visible spectroscopy has been used, not so much for characterization purposes, as for studying the influence of various substituents upon the basicity of amidines.

Statistical analysis was applied to the straight lines obtained from a function of the spectrophotometric titration data[22]. An equation of the form

$$\log \left[(A_{\max} - A)/(A - A_{\min}) \right] = pK - pH$$

where A is the observed absorbance, was used. pH was the independent variable and the pK evaluated from the intercept on the abscissa.

The equation was applied to amidines which have pK values in the range 11–12. Photometric methods were also employed in an investigation of the pK values of a series of N-monosubstituted benzamidines[23]. It was observed that the basicity depended on the nature and position of the substituent and that the pK values were related to the Hammett σ values.

In a more detailed study of protonated cations using electronic, fluorescence and infrared spectroscopic techniques[24] evidence has been

presented for the existence of cyclic amidine structures for singly protonated (at the heterocyclic nitrogen atom) species derived from
2-aminoquinoline and 4-aminoquinaldine. Evidence is presented to demon-
strate that these molecules have protonated amidine electronic structures
in the ground and lowest excited electronic singlet states.

C. Nuclear Magnetic Resonance Spectra

Nuclear magnetic resonance (n.m.r.) studies have been mainly concerned with amidine salts. This arises principally from the ease with which
salts may be examined by the n.m.r. technique even though solvent effects
may present problems. In one such study of amidine salts[25] it was concluded that rotation about the C$\doteq$N bonds was restricted. The dashed
line is used here and subsequently to denote delocalization of electrons

$$CH_3 - C \begin{matrix} R \\ \diagdown \\ N-H \\ + \\ N-R \\ \diagup \\ H \end{matrix} \quad X^-$$

(1)

across the bond. Molecules of the type **1** where R is methyl and X is
chlorine were concluded to exist in the configuration indicated with magnetically non-equivalent pairs of hydrogen and methyl groups and with
nitrogen atoms which differed in chemical reactivity.

The presence of magnetically non-equivalent pairs of hydrogen atoms
and methyl groups was concluded from the non-simultaneous collapse of
the two N—CH$_3$ doublets, at intermediate concentrations of sulphuric acid.
The high field doublet was observed to collapse at lower acidities than the
low field doublet, indicating that the basicities of the two nitrogen atoms
were different.

It was further suggested that in media of high acidity hydrogen exchange
involved the formation of a second conjugate acid and decreased the
residence time of N—H protons in unique spin states. Finally the lifetimes
of the second conjugate acid became long enough to cause magnetic equiv-
alence of the N—CH$_3$ groups, by rotation about C—N$^+$NH$_2$CH$_3$ bond.

The exchange reactions of protons attached to nitrogen in amidinium
ions formed in dilute hydrochloric acid and various water–sulphuric acid
mixtures have been studied[26]. Since the exchange rate is inversely proportional to the [H$^+$] and independent of [AH$^+$] it was concluded that
the dominant mechanism involved reaction with hydroxyl ions. The
mechanism was different in strong acids.

An examination of unsubstituted and symmetrically substituted amidinium salts in solution in dimethyl sulphoxide and water again suggested that rotation about the carbon–nitrogen bonds was hindered[27]. The barrier to rotation about the carbon–nitrogen partial double bonds in acetamidine hydrochloride in dimethyl sulphoxide solution were estimated. The value was expected to lie between 9 ± 2 and 25 ± 8 kcal mol^{-1} (36 ± 8 and 100 ± 32 kJ mol^{-1}). Although the barrier was imprecisely known the value was similar to that in amides (7–18 kcal mol^{-1}, 30–75 kJ mol^{-1}). It was further concluded that the barrier would be higher in amidinium ions due to the greater double bond character associated with an amidinium C$\doteq$N bond.

The barrier to rotation in N,N-dimethyl acetamidinium-d_3 chloride and nitrate has also been studied[28].

The temperature dependence of rotation about the carbon–nitrogen bond in substituted N,N-dimethylformamidines has been investigated[29]. The rotational barrier was attributed to the zwitterionic structure **2**.

$$R\overset{\ominus}{-}N-\underset{\underset{\overset{|}{N}}{\overset{||}{}}}{C}-R'$$

CH$_3$ CH$_3$

(2)

The resonance absorption due to the two methyl groups was a singlet at room temperature. The coalescence temperature was concluded to lie below ambient temperature in dilute solution in chloroform. N,N-Dimethyl-benzamidine (**3**), N,N-dimethyl-N-(O,O-dimethyl thiophosphoryl)-benzamidine (**4**) and N,N-dimethyl-(O,S-dimethyl thiophosphoryl)-benzamidine (**5**) were also studied[29]. The two methyl groups of **4** and **5** exhibited non-equivalence at room temperature in both aromatic and non-aromatic solvents whereas in **3** the resonance was a singlet in chloroform and benzene at room temperature. Since **4** and **5** possess the P—N=C moiety it was suggested that the methyl group doublet could be attributed to the presence of *syn* and *anti* isomers.

N,N-Dimethyl formamidines of the type (X)C$_6$H$_4$—N=CHN(CH$_3$)$_2$ and N'-aralkyl-N,N-dimethyl formamidine, C$_6$H$_5$(CH$_2$)$_n$N:CH·N(CH$_3$)$_2$

(3)   (4)   (5)

were studied and it was observed[30] that the spectra of all the hydro-chlorides in D_2O exhibited the $N(CH_3)_2$ group resonance as two sharp singlets of approximately equal intensity. It was concluded that the non-equivalence existed because of cis/trans isomerism. In deuterochloroform solution the $N(CH_3)_2$ resonance of all the formamidines (as free bases) appeared as a six-hydrogen sharp singlet. However, in hexadeuterobenzene the resonance appeared as a broad singlet the appearance of which was markedly dependent on the substituent in the N'-phenyl group. With a nitro or cyano group in the para position two distinct singlets were observed, whereas with good electron donating groups in the para position of the phenyl group the spectra differed little from those observed in deuterochloroform.

In an examination of nitrogen-15 labelled N,N,N'-trisubstituted amidines by proton magnetic resonance[31] the existence of long range nuclear spin–spin interaction between the nitrogen-15 and the hydrogen atoms has been demonstrated. Nitrogen-15 studies have also been made on N,N-dimethyl-^{15}N-phenylformamidine[32] which demonstrated that the singlet observed at τ 2·69 became a doublet with $J = 2·4 \pm 0·1$ Hz for the $^{15}N{=}C{-}H$ coupling constant.

D. Mass Spectra

The mass spectrum of N,N-dimethyl-N'-phenylformamidine has been examined[33] and a large peak due to the $[M-H]^+$ ion observed. As demonstrated by deuterium labelling, one of the ortho hydrogen atoms of the phenyl ring was lost. The same result had been observed for the corresponding fragmentation of thioformanilide. The mechanism was explained by the formation of benzimidazolium ions (a). The effects of substituents at the benzene ring were also examined.

The proposed mechanism is outlined in Scheme 1.

Detailed information has been obtained from the mass-spectra of N,N,N'-trisubstituted amidines under electron impact, nitrogen-15 and deuterium labelling[31].

E. Chromatography

Amidines have not been extensively investigated by chromatographic techniques and consequently few studies are worth more than passing interest. A method of characterizing amidines as their picrates by the thin layer technique has been described[34]. An interesting series of experiments[35] has demonstrated the separation of some N,N,N'-trifluoroamidines by liquid column chromatography. The explosion hazard was reduced by

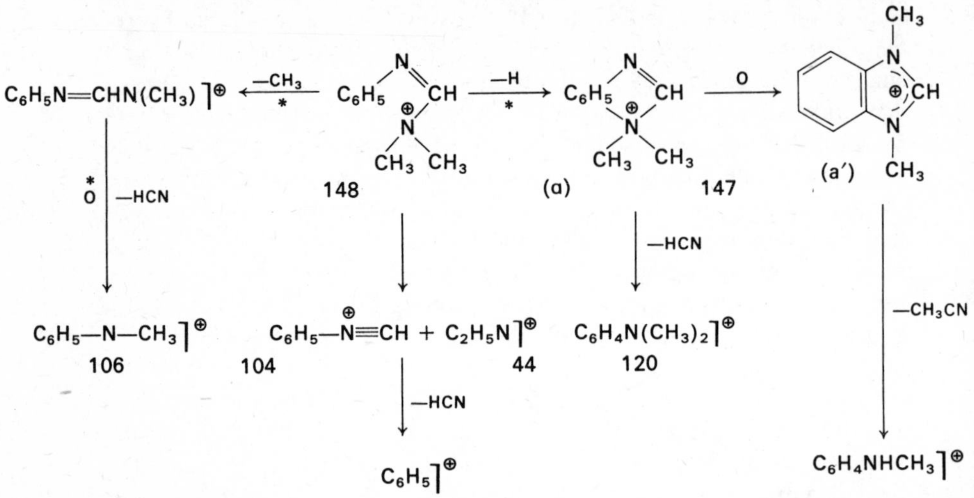

$C_6H_5N=CHN(CH_3)]^\oplus$ $\xleftarrow[*]{-CH_3}$ C_6H_5 ... CH (148) $\xrightarrow[*]{-H}$ C_6H_5 ... CH (147) $\xrightarrow{O}$ (a')

(a)

C_6H_5—N—CH$_3]^\oplus$ 106

C_6H_5—N≡CH + $C_2H_5N]^\oplus$ 104   44

$C_6H_4N(CH_3)_2]^\oplus$ 120

$C_6H_4NHCH_3]^\oplus$

$C_6H_5]^\oplus$

SCHEME 1

using inert solvents and separations were achieved on macro and micro-scales by percolating a 2–4% (w/w) solution of the fluoronitrogen mixture in an inert fluorochemical over a bed of silica gel. Elution was achieved with the solvent and 65% trifluoroacetic acid in the inert fluorocarbon. The separated fractions were subsequently analysed by gas chromatography.

III. AMIDOXIMES

A. Infrared Spectra

Infrared spectroscopy has been the most frequently employed technique for investigating the amidoxime structure. An early and thorough investigation of formamidoxime[36] resolved the previously held supposition that the molecule existed as a tautomeric mixture of the two forms

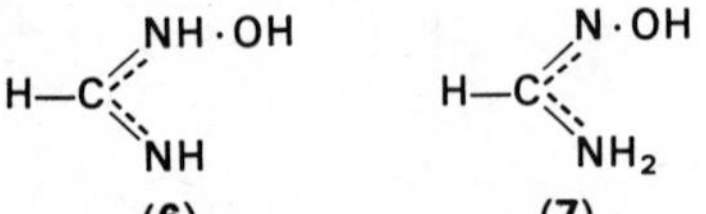

     H—C (NH·OH)(NH)   H—C (N·OH)(NH$_2$)

         (6)   (7)

Examination of the compound as a melt and dissolved in carbon tetrachloride or in methyl cyanide revealed that the molecule possessed exclusively the amino structure. Strong evidence was afforded for this structure by the appearance of a strong band at 1618 cm^{-1} in the crystalline state which moved to 1572 cm^{-1} in methyl cyanide solution. Deuteration of labile hydrogen atoms greatly reduced the intensity of this band indicating that the band arose principally from the motions of the amino hydrogen atoms.

For structural determinations by infrared spectroscopy the amidoximes possess the hydroxyl group which is attached to the carbon–nitrogen double bond and the amino group. All three groups give rise to characteristic infrared absorptions. It should be carefully noted, however, that the hydroxyl and amino groups are subject to extensive hydrogen bonding which shifts the absorptions by up to 200 cm^{-1}. Formamidoxime[36] in the solid state has its ν(OH) at 3410 cm^{-1}, and at 3628 cm^{-1} as a dilute solution in carbon tetrachloride. The ν(NH$_2$) bands are observed at 3300 and 3254 cm^{-1} in the solid and at 3530 and 3424 cm^{-1} respectively in dilute solution in carbon tetrachloride. It should be noted carefully that the choice of solvent is critical (the ν(NH$_2$) absorptions appear at 3496 and 3397 cm^{-1} in solution in methyl cyanide) and also that the solvents employed must be rigorously purified and dried. The interaction of functional groups with solvents is outside the scope of this discussion, but excellent review articles are available[37].

The carbon–nitrogen double bond is not as sensitive to changes of phase.

Although not directly relevant to a study of amidoximes it should be noted that in an examination of the association of oximes[38] the predominant species was stated to be a dimer or trimer. It was also recorded that shifts as great as approximately 500 cm^{-1} occurred, between the monomeric and associated species, for the ν(OH) mode.

An examination of 3-amino amidoximes[39] showed that the compounds existed in a chelated form in solution in carbon tetrachloride and that the molecules were less basic than the simple amidoximes. Whereas simple amidoximes showed three bands near 3620, 3515 and 3410 cm^{-1}, the 3-amino amidoximes gave very complex absorptions in the 3600–3300 cm^{-1} region. Spectra of solutions in carbon tetrachloride or carbon tetrachloride/chloroform (4%) mixtures exhibited a ν(OH) near 3610 cm^{-1}. It was suggested that the molecules existed in the equilibrium

(8)   (9)

Oxamidoxime[40] has been examined by infrared and n.m.r. methods and assigned the diaminoglyoxime structure. Other studies of amidoximes have been made[41, 42, 43]. From one such study[44] it was concluded that amidoximes possessed the $RC(NH_2){=}NOH$ structure. The limited solubility of the amidoximes examined necessitated the use of cells of length between 2·5 and 7 cm for solutions in carbon tetrachloride. The $\nu(OH)$ was observed near 3615 cm^{-1} which is close to the value observed for monomeric phenols. The positions of the $\nu(NH_2)$ bands (Table 2) were near to the values observed for amides. The intensity of the bands was also stated to be close to that of amides and slightly greater than that of arylamines.

TABLE 2. The hydroxyl and amino stretching mode bands* in some amidoximes[44]

Compound	$\nu(OH)$	$\nu_{as}(NH_2)$	$\nu_s(NH_2)$
Formamidoxime	3620	—	—
Acetamidoxime	3610	3516	3412
Propionamidoxime	3620	3516	3410
Butanamidoxime	3610	3510	3400
Benzamidoxime	3615	3515	3410

* All values recorded as very dilute solutions in carbon tetrachloride.

The $\nu(C{=}N)$ mode has not been so extensively examined presumably because of the very definite character of the modes discussed above. One study of amidoximes[42] suggested that the strong band near 1660 cm^{-1} (solid state) was due to the $\nu(C{=}N)$ mode. In an examination of compounds containing the carbon–nitrogen double bond[45] by Raman effect studies it was observed that the shift for the $\nu(C{=}N)$ mode occurred between 1663 and 1623 cm^{-1} for oximes. This is somewhat below the value observed for formamidoxime[36], 1672 cm^{-1} in solution in methyl cyanide but it should be noted that the first member of any series is slightly atypical.

B. Nuclear Magnetic and Electron Spin Resonance Spectra

N.m.r. spectroscopy has been applied to determine the structure of oxamidoxime[40]. Amidoximes of the type **10** have been studied in benzene solution for chemical shifts, solvent shifts and for correlations with respect

$$C_6H_5\diagdown$$
$$C{=}NOH$$
$$|$$
$$N$$
$$XC_6H_4 \quad Y$$

(10)

to their configurations[46]. Analyses of this type are rendered a little difficult in that the resonances of hydrogen atoms attached to nitrogen may be very broad and the positions of the hydroxyl and nitrogen protons are extremely sensitive to the solvent employed.

In the study of molecules of the family given above it was observed that the resonance signals for the N-phenyl protons always occurred at higher field than the C_6H_4 protons, for solutions in carbon disulphide. The N-phenyl protons always gave a multiplet centred near τ 3·1, whereas the C_6H_4 protons occurred as a singlet (unresolved multiplet) when Y = H and as a multiplet when Y = CH_3 or C_2H_5, all very near τ 2·6. These observations were interpreted as reflecting a change in the sterochemical form of the amidoximes from a *cis*-position of the two phenyl rings when Y = H to a *trans* position for derivatives where Y = CH_3 or C_2H_5. In this latter case it was further suggested that the conjugation of the C_6H_4 ring with the C=N bond was more effective and caused the splitting of the C_6H_4 protons.

The solvent shifts of the C_6H_4 protons induced by benzene are paramagnetic for the *ortho* protons ($\Delta \sim 20$ Hz downfield) and diamagnetic for the *meta* and *para* protons ($\Delta \sim 15$ Hz upfield from the position in solution in carbon disulphide). The behaviour of the complexed benzene on the N-phenyl protons is analogous but the shifts are smaller.

A particularly interesting study of the amidoxime structure has been reported[47]. Amidoximes which may exhibit geometrical isomerism and tautomerism appear to exist from physical measurements, in the *syn*-hydroxymino form **11**, which is stabilized by intramolecular hydrogen bonding.

$$R-\underset{\substack{\|\\N^1-OH}}{C}-N^2H_2 \qquad C_6H_5-\underset{\substack{|\\CH_3-N^1-OH}}{C}=N^2H \qquad C_6H_5-\underset{\substack{\|\\N^1-OH}}{C}-N^2H \cdot C_6H_5$$

$$\textbf{(11)} \qquad\qquad \textbf{(12)} \qquad\qquad \textbf{(13)}$$

However, analogues of N-alkyldroxylamines must exist in the imino form **12**. The three compounds **11–13** are amphoteric but in water and aqueous alcohol they behave as bases.

By flow techniques strong electron spin resonance (e.s.r.) signals were obtained by oxidizing the free bases with potassium ferricyanide in 0·1M-sodium hydroxide solution. The e.s.r. spectra were symmetrical and since they did not change on variation of the flow rate were concluded to be the immediately formed radical ions of the amidoximes themselves. The spectra were all of the same type having 18 lines of approximately equal intensity except when obvious overlapping occurred between four pairs of lines towards the centre of the spectrum. The pattern indicated the interaction of the odd electron with two nonequivalent nitrogen atoms and one hydrogen atom as in **14** and **15**. The suggested *syn*-structure was supported by

$$
\begin{array}{ccc}
R-C{=}N^2H^\bullet & & R-C{=}N^2H^{\,-} \\
| & \rightleftharpoons & || \\
N^1{-}O^{\,-} & & N^1{-}O^\bullet \\
\textbf{(14)} & & \textbf{(15)}
\end{array}
$$

the fact that the radical from phenylacetamidoxime, $C_6H_5CH_2\text{-}C(NH_2){=}NOH$ showed no further splitting due to the methylene group. The measured splitting constants for six radicals were determined, the values ranged from 8·05–7·55 and 3·53–3·10 Oe for the nitrogen atoms and 5·88–5·13 for the hydrogen.

The larger value was assigned to the nitrogen atom (N^1) of the oxyimino group partly because this nitrogen is nearer to the oxy radical centre $N^1{-}O$, but more particularly because the assignment accords with the spectrum obtained by oxidation of **12** in alkaline solution.

IV. AMIDRAZONES

A. Miscellaneous Data

Optical rotatory dispersion (o.r.d.) measurements made on the compounds

$$
o\text{-}RC_6H_4 \cdot CH(OH) \cdot C \cdot (NH_2){=}NHNHC_6H_5{}^+ X^-
$$
$$
\textbf{(16)}
$$

(R = H or CH_3O, X = Cl) which exhibited plain Cotton effect curves down to 285 nm[8]. However, the (+) amidrazone(−)mandelate

$$
o\text{-}CH_3OC_6H_4CH(OH)C(NH_2){=}NHNHC_6H_5{}^+ C_6H_5CH(OH)COO^-
$$

exhibited a Cotton effect curve of high amplitude probably due to the presence of the various aromatic chromophores[8].

Molecular orbital calculations[48] have indicated that pK values and coupling activities of vinylogous amidrazones and light stabilities of the resulting dyes were related to the electron densities on the nitrogen atom of the amidrazones.

B. Infrared Spectra

Data pertaining to the amidrazone structure are few. A combined infrared and n.m.r. spectroscopic study of N^1,N^1-disubstituted and N^1,N^1,N^3-trisubstituted amidrazones[49] showed that these compounds exist only in the amidrazone configuration **18** with an intramolecular hydrogen bond **19** and no contribution from **17**.

$$R^1{-}C\underset{N^3{-}R^4}{\overset{N^2H{-}N^1{\diagup}^{R^2}_{R^3}}{}} \;\rightleftharpoons\; R^1{-}C\underset{N^3{\cdot}HR^4}{\overset{N^2{-}N^1{\diagup}^{R^2}_{R^3}}{}} \qquad R^1{-}C\overset{N{-}N{\diagup}^{R^2}}{\underset{N{\diagdown}_{R^4}}{}}{\overset{|}{H}}R^3$$

(17)   **(18)**   **(19)**

The study embraced a wide range of amidrazones. For amidrazones with R^1, R^2 and R^3 alkyl and aryl and $R^4 =$ H (D) two bands were observed in the ranges 3493–3475 cm^{-1} and 3424–3375 cm^{-1} for dilute solutions in carbon tetrachloride. Moreover, when $R^1 =$ H or C_6H_5 and R^2, R^3 and R^4 were alkyl and aryl two bands were again observed in most instances lying in the ranges 3475–3410 and 3395–3297 cm^{-1}, also for dilute solutions in carbon tetrachloride. The band in the lower wavenumber region was of considerable breadth indicating the presence of hydrogen bonding.

In an earlier study[50] amidrazone salts were prepared. It was observed that with $R^1 = C_6H_5$ and with $R^2 = C_6H_5$ and $R^3 = CH_3$ that the free base was stable and soluble in organic solvents. The infrared spectra of solutions of the free bases in carbon tetrachloride showed two bands of approximately equal intensity at 3490 and 3380 cm^{-1}. These results were confirmed by studying the 100 MHz n.m.r. spectrum which possessed a resonance at τ 4·78 (from TMS) and which integrated at two protons. When the spectrum was recorded after shaking the sample with D_2O the τ 4·78 resonance disappeared.

In a study primarily concerned with imidate systems the spectra of a number of amidrazones were recorded for comparison purposes[89]. It was observed that molecules of the type $C_6H_5C(NH_2){=}NN(CH_3)C_6H_5$ in the solid state possessed a very intense asymmetrical band in the infrared spectrum between 1600 and 1595 cm^{-1} and an intense band at 1575 cm^{-1}. In the $\nu(NH)$ region of the infrared spectrum for the solid state four bands were observed near 3460 and 3355 cm^{-1} which were of almost equal intensity and two more intense bands near 3400 and 3300 cm^{-1}. In dilute solutions in carbon tetrachloride only absorptions corresponding to a free NH_2 group were observed at 3500 and 3375 cm^{-1}.

In the spectra of the picrate and hydrochloride salts a large broad peak typical of salts was observed. The $\nu(C{=}N^+)$ mode moved to higher wavenumber (1665–1650 cm^{-1}) than those observed for the free base.

The preparation of perfluoroalkylamidrazones has been reported[90].

Elemental analyses and the neutralization equivalent of the hydrochloride confirmed the compound and the molecular weight. Infrared spectroscopy yielded considerable information but did not lead to the differentiation between the tautomers **20** and **21**.

$$\underset{(20)}{R_F-\overset{\overset{\displaystyle NH}{\|}}{C}\cdot NH\cdot NH_2} \qquad\qquad \underset{(21)}{R_F\cdot\overset{\overset{\displaystyle NH_2}{|}}{C}=N\cdot NH_2}$$

For unsubstituted amidrazones three bands were recorded in the 3500–3100 cm^{-1} region[90]. In the 1700–1600 cm^{-1} region two absorption bands were noted. The band lying in the range 1690–1670 cm^{-1} was assigned to the ν(C=N) mode and that in the range 1660–1655 cm^{-1} to the δ(NH$_2$). No indication was given but it is suspected that the data refer to the pure compound. Deuteration studies would obviously confirm any suspected ambiguities in the assignment.

C. Nuclear Magnetic Resonance Spectra

The n.m.r. spectra of many amidrazones were also recorded during the study mentioned previously[49]. When $R^4 = H$ and R^1, R^2 and R^3 are aryl in structure **18** a two proton resonance was observed in the n.m.r. spectrum in the range τ 5·3–4·6 p.p.m. from TMS. In those systems where $R^1 = H$ or C_6H_5 and R^2, R^3 and R^4 were alkyl or aryl a single proton resonance was observed in the range τ 4·3–2·1 with the majority of the values lying at the lower end of the range.

The syntheses of silyl amidrazones and hydrazidines from alkyl-ω-(tri-methylsilyl) alkaneimidates and 1,1-dialkylhydrazines have been reported[19]. The infrared spectra of ω-(trimethylsilyl) alkanamide dialkylhydrazones (silyl amidrazones) of the structure $(CH_3)_3Si(CH_2)_nC(NH_2)=NNR$ where $R = CH_3$, C_2H_5 and C_4H_9 and $n = 2$, and $R = CH_3$ and $n = 3$ contained absorption bands due to the ν(NH$_2$) modes 3330–3150 cm^{-1}, the ν(CH$_3$—N) group at 2770 cm^{-1} and two absorptions at 1650 and 1610 cm^{-1} due to the overlapping ν(C=N) and δ(NH$_2$) bands. The infra-red spectra were recorded for suspensions of solid material in mineral oil. The butyramide dimethylhydrazone contained a resonance at τ 5·07 attributed to the two hydrogen atoms attached to nitrogen.

V. FORMAZANS

A. Miscellaneous Data

The acid–base properties of several substituted formazans of the benzimidazole series have been examined[51]. Studies of pH changes showed that formazans exist in three forms, cationic, anionic and neutral (pH 5·0–

9·5). pK values for 1,5-dibenzimidrazoyl formazans are in the 8·4–10·5 range for the acid ionization form and 4·0–5·0 for the basic ionization form. The dependence of the acid–base properties of formazans is explained as being dependent upon the molecular configuration.

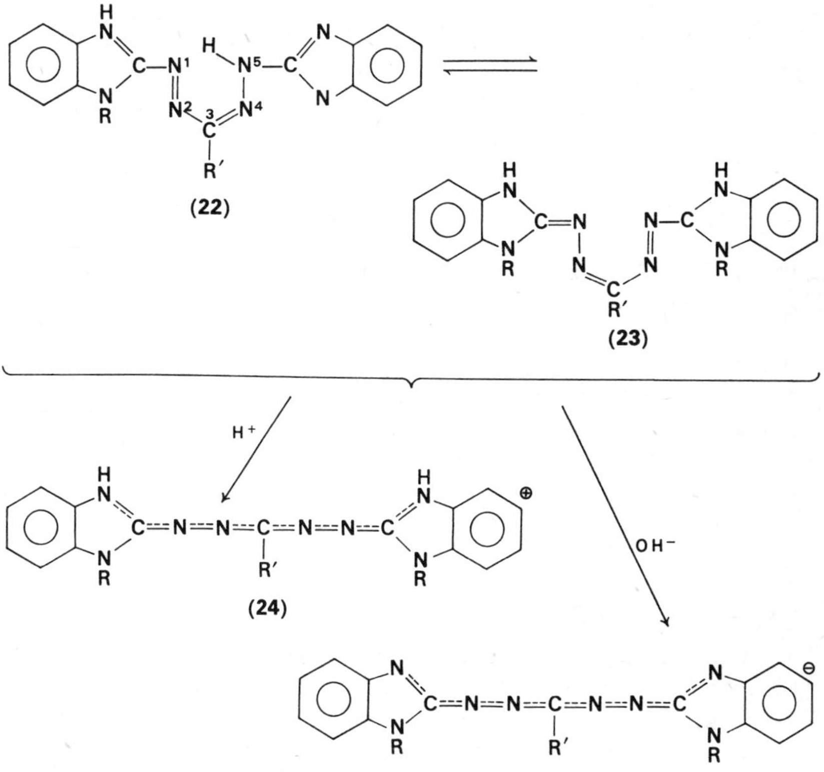

(22)

(23)

(24)

(25)

R = $C_6H_5CH_2$; $C_6H_5CH_2$; $C_6H_5CH_2$; CH_3; C_2H_5; C_2H_5; H

R′ = CH_2; n-C_3H_7; H; CH_3; CH_3; n-C_3H_7; CH_3

Simple L.C.A.O.–M.O. methods have been used to calculate models of substituted 1,5-diphenylformazans[52]. Experimental and calculated dipole moments were compared and discussed in relation to the structure of compound 26.

B. Infrared and Raman Spectra

Infrared spectroscopy has been used[53] to study the red and yellow formazans of the type 26. Solid and dilute solutions in carbon tetrachloride of yellow formazans, structure **27**, show, in most cases, a strong ν(N—H) band in the range 3500–3300 cm^{-1}. The ν(N—H) of the red formazans, structure **28** is very weak (ε less than 1 l mol^{-1} cm^{-1}) and is only observed

(26)   (27)   (28)

in solution with very long path lengths and in the region 3500–3000 cm^{-1}. N,N'-Diphenyl-C-(2,6-dimethoxyphenyl)-formazan in solution in carbon tetrachloride had a band at 3315 cm^{-1}, ε 65 l mol^{-1} cm^{-1}, ν(N—D) = 2465 cm^{-1}. In C-ethoxycarbonyl-N,N'-diphenyl formazan it was observed that the intramolecular N—H---N hydrogen bond was opened by light absorption and gave rise to a new intramolecular N—H---O hydrogen bond.

The infrared spectra of substituted triphenyl formazans[54] in solution in chloroform or in the solid state exhibit no ν(N—H) absorption bands in the 3400–3100 cm^{-1} region. This is taken as being indicative of the formation of chelate structures. The bands lying in the range 1445–1440 cm^{-1} were assigned to the nitrogen–nitrogen double bond because of the similarity to azo dyes. The ν(C=N) mode was assigned to the band between 1517 and 1510 cm^{-1}.

Using the methods of group theory the number of vibrations for all possible types of symmetry have been determined for diphenyl formazan and its nickel complex[55]. The infrared spectra and Raman effect of ten formazans were measured in the solid state (potassium bromide disc) and as solutions in carbon tetrachloride. Formazans in solution were concluded not to possess C_{2v} symmetry and the degree of conjugation in the formazans decreased in carbon tetrachloride solution as compared with the solid.

C. Ultraviolet and Visible Spectra

An extremely comprehensive coverage of the literature to 1954 of the ultraviolet and visible region spectra of formazans and tetrazolium salts is given in reference 9.

Protonation of p-RC$_6$H$_4$N=NCR^1=NNH C$_6$H$_4$R^2-p, (**29**) in sulphuric acid solution shifted the ultraviolet spectrum maximum by 125–250 nm to longer wavelengths as compared to the spectrum of 29 in alcohol[56]. When substituents of different electronic nature were introduced into the phenyl nucleus the values of λ_{max} for unsymmetrical formazans in alcohol and sulphuric acid solutions were not equal to the arithmetic mean of the values of λ_{max} of the two corresponding symmetrically substituted compounds.

The stable chelate ring has been given as the reason for the deep colour of formazans[54]. Formazans were recorded as possessing absorption bands between 255–270 nm and 298–310 nm, which were absent in the case of triphenyl formazans with a nitro group in the three phenyl groups, and having no absorption bands in the visible region. It was also observed that solutions in polar nitrobenzene or non-polar benzene exhibited no solvatochromy (that is they were the same colour). The long wavelength bands were most sensitive to the effects of substituents.

Microspectrophotometric determinations of formazans from the reaction of neotetrazolium chloride (30) with isolated chick embryo kidney have been studied at 530 nm over a wide range of pH and temperature[57].

(30)

D. Nuclear Magnetic Resonance Spectra

The n.m.r. spectrum of the red form of formazans exhibited a resonance attributed to the nitrogen–hydrogen group at low field, approximately τ 14 from TMS[53]. When R = 2,6-dimethoxyphenyl, methyl, phenyl, 4-methoxyphenyl, t-butyl- and ethoxycarbonyl in structure 26 above, the proton attached to nitrogen resonated at τ 9·68, 10·22, 14·54, 14·00, 13·69 and 14·67 respectively, from TMS all measurements in dimethyl sulphoxide-d_6 solution.

In sugar and non-sugar formazans[58] the nitrogen–hydrogen resonates at very low field. It was observed at τ 15·4 for triphenyl formazan and at τ 15·9 for acetyl formazan both measurements as solutions in deuterochloroform, resonances from TMS. The n.m.r. spectrum of penta-O-acetyl-D-galactose-diphenyl formazan was markedly different when a nitrogen-15 atom was substituted in the α_1 position.

Instead of a single peak as observed in the nitrogen-14 compound, a doublet with $J = 46·5$ Hz was observed. This was interpreted as indicating rapid tautomerism in sugar formazans.

E. Chromatography

The application of chromatographic techniques to the determination of formazans is mainly in biological and histological chemistry. Formazans of nitroblue tetrazolium **31** and tetranitroblue tetrazolium **32** are insoluble in common solvents at room temperatures. The most practical way of obtaining formazans for t.l.c. has been suggested[59] as spotting tetrazolium salts onto chromatographic plates and reducing the spots *in situ*. Ammonium sulphide is satisfactory for the reduction when dropped onto the tetrazolium salt on the plates. The plates were warmed, when the ammonium sulphide evaporated off quickly and left no residue. The conditions of the tetrazolium reduction determine the nature of the reaction product. The majority of formazan products of nitroblue tetrazolium and tetranitroblue tetrazolium gave R_f values of zero but even with solvents at the polar end of the eluotropic series it was impossible to get these formazans to run on silica gel or alumina.

An alternative method involves the quantitative elution of nitroblue formazan from tissue sections which is subsequently measured spectrophotometrically[60].

(31) and (32)

	R_2R_2'	R_3R_3'	R_5R_5'
31, Nitroblue tetrazolium	$p\text{-NO}_2\text{C}_6\text{H}_4$	$-\text{C}_6\text{H}_3(\text{OCH}_2)-$	C_6H_5
32, Tetranitoblue tetrazolium	$p\text{-NO}_2\text{C}_6\text{H}_4$	$-\text{C}_6\text{H}_3(\text{OCH}_3)-$	$p\text{-NO}_2\text{C}_6\text{H}_4$

VI. HYDRAZIDINES (HYDRAZIDE-HYDRAZONES OR DIHYDROFORMAZANS)

Literature pertaining to the hydrazidine structure is extremely sparse.

As mentioned in the section dealing with amidrazones the syntheses of silyl amidrazones and hydrazidines have been reported[19]. If one molecular proportion of an alkyl ω-(trimethylsilyl)-alkaneimidate and two or more (instead of one) molecular proportions of the dialkylhydrazine are reacted in the presence of a catalyst (an ammonium salt) the corresponding hydrazidine is formed.

The infrared spectra of the 2′,2′-dialkyl-ω-(trimethylsilyl)-alkanohydrazide dialkylhydrazones[19] contained strong absorption peaks correspond-

$$(CH_3)_3Si(CH_2)_n C \overset{NH}{\underset{OR}{\diagdown}} + 2NH_2 \cdot NR_2' \longrightarrow$$

$$(CH_3)_3Si(CH_2)_n C \overset{NNR_2'}{\underset{NH \cdot NR_2'}{\diagdown}} + NH_3 + R \cdot OH$$

$R = C_2H_5, \; iso\text{-}C_3H_7, \; C_4H_9; \; R' = CH_3, C_2H_5; \; n = 2,3.$

ing to the stretching vibrations of NH (3230 cm^{-1}), CH$_3$—N (2770 cm^{-1}) and C=N (1630 cm^{-1}), all spectra recorded as thin films of pure liquid. These values are at considerable variance with those recorded above even allowing for possible phase changes, and would suggest that more evidence from other sources is required before unambiguous assignments could be made.

The p.m.r. spectrum of 2',2',-dimethyl-4-(trimethylsilyl)-butyrohydrazide dimethylhydrazone[19] contains five resonances. The resonance with a chemical shift of δ 6·45 (τ 3·55) was attributed to the proton attached to nitrogen. The resonance attributed to the methyl groups attached to nitrogen δ 2·27 (τ 7·73) were similar to the value observed for the amidrazone δ 2·18 (τ 7·82).

VII. IMIDATES

A. Miscellaneous Data

Aliphatic and aromatic imidates are very prone to decomposition, especially in the presence of moisture. It is the products of these hydrolyses which are the most likely contaminants in any imidate system. It is therefore relevant that the factors influencing such decompositions are understood. In one such study[61] the influence of the aryl substituent, pH, temperature, general acid base catalysts and solvent polarity on the kinetics of hydrolysis of ethyl N-arylformimidates and ethyl N-arylacetimidates were examined in water and aqueous dioxan solutions to determine the mechanism of formation and breakdown of the intermediates involved in imidate hydrolysis. This information is in turn relevant to the mechanisms of important acyl transfer reactions.

Another investigation of the imidate formation mechanism[62] involved the detection of some formimidate esters and the measurement of their rates of formation using n.m.r. techniques. The results for the rate measurement for methyl formimidate at −10°C (263 K) were the most complete and showed a close proportionality to the hydrogen cyanide concentration.

Imidate–enamine tautomerism **33a** has been the subject of many studies. A recent report[63] describes several experiments conducted upon a number of complex imidates **33b**

iminonitrile   enaminonitrile

(33a)

(33b)

The n.m.r. spectra of solutions in dimethyl sulphoxide showed 15% enamine and 85% imidate whereas for solutions in dimethyl formamide the proportions were 36% enamine and 64% imidate.

In the ultraviolet difference spectra of solutions of **33b** in methanol the characteristic enamine chromophore[64] (253 nm, $\varepsilon = 6630$) was observed. In the presence of acid the absorption disappeared while one at 226 nm, originally $\varepsilon = 7200$ was enhanced.

The thermal behaviour of cellulose trichloroacetimidate has been investigated by differential thermal, dynamic thermogravimetric and isothermal thermogravimetric analyses[65]. Kinetic data were calculated from the curves and explanations of the observed transitions and their relationship to imidate hydrolysis were discussed.

B. Infrared and Raman Spectra

Spectroscopy in the infrared region for the study of the imidate system has a very early history. Several workers have attempted to study the very simple aliphatic imidate systems. A study of methyl acetimidate and its hydrochloride[66] was curtailed by the fairly rapid decomposition of the materials being studied. The study did reveal that the $\nu(C{=}N)$ mode was almost unaffected upon dilution, occurring in the pure liquid and as dilute solution in carbon tetrachloride at 1660 and 1661 cm^{-1} respectively. However, the vapour phase showed a doublet at 1682 and 1668 cm^{-1}. The $\nu(C{=}N^+)$ mode in methyl acetimidate hydrochloride was assigned to a band at 1649 cm^{-1} (mull in Nujol). The observation led the authors to conclude that the more ionic the C=N bond became, the lower was the value of the stretching mode wavenumber. This observation was supported by studies of hydrochlorides of compounds containing the carbon–nitrogen double bond[13].

The hydrochlorides of methyl formimidate and methyl acetimidate have been studied as solutions in concentrated hydrochloric acid by their Raman effect[67]. The hydrolysis was sufficiently slow to allow the measurements to be made. It was concluded that the structure of the alkyl formimidate hydrochlorides, given as $RO \cdot CHCl \cdot NH \cdot CH(OR)NH_2$, HCl could not be reconciled with their infrared spectra. Two strong bands were observed in the 1750–1600 cm^{-1} region. Replacement of R = H by R = CH_3 lowered the band near 1720 cm^{-1} slightly (1720 to 1708 cm^{-1}) and that near 1675 cm^{-1} substantially (1678 to 1623 cm^{-1}). Examination by the Raman effect of solutions indicated that the 10% reaction time for the hydrolysis in concentrated hydrochloric acid at 20°C (293 K) and $\sim 25\%$ cation concentration (estimated by weighing successive precipitates of ammonium chloride) was ~ 20 h for methyl acetimidate. This result paralleled the findings for the methyl benzimidate cation[68].

In infrared spectroscopy the imidate system is characterized[66] by the absorptions due to the $=NH$ and $C=N$ stretching modes, the former near 3330 cm^{-1}, the latter near 1650 cm^{-1}.

Several studies have been devoted to the effects of different solvent environments upon the $\nu(N-H)$ wavenumber. In one such study[69] ethyl butyrimidate, benzimidate and phenylacetimidate were examined in a diverse range of solvents. The values observed for the solutes varied by as much as 60 cm^{-1} from the values observed in hexane. The largest shifts to low wavenumbers were observed for the highly polar solvents pyridine and dimethyl sulphoxide. The situation was complicated by the observation of a double peak for ethyl phenylacetimidate. It is well established that this molecule is not very stable and rearranges slightly upon distillation[70]. A further complication arises because the product of the rearrangement is N-ethyl phenylacetamide which has an identical molecular formula and one N—H group (albeit sp^3 hybridized). Great care must be exercised to ensure that any spectra are not complicated by the presence of this impurity which does not influence a carbon, hydrogen, nitrogen analysis. The doublet structure was explained in terms of the two conformations **34** and **35**. In conformation **35** the NH group can interact with the π-electron system of the aromatic ring.

A similar study was devoted to the influence of substituents in the benzene ring of ethyl benzimidate upon the position of the $\nu(N-H)$ mode in a diverse range of solvents[71]. It was observed that although the maximum shift from the value observed in hexane was ~ 60 cm^{-1} (pyridine) the substituents in the ring had very little influence upon the $\nu(N-H)$

mode, the values for the five compounds studied varying very little from one another when recorded in the same solvent.

(34)   (35)

Measurements of the wavenumber and absolute intensities of the absorption bands $\nu_{0-1}(N-H)$, $\nu_{0-2}(N-H)$ and $\nu(C=N)$ of the $C=NH$ group in a series of phenylacetimidic and benzimidic acid esters have been made[72]. In phenylacetimidates it is suggested that the doublet structure of the fundamental stretching absorption band and its first overtone indicate the existence of two conformations 34 and 35 for these compounds.

It was also concluded that the N—H group in imidates is weakly acid and that it is slightly more acid in phenylacetimidates than in benzimidates. In benzimidates the $\nu(C=N)$ mode was observed in the range 1645–1630 cm^{-1} for the pure materials, the values for solution in carbon tetrachloride being nearly identical (where recorded). For the phenyl-acetimidates the mode was at 1655 ± 1 cm^{-1} for the pure materials with no shift for the solution value.

The influence of substituents attached to the nitrogen atom of the $C=N$ bond has also been investigated[89]. As previously indicated the value of the $\nu(C=N)$ mode moves to lower wavenumber for the hydrochloride salt. It was observed that when groups such as COR, COOR, CH_2CN and CH_2CH_2Br were attached to the nitrogen atom the observed wavenumber lay in the range 1666–1660 cm^{-1}. In order to make meaningful comparisons the spectra of some amidines and amidrazones were also recorded.

In an investigation of acet-, propion-, butyr-, isovaler-, capr-, and isocaprimidate systems[73] the $\nu(N-H)$ mode was observed in the range 3340–3275 cm^{-1} (pure liquids) and only slightly shifted (~ 10 cm^{-1}) to higher wavenumber for capr- and isocaprimidate in solution in heptane and carbon tetrachloride. The $\nu(C=N)$ was only slightly influenced by the solvent and lay in the range 1658–1653 cm^{-1} for the pure liquids. The other absorption bands observed in the 3400–1300 cm^{-1} region were also recorded.

The influence of substituents in the ether group of various trichloro-acetimidates has been observed[74]. For molecules of the type

$CCl_3C(=NH)OCH_2R$ the $\nu(C=N)$ mode was observed at 1679, 1675 and 1680 cm^{-1} when R $= CF_3$, CCl_3 and $C(NO_2)_2CH_3$ respectively.

For a series of trichloroacetimides it has been observed[70] that the shift between pure liquid and vapour phase values for the $\nu(N-H)$ mode was ~ 25 cm^{-1}, the liquid value recorded in the range 3347 $\pm$ 1 cm^{-1}. A much more striking result is obtained for the influence of substituents attached to the nitrogen atom. For methyl trichloroacetimidate the $\nu(C=N)$ mode occurs at 1673 cm^{-1} whereas for the N-deuterated compound the value is 1656 cm^{-1}. The $\nu(C=N)$ mode occurs at slightly higher wavenumbers for the trichloroacetimidates than for the acetimidates. The effect of attaching chlorine to the nitrogen atom in acetimidates is to drop the $\nu(C=N)$ mode to ~ 1620 cm^{-1}, whereas a hydroxyl group attached to the nitrogen causes little if any shift.

C. Ultraviolet and Visible Spectra

An examination has been made of the ultraviolet spectra of a series of imidates of the type[75]:

$$R-C\begin{matrix}\nearrow NH \\ \searrow OCH_3\end{matrix}$$

R $= C_6H_5C\equiv C$ **36**; $(CH_3)_3 CC\equiv C$ **37**;
$CH_3(CH_2)_4C\equiv C$ **38**; $CH_3(CH_2)_5C\equiv C$ **39**.

Absorption maxima were observed at 260 nm for compound **36**, 205 nm for compound **37** and at 207 nm for compounds **38** and **39**. The molar extinction coefficients lay between $(1-2\cdot6) \times 10^4$ for these compounds.

D. Nuclear Magnetic Resonance Spectra

N.m.r. studies of imidates are not as comprehensive as those devoted to infrared spectroscopy. The n.m.r. spectra of benzimidates and phenylacetimidates have been recorded[72]. The NH group proton resonated near τ 1·75 for the benzimidates and near τ 2·85 for phenylacetimidates. In both cases these values for the pure liquids were little affected by substituents in the benzene ring. For solutions in carbon tetrachloride (concentration not recorded) the values shifted to τ 2·3 and τ 3·28 respectively. For the series of aliphatic imidates mentioned previously[73] all the compounds exhibited a broad resonance in the range τ 3·06–2·90 for solutions in deuterochloroform.

In the trichloroacetimidates[70] the proton attached to nitrogen resonates in the range τ 1·74–1·52, values for pure liquid samples.

E. Chromatography

Chromatographic separations and determinations are not usually practised upon simple imidate systems. The technique has been applied to determination of complex imidates, such as methomyl[76]. Methomyl (S-methyl-N-[(methylcarbamoyl)oxy]thioacetimidate) was hydrolysed to the oxime, methyl N-hydroxythioacetimidate, which was extracted with

$$CH_3SC(CH_3)=NOCONHCH_3 \xrightarrow{OH^-} CH_3SC(CH_3)=NOH$$

an organic solvent and selectively measured by micro-coulometric gas chromatography. The method is claimed to be sensitive to 0·02 p.p.m. methomyl on a 25-g sample.

F. Dipole Moments

Several dipole moment studies have been made of imidate systems in attempts to resolve their structure. The situation is complicated by the possibility of the system existing as one of four isomers.

| syn (s-cis) | syn (s-trans) | anti (s-cis) | anti (s-trans) |
| (40) | (41) | (42) | (43) |

In any attempt to use dipole moment measurements for structural studies the final result hangs on the accuracy of the calculated moments with which the measured moment is to be compared. For imidate and other systems containing the carbon–nitrogen double bond the bond moment assigned to the double bond is of paramount importance. The actual moment used varies widely. Another important factor is the solvent used to measure the experimental moment. The formation of complexes between solvent and solute is obviously highly undesirable.

The earliest use of dipole moments to study the imidate system was in 1934[77] and this dealt with ethyl acetimidate, the measured moment being 1·33 debyes for solutions in dioxan.

From a study of ethyl acetimidate and ethyl benzimidate[78] it was concluded that structure 42 was the only one consistent with the measured moments of 1·42 and 1·54 debyes respectively. These values were recorded for solutions in benzene, the values were 1·44 and 1·52 debyes respectively for solutions in dioxan.

A similar study[79] concluded that the *anti* (s-cis) isomer was that which was taken by ethyl acetimidate, benzimidate, p-nitrobenzimidate and p-propoxybenzimidate.

In a study of ethyl benzimidate[80] the conformation was determined. In a later study[81] on the basis of a qualitative comparison of its dipole moment with the moment of ethyl p-nitrobenzimidate it was concluded that the alkoxy group possessed free rotation around the carbon–oxygen axis in contradistinction to a stable ester conformation. The study aimed at determining the conformation formed by rotation around the carbon–oxygen axis on the one hand and of the unstable configuration on the carbon–nitrogen double bond on the other. The s-cis conformation was taken as impossible on steric grounds which left as the only possible interpretation of the result a mixture of both unstable configurations on the C=N bond with simultaneous free rotation around the C—O bond. The maximum uncertainty of this study lay in the choice of value of the bond moment ascribed to the easily polarized bonds, e.g. C=N. The dipole moments were measured for solutions in benzene and dioxan.

G. X-ray Studies

Methyl p-bromobenzimidate has been studied by X-ray crystallography[82].

The needle like crystals elongated in the [001] direction (m.p. 65–66°C) were grown from n-heptane solution. Laue photographs showed that the crystals belonged to the orthorhombic system and that the unit cell dimensions were

$$a = 1.864 \pm 0.0015 \text{ nm}$$
$$b = 1.547 \pm 0.0010 \text{ nm}$$
$$c = 0.582 \pm 0.0005 \text{ nm}$$

The space group was $D_{2h}^{15} - P_{bca}$ and the X-ray density calculated from the measured unit cell dimensions and assuming eight molecules per unit cell was 1.695 compared with 1.68 g cm^{-3} measured by the flotation method. The volume of the unit cell was 1.678 nm^3.

VIII. IMIDOYL HALIDES

A. Infrared and Raman Spectra

Imidoyl halides are reactive organic compounds characterized by the presence of a halogen atom attached to the carbon atom of a C=N bond.

The majority of spectroscopic studies of imidoyl halides have concentrated on the position of the $\nu(C=N)$ mode. N-Chloro-chloroformimidoyl chloride has been studied in the vapour, liquid and solid phases by infra-red spectroscopy[83]. The $\nu(C{=}N)$ mode was observed at 1564 cm^{-1} in the solid, 1565 cm^{-1} in the liquid, as a doublet at 1580 and 1569 cm^{-1} in the vapour phase and as a weak line at 1565 cm^{-1} in the Raman effect. The weak Raman effect for the $C{=}N$ bond was taken to be indicative of a weaker, more polar bond than in similar compounds. The $\nu(N{-}Cl)$ mode was observed at 746 cm^{-1} in the vapour and 642 cm^{-1} in the liquid phase.

In contrast the Raman effect of N-chloro-chlorothioformimidoyl chloride, $ClSC(Cl){=}NCl$, exhibited two very intense shifts at 1587 and 1604 cm^{-1} which were ascribed to the $C{=}N$ bond[84]. The $\nu(C{=}N)$ mode has also been assigned to bands at 1689 and 1672 cm^{-1} in the infrared spectra of $HC(Cl){=}NCH_3$ and $C_6H_5C(Cl){=}NC_6H_5$ respectively[85]. The molecules $CH_3C(Br){=}NH_2^+ Br^-$ and $CH_3C(Br){=}ND_2^+ Br^-$ have been demonstrated to possess bands at 1664 and 1626 cm^{-1} respectively[86].

B. Nuclear Magnetic Resonance Spectra

N.m.r. studies of the imidoyl halides are few. The n.m.r. parameters of 4-chlorohexafluoro-2-azabut-1-ene (**44**), and 4-chlorohexafluoro-2-azabut-2-ene (**45**), have been obtained by analysis on a first order basis[87]. At low temperatures the ^{19}F nuclei located near to a nitrogen atom gave relatively sharp lines as compared to the broad absorption observed at room temperature. This result was interpreted, at least for **44** in terms of stereoisomerism about the $C{=}N$ bond. The effects were particularly evident for fluorine in $CF_2{=}N$. At room temperature they gave a very broad AB

(44)

type spectrum which sharpened at $-35°C$. It was difficult to apply the same argument to **45**. At room temperature the $CF{=}N$ absorption was very broad and at $-70°C$ splittings due to CF_3 and CF_2Cl appeared. If the effect was due to the freezing out of the nitrogen inversion, signals related to the *syn* and *anti* stereoisomers should be observed at low temperatures. The n.m.r. spectrum was consistent with the existence of only one isomer which was probably *syn*.

syn   anti

(45)

C. Dipole Moments

Dipole moment studies have been made in an attempt to confirm the configuration of imidoyl chlorides[88]. Six imidoyl chlorides were studied and the dipole moment for the Z and E configurations of each compound were calculated by vector addition of the bond moments.

Z (*trans*)   E (*cis*)

(46)   (47)

The calculated moments were consistent with the Z configuration. This was made more evident by the use of Exner's graphical method[81]. The Z configuration was assumed proved for the chloride of aromatic amidic acids. The stability appeared to be controlled by strong steric effects which do not favour the E form. Evidence was presented which suggested that the C=N bond moment could be given the constant value (1·8 debye) with the same standard accuracy and range of validity as for other bond moments.

IX. REFERENCES

1. F. C. Schaefer and G. A. Peters, *J. Org. Chem.*, **26**, 2778 (1961).
2. F. C. Schaefer and G. A. Peters, *J. Org. Chem.*, **26**, 2784 (1961).
3. E. M. Smolin and L. Rapoport, *s-Triazines and Derivatives*, Interscience, New York (1959).
4. N. D. Cheronis and T. D. Ma, *Organic Functional Group Analysis by Micro and Semi-micro Methods*, Wiley, New York (1964), p. 278.
5. K. R. Manolov, *Fresenius's Z. Anal. Chem.*, **234**, 37 (1968).
6. W. H. Prichard, *Ph.D. Thesis*, University of Wales, 1962.
7. F. Eloy and R. Lenaers, *Chem. Rev.*, **62**, 155 (1962).
8. D. G. Neilson, R. Roger, J. W. M. Heatlie and L. R. Newlands, *Chem. Rev.*, **70**, 151 (1970).
9. A. W. Nineham, *Chem. Rev.*, **55**, 355 (1955).
10. R. Roger and D. G. Neilson, *Chem. Rev.*, **61**, 179 (1961).
11. H. Ulrich, *The Chemistry of Imidoyl Halides*, Plenum Press, New York (1968).
12. M. Davies and A. E. Parsons, *Z. Phys. Chem.*, **20**, 34 (1959).
13. D. N. Shigorin, *Zh. Fiz. Chim.*, **25**, 789 (1951).
14. D. Pevorsek, *C. R. Acad. Sci. (Paris)*, **244**, 2599 (1957).
15. D. Pevorsek, *Bull. Soc. Chim. France*, 788 (1958).
16. J. Fabian, V. Delaroff and M. Legrand, *Bull. Soc. Chim. France*, 287 (1956).

17. J. C. Grivas and A. Taurins, *Canad. J. Chem.*, **37**, 795 (1959).
18. P. Söhar, *Acta Chim. Acad. Sci. Hung.*, **54**, 91 (1967).
19. G. S. Gol'din, V. G. Poddubnyi, A. A. Simonova, A. B. Kamenskii and G. S. Shor, *Zh. Obshch. Khim.*, **40**, 1288 (1970); Engl. transl., **40**, 1281 (1970).
20. J. Fabian, M. Legrand and P. Poirier, *Bull. Soc. Chim. France*, 1499 (1956).
21. D. N. Shigorin and Yu. K. Syrkin, *Izvest. Akad. Nauk SSSR Ser. Fiz.*, **9**, 225 (1945); *Zh. Fiz. Khim.*, **23**, 241 (1949).
22. M. Mares-Guia and E. Rogana, *Ann. Acad. Brasil Cienc.*, **43**, 563 (1971).
23. J. Sevcik, *Chem. Zvesti*, **26**, 49 (1972).
24. P. J. Kovi, A. C. Capomacchia and S. G. Schulman, *Anal. Chem.*, **44**, 1611 (1972).
25. R. C. Neuman, G. S. Hammond and T. J. Dougherty, *J. Amer. Chem. Soc.*, **84**, 1506 (1962).
26. R. C. Neuman and G. S. Hammond, *J. Phys. Chem.*, **67**, 1659 (1963).
27. G. S. Hammond and R. C. Neuman, *J. Phys. Chem.*, **67**, 1655 (1963).
28. V. Jonas, *Diss. Abstr.*, **B32**, 830 (1971).
29. C. K. Tseng and F. M. Pallos, *Spectros. Letters*, **5**, 43 (1972).
30. J. P. Marsh and L. Goodman, *Tetrahedron Letters*, **8**, 683 (1967).
31. I. Kugajevsky, *Diss. Abstr.*, **B27**, 1817 (1966).
32. A. K. Bose and I. Kugajevsky, *Tetrahedron*, **23**, 1489 (1967).
33. H. F. Grutzmacher and H. Kuschel, *Org. Mass. Spectr.*, **3**, 605 (1970).
34. S. Veibel, *Anal. Chim. Acta*, **51**, 309 (1970).
35. R. L. Rebertus, K. R. Fiedler and G. W. Kottong, *Anal. Chem.*, **39**, 1867 (1967).
36. W. J. Orville-Thomas and A. E. Parsons, *Trans. Faraday Soc.*, **54**, 460 (1958).
37. H. E. Hallam, *Infra-red Spectroscopy and Molecular Structure*, Ed. M. Davies, Elsevier, London 1963, p. 405.
38. S. Califano and W. Lüttke, *Z. Phys. Chem.* (Frankfurt), **5**, 240 (1955).
39. H. Goncalves and A. Secches, *Bull. Soc. Chim. France*, 2589 (1970).
40. H. E. Ungnade, L. W. Kissinger, A. Narath and D. C. Barham, *J. Org. Chem.*, **28**, 134 (1963).
41. H. E. Ungnade and L. W. Kissinger, *J. Org. Chem.*, **23**, 1794 (1958).
42. J. Barrans, R. Mathis-Noël and M. F. Mathis, *C. R. Acad. Sci.* (*Paris*), **245**, 419 (1957).
43. D. Pevorsek, *C. R. Acad. Sci.* (*Paris*), **247**, 1333 (1958).
44. J. Barrans, T. Marty and R. Mathis, *C. R. Acad. Sci.* (*Paris*), Ser. B, **254**, 2736 (1962).
45. K. W. F. Kohlrausch and R. Seka, *Z. Phys. Chem.*, **B38**, 72 (1938).
46. N. E. Alexandrou and D. N. Nicolaides, *Org. Mag. Resonance*, **4**, 229 (1972).
47. M. H. Millen and W. A. Waters, *J. Chem. Soc.*, B, 408 (1968).
48. H. J. Hofmann and M. Scholz, *J. Prakt. Chem.*, **313**, 349 (1971).
49. W. Walter and H. Weiss, *Liebigs Ann. Chem.*, **758**, 162 (1972).
50. B. G. Baccar and J. Barrans, *C. R. Acad. Sci.* (*Paris*), Ser. C., **263**, 743 (1966).
51. G. A. Sereda, R. I. Ogloblina, V. N. Podchainova, and N. P. Bednyagina,

Zh. Anal. Khim., **23**, 1224 (1968); J. Anal. Chem. U.S.S.R. (Engl. transl.), **23**, 1076 (1968).
52. E. N. Yurchenko and I. I. Kukushkina, Zh. Strukt. Khim., **10**, 706 (1969); J. Struct. Chem. (Engl. trans.), **10**, 600 (1969).
53. W. Otting and F. A. Neugebauer, Z. Naturforsch., **P23**, 1064 (1968).
54. N. P. Bednyagina, A. P. Novikova, N. V. Serebryakova, I. I. Mudretsova and I. Ya. Postovskii, Zh. Org. Khim., **4**, 1613 (1968); J. Org. Chem. U.S.S.R., **4**, 1551 (1968).
55. I. I. Kukushkina, E. N. Yurchenko, D. K. Arkhipenko and B. A. Orekhov, Zh. Fiz. Khim., **46**, 1677 (1972).
56. R. G. Dubenko and P. S. Pel'kis, Zh. Org. Khim. **6**, 1101 (1970); J. Org. Chem., U.S.S.R., **6**, 1103 (1970).
57. H. Mujamoto, H. Sone, S. Moori and Y. Oka, Tokushima J. Exp. Med., **7**, 57 (1970).
58. L. Mester, A. Stephen and J. Parello, Tetrahedron Letters, **38**, 4119 (1968).
59. J. H. Tyrer, M. J. Eadie and W. D. Hooper, J. Chromatographie, **39**, 312 (1969).
60. F. P. Attman, Histochemie, **17**, 319 (1969).
61. R. H. DeWolfe, J. Org. Chem., **36**, 162 (1971).
62. J. Nematollahi and L. D. Tuck, J. Pharm. Sci., **56**, 684 (1967).
63. L. Töke, G. Blasko, L. Szabo and Cs. Szantay, Tetrahedron Letters, **24**, 2459 (1972).
64. S. Baldwin, J. Org. Chem., **26**, 3288 (1961).
65. T. L. Vigo, D. J. Daigle and C. H. Mack, J. Appl. Polymer Sci., **15**, 2051 (1971).
66. W. H. Prichard and W. J. Orville-Thomas, J. Chem. Soc., A, 1102 (1967).
67. E. Spinner, Aust. J. Chem., **19**, 2153 (1966).
68. J. T. Edward and S. C. R. Meacock, J. Chem. Soc., 2009 (1957).
69. R. Pujol and F. Mathis, C. R. Acad. Sci. (Paris), Ser. B, **269**, 975 (1969).
70. J. C. Hughes and W. H. Prichard, Unpublished results.
71. R. Pujol and F. Mathis, C. R. Acad. Sci. (Paris), Ser. B, **269**, 1049 (1969).
72. R. Mathis, B. Baccar, J. Barrans and F. Mathis, J. Mol. Structure, **7**, 355 (1971).
73. P. Reynaud and R. C. Moreau, Bull. Soc. Chim. France, 2997 (1964).
74. G. A. Shevkhgeimer and M. L. Shul'man, Dokl. Akad. Nauk SSSR, **173**, 378 (1967).
75. C. A. Grob and B. Fischer, Helv. Chim. Acta, **38**, 1794 (1955).
76. H. L. Peare and J. J. Kirkland, J. Agric. Food Chem., **16**, 554 (1968).
77. W. D. Kumler and C. W. Porter, J. Amer. Chem. Soc., **56**, 2549 (1934).
78. H. Lumbroso, D. M. Bertin and P. Reynaud, C. R. Acad. Sci. (Paris), **261**, 399 (1965).
79. H. Lumbroso and D. M. Bertin, Bull. Soc. Chim. France, 1728 (1970).
80. O. Exner, V. Jehlička and A. Reiser, Coll. Czech. Chem. Comm., **24**, 3207 (1959).
81. O. Exner and V. Jehlička, Coll. Czech. Chem. Comm., **30**, 639 (1965).
82. B. Kolakowski, Acta Phys. Polon., **36**, 713 (1969).
83. J. M. Burke and R. W. Mitchell, Spectrochim. Acta, **A28**, 1649 (1972).
84. F. Feher and H. Weber, Chem. Ber., **91**, 2523 (1958).

85. H. M. Bosshard and H. Zollinger, *Helv. Chim. Acta*, **42**, 1659 (1959).

86. E. Allenstein and P. Quis, *Chem. Ber.*, **97**, 3162 (1964).

87. L. Cavalli and P. Piccardi, *J. Chem. Soc.*, **D**, 1132 (1969).

88. A. Dondoni and O. Exner, *J. Chem. Soc., Perkin II*, 1908 (1972).

89. B. Baccar, R. Mathis, A. Secches, J. Barrans and F. Mathis, *J. Mol. Structure*, **7**, 369 (1971).

90. H. C. Brown and D. Pilipovich, *J. Amer. Chem. Soc.*, **82**, 4700 (1960).

91. P. Bassignana, C. Cogrossi, G. Polla-Mattiot and S. France, *Ann. Chim. Rome*, **53**, 1212 (1963).

CHAPTER **4**

Rearrangements involving imidic acid derivatives

C. G. McCarty and L. A. Garner

West Virginia University, Morgantown, U.S.A.

I. INTRODUCTION

The largest portion of this chapter is devoted to a review of the thermal and catalysed rearrangements of the derivatives of imidic acid which may generally be classified as imidates. Examples given include both open-chain and cyclic imidates. For comparison purposes, short sections have been included on studies of thioimidates, hydrazonates and a few other compounds closely related to imidates. In all cases involving imidates, the rearrangements involve an aryl, alkyl or allyl group migrating from the imidate oxygen to the azomethine nitrogen. Examples of hydrogen migration from a carbon to the azomethine nitrogen (tautomerism) have not been included since they have been dealt with in another chapter in this volume.

There are many fewer references to rearrangements of amidines. A short summary of the literature on this subject is followed by the concluding section on some miscellaneous rearrangements of compounds which are structurally related to imidic acid.

II. REARRANGEMENTS OF IMIDATES AND RELATED COMPOUNDS

A. Aryl Imidates (The Chapman Rearrangement)

I. Introduction

The thermal, uncatalysed rearrangement of an aryl imidate (**1**) to an N-aroyldiarylamine (**2**) via a 1,3-O to N migration was first reported by Mumm, Hesse and Volquartz in 1915[1]. Several years passed before

$$\underset{\textbf{(1)}}{\overset{\displaystyle \overset{OAr^2}{\underset{|}{\,}}}{Ar^1C{=}NAr^3}} \quad \xrightarrow{200\text{-}300^\circ C} \quad \underset{\textbf{(2)}}{Ar^1C\overset{\displaystyle O}{\overset{\|}{\,}}N\!\!\begin{array}{l} {}^{Ar^2}\\ {}_{Ar^3} \end{array}} \tag{1}$$

Chapman began his study of the mechanism and scope of this reaction. Because of Chapman's extensive work on the subject[2-7], this rearrangement now usually bears his name.

Reviews of the Chapman rearrangement by Schulenberg and Archer[8] and by McCarty[9] have been published in the past decade. With these two reviews being so recent and readily accessible, most of the emphasis in this chapter is on some of the more recent applications and mechanistic studies. No attempt has been made to be all-inclusive, but it is hoped that most of the pertinent references to this rearrangement from the late 1960's through most of 1973 have been included. For the sake of comparison, some examples of O to N migration of aryl groups in related compounds are included at the end of this section.

2. Some recent applications

The main synthetic utility of the Chapman rearrangement lies in the fact that the initially formed amides (**2**) can be hydrolysed to diarylamines

(**3**, equation 2). Often this provides a route to diarylamines which may be difficult or impossible to prepare by other methods[8].

$$\underset{\substack{\displaystyle \text{(2)}}}{Ar^1CN\overset{\displaystyle \overset{O}{\|}}{}\!\!\overset{\displaystyle Ar^2}{\underset{\displaystyle Ar^3}{\diagdown}}} \xrightarrow{\text{alc. KOH}} \underset{\substack{\displaystyle \text{(3)}}}{Ar^2NHAr^3} \qquad (2)$$

A recent example can be found in the study of the effect of some tri-fluoromethoxy dibenz[*b,e*][1,4]diazepines (**7**) on the central nervous system. A necessary intermediate in the reaction sequence employed by McEvoy and co-workers[10] was diarylamine **6** which could not be satis-factorily prepared by the copper-catalysed condensation of *p*-trifluoro-methoxyaniline with *o*-nitrochlorobenzene. However, the Chapman rearrangement of imidate **4** to amide **5** in 80% yield was carried out in refluxing *o*-dichlorobenzene and subsequent hydrolysis of **5** gave **6** in 99% yield.

$$p\text{-}CF_3OC_6H_4N\!\!\!=\!\!\!\underset{\underset{\displaystyle C_6H_5}{|}}{COC_6H_4NO_2\text{-}o} \xrightarrow{\Delta} p\text{-}CF_3OC_6H_4\underset{\underset{\displaystyle COC_6H_5}{|}}{N}C_6H_4NO_2\text{-}o$$

(4)   (5)

alc. KOH

(3)

$$\xleftarrow{} p\text{-}CF_3OC_6H_4NHC_6H_4NO_2\text{-}o$$

(6)

(7)

Mukherjee and Block[11] used the Chapman rearrangement to prepare some iodinated diphenylamine analogs of thyroxine (**10**, equation 4) which were expected to show marked physiological activity. They observed a marked rate enhancement from the *ortho* electron withdrawing substituents on the migrating ring in conformity with recent views on the mechanism of this rearrangement (see next section, Mechanism). The formation of **9** from imidate **8** was almost quantitative after only 10 min in refluxing *o*-dichlorobenzene.

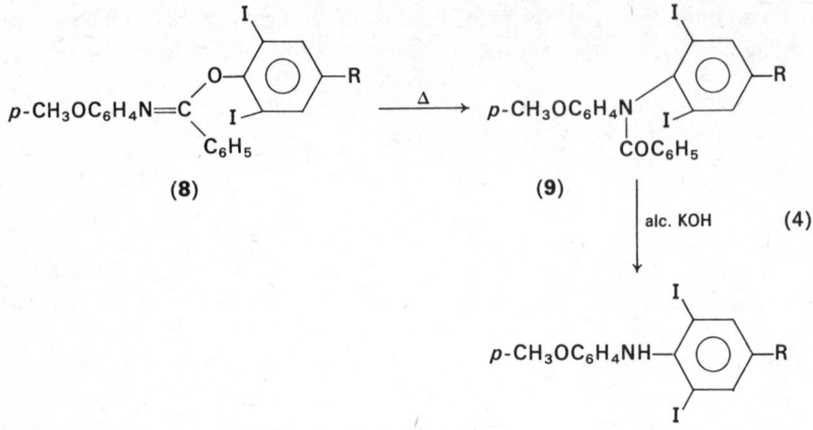

(8)   (9)

alc. KOH   (4)

(10)

Chlorinated diphenylamines have also been prepared in high yields by the Chapman rearrangement of the appropriate imidate precursors. 3,5,3′,5′-Tetrachlorodiphenylamine prepared in this way by Fritsch and co-workers[12] was subjected to 'mild nitration' to yield 2,6,2′,6′-tetranitro-3,5,3′,5′-tetrachlorodiphenylamine—a compound with excellent anti-bacterial activity.

In an attempt to assess the importance of the relative planarity of the aryl rings of N-arylanthranilic acid analogs on anti-inflammatory activity, Westby and Barfknecht[13] prepared 11 and 12 via Chapman rearrangements. These diarylamines proved to be anti-inflammatory agents and

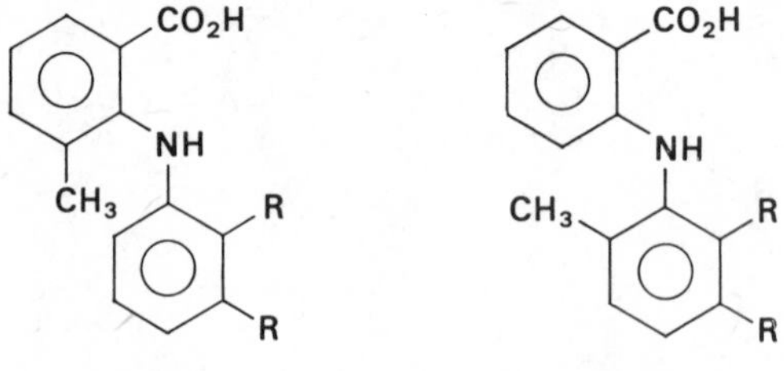

(11)   (12)

R = CH₃,H

comparisons with other compounds revealed that the degree of planarity of the aryl rings was apparently not an important factor in determining anti-inflammatory behaviour.

Sometimes the amides formed from the rearrangement of imidates are valuable because of their ability to undergo subsequent cyclization. Imidate 13 was converted to 15 in 65% yield, probably via the initial formation of the Chapman product **14**[14]. Reduction of **15** led to one of a series of 1,2-diaryltetrahydroquinolines tested by Bell and co-workers[14] for antifertility, estrogenic and antiestrogenic activities in rats. They also employed another route (equation 6) which involved a Chapman rearrangement of a phenoxy-quinoline.

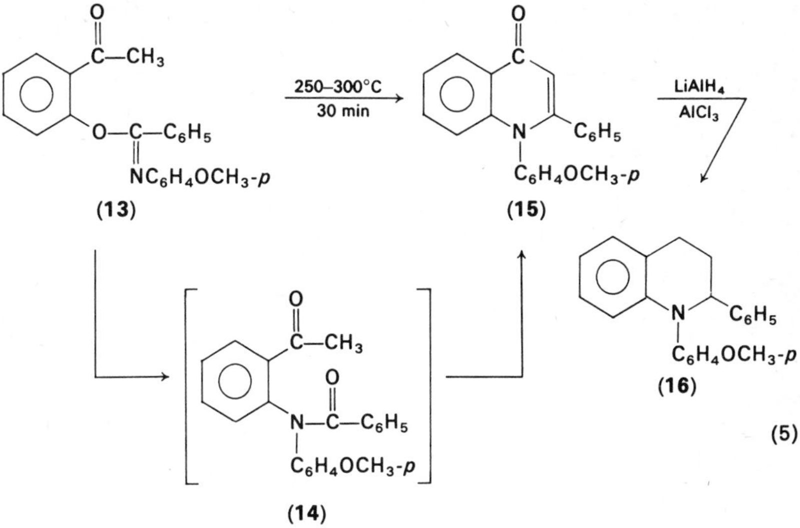

(5)

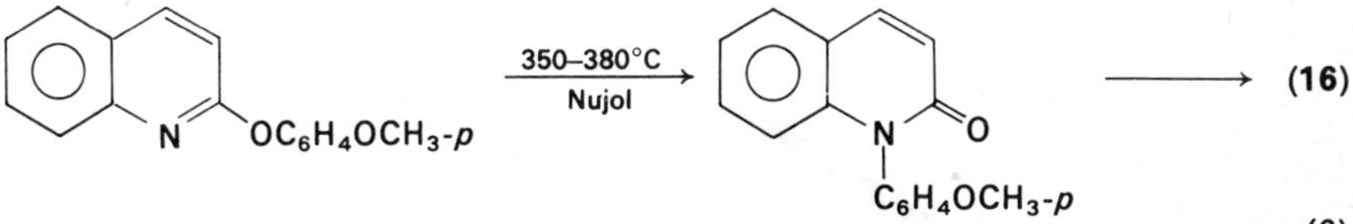

(6)

Schulenberg[15] found that amide **18**, which was formed in 91% yield from **17**, could be cyclized with excess sodium methoxide in benzene to an oxindole or, with slightly less than one equivalent of the same base, to an indole diester.

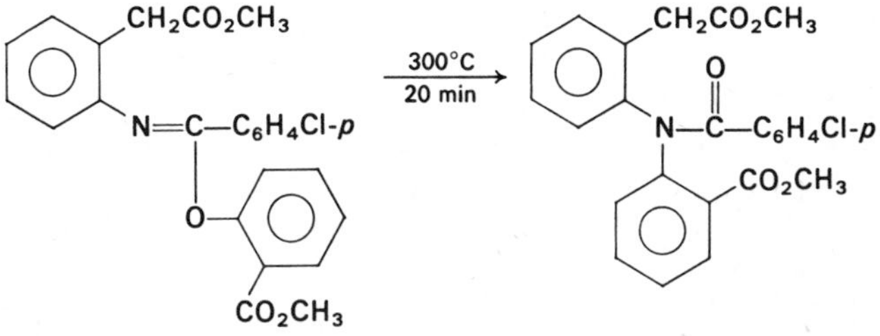

(7)

It has been known for many years that the products of the Chapman rearrangement can be used for the synthesis of acridones[8]. Often, appropriately substituted benzimidates can be converted directly to acridones without isolation of the intermediate N-benzoyldiarylamines. Goldberg and Harris[16] recently made use of this fact in their synthesis of some benz- and dibenzacridones by the method of Cymerman-Craig and Loder[17]. For example, benzimidate **19** underwent rearrangement to dibenzacridone **20** when it was heated to 280–310°C. However, when the temperature was kept just below 280°C, the ester **21** was obtained in excellent yield.

(8)

For the Chapman rearrangement to be useful in a synthetic scheme it should, of course, give a high yield of product which is easily isolated. Many literature references to this reaction involve procedures describing the use of the fused imidate instead of the imidate dissolved in a high boiling solvent. In the absence of activating substituents on the migrating ring (see next section) the yields from the neat rearrangements are often low and isolation of the amide rearrangement product is sometimes difficult. Wheeler and co-workers[18] studied the rearrangement of several aryl imidates in a variety of high boiling solvents and found that they obtained the highest yields (up to 96%) in boiling tetraglyme. The products were easily isolated by the addition of water.

3. Mechanism

The mechanism of the thermal, uncatalysed rearrangement of aryl imidates to amides is now well established. It is known to be unimolecular, intramolecular, and to proceed by an S_N-like nucleophilic attack by the azomethine nitrogen on the migrating aryl group *via* a 4-membered transition state (equation 9). The rest of this section is devoted to a summary of some of the studies which led to this present view of the mechanism of the Chapman rearrangement.

$$
\begin{array}{ccc}
\underset{\textbf{(1)}}{\overset{\displaystyle O\!-\!Ar^2}{\underset{\displaystyle Ar^1\!-\!C\!=\!NAr^3}{\big|}}}
& \longrightarrow &
\left[\;
\begin{array}{c}
O\text{-----}Ar^2 \\
\vdots\qquad\ \vdots \\
Ar^1\!-\!C\text{=====}N\!-\!Ar^3
\end{array}
\;\right]^{\ddagger}
& \longrightarrow &
\underset{\textbf{(2)}}{Ar^1\!-\!\overset{\displaystyle O}{\overset{\|}{C}}\!-\!N\overset{\displaystyle Ar^2}{\underset{\displaystyle Ar^3}{<}}}
& (9)
\end{array}
$$

Chapman first studied the kinetics of this reaction by following the conversion of phenyl N-phenylbenzimidate (**22**) to N-phenylbenzanilide (**23**) at temperatures varying from 228 to 292°C[2]. His method for following

$$
\underset{\textbf{(22)}}{\overset{\displaystyle OC_6H_5}{\underset{\displaystyle C_6H_5C\!=\!NC_6H_5}{\big|}}}
\xrightarrow[\text{neat}]{228\text{--}292^\circ C}
\underset{\textbf{(23)}}{\overset{\displaystyle O}{\overset{\|}{C_6H_5CN(C_6H_5)_2}}}
\qquad (10)
$$

the kinetics would be considered crude compared to modern techniques but many of the ideas he proposed from his early studies have later been shown by other workers to be correct. The conversion of **22** to **23** was found by him to obey first-order kinetics although parameters such as activation energy, entropy and enthalpy of activation were not calculated.

Wiberg and Rowland[19], realizing the crudeness of Chapman's kinetic data, repeated the kinetic study on **22** plus several other aryl imidates substituted in ring 2 (equation 9). Diphenyl ether was used as the solvent and the temperature varied from 203 to 279°C. The progress of the rearrangements was followed by potentiometric titration of a glacial acetic acid solution of the reaction mixture with 0·1M-perchloric acid in glacial acetic acid. Their results supported the unimolecularity of the rearrangement and gave further insight into the nature of substituent effects earlier noted by Chapman. Some rate constants at 255°C are listed in Table 1 and some ΔH^* and ΔS^* values are shown in Table 2.

Several interesting conclusions can be drawn from the data in Tables 1 and 2. Electron withdrawing substituents on the migrating ring facilitate the reaction and substitution in the *ortho* position is especially effective. The effect of electron withdrawing substituents in Ar^2 (**1**) had, in fact, been noted by Chapman earlier[3]. He reported, for example, that when Ar^2 was *o*-nitrophenyl the conversion to amide was 40% complete in 90 min at 163°C whereas when Ar^2 was phenyl the same percentage conversion in

TABLE 1. Rates of rearrangement of some aryl benzimidates (1) at 255°C in diphenyl ether[19]

Substituent on Ar^2 (1)	$k \times 10^5 (sec^{-1})$
H	7·66
p-Methyl	3·55
o-Methyl	8·87
m-Methyl	7·11
p-Chloro	13·8
o-Chloro	64·5
p-Methoxy	1·60
o-Methoxy	8·87
p-i-Propyl	3·79
o-i-Propyl	6·75
p-Ethyl	3·69
p-Bromo	18·1
p-t-Butyl	3·82
o-t-Butyl	2·30

the same time required temperatures of about 255°C. Chapman also studied the effect of substituents in Ar^1 and Ar^3. He found that when Ar^3 was p-methoxyphenyl the rearrangement proceeded faster than when Ar^3 was phenyl[2]. When Ar^3 was 2,4,6-trichlorophenyl the rearrangement was much slower than with the unsubstituted compound. Decomposition occurred before rearrangement took place when Ar^3 was p-nitrophenyl. Similar substituent effects were found for Ar^1 although they seemed to be lesser in magnitude[2].

The effect of substituents in the *ortho* positions of Ar^2 deserves further discussion. The general trend of *ortho* rate > *para* rate for a given substituent is readily apparent from Table 1. It was assumed by Wiberg and Rowland[19] that the *ortho* substituent hindered free rotation of Ar^2; this

TABLE 2. Enthalpies and entropies of activation for the rearrangement of some aryl benzimidates (1)[19]

Substituent on Ar^2 (1)	ΔH^*(kcal/mol)	ΔS^*(eu)
H	37·7 ± 0·9	− 7·2 ± 2
p-Methyl	37·2	− 9·7
o-Methyl	39·5	− 3·6
p-Chloro	37·4	− 6·7
o-t-Butyl	36·7	−11·6

restriction being exactly what was required at the point of the 4-membered transition state. Thus, the *ortho* substituent should raise the entropy of the ground state of **1** and lessen the decrease in entropy in going from **1** to the transition state. Indeed, as can be seen from Table 2, the entropy of activation for *o*-CH$_3$ is less negative than that for *p*-CH$_3$. In the case of *t*-butyl the rate ratio k_{ortho}/k_{para} was 0·60 and it was suggested that the *t*-butyl group was so bulky that in the *ortho* position its steric requirements outweighed any rate enhancement resulting from hindered rotation.

Relles[20] reasoned that if the argument of Wiberg and Rowland was correct then the rate of Chapman rearrangements of aryl imidates with two *ortho* substituents on the migrating ring should be even greater than those with one *ortho* substituent (in cases where steric compression does not negate steric acceleration). His results are combined with some of Wiberg's and Rowland's results for comparison in Table 3. It is apparent from Table 3 that steric acceleration due to hindered rotation (SAHR[20]) predominates over steric and inductive decelerating effects when a second *o*-CH$_3$ group is added to Ar^2. On the other hand, the introduction of a second *o*-*t*-butyl group results in a greater deceleration than that observed for one such group. Clearly the steric compression in proceeding from imidate to transition state is far more important than SAHR when *o*-*t*-butyl groups are present. The interpretation of the rate increase observed with *o*-phenyl groups cannot be based on steric effects alone since resonance and inductive effects must also play a role.

Relles recognized that there are similarities between the Chapman and the Newman–Kwart (equation 11) rearrangements. In a paper[21] giving a detailed comparison of these two reactions he calculated steric substituent constants and found the values for $\sigma^{-\,steric}_{\quad ortho}$ in both rearrangements to be in agreement with the general ideas concerning competition between SAHR and steric rate depression. The rate enhancing effect of one *o*-methyl group is almost the same in both rearrangements as is the percentage increase in

TABLE 3. Rate constants for some Chapman rearrangements at 255°C in diphenyl ether[19, 20]

Substituent(s) on Ar^2 (**1**)	$k \times 10^5 (\text{sec}^{-1})$	k/k_{H}
H	7·66	1·00
2-CH$_3$	8·87	1·16
2,6-di-CH$_3$	12·3	1·61
2-*t*-C$_4$H$_9$	2·30	0·30
2,6-di-*t*-C$_4$H$_9$	0·31	0·040
2,6-(C$_6$H$_5$)$_2$	18·8	2·45

rate obtained when a second *o*-methyl group is added. The ratio of the rate values for two *o-t*-butyl groups and one *o-t*-butyl group is also almost the same in the two rearrangements. Steric compression due to the *o-t*-butyl group(s) wins out over SAHR in the Newman–Kwart reaction just as it does in the Chapman reaction.

$$\qquad\qquad\qquad\qquad\qquad\qquad\qquad\qquad\qquad\qquad\qquad\qquad (11)$$

$$\text{(24)} \qquad\qquad\qquad \text{(25)}$$

The question of intramolecularity vs. intermolecularity was first raised by Chapman[2]. He thought that perhaps the mechanism might involve ionization of the Ar^2—O bond. He did find an increase in electrical conductivity during the rearrangement of **22** to **23**. Evidence for an ionic mechanism or possibly just an increase in polarity in the transition state was suggested by his data. Chapman concluded that the latter interpretation was the more probable one when he observed no cross product **27** (based on melting-point determinations) from a rearrangement of a mixture of equimolar amounts of **22** and **26**. Thus, the rearrangement appeared to be intramolecular in nature.

$$\text{(22)} \qquad\qquad\qquad \text{(26)} \qquad\qquad\qquad \text{(27)}$$

Wiberg and Rowland also investigated the intramolecularity of the Chapman rearrangement[19]. An equimolar mixture of **22** and **28** was heated until transformation to the amides **23** and **29** was complete. The X-ray powder pattern and infrared spectrum of the mixture of amides from the rearrangement described above were identical to the X-ray powder pattern and infrared spectrum of a mixture of amides from separate rearrangements of **22** and **28**. Thus, no cross product **30** was formed and they concluded that the rearrangement was intramolecular.

$$\underset{\textbf{(28)}}{\overset{\displaystyle OC_6H_4Cl\text{-}p}{\underset{|}{C_6H_5C}}=NC_6H_4Cl\text{-}p} \qquad \underset{\textbf{(29)}}{\overset{\displaystyle O}{\overset{\|}{C_6H_5CN(C_6H_4Cl\text{-}p)_2}}} \qquad \underset{\textbf{(30)}}{\overset{\displaystyle O}{\overset{\|}{C_6H_5CN}}\underset{C_6H_4Cl\text{-}p}{\overset{C_6H_5}{\diagdown}}}$$

In a more recent study, Wheeler and co-workers[22] have applied ^{14}C labeling to the question of intramolecularity. An equimolar mixture of ^{14}C-4-bromophenyl-N-phenylbenzimidate (**31**) and 4-bromophenyl-N-4-tolylbenzimidate (**32**) dissolved in tetraglyme was heated for 4 h at 275–280°C. Four products would be possible if the reaction was intermolecular. Compounds **33** and **34** would be the expected rearrangement products and **35** and **36** would be the cross products. The rearranged mixture was treated with potassium permanganate in pyridine to oxidize the methyl groups in **34** and **36** (if they existed) to the corresponding substituted benzoic acids. Analysis for ^{14}C in the non-acidic part of the reaction mixture (which

$$\underset{\textbf{(31)}}{\overset{\displaystyle O-\overset{*}{\bigcirc}\!-Br}{\underset{|}{C_6H_5C}=NC_6H_5}} \qquad \underset{\textbf{(32)}}{\overset{\displaystyle OC_6H_4Br\text{-}p}{\underset{|}{C_6H_5C}=NC_6H_4CH_3\text{-}p}} \qquad \underset{\textbf{(33)}}{\overset{\displaystyle O}{\overset{\|}{C_6H_5CN}}\!-\!\overset{*}{\bigcirc}\!-Br}$$

would contain **33** and **35**) gave a specific activity, within experimental error, of the original compound (**31**). The acidic part of the reaction mixture showed very little activity. Thus, no cross products were indicated.

$$\underset{\textbf{(34)}}{\overset{\displaystyle O}{\overset{\|}{C_6H_5CNC_6H_4Br\text{-}p}}\atop{\underset{C_6H_4CH_3\text{-}p}{|}}} \qquad \underset{\textbf{(35)}}{\overset{\displaystyle O}{\overset{\|}{C_6H_5CNC_6H_4Br\text{-}p}}\atop{\underset{C_6H_5}{|}}} \qquad \underset{\textbf{(36)}}{\overset{\displaystyle O}{\overset{\|}{C_6H_5CN}}\!-\!\overset{*}{\bigcirc}\!-Br}\atop{\underset{C_6H_4CH_3\text{-}p}{|}}$$

All of the aforementioned studies on substituent effects and intramolecularity have produced results which, when taken together, lend very strong support to the mechanism described at the first of this section and shown in equation 9. There seem to be no studies which are at odds with this being the major pathway for the thermal, uncatalysed conversion of an aryl imidate to an amide.

Nothing has been mentioned yet about the reversibility of the Chapman rearrangement. Chapman himself thought that the rearrangement might be reversible[3]. He heated some N-benzoyldiphenylamines but was not successful in isolating any benzimidates and concluded that the equilibrium between imidate and amide lies far toward the amide. Apparently, the only report of anything approaching a reverse Chapman rearrangement is in the mass spectral study by Goldberg and Harris[16]. In the mass spectra of some N-benzoyldiarylamines, $Ar^1Ar^2NCOC_6H_5$, they observed signals corresponding to $(M-OAr^1)^+$ and $(M-OAr^2)^+$. These could be fragments from the corresponding imidates formed by a reverse Chapman rearrangement but, as the authors point out, it would not be necessary for the reverse rearrangement to proceed beyond the formation of a transition complex for the fragment ions to be formed and observed.

There have been several recent studies of the Beckmann–Chapman rearrangement of ketoxime picryl ethers. The thermal rearrangement of benzophenone oxime picryl ether (37a) to N-picrylbenzanilide (39a) was found by Chapman and Howis[23] to proceed readily at moderate temperatures in either polar or nonpolar solvents. This same picryl ether was found by them to rearrange 'almost explosively' at its melting point. Presumably the rearrangement proceeds via the imidate 38a. Chapman

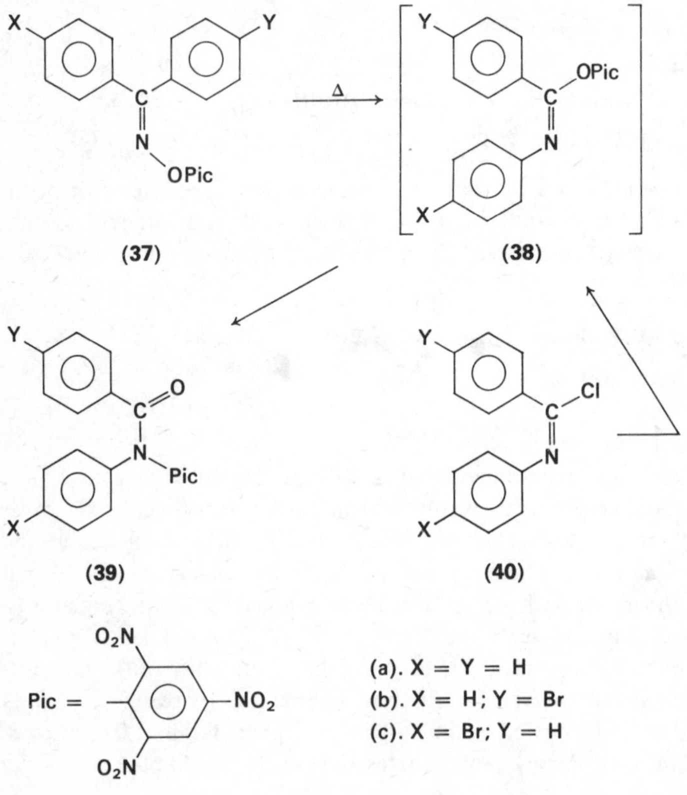

(12)

(a). X = Y = H
(b). X = H; Y = Br
(c). X = Br; Y = H

and Howis[23] tried to isolate **38a** from the reaction of **40a** with sodium picrate in ether but they found only the picramide **39a** upon evaporation of the ether. Other investigators have been equally unsuccessful in attempts to isolate this presumed picryl imidate intermediate in the Beckmann–Chapman rearrangement of oxime picryl ethers[24, 25]. The extremely rapid rearrangement of such an imidate would be expected from a consideration of the rate enhancing effect of three electron withdrawing groups and two *ortho* substituents in the migrating ring.

Curtin and Paul[25] reinvestigated the solid state behaviour of **37a**. They found that it does indeed undergo rearrangement without apparent melting when heated to 70°C or even when standing at room temperature for several months. To facilitate an X-ray study of oxime picryl ethers and to check on the stereospecificity of this rearrangement, Curtin and Paul prepared **37b,c**. In the course of their work they tried to isolate imidates **38b,c** from treatment of the imino chlorides **40b,c** with silver picrate in diethylcarbitol. Even short reaction times at temperatures as low as −78°C failed to reveal any imidate. The only products were the amides **39b,c**. Furthermore, they found no evidence for the accumulation of **38b,c** in their study of the solid state rearrangement of **37b,c** to **39b,c**.

Kukhtenko[26] has used ^{18}O-labeling in a study of intramolecular vs. intermolecular paths in the Beckmann–Chapman rearrangement of **37a**. He studied the rearrangement of **37a** in acetone solution in the presence of picric acid labeled with ^{18}O in the hydroxyl group and noted that ^{18}O exchange accompanied rearrangement. When the mixture of reactants was heated to 47°C for 1 h it was found that the picramide isolated contained 92% of the total amount of ^{18}O and the percentage of ^{18}O in the picric acid had fallen by a corresponding amount. Under these reaction conditions, the initial compound (**37a**) did not exchange —OPic residues and contained no ^{18}O.

To explain these results, Kukhtenko proposed a mechanism involving a slow, irreversible cleavage of the N—OPic bond in **37a** to give an ion-pair

(13)

as shown in equation 13. Intramolecular collapse of this ion-pair would give unlabeled picryl imidate while an intermolecular path would give labeled picryl imidate. The picryl imidates would rapidly rearrange to the corresponding labeled and unlabeled picramides. The ratio of the products formed (labeled and unlabeled) would depend on the reaction conditions. Kukhtenko proposed that if the more nucleophilic $H_2^{18}O$ was in the reaction mixture no picramide would be formed. Instead, the product should be benzanilide containing as much ^{18}O as the labelled water. This is exactly what he found when this hypothesis was tested.

Kukhtenko's ideas on the mechanism of the rearrangement of picryl ethers are essentially the same as those expressed several years earlier by Fischer[27, 28] who was studying the relationships among the Beckmann, Chapman, and Beckmann fragmentation reactions. Fischer's scheme (equation 14) depicts the possible fates of an alkyl ketoxime ether. The Chapman path would be favoured by non-nucleophilic solvents. A good nucleophile such as water should intercept the intermediate nitrilium ion to give the Beckmann product. If R is a good electrofugic group, fragmentation to nitrile and carbonium ion can result. These and other factors which influence the product ratio along the three paths are discussed by Fischer.

$$
\begin{array}{c}
\underset{R^1}{\overset{R}{\diagdown}}C=N\diagup_{OPic}
\xrightarrow{\ k\ }
\left[R^1C\overset{+}{\equiv}NR \quad {}^-OPic \right]
\rightleftharpoons
\underset{PicO}{\overset{R^1\quad R}{C=N}}
\end{array}
$$

$$
\underset{O}{\overset{R^1\quad R}{C-N}}\diagup_{Pic}
$$

$$
R^1C\equiv N + R^+ \qquad\qquad R^1C\overset{+}{\equiv}NR + {}^-OPic
$$

$$
\text{olefin} \qquad \xrightarrow{-H^+}\quad \xleftarrow{H_2O}\quad ROH \qquad\qquad \downarrow H_2O
$$

$$
\underset{\overset{\|}{O}}{R^1CNHR}
$$

(14)

4. Some other O to N migrations of aryl groups

In 1972, Scherrer and Beatty reported a new procedure for the conversion of phenols to anilines[29]. A key step in the reaction sequence (equation 15) is the thermal rearrangement of a 4-aryloxy-2-phenylquinazoline (**41**) to a 3-aryl-2-phenyl-4(3H)-quinazolinone (**42**). Hydrolysis of the resulting quinazolinone gives the aniline substituted in the same manner as the starting phenol. Since this type of rearrangement of an aryl group to a heterocyclic nitrogen was first observed many years ago by Chichibabin[30], these authors propose to call this the Chichibabin rearrangement. However it is essentially a Chapman rearrangement; the factors which influence the conversion of **41** to **42** are the same as those discussed earlier for the

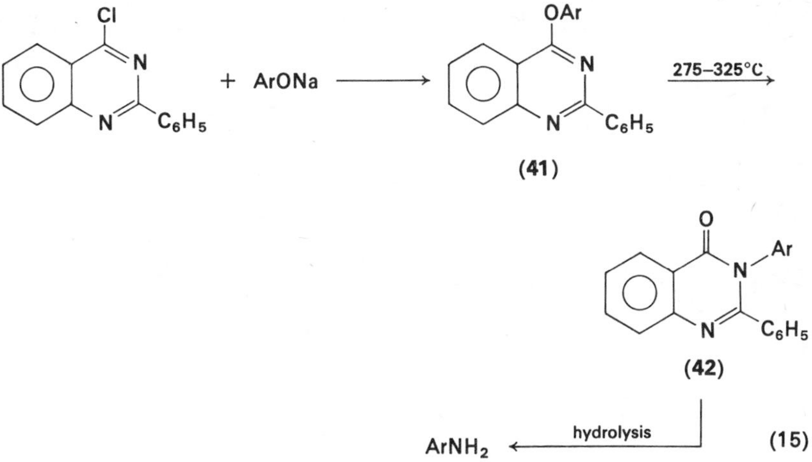

(41)

(42)

(15)

rearrangement of aryl imidates. In this and other studies[31–33], Scherrer and co-workers found the rearrangement of **41** to **42** to be aided by electron withdrawing groups and *ortho* substitution on the migrating ring. They also found the reaction to follow first-order kinetics and proposed a mechanism involving a four-membered transition state analogous to the one in equation 9. Apparently the Chichibabin rearrangement is quite general[31–33] and even occurs in tandem in the case of **43** rearranging to **44**[29, 34].

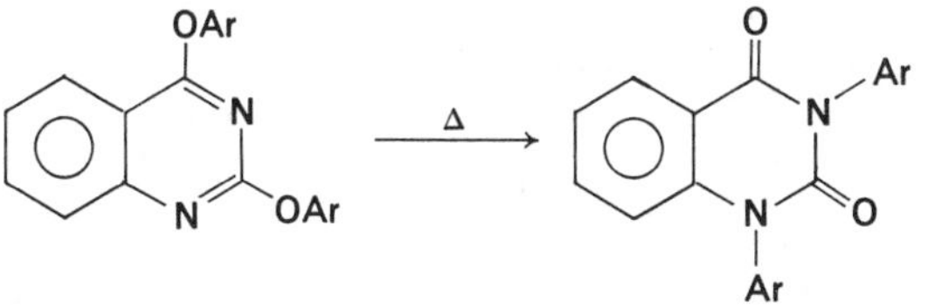

(43)   (44)

(16)

In 1972, Shawali and Hassaneen[35] reported the synthesis and rearrangement of some hydrazones of aryl benzoates. Treatment of hydrazonyl bromides with sodium phenolate in ethanol gave the arylhydrazonates (**45**) in yields of 60–80%. In contrast to the aryl imidates with which they bear a structural relationship, the hydrazonates were reported to be quite stable when stored for extended periods of time at room temperature and they were not hydrolysed in refluxing aqueous dioxane. Heating to 210°C, however, resulted in a rearrangement, which, according to this early communication by Shawali and Hassaneen, paralleled the Chapman rearrangement of aryl imidates and afforded the corresponding aroyl diarylhydrazines (46). Unfortunately, these authors gave no data to support the assigned structure and one must conclude that they were strongly influenced by the analogy they were drawing with aryl imidates and the Chapman rearrangement.

$$\underset{(45)}{\text{Ar}^1\text{C}\overset{\overset{\displaystyle O\text{Ar}^2}{|}}{=}\text{NNHAr}^3} \xrightarrow{\;\;\Delta\;\;}\!\!\!\!/\!\!\!/ \quad \underset{(46)}{\text{Ar}^1\overset{\overset{\displaystyle O}{\|}}{\text{C}}\underset{\underset{\displaystyle \text{Ar}^2}{|}}{\text{N}}\text{NHAr}^3} \tag{17}$$

$$\xrightarrow{\;\Delta\;} \quad \underset{(47)}{\text{Ar}^1\overset{\overset{\displaystyle O}{\|}}{\text{C}}\text{NHNAr}^2\text{Ar}^3}$$

Later the same year (1972), Shawali and Hassaneen[36] acknowledged, in a more detailed paper on this subject, that their previous conclusions had been incorrect. Hegarty and co-workers[37] had already reported in 1971 that the products of rearrangements of arylhydrazonates were $N'N'$-diarylhydrazides (47) instead of aroyl diarylhydrazines (46). However, their hydrazonates did not have the same substituents as those studied by Shawali and Hassaneen so they repeated their work using the same hydrazonates and found, in all cases, that the rearrangements yielded 47 as a product of 1,4-aryl migration[38]. Both research groups were then in agreement that the similarities between this rearrangement (equation 17) and the Chapman rearrangement had limitations

Kinetic studies[39] showed the rearrangement to be first-order and crossover experiments[36, 39] revealed that it is intramolecular. Also, the migrating group (Ar^2 in 45) retained its configuration during migration, i.e. the ring carbon originally attached to oxygen in 45 was attached to nitrogen in 47. The results of studies on substituent effects are in contrast, though, with what is known about the effect of substituents on the rates of rearrangements of aryl imidates. Hegarty and co-workers[39] found that electron-withdrawing groups in the migrating ring (Ar^2) diminished the tendency of the hydrazonates (45) to rearrange to 47. In fact, an electron-withdrawing group in any of the three rings retarded the rearrangement; the effect being the greatest in Ar^3 where $\rho = -2.1$. Problems with reproducing rate constants in different batches of solvent and the observation of apparent autocatalysis in optical density vs. time plots led Hegarty and co-workers to suspect that the rearrangement of **45** to **47** was free radical initiated when carried out in the solvents they were using. Their experiments which showed catalysis by benzoyl peroxide and azobisisobutyronitrile confirmed this suspicion. They proposed a mechanism (equation 18) involving initiation by abstraction of the amino hydrogen to give the hydrazonate radical (**48**). This radical could then undergo an intramolecular rearrangement, possibly via a bridged radical species (**49**), to give the more stable hydrazyl radical (**50**). Hydrogen abstraction from the solvent or from another molecule of the substrate would give the observed product (**47**). They further concluded that the thermal rearrangement of neat hydrazonates could also involve a free radical pathway[39].

$$
\underset{\textbf{(45)}}{\overset{\overset{\displaystyle OAr^2}{|}}{Ar^1-C}=NNHAr^3} \xrightarrow{R\cdot} \underset{\textbf{(48)}}{\overset{\overset{\displaystyle O-Ar^2}{|}}{Ar^1-C}=N\dot{N}Ar^3} \longrightarrow
$$

$$(18)$$

$$
\underset{\textbf{(49)}}{\overset{\overset{\displaystyle O^{\text{\tiny IIIIIII}}Ar^2}{\text{\rotatebox{90}{$\mathbb{I}$}}\quad\text{\rotatebox{90}{$\mathbb{I}$}}}}{Ar^1-\overset{}{\underset{}{C}}^{\text{\tiny IIII}}\dot{N}-NAr^3}} \longrightarrow \underset{\textbf{(50)}}{\overset{\overset{\displaystyle O}{\|}}{Ar^1-C-\dot{N}NAr^2Ar^3}} \xrightarrow{RH} \textbf{(47)}
$$

The base catalysed rearrangement of **45** to **47** would represent a Smiles rearrangement[40]. The facile transformation of hydrazonates with *ortho* or *para* nitro groups in Ar^2 was observed by Elliott and co-workers[41] when they employed refluxing ethanol–triethylamine. Electron withdrawing groups in Ar^3 retarded the reaction, reflecting the decreased nucleophilicity of the amino nitrogen in such compounds. Crossover experiments revealed that these base catalysed conversions of **45** to **47** were intramolecular and a mechanism involving a five-membered cyclic transition state was proposed.

B. Alkyl and Allyl Imidates

I. Introduction

In comparison to the rearrangement of aryl imidates, the rearrangement of alkyl imidates (**51**, R^2 = alkyl) requires more stringent conditions, usually gives poorer yields of the rearrangement product (**52**), proceeds by a different mechanism, and may be very effectively catalysed by a variety of electrophilic species. Allyl imidates (**51**, R^2 = allyl), on the other hand, undergo rearrangement fairly readily and given an excellent yield of re-arrangement product. Here again, though, the mechanism is different from that of the rearrangement of aryl imidates and the process may be effectively catalysed by a variety of compounds.

$$
\underset{(51)}{R^1C\!\!\overset{\displaystyle OR^2}{=}\!\!NR^3} \longrightarrow \underset{(52)}{R^1\overset{\displaystyle O}{\overset{\|}{C}}\!\!-\!\!NR^2R^3} \tag{19}
$$

The rearrangements of alkyl and allyl imidates have been briefly treated in the two previously mentioned reviews of the Chapman rearrangement[8,9]. The following summary of the literature in this area is somewhat more detailed and includes some more recent examples. Once again, a section has been included on rearrangements in related compounds for sake of comparison.

2. Thermal rearrangements

As stated earlier, the temperatures required for the rearrangement of alkyl imidates are generally higher than those required for the Chapman rearrangement of aryl imidates and the yields are lower. The two examples shown in equations 20 and 21 should suffice as illustrations. The imidate **53** gave the amide **54** in 25% yield when heated to 300–330°C[3] and the conversion of **55** to **56** occurred in 20–40% yields at about 300°C[42].

$$
\underset{(53)}{C_6H_5\overset{\displaystyle OCH_3}{C}\!\!=\!\!NC_6H_5} \overset{\Delta}{\longrightarrow} \underset{(54)}{C_6H_5\overset{\displaystyle O}{\overset{\|}{C}}N(CH_3)C_6H_5} \tag{20}
$$

$$
\underset{(55)}{H\overset{\displaystyle OCH_3}{C}\!\!=\!\!NC_6H_4R\text{-}p} \overset{\Delta}{\longrightarrow} \underset{(56)}{H\overset{\displaystyle O}{\overset{\|}{C}}N(CH_3)C_6H_4R\text{-}p} \tag{21}
$$

$$(R = H, CH_3, C_2H_5)$$

More recently, Pilotti and co-workers[43], in a study of several methods of preparing alkyl imidates (**51**, all three R groups alkyl), reported that the various compounds they prepared did not rearrange in their pyrolysis studies. Unfortunately, they gave no conditions except for an O-cyclohexyl derivative which was reported to be stable at temperatures below 250°C. Paquette and co-workers[44] observed a thermal rearrangement of some methoxy azetines but these were vapour phase pyrolyses at temperatures of 600°C or greater. They found, for example, that azetine **57** yielded **58, 59,** and **60** (equation 22) in a ratio of 14:78:8 at 600°C. When the temperature was increased to 700°C compound **60** predominated at the expense of **59** showing it to be the Chapman type rearrangement product of imidate **59**. Apparently, imidate **58** did not rearrange under these conditions.

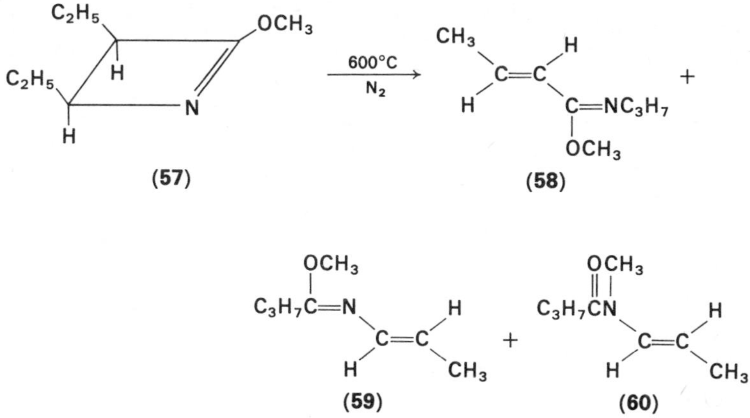

(57) (58) (59) (60)

(22)

A rearrangment of a cyclic imidate formed at the end of nylon-6 chains during pyrolysis studies at 250–290°C has been proposed to account for some of the ammonia evolved and the di-(ω-carboxypentyl)amine (**63**) found upon subsequent hydrolysis with base (equation 23)[45]. There was no spectral or other evidence presented, however, for the presence of imidate **61** or lactam **62**.

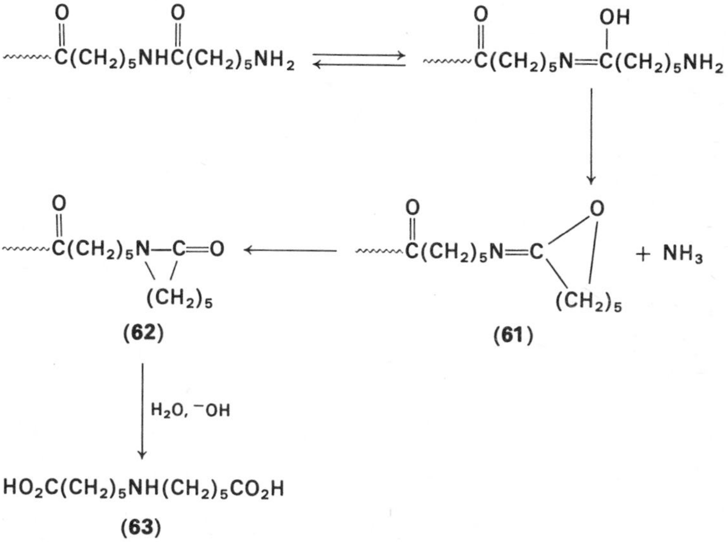

(62) (61) (63)

(23)

When the O-alkyl group in an alkyl imidate has a β-hydrogen, olefin formation seems to be a common pyrolysis pattern[8]. Marullo and co-workers[46] studied the stereochemistry and kinetics of the vapour phase (375–560°C) pyrolysis of *cis*- and *trans*-2-phenylcyclohexyl *N*-phenyl-benzimidates and found the elimination to be a unimolecular, *cis* process. It was suggested that this reaction could be used as a synthesis of olefins since lower temperatures are required relative to acetate pyrolysis.

Benzyl imidates cannot undergo elimination and have been observed to rearrange (equation 24)[47]. Cyclic benzyl imidates also rearrange as illustrated in the last step of equation 25. Hauser and co-workers[48] had

$$
\begin{array}{ccc}
\overset{\displaystyle OCH_2C_6H_5}{\underset{\displaystyle Cl_3CC\!=\!\!NH}{|}} & \overset{\Delta}{\longrightarrow} & \overset{\displaystyle O}{\underset{\displaystyle Cl_3CCNHCH_2C_6H_5}{\|}} \\
\text{(64)} & & \text{(65)}
\end{array}
\tag{24}
$$

$$R^1 = C_6H_5$$
$$R^2 = CH_3$$

reported in 1969 that the products of acid-catalysed cyclodehydration of γ-hydroxyamides (66) were phthalimidines (68). However, Bailey and De-Grazia[49] repeated this work and found that the products were really cyclic imidates (67). To check for the possibility of thermal rearrangement of 67 to 68, imidate 67 was kept at 100°C for 1 h. No lactam (68) could be detected by g.l.c. analysis. However, when 67 was heated at 220–245°C for 1 h, 33% conversion to 68 was observed by Bailey and DeGrazia.

Wiberg and Rowland[19] did not study alkyl imidates in their paper on the mechanism of the rearrangement of aryl imidates but they did conclude that the different conditions required for the rearrangement of alkyl imidates (as reported by others) and the poor yields obtained indicated a free radical process for alkyl imidates. They felt that the inability of an alkyl group to undergo a front side S_N2 attack would make a cyclic transition state (as in the rearrangement of aryl imidates, equation 9) unlikely. A couple of years later Wiberg and co-workers[42] reported their results of a mechanistic study of the rearrangement of alkyl imidates. The mode of rearrangement—intramolecular vs. intermolecular—was decided by a crossover experiment using a ^{13}C-labeled compound (69, 17·8% excess

[13]C) and an unlabeled one (70). An equimolar mixture of 69 and 70 was heated at 300°C for 4 h. The product formamides (71, 72) were reduced with LiAlH$_4$ and analysed by mass spectroscopy. The N,N-dimethyl-p-ethylaniline from 72 was found to contain 7·5% excess ^{13}C indicating the rearrangement to be intermolecular. In a similar fashion they established

$$O^{13}CH_3$$
$$HC{=}NC_6H_4CH_3\text{-}p$$
(69)

$$OCH_3$$
$$HC{=}NC_6H_4C_2H_5\text{-}p$$
(70)

$$\overset{O}{\overset{\|}{HCNC_6H_4CH_3\text{-}p}}$$
$$^{13}CH_3$$
(71)

$$\overset{O}{\overset{\|}{HCNC_6H_4C_2H_5\text{-}p}}$$
$$CH_3$$
(72)

(26)

(73)   (74)

the rearrangement of α-methoxypyridine (73) to N-methylpyridone (74) to also be intermolecular. This latter rearrangement (equation 26) required lower temperatures (200°C) and gave essentially 100% conversion in contrast to the low yields (20–40%) of amides from the rearrangement of 69 and 70. Although Wiberg and co-workers[42] apparently did not add a radical initiator to 69 or 70, they did note that the conversion of 73 to 74 is catalysed by benzoyl peroxide as are several other conversions which involve 1,3-shifts of alkyl groups.

The rearrangement of allyl imidates was first observed by Mumm and Möller[50] when the allyl imidate 75 was found to be quantitatively converted to 76 in 3 h at 210–215°C. They realized that allyl imidates resemble allyl phenyl ethers which readily undergo the Claisen rearrangement to

$$OCH_2CH{=}CH_2$$
$$C_6H_5C{=}NC_6H_5 \quad \xrightarrow{\Delta} \quad C_6H_5\overset{O}{\overset{\|}{C}}NCH_2CH{=}CH_2 \quad (27)$$
$$\phantom{C_6H_5CNCH_2CH{=}CH_2}C_6H_5$$

(75)   (76)

o-allylphenols[51] and designed experiments to distinguish between Chapman and Claisen products. They found that imidates 77 and 78 rearranged with complete inversion of the allyl chain to amides **79** and **80**, respectively. These results suggested a cyclic, concerted process as shown in equation 28.

$$\begin{matrix} \text{CH}_3 \\ | \\ \text{OCHCH}=\text{CH}_2 \\ | \\ \text{C}_6\text{H}_5\text{C}=\text{NC}_6\text{H}_5 \\ \text{(77)} \end{matrix} \quad \xrightarrow{\Delta} \quad \left[\begin{matrix} \text{CH}_3 \\ | \\ \text{CH} \\ O \diagup \quad \diagdown \text{CH} \\ \text{C}_6\text{H}_5\text{C} \quad \quad \text{CH}_2 \\ \diagdown N \diagup \\ | \\ \text{C}_6\text{H}_5 \end{matrix} \right] \quad \quad (28)$$

$$\begin{matrix} O \\ \| \\ \text{C}_6\text{H}_5\text{CCNCH}_2\text{CH}=\text{CHCH}_3 \\ | \\ \text{C}_6\text{H}_5 \\ \text{(79)} \end{matrix}$$

$$\begin{matrix} \text{OCH}_2\text{CH}=\text{CHCH}_3 \\ | \\ \text{C}_6\text{H}_5\text{C}=\text{NC}_6\text{H}_5 \\ \text{(78)} \end{matrix} \quad \xrightarrow{\Delta} \quad \begin{matrix} O \quad \text{CH}_3 \\ \| \quad | \\ \text{C}_6\text{H}_5\text{CNCHCH}=\text{CH}_2 \\ | \\ \text{C}_6\text{H}_5 \\ \text{(80)} \end{matrix} \quad \quad (29)$$

Further information on the mechanism of the rearrangement of allyl imidates was generated by Wheeler and co-workers[22]. They prepared **75** with tritium in the α-position of the allyl chain and then rearranged it by heating it for 3 h at 210–215°C. Amide **76** was converted to a glycol and then cleaved with periodic acid. The formaldehyde (from the terminal carbon) was converted into the dimedone derivative which analysed for 99·0 ± 1·5% of the activity originally in **75**. Thus, they found that the re-arrangement of **75** to **76** must be entirely intramolecular and must proceed with inversion of the allyl chain in a manner analogous to that of a Claisen rearrangement[52].

3. Catalysed rearrangements

The catalysed rearrangement of alkyl imidates was reported by several researchers around 1900. Wislicenus and Goldschmidt[53], for example, observed the facile transformation of ethyl benzimidate (**81**) to N-ethyl-benzamide (**82**) at 100°C in the presence of ethyl iodide. From that point on catalysis by a variety of substances has been reported by numerous research groups.

$$\underset{\textbf{(81)}}{\overset{\overset{\displaystyle OC_2H_5}{|}}{C_6H_5C{=\!\!=}NH}} \xrightarrow[\Delta]{C_2H_5I} \underset{\textbf{(82)}}{\overset{\overset{\displaystyle O}{\|}}{C_6H_5CNHC_2H_5}} \qquad (30)$$

Alkyl halides have been by far the most frequently studied catalysts. Lander[54] was the first person to do an extensive study of the alkyl halide catalysed alkyl imidate rearrangement and this reaction has since been referred to as the Lander rearrangement[8]. He found that methyl iodide, ethyl iodide, ethyl chloride, and ethyl bromide in amounts varying from 1·0 mol to trace amounts catalysed the rearrangements of alkyl N-aryl-acetimidates (83, R = CH_3, C_2H_5), alkyl N-arylbenzimidates (84, R = CH_3, C_2H_5), alkyl N-alkylbenzimidates (85, R = CH_3, C_2H_5), and alkyl N-benzylbenzimidates (86, R = CH_3, C_2H_5). The transformation to the

(83)

(84)

(85)

(86)

corresponding amides was carried out on neat samples at temperatures of 160°C or less (in considerable contrast to the temperatures of 300°C or greater required for thermal, uncatalysed rearrangements). Although the yields were not stated, the implication was that they were good. It was found that compounds which had a methyl group attached to the oxygen rearranged somewhat faster than the O-ethyl analogs. The compounds which had an *ortho* substituent in the N-phenyl ring rearranged more slowly than the *para* substituted analogs. The displacement of a higher by a lower alkyl group was also observed (equation 31) and Lander concluded that the mechanism of this rearrangement must involve addition followed by elimination (equation 32).

Arbuzov and Shishkin[55, 56] agreed, in principle, with the Lander mechanism. From the results of their studies they concluded that the alkyl halide catalysed rearrangement of alkyl imidates proceeds in two steps of which the first involves the addition of alkyl halide to form an ionic adduct

$$\underset{CH_3C=NC_6H_5}{\overset{OC_2H_5}{|}} \quad \xrightarrow[\Delta]{CH_3I} \quad \underset{CH_3CN(CH_3)C_6H_5}{\overset{O}{\parallel}} \tag{31}$$

$$\underset{R^1C=NR^3}{\overset{OR^2}{|}} \quad \xrightarrow{R^4I} \quad \underset{\overset{|}{I}}{\overset{OR^2}{\underset{R^1C-NR^3R^4}{|}}} \quad \xrightarrow{-R^2I} \quad \underset{R^1CNR^3R^4}{\overset{O}{\parallel}} \tag{32}$$

(87)

(usually $R^4 = R^2$)

(a halide salt rather than as in **87**) which undergoes cleavage into the final products. A more quantitative and more recent study was carried out by Challis and Frenkel[57]. They used n.m.r. to follow the catalysed rearrangement of isopropyl N-methylbenzimidate (**88**) at 138°C in nitrobenzene. Their data for this reaction catalysed by isopropyl iodide show that the kinetics follow equation 33, implying an S_N2 mechanism. Catalysis by the

$$\underset{C_6H_5C=NCH_3}{\overset{OPr\text{-}i}{|}}$$

(88)

$$\text{rate} = k_2 \text{ [88] } [i\text{-}PrI] \tag{33}$$

various isopropyl halides decreased sharply in the order i-PrI > i-PrBr > i-PrCl for the rearrangement of **88** which is consistent with an S_N2 mechanism (equation 34) with either the first or second step being rate-determining. The further observation that catalysis decreased with increased branching of the alkyl halide (i.e. CH_3I > i-PrI) regardless of the O-alkyl substituent in other N-methylbenzimidates established the first step as being the slow step for these rearrangements.

$$\underset{(89)}{\underset{C_6H_5C=NCH_3}{\overset{OR}{|}}} + RX \quad \rightleftharpoons \quad \left[\underset{C_6H_5C\text{-----}N(CH_3)R}{\overset{OR}{\vdots}} \right]^+ X^- \tag{34}$$

(90)

$$\downarrow -RX$$

$$\underset{C_6H_5CN(CH_3)R}{\overset{O}{\parallel}}$$

(91)

In equation 34, the intermediate (**90**) may be considered to be a common alkylated derivative of the imidate (**89**) and the amide (**91**). The equilibration of alkyltropic[58] isomers such as **89** and **91** by the use of their common alkylated derivative (**90**) is a technique which has been used by Beak and co-workers on several occasions[58–61]. Beak pointed out in his first paper on this subject[59] that this technique seems to be general and potentially applicable to any alkylated isomers having a common alkylated derivative. In their work on imidate–amide equilibria they studied the isomer pairs **92, 93** (equation 35) and **94, 95** (equation 36). Equilibration in each case was carried out in the liquid phase at 130°C with the use of catalytic amounts of the appropriate common alkylated derivative. Reaction was observed only in experiments starting with the imidates (**92, 94**). No imidates could be detected in experiments starting with the amides (**93, 95**). Thus, the equilibrium constants for these systems in the liquid phase lie far toward the amide ($K_{eq} > 10^3$) and this information was used in their process of arriving at relative chemical binding energies for the imidate–amide isomer pairs.

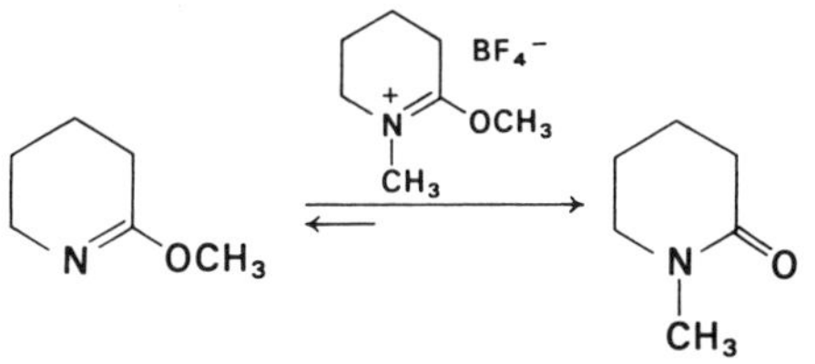

(35)

(**92**) (**93**)

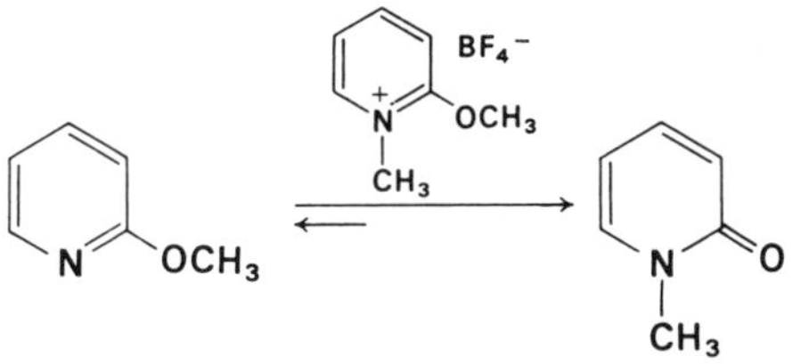

(36)

(**94**) (**95**)

Catalysis by Lewis and Bronsted acids has also been observed. Cramer and Hennrich[62] showed that imidate **96** could be rearranged to amide **97** in 96% yield in refluxing benzene with catalytic amounts of BF_3 added.

$$CCl_3C(OC_2H_5){=}NH \xrightarrow[C_6H_6,\ \Delta]{BF_3} CCl_3C({=}O)NHC_2H_5 \qquad (37)$$

(**96**) (**97**)

Challis and Frenkel[57] also found BF_3 to be very effective in catalysing the rearrangement of **88**. The work of Roberts and Vogt[63] showed that sulfuric acid can serve as a catalyst but the concentration is critical—too little limits the rate of rearrangement and excess results in the formation of tarry products. The 'catalytic quantities' of sulfuric acid used by Challis and Frenkel with **88** were apparently too little since they found no rearrangement at 138°C in nitrobenzene. The concentration of HBr added as a catalyst is also critical. When HBr was added to **88** in 0·22 molar equivalent the catalytic effect was the same as that found for isopropyl bromide (which is presumably formed from the addition of HBr)[57]. However, when HBr was added in equimolar quantities with **88**, no rearrangement was observed at 138°C in nitrobenzene.

Lander[54] reported iodine to be an effective catalyst for alkyl imidate rearrangements and this was confirmed by Challis and Frenkel[57]. Their results showed iodine to be much more effective than isopropyl iodide in catalysing the rearrangement of **88**.

There are several examples in the literature of catalysed rearrangements of cyclic imidates. Bailey and DeGrazia[49] noted that refluxing a trifluoroacetic acid solution of **67** for a few minutes afforded a 92·4% yield of **68**. Singer and co-workers[64, 65] studied the photocycloaddition of fluorenone to ketenimines. Some of the initially formed α adducts (**98**; R = R^1 = C_6H_6; R = C_2H_5, R^1 = C_6H_5) were found to be readily isomerized to the corresponding β-lactams (**99**) during Florisil chromatography. Subsequent work also revealed that another α adduct (**98**, R = R^1 = CH_3) rearranged in a few minutes to a β-lactam at room temperature in acetonitrile in the presence of lithium perchlorate. This latter rearrangement plus the ones observed from Florisil chromatography were formulated as proceeding *via* a zwitterion intermediate (equation 38). Ishibe and Yamaguchi[66] have proposed a similar rearrangement of an α adduct to a cyclic β-lactam in the photoaddition of N-(cyclohexyl)dimethylketenimine to p-benzoquinone. However, they did not have direct evidence for either the α adduct or a β-lactam since the product isolated was a rearrangement product which could have been formed from either the α adduct or the β-lactam.

Mercuric salts dramatically catalyse the rearrangement of allylic trichloroacetimidates[67]. For example treatment of **100** (R = n-C_3H_7) with 0·1 equivalent of $Hg(OCOCF_3)_2$ in THF at 0°C resulted in immediate rearrangement to **101**. On the other hand, the thermal uncatalysed rearrangement of **100** required temperatures of 140–180°C. In another study of allyl imidates it was found that the mode of rearrangement of the allyl group can be affected by the particular catalyst employed. Stewart and

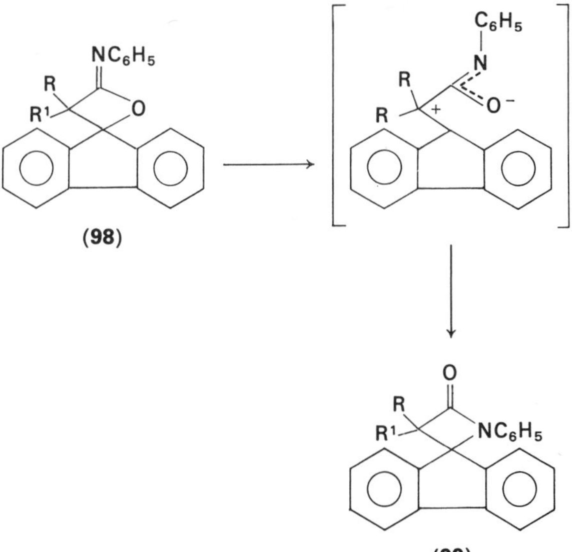

(38)

(98)

(99)

(39)

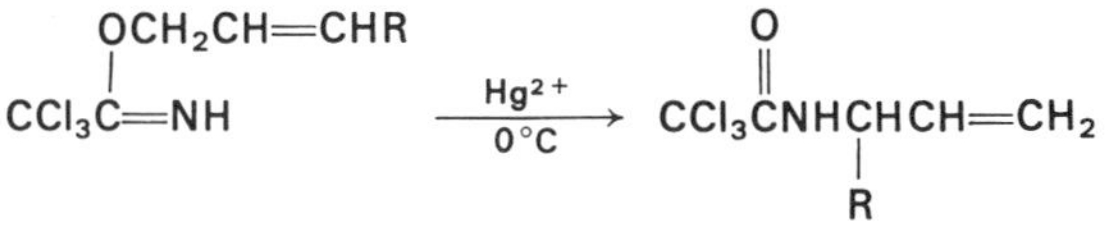

(100)   **(101)**

Seibert[68] found that chloroplatinic acid in isopropyl alcohol quantitatively converted 2-crotoxypyridine (**102**) to 1-(1-methylallyl)-2-pyridone (**103**) at 125°C whereas, at the same temperature, boron trifluoride etherate yielded a mixture of two abnormal Claisen products: 82% 1-crotyl-2-pyridone (**104**) and 18% 3-crotyl-2-pyridone (**105**).

Tin(IV) chloride is another effective catalyst for the rearrangement of allyl imidates. The thermal, uncatalysed conversion of 2-allyloxypyridine (**106**) to 1-allyl-2-pyridone (**107**) requires temperatures of 240°C or higher and gives a mixture of products[69]. The addition of 1% tin(IV) chloride allows the conversion to be carried out at 140°C with yields of 85% or higher[68]. Other equally effective catalysts for this conversion are H_2PtCl_6, $NaPtCl_4$, and $BF_3 \cdot OEt_2$.

Copper(I) chloride may have catalysed the rearrangement of **108** to **109** in the reaction shown in equation 42[70]. The authors apparently did not consider this possibility as they attributed the presence of **109** to a thermal rearrangement of **108** (possibly during the isolation of the components of the mixture by g.l.c. or vacuum distillation). The ratio of **108** to **109** was about 10:1 but with benzyl alcohol in place of allyl alcohol the ratio of the corresponding imidate to amide was about 1:3.

4. Some other O to N migrations of alkyl groups

McCarty and Garner[71] have recently completed a study of the kinetics and mechanism of the thermal uncatalysed Chapman-like rearrangements of **110** and **111**. At 120°C in bromobenzene where **110** and **111** smoothly rearranged, the related methyl N-cyanoacetimidate (**112**) underwent no conversion to the corresponding amide. In fact **112** was unchanged after several hours at 150°C in bromobenzene and after being stored neat for 6 months at room temperature in a container exposed to light. Previously, however, Huffman and Schaefer[72] had given i.r. evidence for the slow rearrangement of **112** (neat) at 165°C.

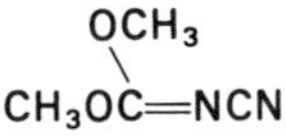
OCH3
CH3OC=NCN
(110)

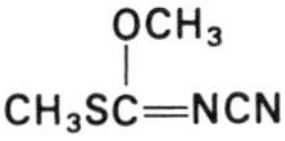
OCH3
CH3SC=NCN
(111)

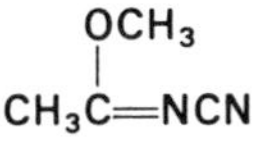
OCH3
CH3C=NCN
(112)

Two interesting examples of methyl migrations in heterocyclic systems are shown in equations 43 and 44. The thermal rearrangement of **113** to **114** can be effected in 70 min at 200°C but the addition of lithium iodide to an acetone solution of **113** allows the rearrangement to proceed smoothly at much lower temperatures[73]. When ethyl iodide plus lithium iodide were added to an acetone solution of **113**, a mixture of *N*-methyl and *N*-ethyl saccharine was produced. The methyl group in **115** migrates to both the adjacent nitrogen atom and the nitrogen in the five-membered ring[74].

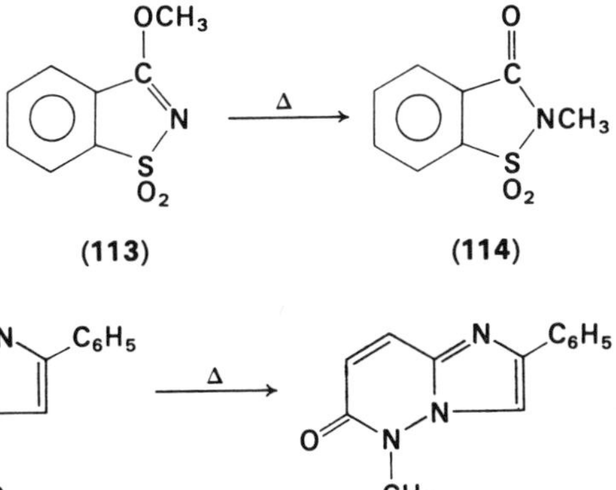
OCH3
C
N
S
O2
Δ
O
C
NCH3
S
O2
(43)

(113) **(114)**

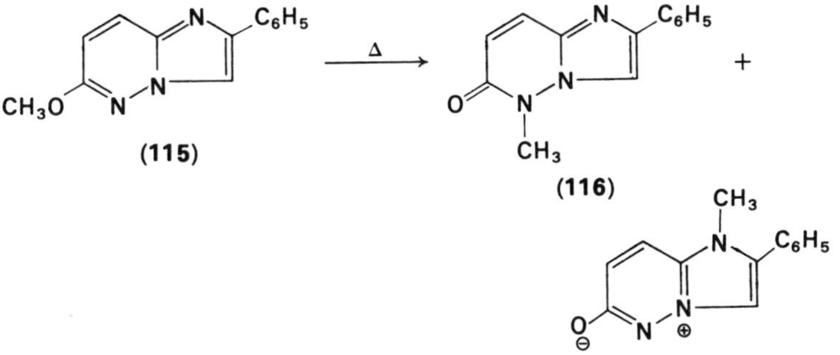
N
C6H5
N
CH3O
N
(115)
Δ
N
C6H5
N
O
N
CH3
(116)
+
CH3
N
C6H5
N
O
N
⊖
⊕
(44)

(117)

The ratio of compounds **115**:**116**:**117** after heating **115** for 2 h at 240°C was found to be 25:61:14.

The migration of the chloroethyl group in **118** was recently reported by Pring and Swahn[75]. Refluxing **118** in dimethylformamide for 4 h gave an 85% yield of **120** while, conversely, refluxing **120** in dimethylformamide led to some **118**. An oxazolinium ion (**119**, equation 45) was postulated as an intermediate in this equilibrium. Quite likely a similar O to N migration of a chloroethyl group is involved in the sequence shown in equation 46. Amide **122** was the only product isolated but imidate **121** was proposed as a likely precursor[76].

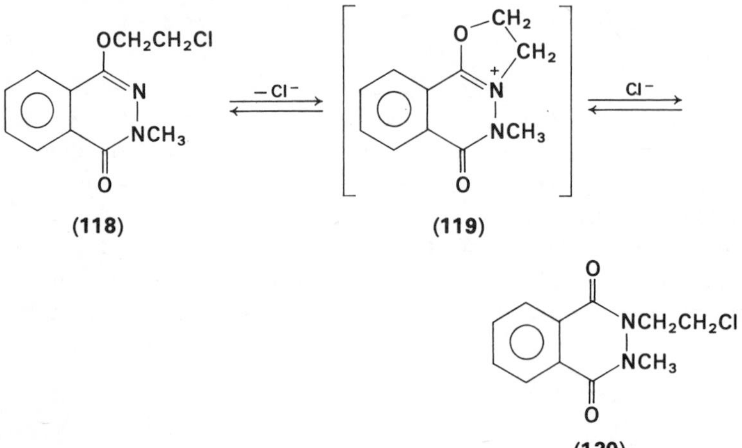

(45)

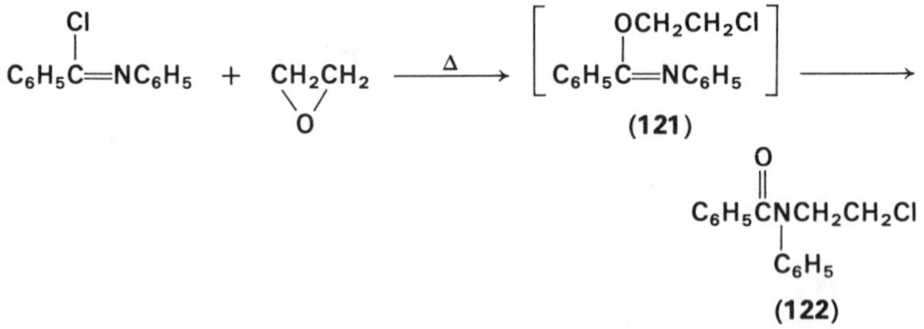

(118)   **(119)**

(120)

(46)

(121)

(122)

Ulbricht and co-workers have published a series of papers on O to N glycosyl rearrangements in heterocyclic systems. An example from the purine series is shown in equation 47[77].

In contrast to the behaviour mentioned earlier for aryl hydrazonates (equation 17), the alkyl analogs (**125**) showed no rearrangement after 2 h at 200°C and no decomposition after 2 years of storage at room temperature[78].

Johnson and co-workers[79] studied the effect of various alkylating agents on the distribution of alkylation products from some alkyl benzohy-

(47)

(123)   **(124)**

(AcGlu = 2,3,4,6-tetra-O-acetyl-D-glucopyranosyl)

(125, R = CH$_3$, C$_2$H$_5$, n-C$_3$H$_7$)

droxamates (**126**). Since the N-alkylated products (**128**) could conceivably arise from the rearrangement of the O-alkylated derivatives (**127**), the latter were subjected alone to the reaction conditions (R^{1}X and K$_2$CO$_3$ in CH$_3$OH/H$_2$O at 38°C for 15 h). No rearrangement of **127** to **128** was observed under these conditions.

(48)

(126)   **(127)**   **(128)**

The last two examples to be included in this section may not be as closely related as some of the preceding examples to alkyl imidate re-arrangements but they do involve migrations of saturated carbon atoms from oxygen to nitrogen (equation 49) or *vice versa* (equation 50). The addition of chlorosulfonyl isocyanat (CSI) to cycloheptatriene (**129**) was originally thought to yield **131** as the thermodynamically controlled pro-duct[80]. Malpass[81] has found that **131** is indeed formed from **129** and CSI in CH$_2$Cl$_2$ at 25°C but it slowly rearranges to **132**, probably by way of a di-polar intermediate (**130**) since the conversion of **131** to **132** is more effi-ciently achieved in a more polar solvent such as nitromethane.

Mackay and co-workers[82] found that the Diels–Alder adduct of azodi-benzoyl and cyclopentadiene (**133**) was labile. It isomerized irreversibly on heating near its melting point or in solution to **135**. This thermal rearrangement was insensitive to solvent character and possessed a large negative entropy of activation. A possible transition state for this sigma-tropic rearrangement is shown by **134**. Further studies on the acid catalysis of this rearrangement have been carried out[83, 84].

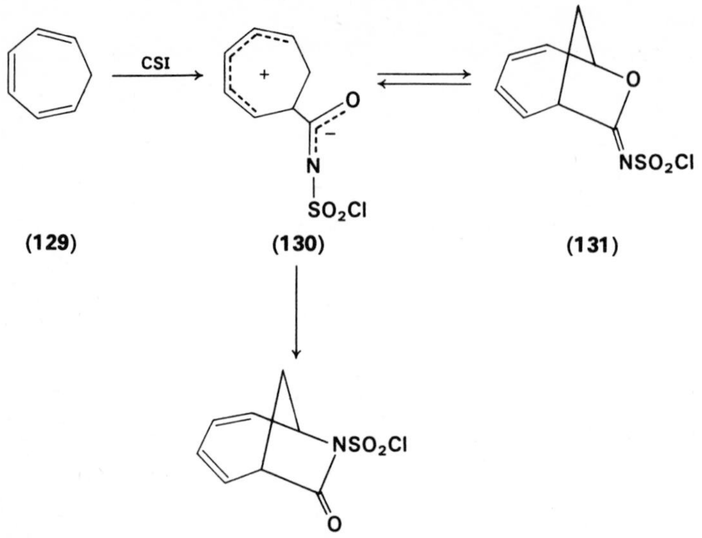

(129)   (130)   (131)

(132)

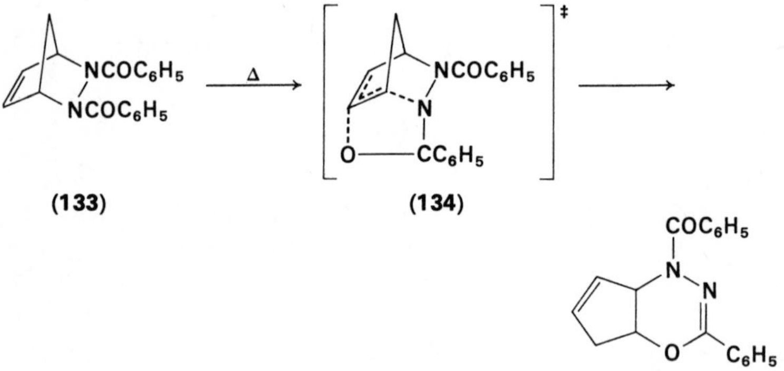

(133)   (134)

(50)

(135)

C. Acyl Imidates (Isoimides)

I. Introduction

Acyl imidates or isoimides are mixed anhydrides of an imidic and carboxylic acid. They are included in this section under imidates because of the relationship of some of their rearrangements to the rearrangements of some of the aforementioned imidates. Since the initial report of the isolation of an isoimide in 1893[85], there have been numerous proposals of isoimides existing as labile intermediates in a variety of reactions. Yet, little attention seems to have been paid to the synthesis and study of stable isoimides until about 1960. References to much of the literature on transient isoimide intermediates can be found in the papers by Roderick[86] and Ernst[87].

Because of the differences in conditions and mechanisms for their rearrangement, it is convenient to separate the acyclic (**136**) and cyclic (**137**) isoimides in the following discussion.

$$
\begin{array}{cc}
\begin{array}{c}
\text{O} \\
\parallel \\
\text{OCAr}^2 \\
\mid \\
\text{Ar}^1\text{C}\!=\!\text{NAr}^3 \\
\textbf{(136)}
\end{array}
&
\begin{array}{c}
\text{(137 structure)} \\
\textbf{(137)}
\end{array}
\end{array}
$$

2. Acyclic acyl imidates

The 1,3-0 to N transfer of an acyl group in an acyclic acyl imidate is sometimes called the Mumm rearrangement[8]. Mumm, Hesse, and Volquartz[1] reported in 1915 on their failure to obtain the expected isoimides from the treatment of imidoyl chlorides with benzoate salts (equation 51). The isoimides were presumably intermediates in these reactions but the only products isolated were the imides.

$$
\text{Ar}^2\text{CO}_2{}^- + \underset{\substack{| \\ \text{Ar}^1\text{C}=\text{NAr}^3}}{\overset{\text{Cl}}{}} \longrightarrow \left[\underset{\substack{| \\ \text{Ar}^1\text{C}=\text{NAr}^3}}{\overset{\substack{\text{O} \\ \parallel \\ \text{OCAr}^2}}{}} \right] \longrightarrow \underset{\substack{| \\ \text{Ar}^3}}{\overset{\substack{\text{O} \ \ \text{O} \\ \parallel \ \ \parallel}}{\text{Ar}^1\text{CNCAr}^2}}
$$

$$(51)$$

Many years later, Curtin and Miller[88, 89] were successful in preparing, isolating, and studying the properties of acyclic isoimides containing nitro groups in Ar³ (**136**). Apparently the electron withdrawing groups in Ar³ are necessary to depress the nucleophilicity of the imido nitrogen atom and thus reduce the tendency for rearrangement to imides. They prepared a series of *N*-(2,4-dinitrophenyl)benzimidoyl benzoates (**138**) containing *para* substituents in Ar² and studied the kinetics and mechanism of their rearrangement to **139**. These acyl imidates rearranged to **139** in benzene or acetonitrile solution at temperatures of 40–65°C and kinetic measurements showed that the reactions were first-order and that the rates were not affected by the addition of small amounts of acetic acid or calcium hydride. The rates were somewhat faster in acetonitrile for a given compound and, in either solvent, the presence of an electron withdrawing

$$
\underset{\textbf{(138)}}{\underset{C_6H_5C=NC_6H_3(NO_2)_2\text{-}2,4}{\overset{\overset{\displaystyle O}{\overset{\displaystyle \|}{OCC_6H_4X\text{-}p}}}{\vert}}} \longrightarrow \underset{\textbf{(139)}}{\underset{C_6H_3(NO_2)_2\text{-}2,4}{\underset{\vert}{\overset{\overset{\displaystyle O\ \ O}{\overset{\displaystyle \|\ \ \|}{C_6H_5CNCC_6H_4X\text{-}p}}}{}}}} \tag{52}
$$

$$(X = H,\ CH_3O,\ Br,\ NO_2)$$

substituent in the migrating ring accelerated the rearrangement. Although evidence was presented to support a *trans* structure for the predominant isomer of **138**, Curtin and Miller assumed that *trans–cis* interconversion (equation 53) would be fast compared to the subsequent rearrangement which should take place through the *cis* isomer by way of a carbonyl addition mechanism. A dipolar transition state (or intermediate) was proposed. An extensive review of acyl migrations in other systems was also included in this paper by Curtin and Miller[89].

$$\tag{53}$$

Schwarz[90] has prepared some acyclic isoimides with $Ar^2 = C_6H_5$ (**136**) and various *para* substituents in Ar^1 and Ar^3. In no case did Ar^3 have two nitro substituents so the isoimides were too labile for easy isolation. The rearrangements to imides were monitored at 0°C by an i.r. method and half-lives were calculated. His results lent support to the mechanism proposed by Curtin and Miller. Schwarz found that, in contrast to the rearrangements of cyclic isoimides (next section), added carboxylate ion did not catalyse the rearrangements. He also looked for intermolecular acylation by an isoimide by adding aniline to one of the reaction mixtures and checking for the formation of benzanilide. There was no evidence for intermolecular acylation although this possibly could take place with a less labile isoimide and a more reactive amine.

That imides may be converted to isoimides, at least in the cases of *N*-alkyl- or *N*-aryl-*N*-formylamides, is a consideration dealt with by Hoy and Poziomek[91]. They reviewed the many reports in the older literature where pyrolysis of *N*-formylamides have yielded such products as isocyanides, nitriles, carbon monoxide, carboxylic acids, and amides. Their own studies of the thermal decomposition of some *N*-substituted *N*-formylacetamides led them to conclude that the best scheme which can account for their data and those of earlier workers is the one shown in equation 54. Although the imides could undergo decarbonylation to explain some of the products, the remainder of the products would be difficult to explain without invoking an imide–isoimide rearrangement (possibly by way of the cyclic intermediate proposed by Curtin and Miller).

$$
\begin{array}{c}
\underset{\substack{\text{R}}}{\overset{\text{O}\ \ \text{O}}{\overset{\|\ \ \|}{\text{HCNCCH}_3}}}
\ \rightleftarrows\
\underset{}{\overset{\overset{\text{O}}{\overset{\|}{\text{OCCH}_3}}}{\underset{}{\overset{|}{\text{HC}=\!\text{NR}}}}}
\ \rightleftarrows\
\underset{}{\overset{\text{O}}{\overset{\|}{\text{CH}_3\text{COH}}}} + \text{RNC} \longrightarrow \text{RCN}
\end{array}
$$

$$\downarrow$$

$$\tag{54}$$

$$
\tfrac{1}{2}\underset{}{\overset{\text{O}\ \ \text{O}}{\overset{\|\ \ \|}{\text{CH}_3\text{COCCH}_3}}} + \text{RNHCHO}
$$

3. Cyclic acyl imidates

There have been many reports (mostly since 1960) on the synthesis, chemistry and physical properties of cyclic acyl imidates or isoimides. Most of the work has been on the isoimides derived from phthalic and maleic acids although there are a few reports on cyclic isoimides derived from some saturated diacids. Extensive references to the literature in this area can be found in the papers by Ernst[87], Sauers[92-94] and Hedaya[95, 96].

It has been suggested[89] that cyclic isoimides are particularly stable because the carbonyl addition mechanism for rearrangement via a four-membered cyclic transition state would be difficult or impossible. Curtin and Miller[89] reinvestigated the rearrangement of **140** to **141** which had been initially reported many years earlier[85]. They found the half-time for the reaction (equation 55) in chlorobenzene at 250°C to be about 24 h. Reactions in dioxane and nitrobenzene at 178°C appeared to be first-order. It is unlikely, however, that they were observing unimolecular internal acyl migration. Probably in this case and other supposed thermal rearrangements the basic solvents or impurities in them were serving as catalysts or acyl transfer agents.

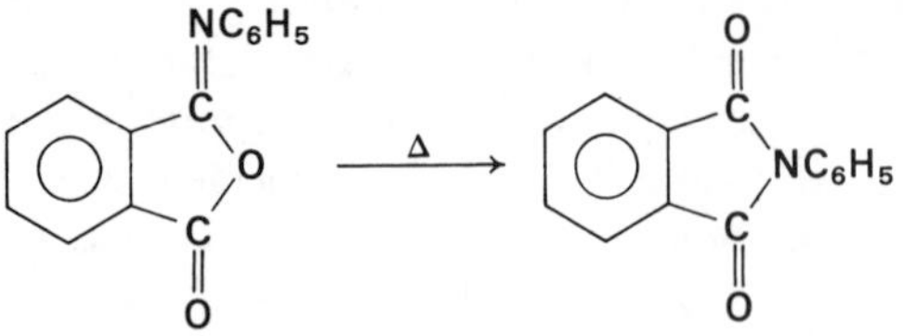

$$(55)$$

(140)   (141)

The base-catalysed rearrangements of isomaleimides (142) to maleimides (143) has been extensively investigated. Reaction conditions employed include the use of acetic anhydride–sodium acetate[97], benzene–triethylammonium acetate[97], and, most recently, ether–aziridine[98]. A variety of

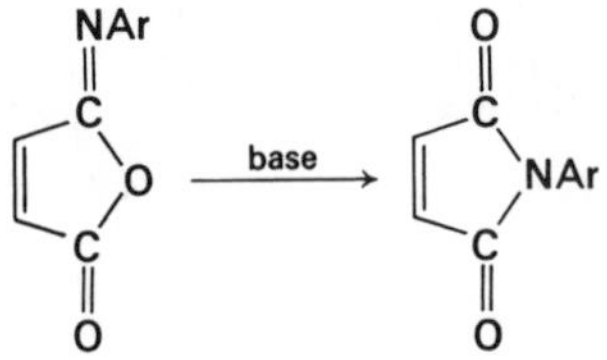

$$(56)$$

(142)   (143)

conditions for the analogous rearrangements of N-arylphthalisoimides (144)[87], N-arylsuccinisoimides (145)[94], and N,N'-biisomaleimides (146)[96] have also been reported.

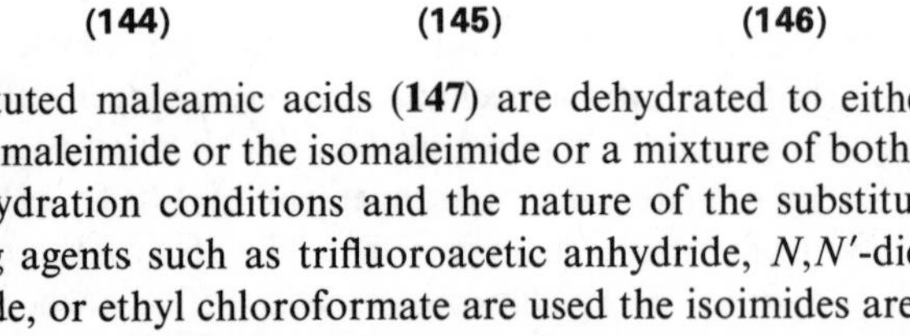

(144)   (145)   (146)

N-Substituted maleamic acids (147) are dehydrated to either the corresponding maleimide or the isomaleimide or a mixture of both depending on the dehydration conditions and the nature of the substituent. When dehydrating agents such as trifluoroacetic anhydride, N,N'-dicyclohexylcarbodiimide, or ethyl chloroformate are used the isoimides are formed as the kinetically controlled products[99]. Rearrangement to the thermodynamically more stable imides is avoided because the relatively weak bases formed as by-products from these dehydrating agents are not effective catalysts. In some cases, dehydration of maleamic acids to isoimides may be effected by thionyl chloride[100, 101], acetyl chloride[99], acetic anhydride[92], acetic anhydride–sodium acetate[92], acetyl chloride-triethylamine[99], or acetic anydride–triethylamine[99]. The results obtained seem to depend heavily on the conditions employed.

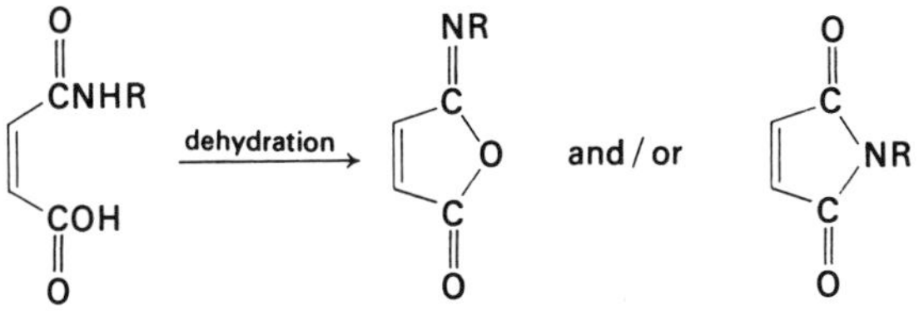

$$\text{(57)}$$

(147)

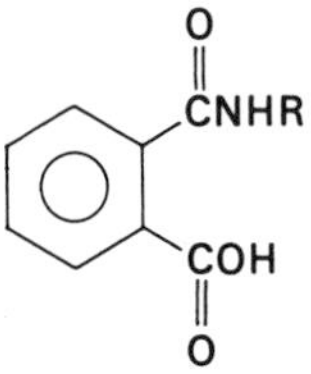

(148)

Much the same is true for the dehydration of N-substituted phthalamic acids (**148**); the product ratios depend on the reaction conditions and substituent[93]. In both series, there has been some concern about whether the isomerization of isoimides to imides is the major source of the latter or whether imides are formed directly from the maleamic or phthalamic acids. Sauers and co-workers[92,93] have reported detailed kinetic results which seem to answer this question, at least for the systems they studied and the conditions they used. For the dehydrations of N-arylmaleamic acids with acetic anhydride they found that isoimides predominated over imides[92]. In the presence of acetate ion more imide was formed initially and the rates of the rearrangements of isoimides to imides were not high enough to account for all of the imide produced in a given time. While the isomerization path was the major source of imides in the presence of acetate ion, some imide was clearly being formed directly from the N-arylmaleamic acid. Acetate ion in acetic anhydride was also found to enhance the direct formation of N-arylphthalimides from N-arylphthalamic acids in addition to increasing the yield of imide produced by the rearrangement of the isoimide[93]. Here again, then, competing paths were found to be involved. A mechanism involving initial mixed anhydride formation was proposed for both the N-arylmaleamic acid and the N-arylphthalamic acid dehydrations. The scheme for the former is shown below (equation 58)[92].

(58)

4. Some other O to N migrations of acyl groups

Curtin and Engelmann commented in 1968 on the scarcity of data available on pairs of isomeric N- and O-acyl derivatives of simple heterocyclic systems[102]. They found that the sodium salt of 6(5H)-phenanthridone (149) reacted with benzoyl chloride under kinetically controlled conditions ($-20°C$) to give the O-acylated product (150). Benzoylation at room temperature, on the other hand, led to 151. When heated alone, 150 rearranged to 151 to the extent of 99%[103]. This rearrangement also occurred in hexane or tetrahydrofuran solution to give an equilibrium mixture with a ratio of 151 to 150 of about 5:1. The approach to equilibrium was observed to be first-order in 150 or 151 and relatively insensitive to solvent polarity. It was concluded that the mechanism for the rearrangement of 150 to 151 involved an intramolecular nucleophilic replacement at the carbonyl group as suggested earlier for acyclic acyl imidates[89].

Curtin and Engelmann pointed out that N-acetylation of 2-pyridone had not been reported[102]. In some of their own preliminary studies, however, spectral evidence suggested a mixture of N- and O-acyl products from the acylation of the sodium salt of 2-pyridone in benzene but they did not pursue this further. Within a few months after that, McKillop, Zelesko, and Taylor[104] reported the acetylation of the thallous salt of 2-pyridone in chloroform at $-40°C$. N.m.r. spectra of the resulting solution strongly suggested that a mixture of 152 and 153 had been formed in a ratio of 3:2

(**152**:**153**). Spectra showed the *N*-acetyl compound to be converted to the *O*-acetyl isomer on standing at room temperature.

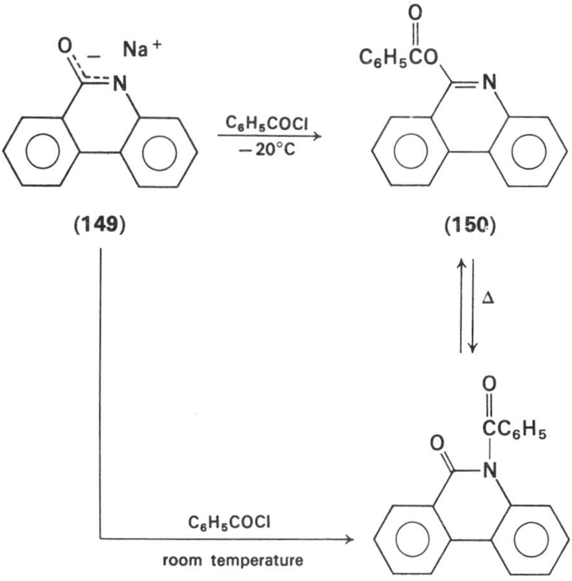

(149)   (150)

(59)

(151)

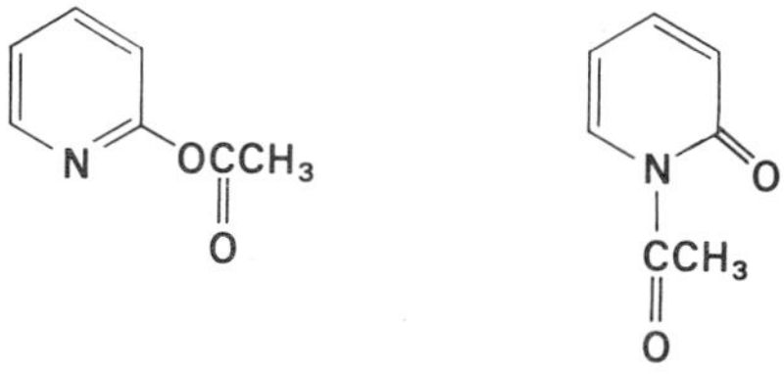

(152)   (153)

Another rearrangement in the heterocyclic series is the one shown in equation 60. The *O*-acylated derivatives of 3-hydroxy-1,2-benzisoxazoles (**154**) were observed to undergo a thermal rearrangement at 225°C to the isomeric acylated benzoxazol-2-ones (**156**) by way of the *N*-acylated

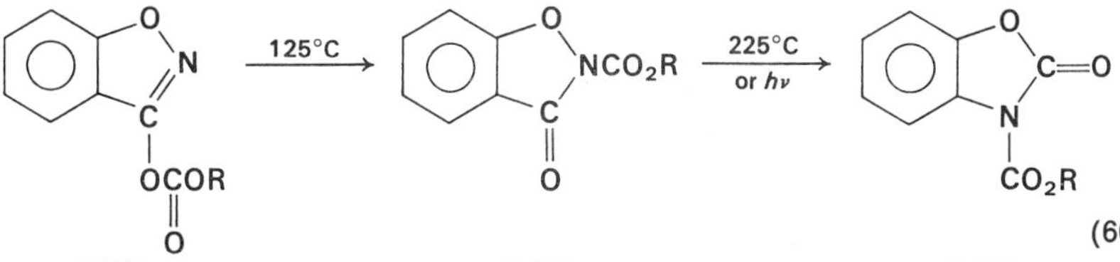

(154)   (155)   (156)

(60)

derivatives (155)[105]. At 125°C the *N*-acylated compounds could be obtained directly from 154 and then thermolytically or photolytically converted to 156.

Rubenstein and co-workers[106] recently prepared a series of aminoisomaleimides (157) and found that they undergo acid catalysed rearrangements to aminomaleimides (158) or pyridazinones (159) depending upon the conditions of the rearrangement and the nature of the substituent.

$$(157) \xrightarrow{\;H^+\;} (158) \quad \text{or} \quad (159) \tag{61}$$

(157)   (158)   (159)

The azoacetates (160) obtained by the oxidation of aldehyde arylhydrazones with lead tetra-acetate should be capable of undergoing a prototropic shift to give hydrazonyl acetates (161) which undergo 1,4-acyl migration to give diacylhydrazines (162)[107]. The sequence shown in equation 62 was observed by Gladstone, Aylward, and Norman[108] but they also provided evidence that a nitrilimine intermediate (163) lies on the main reaction path leading to 162.

$$(160) \longrightarrow (161) \longrightarrow (162) \tag{62}$$

(160)   (161)   (162)

$$\uparrow \; CH_3CO_2H$$

$$RC\equiv\overset{+}{N}-\overset{-}{N}Ar$$
(163)

The reaction of methyl isocyanide with chloroacetic anhydride or trifluoroacetic anhydride in chloroform gives the pyruvamide (165) as a product[109]. Although 164 was not isolated or observed spectroscopically it was proposed by Krivinka and Honzl[109] as an intermediate which should be formed directly from the reactants and which should rearrange to 165.

$$CH_3NC + (RCO)_2O \longrightarrow \left[\begin{matrix} O \\ \| \\ CH_3N{=}CCR \\ | \\ OCR \\ \| \\ O \end{matrix} \right] \longrightarrow \begin{matrix} O \quad O \\ \| \quad \| \\ CH_3NC{-}CR \\ | \\ CR \\ \| \\ O \end{matrix} \quad (63)$$

$$\qquad\qquad\qquad\qquad (164) \qquad\qquad\qquad (165)$$

D. Thioimidates

Chapman[110] examined the effect of heat on thioimidate **166** and found that isomerization to **167** took place to only a small extent at 280–290°C. At temperatures of 320°C or higher, **166** gave a mixture of diphenyl sulfide, benzonitrile, thiophenol and the benzthiazole **168**. The same mixture of products was obtained by heating **167** to 320°C or higher so Chapman concluded that the rearrangement of **166** to **167** is reversible.

$$\begin{matrix} SC_6H_5 \\ | \\ C_6H_5C{=}NC_6H_5 \end{matrix} \xrightarrow[\text{2 h}]{\text{280–290°C}} \begin{matrix} S \\ \| \\ C_6H_5CN(C_6H_5)_2 \end{matrix} \qquad (64)$$

$$\qquad (166) \qquad\qquad\qquad\qquad (167)$$

$$(168)$$

Apparently nobody has obtained results with simple aryl or alkyl thioimidates which would dispute Chapman's findings. Thioimidates simply do not readily undergo a thermal S to N migration to give thioamides unless some structural feature is built in to enhance the rearrangement. Such enhancement was observed by Walter and Krohn[111] with thioimidates which have a benzhydryl group on sulfur. Rearrangement of **169** to **170** in 80–90% yield occurred in refluxing benzene. Elimination of a mercaptan from **169** was not observed so a rearrangement mechanism involving ionization of **169** to a benzhydryl carbonium ion was proposed. A crossover experiment gave results which supported the intermolecularity of this rearrangement. They also found that some of the thioamides could be converted back to thioimidates by heating them for several hours in ether with HCl added.

$$\begin{matrix} SCHAr^2Ar^3 \\ | \\ Ar^1C{=}NH \end{matrix} \xrightarrow{\Delta} \begin{matrix} S \\ \| \\ Ar^1CNHCHAr^2Ar^3 \end{matrix} \qquad (65)$$

$$\qquad (169) \qquad\qquad\qquad (170)$$

Beak and Lee[61] used the common alkylated derivative of N-methyl-2-thiopyridone (171) and 2-methylthiopyridine (172) to effect equilibration of the two isomers at 190°C (equation 66). In the liquid phase the equilibrium ratio of 172 to 171 was about 9:1 starting with either 172 or 171.

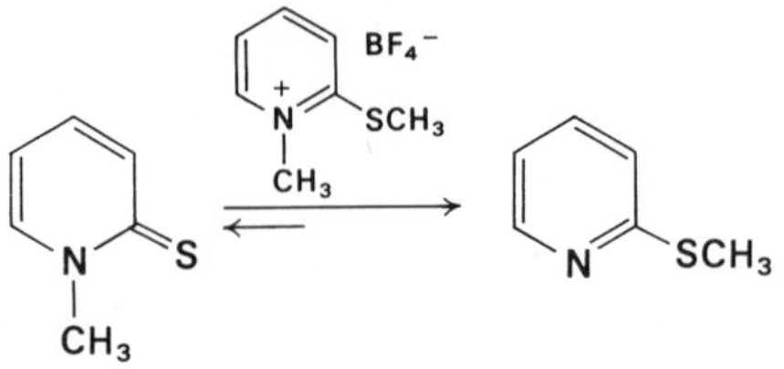

(66)

(171)   (172)

Attempts have been made to rearrange thiohydrazonates (173) but no conversion to 174 has been observed in refluxing xylene[36] or in refluxing dioxane[39].

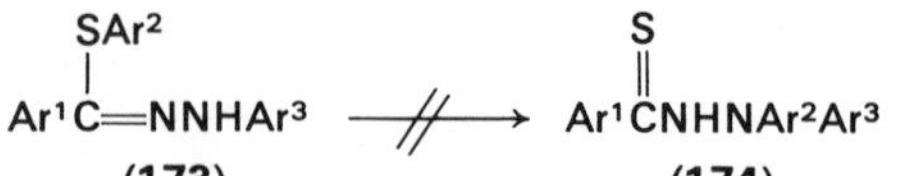

(67)

(173)   (174)

A recent review[112] of S to N rearrangements in heterocycles contains many examples which fit the general scheme shown in equation 68. However, in practically all cases, the examples shown are thiazoles or benzthiazoles (X = S) and not cyclic thioimidates.

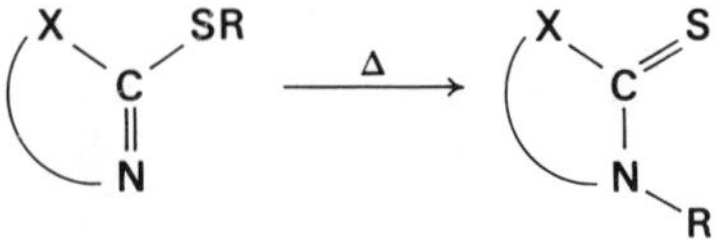

(68)

III. REARRANGEMENTS OF AMIDINES AND RELATED COMPOUNDS

The migration of a group from N to N in an amidine has received scant attention compared to the O to N migrations in imidates. After his intensive study of imidates, Chapman proceeded to look for similar rearrangements in amidines. In a series of papers between 1929 and 1932[5-7, 113,114] he reported his results on aryl amidines. Amidine 175 was found to be stable at temperatures below 300°C but partially rearranged to 176 at 330–340°C. The ratio of 176 to 175 was 2:1 and this same ratio was obtained when heating 176 at 330–340°C[113]. Thus, as one would expect, Chapman found the thermal rearrangement of aryl amidines to be reversible. In further studies he found from heating mixtures of different amidines that no crossover products were formed[5]. Thus, the rearrangement appears to be intramolecular. Kinetic studies showed the approach

$$
\underset{\textbf{(175)}}{\overset{\displaystyle N(C_6H_5)_2}{C_6H_5C=NC_6H_4CH_3\text{-}p}} \quad \underset{\longleftarrow}{\overset{\Delta}{\rightleftharpoons}} \quad \underset{\textbf{(176)}}{\overset{\displaystyle NC_6H_5}{\underset{\displaystyle C_6H_4CH_3\text{-}p}{C_6H_5CNC_6H_5}}} \tag{69}
$$

to equilibrium to be first-order[114]. The effects of various substituents were studied in amidines such as **177** and **178** (equation 70). The value of k_1 was essentially constant as Ar was varied from p-tolyl to p-chlorophenyl to 3,5-dichlorophenyl but k_{-1} varied, being greatest with p-tolyl and least with 3,5-dichlorophenyl[7]. These results would be consistent with a mechanism similar to the one proposed for the thermal rearrangements of aryl imidates (equation 9).

$$
\underset{\textbf{(177)}}{\overset{\displaystyle NC_6H_5}{\underset{\displaystyle Ar}{C_6H_5CNC_6H_5}}} \quad \underset{k_{-1}}{\overset{k_1}{\rightleftharpoons}} \quad \underset{\textbf{(178)}}{\overset{\displaystyle N(C_6H_5)_2}{C_6H_5C=NAr}} \tag{70}
$$

There are few examples of the migration of alkyl groups in amidines. Schwenker and Kolb[115] were studying the reaction of N,N'-dimethyl-benzamidine (**179**) with thiocyanates and to account for the benzonitrile formed from the pyrolysis of **179** in the presence of phenyl isothiocyanate they proposed the scheme shown in equation 71. Under proper conditions,

$$
\underset{\textbf{(179)}}{\overset{\displaystyle NCH_3}{\underset{\displaystyle H}{C_6H_5CNCH_3}}} \quad \xrightarrow{290°C} \quad \underset{}{\overset{\displaystyle N(CH_3)_2}{C_6H_5C=NH}} \quad \xrightarrow{-HN(CH_3)_2} \quad C_6H_5CN \tag{71}
$$

cyclic imidates can be reacted with aziridine to give interesting amidines[116]. An aziridinyl tetrahydroazepine (**180**) thus formed rearranged smoothly to **181** in refluxing acetone containing a small amount of iodine.

$$\tag{72}$$

(180)   **(181)**

Another example of alkyl migration was observed in the transformation of **182** to **183** at 260°C (equation 73)[117]. On the other hand, neither alkyl nor aryl migration occurred in the nitrogen analogs of the aryl hydrazonates studied by Hegarty and co-workers. For example, **184** did not rearrange in the presence of radical initiators or on heating in dioxane under reflux[39].

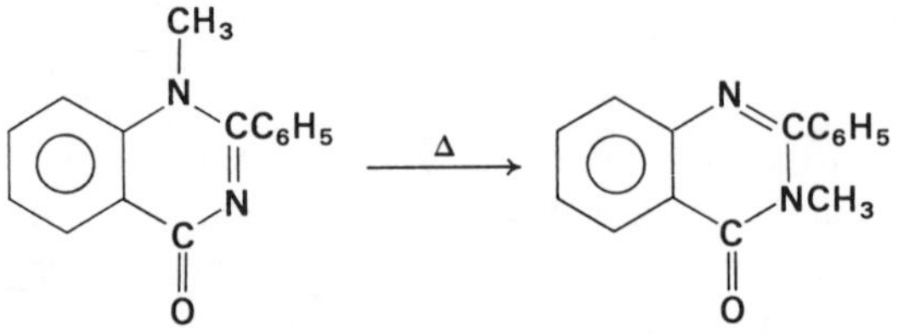

(73)

(182)   (183)

$$N(CH_3)C_6H_5$$
$$|$$
$$C_6H_5C=NNHC_6H_5$$

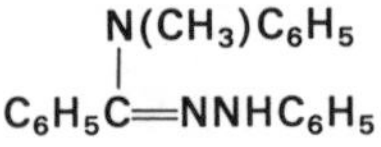

(184)

Migration of an acyl group in acyclic amidines is more facile than the previously described aryl and alkyl rearrangements. *N*-Benzoyl-*N*-phenylbenzamidine (**185**) smoothly rearranges in solution to **186** (or its tautomer) at low temperatures. This first-order conversion of **185** to **186** was also found to take place in the melt and in the solid state[103, 118]. The mechanism is undoubtedly similar to the one proposed for acyclic acyl imidates or isoimides (equation 53). Such a mechanism would not be

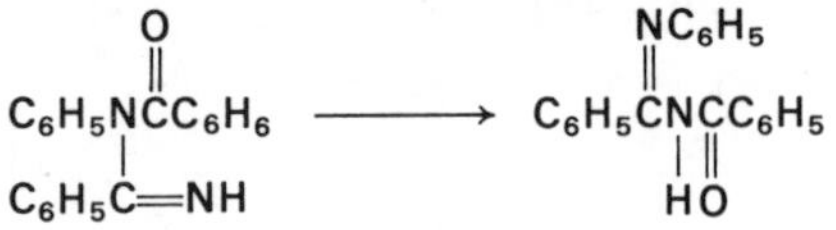

(74)

(185)   (186)

(75)

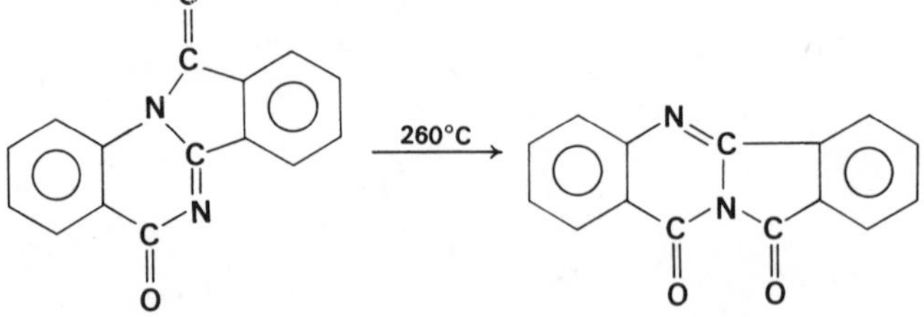

(187)   (188)

possible for the rearrangement of **187** to **188**. Heating **187** to 260°C for 30 min did, however, give an almost quantitative conversion to **188**[119].

N-Chloroamidines (**189**) having a hydrogen on the other nitrogen atom can rearrange upon dehydrohalogenation with base. The mechanism suggested by Fuchigami and co-workers[120] is shown in equation 76. Although they favoured a simultaneous loss of chloride ion and aryl migration (**190** to **191**) a nitrene intermediate was not ruled out. The carbodiimides (**191**) were trapped with water or alcohol as the urea or isourea derivatives. Treatment of **189** with silver oxide in ligroin also led to carbodiimides and, in this case, a nitrenium cation intermediate was proposed (equation 77).

$$
\begin{array}{c}
\underset{\displaystyle \text{ArC}=\text{NCl}}{\overset{\displaystyle \text{NHR}}{|}} \xrightarrow{\ ^{-}\text{OH}\ } \left[\underset{\displaystyle \text{ArC}=\text{NCl}}{\overset{\displaystyle \bar{\text{N}}\text{R}}{|}}\right] \longrightarrow \left[\underset{\displaystyle \text{ArC}-\text{NCl}}{\overset{\displaystyle \text{NR}}{\|}}\right] \\
\text{(189)} \hspace{6cm} \text{(190)}
\end{array}
$$

$$
\Big\downarrow {}_{-\text{Cl}^-}
$$

$$
\text{RN}=\text{C}=\text{NAr} \tag{76}
$$
$$
\text{(191)}
$$

$$
\text{(189)} \xrightarrow{\ \text{Ag}_2\text{O}\ } \left[\underset{\displaystyle \text{ArC}=\overset{+}{\text{N}}}{\overset{\displaystyle \text{NHR}}{|}}\right] \longrightarrow \left[\text{RNHC}=\overset{+}{\text{N}}\text{Ar}\right] \xrightarrow{\ -\text{H}^+\ } \text{(191)} \tag{77}
$$

N-Hydroxyamidines or amidoximes rearrange when treated with benzenesulfonyl chloride and base (equation 78). The reaction is sometimes referred to as the Tiemann rearrangement[121] since it was first reported by Tiemann in 1891[122]. This conversion attracted little more attention until

$$
\underset{\displaystyle \text{C}_6\text{H}_5\text{C}=\text{NOH}}{\overset{\displaystyle \text{NH}_2}{|}} \xrightarrow[\text{base}]{\ \text{C}_6\text{H}_5\text{SO}_2\text{Cl}\ } \text{C}_6\text{H}_5\text{NHCONH}_2 \tag{78}
$$
$$
\text{(192)} \hspace{4cm} \text{(193)}
$$

Partidge and Turner[123] investigated the mechanism some 60 years later. Benzamidoxime (**192**) does not rearrange when heated alone. The purpose of the benzensulfonyl chloride is to form the benzenesulfonyl ester of the amidoxime which then decomposes to phenylcyanamide (most likely *via* its carbodiimide tautomer) and benzenesulfonic acid. Although Partridge and Turner could not isolate the intermediate ester from benzamidoxime (since it rearranged too readily) they did isolate the ester of phenylacetamidoxime (**194**) and showed that in inert solvents it formed benzylcyanamide and benzenesulfonic acid (equation 79). Amidoximes can be easily O-phosphorylated and the resulting esters also rearrange to cyan-amides[124]

$$C_6H_5CH_2\overset{\overset{\displaystyle NH_2}{|}}{C}=NOSO_2C_6H_5 \quad \overset{\Delta}{\longrightarrow} \quad C_6H_5CH_2NHCN + C_6H_5SO_3H \quad (79)$$

(194)

Further information on this rearrangement of amidoximes was provided by Partridge and Turner in 1958[125] when they reported that N-aryl amidoximes give 2-substituted benzimidazoles (**195**) as the major products under certain conditions. Dissociation of the initially formed benzene-

$$R\overset{\overset{\displaystyle NHC_6H_5}{|}}{C}=NOH \quad \overset{C_6H_5SO_2Cl}{\underset{pyridine}{\longrightarrow}} \quad (80)$$

(195)

sulfonyl ester into an azomethine nitrenium ion has been proposed[125] to precede ring closure. The nitrenium ion could inert in an appropriate aromatic C—H bond to give the benzimidazole. However, elimination of benzenesulfonic acid from the ester to leave a nitrene is also possible since a nitrene could also effect ring closure via insertion in an aromatic C—H bond. Evidence for a nitrene intermediate in the analogous reactions of N-alkyl amidoximes with benzenesulfonyl chloride and pyridine has been provided by Boyer and Frints[126]. In addition to obtaining the expected carbodiimide (**197**) from N-cyclohexylbenzamidoxime (**196**) they also observed an amidine (**198**) which could have been formed from a nitrene intermediate (**199**) by hydrogen abstraction from the solvent.

$$C_6H_5\overset{\overset{\displaystyle NHC_6H_{11}}{|}}{C}=NOH \qquad C_6H_5N=C=NC_6H_{11} \qquad C_6H_5\overset{\overset{\displaystyle NHC_6H_{11}}{|}}{C}=NH \qquad C_6H_5\overset{\overset{\displaystyle NC_6H_{11}}{\|}}{C}-N$$

$$\textbf{(196)} \qquad\qquad \textbf{(197)} \qquad\qquad \textbf{(198)} \qquad\qquad \textbf{(199)}$$

IV. SOME RELATED REARRANGEMENTS

Imidocarbonates, thioimidocarbonates, and isoureas are structurally related to the imidates discussed earlier and there are many examples of O to N rearrangements in these systems. Only a few examples are included here for sake of comparison with the imidates. McCarty and Garner[71] have carried out an extensive investigation of the Chapman-like rearrangements of **200**, **202**, and **204**. Imidocarbonate **200** and thioimidocarbonate

202 underwent clean, first-order rearrangements in bromobenzene at temperatures far below those normally observed for alkyl imidates. The unimolecular reactions (equations 81, 82) had large negative entropies of activation, were not accelerated by benzoyl peroxide or iodine, and showed no exchange of alkyl groups in crossover experiments, i.e. they behaved very much like aryl imidates even though it would be difficult to imagine a S_N type mechanism for these alkyl migrations. Although **205** was the main product from the thermal rearrangement of **204**, this reaction was not clean. Other products were identified as oxadiazenes which might have been formed by initial methyl migration to the cyano nitrogen to form a reactive carboiimide intermediate.

$$
\begin{array}{ccc}
\overset{\displaystyle OCH_3}{\underset{\displaystyle}{CH_3O\overset{|}{C}\!=\!NCN}} & \xrightarrow{\;100°C\;} & CH_3O\overset{\displaystyle O}{\overset{\displaystyle \|}{C}}\overset{|}{\underset{\displaystyle CN}{N}}CH_3 \\
\textbf{(200)} & & \textbf{(201)}
\end{array} \qquad (81)
$$

$$
\begin{array}{ccc}
\overset{\displaystyle OCH_3}{\underset{\displaystyle}{CH_3S\overset{|}{C}\!=\!NCN}} & \xrightarrow{\;120°C\;} & CH_3S\overset{\displaystyle O}{\overset{\displaystyle \|}{C}}\overset{|}{\underset{\displaystyle CN}{N}}CH_3 \\
\textbf{(202)} & & \textbf{(203)}
\end{array} \qquad (82)
$$

$$
\begin{array}{ccc}
\overset{\displaystyle OCH_3}{\underset{\displaystyle}{(CH_3)_2N\overset{|}{C}\!=\!NCN}} & \xrightarrow{\;120°C\;} & (CH_3)_2N\overset{\displaystyle O}{\overset{\displaystyle \|}{C}}\overset{|}{\underset{\displaystyle CN}{N}}CH_3 \\
\textbf{(204)} & & \textbf{(205)}
\end{array} \qquad (83)
$$

A cyclic system related to acyclic imidocarbonates is shown in equation 84[127]. These thermal rearrangements were carried out at 250°C and were found by crossover experiments to be intermolecular. Elimination of the alkyl group as the corresponding alkene occurred when R^2 was secondary or tertiary.

$$
\begin{array}{ccc}
\textbf{(206)} & \xrightarrow{\;250°C\;} & \textbf{(207)}
\end{array} \qquad (84)
$$

The 3-alkoxy-1,2,4-benzothiadiazine-1,1-dioxides (**208**) shown in equation 85 bear a structural relationship to acyclic isoureas. Alkyl migration from O to N was studied in this system recently[128]. α,β-Unsaturated groups migrated with inversion as in the Claisen rearrangement.

(208)   **(209)**   (85)

No S to N rearrangements have been observed for **210** or **211** under conditions where **200**, **202**, and **204** readily rearranged [71]. The addition of alkyl halides, iodine, and peroxides as possible catalysts was to no avail. The N-methyl and N-phenyl analogs of **210** could not be isomerized even at 250–270°C [129]. However, some cyclic dithioimidocarbonates (**212**) have been isomerized at temperatures of about 200°C (equation 86) [130].

$$(CH_3S)_2C{=}NCN$$

(210)   **(211)**

(212)   **(213)**   (86)

$$R = CH_3, C_2H_5$$

V. REFERENCES

1. O. Mumm, H. Hesse and H. Volquartz, *Ber.*, **48**, 379 (1915).
2. A. W. Chapman, *J. Chem. Soc.*, 1992 (1925).
3. A. W. Chapman, *J. Chem. Soc.*, 1743 (1927).
4. A. W. Chapman, *J. Chem. Soc.*, 569 (1929).
5. A. W. Chapman, *J. Chem. Soc.*, 2458 (1930).
6. A. W. Chapman and C. H. Perrott, *J. Chem. Soc.*, 1770 (1932).
7. A. W. Chapman and C. H. Perrott, *J. Chem. Soc.*, 1775 (1932).
8. J. W. Schulenberg and S. Archer, *Org. Reactions*, **14**, 1 (1965).
9. C. G. McCarty in *The Chemistry of the Carbon–Nitrogen Double Bond*, (Ed. S. Patai), Interscience, New York, 1970, Chap. 9, pp. 439–447.
10. F. J. McEvoy, E. N. Greenblatt, A. C. Osterberg and G. R. Allen, Jr., *J. Med. Chem.*, **13**, 295 (1970).
11. R. Mukherjee and P. Block, *J. Chem. Soc. (C)*, 1596 (1971).
12. A. J. Fritsch, C. E. Moore, T. S. Meyer, I. L. Shklair, and L. F. Devine, *Appl. Microbiol.*, **18**, 684 (1969).

13. T. R. Westby and C. F. Barfknecht, *J. Med. Chem.*, **16**, 40 (1973).
14. M. R. Bell, A. W. Zalay and R. Oesterlin, *J. Med. Chem.*, **13**, 664 (1970).
15. J. W. Schulenberg, *J. Amer. Chem Soc.*, **90**, 7008 (1968).
16. V. D. Goldberg and M. M. Harris, *J. Chem. Soc., Perkin Trans. 2*, 1303 (1973).
17. J. Cymerman-Craig and J. W. Loder, *J. Chem. Soc.*, 4309 (1955).
18. O. H. Wheeler, F. Roman, M. V. Santiago, and F. Quiles, *Can. J. Chem.*, **47**, 503 (1969).
19. K. B. Wiberg and B. I. Rowland, *J. Amer. Chem. Soc.*, **77**, 2205 (1955).
20. H. M. Relles, *J. Org. Chem.*, **33**, 2245 (1968).
21. H. M. Relles and G. Pizzolato, *J. Org. Chem.*, **33**, 2249 (1968).
22. O. H. Wheeler, F. Roman, and O. Rosado, *J. Org. Chem.*, **34**, 966 (1969).
23. A. W. Chapman and C. C. Howis, *J. Chem. Soc.*, 806 (1933).
24. P. A. S. Smith in *Molecular Rearrangements*, (Ed. P. de Mayo), Interscience, New York, 1963, Vol. 1, Chap. 8.
25. J. D. McCullough, Jr., D. Y. Curtin, and I. C. Paul, *J. Amer. Chem. Soc.*, **94**, 874 (1972).
26. I. I. Kukhtenko, *Zh. Org. Khimii*, **7**, 333 (1971).
27. H. P. Fischer, *Tetrahedron Letters*, 285 (1968).
28. H. P. Fischer and F. Funk-Kretschmar, *Helv. Chim. Acta*, **52**, 913 (1969).
29. R. A. Scherrer and H. R. Beatty, *J. Org. Chem.*, **37**, 1681 (1972).
30. A. E. Chichibabin and N. P. Jeletsky, *Chem. Ber.*, **57**, 1158 (1924).
31. R. A. Scherrer, C. V. Winder, and F. W. Short, *Abstracts, Ninth Annual Medicinal Chemistry Symposium*, Minneapolis, Minn., June, 1964, p. 11i.
32. R. A. Scherrer, *German Patent* 1,190,951 (1965); *Chem. Abstr.*, **63**, 4209d (1965).
33. R. A. Scherrer, *U.S. Patent* 3,238,201 (1966); *Chem. Abstr.*, **64**, 17614b (1966).
34. P. F. Juby, T. W. Hudyma, and M. Brown, *J. Med. Chem.*, **11**, 111 (1968).
35. A. S. Shawali and H. M. Hassaneen, *Tetrahedron Letters*, 1299 (1972),
36. A. S. Shawali and H. M. Hassaneen, *Tetrahedron*, **28**, 5903 (1972).
37. A. F. Hegarty, J. A. Kearney, M. Cashman, and F. L. Scott, *Chem. Commun.*, 689 (1971).
38. A. F. Hegarty, J. A. Kearney, and F. L. Scott, *Tetrahedron Letters*, 3211 (1972).
39. A. F. Hegarty, J. A. Kearney, and F. L. Scott, *J. Chem. Soc., Perkin Trans. 2*, 1422 (1973).
40. T. S. Stevens and W. E. Watts, *Selected Molecular Rearrangements*, Van Nostrand Reinhold Co., New York, 1973, p. 120.
41. A. J. Elliott, M. S. Gibson, M. M. Kayser, and G. A. Pawelchak, *Can. J. Chem.*, **51**, 4115 (1973).
42. K. Wiberg, T. Shryne and R. Kinter, *J. Amer. Chem. Soc.*, **79**, 3160 (1957).
43. A. Pilotti, A. Reuterhall and K. Torssell, *Acta Chem. Scand.*, **23**, 818 (1969).
44. L. A. Paquette, M. J. Wyvratt and G. R. Allen, Jr., *J. Amer. Chem. Soc.*, **92**, 1763 (1970).

45. H. K. Reimschuessel and G. J. Dege, *J. Pol. Sci.-Al.* **8**, 3265 (1970).
46. N. Marullo, C. Smith and J. Terapane, *Tetrahedron Letters*, 6279 (1966).
47. F. Cramer, K. Pawelzwich, and J. Kupper, *Angew. Chem.*, **68**, 649 (1956).
48. C. Mao, I. T. Barnish, and C. R. Hauser, *J. Heterocycl. Chem.*, **6**, 475 (1969).
49. D. M. Bailey and C. G. DeGrazia, *J. Org. Chem.*, **35**, 4093 (1970).
50. O. Mumm and F. Möller, *Chem. Ber.*, **70**, 2214 (1937).
51. D. S. Tarbell, *Org. Reactions*, **2**, 1 (1944).
52. D. Y. Curtin and H. W. Johnson, *J. Amer. Chem. Soc.*, **76**, 2276 (1954).
53. W. Wislicenus and M. Goldschmidt, *Chem. Ber.*, **33**, 1470 (1900).
54. G. Lander, *J. Chem. Soc.*, **83**, 406 (1903).
55. A. Arbuzov and V. Shishkin, *Dokl. Akad. Nauk, SSSR*, **141**, 349 (1961); *Chem. Abstr.*, **56**, 11491 (1962).
56. A. Arbuzov and V. Shishkin, *Dokl. Akad. Nauk. SSSR*, **141**, 611 (1961); *Chem. Abstr.*, **56**, 15424 (1962).
57. B. C. Challis and A. D. Frenkel, *Chem. Commun.*, 303 (1972).
58. P. Beak, J. Bonham, and J. T. Lee, Jr., *J. Amer. Chem. Soc.*, **90**, 1569 (1968).
59. P. Beak, *Tetrahedron Letters*, 863 (1963).
60. P.. Beak and J. Bonham, *Chem. Commun.*, 631 (1966).
61. P. Beak and J. T. Lee, Jr., *J. Org. Chem.*, **34**, 2125 (1969).
62. F. Cramer and N. Hennrich, *Chem. Ber.*, **94**, 976 (1961).
63. R. Roberts and P. Vogt, *J. Amer. Chem. Soc.*, **78**, 4778 (1956).
64. L. A. Singer and G. A. Davis, *J. Amer. Chem. Soc.*, **89**, 941 (1967).
65. L. A. Singer, G. A. Davis, and R. L. Knutsen, *J. Amer. Chem. Soc.*, **94**, 1188 (1972).
66. N. Ishibe and Y. Yamaguchi, *J. Chem. Soc., Perkin Trans. 1*, 2618 (1973).
67. L. E. Overman, *J. Amer. Chem. Soc.*, **96**, 597 (1974).
68. H. F. Stewart and R. P. Seibert, *J. Org. Chem.*, **33**, 4560 (1968).
69. F. J. Dinan and H. Tieckelman, *J. Org. Chem.*, **29**, 892 (1964).
70. T. Saegusa, Y. Ito, S. Kobayshi, K. Hirota and N. Takeda, *Can. J. Chem.*, **47**, 1217 (1969).
71. C. G. McCarty and L. A. Garner, unpublished results.
72. K. R. Huffman and F. C. Schaefer, *J. Org. Chem.*, **28**, 1812 (1963).
73. H. Hettler, *Tetrahedron Letters*, 1793 (1968).
74. M. Furlan, B. Stanovnik, and M. Tišler, *J. Org. Chem.*, **37**, 2689 (1972.)
75. B. G. Pring and C. Swahn, *Acta Chem. Scand.*, **27**, 1891 (1973).
76. F. Nerdel, P. Weyerstahl, and R. Dahl, *Liebigs Ann. Chem.*, **716**, 127 (1968).
77. J. A. Elvidge, G. T. Rogers, and T. L. V. Ulbricht, *J. Heterocycl. Chem.*, **8**, 1039 (1971).
78. J. B. Aylward and F. L. Scott, *J. Chem. Soc. (C)*, 968 (1970).
79. J. E. Johnson, J. R. Springfield, J. S. Hwang, L. J. Hayes, W. C. Cunningham, and D. J. McClaugherty, *J. Org. Chem.*, **36**, 284 (1971).
80. E. J. Moriconi, C. F. Hummel, and J. F. Kelly, *Tetrahedron Letters*, 5325 (1969).
81. J. R. Malpass, *Chem. Commun.*, 1246 (1972).

82. D. Mackay, J. A. Campbell, and C. P. R. Jennison, *Can. J. Chem.*, **48**, 81 (1970).
83. C. Y-J. Chung, D. Mackay and T. D. Sauer, *Can. J. Chem.*, **50**, 1568 (1972).
84. C. P. R. Jennison and D. Mackay, *Tetrahedron*, **29**, 1255 (1973).
85. S Hoogerwerff and W. A. van Dorp, *Rec. Trav. Chim.*, **12**, 12 (1893).
86. W. R. Roderick and P. L. Bhatia, *J. Org. Chem.*, **28**, 2018 (1963).
87. M. L. Ernst and G. L. Schmir, *J. Amer. Chem. Soc.*, **88**, 5001 (1966).
88. D. Y. Curtin and L. L. Miller, *Tetrahedron Letters*, 1869 (1965).
89. D. Y. Curtin and L. L. Miller, *J. Amer. Chem. Soc.*, **89**, 637 (1967).
90. J. S. P. Schwartz, *J. Org. Chem.*, **37**, 2906 (1972).
91. D. J. Hoy and E. J. Poziomek, *J. Org. Chem.*, **33**, 4050 (1968).
92. C. K. Sauers, *J. Org. Chem.*, **34**, 2275 (1969).
93. C. K. Sauers, C. L. Gould and E. S. Ioannou, *J. Amer. Chem. Soc.*, **94**, 8156 (1972).
94. C. K. Sauers, C. A. Marikakis, and M. A. Lupton, *J. Amer. Chem. Soc.*, **95**, 6792 (1973).
95. E. Hedaya, R. L. Hinman, and S. Theodoropulos, *J. Org. Chem.*, **31**, 1311 (1966).
96. E. Hedaya, R. L. Hinman, and S. Theodoropulos, *J. Org. Chem.*, **31**, 1317 (1966).
97. R. J. Cotter, C. K. Sauers, and J. M. Whelan, *J. Org. Chem.*, **26**, 10 (1961).
98. P. Joseph-Nathan, V. Mendoza, and E. Garcia G., *Can. J. Chem.*, **52**, 129 (1974).
99. T. M. Pyriadi and H. J. Harwood, *J. Org. Chem.*, **36**, 821 (1971).
100. H. Feuer, *J. Org. Chem.*, **36**, 3372 (1971).
101. T. M. Pyriadi, *J. Org. Chem.*, **37**, 4184 (1972).
102. D. Y. Curtin and J. H. Engelmann, *Tetrahedron Letters*, 3911 (1968).
103. D. Y. Curtin and J. H. Engelmann, *J. Org. Chem.*, **37**, 3439 (1972).
104. A. McKillop, M. J. Zelesko, and E. C. Taylor, *Tetrahedron Letters*, 4945 (1968).
105. H. Böshagen and W. Geiger, *Chem. Ber.*, **103**, 123 (1970).
106. H. Rubinstein, M. Parnarouskis, and H. Feuer, *J. Org. Chem.*, **38**, 2166 (1973).
107. J. M. Burgess and M. Gibson, *J. Chem. Soc.*, 1500 (1964).
108. W. A. F. Gladstone, J. B. Aylward and R. O. C. Norman, *J. Chem. Soc.* (*C*), 2587 (1969).
109. P. Krivinka and J. Honzl, *Coll. Czech. Chem. Commun.*, **37**, 4035 (1972).
110. A. W. Chapman, *J. Chem. Soc.*, 2296 (1926).
111. W. Walter and J. Krohn, *Chem. Ber.*, **102**, 3786 (1969).
112. M. Chanon, M. Conte, J. Micozzi and J. Metzger, *Int. J. Sulfur Chem.*, *C.*, **6**, 85 (1971).
113. A. W. Chapman, *J. Chem. Soc.*, 2133 (1929).
114. A. W. Chapman and C. H. Perrott, *J. Chem. Soc.*, 2462 (1930).
115. G. Schwenker and R. Kolb, *Tetrahedron*, **25**, 5549 (1969).
116. D. Bormann, *Angew. Chem. Internat. Edit.*, **12**, 768 (1973).
117. Y. Hagiwara, M. Kurihara, and Y. Yoda, *Tetrahedron*, **25**, 783 (1969).

118. D. A. R. Thompson, *Diss. Abst. Int.* (*B*), **31**, 5274 (1971).
119. M. Kurihara, *J. Org. Chem.*, **34**, 2123 (1969).
120. T. Fuchigami, E. Ichikawa, and K. Odo, *Bull. Chem. Soc. Jap.*, **46**, 1765 (1973).
121. P. A. S. Smith in *Molecular Rearrangements*, Part 1 (Ed. P. de Mayo), Interscience, New York, 1962, pp. 564–565.
122. F. Tiemann, *Ber.*, **24**, 4162 (1891).
123. M. W. Partridge and H. A. Turner, *J. Pharm. and Pharmacol.*, **5**, 103 (1953).
124. R. F. Hudson and R. Woodcock, *Chem. Commun.*, 1050 (1971).
125. M. W. Partridge and H. A. Turner, *J. Chem. Soc.*, 2086 (1958).
126. J. H. Boyer and P. J. A. Frints, *J. Org. Chem.*, **35**, 2449 (1970).
127. M. Golfier and R. Milcent, *Tetrahedron Letters*, 4465 (1973).
128. P. Pecorari, A. Albasini, and L. Raffa, *Farm. Ed. Sci.*, **28**, 203 (1973).
129. Y. Ueno, T. Nakai, and M. Okawara, *Bull. Chem. Soc., Jap.*, **44**, 1933 (1971).
130. Y. Ueno, T. Nakai, and M. Okawara, *Bull. Chem. Soc. Jap.*, **43**, 162 (1970).

CHAPTER **5**

The electrochemistry of imidic esters and amidines

HENNING LUND

University of Aarhus, Denmark

I. INTRODUCTION

The electrochemistry of imidic acid derivatives has not been studied extensively; results from electroanalytical and from electrosynthetic investigations in hydroxylic solvents are scarce, and data from work in aprotic media are nearly absent. This chapter will deal with electrolytic reactions which involve imidic acid derivatives either as starting material or product.

Organic chemists have generally been reluctant to consider using electro-chemical reactions, mainly because the apparatus was unfamiliar. There exist now books written for organic chemists which treat both the theo-retical and the practical problems of organic electrochemistry (e.g. Ref. 1); the general aspects will, therefore, not be treated here.

The electrochemistry of imidic esters and amidines resembles in many respects that of the corresponding oxygen analogues, but generally the

241

carbon–nitrogen double bond[2] is more easily reduced than the carbon–oxygen double bond; the easier reduction of the nitrogen analogues is, among other things, connected with the easier protonation of these compounds.

II. ELECTROCHEMICAL PREPARATION OF IMIDIC ACID DERIVATIVES

A. Imidic Esters

One of the very few reported electrolytic preparations of an imidic ester is the reductive ring opening of some alkoxy-substituted phthalazines and dihydrophthalazines[3]. The reduction of 4-methoxy-1-phenylphtalazine (1) proceeds according to

(1)   (2)   (3)

The electrolytic reaction is performed in N-hydrochloric acid containing 30% alcohol at 0°C at a potential controlled at -0.82 V (SCE); the control of the cathode potential is essential for the reduction as 2 is further reducible at a potential 200–300 mV more negative than that of 1. The intermediate 2 has not been isolated in this case; extraction at a higher pH is not feasible since the amino group attacks the azomethine group with ring closure and loss of ammonia, and hydrolysis of the imidic ester to the carboxylic ester makes isolation by evaporation of solvent difficult.

1,2-Dihydro-4-methoxy-1-phenylphthalazine (4) is also reduced with ring opening to 2: the reduction of 1 seems, however, at low pH to pass through the imine (5) rather than through 4.

The reduction of 1 and 4 thus follows the apparently rather general rule[4] that compounds of the type $\diagdown C{=}N{-}Y$, where Y is a heteroatom, are reduced in acid solution to $\diagdown C{=}N^{+}H_2$ and HY. This reduction and its mechanism have previously been discussed[2,3] and the critical step seems to be the protonation of the radical formed after the uptake of one electron and one proton.

It would thus be expected that the reduction at controlled cathode potential of compounds of the general formula $RC(OR'){=}N{-}Y$, where Y is nitrogen or oxygen, would produce the corresponding imidic ester.

$$
\mathbf{1} \xrightarrow[\text{pH} < 1]{2e^- + 4H^+}
$$

C₆H₅ structure (5):

$$
\begin{array}{c}
\text{C}_6\text{H}_5 \\
\diagdown \\
\text{C}=\overset{+}{\text{N}}\text{H}_2 \\
\\
\text{C}=\overset{+}{\text{N}}\text{H}_2 \\
| \\
\text{OCH}_3
\end{array}
$$

(5)

$$
\xrightarrow{2e^- + 2H^+} \mathbf{2} \xleftarrow{2e^- + 4H^+} \mathbf{4}
$$

$$
2e^- + 2H^+ \downarrow \quad \text{pH} > 5
$$

Structure (4):

$$
\begin{array}{c}
\text{C}_6\text{H}_5 \quad \text{H} \\
\diagup \quad \diagdown \\
\text{NH} \\
| \\
\text{N} \\
\parallel \\
\text{C} \\
| \\
\text{OCH}_3
\end{array}
$$

(4)

B. Amidines

I. By oxidation

Anodic oxidation of ethyl alcohol in an aqueous solution of ammonium carbonate at a platinum anode yields acetamidine (**6**), isolated as the nitrate[5]. During the oxidation some ammonia is oxidized to nitrate, thus providing the anion for the isolated product. The reaction probably

$$
\text{CH}_3\text{CH}_2\text{OH} \xrightarrow{[-2e^- - 2H^+]} \text{CH}_3\text{CHO} \xrightarrow{+NH_3}
$$

$$
\text{CH}_3\text{CH(OH)NH}_2 \xrightarrow[+NH_3]{[-2e^- - 2H^+]} \text{CH}_3\text{C}\begin{array}{c}\diagup \text{NH} \\ \diagdown \text{NH}_2\end{array}
$$

$$
\mathbf{(7)} \qquad\qquad\qquad\qquad \mathbf{(6)}
$$

passes through acetaldehyde and 'aldehydammonia' (**7**); this has been made plausible by the finding that **7** gives a better yield of **6** than ethyl alcohol by anodic oxidation. The mechanism of the further oxidation is less clear; Fichter and co-workers[5] suggest that acetamide, which then should add ammonia, is an intermediate but no evidence to substantiate it has been presented. No control of potential has been made during the oxidation and several reactions seem to be occurring simultaneously, so it is difficult to suggest a mechanism for the reaction.

Electrolytic oxidation at a platinum anode of thiourea in sulphuric acid gives the sulphate of formamidine disulphide in a good yield[6]. Alkylated derivatives of thiourea behave similarly[7]. The reaction may be formulated as a loss of one electron from the sulphur atom.

$$2\ (H_2N)_2CS \xrightarrow{-2e^-} \underset{H_2N}{\overset{\overset{+}{H_2N}}{\diagdown}}C-S-S-C\underset{NH_2}{\overset{\overset{+}{NH_2}}{\diagup}}$$

2. By reduction

Reduction of compounds of the general formula $RC(NR'_2){=}N{-}Y$, where Y is a heteroatom would be expected to produce amidines. 4-Dimethylamino-1-phenylphthalazine (8) and 1,2-dihydro-4-dimethyl-amino-1-phenylphthalazine (9) are both reduced in acid solution to the amidine (10)[3], which may lose dimethylamine and form the cyclic amidine (11). The amidine 10 is more stable than the imidic ester 2 and can be isolated as the dihydrochloride; at higher pH 10 forms 11

Another example of the cleavage of $RC(NH_2){=}NY$ is the reduction of amidoximes to amidines[4]. The reduction of benzamidoxime (12) to benzamidine (13) was the first example where a product of the type $RR'C{=}N^+H_2$ was isolated from the reduction of $RR'C{=}NOH$; later other examples have been found[2].

$$C_6H_5C\overset{NOH}{\underset{NH_2}{\diagdown}} \xrightarrow{2e^- + 3H^+} C_6H_5C\overset{\overset{+}{NH_2}}{\underset{NH_2}{\diagdown}}$$

$$(12) \qquad\qquad\qquad (13)$$

C. Amidrazones

Amidrazones $(RC(NR'_2)=NNHR'')$ may be synthesized electrolytically by partial reduction of oximehydrazones and possibly from similar compounds such as hydrazidines; cyclic amidrazones may be formed by reductive ring closure or by partial reduction of certain heterocyclic compounds. In the first case, the reaction involves a cleavage of a simple bond between nitrogen and another heteroatom in a compound of the general type $RC(NHNR'_2)=N-Y$.

Whereas aliphatic oximehydrazines, such as $CH_3C(NHNHC_6H_5)=NOH$ are generally not polarographically reducible[8], the aromatic analogues

(14)   (15)

are[9, 10]. The oximehydrazine (14) has been reduced to the amidrazone (15) in an acetate buffer at a potential where the nitro group is not reduced.

Further reduction of 15 at a more negative potential where the nitro group is reduced to an amino group gives compound (16) which cyclizes to (17); 17 in turn loses ammonia to the cyclic amidrazone (18).

(16)

(17)   (18)

It has not been proved conclusively whether 18 is a 1,4- or a 1,2-dihydrobenzotriazine. 18 may also be obtained by two-electron reduction of 3-phenylbenzo-1,2,4-triazine $(19)^{11}$; 19 and 18 form a nearly reversible system at the dropping mercury electrode.

D. Hydrazidines

Triphenyltetrazolium chloride (20) has been investigated polarographically[12-16]; in acid solution a six-electron reduction is observed at

$t > 40°C$, whereas a four-electron wave is found at $t < 20°C$. This may be explained by the following reaction scheme[10]:

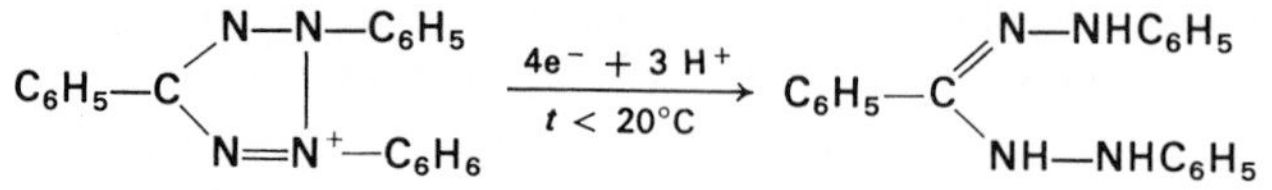

$$C_6H_5-C \underset{N=N^+-C_6H_6}{\overset{N-N-C_6H_5}{\diagdown}} \xrightarrow[t < 20°C]{4e^- + 3 H^+} C_6H_5-C \underset{NH-NHC_6H_5}{\overset{N-NHC_6H_5}{\diagdown}}$$

$$\text{(20)} \hspace{6cm} \text{(21)}$$

At higher temperatures, the triphenylbenzhydrazidine (**21**) disproportionates to phenylbenzamidrazone, aniline, and the easily reducible triphenylformazane (**22**); **22** is then reduced to **21** which disproportionates etc.

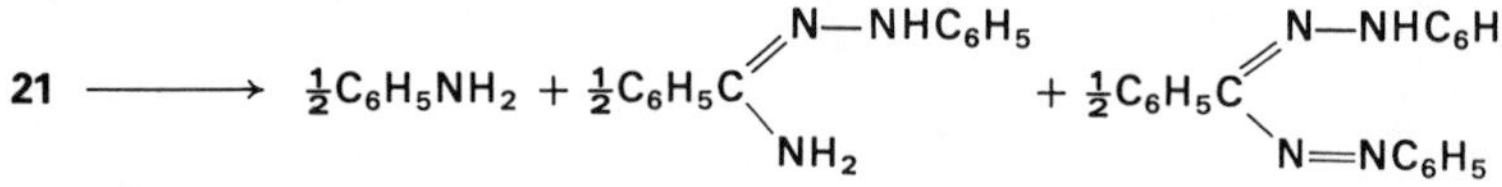

$$\textbf{21} \longrightarrow \tfrac{1}{2}C_6H_5NH_2 + \tfrac{1}{2}C_6H_5C \underset{NH_2}{\overset{N-NHC_6H_5}{\diagdown}} + \tfrac{1}{2}C_6H_5C \underset{N=NC_6H_5}{\overset{N-NHC_6H_5}{\diagdown}}$$

$$\text{(22)}$$

III. ELECTROCHEMICAL REACTIONS OF IMIDIC ACID DERIVATIVES

A. Imidic Esters

Aromatic imidic esters (**23**) have been reduced[17] in 2 N-sulphuric acid solution to amines in good yields (Table 1) at lead cathodes at 0°C; the yield of amines in the reduction of aliphatic imidic esters is lower. The reduction probably follows the scheme

$$\underset{\text{(23)}}{RC(OR')=\overset{+}{N}H_2} \xrightarrow{2e^- + H^+} \underset{\text{(24)}}{RCH(OR')NH_2} \xrightarrow[-R'OH]{H^+} \underset{\text{(25)}}{RCH=\overset{+}{N}H_2}$$

$$\downarrow H_2O \hspace{4cm} 2e^- \downarrow +2 H^+$$

$$RCH_2OH \xleftarrow{2e^- + 2 H^+} RCHO + NH_3 + R'OH \hspace{1cm} RCH_2\overset{+}{N}H_3$$

In acid solution the protonation facilitates the loss of the alcohol from (**24**) to give (**25**) whereas hydrolysis and loss of ammonia is favoured at higher pH. The fate of an intermediate similar to **24** has been discussed in connection with the reduction of oxaziridines[18].

Since imidic esters are more easily reduced than the corresponding nitriles, it has been suggested[17, 19] that nitriles may be reduced electrolytically to amines in ethanolic sulphuric acid, where the first reaction would be an acid catalysed addition of alcohol to the nitrile with formation of a reducible imidic ester.

TABLE 1. Yields of amine in the electrolytic reduction of imidic esters RC(OR′)=NH in 2 N-sulphuric acid at a lead cathode[23]

Imidic ester	Amine	Yield (%)
Benzimidic ester	Benzylamine	76
3-Methylbenzimidic ester	3-Methylbenzylamine	70
4-Methylbenzimidic ester	4-Methylbenzylamine	94
4-Methoxybenzimidic ester	4-Methoxybenzylamine	66
Acetimidic ester	Ethylamine	16
Phenylacetimidic ester	2-Phenylethylamine	14

Aromatic imidic esters are polarographically reducible[20]; in Figure 1 the pH-dependence of benzimidic ester is shown. The reduction of an *o*-substituted benzimidic ester **2** has been studied[3] and the following reduction route proposed:

(26)

(27)   (28)   (29)

The shape of the polarograms of **2** suggested this reaction route rather than one with an initial ring closure to **3**, followed by reduction of the carbon–nitrogen double bond, elimination of alcohol to **(28)** and reduction to **(29)**

B. Amidines

Electrolytic oxidation of acetamidine in liquid ammonia produces among other compounds ethane and cyanamide[21]. The reaction which in a way resembles the Kolbe electrochemical oxidation of acetate to ethane is not simple, and a mechanistic interpretation becomes more difficult to suggest by the finding that the homologues of acetamidine $(R—C(NH_2)=NH)$ on a similar treatment produce methane and ethane

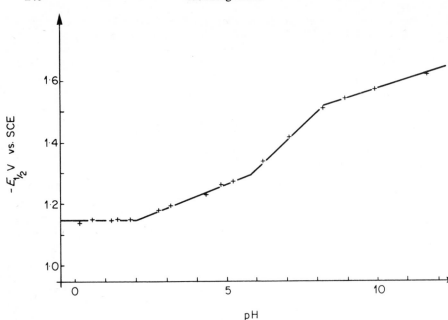

FIGURE 1. Half-wave potentials (vs. SCE) of benzimidic ester at different pH
values in aqueous buffers containing 40% alcohol[20].

and very little, if any, of the 'Kolbe product', the hydrocarbon R—R.

Aliphatic amidines are generally not polarographically reducible in
buffered solutions; aromatic amidines[22] are reducible only at rather
negative potentials at a pH interval from slightly acid to alkaline solution.
The electrode reaction is a four-electron reduction to an amine[22]

$$\overset{+}{RC(NH_2)}{=}\overset{+}{NH_2} \xrightarrow{\ 4e^- + 5H^+\ } RCH_2\overset{+}{NH_3} + \overset{+}{NH_4}$$

Cyclic amidines, such as **11**[3], 3,4-dihydroquinazoline[23] (**30**), and 1,6-
dihydropurine[24-26] (**31**) are reduced in a manner similar to the acyclic
amidines. In Table 2 are given the half-wave potentials of **11** at different
pH.

(30)   **(31)**

Certain aliphatic amidines, such as 2-phenoxyacetamidine (**32**), which
are substituted in the α-position with a heteroatom, are polarographically

TABLE 2. Half-wave potentials (vs. SCE) at different pH of the cyclic amidine 3-iminoisoindoline[3] (11) in aqueous buffers containing 40% ethyl alcohol

pH	4·5–8·5	9·0	10·0	11·0	12·3
$-E_{\frac{1}{2}}$ V(SCE)	1·53	1·57	1·63	1·72	1·80

reducible but the electrode reaction consists of a reductive cleavage of the carbon–heteroatom single bond; 32 thus forms acetamidine and phenol[22].

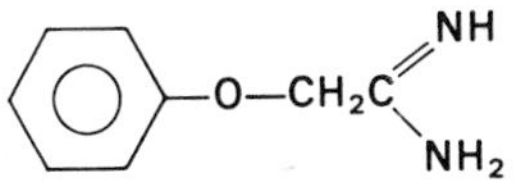

(32)

C. Amidrazones, Hydrazidines

Phenylbenzamidrazone is reducible[10] in approximately the same pH region as benzamidine. A cyclic benzamidrazone, dihydro-3-phenyl-benzo-1,2,4,-triazine (18), is reduced in the following way to phenyl-benzimidazole[11] (33):

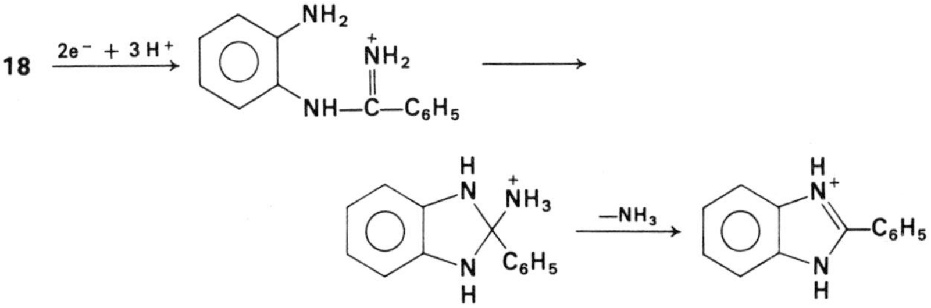

(33)

Hydrazidines have not been investigated much electrochemically; diphenylbenzhydrazidine (34) is polarographically reducible at most pH. In strongly acid solution it may disproportionate and in alkaline solution it is easily oxidized. Anodic oxidation in aqueous-alcoholic solution of 34 would produce the slightly soluble triphenylformazane (35) which in acetonitrile can be oxidized further[10] to the triphenyltetrazolium ion 20:

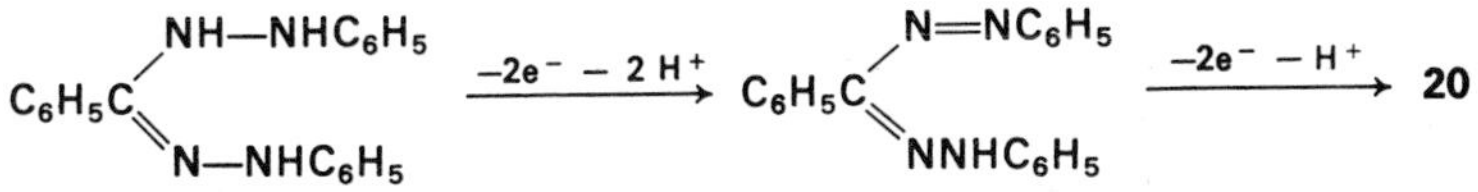

(34)   (35)

D. Amidoximes, Hydroxyamidoximes, Hydrazooximes

The reduction of benzamidoxime[4] in acid solution to benzamidine was discussed in Section II.B.2. Hydroxyamidoximes[27-29] (36) may in neutral and alkaline solution be oxidized to nitrosolates (37) with which they form reversible systems.

$$\begin{array}{ccc} \text{RC—NHOH} & \underset{\longleftarrow}{\overset{-2e^- \; - \; 2\,H^+}{\rightleftarrows}} & \text{RC—NO} \\ \| & & \| \\ \text{NOH} & & \text{NOH} \\ (36) & & (37) \end{array}$$

Hydroxyamidoximes are polarographically reducible in acid solution, and the corresponding amidoximes have been suggested as the products[27-29].

Oximehydrazides, such as (38), are polarographically oxidizable to the oximazo compounds (39) which in turn may be reduced to 38.

$$\begin{array}{ccc} \text{CH}_3\text{C—NHNHC}_6\text{H}_5 & \overset{-2e^- \; - \; 2\,H^+}{\rightleftarrows} & \text{CH}_3\text{C—N=NC}_6\text{H}_5 \\ \| & & \| \\ \text{NOH} & & \text{NOH} \\ (38) & & (39) \end{array}$$

E. Derivatives of Imidic Acid Halides

Imidic acid halides (40) have been investigated very little electrochemically; the simple halides are rather unstable but N,N-disubstituted imidic acid halides are reasonably stable in the absence of nucleophiles. They are, just as carboxylic acid chlorides, very easily reducible; the reduction must take place in an aprotic solvent, such as acetonitrile or N,N-dimethylformamide. The first step would be expected to be an uptake of an electron followed by loss of a halide ion; the radical thus formed may dimerize to an α-diimine (41) or accept an electron further with formation of an aldimine (42), the reaction route depending on many parameters; furthermore, at a more negative potential both 41 and 42 may be reduced further to compounds which may react with 40, 41, or 42.

$$\begin{array}{ccc} \text{R—C—X} & \text{R—C=NR'} & \\ \| & | & \text{R—CH=NR'} \\ \text{N—R'} & \text{R—C=NR'} & \\ (40) & (41) & (42) \end{array}$$

Hydroxamic acid halides have been investigated polarographically[30,31] in aqueous acidic medium, where they are reasonably stable. In Table 3 are given the half-wave potentials of benzhydroxamic acid halides at pH 1. As might be expected from the reduction of other halides, the iodide

TABLE 3. Half-wave potentials (vs. SCE) at pH 1 of some hydroxamic acid halides[30]

Compound	$C_6H_5C(NOH)I$	$C_6H_5C(=NOH)Br$	$C_6H_5C(=NOH)Cl$	$CH_3C(=NOH)I$
$E_{\frac{1}{2}}$V vs. SCE	−0·55	−0·84	−0·86	−0·77

is more easily reduced than the bromide which in turn is more easily reduced than the chloride.

Benzhydroxamic iodide (43) is reduced[30] to benzaldoxime (44) in a two-electron reduction

$$C_6H_5\underset{\underset{NOH}{\|}}{C}-I \xrightarrow{2e^- + H^+} C_6H_5CH{=}NOH + I^-$$

$$\textbf{(43)} \hspace{4cm} \textbf{(44)}$$

44 is more easily reduced than benzhydroxamic bromide (45) or chloride (46), so during the reduction of 45 or 46 the primarily obtained 44 is reduced further in a four-electron reduction as fast as it is formed. The reduction of 45 thus is

$$C_6H_5\underset{\underset{NOH}{\|}}{C}-Br \xrightarrow[-Br^-]{2e^- + H^+} [\textbf{44}] \xrightarrow{4e^- + 5H^+} C_6H_5CH_2\overset{+}{N}H_3 + H_2O$$

$$\textbf{(45)}$$

IV. REFERENCES

1. *Organic Electrochemistry. An Introduction and a Guide* (Ed. M. M. Baizer), Marcel Dekker Inc., New York, 1973.
2. H. Lund, *The Electrochemistry of the Carbon–Nitrogen Double Bond* in *The Chemistry of the Carbon–Nitrogen Double Bond* (Ed. S. Patai), Wiley-Interscience, 1970, p. 505.
3. H. Lund and E. T. Jensen, *Acta Chem. Scand.*, **25**, 2727 (1971).
4. H. Lund, *Acta Chem. Scand.*, **13**, 249 (1959).
5. F. Fichter, C. Stutz, and E. Grieshaber, *Z.Elektrochem.*, **18**, 647 (1912).
6. F. Fichter and W. Wenk, *Ber.*, **45**, 1373 (1912).
7. F. Fichter and F. Braun, *Ber.*, **47**, 1526 (1914).
8. J. Armand, B. Furth, J. Kossayi, and J.-P. Morizur, *Bull. Soc. Chim. France*, 2499 (1968).
9. S. Kwee and H. Lund, *Acta Chem. Scand.*, **23**, 2711 (1969).
10. H. Lund, *Studier over Elektrodereaktioner i Organisk Polarografi og Voltammetri*, Aarhus Stiftsbogtrykkerie, Aarhus 1961.
11. H. Lund, *Disc. Faraday Soc.*, **45**, 193 (1968).
12. B. Jambor, *Acta Chim. Acad. Sci. Hung.*, **4**, 55 (1954).
13. B. Jambor, *J. Chem. Soc.*, 1604 (1958).
14. H. Campbell and P. O. Kane, *J. Chem. Soc.*, 3130 (1956).
15. P. Kivalo and K. K. Mustakallio, *Suomen Kemistilehti* **B29**, 154 (1956).
16. P. Kivalo and K. K. Mustakallio, *Suomen Kemistilehti* **B30**, 214 (1957).
17. H. Wenker, *J. Am. Chem. Soc.*, **57**, 772 (1935).
18. H. Lund, *Acta Chem. Scand.*, **23**, 563 (1969).
19. L. Eberson, Ref. 1, p. 424.
20. H. Lund, to be published.
21. R. A. Fulton and F. W. Bergstroem, *J. Am. Chem. Soc.*, **56**, 167 (1934).

22. P. O. Kane, *Z. Anal. Chem.*, **173,** 50 (1960).
23. H. Lund, *Acta Chem. Scand.*, **18,** 1984 (1964).
24. D. L. Smith and P. J. Elving, *J. Am. Chem. Soc.*, **84,** 1412 (1962).
25. B. Janik and P. J. Elving, *Chem. Rev.*, **68,** 295 (1968).
26. S. Kwee and H. Lund, *Experientia Suppl.*, **18,** 387 (1971).
27. J. Armand, *C. R. Acad. Sci Paris*, **258,** 207 (1964).
28. J. Armand and R.-M. Minvielle, *C. R. Acad. Sci. Paris*, **260,** 2512 (1965).
29. J. Armand, *Bull. Soc. Chim. France*, 1658 (1966).
30. J. Armand, *Bull. Soc. Chim. France*, 882 (1966).
31. J. Armand, P. Souchay, and F. Valentini, *Bull. Soc. Chim. France*, 4585 (1968).

CHAPTER **6**

Biological reactions and pharmaceutical uses of imidic acid derivatives

RAYMOND J. GROUT

Department of Pharmacy, The University, Nottingham, England

ABBREVIATIONS USED

ATP	adenosine triphosphate	P	phosphate (in formulae)
DNA	deoxyribosenucleic acid	PRPP	5-phosphoribose-1-
FIGLU	formiminoglutamate		pyrophosphate
NADP	nicotinamide-adenine-	RNA	ribosenucleic acid
	dinucleotide phosphate	THF	5,6,7,8-tetrahydrofolate

I. INTRODUCTION

It is only relatively recently that imidic acid derivatives have been found in nature. Hydroxamic acids are probably the most common naturally occurring imidic acid derivatives, but since their structure and biochemical functions have been fully reviewed by Emery[1], and their pharmacological aspects have also been reported[2], this group is not discussed in this chapter. Amidines are synthesized by a few micro-organisms; formamidines feature in the biochemical pathways associated with the biosynthesis of imidazoles and purines and in the catabolism of histidine.

The search for chemotherapeutic and pharmacodynamic agents has necessitated the preparation of a large number of synthetic compounds. Many imidic acid derivatives have been investigated as medicinal agents, but relatively few are clinically acceptable. A number of groups of imidic acid derivatives, e.g. the amidrazones, do not occur naturally and have not achieved prominence in the field of medicinal chemistry.

The hetero-atoms of naturally occurring ring systems such as pyrimidine and purines are so arranged that the systems may be regarded as amidines; a discussion of the biological and medicinal chemistry of these and related heterocyclic rings is outside the scope of this review.

An attempt has been made to give an account of the chemical aspects of synthetic and naturally occurring imidic acid derivatives in selected areas; amidines feature very prominently.

II. BIOLOGICAL FORMATION OF THE AMIDINE GROUP

A. Purine Biosynthesis

Amidine formation occurs during the biosynthesis of the purine ring

(1)   (2)

system (1). 5-Phosphoribosylformylglycinamide (2) forms the ring skeleton of the five-membered ring of purine (atoms 9, 4, 5, 7 and 8). The nitrogen atom at 3 is incorporated into the skeleton by amidine formation, the amide group of glutamine being the source of ammonia. This biosynthetic step is reviewed[3] in detail in an earlier volume of this series.

Buchanan and his co-workers[6,7] have clearly established the important steps in this stage. The enzyme, 5-phosphoribosylformylglycinamide:L-glutamate amidoligase requires potassium and magnesium ions and ATP as cofactors. It was suggested that glutamine undergoes reaction with an —SH group in the enzyme, the intermediate behaving as active am-

$$-SH + H_2NCOCH_2CH_2\underset{\underset{^+NH_3}{|}}{CH}CO_2^- \longrightarrow -SCOCH_2CH_2\underset{\underset{^+NH_3}{|}}{CH}CO_2^- + [\dot{N}H_3] \qquad (1)$$

monia (equation 1). Evidence[4,5] in support of this suggestion is provided when the substrate glutamine is replaced by the isosteric azaserine (3) or by 6-diazo-5-oxo-L-norleucine (DON) when S-alkylation of the enzyme occurs (equation 2).

$$-SH + N_2CHCOOCH_2\underset{\underset{^+NH_3}{|}}{CH}CO_2^- \longrightarrow -S-CH_2COOCH_2\underset{\underset{^+NH_3}{|}}{CH}CO_2^- + N_2 \qquad (2)$$

(3)

Ammonia and the glycinamideribotide now react in the presence of ATP (equation 3)[6,7]. Although not suggested by the original workers, presumably this reaction proceeds *via* an imido-phosphate which is susceptible to nucleophilic attack by ammonia (equation 4). In related

$$\underset{NH\text{-ribose-5-phosphate}}{\overset{O}{\overset{\|}{-O_2CCHCH_2CH_2CNH_2}}} \xrightarrow[\text{[NH}_3\text{]}]{ATP, Mg^{2+}, K^+} \underset{NH\text{-ribose-5-phosphate}}{\overset{NH}{\overset{\|}{-O_2CCHCH_2CH_2CNH_2}}} \qquad (3)$$

$$\overset{O}{\overset{\|}{-CNH_2}} \rightleftarrows \overset{OH}{\overset{|}{-C=NH}} \xrightarrow{ATP} \overset{OPO_3H_2}{\overset{|}{-C=NH}} \xrightarrow{\text{[NH}_3\text{]}} \overset{NH_2}{\overset{|}{-C=NH}} + H_2PO_4^-$$
$$(4)$$

reactions it has been shown[8] that with glutamine having ^{18}O-labelled carboxamide the ^{18}O is incorporated into formed phosphate. Mechanistically this reaction then has a close similarity with a reaction, investigated by Oxley, Peak and Short[9], which leads to the formation of amidines from amides. N-Substituted amides, treated with an arenesulphonyl chloride readily form imido-arenesulphonates which with amines form N,N'-disubstituted amidines (equation 5)

$$\overset{O}{\overset{\|}{R^1CNHR^2}} \rightleftarrows \overset{OH}{\overset{|}{R^1C=NR^2}} \xrightarrow[\text{Pyridine}]{ArSO_2Cl} \overset{OSO_2Ar}{\overset{|}{R^1C=NR^2}} \xrightarrow{R^3NH_2}$$

$$\overset{NHR^3}{\overset{|}{R^1C=NR^2}}, ArSO_3H \qquad (5)$$

B. *Histidine Biosynthesis*

Tracer experiments in bacteria have established that the $N_{(1)}$ and $C_{(2)}$ atoms of histidine (4) are derived from the $C_{(2)}$ and $N_{(1)}$ of a purine[10-13] and the $N_{(3)}$ of histidine is derived from the amide group of glutamine[14].

The multistage histidine biosynthesis[15] leading to ring-opening between $C_{(6)}$ and $N_{(1)}$ of the purine nucleus commences with alkylation at the $N_{(1)}$ of ATP by ribose-5-phosphate-1-pyrophosphate to form 5. The enzyme

(4)   (5)

involved ATP–PRPP phosphoribosyl transferase, has been isolated in a substantially pure form from *Salmonella typhimurium*. The molecular weight, determined by equilibrium ultracentrifugation is 215,000, and the enzyme consists of six identical polypeptide units[16]. The alkylation is reversible and is the controlling step in histidine biosynthesis[17].

Following hydrolytic cleavage of the triphosphate to monophosphate (enzyme: phosphoribosyladenosinetriphosphate-pyrophosphohydrolase) the pyrimidine ring of the purine is ring opened (between $N_{(1)}$ and $C_{(6)}$) under the influence of phosphoribosyladenosine monophosphate-1,6-

(6)

cyclohydrolase to form the formamidine (6). Formation of the imidazole ring follows an Amadori rearrangement of 6 to the isomeric formamidine (7) under the influence of the appropriate isomerase[18], cleavage of 7 by ammonia derived from glutamine[18], apparently in two steps, and ring closure to imidazoleglycerol phosphate (8). The later steps in this enzymically controlled sequence are dehydration of 8 to imidazoleacetol

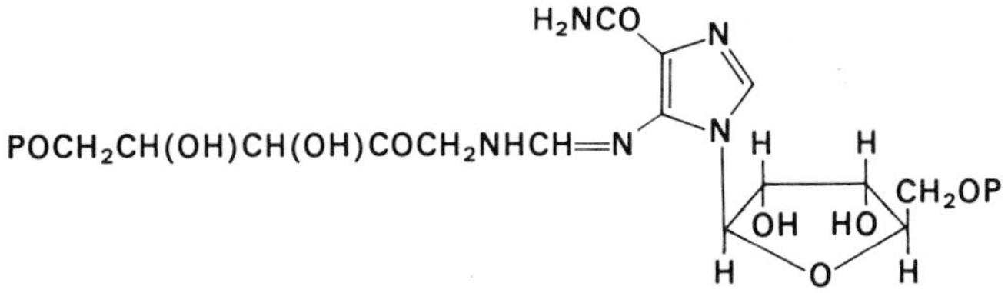

$$H_2NCO \cdots N$$

$$POCH_2CH(OH)CH(OH)COCH_2NHCH=N$$

(7)

$$CH(OH)CH(OH)CH_2OP$$

$$N \quad NH$$

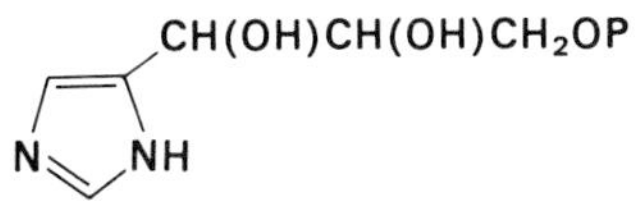

(8)

phosphate, transamination with L-glutamate to L-histidinol phosphate, hydrolysis of the ester and finally oxidation of the L-histidinol to L-histidine (**4**).

Histidine is an essential nutrient for growth of young rats and mice, but in adult man nitrogen balance can be maintained in the absence of histidine.

C. Histidine Catabolism

The main catabolic pathway for histidine in bacteria and mammals results in a one carbon unit being returned to the 'one carbon pool' *via* a

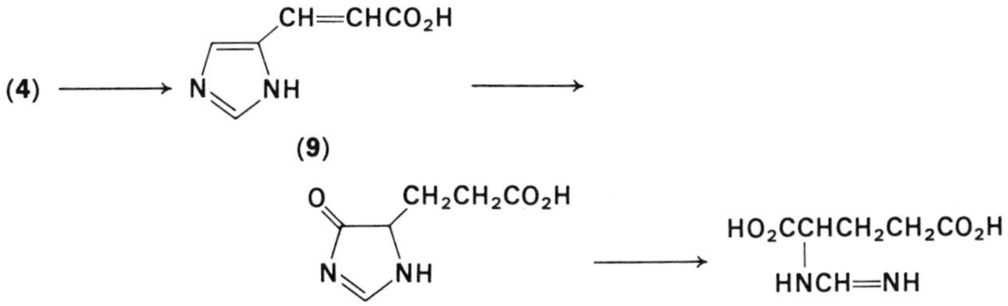

$$CH=CHCO_2H$$

$$(4) \longrightarrow N \quad NH \qquad \longrightarrow$$

(9)

$$O \quad CH_2CH_2CO_2H$$

$$N \quad NH$$

(11)

$$\longrightarrow$$

$$HO_2CCHCH_2CH_2CO_2H$$
$$|$$
$$HNCH=NH$$

(10)

formamidine intermediate. Histidine (**4**), by elimination of ammonia (L-histidine ammonia lyase) is converted into urocanic acid (**9**)[19, 20] which with crude liver extracts and with bacterial enzyme preparations is transformed into *N*-formimino-L-glutamate (FIGLU) (**10**)[21, 22]. The crude liver preparation has been shown to contain two enzyme systems which can be separated[22]. The first enzyme, urocanase, catalyses the addition of water to (**9**) to form 4(5)-imidazolone-5(4)-propionic acid (**11**); the latter compound has not been fully characterized, however. Evidence for the structure is based on the similarity of the spectral characteristics of the isolated and authentic synthetic samples of the acid[23, 24]. The second enzyme, imidazolonepropionic acid hydrolase, has been obtained in a partially purified form from rat liver[25] and *Pseudomonas fluorescens*[26] extracts and hydrolytically ring opens the cyclic acylformamidine (11) and gives optically active FIGLU (10).

The formimino group is transferred from 10 to 5,6,7,8-tetrahydrofolate under the influence of formiminoglutamate formiminotransferase to form the open-chain amidine (12) and glutamate[27–29].

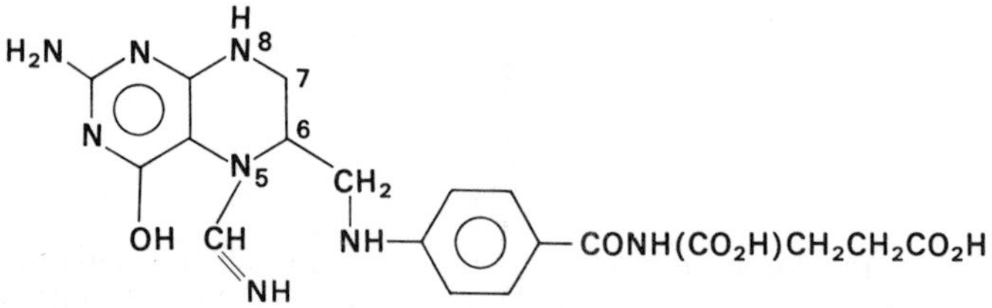

(12)

Formiminoglutamate is excreted in individuals with a folic acid-deficient diet; it manifests as megaloblastic anaemias. For clinical evaluation FIGLU can be estimated microbiologically[30], enzymically[31,32], by electrophoresis[33,34] and by paper chromatography[35,36], but these methods are of low sensitivity. Tests with increased sensitivity include the separation of FIGLU and its precursor, urocanic acid (9), by t.l.c. on cellulose with *n*-butanol–glacial acetic acid–water (114:38:60) as developing solvent. FIGLU is detected (limit 0·125 µg) by its ammonolysis to glutamate

$$\text{FIGLU (10)} \xrightarrow{\ \text{NH}_3\ } {}^-O_2CCH(NH_3{}^+)CH_2CH_2CO_2H + HC\,(=NH)NH_2 \qquad (6)$$

(equation 6) and its location with ninhydrin reagent; urocanic acid (9) (limit 0·0625 µg) which is faster running is detected by Pauly's reagent[37].

A second method[38] depends on the estimation of a colour produced by interaction of urine with nitroprusside and ferricyanide.

D. Folic Acid Derivatives

Brief mention has already been made to 5-formimino-5,6,7,8,-tetra-hydrofolate (12). This formamidine undergoes ring-closure with loss of

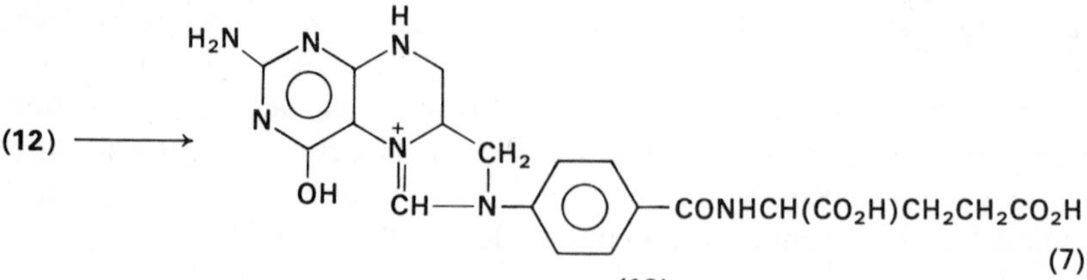

(12) ⟶

CONHCH(CO_2H)CH_2CH_2CO_2H   (7)

(13)

ammonia to give the cyclic formamidinium derivative, 5,10-methenyl-5,6,7,8-tetrahydrofolate [5,10-methenyl-THF (13)] (equation 7). The enzyme facilitating this ring-closure, 5-formiminotetrahydrofolate ammonia lyase (cyclizing) has been isolated from a number of sources including *Clostridium cylindrosporum*[39,40], rabbit and hog[41] liver. The cyclization is readily followed since 12 has an absorption maximum at 285 nm, associated with the formamidine structure, which is lost as 13 is formed; 13 has an absorption maximum at 356 nm[42,43]. The same methenyltetrahydrofolate is formed in a reversible oxidation (NADP) of 5,10-methylenetetrahydrofolate (14), the latter being formed in a two-step, enzymically controlled, sequence from tetrahydrofolate and serine followed by dehydration.

$$\text{CONHCH(CO}_2\text{H)CH}_2\text{CH}_2\text{CO}_2\text{H} \underset{\text{NADP}^+}{\overset{}{\rightleftharpoons}} (13)$$

(14)

The reductase responsible for the conversion 14 → 13 has been isolated from many sources including yeast[44], calf thymus[45] and *Clostridium cylindrosporum*[46].

The methenyl-THF (13) plays an important role in one carbon transfers in purine biosynthesis[47]. It is required for the insertion of $C_{(8)}$ and indirectly for the insertion of $C_{(2)}$ into inosinic acid [see purine numbering (1)].

13 is ring opened by water to form 10-formyl-THF (equation 8) but the reaction is suppressed in maleate buffer. Using an avian liver enzyme preparation, Hartmann and Buchanan[48] were able to establish that 13

$$\text{5,10-methenyl-THF}^+ + \text{H}_2\text{O} \rightleftharpoons \text{10-formyl-THF} + \text{H}^+ \qquad (8)$$
(13)

was the active formyl donor in the enzyme controlled formylation of 5-phosphoribosylglycinamide to provide ultimately the $C_{(8)}$ of the purine (equation 9); 10-formyl-THF was inactive. For the formylation of 5′-phosphorisbosyl-5-aminoimidazole-4-carboxamide, 10-formyl-THF was the active donor, 5,10-methenyl-THF (13) being inactive.

A large number of chemotherapeutic agents indirectly prevent the formation of 13 by blocking either the biosynthesis of 7,8-dihydrofolate

$$\text{H}_2\text{NCH}_2\text{CONH-ribose-5-phosphate} \xrightarrow{\text{(13)}}$$

$$\text{O}{=}\text{CHNHCH}_2\text{CONH-ribose-5-phosphate} \qquad (9)$$

(e.g. the sulphonamides), or, by competitive inhibition of dihydrofolate reductase, prevent the reduction of the dihydrofolate to 5,6,7,8-tetra-hydrofolate (e.g. trimethoprim, pyrimethamine and amethopterin)[49].

III. BIOLOGICAL ACTIVITY OF IMIDATE DERIVATIVES

A. Antiviral Activity of Naturally-occurring Amidines

Unsubstituted amidines have been isolated from a number of micro-organisms; they possess antiviral properties and are thus of considerable chemotherapeutic interest.

Amidinomycin (15), probably the simplest amidine isolated[50], was obtained from a species related to *Streptomyces flavochromogenes.*

$$H_2N$$

$$CONHCH_2CH_2C(=NH)NH_2$$

(15)

The β-amidinoethylamine substituent is retained in noformicin (16), isolated from *Nocardia formica*[51,52]. This microbial metabolite is active against a broad spectrum of viruses, including parainfluenza, type 3, and the causative viruses of a number of animal infections[53]. A number of synthetic analogues of 16 [(17), where $n = 0$ or 1 and $m = 0–3$] have been

$$H_2N \quad N \quad CONHCH_2CH_2C(=NH)NH_2$$

(16)

$$(CH_2)_n$$

$$H_2N \quad N \quad CONH[CH_2]_mC(=NH)NH_2$$

(17)

prepared[53], and although some antiviral activity has been found, all are significantly less active than noformicin. It has been speculated[53] that the noformicin molecule fulfils a very rigid structural requirement for activity.

Netropsin (18), isolated from *Streptomyces netropsis* and other *Streptomyces* sp., is one of a number of naturally occurring compounds which are polypyrrole derivatives[54]. 18 increases the survival time of mice infected with influenza A and B, swine 'flu' and mouse-adapted neutrotropic vaccinia viruses.

The antibiotic distamycin A [(19), $n = 3$] isolated[55] from *Streptomyces distallicus*, has received much attention as a viricide[55,56,57]; it apparently acts on the reverse transcription process. In most organisms DNA is transcribed into complementary RNA or self transcribed into complementary DNA. Some viruses, e.g. polio virus, can produce RNA to DNA reverse transcription and the enzyme RNA-dependent DNA polymerase

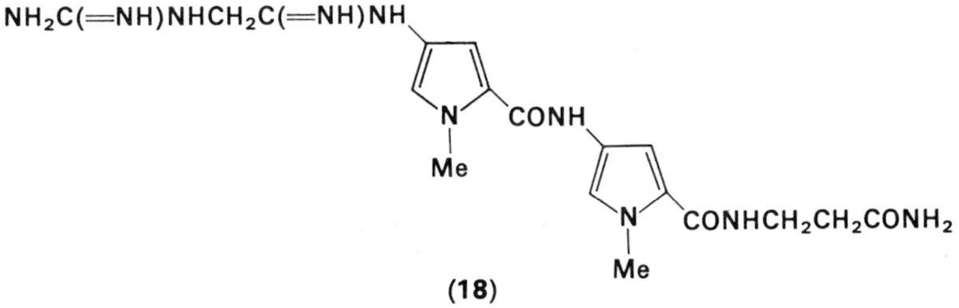

$NH_2C(=NH)NHCH_2C(=NH)NH$

(18)

(reverse transcriptase) obviously supplies a point where selective action against viruses can occur.

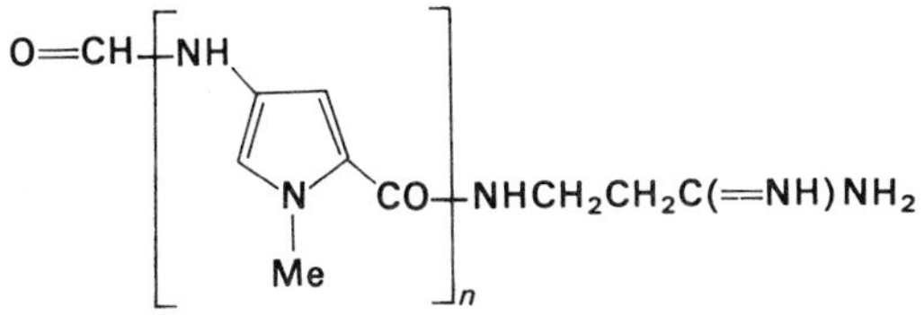

(19)

A number of polypyrrole analogues of distamycin have been examined for reverse transcription inhibitory properties[58]. The polypyrrole [(19, $n = 5$] is reported to be 16 times as active as distamycin A; it reduces by half leukaemia virus and Moloney sarcoma virus reverse transcription at a dose level of 20 μg per ml. It appears to affect attachment between DNA and DNA → RNA polymerase and stops initiation of new RNA. Chain elongation is not affected. Other structural variants of (19) involving different numbers of pyrrole rings and modification of the formamido group are less active compounds.

B. Antibacterial, Antifungal and Antiprotozoal Drugs

Amidines have been the subject of many studies in the search for antibacterial and antiprotozoal drugs in man and domestic animals. A recent report by Kreutzberger[59] documents many of the active compounds in this area.

Large numbers of active compounds possess two benzamidine residues which are separated by a structural unit, X, containing one or several

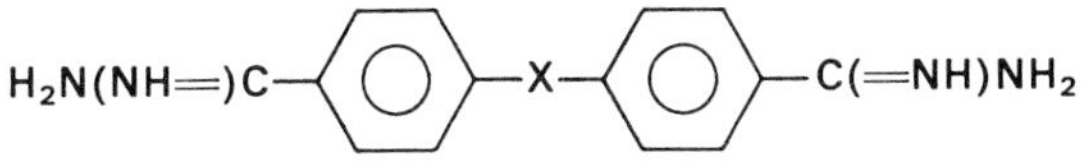

(20)

where X may be (a) —CH=CH— (stilbamidine); (b) —O[CH$_2$]$_n$O— (when $n = 3$ is propamidine, $n = 5$ is pentamidine); (c) —NH—N=N— (diminazene); (d) —S—S—; (e) —NHCO—; (f) —NH—

atoms (20). Some of the common variants are illustrated (a–f). A striking feature is that all of these compounds contain an unsubstituted amidine group.

Recently, 20a and more particularly 2-hydroxystilbamidine, which is less toxic, has been particularly successful in the treatment of blastomycosis[60]; a success rate of 90% has been claimed[61]. Pentamidine (20b) is used in the treatment of pneumonia due to *Pneumocystis carinii*. This is a serious disease in patients receiving immunosuppressive therapy for leukaemia, lymphoma or transplant rejection[62, 63].

In an investigation into the effects of *N*-substitution into the amidine, Cooper[64] and Partridge[65] prepared a series of α,ω-di(*p*-*N*-phenylamidinophenoxy)alkanes (21). *In vitro* activity was obtained against *Mycobacterium*

$$H_2N(NH{=})C\text{—}\langle\bigcirc\rangle\text{—}O[CH_2]_nO\text{—}\langle\bigcirc\rangle\text{—}C({=}NH)NHPh$$

(21)

tuberculosis in the members containing 1, 3 or 5 methylene groups, with a maximum at $n = 5$: homologues possessing an even number of methylene groups ($n = 2$, 4 or 6) were inactive. The related *p*-alkoxy-*N*-phenylbenzamidines (22) had greater activity, with a maximum when alk = n-hexyl, but they showed no alternation in activity with increase in the

$$AlkO\text{—}\langle\bigcirc\rangle\text{—}C({=}NH)NHPh$$

(22)

length of the alkyl chain[66]. Introduction of a further ether link, giving *p*-(ω-alkoxyalkoxy)-*N*-phenylbenzamidines (23), considerably reduced *in*

$$Alk\text{—}OCH_2CH_2O\text{—}\langle\bigcirc\rangle\text{—}C({=}NH)NHPh$$

(23)

vitro activity[67]. All of the substituted benzamidines showed high toxicity and *in vivo* activity could not be demonstrated.

Bisbenzamidines (24, 25) linked through a *meta* bridge have found use in the treatment of babesiasis in domestic animals[68].

(24)   (25)

C. Cancer Chemotherapeutic Agents

I. Terephthalanilide derivatives

Random testing of synthetic compounds as potential cancer therapeutic agents highlighted amidine derivatives of terephthalanilide[69]. In active compounds the amidine function may be substituted by alkyl groups or may be cyclic as in imidazoline and 1,4,5,6-tetrahydropyrimidine. The imidazoline derivative (26) is typical of the class; this compound has been

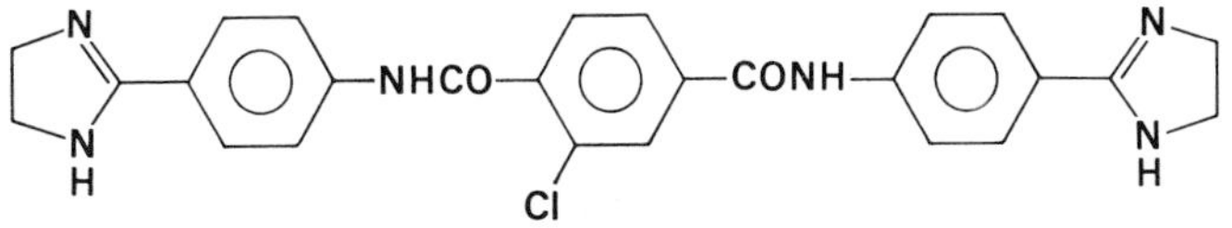

(26)

shown to be active against transplantable leukaemia L1210[70]. It causes inhibition of DNA synthesis at therapeutic levels in *Escherichia coli* ATCC 9637; the precursors of pyrimidine nucleotides, ureidosuccinic acid, dihydro-orotic acid and orotic acid have been shown to accumulate in the bacterial cell[71]. RNA synthesis is not affected[72].

The terephthalanilide (26) interferes with lipid synthesis, [1–^{14}C]acetate is not incorporated *in vitro* into lipids of mouse ascites cells[73]. With P388 lymphocytic leukaemia cells in culture, the terephthalanilide does not inhibit protein synthesis at chemotherapeutic levels[72].

The open chain amidine (27) is 58% as effective as amethopterin against

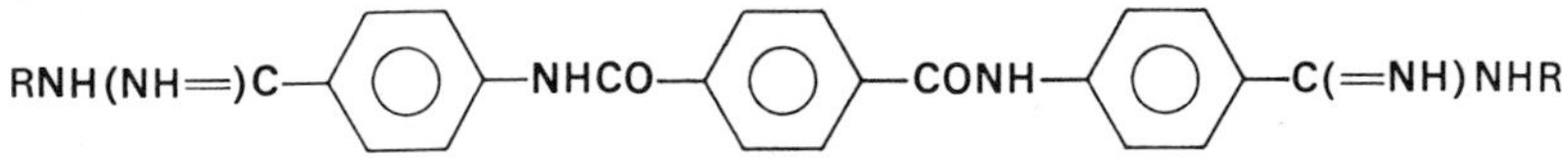

R= —$CH_2CH_2CH_2OMe$

(27)

transplantable mouse leukaemia L1210 and the cyclic amide (28) which has one of the terephthalanilide amide links reversed has antileukaemic activity similar to that of amethopterin[74].

In a clinical trial using the 2-chloroterephthalanilide (26) tumour regression was observed in six out of eighteen children suffering from malignant lymphoma of the jaw. Regression lasted for more than one

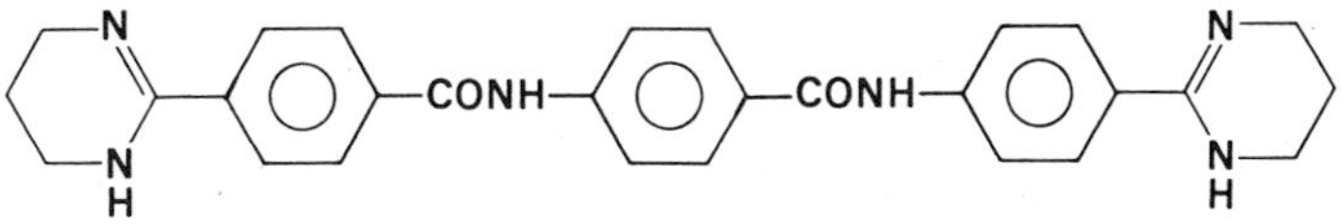

(28)

month in only one patient[75]. The 2-aminoterephthalanilide analogue was not effective in treatment of far advanced cancer cases[76].

In studies of the binding which occurs with terephthalanilide derivatives it has been shown that DNA–phthalanilide complexes are formed *in vitro* and there is some evidence that complexation may occur *in vivo*[77]. The complexes formed between dog-brain lipid and the tetrahydropyrimidine (29) have led to the discovery of new phosphatide fractions. The complexes

(29)

after treatment with acid or calcium salts gave lipid which was resolvable by chromatography into four subfractions, three of which were ninhydrin positive; all four, after hydrolysis, were ninhydrin positive and were previously unreported phosphatides[78].

2. Amidines as potentiators of cancer chemotherapeutic agents

Methylglyoxal bis(guanylhydrazone) (30) is active against transplanted mouse leukaemia L1210[79] and human granulocytic leukaemia[80]. The compound has severe toxicity, but stilbamidine (20a) or its 2-hydroxy analogue in conjunction with the hydrazone is much more effective than the optimally tolerated doses of either drug alone[81]. Stilbamidine also potentiates the action of the terephthalanilide derivative (26)

$$H_2N(NH{=})CNH \cdot N{=}C(Me)CH{=}NNHC({=}NH)NH_2$$
(30)

D. Protease Inhibitors

Extensive studies have been carried out on selective inhibitors of the digestive type proteases, trypsin, chymotrypsin and pepsin. Inhibitors of serum proteases are potentially useful in the chemotherapy of cardio-vascular disease and organ transplant.

I. Trypsin inhibitors

Trypsin assists in the hydrolysis of amide and ester bonds formed through the carboxyl group of arginine and lysine. Esters are more readily split than amides and simple amides are more readily split than peptide bonds. Since acylation of the terminal basic group of substrates prevents hydrolysis at the carboxyl group, it is obvious that there is ionic binding between the substrate and enzyme and accordingly a number of basic compounds have been investigated as potential reversible inhibitors of trypsin.

Benzamidine (**31**) was the first potent inhibitor of trypsin to be reported[82]. It binds more effectively than the substrate, DL-benzoylarginine-*p*-nitroanilide[83]. Cyclohexylcarboxamidine (**32**) is an 84-fold better inhibitor than acetamidine but only 1/24 as effective as benzamidine; it appears that hydrophobic interactions occur between the amidine and enzyme; further *p*-ethoxycarbonylbenzamidine (**33**) is a substrate for trypsin[82], adding weight to the argument that there are hydrophobic interactions between the active site and the binding site at the cationic centre of the enzyme.

$$PhC(=NH)NH_2 \qquad cyclo\text{-}C_6H_{11}C(=NH)NH_2 \qquad p\text{-}EtO_2CC_6H_4C(=NH)NH_2$$
$$\textbf{(31)} \qquad\qquad\qquad \textbf{(32)} \qquad\qquad\qquad \textbf{(33)}$$

Baker and his co-workers in a long series of papers have reported their investigations into active-site-directed irreversible enzyme inhibitors. Such inhibitors may possess as structural features a group which can compete for the active site of the enzyme and a second group which is usually electrophilic which allows a covalent link to be made to a second site in an area adjacent to or within the active site. The second site will obviously possess a nucleophilic group[84].

When covalent bonding is within the active site (*endo* type) there is little tolerance for bulky groups, but with bonding to an adjacent area (*exo* type) there may be sufficient area away from the active site for bulky groups to be placed so that they do not come into contact with the enzyme. For a molecule to fulfil the conformational requirements of both the active site and the second covalent binding site obviously there are more severe limitations on structure than when each site is considered separately.

Initial studies[85] into the bulk tolerance of benzamidine derivatives showed that a number of bulky groups can be placed *meta* or *para* in benzamidine without interfering with formation of a reversible enzyme-inhibitor complex. Particularly interesting is the amidine series (**34**); these compounds are active inhibitors of trypsin with a slight increase in binding over benzamidine (**31**)[85]. Introduction of a terminal fluorosulphonyl containing group into **34** to give (**35a** and **b**) gave amidines which caused

$$m \text{ or } p\text{-}\{PhO[CH_2]_3O\}C_6H_4C(=NH)NH_2$$
$$\textbf{(34)}$$

$$p\text{-}\{RC_6H_4O[CH_2]_nO\}C_6H_4C(=NH)NH_2$$
$$\textbf{(35)}$$

where R = (a) —NHCOC$_6$H$_4$SO$_2$F (*m*)

(b) —NHCOC$_6$H$_4$SO$_2$F (*p*)

(c) —NHCOCH$_2$Br

irreversible inhibition of trypsin. Reversible complex formation preceeds irreversible complex formation with **35b** ($n = 3$) at a concentration to complex 88% of trypsin, total inactivation occurs in 15 min[86].

Because of the differences in the dimensions of the fluorosulphonyl subsituted amidines (**35a**: $n = 3$ or 4) it is unlikely that a covalent link is formed to the same amino acid in each case. It has been suggested that a serine or threonine residue is attacked[87, 88].

The severe structural requirements for active compounds are illustrated by the amidines possessing a terminal bromacetamide residue (**35c**). The phenoxypropyloxyamidine (**35c**, $n = 3$) is a reversible inhibitor of trypsin, whereas the amidine (**35c**, $n = 4$) shows slow irreversible inhibition, with a half life of 5 hours[86].

2. Inhibition of guinea pig complement

Serum complement contains 11 distinct proteins; all are required for cell lysis. One of the functions of complement is the destruction of foreign cells, be they bacteria, protozoa or foreign mammalian cells, therefore it is obviously involved in rejection of organ transplants. Since some of the proteins of complement are proteases, amidine inhibitors of trypsin have been investigated as inhibitors of complement.

Benzamidine is a weak inhibitor of complement, inhibition is increased by a factor of six with a m-phenoxypropoxy substituent and 400-fold by further substitution of m-(p-nitrophenylurea) (**36a**)[89] and about a thousandfold by (**36b**); the latter is the most potent inhibitor of whole complement so far reported.

(36)

(a) R = m-NHCONHC$_6$H$_4$NO$_2$ (p)
(b) R = m-NHCONHC$_6$H$_4$SO$_2$F (p)
(c) R = o-NHCOC$_6$H$_4$SO$_2$F (m)

To investigate which component of complement was involved, Baker and his co-workers separated zymogen C′1 by dialysis of guinea pig complement. The zymogen was activated by incubation at 37°C[90, 91], potential inhibitor was added to component C′1a and the mixture incubated. Complement was reconstituted. If inhibition occurs the rate of lysis of sheep red blood cells is reduced.

The fluorosulphonyl amidine (**36c**) is an excellent irreversible inhibitor of component C'1a of complement but in the inhibition of whole complement by (**36c**) and other related amidines it is probable that inhibition of component C'1a does not occur since the amidine without the fluorosulphonyl group is as effective an inhibitor of whole complement as (**36c**).

3. Thrombin and kallikrein inhibitors

Thrombin does not occur in circulating blood but is formed at the time of blood coagulation by restricted proteolysis of the zymogen, prothrombin. Thrombin cleaves the N-terminal residues from fibrinogen to produce fibrin. Inhibition of the proteolytic activity of thrombin may be of value in the control of coagulation of blood[92] in, for example, thrombosis.

Pancreatic kallikrein releases vasoactive kinins from plasma globulins and thus contributes to vasodilatation of inflamed tissue. Kallikrein and thrombin are similar to trypsin in that each hydrolyse bonds involving the carboxyl group of lysine and arginine.

The action of potential inhibitors for these enzymes can be estimated in rate assays using N^α-benzoyl-DL-arginine-p-nitroanilide (BANA) as substrate[83]. In such estimations, m- and p-alkyl substituted benzamidines are poorer inhibitors for thrombin than trypsin[93, 94]. Diamidines such as **20b** behave as active-site-directed reversible inhibitors although only one amidine group is involved. The most efficient inhibition of thrombin is shown by **20b** ($n = 8$) and of kallikrein and trypsin by **20b** ($n = 12$). Introduction of an iodine atom *ortho* to the amidine group makes 2',2''-diiodo-4',4''-diamidino-1,5-diphenoxypentane the most effective inhibitor of kallikrein and trypsin; the analogous diphenoxyoctane is the most effective inhibitor of bovine thrombin, although the former compound is the most effective in blocking the clotting activity of human thrombin[95].

4. Correlation of activity of substituted benzamidines as proteolytic enzyme inhibitors

Coats, in a detailed study[96] has correlated the inhibition of thrombin, plasmin, trypsin and complement activity by the use of substituent constants and regression analysis (Hansch type analysis[97]).

In this type of analysis it may be possible to relate biological activity with structure in a series of compounds by the appropriate use of substituent constants. A generalized equation suitable for the evaluation of a number of structure-activity relationships is

$$\log \frac{1}{c} = -k\pi^2 + k'\pi + \rho\sigma + k'' \tag{10}$$

where $1/c$ is a fixed term, e.g. isotoxic concentration, LD_{50}, % growth etc., σ is Hammett's polar substituent constant, ρ is the reaction constant, π is a term derived in a manner similar to the Hammett constants and is defined as $\log P_x - \log P_H$ where P_x and P_H are the n-octanol–water partition coefficients of the substituted and unsubstituted compounds. The π constant reflects hydrophobic interactions drugs encounter in their 'random walk' to a receptor and their interactions at the receptor. The constants (K's) are generated by regressional analysis. Other parameters which reflect substituent effects may replace σ.

Coats, using π constants and the polarizability parameter $P_E{}^{98}$, showed that hydrophobic and electronic factors contribute to the binding in each system but to different degrees. Thrombin and complement seem to have similar binding sites and these are different from those of plasmin and trypsin. The overall results suggest that an increase in lipophilicity in the substituted amidine should result in stronger inhibition of the enzyme systems. Electronic effects appear to be different; thrombin and complement inhibition increases with electron-donating groups whereas plasmin and trypsin inhibition increases with electron-withdrawing substituents in the amidine.

E. Anthelmintic Drugs

The late 1960's saw the introduction of a cyclic amidine with broad spectrum anthelmintic activity into veterinary and human use. This drug, pyrantel (**37**, R = H), *trans*-1,4,5,6,-tetrahydro-1-methyl-2-[2-(2-thienyl)-vinyl]pyrimidine, was developed from an observation that while 2-(2-thienylmethylthio)imidazoline (**38**) exhibited nematodicidal activity in mice infected with *Nematospiroides dubius*, it had low activity in sheep,

(37)   (38)

the low activity being attributed to the hydrolysis of **38** into 2-thienyl-methylthiol and imidazolidin-2-one. Structural modification by placing an ethylene bridge or vinyl link between the two rings produces stable and active compounds[99].

The size of the N-heterocycle is important; in the series **39** the tetra-hydro-1,3-diazepine (**39**, $n = 4$) has no activity at a high dose level.

(39)

In **39** ($n = 2$ or 3) there was no marked difference in activity but when the ethylene link was replaced by vinyl in the tetrahydropyrimidine series the activity was greater in *trans* vinylene than in *cis* vinylene compounds. $N^{(1)}$ methylation of the tetrahydropyrimidine increases the activity but larger groups lead to inactivation. The thiophene moiety can be replaced by the isoelectronic benzene or the analogous furan with retention of activity.

McFarland has correlated[100] the biological effects of a substituent R in the structure **40** using a Hansche type analysis (see III, D, 4.). To relate the benzene and thiophene series a term δ was introduced to allow for differences such as the presence of sulphur 'd' orbitals in the thienyl series; with this constant a statistically significant correlation (equation 11, see also equation 10) was obtained.

$X = -S-$ or $-CH=CH-$

(40)

$$\log \left(\frac{1}{ED_{90}} \right) = -1 \cdot 64\pi^2 + 1 \cdot 93\pi + 0 \cdot 66\delta + 0 \cdot 88 \qquad (11)$$

The fixed term in equation 11 is $1/ED_{90}$, the dose to reduce by 90% the *N. dubius* population in infected mice.

Hydrophobic interactions are of supreme importance in these compounds, polar substituent effects are minimal. Compounds with more hydrophobic groups such as methyl and halogen (bromine or chlorine) are more active than the unsubstituted compounds.

Pyrantel exerts persistent nicotinic action which produces spastic paralysis in *Ascaris* sp., and although in cat-muscle preparations the drug produces transient neuromuscular block and some properties of compounds acting like an excess of acetylcholine, the dose to produce an anthelmintic effect is sufficiently low for negligible effects on the host[101]. Pyrantel is formulated as its tartrate for veterinary use and as its pamoate [pamoic acid is 4,4′-methylenedi-(3-hydroxy-2-naphthoic acid)] for human use[102].

The cyclic amidine system is not essential for anthelmintic activity since analogues such as the thiophenepropamidine (**41**) and the thiophenacryl-

(**41**)   (**42**)

amidine (**42**) are active compounds[103]. In these series the substitution pattern is critical. High activity is associated with N,N-substitution; one N-substituent must be methyl, the other may be methyl, ethyl, allyl, methoxy (an O-methylamidoxime) or methylamino (an amidrazone); $N,N′$-disubstitution is unfavourable for activity. Steric factors and hydrophobic interactions appear to be important; with no substituents the compound is too hydrophilic and with substituents larger than N-allyl-N-methyl the compounds are too lipophilic.

Thioimidates [e.g. (**43**)], related to pyrantel, have been screened for

(**43**)

their anthelmintic activity; some are highly potent[104]. The difference in basic strength in the two series is striking; thioimidates are weaker by a factor of 10^6–10^8.

Pyrantel is inactive against adult whipworms (*Trichuris* sp.) but an analogue *trans*-1,4,5,6-tetrahydro-2-(3-hydroxystyryl)-1-methylpyrimidine (**44**) and its open-chain analogue (**45**) are active against *T. muris* and *T. vulpis*[105].

(**44**)   (**45**)

N,N-Dialkyl-4-alkoxynaphthamidines possess anthelmintic activity, one, N,N-dibutyl-4-hexyloxynaphthamidine (bunamidine) (**46**) is outstandingly effective against a variety of cestodes in animals but is not good enough for human pinworm treatment. Recently[106, 107], a large group of analogues of bunamidine have been prepared and examined in the search for a superior compound. Compounds possessing activity against *Taenia pisiformis* in the dog and *Hydatigera taeniaeformis*, *Spirometra mansonoides* or *Dipylidium caninum* in the cat have N-alkyl groups butyl or smaller and an O-alkyl group butyl or higher. Replacement of N,N-dialkyl by N-alkyl-N-aryl, by morpholine or by 4-methylpiperazine abolishes activity against *Hymenslepis nana* and *Oochoristica symmetrica* in mice.

$$C(=NH)N(Bu\text{-}n)_2$$

$$OC_6H_{13}\text{-}n$$

(**46**)

F. Antihypertensive Agents

A number of azacycloheptane and azacyclooctane derivatives possessing an ethylene side chain which terminates with a basic group [structure (**47**)] have been examined for their antihypertensive properties[108]. The correla-

$$(CH_2)_n \quad NCH_2CH_2 \text{ basic group}$$

(**47**)

tion between ring size and nature of the basic group has been investigated. With guanidine on the basic group, the seven-membered ring is optimal for activity (guanethidine). In the amidine series, seven- or eight-membered rings (**48**) are optimal. With the amidoximes a seven-membered ring (**49**)

$$(CH_2)_{6\,or\,7} \quad NCH_2CH_2C(=NH)NH_2$$

(**48**)

$$(CH_2)_6 \quad NCH_2CH_2C(=NOH)NH_2$$

(**49**)

is necessary for pharmacological action[109, 110]; O-acylation and O-alkylation markedly reduces or abolishes activity.

In all the series the cyclic structure is necessary for retention of activity. All the open chain analogues examined were inactive.

The imidazoline derivatives, tolazoline (**50**) and phentolamine (**51**) are vasodilator drugs which are used for the treatment of peripheral circulatory disorders. Tolazoline exerts its action by dilating the blood vessels[111]; phentolamine blocks the pressor action of noradrenaline and adrenaline[112].

Structural modification of these drugs has led to the introduction of clonidine (**52**) for the treatment of hypertension. The amidine analogues

(**50**)	(**51**)	(**52**)

(**53**)[113, 114] of clonidine where Ar is phenyl, o- or m-tolyl, 2,3- or 2,4-xylyl or 2,6-dichlorophenyl are active antihypertensive agents at 10 mg/kg administered intragastrically to rats pretreated with deoxycortone acetate[113.]

(**53**)

G. Cyclic Amidines in the Control of Cardiac Arrhythmias

The cyclic amidine, antazoline (**54**), exhibits a wide range of pharmacological action. It has the properties and uses of an antihistamine drug, but it is one of the least active of the common antihistamines[115]. It also possesses local anaesthetic and anticholinergic properties. The successful use[116, 117] of antazoline in the control of cardiac arrhythmias prompted the Ciba research group to modify the structure of antazoline with the specific aim of improving the pharmacological action. One of the simplest

(**54**)	(**55**)

compounds prepared[118] has the N-benzyl-N-phenyl residue linked by a methylene bridge [5-(2-imidazolylmethyl)-5,6-dihydromorphanthridine (55)]; this compound had interesting antifibrillatory effects on aconitine-induced cardiac arrhythmias.

In a long series of compounds[119] the bicyclic compound [Su-13197 (56)] was examined closely[120]. High antifibrillatory action is associated

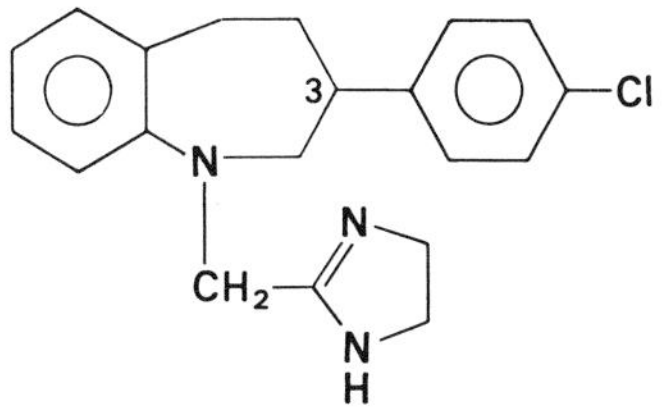

(56)

with o-, m- and p-chlorophenyl substituents in the 3 position of the benzazepine. Good activity is retained with a 3-phenyl substituent and with 1,4,5,6-tetrahydropyrimidin-2-ylmethyl and 4-methylimidazolin-2-ylmethyl as the $N_{(1)}$ substituent.

H. Tranquillizing Drugs

The semi-cyclic amidine chlordiazepoxide (57), widely used as a mild tranquillizing drug in neurotic patients, is one of a number of benzo-

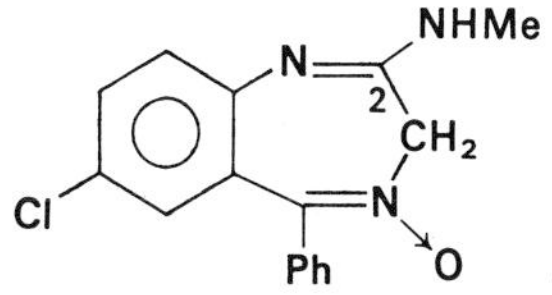

(57)

diazepines with tranquillizing properties[121]. The metabolic fate of this amidine is well established; the main metabolites, in man, are desmethyl-chlordiazepoxide and demoxepam, the latter arising from hydrolytic fission of the 2-methylamino substituent. Further hydrolytic cleavage of demoxepam to N-(2-amino-5-chloro-α-phenylbenzylidene)glycine-N-oxide (58), also occurs[122].

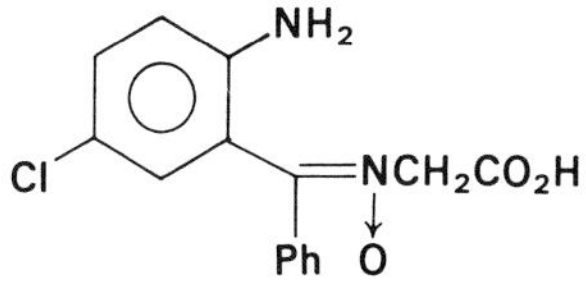

(58)

In contrast with the metabolism of chlordiazepoxide the dibenzo-diazepine, clozapine (**59**), which has been effective in the treatment of psychotic patients[123], has an amidine group which is metabolically stable. The bio-transformations occurring are N-4' demethylation and 4'-N-oxide formation.

(59)

I. Anti-inflammatory and Antipyretic Agents

A number of substituted phenylacetic acids, e.g. ibuprofen (**60**) exhibit anti-inflammatory and antipyretic activity. The carboxylic acid function may be replaced by a hydroxamic acid residue with retention of activity, e.g. p-butoxyphenylacetohydroxamic acid [bufexamac (**61**)]

$p\text{-}i\text{-}BuOC_6H_4CH(Me)CO_2H$ $p\text{-}n\text{-}BuOC_6H_4CH_2CONHOH$
          (60)   **(61)**

This acid is metabolized by reduction to the amide or by hydrolysis to the acid or by hydroxylation in the 2 or 3 positions. The phenols are metabolically conjugated with glucuronic acid and are excreted as their β-glucuronides (**62** and **63**)[125].

(62)   **(63)**

Clinical trials with the amidine, paranylene [α-fluoren-9-ylidene-*p*-toluamidine (**64**)] have shown that the drug is beneficial in various types

(64)

of arthritis, without side effects. Resistance to the compound occurs sooner than with other recognised drugs such as phenylbutazone and cortisone[126]. This amidine also possesses antiviral properties[127].

IV. REFERENCES

1. T. Emery, in *Advances in Enzymology*, Vol. 35 (Ed. A. Meister), Interscience, New York, 1971, pp. 135–185.
2. J. H. Weisburger and F. K. Weisburger, *Pharmacol. Rev.*, **25**, 1 (1973).
3. J. E. Reimann and R. U. Byerrum, in *The Chemistry of Amides* (Ed. J. Zabicky), Interscience, London, 1970, pp. 627–631.
4. I. B. Dawid, T. C. French and J. M. Buchanan, *J. Biol. Chem.*, **238**, 2178 (1963).
5. T. C. French, I. B. Dawid and J. M. Buchanan, *J. Biol. Chem.*, **238**, 2186 (1963).
6. K. Mizobuchi and J. M. Buchanan, *J. Biol. Chem.*, **243**, 4853 (1968).
7. K. Mizobuchi, G. L. Kenyon and J. M. Buchanan, *J. Biol. Chem.*, **243**, 4863 (1968).
8. S. C. Hartman and J. M. Buchanan, *J. Biol. Chem.*, **233**, 456 (1958).
9. P. Oxley, D. A. Peak and W. F. Short, *J. Chem. Soc.*, 1618 (1948).
10. B. Magasanik, H. S. Moyed and D. Karibian, *J. Amer. Chem. Soc.*, **78**, 1510 (1956).
11. B. Magasanik, *J. Amer. Chem. Soc.*, **78**, 5449 (1956).
12. C. Mitoma and E. E. Snell, *Proc. Nat. Acad. Sci. U.S.A.*, **41**, 891 (1955).
13. A. Neidle and H. Waelsch, *Fed. Proc.*, **16**, 968 (p. 225) (1957).
14. H. S. Moyed and B. Magasanik, *J. Biol. Chem.*, **235**, 149 (1960).
15. D. M. Greenberg in Metabolic Pathways, Vol. III (Ed. D. M. Greenberg), Academic Press, London, 1969, pp. 268–277.
16. M. J. Voll, E. Appella and R. G. Martin, *J. Biol. Chem.*, **242**, 1760 (1967).
17. B. N. Ames, R. G. Martin and B. J. Garry, *J. Biol. Chem.*, **236**, 2019 (1961).
18. D. W. E. Smith and B. N. Ames, *J. Biol. Chem.*, **239**, 1848 (1964).
19. A. Peterkofsky, *J. Biol. Chem.*, **237**, 787 (1962).
20. V. R. Williams and J. M. Heroms, *Biochim. Biophys. Acta*, **139**, 214 (1967).
21. R. H. Feinberg and D. M. Greenberg, *Nature*, **181**, 897 (1958).

22. R. H. Feinberg and D. M. Greenberg, *J. Biol. Chem.*, **234**, 2670 (1959).
23. K. Freter, J. C. Rabinowitz and B. Witkop, *Annalen*, **607**, 174 (1957).
24. H. Kny and B. Witkop, *J. Amer. Chem. Soc.*, **81**, 6245 (1959).
25. S. Snyder, O. L. Silva and M. W. Kies, *Biochem. Biophys. Res. Commun.*, **5**, 165 (1961).
26. D. R. Rao and D. M. Greenberg, *J. Biol. Chem.*, **236**, 1758 (1961).
27. A. Miller and H. Waelsch, *J. Biol. Chem.*, **228**, 383 (1957).
28. A. Miller and H. Waelsch, *J. Biol. Chem.*, **228**, 397 (1957).
29. H. Tabor and L. Wyngarden, *J. Biol. Chem.*, **234**, 1830 (1959).
30. M. Silverman, R. C. Gardiner and P. T. Condit, *J. Nat. Cancer Inst.*, **20**, 71 (1958).
31. I. Chanarin and M. C. Bennett, *Brit. Med. J.*, (*i*), 27 (1962).
32. H. Tabor and L. Wyngarden, *J. Clin. Invest.*, **37**, 824 (1958).
33. J. Kohn, M. L. Mollin and L. M. Rosenbach, *J. Clin. Path.*, **14**, 345 (1961).
34. J. P. Knowles, T. A. J. Prankerd and R. G. Westall, *Lancet* (*ii*), 347 (1960).
35. A. L. Luhby, J. M. Cooperman and D. N. Teller, *Amer. J. Clin. Nutr.*, **7**, 397 (1959).
36. A. L. Luhby, J. M. Cooperman and D. N. Teller, *Proc. Soc. Exp. Biol. Med.* (*N. Y.*), **101**, 350 (1959).
37. M. Roberts and S. D. Mohamed, *J. Clin. Path.*, **18**, 214 (1965).
38. J. M. Johnstone, J. H. Kemp and E. D. Hibbard, *Clin. Chem. Acta.*, **12**, 440 (1965).
39. J. C. Rabinowitz and W. E. Pricer, Jr., *J. Amer. Chem. Soc.*, **78**, 5702 (1956).
40. K. Uyeda and J. C. Rabinowitz, *J. Biol. Chem.*, **242**, 24 (1967).
41. H. Tabor and J. C. Rabinowitz, *J. Amer. Chem. Soc.*, **78**, 5705 (1956).
42. R. L. Blakley, in *The Biochemistry of Folic Acid and Related Pteridines*, North Holland, London, 1969, p. 92.
43. J. C. Rabinowitz and W. E. Pricer, Jr., *J. Amer. Chem. Soc.*, **78**, 4176 (1956).
44. B. V. Ramasastri and R. L. Blakley, *J. Biol. Chem.*, **237**, 1982 (1962); **239**, 106 (1964).
45. K. O. Donaldson, V. F. Scott and W. Scott, *J. Biol. Chem.*, **240**, 4444 (1965).
46. K. Uyeda and J. C. Rabinowitz, *J. Biol. Chem.*, **242**, 4378 (1967).
47. R. L. Blakley, in *The Biochemistry of Folic Acid and Related Pteridines*, North Holland, London, 1969, pp. 223–227.
48. S. L. Hartmann and J. M. Buchanan, *J. Biol. Chem.*, **234**, 1812 (1959).
49. A. Albert, in *Selective Toxicity*, 5th ed., Chapman and Hall, London, 1973, Chapter 8, pp. 258–270.
50. S. Nakamura, K. Karasawa, H. Yonehara, N. Tanaka and H. Umezawa, *J. Antibiotics*, **14**, 103 (1961); *Chem. Abstr.*, **56**, 10677 (1962).
51. R. A. Gray, *Phytopathology*, **45**, 281 (1955).
52. R. L. Peck, H. M. Shafer and F. J. Wolf, *U.S. Patent*, 2,804,463 (1957); *Chem. Abstr.*, **52**, 8474 (1958).
53. G. D. Diana, U. J. Salvador, E. S. Zalay and F. Pancic, *J. Medicin. Chem.*, **16**, 1050 (1973).

54. C. W. Waller, C. F. Wolf, W. J. Stein and B. L. Hutchins, *J. Amer. Chem. Soc.*, **79**, 1265 (1957).
55. A. DiMarco, M. Gaetani, P. Orezzi, T. Scotti and F. Arcamone, *Cancer Chemotherapy Reports*, **18**, 15 (1962).
56. P. Chandra, F. Zunino, A. Götz, A. Wacker, D. Gerichi, A. DiMarco., A. M. Casazza and F. Guiliana, *FEBS Letters*, **21**, 154 (1972).
57. B. Puschendorf, E. Petersen, H. Wolf, H. Werchau and H. Grunicke, *Biochem. Biophys. Res. Commun.*, **43**, 617 (1971).
58. F. Arcamore, V. Nicolella, S. Penco and S. Radaelli, *Gazz. Chim. Ital.*, **99**, 632 (1969).
59. A. Kreutzberger, in *Progress in Drug Research*, Vol. II (Ed. E. Jucker), Birkhäuser, Basel, 1968, pp. 356–445.
60. *The Extra Pharmacopoeia*, 26th ed., Pharmaceutical Press, London, 1972, pp. 775–776.
61. E. Grunberg, *Chemother.*, **12**, 272 (1967).
62. J. Ruskin and J. S. Remington, *J. Amer. Med. Assoc.*, **203**, 604 (1968).
63. W. B. Hamlin, *J. Amer. Med. Assoc.*, **204**, 173 (1968).
64. F. C. Cooper and M. W. Partridge, *J. Chem. Soc.*, 5036 (1952).
65. M. W. Partridge, *J. Chem. Soc.*, 2683 (1949).
66. M. W. Partridge, *J. Chem. Soc.*, 3043 (1949).
67. F. C. Cooper and M. W. Partridge, *J. Chem. Soc.*, 459 (1950).
68. G. Schmidt, R. Hirt and R. Fischer, *Res. Vet. Sci.*, **10**, 530 (1969).
69. S. A. Schepartz, I. Wodinsky and J. Leiter, *Cancer Chemotherapy Reports*, **19**, 1 (1962).
70. J. M. Venditti, A. Goldin and I. Kline, *Cancer Chemotherapy Reports*, **19**, 5 (1962).
71. R. F. Pittillo, L. L. Bennett, Jr., W. A. Short, A. J. Tomisek, G. J. Dixon, J. R. Thompson, W. R. Laster, Jr., M. Trader, L. Mattil, P. Allan, B. Bowdon, F. M. Schabel, Jr. and H. E. Skipper, *Cancer Chemotherapy Reports*, **19**, 41 (1962).
72. D. S. Yesair, W. I. Rogers, P. E. Baronowsky, I. Wodinsky, P. S. Thayer and C. J. Kensler, *Cancer Res.*, **27**, 314 (1967).
73. J. Booth, E. Boyland and A. Gellhorn, *Cancer Chemotherapy Reports*, **43**, 11 (1964).
74. I. Kline, M. Ganz, D. J. Tyrer, J. M. Venditti, E. W. Artis and A. Goldin, *Cancer Chemotherapy Reports, Part 2*, **2**, 65 (1971).
75. H. F. Oettgen, P. Clifford and J. H. Burchenal, *Cancer Chemotherapy Reports*, **27**, 45 (1963).
76. J. Van Dyk, W. Kreis, I. H. Krakoff, C. T. C. Tan, N. Cevik and J. H. Burchenal, *Cancer Res.*, **28**, 1566 (1968).
77. A. Sivak, W. I. Rogers and C. J. Kensler, *Biochem. Pharmacol.*, **12**, 1056 (1963).
78. D. W. Yesair, W. I. Rogers, J. T. Funkhouser and C. J. Kensler, *J. Lipid. Res.*, **7**, 492 (1966).
79. B. L. Freedlander and F. A. French, *Cancer Res.*, **18**, 360 (1958).
80. E. J. Freireich, E. J. Frei III and M. Karon, *Cancer Chemotherapy Reports*, **16**, 189 (1962).
81. J. H. Burchenal, J. R. Purple, E. Bucholz and P. W. Staub, *Cancer Chemotherapy Reports*, **29**, 85 (1963).

82. M. Mares-Guia and E. Shaw, *J. Biol. Chem.*, **240**, 1579 (1965).
83. B. F. Erlanger, N. Kokowsky and W. Cohen, *Arch. Biochem. Biophys.*, **95**, 271 (1961).
84a. W. T. Ashton, L. L. Kirk and B. R. Baker, *J. Medicin. Chem.*, **16**, 453 (1973).
84b. B. R. Baker, in *Design of Active-Site-Directed Irreversible Enzyme Inhibitors. The Organic Chemistry of the Enzymic Active Site*, John Wiley & Sons, New York, 1967, pp. 172–190.
85. B. R. Baker and E. H. Erickson, *J. Medicin. Chem.*, **10**, 1123 (1967).
86. B. R. Baker and E. H. Erickson, *J. Medicin. Chem.*, **11**, 245 (1968).
87. B. R. Baker and J. A. Hurlbut, *J. Medicin. Chem.*, **11**, 241 (1968).
88. A. M. Gold and D. Fahrney, *Biochemistry*, **3**, 783 (1964).
89. B. R. Baker and M. Cory, *J. Medicin. Chem.*, **12**, 1053 (1969); **14**, 119 (1971).
90. E. E. Ecker and S. Seifter, *Proc. Soc. Exp. Biol. Med.*, **47**, 18 (1943).
91. E. H. Lepow, O. D. Ratnoff, F. N. Rosen and L. Pillemer, *Proc. Soc. Exp. Biol. Med.*, **92**, 32 (1956).
92. L. Lorand and J. L. G. Nilsson, in *Drug Design*, Vol. III, (Ed. E. J. Ariens), Academic Press, New York, 1972, Ch. 8.
93. F. Markwardt, P. Walsmann and H. Landmann, *Pharmazie*, **25**, 551 (1970).
94. F. Markwardt, H. Landmann and P. Walsmann, *European J. Biochem.*, **6**, 502 (1968).
95. J. D. Geratz, A. C. Whitmore, M. C.-F. Cheng and C. Piantadosi, *J. Medicin. Chem.*, **16**, 970 (1973).
96. E. A. Coats, *J. Medicin. Chem.*, **16**, 1102 (1973).
97. C. Hansch and T. Fujita, *J. Amer. Chem. Soc.*, **86**, 1616 (1964).
98. A. Leo, C. Hansch and C. Church, *J. Medicin. Chem.*, **12**, 766 (1969).
99. J. W. McFarland, L. H. Conover, H. L. Howes, Jr., J. E. Lynch, D. R. Chisholm, W. C. Austin, R. I. Corwell, J. C. Danilewicz, W. Courtney and D. H. Morgan, *J. Medicin. Chem.*, **12**, 1066 (1969).
100. J. W. McFarland, in *Progress in Drug Research*, Vol. 15, (Ed. J. Jucker), Birkhäuser, Basel, 1971, p. 123.
101. M. L. Aubry, P. Cowell, M. J. Davey and S. Shevde, *Brit. J. Pharmacol.*, **38**, 332 (1970).
102. *The Extra Pharmacopoeia*, 26th ed., Pharmaceutical Press, London, 1972, p. 2036.
103. J. W. McFarland and H. L. Howes, Jr., *J. Medicin. Chem.*, **13**, 109 (1970).
104. J. W. McFarland, H. L. Howes, Jr., L. H. Conover, J. E. Lynch, W. C. Austin and D. H. Morgan, *J. Medicin. Chem.*, **13**, 113 (1970).
105. J. W. McFarland and H. L. Howes, Jr., *J. Medicin. Chem.*, **15**, 365 (1972).
106. R. B. Burrows, C. J. Hatton, W. G. Lillis and G. R. Hunt, *J. Medicin. Chem.*, **14**, 87 (1971).
107. M. Harfenist, R. B. Burrows, R. Baltzly, E. Pedersen, G. R. Hunt, S. Gurbaxani, J. E. D. Keeling and O. D. Standen, *J. Medicin. Chem.*, **14**, 97 (1971).
108. E. Schlittler, J. Druey, A. Marxer, *Progress in Drug Research*, Vol. IV., (Ed. E. Jucker), Birkhäuser, Basel, 1962, pp. 341–348.

109. R. P. Mull, P. Schmidt, M. R. Dapero, J. Higgins and M. J. Weisbach, *J. Amer. Chem. Soc.*, **80**, 3769 (1958).
110. R. A. Maxwell, S. D. Ross and A. J. Plummer, *J. Pharmacol. Exptl. Therap.*, **123**, 128 (1958).
111. *The Extra Pharmacopoeia*, 26th ed., Pharmaceutical Press, London, 1972, p. 1790.
112. *The Extra Pharmacopoeia*, 26th ed., Pharmaceutical Press, London, 1972, p. 1801.
113. F. M. Hershenson and L. F. Rozek, *J. Medicin. Chem.*, **14**, 907 (1971).
114. H. Wollweber, R. Hiltman and K. Stoepel, *S. African Patent* 6,802,296 (1968); *Chem. Abstr.*, **70**, 68177p (1969).
115. *The Extra Pharmacopoeia*, 26th ed., Pharmaceutical Press, 'London, 1972, p. 1542.
116. S. R. Kline, L. S. Dreifus, Y. Watanabe, T. F. McGarry and W. Likoff, *Amer. J. Cardiol.*, **9**, 564 (1962).
117. E. W. Reynolds, W. M. Baird and M. E. Clifford, *Amer. J. Cardiol.*, **14**, 513 (1964).
118. L. H. Werner, S. Ricca, E. Mohacsi, A. Rossi and V. P. Arya, *J. Medicin. Chem.*, **8**, 74 (1965).
119. L. H. Werner, S. Ricca, A. Rossi and G. DeStevens, *J. Medicin. Chem.*, **10**, 575 (1967).
120. W. E. Barrett, T. Garces, R. Rutledge and A. J. Plummer, *Pharmacologist*, **7**, 229 (1965).
121. *The Extra Pharmacopoeia*, 26th ed., Pharmaceutical Press, London, 1972, p.1808.
122. B. A. Koechlin, M. A. Schwartz, G. Krol and W. Oberhansli, *J. Pharmacol. Exp. Ther.*, **148**, 399 (1965).
123. J. Angst, U. Jaenicke, A. Padrutt and C. Scharfetter, *Pharmakopsychiat./ Neuro-psychopharmakol*, **4**, 192 (1971); *Chem. Abstr.*, **76**, 196n (1972).
124. D. E. Hathway, in *Foreign Compound Metabolism in Mammals*, Vol. 2. The Chemical Society, London, 1972, p. 222.
125. R. Roncucci, M. J. Simon, G. Lambelin, N. P. Buu-Hoi and J. Thiriaux, *Biochem. Pharmacol.*, **15**, 1563 (1966); **17**, 187 (1968).
126. W. J. Poznanski and J. D. Wallace, *Canad. Med. Assoc. J.*, **83**, 1302 (1960).
127. K. A. Ludwig, I. Ruchman and F. J. Murray, *Proc. Soc. Exp. Biol. Med.*, **100**, 495 (1959).

CHAPTER **7**

Preparation and synthetic uses of amidines

JEAN-ALBERT GAUTIER, MARCEL MIOCQUE AND
CLAUDE COMBET FARNOUX

U.E.R.* de Chimie Thérapeutique, Université Paris XI, France

* U.E.R. "Unité d'Enseignement et de Recherche" means: Teaching and Research Unit.

I. INTRODUCTION

This review considers the amidines from a practical point of view and intends to give a survey of their methods of preparation and of their utilization as starting materials for other syntheses.

Some reviews have already been published in the field, the best known being that of Shriner and Newmann[1]. References 2 and 3 discuss in detail methods of synthesis and cyclization reactions of amidines.

Discussion of earlier work in the field will be only perfunctory since it has already been amply covered in references 1 and 2. We shall, however, refer to those early works which, although of limited applicability, may in our opinion open new possibilities.

We have tried to give a general survey of the subject in as modern and complete a way as possible in order to provide guide-lines for future investigators.

II. PREPARATION OF AMIDINES

The synthesis of amidines proceeds in general through starting materials having an unsubstituted carbon–nitrogen bond; the introduction of the second nitrogen is realized by action of ammonia or of primary or secondary amines.

This simple general scheme may proceed by two routes:

(1). Transformation of nitriles by addition of amines:

$$
\underset{\substack{\diagdown \\ \text{NH} \diagup}}{\text{R—C}\!\equiv\!\text{N}} \;\rightleftarrows\; \underset{\substack{\diagdown \\ \overset{\oplus}{\text{NH}} \diagup}}{\text{R—C}\!=\!\text{N}^{\ominus}} \;\longrightarrow\; \underset{\substack{\diagdown \\ \text{N} \diagup}}{\text{R—C}\!=\!\text{NH}}
$$

(2). Substitution by nucleophilic attack on the carbon atom of amides or their derivatives:

$$
\underset{}{\overset{\overset{\text{A}}{\|}}{\text{R—C—NHR}}} \;\rightleftarrows\; \underset{\substack{\diagdown \\ \text{NH} \diagup}}{\overset{\overset{\text{AH}}{|}}{\text{R—C}\!=\!\text{NR}}} \;\rightleftarrows\;
$$

$$
\underset{\substack{\diagdown \\ \text{NH} \diagup \\ +}}{\overset{\overset{\text{AH}}{|}}{\text{R—C—NR}}} \;\longrightarrow\; \underset{\substack{\diagdown \\ \text{N} \diagup}}{\text{R—C}\!=\!\text{NR}} + \text{AH}_2
$$

$$[\text{A} = \text{O or S}]$$

Unfortunately, these hypothetical equations hold only for a few particular cases so that while nitriles and amides are indeed the most frequent starting materials for the syntheses of amidines, they must generally first be tranformed into more reactive intermediates such as imido esters (X = OR) or imidoyl halides (X = Hal):

$$
\text{R—C}\underset{\diagdown}{\overset{\diagup \text{X}}{\big\langle}}\underset{\text{N—R}'}{}
$$

The large variety of possible substituents on the carbon or on the nitrogen atoms complicates the study. According to the nature of the substituents, methods of preparation are more or less successful. Furthermore, in some cases, although based on the fundamental schemes described above, precursors or derivatives of the reactants are used. Finally, some processes that are interesting but have only limited applications, are based on completely different reactions.

In order to simplify the presentation, we shall deal with the various ways of preparing amidines in three groups based on the nature of the starting material: (A) Preparations from nitriles; (B) Preparations from amides or thioamides; (C) Preparations from miscellaneous starting materials.

A. Preparations from nitriles

Simple additions of ammonia and amines to nitriles are, unfortunately, only observed in the case of nitriles activated by electron-attracting substituents in the position α of the C≡N bond.

In practice three types of reactions are used:

(1). Addition of metal amides to nitriles:

$$R-C\equiv N + R'-NH^{\ominus} \longrightarrow R-C{\overset{N^{\ominus}}{\underset{NH-R'}{}}} \xrightarrow{H^{\oplus}} R-C{\overset{NH}{\underset{NHR'}{}}}$$

(2). Addition of salts of ammonia or amines to nitriles. Theoretically the scheme is:

$$R-C\equiv N \atop {\underset{>NH}{}} \xrightarrow{H^{\oplus}} R-C{\overset{=NH}{\underset{\underset{\oplus}{>NH}}{}}} \longrightarrow R-C{\overset{=NH}{\underset{>N}{}}} + H^{\oplus}$$

The anion of the salt may play a role in the reaction process.

(3). Formation of an imidoester which subsequently reacts with ammonia or an amine (Pinner method):

$$R-C\equiv N \xrightarrow[R'OH]{HCl} R-C{\overset{=NH, HCl}{\underset{OR'}{}}} \xrightarrow{HN<} R-C{\overset{=NH, HCl}{\underset{N<}{}}} + R'OH$$

I. Addition of metal amides to nitriles

Three types of reactions are known which use metal derivatives of ammonia or of amines as reactive nucleophiles: (a) Condensation of an alkali amide with nitriles; (b) Reaction of an amine (after its metallation by an alkali amide) with nitriles; (c) Condensation of aminomagnesium derivatives with nitriles.

a. *Reaction of alkali amide anions with nitriles.* Addition of sodium, potassium or calcium amides to nitriles, gives metallic derivatives that are converted into amidines by the action of proton-donors[4-7].

$$RC\equiv N + KNH_2 \longrightarrow RC{\overset{NK}{\underset{NH_2}{}}} \xrightarrow{H^{\oplus}} RC{\overset{NH}{\underset{NH_2}{}}}$$

Liquid ammonia which is the best solvent for the formation of amide anions doesn't always allow the use of a sufficiently high temperature for the condensation to occur, so that ammonia under pressure or anhydrous solvents such as benzene, toluene or xylene are used. During the final hydrolysis, care must be taken to avoid the transformation of the metallated amidine into an amide[1].

The interest in this process is limited for several reasons. It applies only for obtaining unsubstituted amidines. When higher temperatures of condensation are necessary, side reactions are observed with nitriles having an α-hydrogen.

Besides the cases studied before[1] a most recent application of α-substituted acetonitriles is the following[8]:

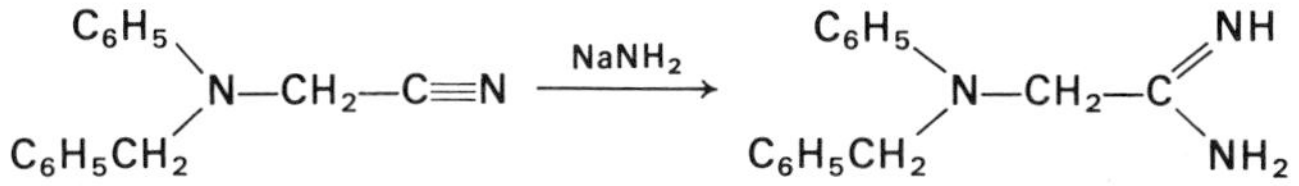

$$C_6H_5 \diagdown N-CH_2-C\equiv N \xrightarrow{NaNH_2} C_6H_5 \diagdown N-CH_2-C \diagup{NH} \diagdown{NH_2}$$
$$C_6H_5CH_2 \diagup \qquad\qquad\qquad C_6H_5CH_2 \diagup$$

b. *Reaction of metallated amines with nitriles.* This is the extension of the previous method for the preparation of monosubstituted amidines.

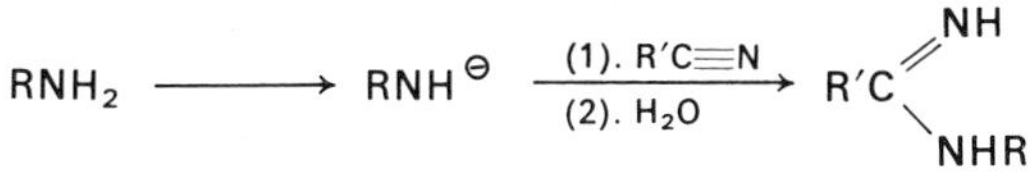

$$RNH_2 \longrightarrow RNH^\ominus \xrightarrow[(2).\ H_2O]{(1).\ R'C\equiv N} R'C\diagup{NH}\diagdown{NHR}$$

In a first method the metallation of amines is brought about by sodium metal; Cooper and Partridge[9] studied in detail this process and verified the equation proposed by Lottermoser[10]:

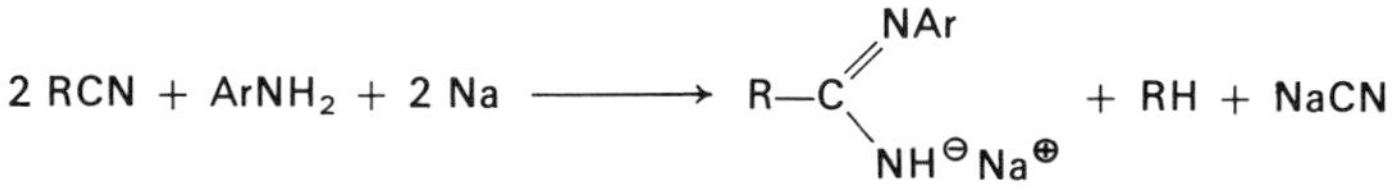

$$2\ RCN + ArNH_2 + 2\ Na \longrightarrow R-C\diagup{NAr}\diagdown{NH^\ominus Na^\oplus} + RH + NaCN$$

According to the following scheme, the hydrocarbon RH which occurs among the products of oxidation is formed by 'reduction of part of the nitrile by free hydrogen', the latter being a by-product of the metallation of the free amine[9].

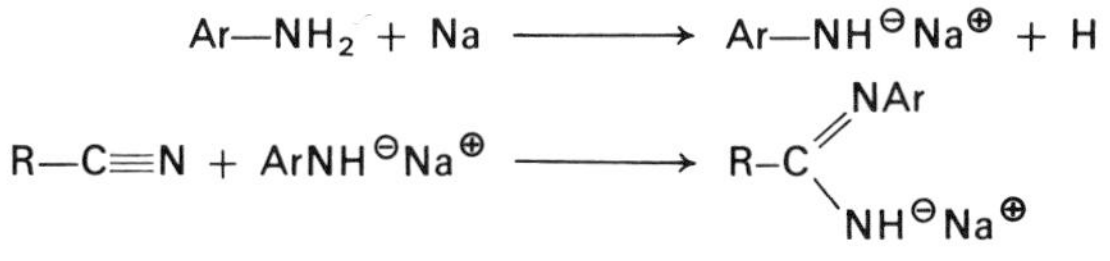

$$Ar-NH_2 + Na \longrightarrow Ar-NH^\ominus Na^\oplus + H$$
$$R-C\equiv N + ArNH^\ominus Na^\oplus \longrightarrow R-C\diagup{NAr}\diagdown{NH^\ominus Na^\oplus}$$

Easy metallation of the amines is achieved by action of metal amides, and side reactions like reduction of the nitrile by sodium are avoided. The reaction is conducted in dry inert solvents at the boiling point. Reference 2 gives the general methods of preparation as well as some applications.

In a new method, nitriles ArC≡N are condensed with aryl amines[11] in liquid ammonia. The products are obtained in good yields and are easily purified since the low temperature minimizes side reactions.

An extension of this method starts with diaryl ketoximes, which after reaction with an excess of sodium in liquid ammonia followed by reaction with benzonitriles gave amidines derived from benzhydrylamine[12].

Benzhydrylamine metallated by sodium amide gives, under identical conditions, the same amidine that results from benzophenone oxime.

A reaction of the same type was developed starting from aromatic nitriles and substituted anilines metallated by sodium hydride in DMSO at room temperature[13].

These methods are often convenient for o-substituted nitriles as well as for naphthonitriles, and complement the Pinner method. Unfortunately they give low yields and cannot be used in the case of nitriles having α-hydrogen atoms.

c. *Condensation of aminomagnesium derivatives with nitriles.* Aminomagnesium intermediates derived from amines and ethylmagnesium halides react with nitriles to give amidines.

Reference 2 gives the experimental conditions for the reaction as well as its various applications.

The best results by this method are obtained with secondary amines which react even with aromatic nitriles substituted in the *ortho* position[14–16].

2. Addition of ammonia and amines to nitriles

a. *Addition of ammonia and free amines.* Ammonia and free amines give a direct addition only with nitriles activated by electron-attracting groups on the α-carbon: trichloroacetonitrile, often mentioned in the literature, gives substituted or unsubstituted amidines[1,2].

$$Cl_3C—C\equiv N \xrightarrow{\ RNH_2\ } Cl_3C—C\overset{\displaystyle NR}{\underset{\displaystyle NH_2}{\diagup\!\!\diagdown}}$$

Perfluoroalkyl nitriles[17] and ethyl cyanotartronate[18] are also subject to direct addition.

$$\begin{array}{c} HO\diagdown\quad\diagup COOEt \\ C \\ N\!\equiv\!C\diagup\quad\diagdown COOEt \end{array} + \begin{array}{c} Pr\diagdown \\ NH \\ Pr\diagup \end{array} \longrightarrow \begin{array}{c} HO\diagdown\quad\diagup COOEt \\ C \\ HN\!=\!C\diagup\quad\diagdown COOEt \\ | \\ NPr_2 \end{array} \quad \text{(ref. 18)}$$

Only one of the nitrile functions of cyanogen reacts with secondary amines to give cyanoformamidines[19,20], while both functions react with primary amines resulting in oxamidines[20-21].

$$R_2NH + (CN)_2 \longrightarrow R_2NC\overset{\displaystyle NH}{\underset{\displaystyle C\equiv N}{\diagup\!\!\diagdown}} \quad \text{(ref. 19)}$$

$$2\,RNH_2 + (CN)_2 \longrightarrow RNH—\underset{\displaystyle \overset{\|}{NH}}{C}—\underset{\displaystyle \overset{\|}{NH}}{C}—NHR \quad \text{(ref. 21)}$$

2-Alkylmercaptoethylamines react like primary amines and give oxamidines while unsubstituted 2-mercaptoethylamine leads to cyclic derivatives[22]. Condensation of cyanogen with o-phenylenediamine leads to diaminoquinoxaline[23]. Substituted formamidines are prepared by condensation of primary and secondary aliphatic amines with hydrocyanic acid[24]

b. *Addition of ammonium salts.* The early studies of Cornell[4] and Bernthsen[25,26] on the condensation of ammonium salts with nitriles have been developed in two main directions:

(i). In a systematic study, Short and co-workers recommend heating of molten mixtures of nitriles with ammonium thiocyanate[27], ammonium arylsulphonates or ammonium alkylsulphonates.

(ii). Schaefer and Krapcho[29] modified the process by decreasing the temperature and working with ammonia under pressure. They suggest heating in an autoclave at 125–150°C a mixture of the nitrile and of an excess of the ammonium salt in liquid ammonia. In some cases, a cosolvent (methanol, ethanol) may be used.

Among the ammonium salts tried, chlorides and bromides were preferred.

c. *Addition of amine salts.*

(i). Amine hydrochlorides give satisfactory results only rarely:

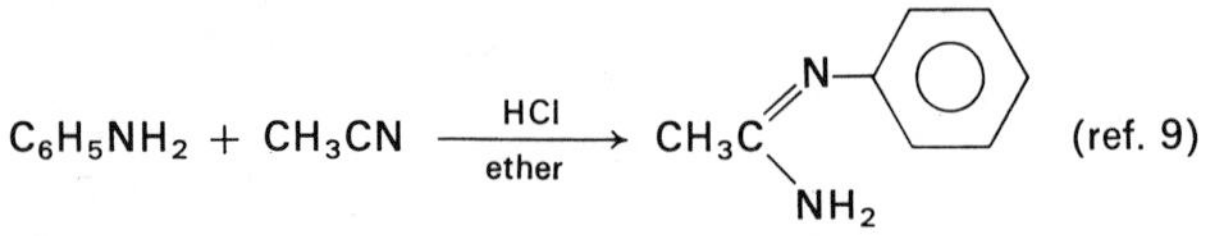

$$C_6H_5NH_2 + CH_3CN \xrightarrow[\text{ether}]{\text{HCl}} CH_3C \begin{subarray}{l} =N\text{-} \\ \\ NH_2 \end{subarray} \text{(ref. 9)}$$

Yield $= 90\%$

$$C_6H_{11}NH_2 \cdot HCl + C_6H_5CH_2CN \xrightarrow{NH_4Cl} C_6H_5CH_2C \begin{subarray}{l} =NC_6H_{11} \\ \\ NH_2 \cdot HCl \end{subarray} \text{(ref. 30)}$$

Yield $= 75\%$.

With *o*-phenylenediamine, this reaction leads to 2-substituted benzimidazoles[31, 32]:

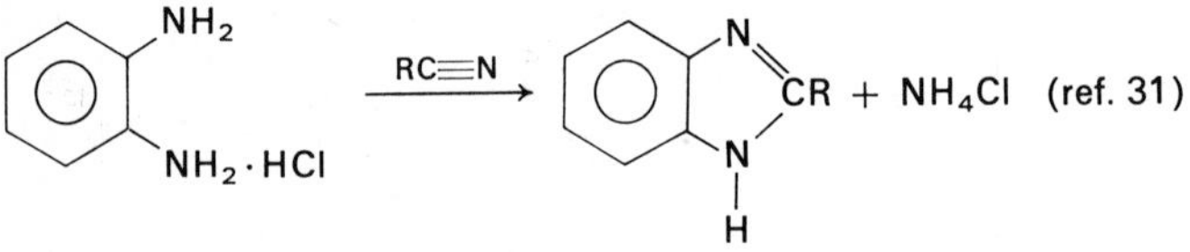

$+ \ NH_4Cl$　(ref. 31)

These reactions and condensations of halogenonitriles with free amines[33,34] are comparable: cyclization by creation of an amidine function is preceded by formation of an amine hydrochloride isolatable only under mild conditions.

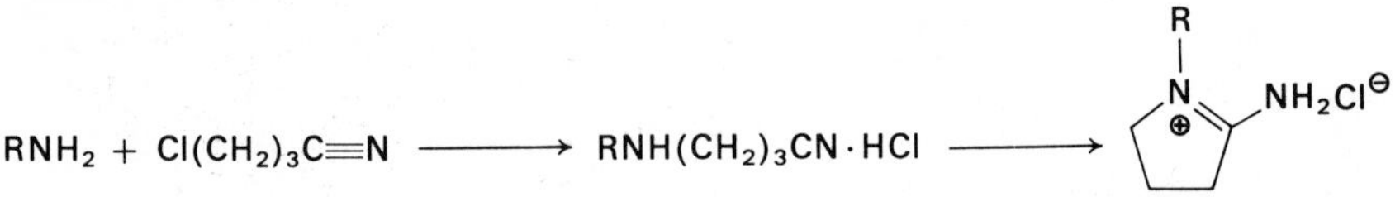

$$RNH_2 + Cl(CH_2)_3C\equiv N \longrightarrow RNH(CH_2)_3CN \cdot HCl \longrightarrow$$

Yield $= 8\%$
(ref. 34)

(ii). Thiocyanates of aliphatic amines sometimes give good yields[27].

(iii). Amine arylsulphonates react according to the following scheme proposed by Oxley and Short[28]:

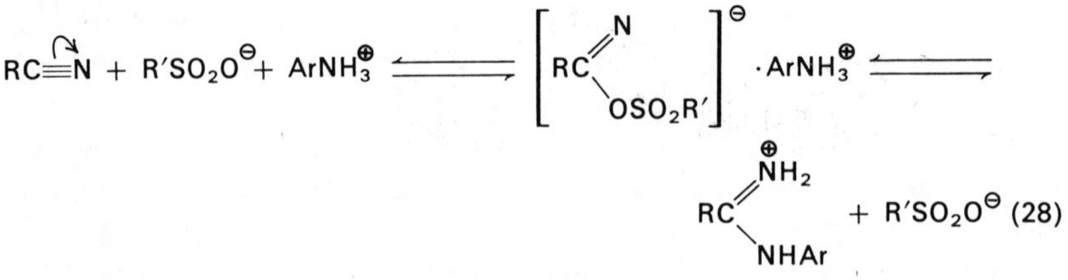

R = alkyl or aryl.

The authors reject for this reaction any mechanism involving dissociation of the ammonium salt and attack on the nitrile by ammonia as well as ammonolysis of intermediate imidoyl sulphonate:

$$RC\underset{\diagdown OSO_2R'}{\overset{\diagup NH}{}}$$

In this reaction, the reagents are heated at temperatures between 180 and 300°C without any solvent. Applications of the method are mainly limited to the preparation of monoarylamidines and of cyclic amidines[35-39] (formation of imidazoles, tetrahydropyrimidines and diazacycloheptenes). Reference 2 gives applications of this reaction. Only arylsulphonates of monoalkylamines[30, 37] and diarylamines[40, 41] may be employed in this reaction. With the arylsulphonates of dialkylamines, dealkylation results under the conditions of the reaction[42]. Delaby and co-workers[43] applied this method to prepare in an indirect way amidines in the pyridine series. Thus, cyano pyridine is converted into the corresponding N-phenylamidine, which after ammonolysis at 140°C gives unsubstituted picolinic amidine.

(a) $RC\equiv N + C_6H_5SO_3H, NH_2C_6H_5 \longrightarrow RC\underset{\diagdown NH-C_6H_5}{\overset{\diagup NH}{}}, C_6H_5SO_3H$

(b) $RC\underset{\diagdown NHC_6H_5}{\overset{\diagup NH}{}}, C_6H_5SO_3H + NH_3 \longrightarrow RC\underset{\diagdown NH_2}{\overset{\diagup NH}{}}, C_6H_5SO_3H + C_6H_5NH_2$

(ref. 43)

R = 2-pyridyl

(iv). Primary and secondary amines in the presence of aluminium chloride react with aliphatic and aromatic nitriles.

$$R-C\equiv N + R'NH_2 \xrightarrow{AlCl_3} R-C\underset{\diagdown NHR'}{\overset{\diagup NH}{}}$$

The nitrile and the amine are simply heated together (variable temperature and time) in an inert solvent[43] or without solvent[2, 42, 44].

At the end of the reaction the amidine–aluminium chloride complex is decomposed by water or very dilute acid. Ref. 2 gives a table that summarizes the amidines prepared in good yields, by this method. Applications are given in references 45 and 46.

3. Pinner synthesis through imido ester intermediates

Nitriles resist condensation with bases containing nitrogen, but with alcohols they easily form imidoesters that are transformed in a second step into amidines.

The Pinner synthesis[47, 48] is a two-step reaction: the first is the transformation of nitriles into imido esters that are generally isolated, and then condensed with ammonia or an amine in a separate operation.

The imido ester is formed according to the scheme:

$$R-C{\equiv}N + R'OH \xrightarrow{\text{HCl}} R-C{\underset{\text{OR', HCl}}{\overset{\text{NH}}{\Big\langle}}}$$

Generally the nitrile is dissolved in an alcohol (usually anhydrous ethanol), cooled to a low temperature and an excess of hydrochloric acid is bubbled through the mixture.

According to Pinner[48] the optimal ratio of the three reagents (nitrile, alcohol, hydrochloric acid) is when they are practically equimolar. However, different ratios were employed on various occasions, and various authors obtained good results with a nitrile/alcohol ratio of 1:1[49, 50] or 1:2 or even up to 1:3[51, 52]. Others report that a large excess of alcohol (1:10) sometimes improves the yield[53, 54], whereas in other cases this ratio must be decreased[55].

Various non-hydroxylic solvents may be added to the medium, such as ether, chloroform, and dioxan.

In the particular case of trichloroacetonitrile its highly electrophilic character, resulting from the effect of the CCl_3 group, facilitates addition of methanol even without the use of an acid, and the formation of the imido ester is much easier.

$$RC{\equiv}N + R'OH \underset{\text{R'ONa}}{\rightleftarrows} RC{\underset{\text{OR'}}{\overset{\text{NH}}{\Big\langle}}}$$

However, the nucleophilic attack on the nitrile by alkoxide ion[56] gives good results only with aliphatic nitriles substituted in the α position by electron-attracting groups, or with nitroaromatic nitriles. From a practical aspect this reaction is of limited interest only.

Usually, the imido esters are isolated as the hydrochlorides and, because of their instability, used immediately. References 1 and 57 list the side reactions caused by this instability. Some imido esters were isolated as the free bases and were studied spectrophotometrically[49, 58, 59].

Other synthetic methods similar to that of Pinner's, were also used for the preparation of amidines through imidoesters. For instance, condensation of ethoxyacetylene with amines in the absence of water yields acetamidines[60]:

$$EtO-C\equiv CH + RNH_2 \longrightarrow \left[H_2C=C \begin{array}{c} OEt \\ \diagdown \\ NHR \end{array} \right] \rightleftarrows$$

$$CH_3C \begin{array}{c} OEt \\ \diagup \\ \diagdown \\ NR \end{array} \xrightarrow{RNH_2} CH_3C \begin{array}{c} NHR \\ \diagup \\ \diagdown \\ NR \end{array}$$

In another method, ethoxyacetylene gives amidines through the intermediate 1,1-diazidoethoxyethane[61].

Heating of an equimolar mixture of an O-alkyl thionester, an amine and its amine hydrochloride in an alcoholic solution gives, after evolution of hydrogen sulphide, the amidine, and again the corresponding imidoester is the assumed intermediate of the reaction[62]:

$$RC \begin{array}{c} S \\ \diagup \\ \diagdown \\ OEt \end{array} + NH_2R' \xrightarrow{(alcohol)} RC \begin{array}{c} N-R' \\ \diagup \\ \diagdown \\ OEt \end{array} + H_2S \xrightarrow[\text{(alcohol)}]{\overset{\oplus}{R'}NH_3\overset{\ominus}{Cl}}$$

$$RC \begin{array}{c} NR' \\ \diagup \\ \diagdown \\ NHR' \cdot HCl \end{array} + EtOH$$

Several types of amidines may be obtained by this method (e.g., R = alkyl, Ph, $PhCH_2$; R' = H, Me, Ph).

In the second step of the Pinner method, ammonia or an amine attacks the imido ester to give the amidine.

Ammonia yields an unsubstituted amidine; e.g. the hydrochloride of the imido ester may react with ammonia in an alcoholic solvent[44, 51, 63–71, 75]:

$$RC \begin{array}{c} NH \\ \diagup \\ \diagdown \\ OR' \end{array} \cdot HCl \xrightarrow{NH_3} RC \begin{array}{c} NH \\ \diagup \\ \diagdown \\ NH_2 \end{array} + NH_4Cl + R'OH$$

Imido esters may also be treated with ammonium salts in an aqueous-alcoholic medium[55, 72–74]:

$$RC \begin{array}{c} NH \\ \diagup \\ \diagdown \\ OR' \end{array} + NH_4X \longrightarrow RC \begin{array}{c} NH \\ \diagup \\ \diagdown \\ NH_2 \cdot HX \end{array} + R'OH$$

Various side reactions were observed during the Pinner reaction, such as ammonolysis of an ester function present in the molecule[75].

$$\text{ROOC(CH}_2)_n\text{—C} \overset{\overset{\oplus}{\text{NH}_2}\overset{\ominus}{\text{Cl}}}{\underset{\text{OEt}}{\diagdown}} \xrightarrow{\text{NH}_3} \text{H}_2\text{NOC(CH}_2)_n\text{—C} \overset{\text{NH}}{\underset{\text{NH}_2, HCl}{\diagdown}}$$

However, ethyl cyanacetate gives a normal Pinner reaction[76] and ammonolysis may be avoided by choosing the appropriate experimental conditions[52,67]. Other side reactions occurring include cyclization of the imido ester[77] or ammonolysis of phthalimidic structures[63]:

$$\text{Cl(CH}_2)_2\text{CH—C} \overset{\text{NH}}{\underset{\text{OEt, HCl}}{\diagdown}} \quad \underset{\text{C}_6\text{H}_5}{} \xrightarrow{\text{NH}_3} \quad \text{C}_6\text{H}_5\text{—}\overset{\text{HN}\diagup\overset{\text{H}}{\text{N}}}{}$$

Imido esters yield monosubstituted amidines with primary amines[1,48,50,53,64,78-84]; they yield disubstituted amidines with secondary amines[1,48,80,85-87], while no reaction takes place with tertiary amines.

Various factors influence the reaction process:

(i). The basicities of the amine and of the imido ester both seem to play an important role, since e.g. aliphatic amines attack more easily than aromatic amines[48,57].

(ii). The presence of alcohol in the medium may displace the equilibrium towards the starting materials. On the other hand, some imido ester chlorhydrates give orthoesters with excess of alcohol. These effects are avoided when the reaction is carried out in a nonhydroxylic solvent like dioxan or ether[39,88,89];

(iii). When working with amine in excess, the equilibrium is displaced to the products increasing the yield of the expected amidine. Unfortunately,

$$\text{RC} \overset{\overset{\oplus}{\text{NH}_2}\ \overset{\ominus}{\text{Cl}}}{\underset{\text{NHR}'}{\diagdown}} \underset{\text{R}'\text{NH}_2}{\overset{\longrightarrow}{\rightleftharpoons}} \text{RC} \overset{\text{NHR}'}{\underset{\overset{\oplus}{\text{NR}'}\ \overset{\ominus}{\text{Cl}}}{\diagdown}}$$

excess of amine at high temperatures and with long reaction times may give N,N'-disubstituted amidines [48, 79]:

Bristow [64] obtained a double reaction involving both NH_2 groups, when reacting ethylenediamine with disubstituted mandelonitriles: one of the NH_2 groups displaces the OEt group, while subsequent reaction of the second one gives a substituted dihydroglyoxaline in a cyclization reaction:

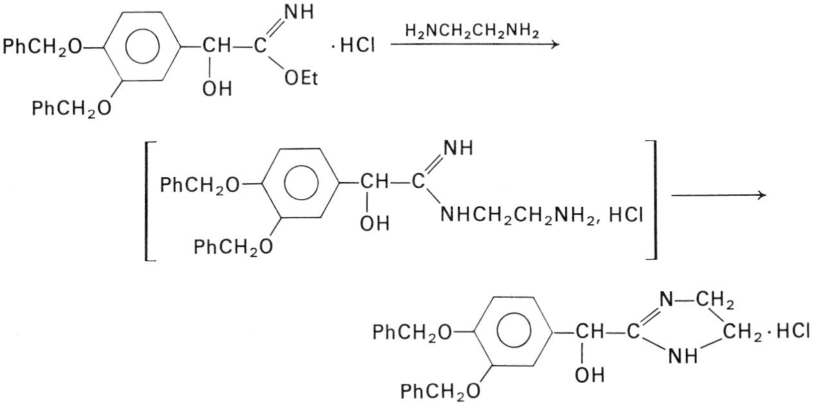

In practice, the Pinner reaction may be used for formation of mainly monosubstituted or mainly N,N'-disubstituted amidines according to the experimental conditions. Interest in the method is limited by the fact that in all cases a mixture of two products is obtained.

Attack of imido esters by sulphonamides, in hot benzene or alcohol gives sulphonylamidines [90, 91]:

p-Aminophenylsulphonamides react selectively with imido esters through their SO_2NH_2 group rather than through their amine function [92, 93]:

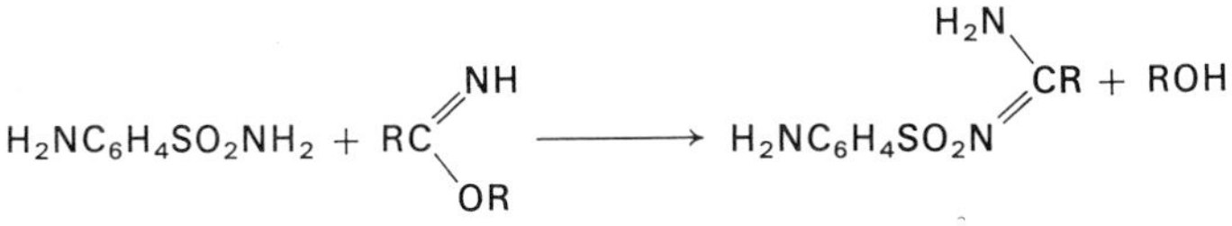

There are some exceptions to the applicability of the Pinner reaction:

Acyl cyanides of the type RCOCN do not give the expected imidoesters [48], and neither do *ortho*-substituted aromatic nitriles and α-naphthonitrile [1]. Steric hindrance by an *ortho* group is not general, e.g. one of the cyano functions of *o*-phthalonitrile gives the reaction and reaction of *o*-ethoxybenzonitrile leads to the imido ester and the amidine[94].

The mechanism of the reaction, according to reference 1, is a nucleophilic attack on the imine bond.

Pinner's method is the most general one for the preparation of unsubstituted amidines. The method is less successful for substituted amidines when various side reactions involving the substituents may occur.

B. Preparation from Amides and Thioamides

In this section we will describe the condensation of amides with amines in presence of halogenating reagents which yield amidines through imido chlorides, as well as other similar methods and also the conversion of thioamides into amidines.

I. Condensations of amides and amines in presence of halogenating agents:

a. *From unsubstituted amides.* The method of acetamidine formation by leading gaseous hydrochloric acid into molten acetamide[95] mentioned in ref. 1 was not developed further because of the formation of by-products in the reaction. The reason is the instability of the imido chlorides generally considered as the intermediates of the reaction.

$$\begin{array}{c} \text{Cl} \\ | \\ \text{R—C}{=}\text{NH} \end{array}$$

This lack of stability is shown especially by *N*-unsubstituted imidochlorides.

Heating of diacetamide with aryl amine hydrochlorides[96, 97] leads also to amidines.

b. *From monosubstituted amides.* Treatment of a monosubstituted amide with an halogenating agent gives an imido chloride which reacts with ammonia or with an amine to yield various substituted amidines.

This method of preparation[46, 98, 99] is widely utilized although it gives side reactions occasionally. According to the scheme opposite proposed by Delaby and co-workers[100,101] the imidochloride (**a**) gives (**b**) and the

$$RC(=O)-NHR' \xrightarrow{PCl_5} RC(Cl)=NR'$$

$$RC(Cl)=NR' \xrightarrow{H_3N} RC(NH_2)=NR'$$

$$RC(Cl)=NR' \xrightarrow{H_2NR''} RC(NHR'')=NR'$$

$$RC(Cl)=NR' \xrightarrow{HNR''_2} R-C(NR''_2)=NR'$$

amidine (**c**). The latter may react with an excess of amide to produce (**d**) and an acid chloride that will give the amide (**e**) which in turn may yield the amidine (**f**).

$$\underset{\textbf{(a)}}{RC(=O)-NHR'} \xrightarrow{PCl_5} \underset{\textbf{(b)}}{RC(Cl)=NR'} \xrightarrow{R''NH_2} \underset{\textbf{(c)}}{R-C(NHR'')=NR'}$$

$$\downarrow RC(=O)-NHR'$$

$$\underset{\textbf{(d)}}{RC(NHR')=N-R'} + \underset{}{RC(=O)Cl}$$

$$\downarrow R''NH_2$$

$$\underset{\textbf{(e)}}{RC(=O)-NHR''} \xrightarrow[R''NH_2]{PCl_5} \underset{\textbf{(f)}}{R-C(NR'')=NHR''}$$

The nature of the amine and of the amide as well as the experimental conditions influence greatly the importance of the side reactions. From a practical point of view, the method causing the least amount of side reactions is preferable. Delaby and co-workers[100] stipulate the best routes to obtain amidines are those at the top of p. 298, between which route (b) is better than route (a).

The synthesis of diaryl benzamidines gives the best results when the amine carrying the most bulky substituents participates in the reaction as an amide[2, 102].

$$CH_3\overset{\overset{\displaystyle O}{\|}}{C}\!-\!NHAr^1 + Ar^2NH_2 \xrightarrow[\text{(a)}]{POCl_3}$$

$$CH_3\overset{\overset{\displaystyle NAr^1}{\|}}{C}\!-\!NHAr^2 + CH_3\overset{\overset{\displaystyle N\!-\!Ar^1}{\|}}{C}\!-\!NHAr^1 + CH_3\overset{\overset{\displaystyle NAr^2}{\|}}{C}\!-\!NHAr^2$$

$$70\% \qquad\qquad 10\%$$

$$CH_3\overset{\overset{\displaystyle O}{\|}}{C}\!-\!NHAr^2 + Ar^1NH_2 \xrightarrow[\text{(b)}]{POCl_3} CH_3\overset{\overset{\displaystyle NAr^1}{\|}}{C}\!-\!NHAr^2 \quad \text{(Yield 93\%)}$$

$$Ar^1 = \beta\text{-naphthyl}; \; Ar^2 = p\text{-isopropoxyphenyl.}$$

The method with amides of aliphatic acids leads mainly to *N*-aryl-substituted amidines [102,105]. Synthesis of trisubstituted amidines by action of a secondary amine on a monosubstituted amide gives the expected amidine and the *N,N'*-symmetrically disubstituted amidine, which is formed from the R—CO—NHR' amide [106]. Starting from *N,N*-disubstituted amides and the primary amines, better yields are obtained: (method b)

$$RC\overset{\overset{\displaystyle O}{\|}}{}\!-\!NHR' + HNR_2'' \xrightarrow[\text{(a)}]{PXn} R\!-\!\overset{\overset{\displaystyle NR_2''}{|}}{C}\!=\!NR' \xleftarrow[\text{(b)}]{PXn} R'NH_2 + R\!-\!\overset{\overset{\displaystyle O}{\|}}{C}\!-\!NR_2''$$

Usually the amide and the amine are heated for several hours in the presence of phosphorous pentachloride, a solvent (usually benzene), at the boiling point [2]. Use of phosphorous oxychloride or trichloride or of thionyl chloride leads to amidines as well, but generally with poorer yields [102, 103, 107–110].

Sometimes the imido chloride intermediate is isolated and reacted with the amine in a second step [111, 112]. This procedure is used when the amine is subject to attack by phosphorous halides and it gives good results for synthesis of *N*-alkylated amidines, but it fails for amidines derived from aliphatic carboxylic acids [2, 113, 114].

c. *From disubstituted amides.*   *N,N*-Disubstituted amides react in same conditions as the monosubstituted amides [102, 110, 115, 116] but in this case the intermediate is a dichloro derivative of the starting amide.

Thus, dialkylformamides react with sulphonamides and yield trisubstituted sulphonylamidines [117]. Reference 118 describes the isolation of the dichloro intermediate.

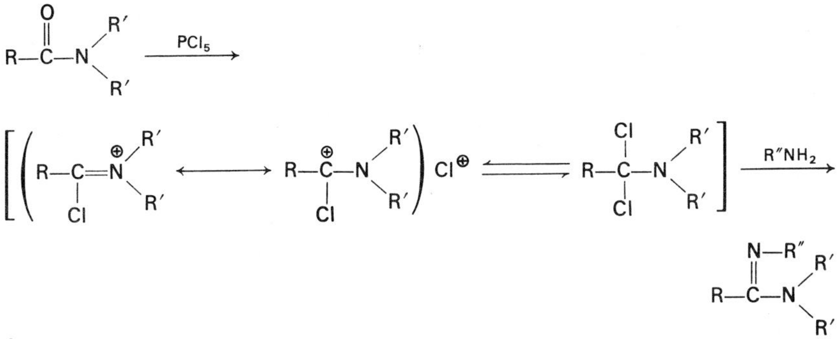

2. Modified procedures and related reactions

Other halogenating agents such as phosgene[99,119] or gaseous hydrochloric acid[120,121] are sometimes used. *m*-Chloroformanilide heated at 160°C under reduced pressure with *m*-chloroaniline gives *N*,*N*'-di(*m*-chlorophenyl) formamidine[122], when the halogenating agent is the hydrochloric acid salt of the amine used.

In some condensations, without use of any halogenating derivative, water elimination takes place such as, for example, with phosphoric anhydride[123].

When there is a possibility of ring closure by dehydration, the formation of the amidine is facilitated[124].

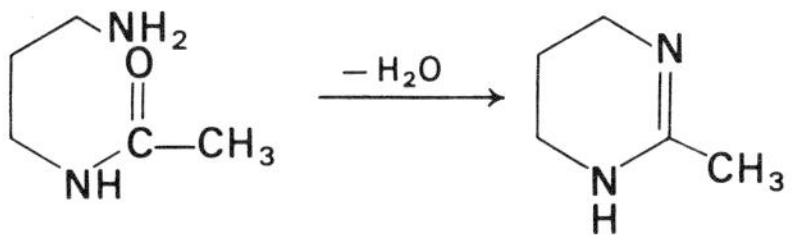

In the absence of an amine, heating of amide above but in the presence of a halogenating agent gives amidines[100]:

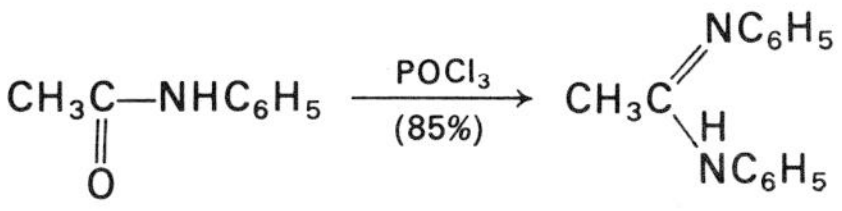

According to the next scheme, monosubstituted formamides with phosgene give dichloromethylformamidines[125]:

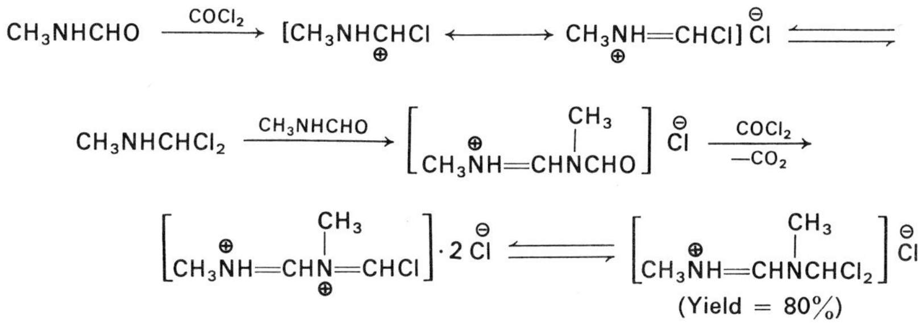

In an aminolysis reaction, these dichloromethylformamidines lose the $CHCl_2$ group[126]. With urea, they give carbamoyl formamidines[127].

The transformation of o-tolylformamide into N,N'-di-o-tolylformamidine by phosgene occurs through a cyclic intermediate[128]:

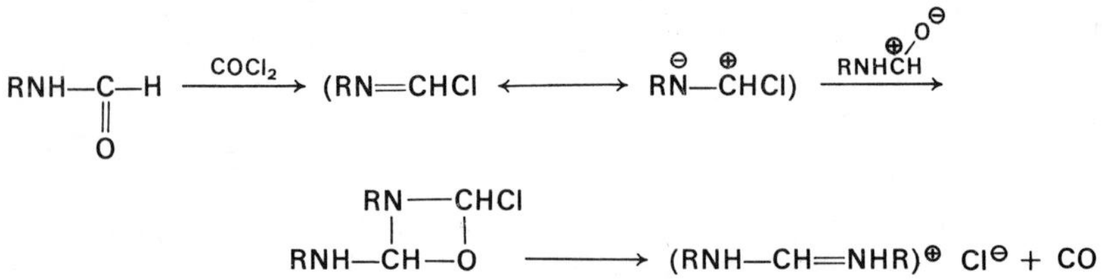

$$RNH-C-H \xrightarrow{COCl_2} (RN=CHCl \longleftrightarrow RN-CHCl) \xrightarrow{RNHCH}$$

$$RNH-CH-O \longrightarrow (RNH-CH=NHR)^{\oplus} Cl^{\ominus} + CO$$

Formation of disulphonyl formamidines from N-formyl sulphonamides and phosphorous oxychloride in the presence of pyridine may be explained by concerted process[129]:

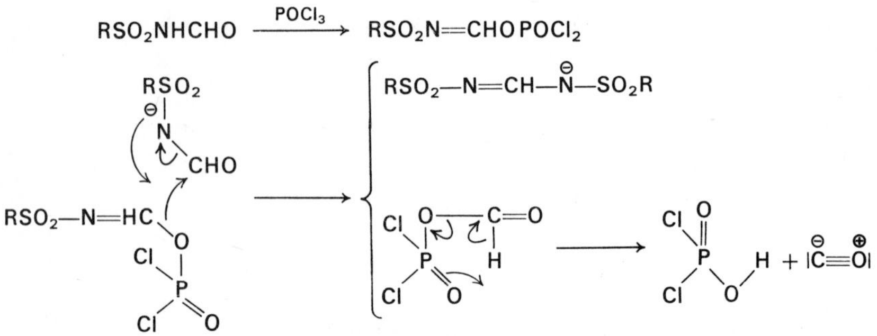

$$RSO_2NHCHO \xrightarrow{POCl_3} RSO_2N=CHOPOCl_2$$

Amidines may be also obtained from mixtures of acids and amines which give amides (for example on heating aniline with formic acid in the presence of boric acid)[130], or from phenolic acids with amines in chlorobenzene in the presence of silicon tetrachloride[131].

Other imido derivatives, in addition to halides, e.g. imido sulphonates or imidoyl fluoborates, may also be intermediates in the syntheses of amidines.

Imido sulphonates are formed when heating together amide, arylsulphonyl chloride and amine, or when an arylsulphonyl amide is treated by an ammonium salt[132] according to the following scheme:

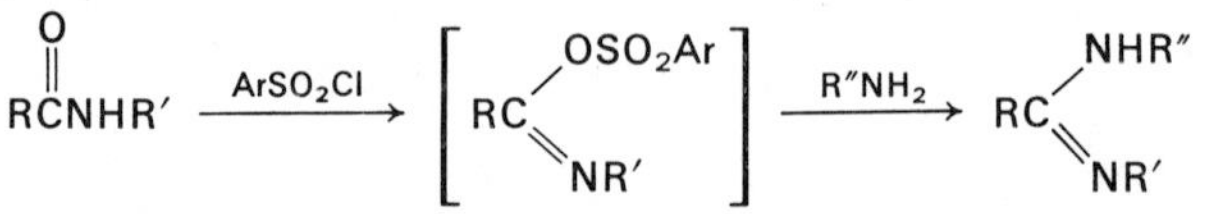

Ammonia reacts as well as primary and secondary amines: for applications see ref. 2.

N-Acylated arylsulphonamides, $ArSO_2NHCOR$, heated with ammonium salts lead to amidines via imido sulphonate intermediates[133].

In a similar synthesis, a Beckmann rearrangement gives, in good yields, amidines from arylsulphonyl ketoximes in the presence of amines[134]:

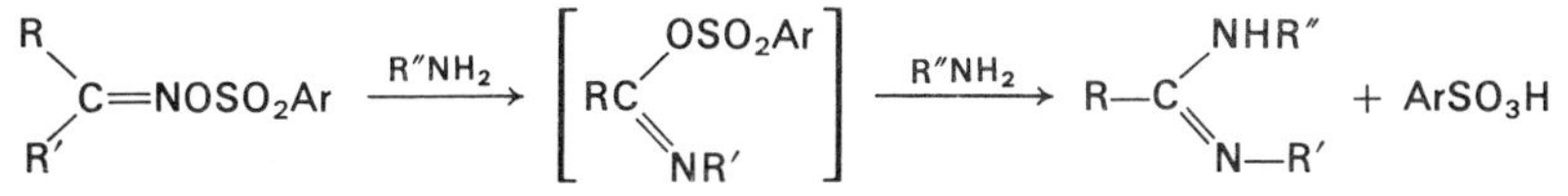

Amides and thioamides may be converted by sultones into imido ester sulphonate salts, which in turn react with amino acids to give amidines[135].

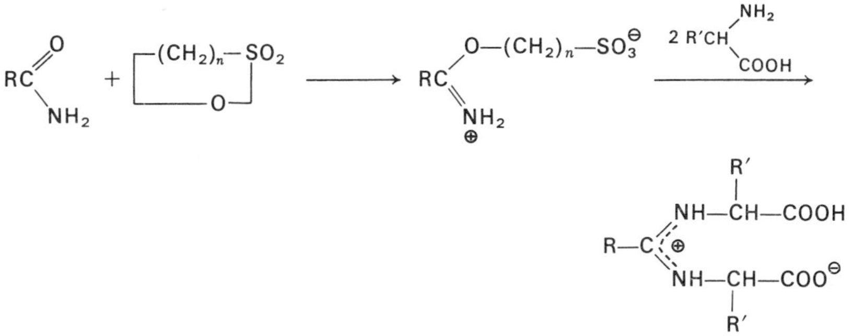

This reaction was studied particularly for the synthesis of multi-functional or cyclic amidines.

Weintraub and co-workers[136] found a new two-step method for the formation of amidines: an amide reacts with triethyloxonium fluoborate to give an easily isolable imido ester fluoborate which, by reaction with an amine yields amidines.

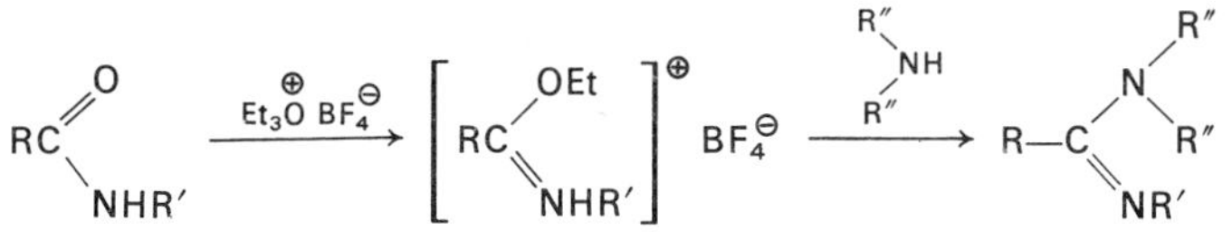

Yields = 71 to 91% for R = Ar or CH_3; R' and R'' = H or CH_3

3. Reactions of miscellaneous mechanisms

Syntheses from amides sometimes occur through particular mechanisms: such cases are methods using titanium complexes, isocyanates, amides in the presence of halogenating agents or reductions of urea derivatives.

Use of titanium complexes: *N*-monosubstituted amides react with tetrakis(dimethylamino)titanium without solvent, or in a benzene–ether mixture or THF according to the general scheme[137]:

$$2\ R^1\!-\!\overset{\overset{\displaystyle O}{\|}}{C}\!-\!NH\!-\!R^2 + Ti\,[N(CH_3)_2]_4 \longrightarrow$$

$$2\ R^1\!-\!\overset{\overset{\displaystyle N\!-\!R^2}{\|}}{C}\!-\!N(CH_3)_2 + TiO_2 + 2(CH_3)_2NH$$

Trisubstituted amidines derived from aliphatic (R^1 = H or alkyl) or aromatic (R^2 = aryl) acids were prepared according to this one-step method. Secondary heterocyclic amides are also transformed by this method into amidines[138].

Action of isocyanates on amides: Aryl isocyanates react with amides and give amidines[139–141]

$$R'NH\!-\!\overset{\overset{\displaystyle O}{\|}}{C}R'' + RN\!\!=\!\!C\!\!=\!\!O \xrightarrow{150°C} R'C\!\!\overset{\nearrow NR''}{\searrow NHR'} + CO_2$$

New reactions were developed, such as action of phenyl isocyanate on dimethylformamide[142] or p-tolylsulphonylisocyanate on dialkylamides[143]. A mechanism involving a cyclic intermediate is proposed[142]:

$$C_6H_5N\!\!=\!\!C\!\!=\!\!O + H\!-\!\overset{\overset{\displaystyle O}{\|}}{C}\!-\!N\overset{\nearrow CH_3}{\searrow CH_3} \longrightarrow \left[\begin{array}{c} C_6H_5\!-\!N\!-\!\overset{\overset{\displaystyle O}{}}{C} \\ CH_3\ \ \ |\ \ \ | \\ \ \ \ N\!-\!C\!-\!O \\ CH_3\ \ H \end{array}\right] \longrightarrow$$

$$CO_2 + C_6H_5\!-\!N\!\!=\!\!CH\!-\!N\overset{\nearrow CH_3}{\searrow CH_3}$$

In the reaction of a substituted urea with an acylating reagent, the acylurea decomposed after heating and gave an amide and an isocyanate that react together to form an amidine[144, 145]:

$$RNH\!-\!\overset{\overset{\displaystyle O}{\|}}{C}\!-\!NHR + R'COCl \xrightarrow{150°C} RNH\!-\!\overset{\overset{\displaystyle O}{\|}}{C}\!-\!N\overset{\nearrow R}{\searrow\underset{\overset{\displaystyle \|}{\underset{\displaystyle O}{C}}\!-\!R'}{}} + HCl \longrightarrow$$

$$RNH\!-\!\overset{\overset{\displaystyle O}{\|}}{C}\!-\!R' + RN\!\!=\!\!C\!\!=\!\!O \longrightarrow R'C\!\!\overset{\nearrow NR}{\searrow NHR} + CO_2$$

Reduction of di-substituted urea by sodium borohydride: The heating of a 1,3-disubstituted urea with approximately equivalent amounts of sodium borohydride gives N,N'-disubstituted amidines[146]:

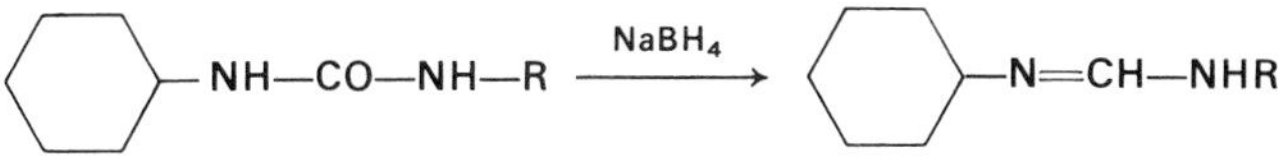

$$\text{Yield} = 66.8\% \text{ for R} = \text{cyclohexyl}$$
$$17.5\% \text{ for R} = \text{phenyl}$$

Under the same conditions, N,N',N'-trisubstituted urea derivatives are cleaved to give formamide and a secondary amine.

4. Synthesis from thioamides

Condensation of ammonia or an amine with a thioamide yields the H_2S salt of the amidine[26, 147-151]. Addition of a mercuric salt may displace the equilibrium by removal of the sulphide ions:

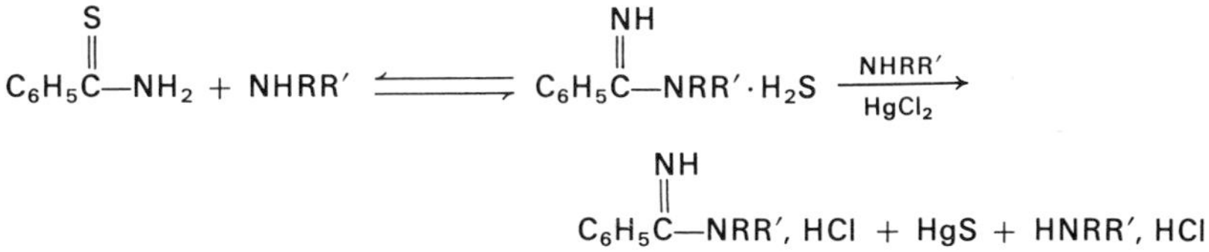

Unsubstituted (R = R' = H) as well as mono and multisubstituted amidines (R and R' = alkyl) may be prepared by this method.

The same amidine may be obtained through two different routes:

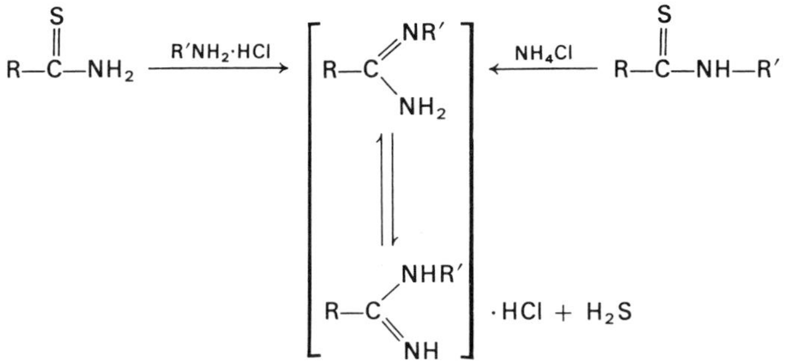

Thioimido ester salts, obtained by the addition of short-chain alkyl iodides to thioamides[152] react easily with molecules containing nitrogen—such as ammonia and primary amines—to form amidines[153-155].

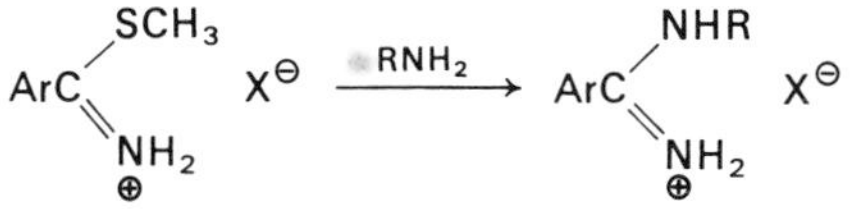

This reaction leads exclusively to amidines if R = C_6H_5, but with bases stronger than aniline, either a mixture of an amidine and a nitrile or the nitrile alone is obtained[155].

N,N'-Dialkylamidines may be formed as well[156] in similar reactions:

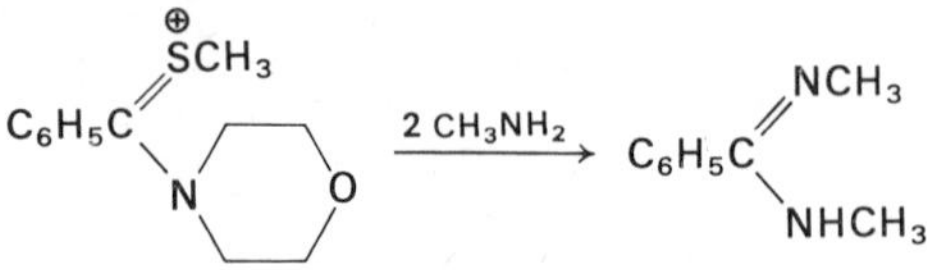

Diarylthiourea reacts with either methylmagnesium iodide[157] or with sodium diethyl malonate[158] to give N,N'-diaryl substituted acetamidines.

C. Miscellaneous Preparations

The following methods are less general than the previous ones. Starting materials contain various organic functions: Schiff bases, hydrazones, amidoximes, carbodiimides, halides, ortho-esters, amines and even amidines.

I. Synthesis from Schiff bases

On heating with sodium amide, Schiff bases give amidines[159]:

$$C_6H_5CH{=\!=}NC_6H_5 + NaNH_2 \xrightarrow[(20\%)]{} C_6H_5\overset{\displaystyle |}{\underset{\displaystyle NH_2}{C}}{=\!=}NC_6H_5 + NaH$$

The mechanism is a nucleophilic substitution of hydrogen, like Tchitchibabin's amination. Yields are poor because of the formation of side products, especially those which result from the reduction of the imino group by the hydride obtained in the reaction.

Action of hypochlorites on some Schiff bases yields addition compounds which with amines give amidines[160].

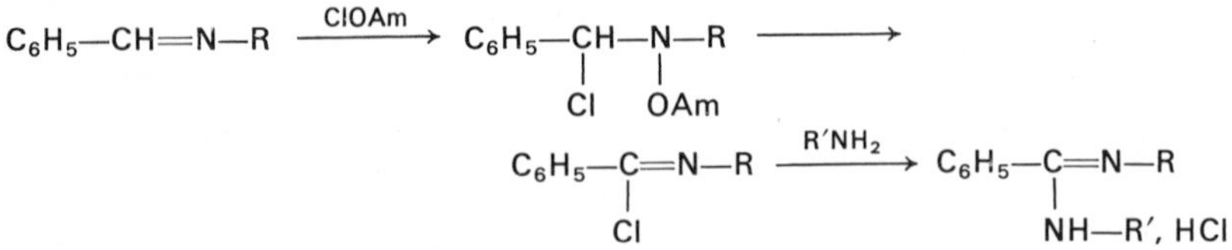

Kröhnke and Steuernagel[161] improved an old method[162] for the preparation of amidines from imines carrying a nitrile function at the imino carbon atom: the imines are heated in pyridine or acetic acid or are melted without any solvent with amine hydrochlorides:

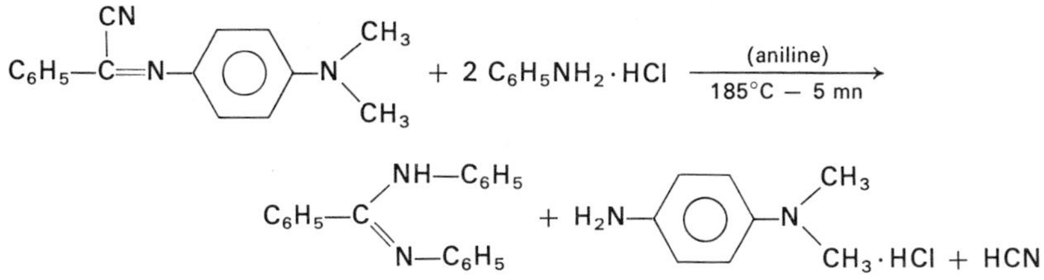

$$C_6H_5-\overset{\overset{\displaystyle CN}{|}}{C}=N-\underset{}{\bigcirc}-N\overset{\displaystyle CH_3}{\underset{\displaystyle CH_3}{\diagdown}} + 2\ C_6H_5NH_2\cdot HCl \xrightarrow[\ 185°C\ -\ 5\ mn\ ]{(aniline)}$$

$$C_6H_5-C\overset{\displaystyle NH-C_6H_5}{\underset{\displaystyle N-C_6H_5}{\diagdown}} + H_2N-\bigcirc-N\overset{\displaystyle CH_3}{\underset{\displaystyle CH_3}{\diagdown}}\cdot HCl\ +\ HCN$$

$$\text{crude yield} = 84\%$$

The proposed mechanism[161] is a nucleophilic attack followed by a double elimination:

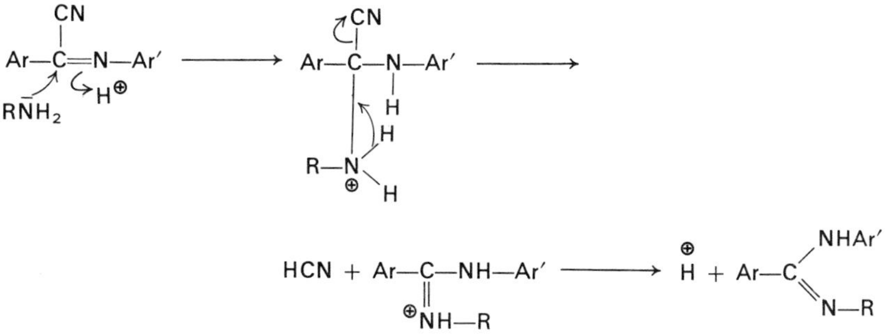

$$\underset{R\overset{\cdot\cdot}{N}H_2}{Ar-\overset{\overset{\displaystyle CN}{|}}{C}}=N-Ar' \longrightarrow Ar-\overset{\overset{\displaystyle CN}{|}}{\underset{\underset{\displaystyle \overset{\oplus}{H}}{R-N}}{C}}-N-Ar' \longrightarrow$$

$$HCN\ +\ Ar-\overset{|}{\underset{\underset{\displaystyle \overset{\oplus}{N}H-R}{\|}}{C}}-NH-Ar' \longrightarrow \overset{\oplus}{H}\ +\ Ar-C\overset{\displaystyle NHAr'}{\underset{\displaystyle N-R}{\diagdown}}$$

A second amine molecule may then substitute Ar'NH$_2$

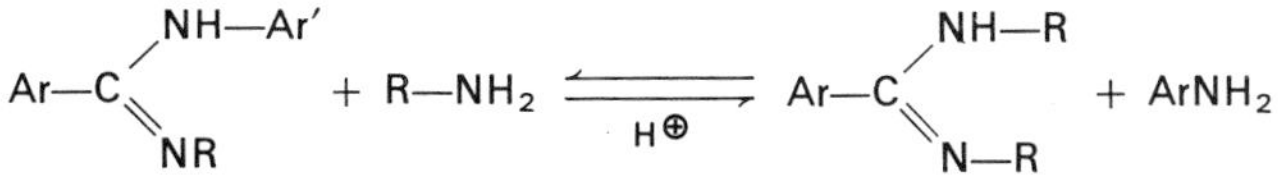

$$Ar-C\overset{\displaystyle NH-Ar'}{\underset{\displaystyle NR}{\diagdown}}\ +\ R-NH_2\ \underset{H^{\oplus}}{\overset{}{\rightleftarrows}}\ Ar-C\overset{\displaystyle NH-R}{\underset{\displaystyle N-R}{\diagdown}}\ +\ ArNH_2$$

2. From hydrazones by transposition

According to Robeff[163], phenylhydrazones of aromatic aldehydes heated with an alkali amide or a Grignard or phenyllithium reagent in xylene lead to monosubstituted amidines[164-166]:

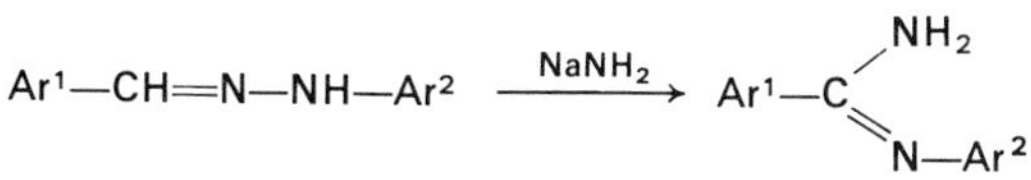

$$Ar^1-CH=N-NH-Ar^2 \xrightarrow{NaNH_2} Ar^1-C\overset{\displaystyle NH_2}{\underset{\displaystyle N-Ar^2}{\diagdown}}$$

This reaction was formerly explained by cleavage of hydrazone into amine and nitrile[167] and subsequent recombination of the two in the presence of a strong base. In fact the reaction does not occur in the absence of oxygen and is inhibited by hydroquinone[164-167], and therefore a free-radical reaction is proposed according to the following scheme[168]:

$$Ar^1-CH=N-NH-Ar^2 \longrightarrow Ar^1-CH=\overset{\cdot}{N} + Ar^2-\overset{\cdot}{N}H$$

$$(Ar^1-CH=\overset{\cdot}{N} \rightleftharpoons Ar^1\overset{\cdot}{C}=NH) + Ar^2-\overset{\cdot\cdot}{N}H \longrightarrow Ar^1-\overset{\overset{\displaystyle NH}{\|}}{C}-NHAr^2$$

$$Ar^1-\overset{\overset{\displaystyle NH_2}{|}}{C}=N-Ar^2$$

The reaction seems to proceed through an intermolecular mechanism[166] and was applied also for substituted hydrazones[169,170].

3. Synthesis from amidoximes

Reduction of amidoximes to amidines[10] has been studied first in 1896 and several applications are given in the literature[171–173].

Recent studies showed that amidoximes may undergo dehydration to nitrenes and subsequent protonation may lead in some cases to amidines in low yields[174].

$$C_6H_5-CH_2-\overset{\overset{\displaystyle N-OH}{\|}}{C}-NH-C_6H_5 \longrightarrow$$

$$C_6H_5-CH_2-C(N)=N-C_6H_5 \xrightarrow[(10\%)]{} C_6H_5CH_2-C\overset{\displaystyle \diagup NH}{\diagdown NHC_6H_5}$$

4. Preparation from cyanamides and carbodiimides

These two compounds having two nitrogens on the same carbon atom may add organometallic reagents and yield amidines:

Normal addition of Grignard reagents to substituted cyanamides leads to amidines:

$$\overset{R}{\underset{R}{\diagup}}N-C\equiv N + R'MgBr \longrightarrow \overset{R}{\underset{R}{\diagup}}N-\overset{\overset{\displaystyle |}{C}-R'}{\underset{\displaystyle N-MgBr}{\|}} \xrightarrow{H_2O} \overset{R}{\underset{R}{\diagup}}N-\overset{\overset{\displaystyle |}{C}-R'}{\underset{\displaystyle NH}{\|}}$$

Arylcyanamides[175] and dibenzylcyanamides[176] have been used as starting materials in this reaction. Beside the normal addition to the nitrile function that gives amidines (reaction 1), some organometallic derivatives lead to mono and dinitriles (reactions 2 and 3)[177].

This reaction is therefore not general for obtaining amidines; sometimes

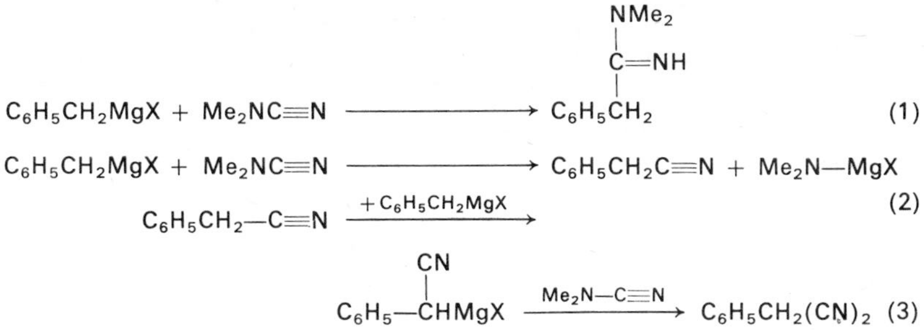

$$C_6H_5CH_2MgX + Me_2NC\equiv N \longrightarrow \underset{\displaystyle C_6H_5CH_2}{\overset{\displaystyle \overset{\displaystyle NMe_2}{|}\ \underset{|}{C}=NH}{}} \tag{1}$$

$$C_6H_5CH_2MgX + Me_2NC\equiv N \longrightarrow C_6H_5CH_2C\equiv N + Me_2N{-}MgX \tag{2}$$

$$C_6H_5CH_2{-}C\equiv N \xrightarrow{+C_6H_5CH_2MgX}$$

$$\underset{\displaystyle C_6H_5{-}\overset{\displaystyle \overset{CN}{|}}{C}HMgX}{} \xrightarrow{Me_2N{-}C\equiv N} C_6H_5CH_2(CN)_2 \tag{3}$$

it may be applied to obtain only nitriles without simultaneous formation of amidines[178].

Amidines are also prepared by action of Grignard reagents on carbodiimides[175]. Phenylacetylene treated with sodium in xylene in the presence of triethylamine gives an acetylide anion which with various disubstituted carbodiimides leads to acetylenic amidines[179]:

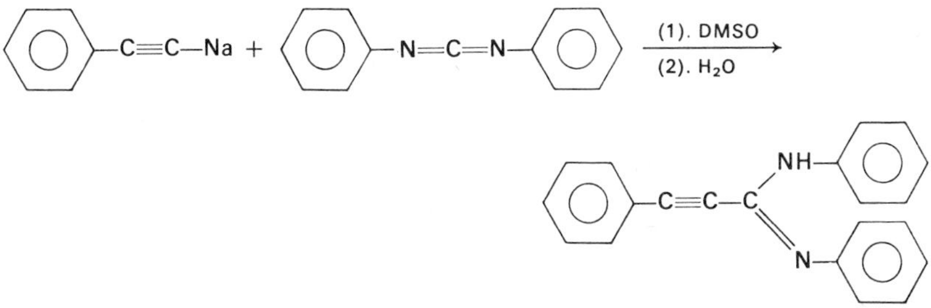

Yield = 50%

Sodium salts of acetylacetone, acetoacetic and malonic esters give substituted amidines through an addition reaction[158, 180]:

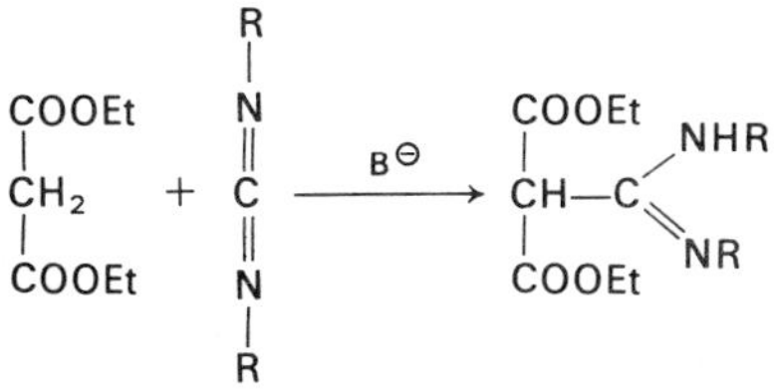

In a similar mechanism but without the need for the formation of a metal derivative, hydrocyanic acid yields α-cyano substituted amidines[180]:

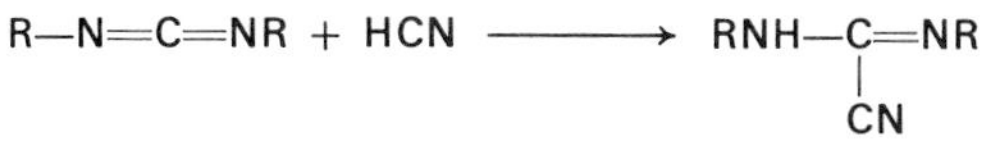

$$R{-}N{=}C{=}NR + HCN \longrightarrow \underset{\displaystyle \underset{|}{CN}}{RNH{-}C{=}NR}$$

5. Synthesis from halogenated compounds

Some compounds having two halogens on the same carbon react with amines to yield amidines. Thus, ethylene derivatives give several types of amidines, e.g. substituted α-amino amidines are obtained from trichlorethylene[181]:

Chlorofluorethylenes lead to α-halogeno amidines[182, 183]

Phenyltrichloromethane reacts with arylamines and yields *N,N'*-disubstituted amidines[183–185] again probably through an imido chloride intermediate[185]:

ortho-Substituted anilines and benzidines do not react as above. *gem*-Dichloroaziridines, easily accessible by addition of dichlorocarbene to Schiff bases, after aminolysis often lead to α-amino amidines[186].

N,N,N'-Trifluoroamidines are obtained by dehydrohalogenation of N,N,N',N'-tetrafluoro *gem*-diamines by pyridine[187].

A similar reaction starts from trichlorovinylamines[188] obtained from trialkylphosphites and α-trihaloacetamides[189].

6. Preparation from orthoesters

Ethyl orthoformate reacts with aromatic amines[190–192]:

This method has been often used for the preparation of N,N'-diaryl-amidines[193–198].

According to one author trianilinomethane was formed by the condensation of aniline with ethyl orthoformate[199], however more recent results[200] proved that an amidine was formed.

The reaction[201, 202] occurs through a two-step mechanism: The initial formation of an imido ester is followed by its transformation by a second aniline molecule into the amidine.

$$H-\underset{\underset{OEt}{|}}{\overset{\overset{OEt}{|}}{C}}-OEt + C_6H_5NH_2 \rightleftharpoons C_6H_5-N{=}CH-OEt + 2\ EtOH$$

$$C_6H_5-N{=}CH-OEt + C_6H_5NH_2 \rightleftharpoons C_6H_5-N{=}CH-NH-C_6H_5 + EtOH$$

Claisen[192] in an early study considered that the amidine is formed first and then reacts with free ethanol in the medium to give an imido ester. This mechanism is discussed in reference 203.

Primary aryl amines are used in this reaction mainly with orthoformic ester or with orthoacetic ester[204, 205]. The same orthoesters and aliphatic amines give similar reactions[204]. Diamines yield cyclic amidines[203].

7. Synthesis from amines

In these amidine syntheses, an amine furnishes the functional carbon atom, which already carries one of the two nitrogens. This reaction cannot be applied to amines which undergo facile dehydrogenation or to unsaturated amines (e.g. ynamines).

Dehydrogenation by mercuric oxide of aliphatic amines leads to amidines[206]. Mercuric salts may be used as dehydrogenating agents to yield cyclic amidines[207]. Hydrogenation by sodium borohydride reforms the starting diamine.

(Theoretical yield)

In similar reactions, sulphonylamidines[208] are obtained by addition of sulphonamides to ynamines.

$$C_6H_5C{\equiv}C-NEt_2 + C_6H_5SO_2NH_2 \longrightarrow$$

$$\left[C_6H_5CH{=}C\underset{\overset{\diagdown}{NHSO_2C_6H_5}}{\overset{\overset{\diagup}{NEt_2}}{}} \right] \xrightarrow{80\%} C_6H_5CH_2C\underset{\overset{\diagdown}{NSO_2C_6H_5}}{\overset{\overset{\diagup}{NEt_2}}{}}$$

8. Syntheses from other amidines

a. *Ammonolysis of amidines.* Displacement of the NH group occurs at high temperatures at which in the following equilibrium the right hand side is favoured:

$$R-C\underset{\overset{\diagdown}{NHR'}}{\overset{\overset{\diagup}{NH}}{}} + R'NH_2 \rightleftharpoons R-C\underset{\overset{\diagdown}{NHR'}}{\overset{\overset{\diagup}{NR'}}{}} + NH_3$$

This substitution is an important side reaction in various amidine syntheses[1,2], and may be preparatively useful when an excess of amine or ammonia displaces the equilibrium or when a more basic amine displaces a less basic one[209,210]. Examples are given in refs. 1, 211 and 212.

$$HC \overset{NC_6H_5}{\underset{NHC_6H_5}{\big<}} + HN \bigcirc \longrightarrow HC \overset{NC_6H_5}{\underset{N}{\big<}} \bigcirc + C_6H_5NH_2$$

b. *Alkylation of amidines.* Amidines having at least one hydrogen atom bound to nitrogen, may undergo alkylation by heating with halides[118,213-220].

$$C_6H_5C \overset{NH}{\underset{NH_2}{\big<}} \xrightarrow{C_2H_5I} C_6H_5C \overset{NC_2H_5}{\underset{NH_2}{\big<}}$$

$$C_6H_5C \overset{NH}{\underset{N(CH_3)_2}{\big<}} \xrightarrow{CH_3I} C_6H_5-C \overset{NCH_3}{\underset{N(CH_3)_2}{\big<}}$$

Amidines may be metallated before their reaction with the halide[221]. The formation and alkylation of amidines may be carried out in a one step reaction[7,222]:

$$R-\overset{R}{\underset{R}{C}}-CN + NaNH_2 \longrightarrow R-\overset{R}{\underset{R}{C}}-C \overset{N-Na}{\underset{NH_2}{\big<}} \xrightarrow{R'Cl} R-\overset{R}{\underset{R}{C}}-C \overset{N-R'}{\underset{NH_2}{\big<}}$$

Interest in these methods is limited by the fact that usually mixtures are obtained[213,222,223]:

$$RC \overset{NR'}{\underset{NHR''}{\big<}} \rightleftarrows RC \overset{NHR'}{\underset{NR''}{\big<}}$$

$$\Big\downarrow CH_3I \qquad\qquad \Big\downarrow CH_3I$$

$$RC \overset{NR'}{\underset{\underset{CH_3}{NR''\cdot HI}}{\big<}} \qquad\qquad R-C \overset{\overset{CH_3}{\underset{}{NR'}}}{\underset{NR''\cdot HI}{\big<}}$$

$$e.g.: \quad 2\,C_6H_5C \overset{NCH_3}{\underset{NH_2}{\big<}} \xrightarrow{CH_3I} C_6H_5C \overset{NCH_3}{\underset{NHCH_3\cdot HI}{\big<}} + C_6H_5C \overset{NH}{\underset{N(CH_3)_2}{\big<}}$$

c. *From a reaction of acetylenic Grignard reagents on chloroformamidines.*
Ried and Weidemann[224] obtained phenylpropiolamidines:

$$C_6H_5C{\equiv}CMgBr + ClC{=}NC_6H_5 \xrightarrow[(90\%)]{} C_6H_5C{\equiv}CC\underset{N}{\overset{NC_6H_5}{\diagup}}\diagdown O + MgBrCl$$

(structures with morpholine rings)

III. SYNTHETIC USES OF AMIDINES

In this Section we shall describe different synthetically useful reactions starting from amidines. We exclude studies of tautomerism as well as the formation of metallic salts, the action of alkylating, acylating and sulphonating reagents, since these reactions have mainly interest for the study of structural problems and hardly for practical purposes.

The discussion is divided into two parts, one dealing with the formation of acyclic derivatives and the second with the formation of heterocyclic compounds.

A. Synthesis of Acyclic Derivatives

In a variety of reactions, amidines may yield thioamides, amidrazones and amidoximes. By reduction of these products aldehydes are formed. Into compounds containing an active methylene group, formamidines may introduce an aminomethylidene group.

I. Synthesis of thioamides

Whereas the acid hydrolysis of amidines to give amides has no synthetic value, the action of hydrogen sulphide on amidines represents a useful route to thioamides.

The early studies of Bernthsen[26] were little utilized[225]. The reaction is carried out at 135–165°C and yields mixtures of thioamides:

$$2\ C_6H_5C\underset{N(C_6H_5)_2}{\overset{NH}{\diagup}} + H_2S \xrightarrow{\Delta} (C_6H_5)_2NH + C_6H_5\overset{S}{\overset{\|}{C}}NH_2 + C_6H_5\overset{S}{\overset{\|}{C}}{-}N(C_6H_5)_2$$

Reynaud and co-workers[226, 227] experimented with a process in which the attack on the amidine by H_2S is facilitated by the presence of pyridine and which is carried out at comparatively low temperatures thus avoiding side reactions. In general a mixture of two thioamides is obtained by this method.

$$\left[RC \underset{NH_2}{\overset{NR'}{<}} \; \rightleftarrows \; R-C \underset{NH}{\overset{NHR'}{<}} \right] \xrightarrow[\text{(pyridine)}]{H_2S}$$

$$R-C \underset{NH_2}{\overset{NHR'}{\underset{|}{-}}} SH \longrightarrow \begin{cases} RC \underset{S}{\overset{NH_2}{<}} + R'NH_2 \\ \\ RC \underset{S}{\overset{NHR'}{<}} + NH_3 \end{cases}$$

In some cases, this reaction yields a single thioamide, e.g. unsubstituted thioamides are obtained in good yields [227]:

$$CH_3C \underset{NH}{\overset{NEt_2}{<}} \xrightarrow{H_2S} \left[CH_3C \underset{NH_2}{\overset{NEt_2}{\underset{|}{-}}} SH \right] \longrightarrow HNEt_2 + CH_3C \underset{S}{\overset{NH_2}{<}}$$

Bernthsen [26] described a similar reaction using carbon disulphide:

$$C_6H_5C \underset{NH-C_6H_5}{\overset{N-C_6H_5}{<}} + CS_2 \xrightarrow{\Delta} C_6H_5 \overset{S}{\overset{\|}{C}}-NHC_6H_5 + C_6H_5N{=}C{=}S$$

2. Syntheses of amidrazones and of amidoximes

Ammonia or amines used in excess and at high temperatures may displace one of the nitrogens of the amidino group [48, 147]:

$$RC \underset{NHR'}{\overset{NH}{<}} + R''NH_2 \rightleftarrows RC \underset{NHR'}{\overset{NR''}{<}} + NH_3$$

Similar equilibria are also established with other derivatives of ammonia such as hydrazines and hydroxylamines, e.g. phenylhydrazine heated with amidines yields imidohydrazides (amidrazones) [218, 228]:

$$ArC \underset{NH_2}{\overset{NC_6H_5}{<}} + C_6H_5NHNH_2, HCl \longrightarrow NH_4Cl + ArC \underset{NHNHC_6H_5}{\overset{NC_6H_5}{<}} \rightleftarrows$$

$$ArC \underset{NNHC_6H_5}{\overset{NHC_6H_5}{<}}$$

Hydroxylamine gives amidoximes[10, 48, 218, 229], but in practice this reaction is not very useful[230].

$$
RC \diagdown{\overset{NR'}{\quad} \atop \underset{NH_2}{\quad}} + NH_2OH \cdot HCl \longrightarrow NH_4Cl + RC \diagdown{\overset{NHR'}{\quad} \atop \underset{NOH}{\quad}}
$$

3. Synthesis of aldehydes

Reduction of amidines by sodium in ethanol[231, 232] or by sodium amalgam in dilute acid[233] yields in some cases aldehydes. Birch, Cymerman-Craig and co-workers[234–236] modified the reaction and carried out the reduction in liquid ammonia in the presence of ethanol as proton donor. The amidine is first reduced to a *gem*-diamine, which is hydrolysed subsequently in an acid medium to yield the aldehyde:

$$
RC{\overset{NH}{\underset{NH_2}{\quad}}} \xrightarrow{2\,Na} RC{\overset{\overset{\ominus}{NH}}{\underset{NH_2}{\quad}}} \xrightarrow{2\,ROH} RCH{\overset{NH_2}{\underset{NH_2}{\quad}}} \xrightarrow[H_2O]{H^{\oplus}} RC{\overset{O}{\underset{H}{\quad}}}
$$

The insolubility of the amidine or the amidine salt in ammonia limits the usefulness of the method; benzamidine or benzamidine hydrochloride lead to benzaldehyde in 100% yield, while the insoluble N,N'-diphenyl-benzamidine does not react.

4. Action of formamidines on active methylene compounds

Dains[198, 237–244] condensed formamidine with active methylene compounds by heating the mixture at 125–200°C:

$$
RN{\underset{RHN}{\overset{\quad}{\diagup}}}CH + H_2C{\overset{X}{\underset{Y}{\quad}}} \longrightarrow RNH_2 + RHNCH{=}C{\overset{X}{\underset{Y}{\quad}}}
$$

The formamidines used should carry two identical substituents. Acetylacetone or acetoacetic, cyanoacetic and malonic esters are used as the active methylene compounds. Benzyl cyanide and deoxybenzoin react with difficulty[237]. This reaction is applicable to heterocyclic compounds with CH_2 groups activated by carbonyl groups and carbon–nitrogen double bonds (pyrazolones, isoxazolones) or carbonyl groups and sulphur (thioimidazolones):

If the active methylene compound is used in excess, the formation of a 'double' product may occur instead of that of an aminomethylidene derivative:

Condensation of diarylformamidines with ethyl malonate yields quinoline derivatives[194].

B. *Syntheses of Cyclic Derivatives*

Cyclizations are the most important reactions of amidines. Different heterocycles containing the —N=C—N= grouping are obtained. We shall describe reactions leading to 3, 4, 5 and 6 membered rings.

I. Synthesis of three-membered rings

Diazirines are obtained from the reaction of amidines with sodium hypobromite or hypochlorite[245].

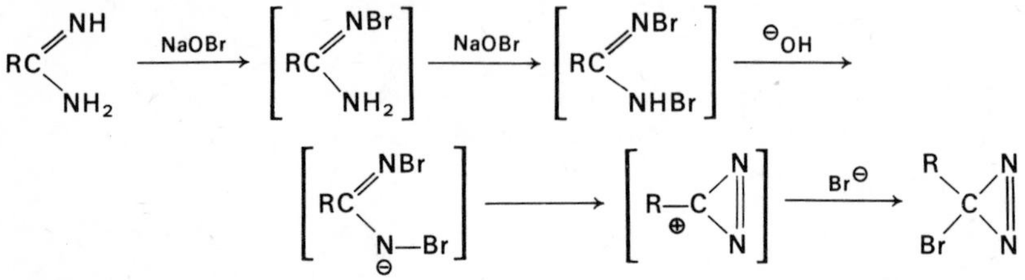

2. Synthesis of four-membered rings

Only a few examples are known, and the products are lactams, e.g. diphenylketen reacts with trisubstituted amidines and yields azetidinone derivatives[246]:

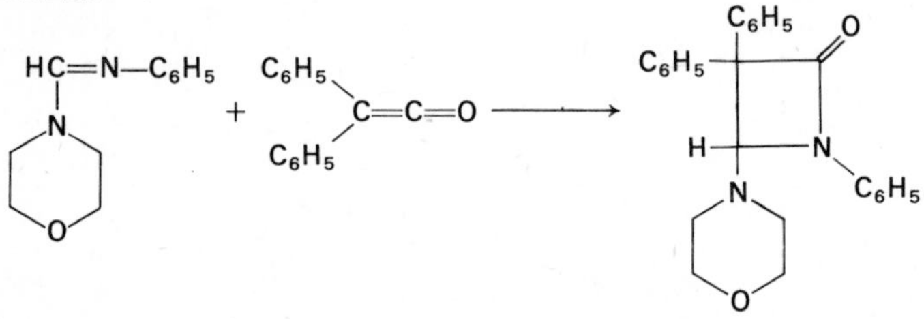

In a similar process, aminoamidines give diazetidinones[247]:

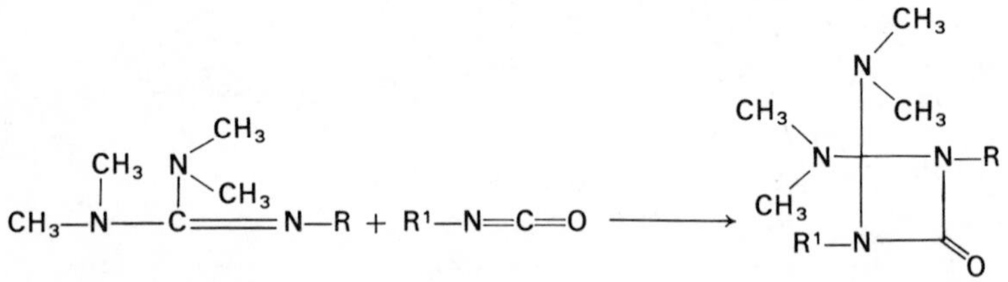

3. Synthesis of five-membered rings

Condensation reactions of amidines may form pyrrole, oxazole, oxadiazole, oxathiadiazole, and especially imidazole ring systems.

a. *Pyrrole derivatives.* One of the rather rare examples of this mode of reaction is when *N'-o*-tolyl-*N*-methyl-*N*-phenyl formamidine reacts on heating with sodium amide to yield indole[248]:

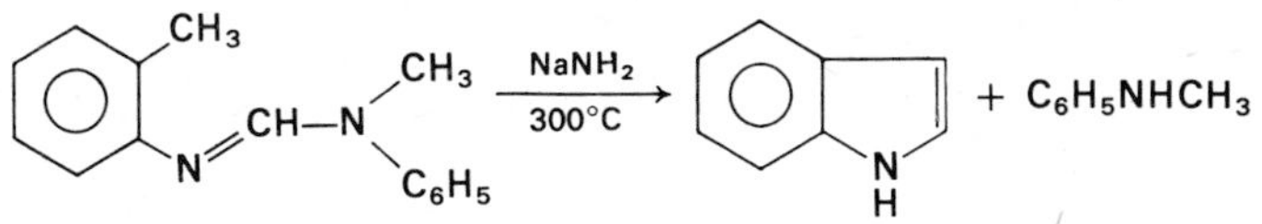

Isatin derivatives are obtained by dehydrogenating N,N'-diphenyl-amidines of phenylglycine[249]:

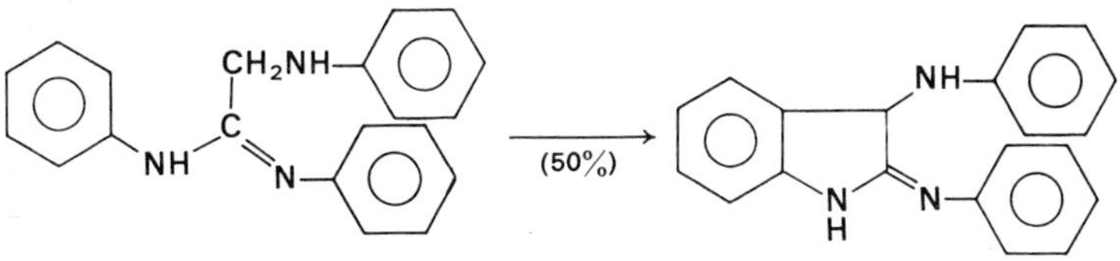

b. *Oxazole derivatives.* Heating of N-phenyl-N'-(o-hydroxyphenyl)-benzamidine leads to 2-phenylbenzoxazole[31].

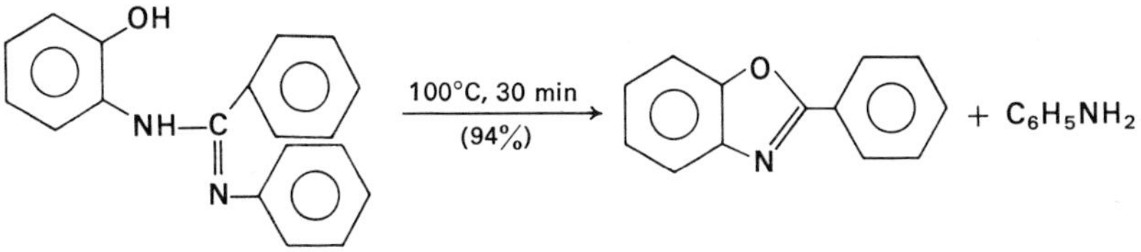

Benzoxazoles are also obtained from the condensation of o-aminophenol with amidines[250]:

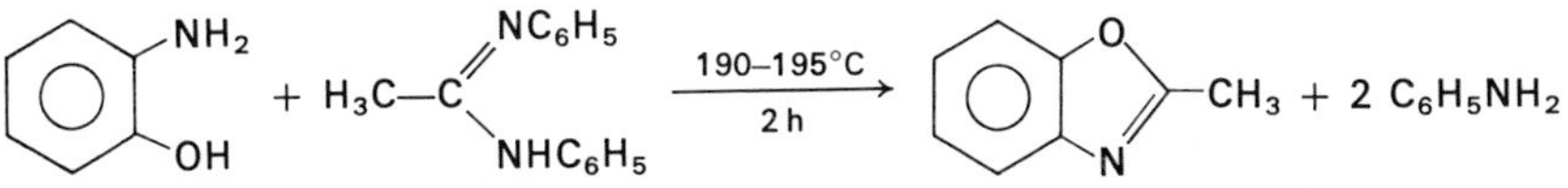

Aminoethanol gives a similar reaction leading to oxazolines[251]:

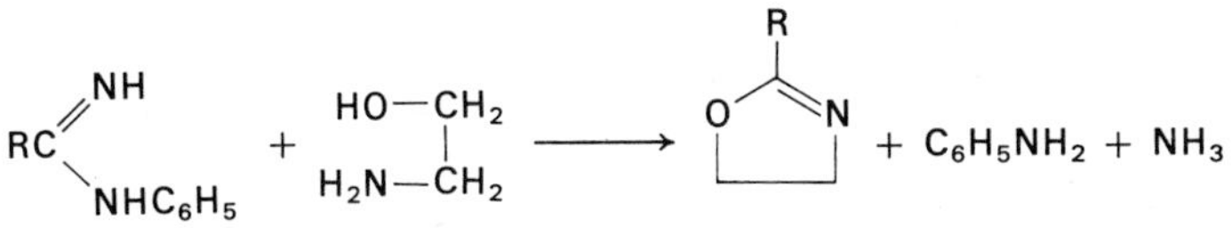

Epoxides react at room temperature with butyramidine: heating of the unstable intermediate addition compound leads to 2-propyloxazolines[252]:

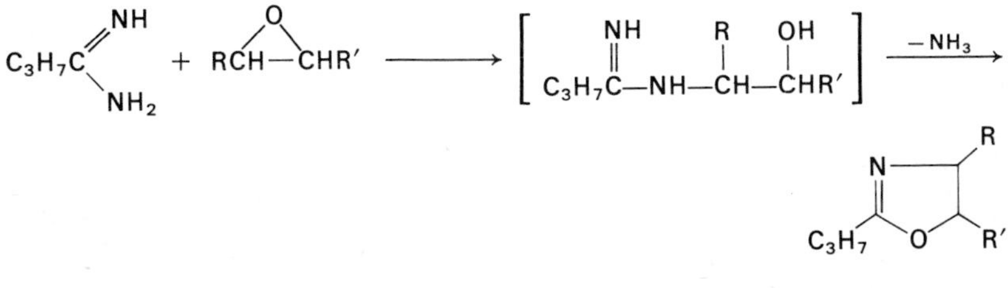

In similar condensations, cyclic amidines (e.g., imidazolines and tetra-hydropyrimidines), give bicyclic systems by action of epoxides[253].

c. *Oxadiazole, thiadiazole and oxathiadiazole derivatives.* 3,5-Diphenyl-1-oxa-2,4-diazole is formed in the condensation of benzamidine with α-chlorobenzaldoxime[254]:

$$C_6H_5C \underset{NOH}{\overset{Cl}{<}} \; + \; \underset{H_2N}{\overset{HN}{>}}CC_6H_5 \xrightarrow[\text{room temp.}]{24\,h} \text{[3,5-diphenyl-1-oxa-2,4-diazole]} + NH_4Cl$$

N-Acyl-*S*-chloroisothiocarbamoyl chlorides react with monosubstituted amidines: the mixtures of thiadiazoles formed are separated by chromatography[255]:

$$ArC(=O)-N=C(SCl)(Cl) \; + \; 3\; \underset{HN-R}{\overset{HN}{>}}CAr' \longrightarrow$$

$$ArC(=O)-N=C\overset{S}{<}N-\text{...}(R)(Ar') \quad + \quad ArC(=O)-N=C\overset{S}{<}N-\text{...}(R)(Ar')$$

Sulphenyl chlorides react with unsubstituted amidines and yield thiadiazoles[256–258]

$$Cl_3C-S-Cl \; + \; H_3C-C\underset{NH_2}{\overset{NH}{<}} \xrightarrow[\text{CH}_2\text{Cl}_2/\text{H}_2\text{O}]{NaOH} \text{[CH}_3\text{-thiadiazole-Cl]} + 4\,NaCl + 4\,H_2O$$

In similar reactions, thiadiazolopyridinic systems are obtained from 2-aminopyridines and trichloromethanethiol[259]. *N*-Chloroamidines with thioamides also yield thiadiazole ring systems[260].

$$C_6H_5C\underset{NHR}{\overset{NCl}{<}} \; + \; C_6H_5-C(=S)-NH_2 \xrightarrow{EtOH} C_6H_5C\overset{N-S}{<}\underset{N=C}{|}(C_6H_5) \; + \; C_6H_5C\underset{NHR}{\overset{NH}{<}} \quad HCl + S$$

d. *Imidazole derivatives.* Non-condensed imidazoles: The condensations of amidines with α-dicarbonyl compounds were widely studied[48, 261–264].

In these condensations, Jacquier and co-workers[265] showed that 4,5-dihydroxyimidazolines are formed, and no open-chain compounds can be isolated. Usually the 4,5-dihydroxyimidazolines undergo dehydration on heating and give imidazolinones as well as other by-products.

It is also possible to obtain imidazole rings by condensation of amidines with α-halo ketones[266], with ethyl phenylpropiolate[267], or with α-hydroxy ketones[268].

Heating at 180–200°C of N-allyl-N'-arylacetamidine hydrochlorides yields imidazolines[46].

If the cyclization occurs in the presence of polyphosphoric acid, dihydroquinazolines sometimes mixed with dihydrobenzodiazepines are obtained.

The reaction of *N*-chlorethylbenzamide with phosphorous pentachloride in the presence of aniline leads to diphenylimidazolines[269], the intermediate seems to be an amidine which however cannot be isolated.

Amidine salts, when heated with ethylene diamine, yield dihydroimidazoles. Unsubstituted amidines[270, 271] react as well as substituted amidines[272].

In the reaction of 2-amino-3-phenacyloxadiazolium halides with benzamidine, a rearrangement occurs leading to 1,2-diaminoimidazole derivatives[273].

Amidines react with oxalyl chloride to yield imidazolinediones[274]:

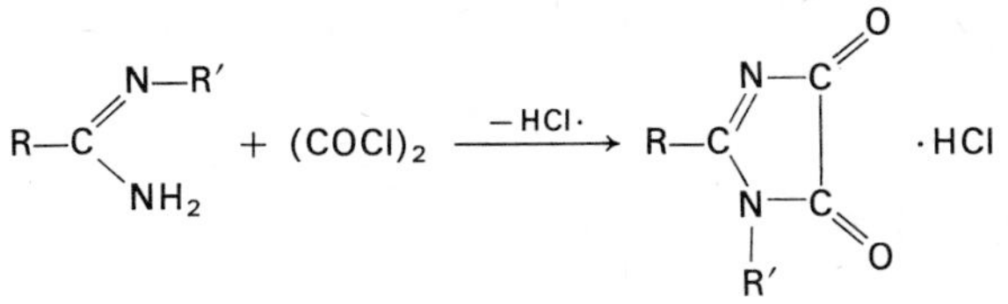

Condensed imidazoles: Benzimidazoles. *N*-arylamidines by action of hypochlorites in basic medium give benzimidazoles; the *N*-chloroamidine intermediates are sometimes isolated[275-277]:

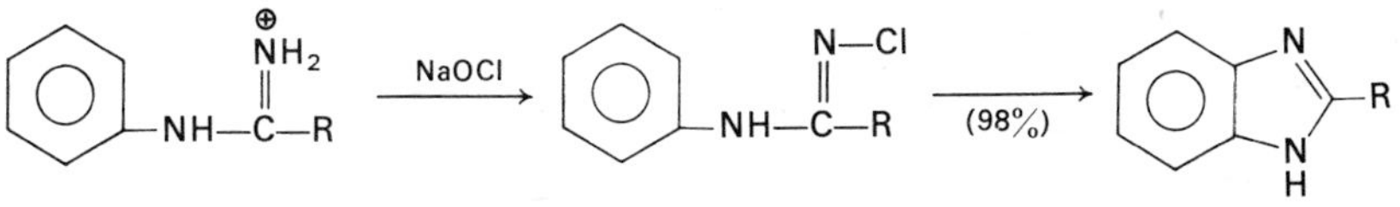

This reaction is applicable for obtaining complex heterocyclic systems, e.g. 2-phenyl(1,2-*d*)naphthoimidazole is obtained by action of *t*-butyl hypochlorite on α or β naphthyl benzamidines[276].

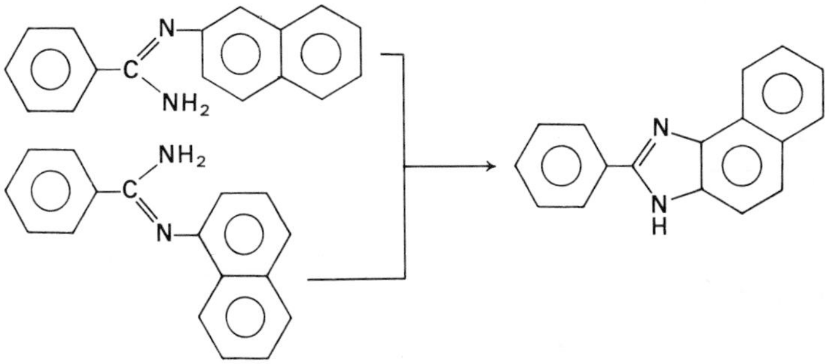

ortho-Phenylene diamine reacts with amidines to yield benzimidazoles[31, 278]:

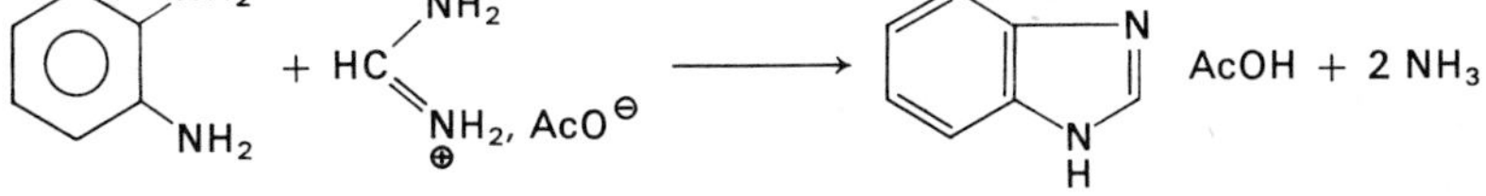

Imidazoles condensed to a non-benzenic cycle: Amino malonic acid diamidine reacts with ethyl orthoformate in DMF and gives a good yield of adenine[279].

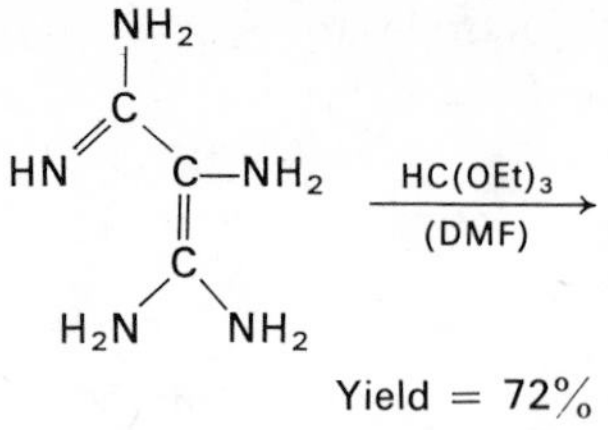

Yield = 72%

Amidines react with 4,5-diaminopyrimidines to yield purines[280]:

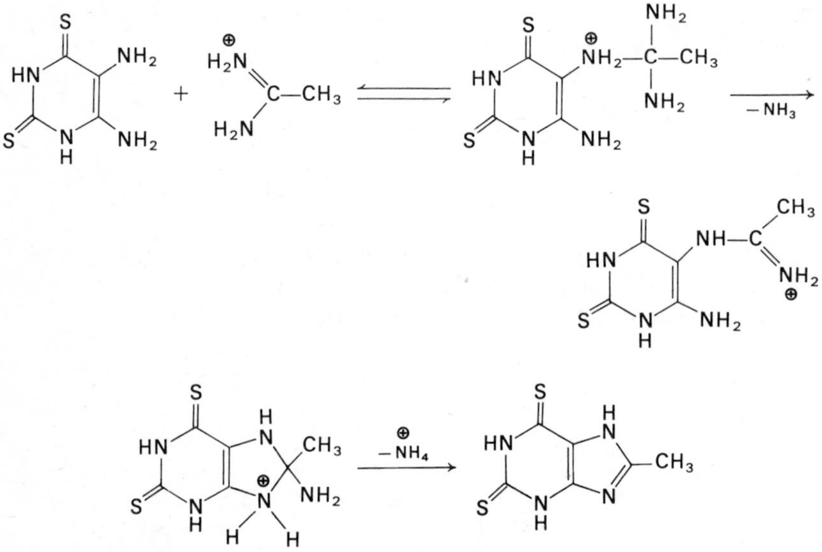

In the condensation of benzamidine with ethyl oxalate the first step leads to an imidazolinedione[281]; the reaction may continue to yield 2,5-diaryl imidazo [4,5-d]imidazoles[282].

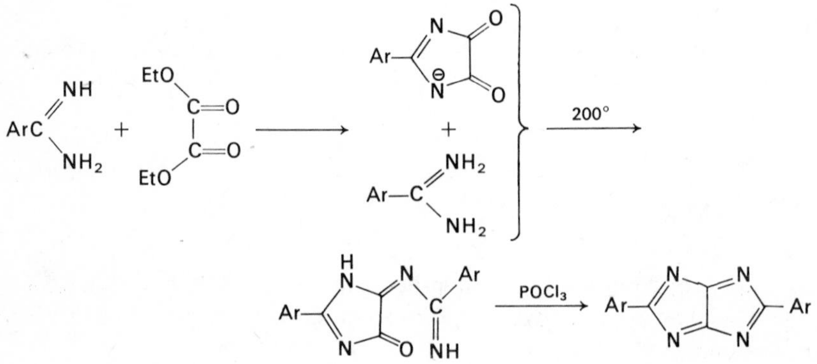

1,5-Diaryl [1,2-*b*] imidazo triazoles are obtained from imidazole substituted amidines[273]:

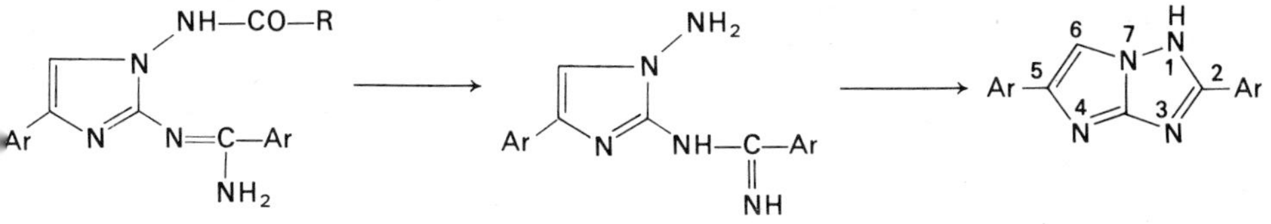

These cyclizations are similar to the formation of [1:3:3*a*]-triazaindenes by dehydrogenation of *N*-2-pyridylamidines[283].

4. Synthesis of six-membered rings

a. *Pyridine derivatives.* The formation of pyridine rings from amidine starting materials is unusual. Some amidines carrying an amide function condense with β-diketones and yield aminopyridines[284].

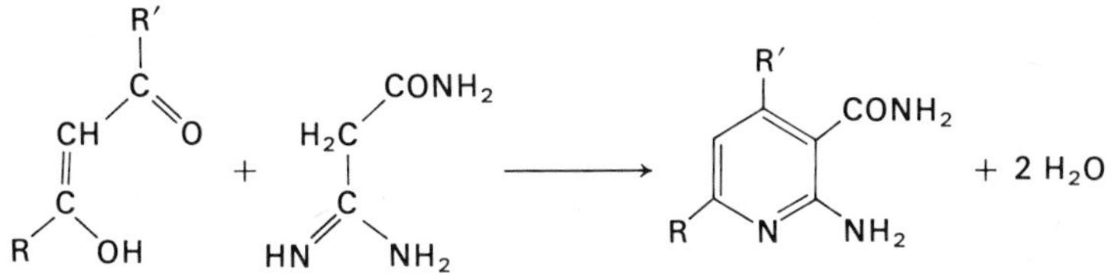

In the particular case of amidines derived from phenylethylamine, dihydroisoquinolines are formed by heating in the presence of phosphorous oxychloride[285].

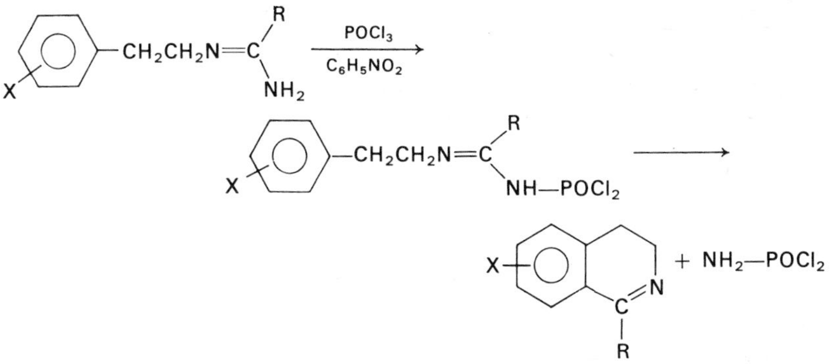

The condensation of chloroformamidines with acetylenic Grignard reagents gives amidines derived from propiolic acid. These amidines may yield 2-aminoquinolines in a cyclization reaction by action of polyphosphoric acid[224].

N-Biphenylylbenzamidines treated by $POCl_3$ yield phenanthridines[286]:

b. *Pyrimidine derivatives.* Condensation of unsubstituted amidines with β-difunctional compounds is a general method leading to pyrimidine derivatives. The reactions will be classified according to the reagents which react with the amidines.

From β-dicarbonyl compounds: β-Diketones are easily accessible and are widely used in pyrimidine syntheses[287-290]:

Yield = 70%; $n = 6$.

β-Dialdehydes may also be used and give similar reactions[291, 292].

From β-keto esters: Unsubstituted amidines easily condense with β-keto esters: the reaction was first described by Pinner[293, 294] and is useful for pyrimidine syntheses[295–297].

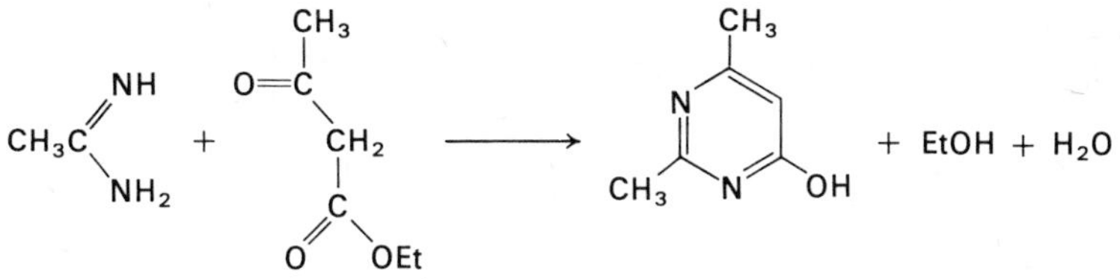

β-Aldehydo esters can also be used in these reactions. Ethyl-2-formyl-3-ethoxy-propionate gives rather poor yields in this condensation[298, 299]:

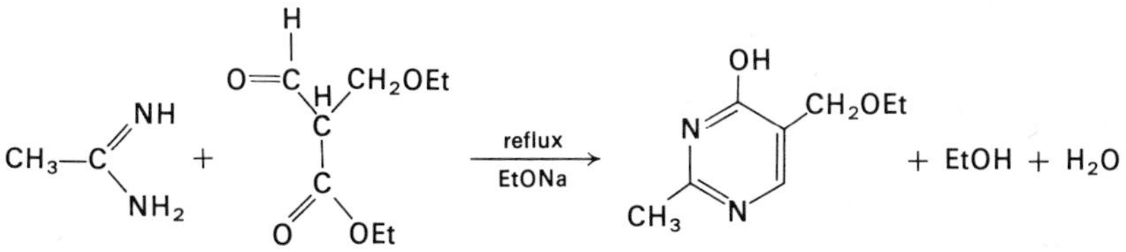

Ethyl-2-formyl succinate gives similar reactions[300–302]:

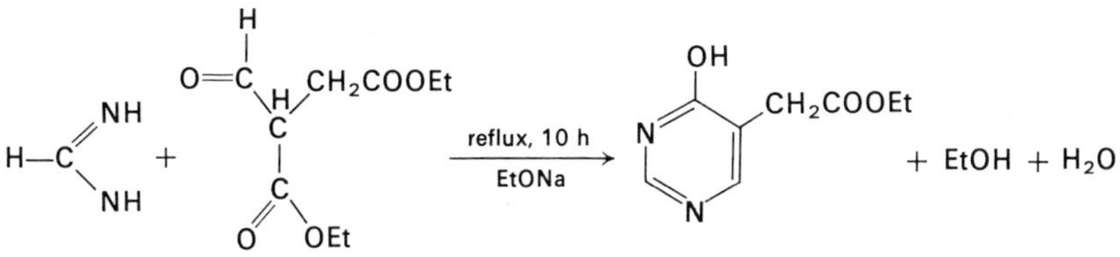

Vitamin B$_1$ was synthetized by a modification of the same method, employing trifluoroacetamidine[303]. Ethyl-N-methyl piperidone carboxylate in a similar reaction yielded a bicyclic compound[304]:

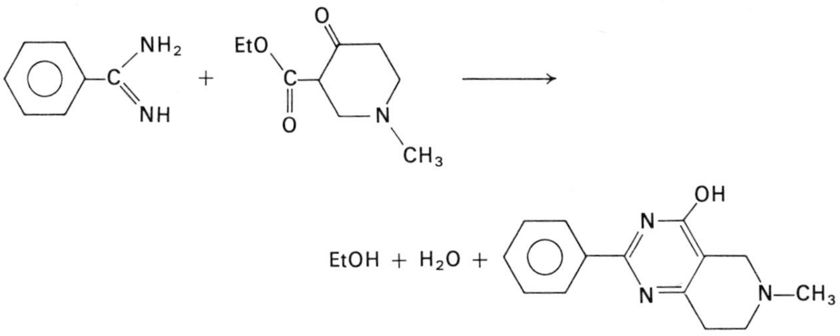

From malonic derivatives: The aptitude of malonic esters for cyclization in basic media is apparent also in their reactions with amidines. Thus formamidine reacts with ethyl malonate[305], yielding 4,6-dihydroxy-pyrimidine:

Various other compounds are obtained in similar reactions.

Substituents in the 2-position of the pyrimidine ring are introduced by amidines carrying alkyl[306–308], aryl[309], p-alkoxybenzyl[310], or amide functions[311].

Substituents in the 5-position are obtained by the choice of suitably substituted alkyl-[312] or cyclobutyl-[313] malonic esters.

A protected functional group in the malonic ester may lead to bicyclic systems[314].

The reaction is more difficult with substituted amidines, but is still sometimes possible by using malonyl chloride[315].

The dithioester $EtOOC-CH_2—CSSCH_3$ is much less reactive than malonic ester itself and did not condense with formamidine[316].

From β-dinitriles, β-cyano esters and β-keto nitriles: The easily accessible cyanoacetic esters are condensed with amidines in the presence of alkali alcoholates to yield 4-hydroxy-6-amino-pyrimidines. Substituents in the

2-position are introduced by using the appropriately substituted starting amidine[305,317–322].

CH$_3$—C(=NH)(NH$_2$) + EtO—C(=O)—CH$_2$CN $\xrightarrow[\text{reflux, 2 h}]{\text{Na/CH}_3\text{OH}}$ (80%) → [pyrimidine product] + EtOH

2-Dialkoxymethyl-3-alkoxypropionitrile which is an acetal of a β-alde-hydo nitrile condenses with acetamidine and yields a dihydropyrimido-pyrimidine; one of the rings of this compound is subsequently split open in the reaction[323].

ROCH$_2$—CH(—CN)—CH(OR′)(OR′) + CH$_3$C(=NH)(NH$_2$) → ROCH$_2$—C(—CN)=CH—N=C(NH$_2$)(CH$_3$) + CH$_3$C(=NH)(NH$_2$) →

[dihydropyrimidopyrimidine] → [pyrimidine with CH$_2$—NH—CO—CH$_3$]

From α, β-unsaturated esters, nitriles and carbonyl derivatives: Any of the following three systems may be used as starting materials:

[three structures: βC=αC systems with R^2, R^1, and C=O (R), C=O (RO), and C≡N]

These structures react with amidines in two steps which may be more or less easy to characterize. In the first step, the amidine usually attacks the β-carbon atom, e.g. R^2 = H[324–326], R^2 = C$_6$H$_5$[327,328] or R^2 = alkyl[329,330] and addition is obtained. If R^2 is a good leaving group, the first step is a substitution reaction, e.g. when R^2 = OR[331–340]; R^2 = Cl[341], R^2 = CH(COOEt)$_2$[342], or R^2 = NRR[343,344].

In the second step, ring closure occurs by attack of the second amidine nitrogen on the ketone, ester, or nitrile function.

The nature of R^1 does not influence the reaction and it appears as a substituent in the pyrimidinic end-product.

Condensation of amidines with acrylonitrile[324], α,β-unsaturated esters[325,330] or ketones[328,345] leads to dihydropyrimidines.

$$C_6H_5C\begin{array}{c}NH\\\\NH_2\end{array} + CH_2{=}CH{-}C{\equiv}N \longrightarrow C_6H_5\text{-dihydropyrimidine-}NH_2$$

The dihydropyrimidine may be easily oxidized into a pyrimidine; e.g. during the condensation of benzalacetophenone with benzamidine, the unsaturated ketone is reduced while the dihydropyrimidine is oxidized[345].

$$C_6H_5C\begin{array}{c}NH\\\\NH_2\end{array} + C_6H_5CH{=}CHCOC_6H_5 \longrightarrow$$

$$\left[\text{dihydropyrimidine ring}\right] \xrightarrow{C_6H_5CH=CHCOC_6H_5} \text{pyrimidine} \quad +$$

$$C_6H_5CH_2CH_2COC_6H_5$$

Alkoxymethylenemalonates condense with amidines and yield 4-hydroxypyrimidine-5-carboxylic esters[340].

$$RC\begin{array}{c}NH\\\\NH_2\end{array} + \begin{array}{c}H\quad OR\\C\\\|\\C\\RO{-}C\quad C{-}OR\\\|\quad\ \|\\O\quad O\end{array} \longrightarrow R\text{-pyrimidine}\begin{array}{c}COOR\\\\OH\end{array}$$

Alkoxymethylene cyanoacetic esters may yield with amidines mixtures of compounds[337]:

$$CH_3C\begin{array}{c}NH\\\\NH_2\end{array} + \begin{array}{c}EtO\quad H\\C\\\|\\C\\N{\equiv}C\quad C{-}OEt\\\ \ \ \|\\\ \ \ O\end{array} \longrightarrow$$

(a) intermediate structure

(b) pyrimidine with COOEt, CH$_3$, NH$_2$

(c) pyrimidine with C$\equiv$N, CH$_3$, OH

The intermediate (**a**) obtained by Todd and Bergel[336] when heated in an alkaline medium gave the cyanopyrimidine (**c**). However, Nishigaki and co-workers[334] later showed that in the same condensation, the derivative (**b**) may also be formed if an excess of the amidine (3:1) is used. Other side reactions may also occur accompanying these cyclizations, e.g. the condensation of amidines with some dihydrofurane derivatives which contain an α,β-unsaturated ester structure, involves the opening of the dihydrofurane ring[344]:

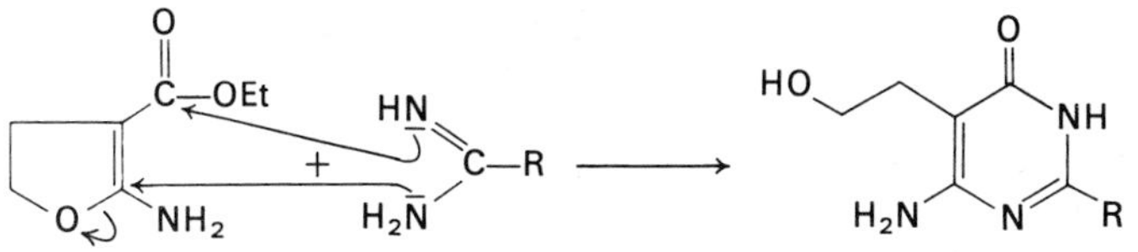

Miscellaneous cyclizations: According to Lacey[346], condensation of *diketene* with acetamidine or with benzamidine leads to the same 4-keto dihydropyrimidines which are also obtained in the condensations of these amidines with acetoacetic ester.

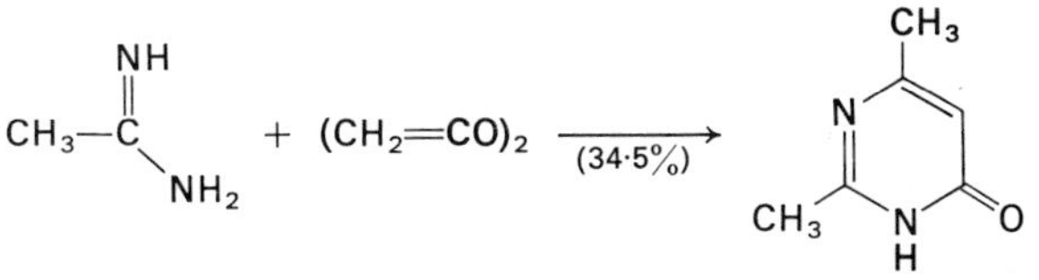

In a similar reaction, benzamidines were condensed with trichloromethylpropiolactone, giving tetrahydropyrimidones[347]. It has been pointed out, that diketene itself can be regarded as a β-lactone.

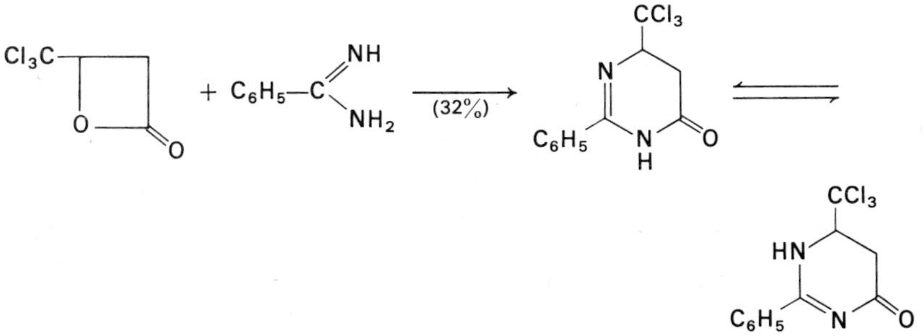

N-(α-Chloroalkenyl)amidines react on heating with phosgene and yield chloro derivatives of pyrimidines or dihydropyrimidines[348]:

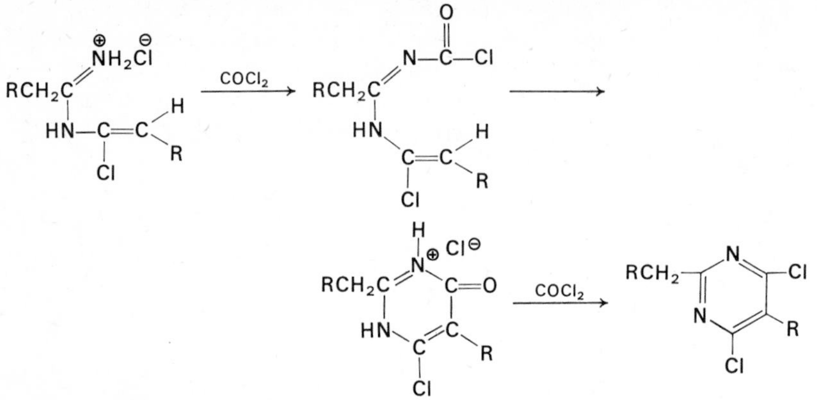

Condensation of mucobromic acid with benzamidine leads to bromo-pyridine carboxylic acid[349].

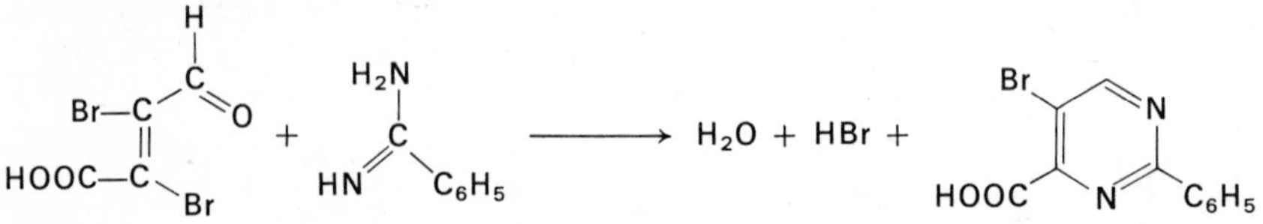

Malonic acid diamidine reacts with some esters or acylating agents and gives 4,6-diaminopyrimidines, but the method is not a general one[350, 351]:

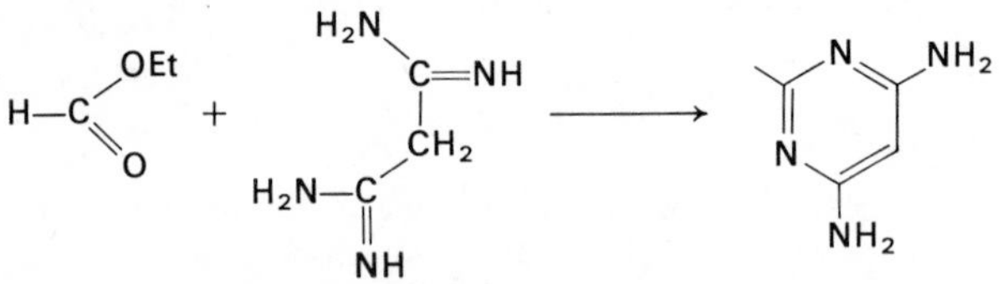

3-Methoxy-2-(dimethoxymethyl) propionitrile and the corresponding esters react with amidines to form pyrimidine derivatives. The nitrile

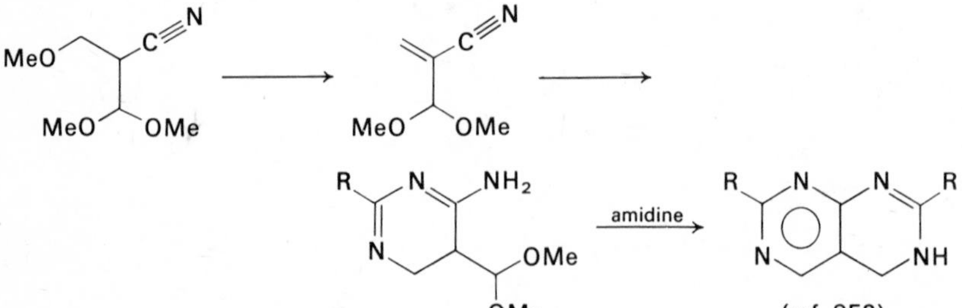

(ref. 352)

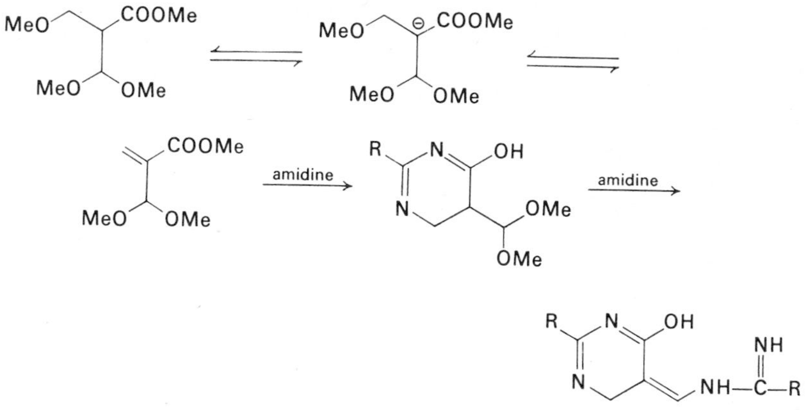

(ref. 354)

reacting with two molecules of amidine yields pyrimidinopyrimidine[352,353]. The ester, through analogous intermediate steps, leads to different products[354].

c. *Triazines: S*-Triazine may be obtained by heating of formamidine hydrochloride (a)[355,356]. The reaction also gives good yields with trichloracetamidine; other amidines give mainly nitriles (b) and only poor yields of triazines[355].

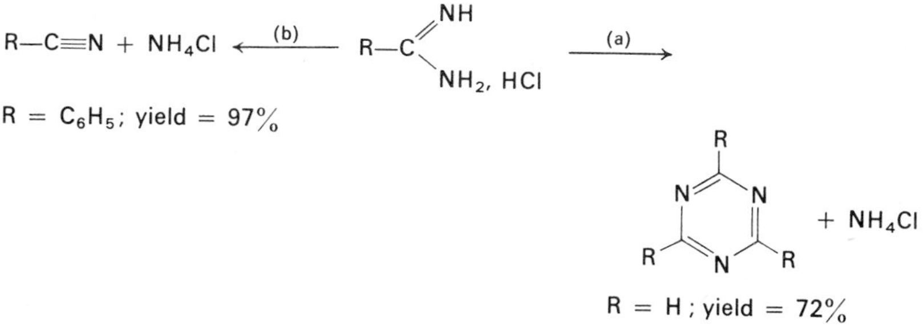

$R = C_6H_5$; yield = 97%

$R = H$; yield = 72%

The first step of this cyclization is the formation of a dimer which is facile if the group R is small or if it is strongly electron-attracting.

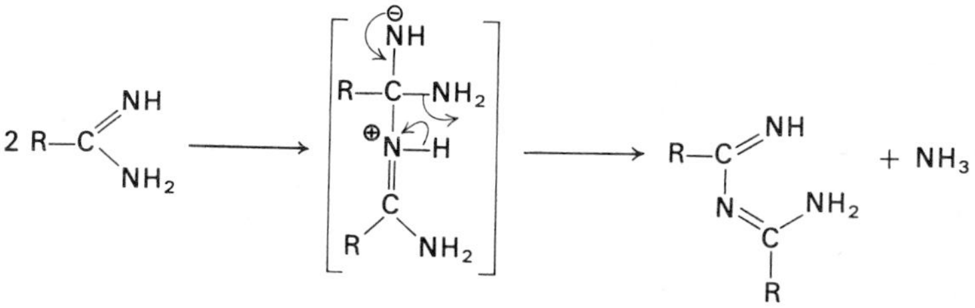

Cotrimerization of two amidines is possible but the reaction has only limited interest because in all such cases mixtures are obtained. In a similar reaction, the attack on sym-triazine by an amidine leads to substituted triazines[357, 358]:

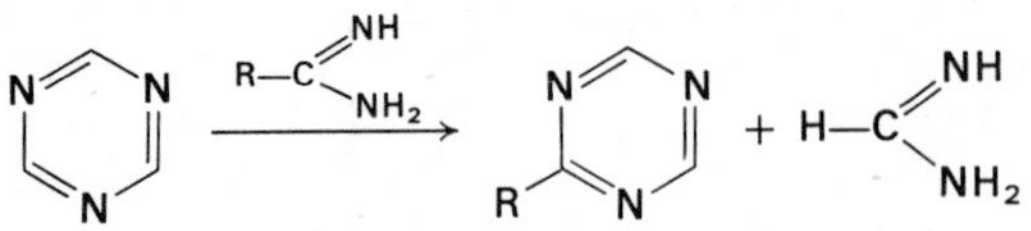

A synthesis of S-triazines was described by Pinner: this starts from arylamidines and phosgene[76,359,360], and symmetrical 2,4-diaryl-6-hydroxy triazines are obtained. Bis-imidoyl urea is an intermediate of the reaction.

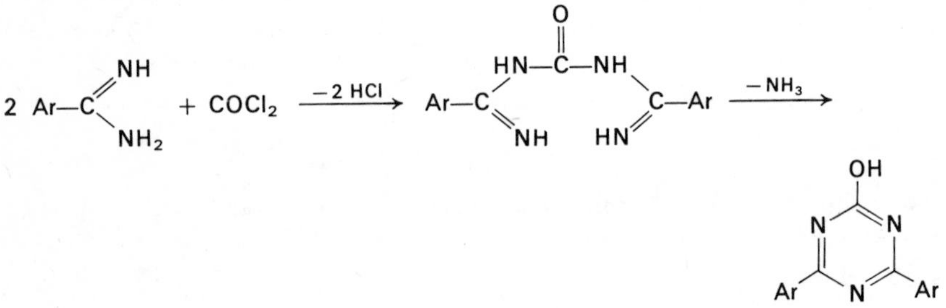

This method was extended for the synthesis of other triazines[361, 362]. Such cyclizations may also occur between amidines and N-(α-chloroalkylidene)-carbamoyl chlorides[363] or polychloroazaalkenes[364]:

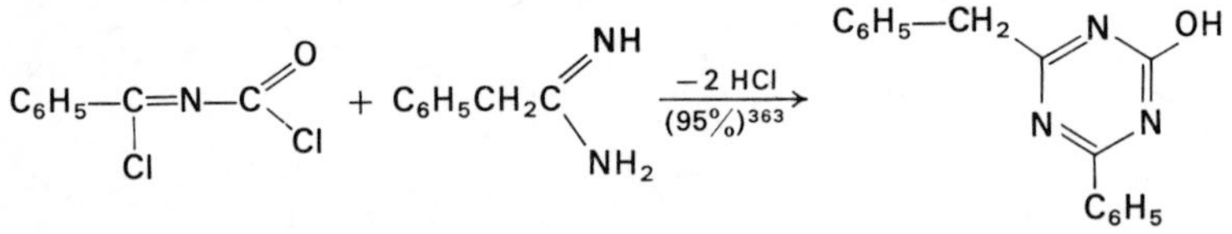

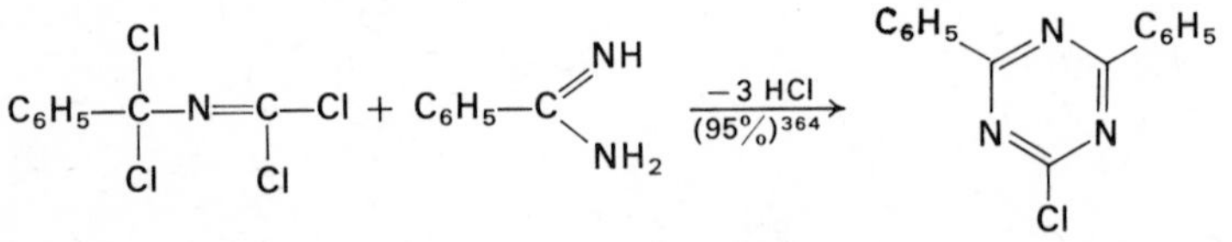

More generally, condensation of two molecules of benzamidine in the presence of various acylating reagents leads to 6-substituted 2,4-diphenyltriazines. Thus, ethyl formate gives 6-unsubstituted triazines[365], ethyl chloroformate reacts like phosgene[360]. In a more complex reaction, ethyl acetylmalonate gives 2,4-diphenyl-6-hydroxy-S-triazine[366].

In the reaction with acetic anhydride, 6-methyl-2,4-diphenyl-S-triazine is formed[367]:

Phenyl salicylate introduces an *o*-hydroxyphenyl group[368] into position 6:

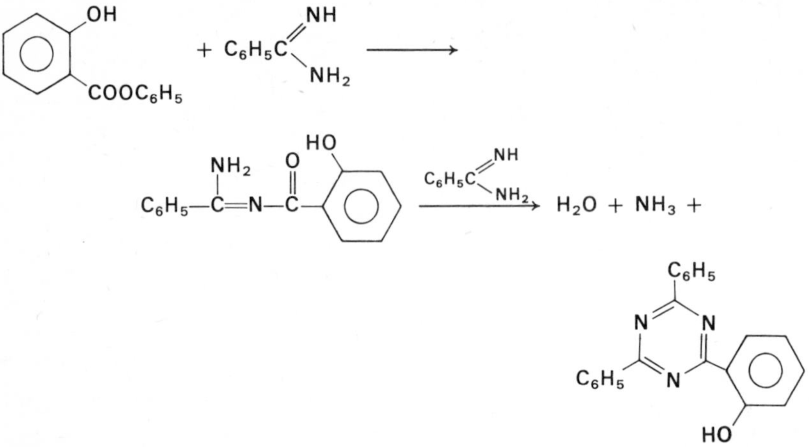

In similar reactions amidines may be condensed with a molecule of acylimidate[369]:

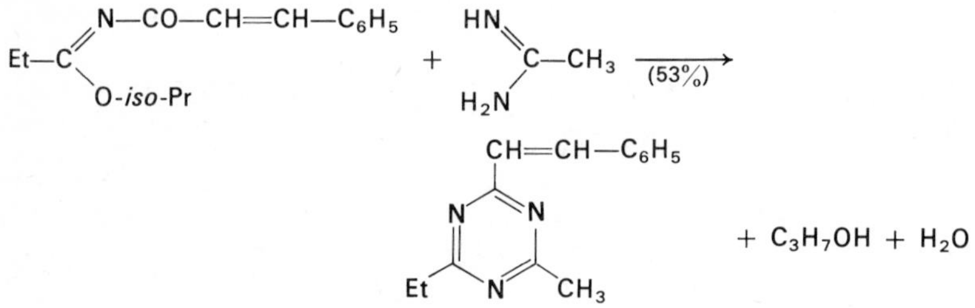

Although this method gives mixtures it nevertheless has preparative interest since it is a route to obtain *S*-triazines differently substituted on the three carbon atoms of the ring[370].

N-Amidinoamidines give by condensation with oxalic or oxamic esters, 4-amino-*S*-triazines having an ester or an amide function in the 6 position[371]:

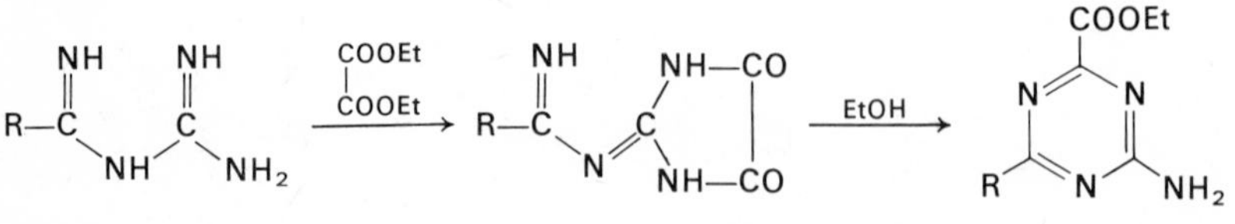

Condensation of benzonitriles with urea in the presence of sodium hydride in DMSO yields 4,6-diaryl-2-hydroxy-S-triazines[13]:

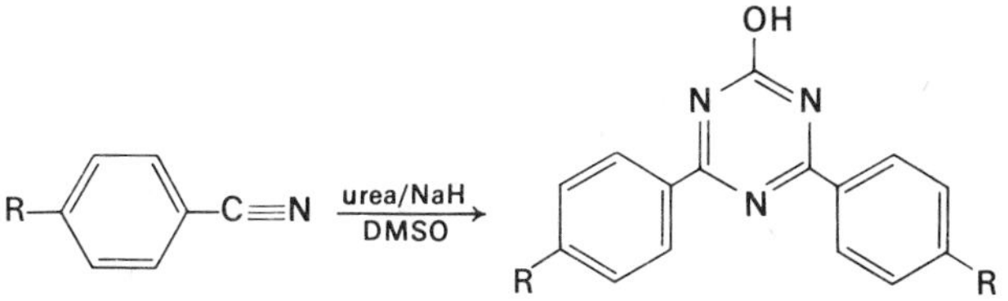

Condensation of various aldehydes with benzamidine in the presence of cyanhydric acid probably occurs through the formation of the corresponding cyanhydrins and N-iminoamidines that react further differently according to the nature of aldehyde. With aliphatic aldehydes, S-triazines are obtained while aromatic aldehydes lead to oxazoles[372].

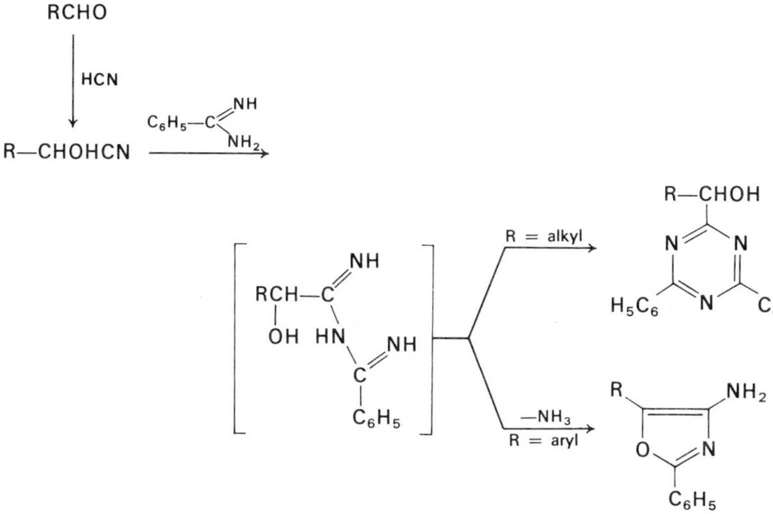

Condensation, at moderate temperatures, of aromatic isocyanates with amidines leads to S-triazine derivatives[373]:

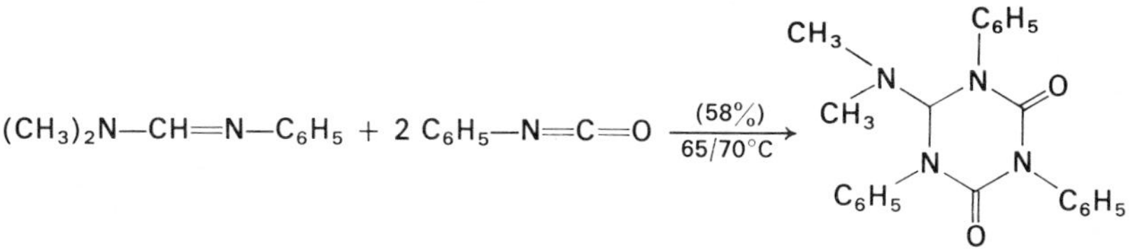

$$(CH_3)_2N-CH=N-C_6H_5 + 2\ C_6H_5-N=C=O \xrightarrow[65/70°C]{(58\%)}$$

At higher temperatures these reactions give abnormal products which may be attributed to cleavage followed by recombination[374]:

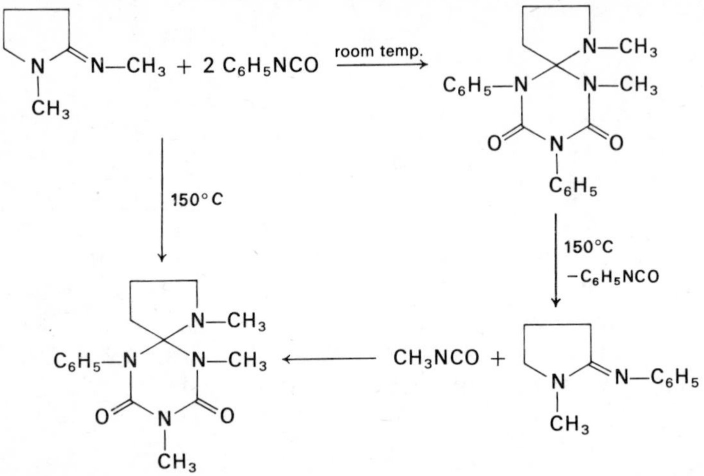

According to Goerdeler and Neuffer[375, 376], amidines give cyclizations with substituted isothiocyanates. Aroyl isothiocyanates combine at room temperature with amidines to form triazinethiones[375]:

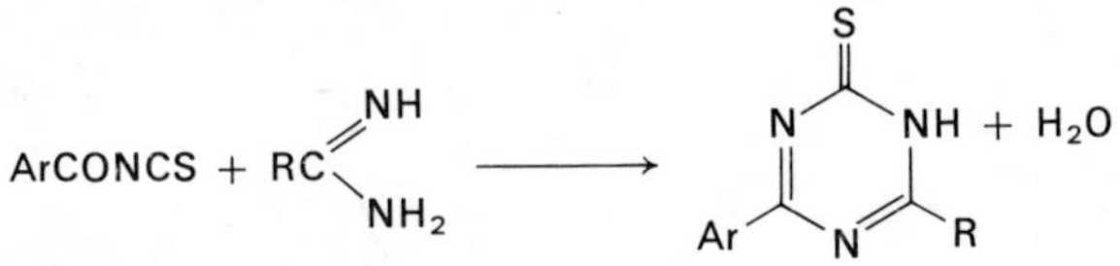

In some cases, a non-cyclic intermediate may be isolated (Ar^1 = mesityl and Ar^2 = phenyl):

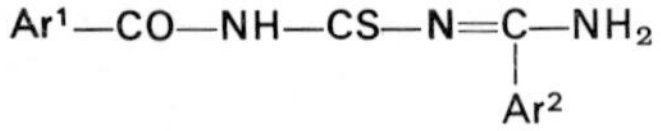

$$Ar^1—CO—NH—CS—N{=}C—NH_2$$
$$\overset{|}{Ar^2}$$

With strongly basic amidines, competition occurs between the formation of triazinethiones and amidine acylation[375]:

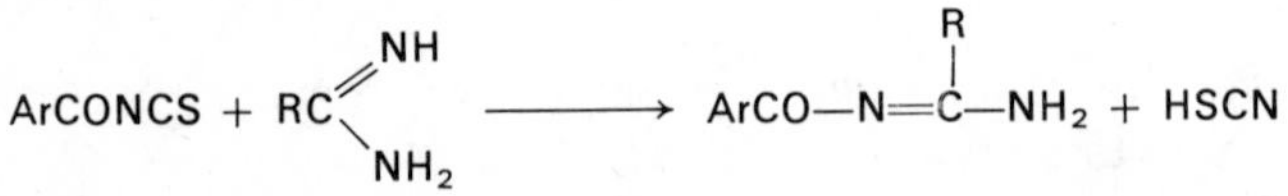

Ethoxycarbonylisothiocyanate yields oxotriazinethiones[376]:

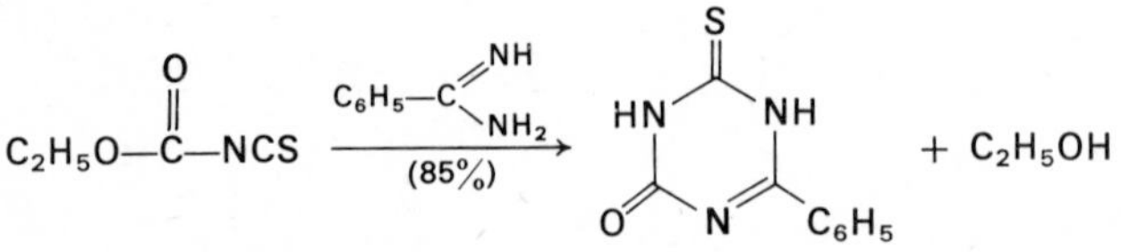

Imidoylisothiocyanates react with amidines to form triazinethiones with elimination of $ArNH_2$ [377]:

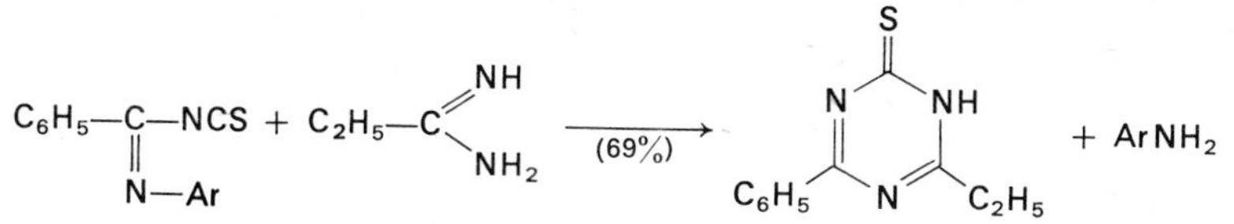

By condensation of cyclic amidines with methyl isothiocyanate, bicyclic systems containing a triazinedithione ring are obtained [378]:

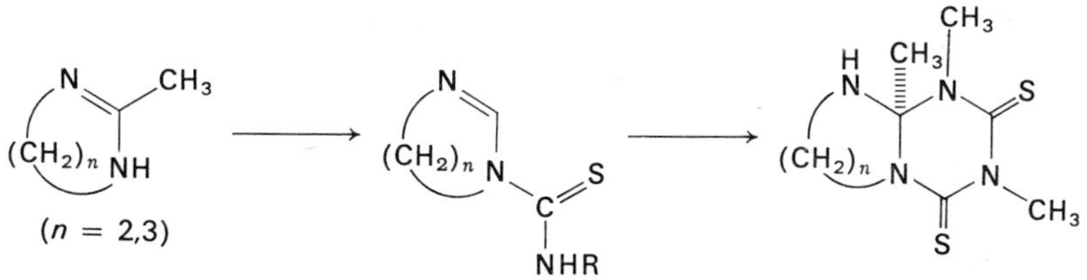

$(n = 2,3)$

d. *Tetrazines:* 1,2,4,5-Dihydrotetrazines were prepared by action of monosubstituted amidines on hydrazine hydrate [379]:

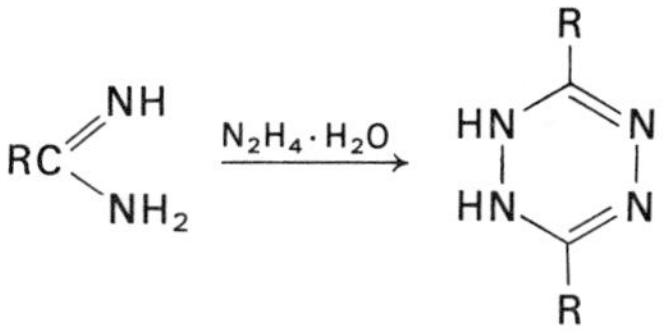

e. *Oxazines and thiadiazines*: Reaction between phenyl salicylate and N-phenylbenzamidine yields a benzoxazinone [380]:

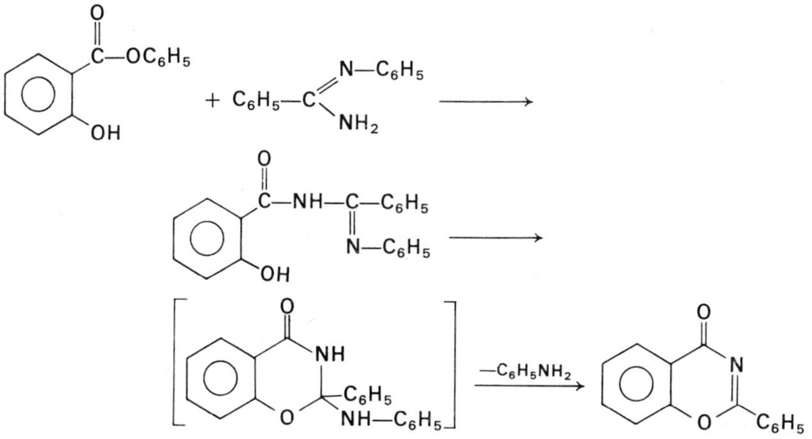

Phenylbenzamidine reacts with *N*-sulphinyl *p*-toluenesulphonamide to form a substituted benzothiadiazine[381]:

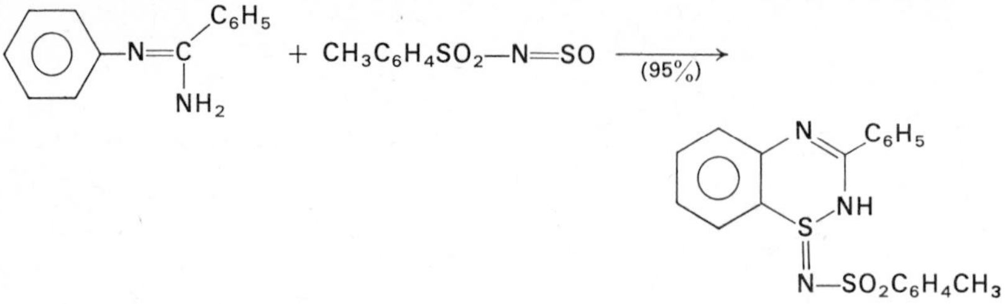

IV. REFERENCES

1. R. L. Shriner and F. W. Neumann, *Chem. Rev.*, **35**, 351 (1944).
2. Houben-Weyl, *Die Methoden der Organischen Chemie*, 4th ed., Vol. XI (2), p. 47, G. Thieme, Leipzig.
3. M. Miocque, C. Fauran and A. Y. Le Cloarec, *Ann. Chim.*, **7**, 89 (1972).
4. E. F. Cornell, *J. Amer. Chem. Soc.*, **50**, 3311 (1928).
5. A. Kirssanoff and I. Poliakowa, *Bull. Soc. chim. France*, **2**, 1600 (1936).
6. *Fr. Patent* 1,451,284 (1965); *Chem. Abstr.*, **66**, 85785v (1967).
7. G. Newberry and W. Webster, *J. Chem. Soc.*, 738 (1947).
8. W. F. Short, D. A. Peak and P. T. Charlton, *Brit. Patent*, 699,644 (1953); *Chem. Abstr.*, **49**, 2487a (1955).
9. F. C. Cooper and M. W. Partridge, *J. Chem. Soc.*, 255 (1953).
10. A. Lottermoser, *J. Prakt. Chem.*, **54**, 113 (1896).
11. J. A. Gautier, M. Miocque, C. Fauran and A. Y. Le Cloarec, *Bull. Soc. chim. France*, 200 (1970).
12. J. A. Gautier, M. Miocque, C. Fauran and A. Y. Le Cloarec, *Bull. Soc. chim. France*, 791 (1969).
13. B. Singh and J. C. Collins, *J. Chem. Soc., D, Chem. Comm.*, 498 (1971).
14. R. P. Hullin, J. Miller and W. F. Short, *J. Chem. Soc.*, 394 (1947).
15. E. Lorz and R. Baltzly, *J. Amer. Chem. Soc.*, **70**, 1904 (1948).
16. The Wellcome Foundation Ltd., *Brit. Patent*, 619,559 (1949); *Chem. Abstr.*, **43**, 5801b (1949).
17. W. L. Reilly and H. C. Brown, *J. Org. Chem.*, **22**, 698 (1957).
18. R. S. Curtiss and L. F. Nickell, *J. Amer. Chem. Soc.*, **35**, 885 (1913).
19. H. M. Woodburn, B. A. Morehead and W. H. Bonner, *J. Org. Chem.*, **14**, 555 (1949).
20. H. M. Woodburn, C. M. Chih and D. H. Thorpe, *J. Org. Chem.*, **22**, 846 (1957).
21. H. M. Woodburn, B. A. Morehead and C. M. Chih, *J. Org. Chem.*, **15**, 535 (1950).
22. H. M. Woodburn and B. G. Pautler, *J. Org. Chem.*, **19**, 863 (1954).
23. D. I. Shiho and S. Tagani, *Ph. Bull. Jap.*, **5**, 45 (1957).
24. J. G. Erickson, *J. Org. Chem.*, **20**, 1569 (1955).
25. A. Bernthsen, *Ann. Chem.*, **184**, 290 (1877).
26. A. Bernthsen, *Ann. Chem.*, **192**, 1 (1878).
27. M. W. Partridge and W. F. Short, *J. Chem. Soc.*, 390 (1947).

28. P. Oxley and W. F. Short, *J. Chem. Soc.*, 147 (1946).
29. F. C. Schaefer and A. P. Krapcho, *J. Org. Chem.*, **27**, 1255 (1962).
30. P. Oxley, M. W. Partridge and W. F. Short, *J. Chem. Soc.*, 303 (1948).
31. E. L. Hölljes, Jr. and E. C. Wagner, *J. Org. Chem.*, **9**, 31 (1944).
32. B. L. Bastic, R. P. Saper and V. B. Golubovic, *Glasnik Khem. Drushtva, Beograd*, **21**, 151 (1956); *Chem. Abstr.*, **52**, 16339i (1958).
33. E. J. Moriconi and A. A. Cevasco, *J. Org. Chem.*, 2109 (1968).
34. R. Kwok and P. Pranc, *J. Org. Chem.*, 738 (1967).
35. P. Oxley and W. F. Short, *J. Chem. Soc.*, 497 (1947).
36. M. W. Partridge and H. A. Turner, *J. Pharm. Pharmacol.*, **5**, 111 (1953).
37. L. Bauer and J. Cymerman, *J. Chem. Soc.*, 2078 (1950).
38. M. W. Partidge, *J. Chem. Soc.*, 2683 (1949).
39. P. T. Charlton, G. K. Maliphant, P. Oxley and D. A. Peak, *J. Chem. Soc.*, 485 (1951).
40. M. W. Partridge, *J. Chem. Soc.*, 2901 (1950).
41. M. W. Partridge, *J. Chem. Soc.*, 3043 (1949).
42. P. Oxley, M. W. Partridge and W. F. Short, *J. Chem. Soc.*, 1110 (1947).
43. R. Delaby, P. Reynaud and T. Tupin, *Bull. Soc. chim. France*, 714 (1957).
44. R. Delaby, P. Reynaud and F. Lilly, *Bull. Soc. chim. France*, 2067 (1961).
45. F. C. Cooper and M. W. Partridge, *Organic Syntheses*, **36**, 64 (1956).
46. M. W. Partridge and A. Smith, *J. Chem. Soc. Perkin Transactions I*, **5**, 453 (1973).
47. A. Pinner and Fr. Klein, *Ber.*, **10**, 1889 (1877).
48. A. Pinner, *Die Imidoäther und ihre Derivate*, Oppenheim, Berlin (1892).
49. R. C. Moreau and P. Reynaud, *Bull. Soc. chim. France*, 2997 (1964).
50. F. C. Cooper and M. W. Partridge, *J. Chem. Soc.*, 5036 (1952).
51. J. N. Ashley, H. J. Barber, A. J. Ewins, G. Newbery and A. D. H. Self, *J. Chem. Soc.*, 103 (1942).
52. F. E. Di Gangi and O. Gisvold, *J. Amer. Pharm. Ass.*, **38**, 154 (1949).
53. R. E. Allen, E. L. Schumann, W. C. Day and M. G. Van Campen Jr., *J. Amer. Chem. Soc.*, **80**, 591 (1958).
54. H. J. Barber and R. Slack, *J. Chem. Soc.*, 82 (1947).
55. H. J. Barber and R. Slack, *J. Amer. Chem. Soc.*, **66**, 1607 (1944).
56. F. C. Schaefer and G. A. Peters, *J. Org. Chem.*, **26**, 412 (1961).
57. R. Roger and D. G. Neilson, *Chem. Rev.*, **61**, 179 (1961).
58. P. Reynaud and R. C. Moreau, *Bull. Soc. chim. France*, 2002 (1960).
59. H. Lumbroso, D. M. Bertin and P. Reynaud, *C.R. Acad. Sci. Paris*, **261**, 399 (1965).
60. J. F. Arens and Th. R. Rix, *Proc. Koninkl. Ned. Akad. Wetenschap*, **275** B 57 (1954).
61. Y. A. Sinnema and J. F. Arens, *Rec. Trav. chim. Pays-Bas*, **75**, 1423 (1956).
62. P. Reynaud, R. C. Moreau and J. C. Tetard, *C.R. Acad. Sci. Paris*, **262**, 665 (1966).
63. A. K. Bose, F. Greer, J. S. Gots and C. C. Price, *J. Org. Chem.*, **24**, 1309 (1959).
64. N. W. Bristow, *J. Chem. Soc.*, 513 (1957).
65. *Brit. Patent*, 1,288,376 (1968); *Chem. Abstr.*, **78**, 3980r (1973).

66. K. Kratzl and E. Meisert, *Monatsh. Chem.*, **88**, 1056 (1957).
67. H. Loewe, J. Urbanietz and H. Mieth, *Int. Congr. Chemother. Proc.*, *5th*, **2**(2) 645 (1967); *Chem. Abstr.*, **70**, 47042a (1969).
68. M. Mengelberg, *Ber.*, **89**, 1185 (1956).
69. R. Delaby, P. Reynaud and F. Lilly, *C.R. Acad. Sci. Paris*, **246**, 2905 (1958).
70. *Ger. Patent*, 2,239,606 (1971); *Chem. Abstr.*, **78**, 110958m (1973).
71. N. S. Drozdov and A. F. Bekhli, *J. Gen. Chem.* (*USSR*), **14**, 480 (1944).
72. H. J. Barber, P. Z. Gregory, F. W. Major, R. Slack and A. M. Woolman, *J. Chem. Soc.*, 84 (1947).
73. P. Z. Gregory, S. J. Holt and R. Slack, *J. Chem. Soc.*, 87 (1947).
74. R. Delaby, P. Reynaud and P. Berçot, *C.R. Acad. Sci. Paris*, **246**, 125 (1958).
75. J. O. Jilek and V. Michajlyszyn, *Chem. Listy*, **48**, 1210 (1954); *Chem. Abstr.*, **49**, 9507 (1955).
76. A. Pinner, *Ber.*, **28**, 473 (1895).
77. F. E. King, K. G. Latham and M. W. Partridge, *J. Chem. Soc.*, 4268 (1952).
78. R. Delaby, J. V. Harispe and F. Bonhomme, *Bull. Soc. chim. France*, **12**, 152 (1945).
79. A. J. Hill and I. Rabinowitz, *J. Amer. Chem. Soc.*, **48**, 732 (1926).
80. J. O. Jilek, M. Borovicka and M. Protiva, *Chem. Listy*, **43**, 211 (1949).
81. H. M. Woodburn, A. B. Whitehouse and B. G. Pautler, *J. Org. Chem.*, **24**, 210 (1959).
82. M. Yamazaki, Y. Kitagawa, S. Hiraki and Y. Tsukamoto, *J. Pharm. Soc. Japan*, **73**, 294 (1953).
83. E. I. Boksiner and A. N. Shchavlinskii, *Izv. Vyssh. Ucheb. Zaved, Khim. i Khim. Tekhnol.*, **15**(2), 248 (1972); *Chem. Abstr.*, **77**, 34102 (1972).
84. *Brit. Patent*, 1,305,550 (1969); *Chem. Abstr.*, **78**, 136291p (1973).
85. *U.S. Patent*, 2,676,968 (1954); *Chem. Abstr.*, **49**, 6307a (1955).
86. P. Benko and L. Pallos, *J. Prakt. Chem.*, **314**, 627 (1972); *Chem. Abstr.*, **78**, 111074 (1973).
87. P. Benko and L. Pallos, *J. Prakt. Chem.*, **314**, 639 (1972); *Chem. Abstr.*, **78**, 110993 (1973).
88. P. G. Bay and O. Gisvold, *J. Amer. Pharm. Assoc.*, **44**, 585 (1955).
89. A. N. Baksheev and N. I. Gavrilov, *Zh. Obshch. Khim.*, **22**, 2021 (1952).
90. *Ger. Patent*, 839,493 (1952); *Chem. Abstr.*, **47**, 1737b (1953).
91. E. B. Knott, *J. Chem. Soc.*, 686 (1945).
92. T. Fujisawa and C. Mizuno, *J. Pharm. Soc. Japan*, **72**, 694, 698 (1952).
93. *Jap. Patent*, 4477 (1931); *Chem. Abstr.*, **48**, 8259h (1954).
94. A. Pinner, *Ber.*, **23**, 2942 (1890).
95. A. Strecker, *Ann. Chem.*, **103**, 321 (1857).
96. K. Brunner, M. Matzler and V. Mössmer, *Monatsh. Chem.*, **48**, 125 (1927).
97. K. Brunner and F. Haslwanter, *Monatsh. Chem.*, **48**, 133 (1927).
98. A. J. Hill and J. V. Johnston, *J. Amer. Chem. Soc.*, **76**, 920 (1954).
99. *Jap. Patent*, 7,103,366 (1966); *Chem. Abstr.*, **74**, 124914y (1971).
100. G. Tsatsas, R. Delaby, A. Quevauviller, R. Damiens and O. Blanpin, *Ann. Pharm. Fr.*, **14**, 607 (1956).

101. G. Tsatsas and R. Delaby, *Ann. Pharm. Fr.*, **14**, 621 (1956).
102. G. Mandel and A. J. Hill, *J. Amer. Chem. Soc.*, **76**, 3978 (1954).
103. A. J. Hill and M. V. Cox, *J. Amer. Chem. Soc.*, **48**, 3214 (1926).
104. J. G. Cannon and G. L. Webster, *J. Amer. Pharm. Assoc., Scientific edition*, **42**, 740 (1953).
105. N. S. Drozdov and A. F. Bekhli, *J. Gen. Chem. (USSR)*, **14**, 472 (1944); *Chem. Abstr.*, **39**, 45904 (1945).
106. N. J. Sintov, J. S. Rodia, J. A. Tursich, H. L. Davis and G. L. Webster, *J. Amer. Chem. Soc.*, **71**, 3990 (1949).
107. *Ger. Patent*, 2,113,978 (1971); *Chem. Abstr.*, **78**, 3981s (1973).
108. *Org. Synth., Coll. Vol.* 4, 383.
109. H. Bredereck and K. Bredereck, *Ber.*, **94**, 2278 (1961).
110. H. Bredereck, R. Gompper, K. Klemm and H. Rempfer, *Ber.*, **92**, 837 (1959).
111. G. L. Webster and J. S. Rodia, *J. Amer. Chem. Soc.*, **75**, 1761 (1953).
112. F. H. S. Curd and C. G. Raison, *J. Chem. Soc.*, 160 (1947).
113. J. V. Braun, F. Jostes and W. Münch, *Ann. Chem.*, **453**, 113 (1927).
114. J. V. Braun, *Angew. Chem.*, **47**, 611 (1934).
115. H. Eilingsfeld, M. Seefelder and H. Weidinger, *Ber.*, **96**, 2671 (1963); *Angew. Chem.*, **72**, 836 (1960).
116. *Jap. Patent*, 7,247,388 (1970); *Chem. Abstr.*, **78**, 124596d (1973).
117. *Ger. Patent*, 949,285 (1956); *Chem. Abstr.*, **53**, 9250e (1959).
118. J. V. Braun, *Ber.*, **37**, 2678 (1904).
119. *Ger. Patent*, 372,842 (1924); *Chem. Abstr.*, **18**, 2176 (1924).
120. B. I. Ardashev, V. I. Minkin and M. B. Minkin, *Nauch. Dokl. Vyssh. Shk., Khim i Khim Tekhnol.*, **3**, 526 (1958); *Chem. Abstr.*, **53**, 3227e (1959).
121. O. Wallach, *Ber.*, **15**, 208 (1882).
122. C. C. Price and R. M. Roberts, *J. Amer. Chem. Soc.*, **68**, 1255 (1946).
123. *U.S. Patent*, 1,384,637; *Chem. Abstr.*, **15**, 3725 (1921).
124. O. Bayer, *Angew. Chem.*, **61**, 236 (1949).
125. W. Jentzsch, *Ber.*, **97**, 1361 (1964).
126. W. Jentzsch, *Ber.*, **97**, 2755 (1964).
127. W. Jentzsch and M. Seefelder, *Ber.*, **98**, 274 (1965).
128. A. A. R. Sayigh and H. Ulrich, *J. Chem. Soc.*, 3146 (1963).
129. I. Hagedorn, H. Etling and K. E. Lichtel, *Ber.*, **99**, 520 (1966).
130. *Fr. Patent*, 717,145 (1931); *Chem. Abstr.*, **26**, 2748 (1932).
131. A. V. Kirsanov, E. S. Levchenko and I. N. Zhmurova, *Ukrain. Khim. Zhur.*, **22**, 498 (1956); *Chem. Abstr.*, **51**, 4333h (1957).
132. P. Oxley, D. A. Peak and W. F. Short, *J. Chem. Soc.*, 1618 (1948).
133. P. Oxley and W. F. Short, *J. Chem. Soc.*, 382 (1947).
134. P. Oxley and W. F. Short, *J. Chem. Soc.*, 1514 (1948).
135. W. Ried and E. Schmidt, *Ann. Chem.*, **676**, 114 (1964).
136. L. Weintraub, S. R. Oles and N. Kalish, *J. Org. Chem.*, **33**, 1679 (1968).
137. J. D. Wilson, J. S. Wager and H. Weingarten, *J. Org. Chem.*, **36**, 1613 (1971).
138. R. I. Fryer, J. V. Earley, G. F. Field, W. Zally and L. H. Sternbach, *J. Org. Chem.* **34**, 1143 (1969).
139. J. L. Neumeyer, *J. Pharm. Sci.*, **53**, 1539 (1964).

140. D. Duerr, H. Aebi and L. Ebner, *U.S. Patent*, 3,284,289 (1966); *Chem. Abstr.* **66**, 28499f (1967).
141. A. Jovtscheff and F. Falk, *J. Prakt. Chem.*, **13**, 265 (1961).
142. M. L. Weiner, *J. Org. Chem.*, **25**, 2245 (1960).
143. C. King, *J. Org. Chem.*, **25**, 352 (1960).
144. F. B. Dains, *J. Amer. Chem. Soc.*, **22**, 188 (1900).
145. F. B. Dains, R. C. Roberts and R. Q. Brewster, *J. Amer. Chem. Soc.* **38**, 131 (1916).
146. Y. Kikugawa and S. Yamada, *Tetrahedron Letters*, 699 (1969).
147. A. Bernthsen, *Ann. Chem.*, **184**, 321 (1876).
148. F. Micheel, Z. Krzeminski, W. Himmelmann and A. Kühlkamp, *Ann. Chem.*, **90**, 575 (1952).
149. F. Micheel and W. Flitsch, *Ann. Chem.*, **577**, 234 (1952).
150. *Neth. Patent*, 6,607,412, (1965); *Chem. Abstr.*, **67**, 3090y (1967).
151. G. R. Pettit and L. R. Garson, *Canad. J. Chem.*, **43**, 2640 (1965).
152. P. Reynaud, R. C. Moreau and Nguyen Hong Thu, *C.R. Acad. Sci. Paris*, **253**, 1968 (1961).
153. H. Bredereck, R. Gompper and H. Seiz, *Chem. Ber.*, **90**, 1837 (1957).
154. P. Reynaud, R. C. Moreau and T. Gousson, *C.R. Acad. Sci. Paris*, **259**, 4067 (1964).
155. P. Reynaud, R. C. Moreau and Nguyen Hong Thu, *C.R. Acad. Sci. Paris*, **253**, 2540 (1961).
156. P. Chabrier and S. H. Renard, *C.R. Acad. Sci. Paris*, **230**, 1673 (1950).
157. N. V. Koshkin, *Zhur. Obshch. Khim.*, **28**, 695 (1958); *Chem. Abstr.*, **52**, 17144i (1958).
158. W. E. Tischtshenko and N. V. Koshkin, *Zhur. Obshch. Khim*, **4**, 1021 (1934).
159. A. Kirssanov and J. Iwastchenko, *Bull. Soc. chim. France*, 2109 (1935).
160. R. Fusco and C. Mustante, *Gazz. Chim. Ital.*, **66**, 258 (1936).
161. F. Kröhnke and H. H. Steuernagel, *Ber.*, **96**, 486 (1963).
162. F. Sachs and E. Bry, *Ber.*, **34**, 118 (1901).
163. St. Robeff, *C.R. Acad. bulg. Sci.*, **7**, (3), 37 (1954).
164. St. Robeff, *Dokl. Akad. Nauk. SSSR*, **101**, 277 (1955); *Chem. Abstr.*, **50**, 3315d (1956).
165. St. Robeff, *C.R. Acad. bulg. Sci.*, **8**, (2), 29 (1955); *Chem. Abstr.*, **50**, 13838g (1956).
166. St. Robeff, *C.R. Acad. bulg. Sci.*, **13**, 159 (1960); *Chem. Abstr.*, **55**, 18676g (1961).
167. I. Grandberg, Y. Naumov and A. Kost, *C.R. Acad. bulg. Sci.*, **17**, 11, 1025 (1964).
168. St. Robeff, *Ber.*, **91**, 244 (1958).
169. St. Robeff and T. Sumerska, *C.R. Acad. bulg. Sci.*, **12**, 137 (1959), *Chem. Abstr.* **54**, 4480d (1960).
170. St. Robeff, *C.R. Acad. bulg. Sci.*, **12**, 141 (1959); *Chem. Abstr.*, **54**, 4480h (1960).
171. *Fr. Patent*, 1.531,264 (1968); *Chem. Abstr.*, **71**, 49620g (1969).
172. *Brit. Patent*, 551,445 (1943); *Chem. Abstr.*, **38**, 2344[8] (1944).
173. R. P. Mull, R. H. Mizzoni, M. R. Dapero and M. E. Egbert, *J. Med. Pharm. Chem.*, **5**, 651 (1962).

7. Preparation and synthetic uses of amidines 343

174. J. H. Boyer and P. J. A. Frints, *J. Org. Chem.*, **35**, 2449 (1970).
175. M. Busch and R. Hobein, *Ber.*, **40**, 4296 (1907).
176. R. Adams and C. H. Beebe, *J. Amer. Chem. Soc.*, **38**, 2768 (1916).
177. L. Vuylsteke, *Bull. Acad. r. Med. Belg.*, *Sciences*, 535 (1926).
178. H. Lettre, P. Jungmann and J. C. Salfeld, *Ber.*, **85**, 397 (1952).
179. H. Fujita, R. Endo, A. Aoyama and T. Ichii, *Bull. Chem. Soc. Jap.*, **45**, 1846 (1972).
180. N. Traube and A. Eyme, *Ber.*, **32**, 3176 (1899).
181. P. Ruggli and I. Marszak, *Helv. chim. Acta*, **11**, 180 (1928).
182. R. L. Pruett, J. T. Barr, K. E. Rapp, C. T. Bahner, J. D. Gibson and R. H. Lafferty Jr., *J. Amer. Chem. Soc.*, **72**, 3646 (1950).
183. D. C. England, L. R. Melby, M. A. Dietrich and R. V. Lindsey Jr., *J. Amer. Chem. Soc.*, **82**, 5116 (1960).
184. O. Döbner, *Ber.*, **15**, 232 (1882); *Ann. Chem.* **217**, 223 (1883).
185. S. P. Joshi, A. P. Khanolkar and T. S. Wheeler, *J. Chem. Soc.*, 793 (1936).
186. M. K. Meilahn, L. L. Augenstein and J. L. McManaman, *J. Org. Chem.*, **36**, 3627 (1971).
187. D. L. Ross, C. L. Coon and M. E. Hill, *J. Org. Chem.*, **35**, 3093 (1970).
188. A. J. Speziale and R. C. Freeman, *J. Amer. Chem. Soc.*, **82**, 909 (1960).
189. A. J. Speziale and R. C. Freeman, *J. Amer. Chem. Soc.*, **82**, 903 (1960).
190. H. Wickelhaus, *Ber.*, **2**, 115 (1869).
191. R. Walther, *J. Prakt. Chem.*, **53**, 472 (1896).
192. L. Claisen, *Ann. Chem.*, **287**, 360 (1895).
193. C. Goldschmidt, *Chem.-Ztg.*, **64**, 743 (1902).
194. R. M. Roberts, *J. Org. Chem.*, **14**, 297 (1949).
195. C. C. Price, N. J. Leonard and H. F. Herbrandson, *J. Amer. Chem. Soc.*, **68**, 1251 (1946).
196. W. Bradley and I. Wright, *J. Chem. Soc.*, 640 (1956).
197. G. F. Duffin and J. D. Kendall, *J. Chem. Soc.*, 408 (1954).
198. F. B. Dains, O. O. Malleis and J. T. Meyers, *J. Amer. Chem. Soc.*, **35**, 970 (1913).
199. A. Giacalone, *Gazz. Chim. Ital.*, **62**, 577 (1932).
200. H. J. Backer and W. L. Van Maker, *Rec. Trav. chim. Pays-Bas*, **67**, 257 (1948); **68**, 247 (1949).
201. R. M. Roberts, *J. Amer. Chem. Soc.*, **71**, 3848 (1949).
202. R. M. Roberts and R. H. De Wolfe, *J. Amer. Chem. Soc.*, **76**, 2411 (1954).
203. E. H. Cordes, in *The Chemistry of Carboxylic Acids and Esters*, S. Patai (Ed.), Intescience, p. 656 (1969).
204. E. C. Taylor and W. A. Ehrhart, *J. Org. Chem.*, **28**, 1108 (1963).
205. R. H. De Wolfe, *J. Org. Chem.*, **27**, 490 (1962).
206. *U.S. Patent*, 3,385,891 (1965); *Chem. Abstr.*, **70**, 3317q (1969).
207. H. Möhrle and S. Mayer, *Tetrahedron Letters*, **51**, 5173 (1967).
208. M. E. Kuehne and P. J. Sheeran, *J. Org. Chem.*, 4406 (1968).
209. J. Oszczapowicz and R. Orlinski, *Rocz. Chem.*, **44**, 2327 (1970).
210. J. Oszczapowicz and R. Orlinski, *Rocz. Chem.* **45**, 103 (1971).
211. J. Oszczapowicz, *Rocz. Chem.*, **44**, 453 (1970).
212. T. Fujigawa and Y. Deguchi, *J. Pharm. Soc. Jap.*, **73**, 225 (1953).

213. E. Beckmann and E. Fellrath, *Ann. Chem.*, **273**, 1 (1893).
214. C. Chew and F. L. Pyman, *J. Chem. Soc.*, **23**, 2318 (1927).
215. A. Pinner and F. Klein, *Ber. Chem.*, **11**, 4 (1878).
216. F. L. Pyman, *J. Chem. Soc.*, 367 (1923).
217. *Brit. Patent*, 528,915 (1940); *Chem. Abstr.*, **35**, 7976[4] (1941).
218. R. Von Walther and A. Grossmann, *J. Prakt. Chem.*, **78**, 478 (1908).
219. A. E. Mil'grom, B. B. Paleev, *USSR*, 256,779 (1968); *Chem. Abstr.*, **72**, P.132703c (1970).
220. A. E. Mil'grom, B. B. Paleev, *USSR*, 256,787 (1968); *Chem. Abstr.*, **72**, P.132752t (1970).
221. J. A. Gautier, M. Miocque, C. Fauran and A. Y. Le Cloarec, *Bull. Soc. chim. France*, 478 (1971).
222. F. L. Pyman, *J. Chem. Soc.*, 3359 (1923).
223. H. Pechmann and B. Heinze, *Ber.*, **30**, 1783 (1897).
224. W. Ried and P. Weidemann, *Ber.*, **104**, 3329(1971).
225. D. Vorländer, *Ber.*, **24**, 803 (1891).
226. P. Reynaud, R. C. Moreau and P. Fodor, *C.R. Acad. Sci. Paris*, **263**, 788 (1966).
227. P. Reynaud, R. C. Moreau and P. Fodor, *C.R. Acad. Sci. Paris*, **264c**, 1414 (1967).
228. H. Pechmann, *Ber.*, **28**, 2362 (1895).
229. H. Müller, *Ber.*, **19**, 1669 (1886).
230. F. Eloy and R. Lenaers, *Chem. Rev.*, **62**, 162 (1962).
231. G. Merling, *Ber.*, **41**, 2064 (1908).
232. L. Ruzicka and H. Schinz, *Helv. chim. Acta*, **23**, 959 (1940).
233. F. Henle, *Ber.*, **35**, 3039 (1902).
234. A. J. Birch, J. Cymerman-Craig and M. Slaytor, *Chem. and Ind.*, 1559 (1954).
235. A. J. Birch, J. Cymerman-Craig and M. Slaytor, *Austral. J. Chem.*, **8**, 512 (1955).
236. H. Smith, *Organic reactions in liquid ammonia*, Interscience, New York (1963), p. 222.
237. F. B. Dains, *Ber.*, **35**, 2496 (1902).
238. F. B. Dains and E. W. Brown, *J. Amer. Chem. Soc.*, **31**, 1148 (1909).
239. F. B. Dains, H. R. O'Brien and C. L. Johnson, *J. Amer. Chem. Soc.*, **38**, 1510 (1916).
240. F. B. Dains and A. E. Stephenson, *J. Amer. Chem. Soc.*, **38**, 1841 (1916).
241. F. B. Dains and R. N. Harger, *J. Amer. Chem. Soc.*, **40**, 562 (1918).
242. F. B. Dains, R. Irvin and C. G. Harrel, *J. Amer. Chem. Soc.*, **43**, 613 (1921).
243. F. B. Dains, R. Thompson and W. F. Asendorf, *J. Amer. Chem. Soc.*, **44**, 2310 (1922).
244. F. B. Dains and E. L. Griffin, *J. Amer. Chem. Soc.*, **35**, 959 (1913).
245. W. H. Graham, *J. Amer. Chem. Soc.*, **87**, 4396 (1965).
246. A. K. Bose and I. Kugajevsky, *Tetrahedron*, **23**, 957 (1967).
247. H. Ulrich, B. Tucker and A. A. R. Sayigh, *Angew. Chem Int. Ed. Engl.*, **7**, 291 (1968).
248. R. R. Lorenz, B. F. Tullar, C. F. Koelsch and S. Archer, *J. Org. Chem.*, **30**, 2531 (1965).

249. E. Zieglek, W. Kaufmann and W. Klementschitz, *Monatsh. Chem.*, **83**, 1334 (1952).
250. E. C. Wagner, *J. Org. Chem.*, **5**, 133 (1940).
251. P. Oxley and W. F. Short, *J. Chem. Soc.*, 1100 (1950).
252. R. F. Lambert and C. E. Kristofferson, *J. Org. Chem.*, **30**, 3938 (1965).
253. K. H. Magosch and R. Feinaner, *Ann. Chem.*, **742**, 128 (1970).
254. C. Musante, *Gazz. chim. ital.*, **68**, 331 (1938).
255. R. Neidlein and H. Reuter, *Tetrahedron*, **27**, 4117 (1971).
256. J. Goerdeler, *Ber.*, **87**, 57 (1954).
257. J. Goerdeler and M. Budnowski, *Ber.*, **94**, 1682 (1961).
258. J. Goerdeler, H. Groschopp and U. Sommerlad, *Ber.*, **90**, 182 (1957).
259. K. T. Potts and R. Armbruster, *J. Org. Chem.*, **35**, 1965 (1970).
260. E. Haruki, T. Inaike and E. Imoto, *Bull. Chem. Soc. Jap.*, **41**, 1361 (1968).
261. J. B. Ekeley and A. R. Ronzio, *J. Amer. Chem. Soc.*, **59**, 1118 (1937).
262. O. Diels and K. Schleich, *Ber.*, **49**, 1711 (1916).
263. J. O. Cole and A. R. Ronzio, *J. Amer. Chem. Soc.*, **66**, 1584 (1944).
264. G. Rio and A. Ranjon, *Bull. Soc. chim. France*, 543 (1958).
265. J. L. Imbach, R. Jacquier, J. M. Lacombe and G. Maury, *Bull. Soc. chim. France*, 1053 (1971).
266. F. Kunckell, *Ber.*, **34**, 637 (1901).
267. S. Ruhemann and A. V. Cunnington, *J. Chem. Soc.*, **75**, 954 (1899).
268. A. Kreutzberger and R. Schücker, *Archiv. der Pharm.*, 935 (1972).
269. M. W. Partridge and H. A. Turner, *J. Chem. Soc.*, 1308 (1949).
270. W. Klarer and E. Urech, *Helv. chim. Acta.*, **27**, 1762 (1944).
271. C. Djerassi and C. R. Scholz, *J. Amer. Chem. Soc.*, **69**, 1688 (1947).
272. P. Oxley and W. F. Short, *J. Chem. Soc.*, 859 (1950).
273. A. Hetzheim and G. Manthey, *Ber.*, **103**, 2845 (1970).
274. L. I. Samarai, V. P. Belaya, V. A. Bondar and G. I. Derkach, *Dop. Akad. Nauk Ukrajin R.S.R.*, *Ser. B*, **30**, 1024 (1968); *Chem. Abstr.*, **70**, 57732f (1969).
275. V. J. Grenda, R. E. Jones, G. Gal and M. Sletzinger, *J. Org. Chem.*, **30**, 259 (1965).
276. E. Haruki, T. Inaïke and E. Imoto, *Bull. chem. Soc. Jap.*, **38**, 1805 (1965).
277. M. Osone, S. Tanimoto and R. Oda, *Yuki Gosei Kagaku Kyokaï Shi*, **24**, 562 (1966); *Chem. Abstr.*, **65**, 10577g (1966).
278. E. C. Taylor and W. A. Ehrhart, *J. Amer. Chem. Soc.*, **82**, 3138 (1960).
279. E. Richter, J. E. Loeffler and E. C. Taylor, *J. Amer. Chem. Soc.*, **82**, 3144 (1960).
280. F. Bergmann and M. Tamari, *J. Chem. Soc.*, 4468 (1961).
281. J. Goerdeler and R. Sappelt, *Ber.*, **100**, 2064 (1967).
282. H. R. Kwasnik, J. E. Oliver and R. T. Brown, *J. heterocycl. Chem.*, **9**, 1429 (1972).
283. J. D. Bower and G. R. Ramage, *J. Chem. Soc.*, 4506 (1957).
284. A. Dornow and E. Neuse, *Ber.*, **84**, 296 (1951).
285. C. I. Brodrick and W. F. Short, *J. Chem. Soc.*, 1343 (1951).
286. J. Cymerman and W. F. Short, *J. Chem. Soc.*, 703 (1949).
287. S. Gabriel and J. Colman, *Ber.*, **32**, 1525 (1899).

288. A. Bowman, *J. Chem. Soc.*, 494 (1937).
289. H. R. Sullivan and W. T. Caldwell, *J. Amer. Chem. Soc.*, **77**, 1559 (1955).
290. D. D. Libman, D. L. Pain and R. Slack, *J. Chem. Soc.*, 2305 (1952).
291. W. J. Hale and H. C. Brill, *J. Amer. Chem. Soc.*, **34**, 82 (1912).
292. P. E. Fanta and E. A. Hedman, *J. Amer. Chem. Soc.*, **78**, 1434 (1956).
293. A. Pinner, *Ber.*, **22**, 2609 (1889).
294. A. Pinner, *Ber.*, **26**, 2122 (1893).
295. H. R. Snyder and H. M. Foster, *J. Amer. Chem. Soc.*, **76**, 118 (1954).
296. F. E. King, T. J. King and I. H. M. Muir, *J. Chem. Soc.*, 5 (1946).
297. S. Ruhemann, *J. Chem. Soc.*, 717 (1903).
298. R. R. Williams and J. K. Cline, *J. Amer. Chem. Soc.*, **58**, 1504 (1936).
299. J. K. Cline, R. R. Williams and J. Finkelstein, *J. Amer. Chem. Soc.*, **59**, 1052 (1937).
300. H. Andersag and K. Westphal, *Ber.*, **70**, 2035 (1937).
301. W. Wislicenus, E. Böklen and F. Reuthe, *Ann. Chem.*, **363**, 340 (1908).
302. G. G. Massaroli and G. Signorelli, *Boll. chim. farm.*, **105** (5), 400 (1966).
303. J. A. Barone, E. Peters and H. Tieckelmann, *J. Org. Chem.*, **24**, 198 (1959).
304. A. H. Cook and K. J. Reed, *J. Chem. Soc.*, 399 (1945).
305. G. W. Kenner, B. Lythgoe, A. R. Todd and A. Topham, *J. Chem. Soc.*, 388 (1943).
306. H. R. Henze, W. J. Clegg and C. W. Smart, *J. Org. Chem.*, **17**, 1320 (1952).
307. H. R. Henze and J. L. McPherson, *J. Org. Chem.*, **18**, 653 (1953).
308. H. R. Henze and S. O. Winthrop, *J. Amer. Chem. Soc.*, **79**, 2230 (1957).
309. E. L. Pinner, *Ber.*, **41**, 3517 (1908).
310. A. A. Aroyan and R. G. Melik-Ogand-Zhanyan, *Arm. Khim. Zh.*, **20**, 314, (1967). *Chem. Abstr.* **68**, 29669g (1967).
311. S. M. McElvain and B. E. Tate, *J. Amer. Chem. Soc.*, **73**, 2760 (1951).
312. A. W. Dox and L. Yoder, *J. Amer. Chem. Soc.*, **44**, 361 (1922).
313. A. W. Dox and L. Yoder, *J. Amer. Chem. Soc.*, **43**, 677 (1921).
314. J. P. Marquet, J. Andre-Louisfert and E. Bisagni, *Bull. Soc. chim.*, 4344 (1969).
315. J. C. Martin, K. C. Brannock and R. H. Meen, *J. Org. Chem.*, **31**, 2966 (1966).
316. D. Isbecque, R. Promel, R. C. Quinaux and R. H. Martin, *Helv. chim. Acta*, **42**, 1317 (1959).
317. A. Maggiolo, A. P. Phillips and G. H. Hitchings, *J. Amer. Chem. Soc.*, **73**, 106 (1951).
318. W. Traube and L. Herrmann, *Ber.*, **37**, 2267 (1904).
319. R. Hull, B. J. Lovell, H. T. Openshaw, L. C. Payman and A. R. Todd, *J. Chem. Soc.*, 357 (1946).
320. Z. Földi, G. V. Fodor, I. Demjen, H. Szekeres and I. Halmos, *Ber.*, **75**, 755 (1942).
321. P. D. Landauer and H. N. Rydon, *J. Chem. Soc.*, 3721 (1953).
322. O. Vogl and E. C. Taylor, *J. Amer. Chem. Soc.*, **79**, 1518 (1957).
323. A. Takamizawa, S. Hayashi and K. Tori, *Yakugaku Zasshi*, **78**, 1166 (1958); *Chem. Abstr.*, **53**, 5276a (1959).

324. S. Pietra, *Boll. Sci. Fac. Chim. Ind. Bologna*, **11**, 78 (1953); *Chem. Abtsr.*, **49**, 13975g (1955).
325. E. I. Boksiner, A. A. Golubyatnikova, and I. Kh. Fel'dman, *Zh. Obshch. Khim.*, **38**, 99 (1968); *Chem. Abstr.*, **69**, 77205 (1968).
326. *Brit. Patent*, 633,353 (1949); *Chem. Abstr.*, **44**, 5924a (1950).
327. S. Ruhemann, *J. Chem. Soc.*, 375 (1903).
328. S. Ruhemann, *J. Chem. Soc.*, 1371 (1903).
329. W. Traube and R. Schwarz, *Ber.*, **32**, 3163 (1899).
330. R. Grewe, *Z. Physiol. Chem.*, **242**, 89 (1936).
331. A. Takamizawa and K. Hirai, *Chem. Pharm. Bull.*, **12**, 393 (1964).
332. A. Takamizawa, K. Tokuyama and H. Satoh, *Yakugaku Zasshi*, **79**, 664 (1959).
333. D. Brutane, A. Y. Strakov, A. M. Moiseenkov and A. A. Akhrem, *Latv. P.S.R. Zinat. Akad. Vestis, Kim. Ser.*, **5**, 610 (1970); *Chem. Abstr.*, **74**, 53703r (1971).
334. S. Nishigaki, K. Aida, K. Senga and F. Yoneda, *Tetrahedron Letters*, **4**, 247 (1969).
335. H. Antaki, *J. Amer. Chem. Soc.*, **80**, 3066 (1958).
336. A. R. Todd and F. Bergel, *J. Chem. Soc.*, 364 (1937).
337. Z. Földi and A. Salomon, *Ber.*, **74**, 1126 (1941).
338. *Neth. Patent*, 52873 (1940); *Chem. Zent.*, **114**, 1912 (1943).
339. G. V. Chelintsev and Z. V. Benevolenskaya, *Z. Gen. Chem. (USSR)*, **14**, 1142 (1944); *Chem. Abstr.*, **40**, 4069[3] (1946).
340. P. C. Mitter and J. C. Bardhan, *J. Chem. Soc.*, **123**, 2179 (1923).
341. *Brit. Patent*, 1,174,165 (1969); *Ger. Appl.* (1966); *Chem. Abstr.*, **72**, 90504f (1970).
342. S. Ruhemann, *Ber.*, **30**, 821 (1897).
343. H. Bredereck, H. Herlinger and E. H. Schweizer, *Ber.*, **93**, 1208 (1960).
344. H. Wamhoff and C. Materne, *Ann. Chem.*, **754**, 113 (1971).
345. R. M. Dodson and J. K. Seyler, *J. Org. Chem.*, **16**, 461 (1951).
346. R. N. Lacey, *J. Chem. Soc.*, 839 (1954).
347. F. I. Luknitskii, D. O. Taube and B. A. Vovsi, *Zr. Org. Khim USSR*, 1851 (1969).
348. S. Yanagida, T. Fugita, M. Ohoka, R. Kumagai and S. Komori, *Bull. Chem. Soc. Jap.*, **46**, 299 (1973).
349. F. Kunckell and L. Zumbusch, *Ber.*, **35**, 3164 (1902).
350. G. A. Howard, B. Lythgoe and A. R. Todd, *J. Chem. Soc.*, 476 (1944).
351. G. W. Kenner, B. Lythgoe, A. R. Todd and A. Topham, *J. Chem. Soc.*, 574 (1943).
352. H. Morimoto, N. Hayashi, T. Naka and S. Kato, *Ber.*, **106**, 893 (1973).
353. T. Nishino, M. Kiyokawa, Y. Miichi and K. Tokuyama, *Bull. chem. Soc. Jap.*, **46**, 253 (1973).
354. T. Nishino, Y. Miichi and K. Tokuyama, *Bull. chem. Soc. Jap.*, **46**, 580 (1973).
355. F. C. Schaefer, I. Hechenbleikner, G. A. Peters and V. P. Wystrach, *J. Amer. Chem. Soc.*, **81**, 1466 (1959).
356. C. Grundmann, H. Schröder and W. Ruske, *Ber.*, **87**, 1865 (1954).
357. F. C. Schaefer and G. A. Peters, *J. Amer. Chem. Soc.*, **81**, 1470 (1959).
358. *U.S. Patent*, 2,845,422 (1958); *Chem. Abstr.*, **52**, 20217e (1958).
359. A. Pinner, *Ber.*, **23**, 2919 (1890).

360. T. Rappeport, *Ber.*, **34,** 1983 (1901).
361. Ch. Grundmann and H. Schroeder, *Ber.*, **87,** 747 (1954).
362. H. Schroeder and Ch. Grundmann, *J. Amer. Chem. Soc.*, **78,** 2447 (1956).
363. E. Degener, H. G. Schmelzer and H. Holtschmidt *Angew. Chem. Ed, Int.*, **78,** 981 (1966).
364. H. G. Schmelzer, E. Degener and H. Holtschmidt, *Angew. Chem.*, **78,** 982 (1966).
365. H. Bredereck, F. Effenberger and A. Hofmann, *Ber.*, **96,** 3265 (1963).
366. A. Pinner, *Ber.*, **23,** 161 (1890).
367. A. Pinner, *Ber.*, **25,** 1624 (1892).
368. A. W. Titherley and E. C. Hugues, *J. Chem. Soc.*, **99,** 1493 (1911).
369. H. Bader, *J. Org. Chem.*, **30,** 707 (1965).
370. B. G. Baggar, *C.R. Acad. Sci. Paris*, **264,** 352 (1967).
371. S. Hayashi, M. Furukawa, Y. Fujino and H. Morishita, *Chem. Pharm. Bull.*, **19,** 1789 (1971).
372. E. Haruki, H. Imanaka and E. Imoto, *Bull. chem. Soc. Jap.*, **41,** 1368 (1968); *Chem. Abstr.*, **69,** 96530w.
373. R. Richter, *Chem. Ber.*, **101,** 3002 (1968).
374. R. Richter and W. P. Trautwein, *Ber.*, **102,** 931 (1969).
375. J. Goerdeler and J. Neuffer, *Ber.*, **104,** 1580 (1971).
376. J. Goerdeler and J. Neuffer, *Ber.*, **104,** 1616 (1971).
377. J. Neuffer and J. Goerdeler, *Ber.*, **104,** 3498 (1971).
378. A. C. Veronese, C. Di Bello, F. Filira and F. D'Angeli, *Gazz. chim. ital.*, **101,** 569 (1971).
379. J. L. Fahey, P. A. Foster, D. G. Neilson, K. M. Watson, J. L. Brokenshire and D. A. V. Peters, *J. Chem. Soc.*, 719 (1970).
380. A. W. Titherley, *J. Chem. Soc.*, 200 (1910).
381. G. Kresse, C. Seyfried and A. Trede, *Tetrahedron Letters*, 3933 (1965).

CHAPTER 8

Kinetics and mechanisms of reactions of amidines

ROBERT H. DE WOLFE

University of California at Santa Barbara, U.S.A.

I. INTRODUCTION

The most powerful tool for the study of reaction mechanisms is chemical kinetics. No proposed reaction mechanism can be more than a temporary working hypothesis until it is supported by kinetic data[1]. The literature abounds with proposed mechanisms for most of the reactions of amidines, based on little more than a knowledge of the reaction conditions and products. With few exceptions, the only amidine reactions whose mechanisms are supported by solid experimental evidence are those whose kinetics have been studied.

These reactions are few in number. They include hydrolysis, alkylations and acylations, and thermal isomerizations. Hydrolysis reactions have been the most thoroughly studied. Another process (not strictly a reaction) whose kinetics have been studied involves rotation about the C—N bonds of amidines. The kinetics of these chemical reactions and conformational transformations are the subject of this chapter.

II. HYDROLYSIS REACTIONS OF AMIDINES

A. General Characteristics

Amidines are hydrolysed under milder conditions than the corresponding nitriles, amides or esters[2]. Amidine hydrolysis occurs in two steps, with the first step usually being faster than the second:

$$RC\underset{NR^2R^3}{\overset{NR^1}{<}} \quad \xrightarrow{H_2O} \quad RCONHR^1 + RCONR^2R^3 + R^1NH_2 + R^2R^3NH \quad (1)$$

$$RCONHR^1 + RCONR^2R^3 \quad \xrightarrow{H_2O} \quad RCO_2H + R^1NH_2 + R^2R^3NH \quad (2)$$

The composition of the mixture of products formed in the first step of hydrolysis of unsymmetrically substituted amidines depends on reaction conditions and the nature of the N-substituents.

Until recently, only qualitative information existed concerning the effects of amidine structure and reaction conditions on rate of hydrolysis. Amidines often hydrolyse on standing in the presence of water, or when they are dissolved in water or an organic solvent containing water. These reactions occur under alkaline conditions, since amidines are relatively strong organic bases. Usually hydrolysis occurs more rapidly in alkaline solutions than in acidic solutions. Amidines generally hydrolyse more slowly in moderately concentrated than in dilute solutions of strong acids.

The hydrolytic reactivity of amidines is very sensitive both to substituents on acyl carbon and substituents on the nitrogen atoms. Substituents appear to influence reactivity by both steric and inductive effects. Unsubstituted amidines are much more reactive than N-substituted amidines. For example, acetamidine hydrolyses rapidly in aqueous solutions at room temperature[3], and α-phenylacetamidine hydrolyses when its aqueous solution is warmed[2]. In contrast, N,N'-diphenylformamidine survives steam distillation[4], and most N-substituted amidines are relatively inert to water at room temperature. All amidines are hydrolysed by sufficiently vigorous treatment with aqueous acid or alkali, but again their reactivity varies markedly with structure. For example, N,N-dimethylbenzamidine is hydrolysed by boiling aqueous 20% sodium hydroxide, but N,N-dimethyl-N'-benzylbenzamidine and N-benzyl-N'-methylbenzamidine are not[5].

In the remainder of this section, experimental observations relevant to amidine hydrolysis reactions are discussed first, and then the most probable mechanisms of these reactions are considered.

B. Kinetics of Hydrolysis of Acyclic Amidines

An investigation of the hydrolysis of N,N'-diphenylformamidine in aqueous dioxane buffer solutions was the first kinetic study of amidine hydrolysis (Equation (3), X = H)[6].

$$XC_6H_4NH—CH\!\!=\!\!N—C_6H_4X + H_2O \xrightarrow{\ H_3O^+,\ HA\ }$$

$$(1) \qquad\qquad\qquad\qquad\qquad XC_6H_4NH_2 + XC_6H_4NHCHO \quad (3)$$

The reaction is general acid–base catalysed. Reaction rate is insensitive to the ionic strength of the reaction solution, but varies with solvent composition, passing through a maximum at about 35% dioxane in acetate buffers. The following rate law was derived from the kinetic data:

$$k_{exp} = \frac{1}{1 + K_b[H_3O^+]}(k_0 + k_H[H_3O^+] + k_{HA}[HA]),$$

where K_b is the basicity constant ($1/K_a$) of the amidine (the first term on the right side of this equation represents the fraction of the amidine which is present as the free base). Hydrolysis rates in p-nitrophenol buffers, in which the amidine is present largely as the free base, permitted evaluation of k_H and K_b in 30·7% dioxane at 35°C: $k_H = 230$ and $K_b = 1\cdot4 \times 10^6$. k_{HA} (buffer acid catalytic coefficients) were $1\cdot3 \times 10^{-2}$ l mol^{-1} sec^{-1} for acetic acid and $1\cdot6 \times 10^{-4}$ l mol^{-1} sec^{-1} for p-nitrophenol. A Broensted catalysis law plot[7] of k_{HA} vs K_{HA} for the hydrolysis reaction under these conditions has a slope $\alpha \simeq 0\cdot6$ ($\beta \simeq 0\cdot4$). Buffer catalysis is also exhibited in hydrolysis of N,N'-di-m-chlorophenylformamidine in 20% dioxane acetate, chloroacetate, and dichloroacetate buffers at 25°C. For this reaction the Broensted catalysis law $\alpha \simeq 0\cdot7$[8].

In dilute solutions of mineral acids, the general rate equation simplifies to:

$$k_{exp} = \frac{k_0 + k_H[H_3O^+]}{K_b[H_3O^+]} \simeq k_H/K_b$$

As predicted by this equation, hydrolysis rate was found to be independent of hydrochloric acid concentration in dilute hydrochloric acid solutions.

The fact that N,N'-diarylformamidines are considerably more reactive than homologous N,N'-diarylamidines makes possible kinetic studies under experimentally convenient reaction conditions. For this reason, N,N'-diarylformamidines were selected as substrates for studies of the dependence of hydrolysis rate on pH, temperature, solvent polarity, water activity, and aryl substituents. In aqueous 20% dioxane solutions 0·415 N in hydrochloric acid, hydrolysis of N,N'-diarylformamidines is strongly accelerated by electron-withdrawing aryl substituents[8]. Plots of k_{exp} vs Hammett's σ-constants for the aryl substituents[9] yielded straight lines of slope $\rho = 3\cdot6$–$3\cdot8$ in the temperature range 25–55°C:

$$\log k_{exp} = \log k_0 + \rho\sigma$$

where $\log k_0 = -4\cdot46$ and $\rho = 3\cdot64$ at 25·0°C; $\log k_0 = -3\cdot79$ and $\rho = 3\cdot78$ at 39·7°C; and $\log k_0 = -3\cdot24$ and $\rho = 3\cdot63$ at 54·6°C. The ρ-values for these reactions are unusually large[10].

Ortho substituents appear to exert little steric effect on reactivity, since N,N'-di-*o*-chlorophenylformamidine is more than three times as reactive as the di-*m*-chlorophenylformamidine.

Energies and entropies of activation derived from rates of hydrolysis of N,N'-diarylformamidines in 20% dioxane-0·415 N-HCl at several temperatures show that the effects of aryl substituents on reactivity are largely due to their influence on activation energies. Entropies of activation ranged between -19 and -25 e.u., and showed no systematic variation with the nature of the aryl substituent. In contrast, energies of activation diminished steadily with increasing electron-withdrawing power of the aryl substituent. (For hydrolysis of N,N'-diphenylformamidine in 20% dioxane, 0·415 N-HCl, $E_a = 18\cdot4$ kcal/mol and $\Delta S^{\ddagger} = -19$ e.u.; for N,N'-di-*m*-chlorophenylformamidine hydrolysis, $E_a = 16\cdot4$ kcal/mol and $\Delta S^{\ddagger} = -20$ e.u.)

In 0·415 N-HCl, the rate of hydrolysis of N,N'-di-*m*-chlorophenyl-formamidine in aqueous dioxane solutions was found to go through a maximum at about 60% dioxane.

In dilute acid solutions, the rate of hydrolysis of N,N'-diphenylform-amidine is nearly independent of acid concentration. In more concentrated acid solutions, hydrolysis rate diminishes rapidly with increasing acidity. For hydrolysis of N,N'-diphenylformamidine in aqueous hydrochloric acid at 25°C, the following rate law is followed approximately:

$$k_{exp} = C[H_3O^+]a_{H_2O}/h_0,$$

where a_{H_2O} is the thermodynamic activity of water, h_0 is Hammett's acidity function, and $C = 2\cdot5 \times 10^{-5}$ l mol^{-1} sec^{-1}. The rate of hydrolysis of N,N'-di-*m*-chlorophenylformamidine at 25°C exhibited a similar sharp decrease with increasing acidity in aqueous 40% dioxane perchloric acid solutions.

The kinetics of hydrolysis of N,N'-diarylformamidines in alkaline aqueous 20% dioxane solutions are complex[11]. The influence of hydroxide ion concentration on hydrolysis rate depends on the nature of the aryl substituent. For **1**, X = H, 2- and 3-CH$_3$, 4-CH$_3$O and 4-NO$_2$, the rate of hydrolysis is nearly independent of hydroxide ion concentration. When X of **1** is 3- or 4-Cl, 4-Br or 3-NO$_2$, there is a pronounced increase in hydrolysis rate with increasing hydroxide ion concentration.

In aqueous alkaline 40% dioxane solutions, hydrolysis of **1**, X = 2-, 3- or 4-Cl, 4-Br, or 3-C$_2$H$_5$O, involves two competing reactions, one independent of hydroxide ion concentration, and one whose rate increases with increasing hydroxide ion concentration. Graphs of k_{exp} vs [OH$^-$] are concave downward for **X** = halogen. Slopes of these plots at [OH$^-$] = 0 show that rates of the hydroxide ion-catalysed reaction increase as X of **1** varies in the order X = 3-C$_2$H$_5$O, 4-Cl, 4-Br, 2-Cl, 3-Cl, 3-NO$_2$. Rates extrapolated to zero hydroxide ion concentration show that the uncatalysed hydrolysis rates for **1**, X = H, 4-C$_2$H$_5$O, 4-CH$_3$ and 3-CH$_3$ are practically the same, while the uncatalysed rates for **1**, X = 3-C$_2$H$_5$O, 3- and 4-Cl, and 4-Br decrease as Hammett's σ-constants of X increase. The slopes of a Hammett plot of log k_0 vs σ (k_0 is the uncatalysed hydrolysis rate) are concave downward, with the slope varying from 0 to -3. This curvature indicates a change in the rate-limiting step of the reaction, since hydrolysis by two competing mechanisms would result in upward, rather than downward curvature of the Hammett plot.

The curvature of the k_{exp} vs [OH$^-$] plots for hydrolysis of diarylformamidines having electron-withdrawing aryl substituents is understandable if the amidines are in equilibrium with unreactive conjugate bases (equation 4):

$$\mathbf{1} + \text{OH}^- \overset{K}{\underset{}{\rightleftharpoons} } (\text{Ar}-\text{N}\cdots\text{CH}\cdots\text{N}-\text{Ar})^- + \text{H}_2\text{O} \qquad (4)$$

The equilibrium constants for this dissociation can be evaluated spectrophotometrically. In aqueous 40% diozane at 25°C, $K = 65$ when Ar = 4-NO$_2$C$_6$H$_4$; $K = 1\cdot 48$ when Ar = 3-NO$_2$C$_6$H$_4$; and $K = 3\cdot 02$ when Ar = 3,4-Cl$_2$C$_6$H$_3$.

Hydrolysis of **1**, X = 4-NO$_2$, in aqueous 20% dioxane, 0·2 N-NaOH at 25°C, is somewhat faster in ordinary water than in deuterium oxide: $k_{\text{H}_2\text{O}}/k_{\text{D}_2\text{O}} = 1\cdot 33$.

Effects of acyl substituents on rates of amidine hydrolysis have not been extensively investigated. *N,N'*-Diarylacetamidines hydrolyse less than a thousandth as fast as the corresponding diarylformamidines in acidic aqueous 20% dioxane[12]. This is in sharp contrast to acid-catalysed ester hydrolysis: acetate esters undergo acid-catalysed hydrolysis about a twentieth as fast as formate esters[13].

The very great sensitivity of acid hydrolysis of amidines to alkyl sub-stitution at the acyl carbon is due to the fact that amidines are much stronger bases than carboxylate esters. Only a minute fraction of an ester is present as the conjugate acid in dilute solutions of strong acids. Replace-ment of the acyl-H of a formate ester with CH_3 should increase the equilib-rium concentration of the conjugate acid of the ester, while simultaneously decreasing its susceptibility to nucleophilic attack by water. For acid-catalysed ester hydrolysis, these opposing substituent effects tend to cancel each other. The opposite effects of acyl substituents on basicity of car-boxylate esters and hydrolytic reactivity of their conjugate acids accounts for the fact that Hammett's ρ for acid hydrolysis of ethyl benzoates is approximately zero (see ref. 9, p. 191).

Amidines are much stronger bases than esters, and are present almost entirely as amidinium ions in dilute solutions of strong acids. Therefore, the most important effect of the acetamidine C-methyl group is its in-fluence on the susceptibility of the amidinium ion to hydrolysis. Since both the polar and steric effects of the acyl methyl substituent decrease hydro-lytic reactivity, it is not surprising that diarylacetamidines hydrolyse so much more slowly than diarylformamidines.

Aryl substituent effects on N,N'-diarylacetamidine hydrolysis and N,N'-diarylformamidine hydrolysis are similar. For reaction (5) in aqueous 20% dioxane, 0·415 N-HCl at 86°C and 100°C, $\rho \simeq 3·1$.

$$(XC_6H_4NH \cdots C(CH_3) \cdots NHC_6H_4X)^+ + H_2O \longrightarrow$$
$$XC_6H_4NH_3^+ + XC_6H_4NHCOCH_3 \quad (5)$$

Within experimental error, the entropies of activation for hydrolysis of N,N'-diarylacetamidines and N,N'-diarylformamidines are the same ($\Delta S^\ddagger \simeq -22$ e.u.). The energies of activation for N,N'-diarylacetamidine hydrolysis are about 4 kcal/mol larger than for hydrolysis of the cor-responding formamidines. Thus, the effect of the acyl methyl substituent of acetamidines is reflected in E_a rather than in $\Delta S^\ddagger$.

The rates of diarylacetamidine hydrolyses, like those of diarylform-amidine hydrolyses, decrease as the acidity of the reaction medium is increased.

Two less detailed studies of amidine hydrolysis kinetics have been reported. Gould and Jameson[14] found that mandelamidine and α-sub-stituted mandelamidines (2, R = H, CH_3, C_2H_5) are quite stable in acidic

$$C_6H_5CR(OH)-C \overset{\displaystyle NH}{\underset{\displaystyle NH_2}{\diagdown}}$$

(2)

solution, but hydrolyse in alkaline solutions at rates which are proportional to hydroxide ion concentration. The specific rates for the hydroxide ion catalysed reaction at 25°C are 1.8×10^{-4} l mol^{-1} sec^{-1} when R = H, 8.4×10^{-5} l mol^{-1} sec^{-1} when R = CH$_3$, and 8.5×10^{-5} l mol^{-1} sec^{-1} when R = C$_2$H$_5$. The effects of α-substituents on reactivity closely parallel substituent effects on alkaline hydrolysis of carboxamides, and are probably steric in origin.

Holy and Zemlicka studied the kinetics of hydrolysis of the N-dimethylaminomethylene nucleosides **3–5**[15]. These compounds are of interest

(3)   (4)   (5)
(R = ribofuranosyl and 2-deoxyribofuranosyl)

because the amino-protecting N-dimethylaminomethylene group is easily introduced by treating the nucleosides with N,N-dimethylformamide acetals[16], and can be removed by hydrolysis under mild acidic or alkaline reaction conditions. The N-dimethylaminomethylene groups of formamidines **3–5** are hydrolysed to N-formyl groups at pH 5–8. The protecting dimethylaminomethylene groups are completely removed by allowing the amidines to stand in aqueous 10% acetic acid for several hours at room temperature. Rates of hydrolysis of **3–5** were found to pass through minima at pH 6–8. At 20°C and pH 4, the riboside derivatives are slightly more reactive than the deoxyriboside derivatives. Half-lives for hydrolysis of the various formamidines under these reaction conditions ranged from 6–120 h. Energies of activation varied with pH, but generally were in the range 10–20 kcal/mol. The pH–rate profiles obtained in this study are only approximate, since the observed hydrolysis rates were not corrected for buffer catalysis.

Alkaline hydrolysis of N,N'-disubstituted amidines is complicated by the existence of pH-dependent equilibria between the free amidines, their conjugate acids, and their conjugate bases. The amidinium ions are the more interesting of these three species, since they are implicated as intermediates under both acidic and alkaline conditions. For this reason, DeWolfe and Cheng studied the kinetics of hydrolysis of a series of N,N'-dimethyl-N,N'-diphenylamidinium ions, **6**[17]. These tetrasubstituted amidinium ions are isoelectronic with the conjugate acids of N,N'-disubstituted amidines, but possess no acidic proton. Their hydrolysis

$$C_6H_5-\overset{|}{\underset{CH_3}{N}}\cdots\overset{+}{\underset{R}{C}}\cdots\overset{|}{\underset{CH_3}{N}}-C_6H_5$$

(6)

thus permits the study of solvent, salt, substituent and pH effects on amidinium ion hydrolysis under alkaline conditions.

In aqueous solutions at 30°C, hydrolysis of the formamidinium and benzamidinium salts (6, R = H, C_6H_5) is approximately first order in hydroxide ion concentration in the pH range 8–14. The rate of hydrolysis of the acetamidinium ion (6, R = CH_3) levels off at high pH, possibly due to reversible formation of the ketene aminal, $CH_2=C[N(CH_3)C_6H_5]_2$. (This ketene aminal is a known compound[18]; its hydrolytic behaviour apparently has not been studied.)

The hydrolysis of these amidinium cations is generally based-catalysed in carbonate, butylamine and borate buffers, with Broensted catalysis law β-values of approximately 0·4. The hydroxide ion and butylamine catalysed reactions have substantial negative entropies of activation. The energies of activation for the hydroxide ion catalysed hydrolysis of 6 increase in the order R = $CH_3 < H < C_6H_5$. Hydrolysis of the formamidinium salt has a substantially less negative entropy of activation ($\Delta S^\ddagger = -8$ e.u.) than hydrolysis of the acetamidinium salt ($\Delta S^\ddagger = -23$ e.u.) or the benzamidinium salt ($\Delta S^\ddagger = -13$ e.u.). For all of the catalysts used, the formamidinium salt is about 100 times as reactive as the acetamidinium salt, which is 3–4 times as reactive as the benzamidinium salt. Hydrolysis of the acetamidinium and benzamidinium salts is insensitive to the ionic strength of the reaction solution.

For hydrolysis of 6, R = $X-C_6H_4$, in aqueous butylamine buffers at 30°C, a linear Hammett plot is obtained with $\rho = 1·6$. The linearity of this plot indicates that electron-donating substituents such as $p-CH_3$ and $p-CH_3O$ do not significantly stabilize 6 by resonance, probably due to steric hindrance to coplanarity of the acyl substituent and the phenyl-methylamino groups.

The hydroxide-ion-catalysed hydrolysis of 6 is somewhat faster in deuterium oxide than in ordinary water: $k_{\text{exp H}}/k_{\text{exp D}} = 0·79$ when R = CH_3, and 0·56 when R = C_2H_5.

C. Kinetics of Hydrolysis of Imidazolines and Imidazolinium Ions

The chemical properties of heterocyclic amidines are similar to those of the acyclic amidines. The only group of heterocyclic amidines whose hydrolysis kinetics have been studied are the imidazolines, **7**. Several studies of hydrolyses of imidazolinium ions, **8**, have also been reported.

(7)   (8)

As in the preceding section, hydrolysis reactions of the amidines **7** are considered first, followed by a review of hydrolysis reactions of the amidinium ions **8**.

Martin and Parcell briefly examined the hydrolysis of 2-methyl-imidazoline (**9**)[19]. This compound is relatively stable in acidic and neutral solutions, and hydrolyses at a significant rate only at high pH:

In alkaline solutions the reaction followed the rate law:

$$k_{exp} = \frac{k[OH^-]}{(K_a/K_w)[OH^-] + 1} \qquad (6)$$

where K_a is the dissociation constant of the conjugate acid of **9**, and K_w is the autoprotolysis constant of water. This dependence of rate on hydroxide ion concentration suggests that the rate limiting step of the reaction involves addition of hydroxide ion to the 2-methylimidazolinium ion.

Harnsberger and Riebsomer studied the alkaline hydrolysis of 1,2-

disubstituted imidazolines[20, 21]. A detailed study of the effects of pH and ionic strength on hydrolysis of 1-(2-hydroxyethyl)-2-pentylimidazoline (equation (7), $R = C_5H_{11}$, $R' = CH_2CH_2OH$)[21] revealed that sodium perchlorate exerts a strong inhibitory effect on hydrolysis rate at pH 11·4, but not at pH 13·7. The rate–pH profile for this reaction was determined in aqueous 1 M-NaClO$_4$ solutions at 25·6°C, in which salt effects should be approximately constant. Hydrolysis rate is approximately first order in hydroxide ion below pH 13, and becomes independent of pH above pH 13. The authors derived a rate law to account for the observed pH–rate profile, which was based on a mechanism which they realized is incompatible with their spectrophotometric data. Actually, this reaction, like 2-methylimidazoline hydrolysis, is adequately described by the rate law of equation (6).

The energy of activation of this reaction is pH-dependent: at pH 12·45, $E_a = 16·7$ kcal/mol, while at pH 11·5, $E_a = 20·2$ kcal/mol.

The effect of acyl substituents on hydrolytic reactivity of 1-(2-hydroxyethyl)-2-alkylimidazolines (**7**, R = alkyl, R' = CH$_2$CH$_2$OH, R^2–R^5 = H) was determined in 95% ethanol–0·0375 M-NaOH at 70·5°C[20]. With the exception of the 2-*t*-butylimidazoline, the effects of 2-alkyl substituents on reactivity correlate reasonably well with the Taft polar and steric parameters of the substituents[22]. The reaction constant, ρ^*, derived from the relationship $\rho^*\sigma^* = \log(k/k_0) - E_s$[23] is approximately 4·0. The 2-*t*-butylimidazoline is less reactive by a power of 10 than predicted from the Taft equation.

Hydrolytic reactivities of a series of 1-alkyl-2-pentylimidazolines (**7**, R = C$_5$H$_{11}$, R^1 = H or alkyl, R^2–R^5 = H) were also determined under these reaction conditions. Interestingly, the imidazoline having no *N*-substituent is less reactive than the *N*-methyl derivative by nearly a factor of 10. The compounds **7** having *N*-alkyl substituents diminished in reactivity in the order: R^1 = CH$_3$ > C$_2$H$_5$ > (CH$_3$)$_2$CH. The *N*-2-hydroxyethyl derivative is more than twice as reactive as the *N*-ethyl derivative. 1-Isopropyl-2-pentyl-4,4-dimethylimidazoline (**7**, R = C$_5$H$_{11}$, R^1 = (CH$_3$)$_2$CH, R^2 = CH$_3$, R^4 = R^5 = H) undergoes alkaline hydrolysis at less than a thousandth the rate of hydrolysis of the 1,2-disubstituted imidazolines. These imidazolines hydrolyse only very slowly in acidic solutions.

The kinetics of alkaline hydrolysis of a series of 1-(3-silylpropyl)-2-imidazolines (**10**, R = C$_3$H$_7$, (CH$_3$)$_3$Si(CH$_2$)$_3$, (CH$_3$O)(CH$_3$)$_2$SiCH$_2$CH-(CH$_3$)CH$_2$, (CH$_3$O)$_2$(CH$_3$)SiCH$_2$CH(CH$_3$)CH$_2$, and (CH$_3$O)$_3$SiCH$_2$CH-(CH$_3$)CH$_2$) were described by Saam and Bank[24]:

$$\text{imidazoline}-H + H_2O \xrightarrow{\ OH^-\ } RNHCH_2CH_2NHCHO + H_2NCH_2CH_2NRCHO$$

(10)

The methoxysilyl derivatives hydrolyse almost instantly to silanols or siloxanes under the reaction conditions used, and it is these derivatives whose hydrolysis was actually followed. **10** did not hydrolyse appreciably in acidic solution. In basic solutions (aqueous 1% isopropyl alcohol) the rate of hydrolysis increases with increasing hydroxide ion concentration, and is independent of phosphate buffer concentration at constant pH. For all of the imidazolines studied except the 3-trimethoxysilyl-2-methyl-propyl derivative (**10**, R = $(CH_3O)_3SiCH_2CH(CH_3)CH_2$), hydrolysis rates level off as the hydroxide ion concentration increases, and the observed rate constants are described by rate law (6).

Hydrolysis of the silanol derived from the 3-trimethoxysilyl compound followed a different rate law:

$$k_{exp} = \frac{k_1[OH^-] + k_2[OH^-]^2}{1 + (K_a/K_w)[OH^-]}$$

A possible explanation for the second-order dependence of hydrolysis rate on hydroxide ion concentration for this compound is discussed in the next section.

The hydrolysis of imidazolinium cations has received much more attention than hydrolysis of uncharged imidazolines. This is in part due to the use of N,N-disubstituted imidazolinium cations as model systems in studies of the chemistry of 5,10-methenyltetrahydrofolic acid, **11**, and its hydrolysis product, **12**, important formate-carrying cofactors[25].

(11)

The structure of this complex imidazolinium ion was first established by Shive and co-workers[26].

11 is stable in acidic solutions, but hydrolyses reversibly to N-10-formyltetrahydrofolic acid, **12**, in alkaline solutions. N-10-Formyltetra-hydrofolic acid is the kinetically controlled hydrolysis product. If **12** is heated, or if the hydrolysis reaction mixture is heated or allowed to stand for long periods of time, N-5-formyltetrahydrofolic acid, **13** is obtained.

Shive and co-workers[26] showed that the hydrolysis of **11** to **12** is base-catalysed, and that **12** reverts to **11** in acidic solutions. Tabor and Wynd-garden[27] demonstrated that both **12** and **13** are converted to **11** in acidic

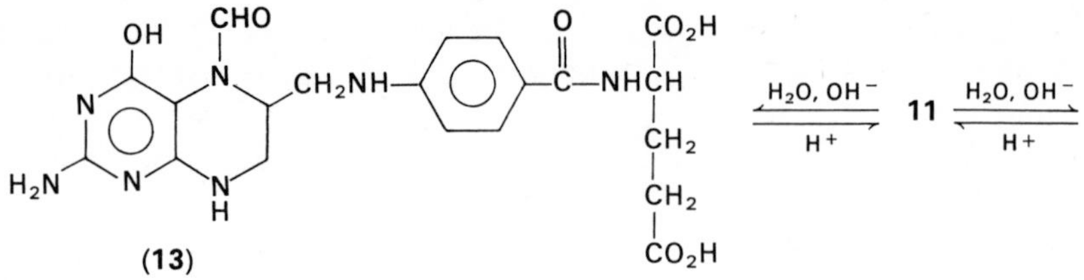

(13)

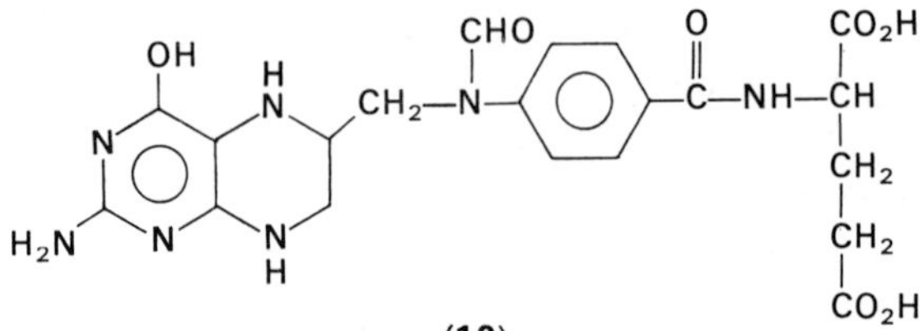

11

(12)

solutions, and that at pH 5·7 the equilibrium constant for the reaction
$12 + H^+ \overset{K}{\rightleftharpoons} 11 + H_2O$ is $K \simeq 4 \times 10^7$. They also reported that the half-life of **11** in maleate buffers at pH 6·8 is approximately 50 minutes.

Hartman and Buchanan[28] found that at pH 7·4 the rate of hydrolysis of **11** to **12** depends on the nature of the buffer base. The reaction is fastest in phosphate buffers, slowest in tris buffers, and occurs at an intermediate rate in maleate buffers (the observed rates at 27°C were approximately 5×10^{-3} sec^{-1} in the phosphate buffer, $2 \cdot 3 \times 10^{-3}$ sec^{-1} in the tris buffer and $4 \cdot 1 \times 10^{-3}$ sec^{-1} in the maleate buffer).

Kay and co-workers measured the equilibrium constants for the acid-catalysed cyclization of **12** and **13** to **11**[29]. These authors reported that for the reaction $12 + H^+ \rightleftharpoons 11 + H_2O$, $K = 9 \times 10^5$, while for $13 + H^+ \rightleftharpoons 11 + H_2O$, $K = 6 \cdot 5 \times 10^2$.

11 is a rather complex molecule, but its hydrolytic reactions are closely paralleled by those of 1,3-diarylimidazolinium salts. Such salts have been used as model compounds to gain insight into the reactions of **11**. Shive and co-workers[26], for example, made a qualitative study of the effect of pH on the rate of hydrolysis of 1,3-diphenylimidazolinium chloride, **14**:

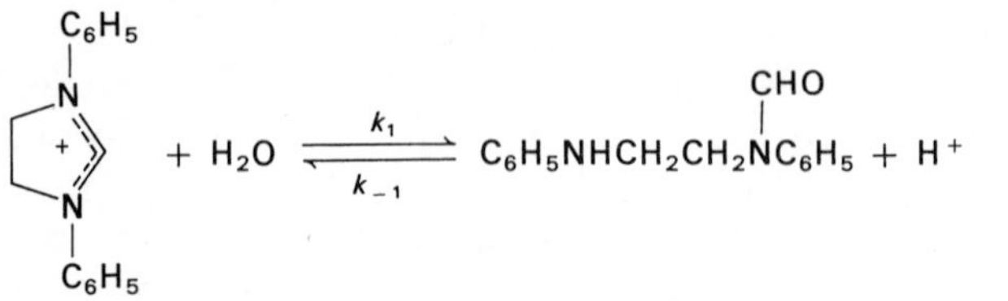

$$+ H_2O \underset{k_{-1}}{\overset{k_1}{\rightleftharpoons}} C_6H_5NHCH_2CH_2\overset{\text{CHO}}{\overset{|}{N}}C_6H_5 + H^+ \qquad (8)$$

(14)   **(15)**

14 is symmetrically substituted, and can yield only one hydrolysis product. The hydrolysis of **14** is general base-catalysed in phosphate buffers. Spectrophotometric data showed that the equilibrium constant for reaction (8) is approximately 1.4×10^{-5}. Jaenicke and Brode[30] also found that 1-formyl-1,3-diarylethylenediamines are nearly completely converted to 1,3-diarylimidazolinium ions in acidic solutions. The fact that formylethylenediamines cyclize almost completely to imidazolines in acidic solutions explains earlier reports that imidazolines appear to be inert in acidic solutions.

Robinson and Jencks thoroughly investigated the kinetics of hydrolysis of 1,3-diphenylimidazolinium chloride (**14**) to 1-formyl-1,3-diphenylethylenediamine, **15**, and the reverse of this reaction, cyclization of **15** to **14** [equation (8)][31, 33]. They found that K for reaction (8) at seven different values of pH averages to 1.14×10^{-5} M^{-1}. Hydroxide-ion-catalysed hydrolysis of **14** follows the rate law:

$$k_{exp} = k_0 + k_2 a_{OH} + k_3 a_{OH^2},$$

where a_{OH} is the thermodynamic activity of hydroxide ion. Above pH 10, the term which is second order in hydroxide ion accounts for most of the reaction.

Hydrolysis of **14** exhibits buffer catalysis. The buffer catalysis increases with increasing pH, indicating that the buffer bases are the effective catalysts. The rate law followed by the buffer-catalysed portion of the reaction is:

$$k_{buffer} = k_2[B] + k_3[B][OH^-], \qquad (9)$$

where B represents the buffer base. The Broensted catalysis law β-value for the $k_3[B][OH^-]$ term of equation (9) is 0.26, while the β-value for the $k_2[B]$ term is 0.44.

The rate law for cyclization of **15** to **14** in acidic solutions is:

$$k_{exp} = \frac{[15]}{[15] + [15 \cdot H^+]} (k_{H^+} a_{H^+} + k_{HA}[HA])$$

This reaction is general acid-catalysed, since the catalytic effectiveness of formate buffers increases with decreasing pH, and since acetate buffers are catalytically less effective than formate buffers.

The general base-catalysed reaction may be subject to bifunctional catalysis by HPO_4^{2-} and HCO_3^-. The catalytic effectiveness of these bases is 8–30 times greater than predicted from their pK values and the Broensted catalysis law equation for this reaction. Monofunctional buffer bases of diverse types, including carboxylate, hydroxide, amines, carbonate, methylphosphate, and hydrazine, are accommodated quite well by the Broensted catalysis law.

In triethylenediamine buffers, the amine-catalysed portion of the reaction exhibits a deuterium solvent isotope effect of $k_{H_2O}/k_{D_2O} = 0.6$. This is a composite of isotope effects on the pre-equilibrium and rate-determining steps, and is similar to solvent isotope effects on other general base-catalysed reactions.

Above pH 11·7, the ultraviolet spectra of solutions of **14** show an initial rapid change, followed by a slower change as hydrolysis products are formed. These spectroscopic results, together with a shift from kinetics second order in hydroxide ion toward kinetics zero order in hydroxide ion at sufficiently high pH, suggest that **14** reacts with hydroxide ion to form a tetrahedral intermediate, whose conjugate base undergoes water-catalysed conversion to the hydrolysis products.

Robinson and Jencks also made the first detailed study of the kinetics of hydrolysis of 5,10-methenyltetrahydrofolic acid, and cyclization of the resulting 10-formyltetrahydrofolic acid $(\mathbf{11} \rightleftarrows \mathbf{12})$[34]. Hydrolysis occurs at convenient rates under mildly alkaline conditions, and cyclization occurs in acidic solutions.

In the pH range 8·80–9·90, the hydroxide-ion-catalysed hydrolysis of **11** is described by the rate law:

$$k_{OH} = \left(\frac{a_{H^+}}{a_{H^+} + K}\right)(k_2 a_{OH^-} + k_3 a_{OH^-}{}^2),$$

where K is the equilibrium constant for: $\mathbf{11} + H_2O \rightleftarrows \mathbf{12} + H^+$ ($K = 1.1 \times 10^{-9}$ in water at 25°C). The first term on the right side of this equation represents the fraction: $[\mathbf{11}]/([\mathbf{11}] + [\mathbf{12}])$.

The hydrolysis of **11** was found to be buffer catalysed, as previously reported **28**. Slopes of plots of hydrolysis rate vs buffer concentration at constant pH decrease with increasing buffer concentration. At low buffer concentrations, the buffer-catalysed hydrolysis rate is described by the rate law:

$$k_{buffer} = \left(\frac{a_{H^+}}{a_{H^+} + K}\right)(k_2[B] + k_3[B]a_{OH^-})$$

In the pH region 7 to 10, most of the hydrolysis is accounted for by the second-order term. The fact that the rate law for this reaction is the same as that for diphenylimidazolinium chloride hydrolysis [equation (8)][31-33] suggests that both reactions have the same mechanism.

5,10-Methenyltetrahydrofolic acid (**11**) differs in two important ways from diphenylimidazolinium chloride (**14**): the nitrogens of the imidazolinium ring of **11** are unsymmetrically substituted, and the imidazolinium ring of **11** is part of a fused-ring system. Benkovic and co-workers selected 2-aryltetrahydroimidazo[1,5-*a*]-quinazolines (**16**) which more closely approximate to the structure of methenyltetrahydrofolic acid than does diphenylimidazolinium chloride, as model compounds for hydrolysis studies[35].

$$(X = CO_2Et, Cl, CH_3)$$

(16)

Hydrolysis of **16** in buffers above pH 6 followed the rate law:

$$k_{exp} = k_{H_2O} + k_{OH^-} a_{OH^-} + k_B[B] + k_{OH^{-2}} a_{OH^{-2}} + k_{B \cdot OH^-}[B]a_{OH^-}$$

where B is the buffer base and $a_{OH^-} = 10^{-14}/a_{H^+}$.

Hydrolysis product composition depends on the nature of X in **16**, and upon reaction conditions:

$$16 + H_2O \xrightleftharpoons{\;\;-H^+\;\;}$$

(17)   **(18)**

Under the conditions (aqueous buffers, pH 6·1–8·5) where the rate law applies, the 1-formylquinoxaline derivative **17** is the sole product when X = Cl or CH_3. When X = $CO_2C_2H_5$, the hydrolysis products are 20% **17** and 80% **18**. At pH = 6, when X = $CO_2C_2H_5$, **18** is slowly converted to **17**, the thermodynamically more stable product. Equilibrium constants were determined for the reaction: $16 + H_2O \rightleftarrows 17 + H^+$. $K \simeq 10^{-3}$ when X = $CO_2C_2H_5$, 5×10^{-5} when X = Cl, and $7·8 \times 10^{-6}$ when X = CH_3. For the *p*-carbethoxy derivative, the equilibrium constant for the reaction: $18 \rightleftarrows 17$, has a value of about 160. In the case of the *p*-carbethoxy derivative, the ratio of **17** to **18** in the hydrolysis products is pH-dependent. The fraction of **17** decreases with increasing pH above pH 6·5. **16**, X = $CO_2C_2H_5$, is a close structural analogue of 5,10-methenyltetrahydrofolic acid, **11**, and it is not surprising that the hydrolysis kinetics of the two compounds are closely similar. The hydrolytic behaviour of **16** supports the hypothesis that the principal features of tetrahydrofolic acid chemistry depend on the difference in basicity between $N_{(5)}$ and $N_{(10)}$, and are essentially independent of the pyrimidine ring and the glutamate residues.

As pointed out in the preceding discussion, imidazolines are relatively inert in weakly acidic solutions, but hydrolyse to acylethylenediamines in alkaline solutions. In moderately concentrated solutions of strong acids, however, imidazolines undergo acid-catalysed hydrolysis[36, 37].

The rate of hydrolysis of 2-methylimidazoline in moderately concentrated (2–8 M) solutions of HCl and H_2SO_4 is roughly proportional to acid concentration. In sulfuric acid solutions, hydrolysis rate goes through a maximum at about 10 M-H_2SO_4, and then decreases rapidly up to 16 M-H_2SO_4.

Up to 12 M-H_2SO_4, the kinetic data yield an excellent Bunnett–Olsen plot[38] of $(\log k_{exp} + H_0)$ vs $(H_0 + \log H^+)$ with a slope ϕ of 1·01. Subject to the limitations of comparing cationic bases with uncharged bases, this ϕ-value correlates empirically with values for reactions in which water functions as a proton transfer agent, rather than as a nucleophilic reagent, in the rate-limiting step.

The hydrolysis of 2-methylimidazoline was found to exhibit a solvent deuterium isotope effect of $k_H/k_D = 0·71$ in 4 M-H_2SO_4. The entropy of activation was -24 e.u. in 4 M-H_2SO_4, and -31 e.u. in 14 M-H_2SO_4.

The ρ-value for hydrolysis of a series of m- and p-substituted 2-arylimidazolines is approximately zero in 9 M-H_2SO_4.

D. Kinetics of Hydrolysis of N,N'-Dihydroxyamidines

Armand and co-workers studied the kinetics of hydrolysis of N,N'-dihydroxyamidines (hydroxamic acid oximes), **19**[39, 40].

$$RC\underset{NHOH}{\overset{NOH}{<}}$$

(19)

In acidic aqueous solutions, in which the dihydroxyamidines are essentially completely protonated, **19** hydrolyses to a hydroxamic acid and hydroxylammonium ion.

$$RC\underset{NHOH}{\overset{NHOH}{<}}{}^{+} + H_2O \longrightarrow RCONHOH + H_3NOH^+$$

The reaction is first order, and is independent of pH and ionic strength. At $30°C$, $10^6 k_{exp}$ (sec^{-1}) are: for $R = H$, 115; $R = CH_3$, 4·5; $R = C_2H_5$, 4·2; $R = C_6H_5$, 2·7; $R = C_6H_5CH_2$, 2·1.

Hydrolysis of **19** in alkaline solutions is first order in hydroxide ion up to pH 10, above which the reaction rate levels off:

$$2\ RC\!\!\begin{array}{c}\nearrow NOH\\[2pt]\searrow NHOH\end{array}\ +\ 2\ OH^-\ \dashrightarrow\ RC\!\!\begin{array}{c}\nearrow NO^-\\[2pt]\searrow NO\end{array}\ +\ R\!-\!C\!\!\begin{array}{c}\nearrow NO^-\\[2pt]\searrow NH_2\end{array}\ +\ 3\ H_2O$$

The pH dependence of reaction rate, the nature of the reaction products, and the insensitivity of the reaction rate to steric substituent effects, indicate that the mechanism of the base-promoted reaction is fundamentally different from that of acid-catalysed hydrolysis.

The mechanism of hydrolysis of N,N'-dihydroxyamidines in acidic solutions is probably essentially the same as that for acid hydrolysis of other amidines (see next section). The products of alkaline hydrolysis of these compounds, however, clearly required a mechanism which is quite different from alkaline hydrolysis of other amidines. Armand[39] proposed the following mechanism for this reaction:

$$\mathbf{19} + OH^-\ \xrightarrow{-H_2O}\ RC\!\!\begin{array}{c}\nearrow NHOH\\[2pt]\searrow NO^-\end{array}\ \longrightarrow\ RC\!\!\begin{array}{c}\nearrow NH\\[2pt]\searrow N{=}O\end{array}\ +\ OH^-$$

$$OH^- + 2\ RC\!\!\begin{array}{c}\nearrow NH\\[2pt]\searrow N{=}O\end{array}\ \xrightarrow{-H_2O}\ R\!-\!\underset{\underset{O^-}{\overset{|}{N{=}N}}}{C}\!\!\begin{array}{c}\nearrow NH\\[2pt]\end{array}\overset{NO^-}{\underset{}{\parallel}}\!C\!-\!R\ \xrightarrow{OH^-}\ RC\!\!\begin{array}{c}\nearrow NO^-\\[2pt]\searrow N{=}O\end{array}\ +\ RC\!\!\begin{array}{c}\nearrow NO^-\\[2pt]\searrow NH_2\end{array}$$

The observed kinetics suggest that the second step of this reaction scheme is rate-limiting.

E. Mechanisms of Hydrolysis of Amidines and Imidazolines

Amidines and imidazolines hydrolyse by similar mechanisms. Imidazolines differ from amidines in that their hydrolysis in acidic solutions is reversible, with equilibrium constants favouring the imidazoline. N,N,N',N'-Tetrasubstituted amidinium ions and 1,3-disubstituted imidazolinium ions also hydrolyse by similar mechanisms.

Imidazolines are nearly inert in acidic solutions. The slow hydrolysis which occurs under drastic conditions probably involves the irreversible hydrolysis of the small amount of monoacylethylenediammonium ion in equilibrium with the imidazoline, or (possibly) hydrolysis of the diprotonated imidazoline.

Acid hydrolysis of acyclic amidines to carboxamides and ammonium ions or substituted ammonium ions is, in contrast, essentially irreversible in dilute solutions. The only amidines whose hydrolysis kinetics have been studied in detail are N,N'-diarylformamidines[6, 8] and N,N'-diarylacetamidines[12].

N,N'-Diarylformamidine hydrolysis is general base-catalysed in buffer solutions, with Broensted catalysis law β-values of ~ 0.4. In dilute aqueous dioxane solutions of strong acids, the rates of hydrolysis of diarylformamidines are independent of hydronium ion concentration and ionic strength, and pass through a maximum at about 60% dioxane. The reactions have unusually large positive Hammett ρ-constants in dilute aqueous dioxane acid solutions. In 20% dioxane, 0.4 N-HCl, hydrolyses of diarylformamidines and diarylacetamidines have large negative entropies of activation (~ -20 e.u.) which are nearly independent of the nature of both the N-aryl and the acyl substituents. The rate of hydrolysis of N,N'-diarylformamidines diminishes sharply with increasing acid concentration in perchloric and hydrochloric acid solutions when the acid concentration exceeds about 0.5 M.

The kinetics of N,N'-diarylamidine hydrolysis support a mechanism involving general base-catalysed breakdown of a tetrahedral hydrate of the conjugate acid of the amidine:

The rate law required by this mechanism is:

$$k_{exp} = \sum_i [B]_i \left(\frac{1}{1 + K_b[H^+]} \right) k_3 K_2 K_b[H^+]$$

Since $K_b \simeq 10^6$, rate of hydrolysis will be independent of pH below about pH 5. The Broensted catalysis law β-value of 0·4 requires that the water-catalysed reaction be faster than the hydroxide-ion-catalysed reaction below pH 6.

This mechanism accounts for the observed buffer catalysis, and for the lack of dependence of hydrolysis rate on hydrogen ion concentration in dilute solutions of strong acids. It also accounts for the effects of aryl substituents on rate, since the destabilizing electrostatic interaction of electron-withdrawing aryl substituents with the positively charged nitrogen in the tetrahedral hydrate is partially relieved by charge dispersal in the rate-limiting transition state.

Salt and solvent effects on the equilibrium formation of the cationic hydrate and its conversion to products should be opposite in sign. Added electrolytes have a negligible effect on reaction rate. Hydrolysis rate is observed to increase somewhat with increasing dioxane concentration up to about 60% dioxane. This increase may be due in part to the effect of dioxane on water structure.

This mechanism accounts for the large negative entropies of activation for diarylamidine hydrolysis, and for the sharp drop-off in rate with increasing acid concentration in moderately concentrated solutions of strong acids, in which water functions as the general base catalyst for the slow step of the reaction. Both of these observations suggest that there is considerable involvement of water in the rate-limiting transition state, a conclusion which is supported by the Bunnett w-value[41] ($+7·75$) and the Bunnett-Olsen ϕ-value[38] ($+1·30$) for N,N'-diphenylformamidine hydrolysis. These values correlate empirically with values for other reactions in which water functions as a proton-transfer agent in the rate-limiting step.

Kinetic experiments can reveal only the composition of the rate-limiting transition state of a reaction. For N,N'-diarylamidine hydrolysis, this transition state contains the amidine, a proton, one or more water molecules, and a general base (which may be an additional water molecule). The mechanism outlined above is a reasonable route to such a transition state, but others can be imagined. Bunnett, for example, suggested that diarylamidine hydrolysis involves rate-limiting general base-catalysed nucleophilic attack by water on the amidinium ion[41].

Hydrolysis of N,N'-diarylformamidines in alkaline aqueous solutions involves two competing reactions, which are zero order and first order with respect to hydroxide ion. In the case of amidines with electron-withdrawing aryl substituents, the hydrolysis reaction is complicated by a parasitic side equilibrium in which the amidine is partially converted to an unreactive conjugate base. The hydroxide ion-catalysed reaction is detectable only for amidines having electron-attracting aryl substituents

The general rate law for all of these reactions is:

$$k_{exp} = \left(\frac{1}{1 + K_a'[OH^-]K_w}\right)(k + k'[OH^-]),$$

where K_a' is the acid dissociation constant of the amidine. The first term on the right side of this equation has values of less than unity only for amidines with one or more strongly electron-withdrawing aryl substituents. The $k'[OH^-]$ term is negligible for amidines having electron-releasing N-aryl substituents. For the more acidic amidines (such as N,N'-di-p-nitrophenylformamidine), the general rate equation simplifies to $k_{exp} \simeq k'K_w/K_a$ at sufficiently high hydroxide ion concentrations.

The experimental results support a mechanism similar to that of diaryl-amidine hydrolysis in acidic solutions, except that products are formed from both the protonated and unprotonated tetrahedral hydrate of the amidine, and that the parasitic ionization influences hydrolysis rate at high pH when the amidine is sufficiently acidic:

$$
\begin{array}{l}
HC\!\!\begin{array}{c}{}^{NAr}\\{}_{NHAr}\end{array} \xrightarrow{\;K_a'\;} HC\!\!\begin{array}{c}{}^{NAr}\\{}_{NAr}\end{array} + H^+ \\[1em]
HC\!\!\begin{array}{c}{}^{NAr}\\{}_{NHAr}\end{array} + H_2O \xrightarrow{\;K_1\;} \begin{array}{c}NHAr\\|\\H-C-OH\\|\\NHAr\end{array} + H_2O \underset{k_{-2}}{\overset{k_2}{\rightleftharpoons}} \begin{array}{c}{}^+NH_2Ar\\|\\H-C-OH\\|\\NHAr\end{array} + OH^- \\[2em]
\hspace{7em} \downarrow {\scriptstyle k_3, OH^-} \hspace{4em} \swarrow {\scriptstyle k_4, OH^-} \\[1em]
\hspace{7em} HC\!\!\begin{array}{c}{}^{O}\\{}_{NHAr}\end{array} + ArNH_2
\end{array}
$$

This mechanism accounts for the observed rate law, and for the effects of substituents on the hydroxide-ion catalysed and the uncatalysed reactions. Amidines with electron-releasing aryl substituents hydrolyse mostly *via* the k_4 route. Since substituent effects on formation of the protonated tetrahedral intermediate and its reaction with hydroxide ion to form products should be opposite in sign, the mechanism accommodates the observed Hammett ρ-value of approximately zero for diarylformamidines having electron-releasing aryl substituents.

Diarylformamidines with electron-attracting aryl substituents hydrolyse by both the k_3 and k_4 pathways. The k_3 path probably predominates because the $ArNH^-$ leaving group is stabilized by electron-attracting substituents. Hammett's ρ for the k_4 hydrolysis of these amidines is -3, indicating that the rate-limiting step for the uncatalysed hydrolysis of these amidines differs from that for the amidines with electron-releasing substituents. Apparently for the amidines with electron-attracting substituents $k_4[OH^-] > k_{-2}$, and formation of the protonated hydrate is rate-limiting. The destabilizing effect of electron-attracting substituents on this cationic intermediate would then account for the negative ρ-value for the uncatalysed reaction.

$N,N,N'N'$-Tetrasubstituted amidinium ions are isoelectronic with the conjugate acids of amidines, but are capable of existing in significant concentrations in alkaline solutions, since they have no acidic proton. The hydrolysis of N,N'-dimethyl-N,N'-diphenylamidinium cations **6**, R = H and C_6H_5, are first order in hydroxide ion in alkaline solution, and are general base-catalysed, with Broensted catalysis law β-values of ~ 0.4[17]. The hydroxide- and butylamine-catalysed reactions have substantial negative entropies of activation. For N,N'-dimethyl-N,N'-diphenylbenzamidinium salts having substituents on the acyl phenyl group, Hammett's ρ-constant for hydrolysis in aqueous butylamine buffers is $+1.6$.

These and other experimental observations are concordant with a mechanism involving rate-limiting general base-catalysed hydrolysis of the tetrahedral hydrate of the amidinium ion:

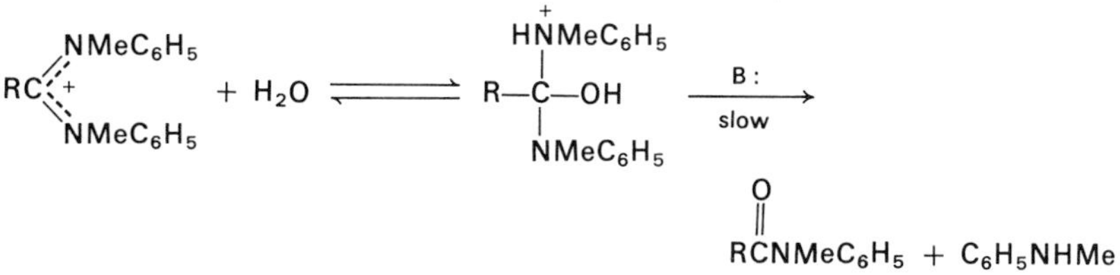

Alkaline hydrolyses of 2-substituted imidazolines[19] and 1,2-disubstituted imidazolines[20, 21] follow a rate law similar to that for alkaline hydrolysis of N,N'-diarylformamidines, except that uncatalysed hydrolysis is negligible. Presumably the imidazolines hydrolyse by the same mechanism as the formamidines. The only compound whose alkaline hydrolysis did not conform to this rate law was 1-(3-trimethoxysilyl-2-methylpropyl)-imidazoline [**10**, R = $(CH_3O)_3SiCH_2CH(CH_3)CH_2$]. The rate law for hydrolysis of this compound contains a term which is second order in hydroxide ion. This second order term was attributed to intramolecular Lewis-acid catalysis of the hydrolysis by the silanol side chain derived from the trimethoxsilylpropyl group by a rapid initial hydrolysis:

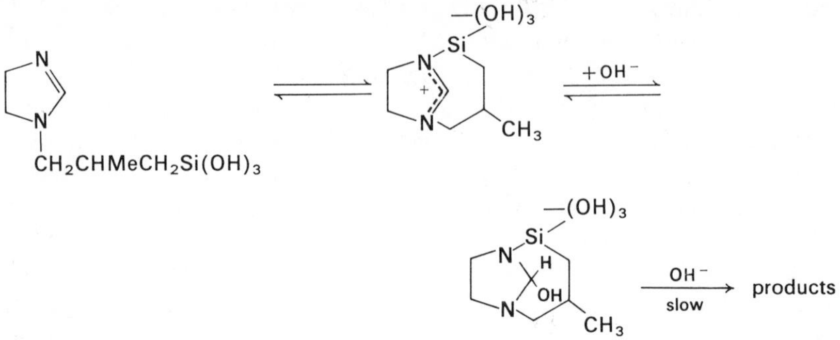

The hydrolysis of 1,3-disubstituted imidazolinium cations has been extensively studied. These reactions, which yield, N,N'-disubstituted-N-acylethylenediamines, occur only under neutral or alkaline conditions. In acidic solutions hydrolysis is reversible, and the equilibria strongly favour the imidazolinium ions:

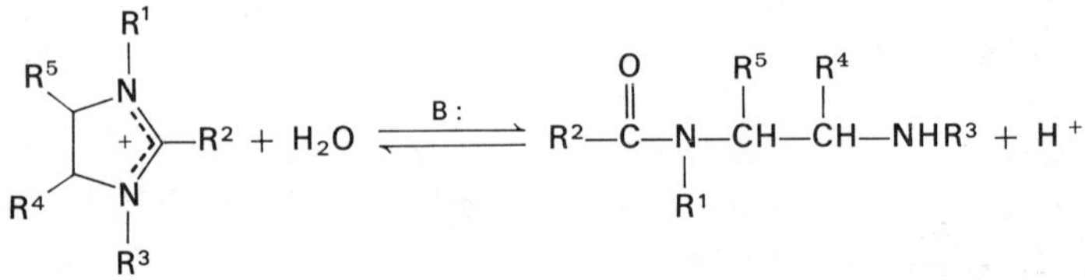

Hydrolysis is general base-catalysed in buffer solutions. Tetrahedral products of addition of hydroxide ion to the acyl carbons of the amidinium ions are usually assumed to be reactive intermediates in hydrolyses of these compounds. Some heterocyclic amidine hydrates are sufficiently stable to be isolable. Examples are 1,3-dibenzoyl-2-hydroxybenzimidazoline (**20**, R = C_6H_5CO)[42], 1,3-dimethyl-2-hydroxybenzimidazoline (**20**, R = CH_3)[43], and 1-*p*-acetamidobenzenesulphonyl-2-methyl-2-hydroxyimidazolidine (**21**)[44].

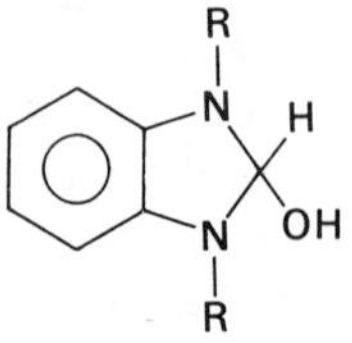

(**20**) 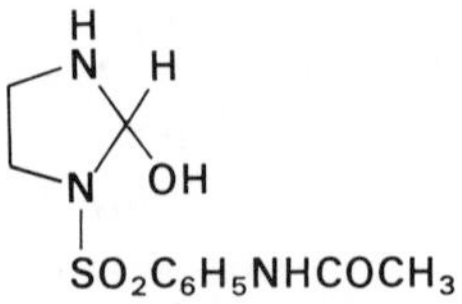   (**21**)

Robinson[32,33] was the first to obtain direct evidence for the intermediacy of 2-hydroxyimidazolidines in hydrolysis reactions.

Above pH 11·7, changes in the ultraviolet absorption spectra of solutions of 1,3-diphenylimidazolinium chloride (14) can most reasonably be accounted for by assuming the following hydrolysis mechanism:

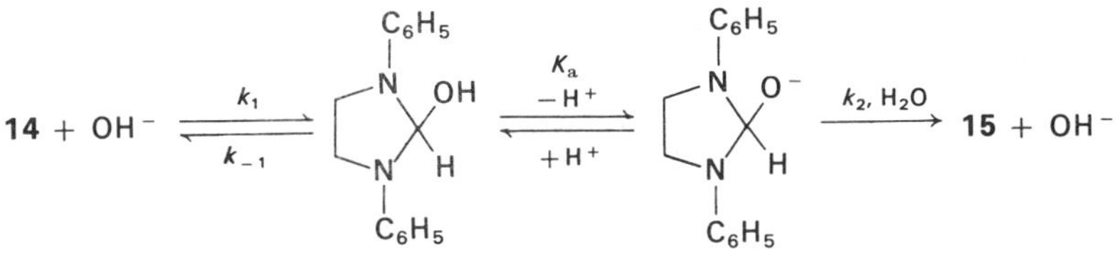

This mechanism involves general acid-catalysed hydrolysis of the conjugate base of the tetrahedral intermediate, and is kinetically indistinguishable from a mechanism involving general base-catalysed hydrolysis of the intermediate. At 25°C, the kinetic and spectrophotometric results are accommodated by this scheme if $K_a = 1·8 \times 10^{-13}$, $k_1 = 1·6 \times 10^4$ M^{-1} sec^{-1}, $k_{-1} = 250$ sec^{-1} and $k_2 = 180$ sec^{-1}. This reaction scheme also accounts for the fact that at pH 10–11, hydrolysis of 14 is nearly second order in hydroxide ion (due to accumulation of the intermediate) whereas at higher pH, where the tetrahedral intermediate is nearly completely converted to its conjugate base, the kinetic order with respect to hydroxide ion approaches zero.

The proposed mechanism for the hydrolysis of 14 is probably applicable to hydrolysis of other imidazolinium ions. The fact that the experimental rate law for hydrolysis of 5,10-methenyltetrahydrofolic acid (11) is the same as that for the hydrolysis of 14 supports this conclusion[34].

The observed decrease in catalytic effectiveness of buffer bases with increasing buffer concentration at constant pH suggests that the rate-limiting step in hydrolysis of 11 shifts from breakdown of the tetrahedral intermediate in dilute buffers to formation of the tetrahedral intermediate in concentrated buffer solutions.

The imidazolinium ring in 5,10-methenyltetrahydrofolic acid (11) differs from that of 14 in an important way: the $N_{(5)}$ and $N_{(10)}$ positions of the imidazolinium ring of 11 are unsymmetrically substituted. One of them is incorporated into the fused pteridine ring system, and the other bears a p-carboxamidophenyl substituent. Cleavage of the hydroxy-imidazolidine ring of the tetrahedral intermediate could give two products, 12 or 13, depending on which C—N bond of the intermediate is broken.

Whether the product-forming step involves general base-catalysed hydrolysis of the 2-hydroxyimidazolidine intermediate, or general acid-catalysed hydrolysis of its conjugate base, the C—N bond cleaved should be that joining the acyl carbon of the imidazoline ring to the more basic of the two nitrogen atoms[35] and this is what is observed. In the hydrolysis of **11**, the formyl group in the kinetically controlled product is located on $N_{(10)}$, which is less basic than $N_{(5)}$ by about two pK units.

Similar results were obtained by Benkovic and co-workers[35], who studied the kinetics of hydrolysis of tetrahydroimidazo[1,5-*a*]quinazolines (**16**), which are structural analogues of 5,10-methenyltetrahydrofolic acid. Kinetic data indicate that these compounds hydrolyse by the same mechanism as **11** and **14**. Of the compounds studied, only that (**16**, X = $CO_2C_2H_5$) in which $N_{(10)}$ differed from $N_{(5)}$ in basicity by about two pK units, formed appreciable amounts of the $N_{(10)}$ formyl product on hydrolysis. In the case of both **11** and **16**, the thermodynamically controlled hydrolysis product is the *N*-formyl derivative in which the formyl group is bonded to the more basic of the two nitrogen atoms of the original imidazolinium ring: $N_{(5)}$ in the case of **11**, and $N_{(10)}$ in the case of **16**.

Although imidazolinium ions are rather stable in dilute acid solutions, they hydrolyse in moderately concentrated solutions of strong acids. Watson and co-workers[36,37] studied the hydrolysis of 2-methylimidazoline and a series of 2-arylimidazolines in moderately concentrated solutions of sulphuric and hydrochloric acid. They interpreted the dependence of hydrolysis rate on solvent acidity, the large negative entropies of activation, and the negligible effect of 2-aryl substituents on reactivity in terms of a mechanism involving nucleophilic attack by water on the diprotonated imidazoline. As supporting evidence for this mechanism, they found that 2-methylimidazoline is significantly diprotonated in 100% sulphuric acid.

The experimental results are rationalized equally well by a mechanism involving acid-catalysed hydrolysis of the small amount of monoacylethyleneammonium ion in equilibrium with the imidazolinium ion, which was not considered by Watson and co-workers:

$$\underset{H}{\overset{H}{\underset{N}{\overset{N}{\diagup}}}}\!\!\diagdown\!\! R + H_2O \underset{}{\overset{K_1}{\rightleftharpoons}} \overset{O}{\overset{\parallel}{R}}CNHCH_2CH_2\overset{+}{N}H_3$$
                                                                    20

$$\textbf{20} + H^+ \overset{K_2}{\rightleftharpoons} \underset{+}{\overset{\overset{OH}{|}}{RC}}-NHCH_2CH_2NH_3^+ \overset{H_2O}{\underset{k_3}{\longrightarrow}} RCO_2H + (CH_2NH_3^+)_2$$

The rate law derived from this reaction scheme is:

$$k_{exp} = K_1 K_2 k_3 a_{H^+} a_{H_2O}{}^2$$

The effects of solvent acidity, water activity, substituents, and temperature on hydrolysis rate predicted by this mechanism are in excellent agreement with experimental observations.

III. AMIDINES AS NUCLEOPHILES: KINETICS OF REACTIONS OF AMIDINES AND AMIDOXIMES WITH ESTERS, ARYL HALIDES, AND ACID HALIDES

Although N-acylation of amidines has been known for over a hundred years, the kinetics of these reactions have received attention only recently. The impetus for studying the detailed kinetics and mechanisms of amidine acylation stems from the fact that these reactions may conceivably involve the amidine as a bifunctional nucleophile. That is, the amidine amino group may transfer a proton to the carbonyl oxygen of the acylating agent at the same time that the imino nitrogen atom of the amidine attacks the carbonyl carbon. Acylation of amidines by reactive esters such as p-nitrophenylacetate in non-polar solvents, by a bifunctional mechanism in which there is little or no charge separation in the transition state of the rate-limiting step, would be relevant to the mechanism of enzyme-catalysed transacylation reactions. If bifunctional catalysis can be demonstrated in the amidine reactions, it becomes more reasonable to assume that biological transacylation reactions may involve bifunctionally-catalysed reactions in hydrophobic regions of enzymes.

This hypothesis was advanced by Menger[45], who studied the benzamidinolysis and n-butylaminolysis of p-nitrophenylacetate (PNPA). Menger did not actually isolate the anticipated product of the reaction, N-acetylbenzamidine, but noted that benzamidine is benzoylated by heating it with phenyl benzoate[46].

The kinetics of amidinolysis of PNPA are compatiable with a bifunctional mechanism. In chlorobenzene at 25°C, the reaction of PNPA with n-butylamine is third-order, first-order in PNPA and second-order in butylamine. In contrast, the reaction of benzamidine with PNPA under the same conditions is second-order, first-order each in PNPA and benzamidine. When the nucleophiles are present in concentrations of 0·0221 M, benzamidinolysis is 2500 times faster than butylaminolysis. It was estimated that benzamidine is at least 15,000 times as reactive as monomeric butylamine in chlorobenzene solutions—this in spite of the fact that benzamidine ($pK_a = 11·6$) is only slightly more basic than n-butylamine ($pK_a = 10·6$). An unspecified aliphatic amidine was found to be even more reactive than benzamidine toward PNPA.

These kinetic results were interpreted as evidence for bifunctional attack by the amidine on the aryl ester, involving a cyclic transition state with little charge separation:

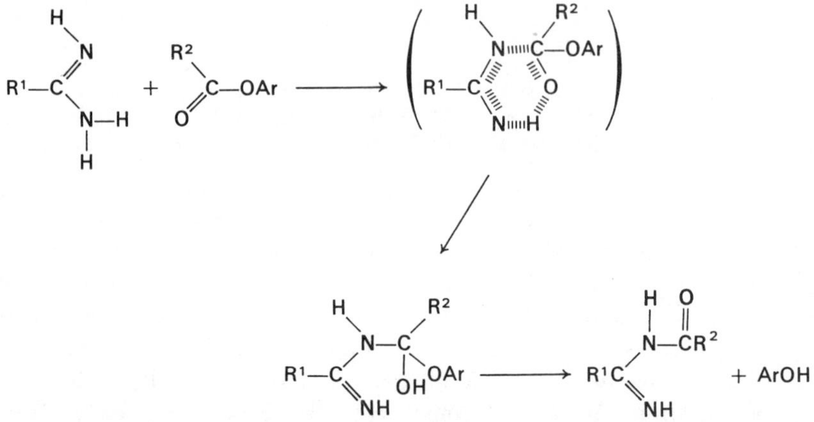

The third-order kinetics observed in butylaminolysis of PNPA might be due to nucleophilic attack by a hydrogen-bonded dimer of the amine, or might be due to involvement of two amine molecules in a cyclic transition state involving little charge separation. The observation that addition of N-methylpiperidine to the n-butylaminolysis reaction mixture has little effect on reaction rate was interpreted as evidence for the cyclic, concerted mechanism, which requires a transferable proton on each of the participating amine molecules. By analogy, this result was also interpreted as supporting evidence for the bifunctional mechanism of amidinolysis of PNPA.

These conclusions were questioned by Anderson, Su, and Watson[47], who found that the acetylation of 3,4,5,6-tetrahydropyrimidine by PNPA (the product, N-acetyltetrahydropyrimidine, apparently was not actually isolated and characterized) is first order with respect to tetrahydropyrimidine:

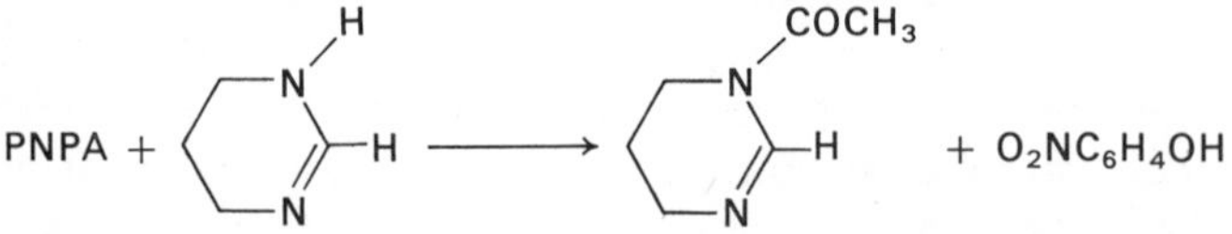

(21)

Tetrahydropyrimidine, which cannot form the kind of cyclic transition state proposed for bifunctional reaction of benzamidine with PNPA, is about 46 times as reactive as benzamidine. This clearly indicates that a bifunctional reaction is not necessarily the correct explanation for the high reactivity of benzamidine with PNPA. Further indications are the facts that 1,3-diaminopropane and N,N-dimethyl-1,3-diaminopropane, whose rate equations for reactions with PNPA contain terms both first-order and second-order in diamine, are about equally reactive. If cyclic transition states of the type proposed for amidinolysis and butylaminolysis of PNPA were involved in these reactions, the diprimary amine would be expected to be much more reactive than the primary-tertiary amine. Further, in third-order butylaminolysis of PNPA, 1,4-diazabicyclooctane (an unhindered tertiary amine) was found to be a slightly more effective catalyst than N-butylamine itself, although it is more than 2 pK units less basic than butylamine.

These observations taken together clearly rule out concerted reactions involving cyclic transition states as the only mechanism of aminolysis of PNPA in non-polar solvents such as chlorobenzene, and imply that the amidinolysis reaction also may occur by a non-cyclic, non-concerted mechanism. The high reactivity of amidines was attributed to high electron-density on the imine nitrogen, and charge dispersal in the transition state for formation of the tetrahedral intermediate[47]:

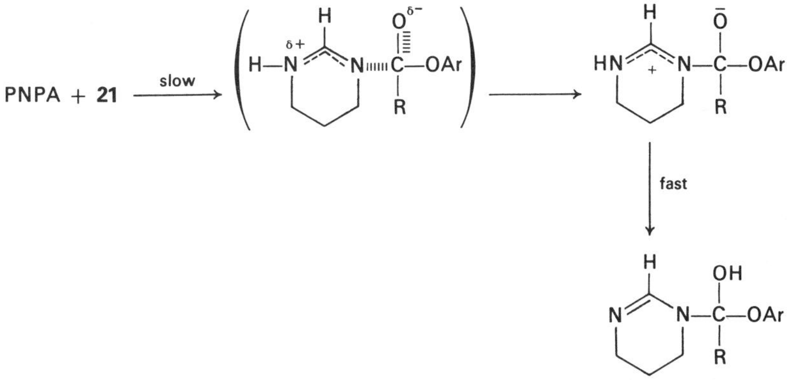

Biggi, Del Cima and Pietra[48,49] attempted to resolve the question of the existence of bifunctional mechanisms for reactions involving amidine as nucleophiles by a comparative study of the kinetics of aminolysis and amidinolysis of activated aryl halides in chlorobenzene solution. The amidine reactions apparently are the first reported examples of N-arylation of amidines with aryl halides.

Nucleophilic substitution of chloride in reactions of amines with 2,4-dinitrochlorobenzene is not general acid–base catalysed, and both butyl-amine and benzamidine react with this aryl halide by kinetically second order processes which are first order in the nucleophile. In contrast to reactions of the same nucleophiles with PNPA, benzamidine is substantially *less* reactive than butylamine with chlorodinitrobenzene. On the grounds that charge dispersal into the amidine system in the transition state for the slow step of the amidinolysis reaction should facilitate arylation as well as acylation, and that such facilitation is not observed in the arylation reaction, Pietra and co-workers[48] argued that transition state charge dispersal is not the explanation for the high reactivity of amidines with PNPA. This may be true, but it does not follow, as Pietra suggests, that the transacylation reactions are therefore bifunctional.

Pietra and co-workers found[49] that the kinetics of aminolysis and amidinolysis of 4-fluoro-1,6-dinitronaphthalene (**22**) in chlorobenzene

resemble the kinetics of reactions of PNPA with the same nucleophiles. The reaction of *n*-butylamine with **22** is second-order in amine. The reaction of **22** with benzamidine is first-order in benzamidine, and, at a particular nucleophile concentration, benzamidinolysis is much faster than butylaminolysis. Since a kinetic term first-order in butylamine could not be detected, it is not possible to compare the reactivities of benzamidine and butylamine in a mechanistically relevant way. The data are consistent with, but do not require, a concerted, bifunctional mechanism for the amidinolysis reaction.

4-Fluoro-1,6-dinitronaphthalene reacts with benzamidine in chlorobenzene solution at 84°C about 6 times as fast as 4-chloro-1,3-dinitrobenzene. 4-Fluoro-1,3-dinitrobenzene is several thousand times as reactive as the chlorodinitrobenzene.

Yet another substrate which undergoes third-order *n*-butylaminolysis and second-order benzamidinolysis in chlorobenzene solutions is *p*-nitrophenyl triphenylmethanesulfenate, **23**[50]:

Benzamidine is several thousand times as reactive as butylamine with **23**.

If, as has been proposed[51, 52, 53] nucleophilic substitution at divalent sulfur involves backside displacement of the leaving group, with formation of an intermediate complex having a trivalent sulfur atom, it would be sterically impossible for benzamidine simultaneously to attack the sulfur atom and transfer a proton to the leaving group. Therefore, a bifunctional mechanism seems unlikely for this amidinolysis reaction.

In fact, it is not necessary to invoke bifunctional mechanisms for amidinolysis of the other substrates (PNPA, aryl fluorides) in chlorobenzene either. All of these amidinolysis and aminolysis reactions involve reaction of the nucleophile and substrate to form intermediate complexes, which may either lose the nucleophile (with reversion to starting materials) or the leaving group (with formation of products). Further, loss of the leaving group may or may not be subject to general acid–base catalysis. The following reaction scheme applies to these reactions:

$$\text{HB} + \text{Substrate} \underset{k_{-1}}{\overset{k_1}{\rightleftharpoons}} \text{Intermediate} \underset{k_3[\text{HB}]}{\overset{k_2}{\rightarrow}} \text{Products} \qquad (10)$$

The rate law for this reaction scheme, assuming that the intermediate does not accumulate is:

$$k_{\text{exp}} = \frac{k_1[\text{HB}](k_2 + k_3[\text{HB}])}{k_{-1} + k_2 + k_3[\text{HB}]} \qquad (11)$$

The observed kinetics for a particular reaction will depend on the relative values of k_{-1} and k_2, and on whether the product-forming step is general acid–base catalysed. As Bunnett has pointed out[54], there are several variants of scheme (10) which lead to the same rate law. These include: reversible transformation of the intermediate complex into its conjugate base, followed by general acid-catalysed detachment of the leaving group: rate-limiting proton removal from the intermediate by HB, followed by rapid expulsion of the leaving group (unlikely, due to the exceedingly fast transfers of protons from relatively strong acids to relatively strong bases); concerted proton removal and leaving-group departure from the intermediate; and simultaneous proton removal from nitrogen and proton transfer to the leaving group by HB. The first of these alternatives, general acid-catalysed conversion of the conjugate base of the intermediate to products, seems the most probable mechanism, although the last, simultaneous proton transfer from nitrogen and proton transfer to the leaving group by HB, deserves serious consideration for reactions in non-polar solvents.

Any of the variants of reaction scheme (10) adequately account for the kinetics of all of the aminolysis and amidinolysis reactions discussed above. The only reactions which involve a good leaving group are the aminolysis and amidinolysis reactions of 2,4-dinitrochlorobenzene. With this substrate, $(k_2 + k_3[HB]) \gg k_{-1}$, and equation (11) simplifies to equation (12):

$$k_{exp} = k_1[HB] \qquad (12)$$

Formation of the intermediate is rate-limiting, and k_1 should parallel the basicity of the nitrogen nucleophile. Hence, the reactions are not only first order in the nucleophile, but the more basic nucleophilic reagent, n-butylamine, is more reactive than benzamidine.

In all of the other displacement reactions, the leaving group (F^- or ArO^-) is a poor one, and the observed kinetics depend on the facility of departure of HB from the intermediate. In reactions of PNPA, activated aryl fluorides, and p-nitrophenyl triphenylmethanesulfonate with n-butylamine, the breakdown of the intermediate to products is rate limiting. That is, $C_4H_9NH_2$ is a better leaving group than F^- or ArO^-, and $k_{-1} > k_2 + k_3[HB])$. For these reactions, equation (11) simplifies to equation (13):

$$k_{exp} = \frac{k_1[HB](k_2 + k_3[HB])}{k_{-1}} \qquad (13)$$

If $k_3[HB] > k_2$, the reactions are second order in butylamine.

In the intermediate for amidinolysis of these same substrates, the positive charge developed in the amidinium portion of the intermediate is delocalized:

$$R-C \overset{\overset{\displaystyle H}{\underset{\displaystyle \;}{N-(substrate)^-}}}{\underset{\displaystyle NH_2}{{}^+}}$$

This makes the amidine a poorer leaving group than the primary amine, which has a full positive formal charge on the amino nitrogen in the intermediate. For the amidinolysis reactions the intermediate-forming step is rate-limiting, and equation (12) is the rate-law for these reactions. The greater reactivity of amidines than primary amines with these substrates is understandable if $(k_2 + k_3[HB])/k_{-1} \ll 1$ for the aminolysis reactions.

Aubort and Hudson[55] studied the O-acylation of a number of amidoximes (N-hydroxyamidines) by PNPA, benzoyl fluoride, and ethyl chloroformate:

$$RC\overset{\displaystyle NOH}{\underset{\displaystyle NR'_2}{\big<}} + R^2COX \longrightarrow RC\overset{\displaystyle NOCOR^2}{\underset{\displaystyle NR'_2}{\big<}} + HX$$

$(R = C_6H_5, R' = H, C_2H_5 ; R = CH_3, R' = H, C_2H_5 ; R = R' = H)$

In water or aqueous acetone, these reactions follow the rate law:

$$k_{exp} = k_a \left[\frac{K_H}{K_H + [H^+]} - \frac{K_A}{K_A + [H^+]} \right] + k_b \left[\frac{K_A}{K_A + [H^+]} \right]$$

where k_a and k_b are the rate constants for acylation of the neutral and anionic forms of the amidoxime, K_H is the equilibrium constant for protonation of the amidoxime, and K_A is its dissociation constant. The neutral amidoximes are 700–900 times more reactive toward ethyl chloroformate and benzoyl fluoride than are aldoximes of similar basicity, but are only slightly more reactive toward PNPA.

These results are accounted for by a reaction scheme which is similar to that proposed for amidinolysis reactions of PNPA and other substrates which form intermediate complexes:

$$\underset{R}{\overset{R'_2N}{\big>}}C{=}NOH + R^2COX \underset{k_{-1}}{\overset{k_1}{\rightleftharpoons}} \underset{R}{\overset{R'_2N}{\big>}}C{=}N{-}O{-}\underset{\underset{R^2}{|}}{\overset{\overset{OH}{|}}{C}}{-}X \overset{k_2}{\longrightarrow}$$

$$R{-}C\overset{\displaystyle NOCOR^2}{\underset{\displaystyle NR'}{\big<}} + HX$$

The product-forming step may involve intramolecular proton transfer from nitrogen to carbonyl oxygen, or preliminary hydrogen bonding between carbonyl oxygen and nitrogen. The amidoximes are much more reactive than aldoximes when $k_2 > k_{-1}$ (when ethyl chloroformate or benzoyl fluoride is the substrate), but the amidoximes and aldoximes are of similar reactivity when the product-forming step is rate limiting (i.e., when $k_{-1} > k_2$, for PNPA).

Arguments based on molecular orbital theory[56] have been advanced to support the view that the anomalous reactivity of the N-hydroxyamidines with certain substrates is due to intramolecular catalysis of formation of the intermediate complex, rather than to the operation of a so-called 'α-effect'.

Haruki, Fujii and Imoto[57] reported that amidines are effective catalysts for hydrolyses of a number of carboxylate esters (ethyl acetate, γ-butyro-lactone, phenyl acetate, p-cresyl acetate, and glycidic esters). The reactions were followed by titrating 'free' amidine (i.e., amidine not hydrolysed or present as N-acylamidine or as amidinium carboxylate), so that there is some uncertainty regarding what reactions were followed. The initial products of these reactions should be N-acylamidines, which would hydrolyse to amidinium carboxylates. The amidinium ions also hydrolyse to amines and carboxamides. In any event, formamidines appear to be more effective catalysts than hydroxide ion for the hydrolysis of carboxylate esters. Acetamidine and benzamidine are less effective as catalysts than formamidines. Glycidic esters hydrolyse in the presence of amidines to amidinium glycidates, without hydrolysis of the epoxide function.

IV. CONFORMATIONAL ISOMERIZATIONS OF AMIDINES AND AMIDINIUM SALTS

A rate process which is not, strictly speaking, a chemical reaction involves rotation of trivalent nitrogen atoms of N,N-dimethylamidines and N,N-dimethylamidinium ions (**24** and **25**) about acyl C—N bonds. These con-

$$R-C \begin{array}{c} \nearrow N(CH_3)_2 \\ \searrow NR \end{array} \qquad R-C \begin{array}{c} \nearrow N(CH_3)_2 \\ \searrow NR^1R^2 \end{array} +$$

$$(\mathbf{24}) \qquad\qquad (\mathbf{25})$$

formational changes can be studied by temperature-dependent n.m.r. spectroscopy. At sufficiently low temperatures, rotation of the dimethylamino groups of these molecules and ions are slow, and the *syn*- and *anti*-methyl groups produce separate peaks of equal intensity in an n.m.r. spectrum. At sufficiently high temperatures, the rotation of the dimethylamino group is rapid on the n.m.r. time scale, and the dimethylamino group appears as a single peak. Coalescence temperatures can be used to calculate the free energy barrier to rotation about the C—N(CH$_3$)$_2$ or C$\cdots$N$^{\delta+}$(CH$_3$)$_2$ bonds.

Energy barriers to rotation (from which rotation rates at a given temperature can be calculated) have been reported for a number of amidines[58-62,69] and amidininium ions[62-64].

Rotational barriers for simple amidines fall mostly in the range of 12–15 kcal mol^{-1}. The relatively large barrier to rotation about the C—N(CH$_3$)$_2$ bond of N,N-diethylamidines is probably due to partial double-bond character of this bond attributable to resonance delocalization of the π-electrons of the imino C=N bond. As expected, the rotational barrier about the N,N-dimethyl-acyl-C bond in amidinium salts is several kilocalories per mole larger than the barrier for the amidines due to the substantially larger double bond character of the C—N bond in the amidinium salts. For discussions of substituent and solvent effects on these rotational barriers, see the original publications.

V. PYROLYSIS AND THERMAL ISOMERIZATION REACTIONS OF AMIDINES

Azobisisobutyramidines and their salts undergo thermal decomposition to nitrogen ammonia and organic products formed from free radicals and the diamidines produced by radical recombinations:

$$\left(=N-C(CH_3)_2-C \overset{NR}{\underset{NHR}{\diagup}} \right)_2 \quad \xrightarrow[\text{slow}]{\Delta} $$

$$N_2 + 2 \cdot C(CH_3)_2 - C \overset{NR}{\underset{NHR}{\diagup}} \quad \xrightarrow{\text{fast}} \quad \text{other products}$$

$$(R,R = H,H \text{ or } -CH_2-CH_2-)$$

The mono- and diamidinium cations undergo similar reactions. The kinetics of these reactions, which do not involve the amidine or amidinium function in the rate-limiting steps, have been studied by Hammond and Neuman[65] and by Dougherty[66].

The kinetics of these reactions in water, dimethylsulfoxide or dimethyl sulfoxide–cumene were studied spectrophotometrically[66] or by measuring the rate of nitrogen evolution. For each azobisamidine, the first conjugate acid decomposes considerably faster than the free base, but at about the same rate as the di-conjugate acid. Electrostatic effects on geminate radical recombinations are small for these compounds.

Chapman studied the thermal isomerization reactions of N,N,N'-triarylbenzamidines[67, 68]:

$$C_6H_5C \overset{NAr}{\underset{NArAr'}{\diagup}} \quad \underset{k_{-1}}{\overset{k_1}{\rightleftharpoons}} \quad C_6H_5C \overset{N(Ar)_2}{\underset{NAr'}{\diagup}}$$

$$(26) \qquad\qquad\qquad (27)$$

These reactions were carried out by heating the melted amidines at 330°C, and were followed by removing samples of the reaction mixtures at intervals, determining the melting-point of the sample, and estimating its composition from melting-point vs composition graphs. When Ar = Ar' = C_6H_5 or p-$CH_3C_6H_4$, heating the amidine at 330°C resulted in no significant melting-point lowering, which indicates that there are no side reactions. When a mixture of **26**, $Ar = Ar' = C_6H_5$, and **26**, $Ar = Ar' = p\text{-}CH_3C_6H_5$, were heated together at 330°C and the mixture then hydrolysed, no $C_6H_5NH\text{—}C_6H_4CH_3$ was found in the hydrolysis products. This indicates that aryl migrations in the isomerization reactions are intramolecular rather than intermolecular. The results of this investigation are summarized in Table 1.

TABLE 1. Thermal isomerizations of N,N,N'-triarylbenzamidines at 330°C

$$(26 \underset{k_{-1}}{\overset{k_1}{\rightleftharpoons}} 27)$$

Ar	Ar'	$10^5 k_1{}^a$	$10^5 k_{-1}{}^a$	k_1/k_{-1}
C_6H_5	$4\text{-}CH_3C_6H_4$	3·2	7·2	0·44
C_6H_5	$4\text{-}ClC_6H_4$	3·2	4·5	0·71
C_6H_5	$3,5\text{-}Cl_2C_6H_3$	2·8	2·5	1·12
$4\text{-}CH_3C_6H_4$	C_6H_5	1·6	1·7	0·94
$4\text{-}ClC_6H_4$	C_6H_5	5·5	13·3	0·41
$3,5\text{-}Cl_2C_6H_3$	C_6H_5	41·7	140	0·30

^{a} first-order rate constants, sec^{-1}.

It is apparent from the data in Table 1 that the rate of isomerization is relatively insensitive to the structure of the aryl group attached to the imino nitrogen of the amidine (the migration terminus). However, electron-withdrawing aryl substituents on the migrating aryl group substantially increase the rate of isomerization. The presence of electron-withdrawing groups on the non-migrating aryl group stabilizes **27** relative to **26**, while the presence of electron-withdrawing substituents on the migrating aryl group has the reverse effect. These results suggest that these isomerizations involve intramolecular nucleophilic replacements, with intermediates resembling:

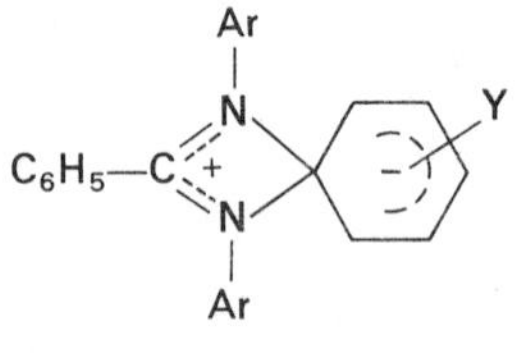

VI. REFERENCES

1. R. Livingston, *Technique of Organic Chemistry*, A. Weissberger, Editor, Interscience Publishers, Inc., New York, 1953, 1st Ed., Vol. 8, p. 208.
2. R. L. Shriner and F. W. Neumann, *Chem. Reviews*, **35**, 351 (1944).

3. M. Davies and A. E. Parsons, *Chem. and Ind.* (*London*), 628 (1958).
4. E. C. Wagner, *J. Org. Chem.*, **5**, 133 (1940).
5. F. L. Pyman, *J. Chem. Soc.*, **123**, 3359 (1923).
6. R. H. DeWolfe and R. M. Roberts, *J. Amer. Chem. Soc.*, **75**, 2942 (1953).
7. J. N. Broensted and K. J. Pedersen, *Z. Phys. Chem.*, **108**, 185 (1924).
8. R. H. DeWolfe, *J. Amer. Chem. Soc.*, **82**, 1585 (1960).
9. L. P. Hammett, *Physical Organic Chemistry*, McGraw-Hill Book Co. Inc., New York, 1940, p. 184.
10. H. H. Jaffe, *Chem. Reviews*, **53**, 191 (1953).
11. R. H. DeWolfe, *J. Amer. Chem. Soc.*, **86**, 864 (1964).
12. R. H. DeWolfe and J. R. Keefe, *J. Org. Chem.*, **27**, 493 (1962).
13. R. P. Bell, A. L. Dowding and J. A. Noble, *J. Chem. Soc.*, 3106 (1955).
14. R. O. Gould and R. F. Jameson, *J. Chem. Soc.*, 296 (1962).
15. A. Holy and J. Zemlicka, *Coll. Czech. Chem. Commun.*, **34**, 2449 (1969),
16. R. H. DeWolfe, *Carboxylic Ortho Acid Derivatives*, Academic Press, New York, 1970, p. 481.
17. R. H. DeWolfe and M. W. Cheng, *J. Org. Chem.*, **34**, 2595 (1969).
18. J. D. Wilson, C. R. Hobbs and H. Weingarten, *J. Org. Chem.*, **35**, 1542 (1970).
19. R. B. Martin and A. Parcell, *J. Amer. Chem. Soc.*, **83**, 4835 (1961).
20. B. G. Harnsberger and J. L. Riebsomer, *J. Heterocyclic Chem.*, **1**, 188 (1964).
21. B. G. Harnsberger and J. L. Riebsomer, *J. Heterocyclic Chem.*, **1**, 229 (1964).
22. R. W. Taft, *J. Amer. Chem. Soc.*, **74**, 3120 (1952).
23. J. Hine, *Physical Organic Chemistry*, McGraw-Hill Book Co., New York, 1956, p. 278.
24. J. C. Saam and H. M. Bank, *J. Org. Chem.*, **30**, 3350 (1965).
25. R. L. Blakeley, *The Biochemistry of Folic Acid and Related Pteridines*, North-Holland Publishing Company, Amsterdam, 1969.
26. M. May, T. J. Bardos, F. L. Barger, M. Lansford, J. M. Ravel, G. L. Sutherland and W. Shive, *J. Amer. Chem. Soc.*, **73**, 3067 (1951).
27. H. Tabor and L. Wyndgarden, *J. Biol. Chem.*, **234**, 1830 (1959).
28. S. C. Hartman and J. M. Buchanan, *J. Biol. Chem.*, **234**, 1812 (1959).
29. L. D. Kay, M. J. Osborn, Y. Hatefi and F. M. Huenshens, *J. Biol. Chem.*, **235**, 195 (1960).
30. L. Jaenicke and E. Brode, *Liebigs Ann. Chem.*, **624**, 120 (1959).
31. D. R. Robinson and W. P. Jencks, *J. Amer. Chem. Soc.*, **89**, 7088 (1967).
32. D. R. Robinson, *Tetrahedron Letters*, 5007 (1968).
33. D. R. Robinson, *J. Amer. Chem. Soc.*, **92**, 3138 (1970).
34. D. R. Robinson and W. P. Jencks, *J. Amer. Chem. Soc.*, **89**, 7098 (1967).
35. S. J. Benkovic, W. P. Bullard and P. A. Benkovic, *J. Amer. Chem. Soc.*, **94**, 7542 (1972).
36. P. Haake and J. W. Watson, *J. Org. Chem.*, **35**, 4063 (1970).
37. S. Limatibul and J. W. Watson, *J. Org. Chem.*, **36**, 3803 (1971).
38. J. F. Bunnett and F. P. Olsen, *Can. J. Chem.*, **44**, 1917 (1966).
39. J. Armand, *Bull. Soc. Chim. France*, 1658 (1966).
40. S. Deswarte, A. Pezzoli and J. Armand, *C.R. Acad. Sci., Ser. C*, **270**, 2062 (1970).

41. J. F. Bunnett, *J. Amer. Chem. Soc.*, **83**, 4968 (1961).
42. E. Bamberger and B. Berle, *J. Liebig's Ann. Chem.*, **273**, 342 (1893).
43. O. Fischer, *Chem. Ber.*, **34**, 930 and 4202 (1901).
44. F. B. Zienty, *J. Amer. Chem. Soc.*, **67**, 1138 (1945).
45. F. M. Menger, *J. Amer. Chem. Soc.*, **88**, 3081 (1966).
46. A. W. Titherly and E. C. Hughes, *J. Chem. Soc.*, **99**, 1493 (1911),
47. H. Anderson, C.-W. Su and J. W. Watson, *J. Amer. Chem. Soc.*, **91**, 482 (1969).
48. G. Biggi, F. Del Cima and F. Pietra, *Tetrahedron Letters*, 2811 (1971).
49. G. Biggi, F. Del Cima and F. Pietra, *J. Chem. Soc. Perkin II*, 188 (1972).
50. E. Ciuffarin, L. Senatore and L. Sagramora, *J. Chem. Soc. Perkin II*, 534 (1973).
51. E. Ciuffarin and A. Fava, *Progr. Phys. Org. Chem.*, **6**, 81 (1968).
52. W. A. Prior and K. Smith, *J. Amer. Chem. Soc.*, **92**, 2731 (1970).
53. E. Ciuffarin and F. Griselli, *J. Amer. Chem. Soc.*, **92**, 6015 (1970).
54. J. F. Bunnett and R. H. Garst, *J. Amer. Chem. Soc.*, **87**, 3879 (1965).
55. J. D. Aubort and R. F. Hudson, *J. Chem. Soc., Chem. Commun.*, 1342 (1969).
56. J. D. Aubort and R. F. Hudson, *J. Chem. Soc., Chem. Commun.*, 937 (1970).
57. E. Haruki, T. Fujii and E. Imoto, *Bull. Chem. Soc. Japan*, **39**, 852 (1966).
58. D. L. Harris and K. M. Wellman, *Tetrahedron Letters*, 5225 (1968).
59. D. J. Bertelli and J. T. Gerig, *Tetrahedron Letters*, 2481 (1967).
60. H. J. Jakobsen and A. Senning, *J. Chem. Soc., Chem. Commun.*, 1245 (1968).
61. G. Schwenker and H. Rosswag, *Tetrahedron Letters*, 2691 (1968).
62. J. S. McKennis and P. A. S. Smith, *J. Org. Chem.*, **37**, 4173 (1972).
63. R. C. Neuman and Y. Jonas, *J. Phys. Chem.*, **75**, 3532 (1971).
64. J. Rauft and S. Dahn, *Helv. Chim. Acta*, **47**, 1160 (1964).
65. G. S. Hammond and R. C. Neuman, *J. Amer. Chem. Soc.*, **85**, 1501 (1963).
66. T. J. Dougherty, *J. Amer. Chem. Soc.*, **83**, 4849 (1961).
67. A. W. Chapman, *J. Chem. Soc.*, 2458 (1930).
68. A. W. Chapman and C. H. Perrot, *J. Chem. Soc.*, 1770 (1932).
69. Z. Rapoport and R. Ta-Shma, *Tetrahedron Letters*, 5281 (1972).

CHAPTER 9

Imidates including cyclic imidates

DOUGLAS G. NEILSON

Chemistry Department, The University, Dundee, Scotland

I. INTRODUCTION, NOMENCLATURE AND SCOPE

Imidates (**1**) are esters of the hypothetical imidic acids or iso-amides (**2**). This chapter will deal with the preparation and properties of both open chain and cyclic imidates. Over the years various names have been given

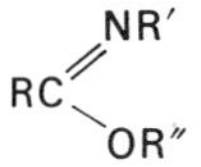

(1) 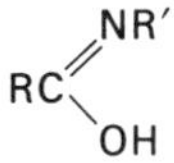   (2)

to these compounds (or their thio-analogues). The most common of these are imino ethers, imido or imidic esters, imidoates and in the case of some *N*-substituted derivatives, alkyl isoanilides. It is proposed here that, except for some cyclic imidates, these compounds will be named from the parent imidic acids; compound (**3a**) is thus called ethyl acetimidate, (**3b**) is methyl *N*-phenylformimidate hydrochloride and (**3c**) is phenyl *N*-phenylbenzthioimidate. Although the synthetically important *iso*-ureas (**3d**) and their thioanalogues are in fact imidates they lie outside the scope of this review as they belong more properly to the volume in this series dealing with urea derivatives.

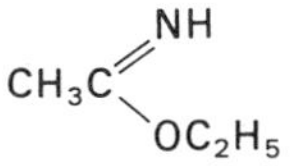

(3a) 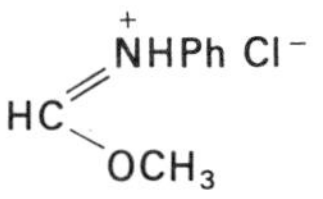   (3b) 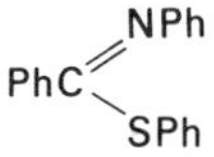   (3c)

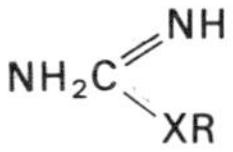

X = O or S

(3d)

Cyclic imidates may be divided into three distinct groups. In the first of these the imidate function (—N=C—O—) lies completely within the ring. Examples of this class of compound include oxazolines (**4a**) and dihydro-oxazines (**4b**). Again these compounds are more naturally dealt

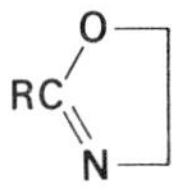

(4a) 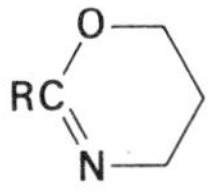   (4b)

with in reviews of heterocyclic chemistry and hence will not be discussed in any great detail here. The other types of cyclic imidates have either the oxygen function or the imino-nitrogen in an exocyclic position as in compounds 5a and 5b. Compound 5a is O-methylcaprolactim and compound 5b is the 2-imine of tetrahydropyran.

(5a)   (5b)

Previous reviews of imidate chemistry have appeared in the literature. The first of these which covered the classical researches of Pinner and his colleagues is to be found in Pinner's book 'Die Imidoäther und ihre Derivate'[1]. Almost 70 years were to pass before a further comprehensive review appeared in the names of Roger and Neilson[2]. The chemistry of lactim ethers (5a) has recently been discussed by Glushkov and Granik[3], and there is also a brief review[4] of cyclic imidates of the type 4. Another review[5] confined to imidate preparations has also recently appeared in the series 'Organic Functional Group Preparation'. In addition short articles and monographs dealing with particular aspects of imidate chemistry have appeared at various times and reference will be made to these in the appropriate section of the text, e.g. under Chapman rearrangement, ortho ester formation etc.

II. TAUTOMERISM

It has been proposed from time to time that amides may exist in part in the iminol form due to amide–iminol (lactam–lactim) tautomerism; however, all early claims to have isolated such iminol forms of simple amides have been discredited[2, 6].

Recently it has been suggested that amide ligands in certain complexes exist in the iminol form[7, 8], e.g. spectral evidence supports related, protonated iminol forms for the two species 6a and 6b. In the case of the complex 6a, however, it should be emphasised that the amide moiety is in its protonated form and hence the formation of a species such as 6a does not prove or disprove the existence of the unprotonated iminol form of the amide (i.e. a free imidic acid). N.m.r. evidence has also been put for-

$$\underset{\text{(6a)}}{HC\!\!\underset{OH}{\overset{\overset{+}{N}H_2}{{<}}}}\quad SbCl_6^-$$ $$\underset{\text{(6b)}}{HC\!\!\underset{OR}{\overset{\overset{+}{N}H_2}{{<}}}}\quad SbCl_6^-$$ $$\underset{\text{(7)}}{R_FC\!\!\underset{OH}{\overset{NF}{{<}}}}$$

ward in support of the iminol forms of N-fluoro-perfluoroamides[9], (7). Work on the structure and nature of the amide group has recently been reviewed and it is obvious that this area of chemistry still presents problems of interpretation[6].

III. SYNTHESIS

A. The Pinner Synthesis

The pioneering work of Pinner in this field is recognised in the naming of the reaction involving a nitrile with an alcohol, phenol or thiol under acid conditions as the Pinner synthesis[1,2,5]. A diluent is often employed and the reaction normally requires equimolar proportions of the reactants, or a very slight excess of the alcohol.

$$RCN + R'OH + HCl \longrightarrow RC\!\!\underset{OR'}{\overset{\overset{+}{N}H_2\ Cl^-}{{<}}}$$

Three side reactions may be troublesome; of these, by far the most important is the action of water on the imidate salt. This necessitates the use of anhydrous reactants and diluents, particular care having to be taken in the case of the lower aliphatic members. Secondly, the temperature

$$RC\!\!\underset{OR'}{\overset{\overset{+}{N}H_2\ Cl^-}{{<}}} + H_2O \longrightarrow RC\!\!\underset{OR'}{\overset{O}{{<}}} + NH_4Cl$$

of the reaction should be controlled around 0–5°C to prevent decomposition of the imidate salt into the amide; this reaction is discussed in greater

$$RC\!\!\underset{OR'}{\overset{\overset{+}{N}H_2\ Cl^-}{{<}}} \overset{\Delta}{\longrightarrow} RC\!\!\underset{NH_2}{\overset{O}{{<}}} + R'Cl$$

detail later (see Section IV, D.). In addition, excess of alcohol over a prolonged period of time may cause ortho ester formation (see Section IV, G.) but this is the least troublesome of the three side reactions except possibly in the case of formimidates.

$$RC\!\!\underset{OR'}{\overset{\overset{+}{N}H_2\ Cl^-}{{<}}} + 2\ R'OH \longrightarrow RC(OR')_3 + NH_4Cl$$

Although diluents are often employed it has been suggested that in some instances it is better to keep the reactants at 0–5°C for 12–48 h before adding the diluent which will then cause the imidate to crystallise within a short period of time[2, 10, 11]. Most commonly anhydrous ether[2] is employed as the diluent but chloroform, nitrobenzene, dioxan, dimethyl cellosolve and even an excess of alcohol have been used[2, 5]. Recent industrial processes have made use of diethylbenzene[12] and esters related to the imidate under synthesis[13] as diluents. Benzene also appears favoured in some older reports[2] but due to its toxicity is not now normally used.

The vast bulk of Pinner preparations involve anhydrous hydrogen chloride[1, 2, 5] but hydrogen bromide can also be used successfully[14, 15]. In a very few instances sulphuric acid has been employed[2, 16].

The Pinner synthesis can make use of a wide variety of primary or secondary but not tertiary alcohols. It is not altogether clear whether tertiary alcohols fail to react or in some instances give rise to unstable products[17]. Methanol and ethanol are by far the most commonly used of the alcohols and it has been suggested that better yields are obtained with methanol than with ethanol. However, most simple alcohols[2], e.g. propyl, iso-propyl, n-butyl, iso-butyl, sec-butyl and benzyl alcohols as well as their higher homologues, e.g. octan-2-ol[18], have been used in specific syntheses. More recently propargyl alcohol[19], cyclohexanol[17], and glycollic acid[20] have been utilised as have the optically active forms of some secondary alcohols[21, 22], e.g. butan-2-ol. In addition, for nitriles with powerful electron-withdrawing substituents (such as trichloro-acetonitrile), 2,2,2-trichloroethanol[21, 23] and some 2-nitroalkanols[24] have been used in attempts to prepare stable imidate salts—the imidate salts of such nitriles with simple alcohols spontaneously decomposing to the

$$CCl_3CN + EtOH + HCl \longrightarrow CCl_3CONH_2 + EtCl$$

amides. Phenols may successfully replace alcohols in the Pinner synthesis although much less work has been carried out using them[1, 2, 5, 25].

Alkyl[2, 14, 26, 27] or aryl thiols[28] give rise in the Pinner synthesis to the corresponding thioimidate salts and t-butyl mercaptan is reported to have been converted into a thioimidate salt[28, 29].

$$RCN + R'SH + HCl \longrightarrow RC\overset{\displaystyle \overset{+}{N}H_2\ Cl^-}{\underset{\displaystyle SR'}{\big\langle}}$$

Glycols[2] and in particular ethylene glycol, the 1,2- and 1,3-propanediols and 2,3-butanediols have all been converted into imidates although the 1,2-, 1,3-, and 1,4-butanediols failed to give imidates with cyanogen[30].

$$2\ RCN + CH_2OHCH_2OH \longrightarrow \underset{\underset{RCOCH_2CH_2OCR}{\parallel}}{\overset{\overset{+}{N}H_2\ Cl^-}{}}\ \underset{}{\overset{\overset{+}{N}H_2\ Cl^-}{}}$$

2,4-Pentanediol condensed with acetonitrile to give a cyclic imidate (8) possibly by electrophilic attack of a carbonium ion on the nitrile[16].

$$(CH_3)_2C(OH)CH_2CH(OH)CH_3 + H^+ \longrightarrow (CH_3)_2\overset{+}{C}CH_2CH(OH)CH_3$$

(8)

The Pinner synthesis utilises a wide variety of nitriles and failure has been noted only under two general circumstances: (a) where the nitrile is severely hindered, or (b) where the nitrile has powerful electron-withdrawing substituents. In this second case the imidate is formed but decomposes spontaneously to the amide.

In the aliphatic series even hydrogen cyanide[2, 12] has been converted into an imidate although there now exists a much more convenient preparation starting from formamide[31]. Simple aliphatic nitriles and dinitriles[2] can be converted readily into imidates although particular care must be taken in these cases to ensure anhydrous conditions. In addition higher homologues are known, e.g. imidates derived from palmito-[32] and stearo-[33] nitriles. Aldehyde and ketone cyanohydrins are also smoothly converted into imidates by the Pinner procedure[2, 10, 11, 34, 35].

$$RR'C(OH)CN + EtOH + HCl \longrightarrow RR'C(OH)C\overset{\overset{+}{N}H_2\ Cl^-}{\underset{OEt}{\diagup\diagdown}}$$

In the case of α-aminonitriles[2], however, the amino group, once protonated, acts as an electron sink and spontaneous decomposition of the imidate salt takes place giving the α-aminoamide[36] often in excellent yield and in a state of high purity. However, tosylation or acylation of the amino function permits the formation of stable imidate salts[37, 38, 39, 40].

$$PhHC(NHTs)CN + i\text{-}PrOH + HCl \longrightarrow PhHC(NHTs)C\overset{\overset{+}{N}H_2\ Cl^-}{\underset{OPr\text{-}i}{\diagup\diagdown}}$$

In the case of certain thioimidates from α-aminonitriles it has been noted that a secondary cyclisation reaction can take place and this appears to happen readily if a benzoyl group has been used to block the amino

$$PhCONHCRR'CN + EtSH + HCl \longrightarrow PhC\overset{N-CRR'}{\underset{O-C=\overset{+}{N}H_2\ Cl^-}{\diagdown}}$$

group[41]. It has also recently been shown that β-aminopropionitrile forms a stable thioimidate dihydrochloride unlike glycinonitrile[42] which gives a stable thioimidate only when the amino group is protected from protonation.

Some aliphatic nitriles with unsaturated centres have been successfully converted into imidate salts[2, 43, 44, 45], but in other instances addition of

$$PhCH=CHCN + MeOH + HCl \longrightarrow PhCH=CHC\overset{\overset{+}{N}H_2\ Cl^-}{\underset{OMe}{\diagup}}$$

hydrogen halide across the unsaturated centre takes place along with imidate formation[2, 46].

$$CH_2=C(OAc)CN + HCl + EtOH \longrightarrow ClCH_2CH(OAc)C\overset{\overset{+}{N}H_2\ Cl^-}{\underset{OEt}{\diagup}}$$

$$PhC\equiv CCN + HCl + EtOH \longrightarrow PhCCl=CHC\overset{\overset{+}{N}H_2\ Cl^-}{\underset{OEt}{\diagup}}$$

Steinkopf and Malinowski[47] found that acetonitriles substituted with either two or more chlorine atoms or with a nitro group tended to give amides under the Pinner synthesis. Free α-aminonitriles act similarly[36] and the author[48] has noted that although mandelimidate salts derived from primary alcohols are stable, decomposition to mandelamide takes place readily when secondary alcohols are employed. Similarly, attempts to convert N-cyanoamidines into imidates gave only the N-carboxamido-amidines[49]. However, it should be noted that the base-catalysed reaction

$$RC\overset{NCN}{\underset{NH_2}{\diagup}} + EtOH + HCl \longrightarrow RC\overset{\overset{+}{N}HCONH_2\ Cl^-}{\underset{NH_2}{\diagup}}$$

of alcohols with nitriles to form imidate bases is most successful with just such nitriles which contain electronegative substituents (see Section III. B) and hence these two processes in acid and basic media complement each other.

A wide variety of aromatic nitriles has also been employed successfully in the Pinner synthesis[1, 2]. The reaction appears quite general except for certain sterically hindered nitriles. For example methyl, nitro, amino, chloro and sulphonamide groups in the *ortho* position of benzonitrile prevent imidate formation although *ortho*-hydroxy or alkoxy groups have less effect[2, 5, 50–55]; in addition α-cyanonaphthalene but not the β-isomer exhibits steric hindrance to imidate formation. When two cyano-groups lie *ortho* to one another only one can be successfully converted into the imidate salt; the other positional isomers give di-imidates[50, 51] without

difficulty. It has proved possible however to obtain aryl imidates with *ortho*-substituents by other routes[56] even in those cases where the Pinner method failed completely (see Section III. D).

Various heterocyclic compounds in which the cyano group is attached either directly to a heterocyclic nucleus or to it via an aliphatic side chain have also been used successfully in Pinner syntheses[2, 45, 57, 58, 59]. Heterocyclic 1,2-dinitriles behave similarly to *o*-phthalonitrile yielding mono-imidates[60].

In certain cases where both the cyano group and the alcohol function are held on one nucleus and so placed that interaction can take place, a cyclic imidate salt is formed[2], e.g. the ester **10** must arise from the intermediate cyclic imidate[60] **9**. Equations (1) and (2) illustrate other similar

$$HOCH_2CH_2OCH_2CH_2CN + HCl \longrightarrow \quad (1)$$

$$HO(CH_2)_nCN + HCl \longrightarrow (CH_2)_n\!\!-\!\!C\!\!=\!\!\overset{+}{N}H_2 \; Cl^- \quad (2)$$

reactions[61, 62]. Secondary cyclisation reactions may also occur when there is a suitably placed reactive functional group on one of the reactants, e.g. as in the case of the unsaturated alcohol[63] **11** or in the case of the formation

$$CH_3CN + HOCH_2CH_2C(CH_3)\!\!=\!\!CH_2 \xrightarrow{H_2SO_4} CH_3C$$

(11)

of the *meso*-ionic compound **12** via the double imidate synthesis[64] illustrated in equation (3).

$$PhC\overset{S}{\underset{NHPh}{\big<}} \xrightarrow{ClCH_2CN} PhC\overset{NPh}{\underset{SCH_2CN}{\big<}} \xrightarrow[HCl]{i\text{-}PrOH} \quad \overset{H\bar{N}\underset{\quad}{\overline{\quad}}NPh\ HCl}{\underset{S}{\bigcirc}}\text{-}Ph \qquad (3)$$

$$(12)$$

Some reagents containing several cyano groups, e.g. the tetracyano compound **13**, have been successfully converted into imidates but other related compounds reacted incompletely due to problems of solubility[65].

$$(p\text{-}NCC_6H_4)_2C{=}C(C_6H_4CN\text{-}p)_2$$
$$(13)$$

The Hoesch[66] reaction is closely related to the Pinner synthesis and can at times produce imidate salts particularly when a simple phenol rather than a polyhydric phenol is employed (equation 4). The more normal product of this reaction is however an aryl ketone, (equation 5).

$$RCN\ +\ PhOH \xrightarrow[ZnCl_2]{HCl} RC\overset{\overset{+}{N}H_2\ Cl^-}{\underset{OPh}{\big<}} \qquad (4)$$

$$RCN\ +\ C_6H_4(OH)_2 \xrightarrow[ZnCl_2]{HCl} RCOC_6H_3(OH)_2 \qquad (5)$$

Although versatile in many ways, the Pinner synthesis has the limitation that only *N*-unsubstituted imidates can be formed directly by it. However, Borch[67] has shown recently that nitriles can react to give *N*-ethylimidates on treatment first with $Et_3O^+BF_4^-$ followed by addition of ethanol.

$$RCN\ +\ Et_3\overset{+}{O}\ \overset{-}{B}F_4 \longrightarrow RC{\equiv}\overset{+}{N}Et\ \overset{-}{B}F_4 \xrightarrow{EtOH} RC\overset{NEt}{\underset{OEt}{\big<}}$$

B. Base Catalysed Reactions of Nitriles with Alcohols

Schaefer and Peters[68] recently reexamined this preparative route to imidates confirming its usefulness in the case of nitriles with powerful electron-withdrawing substituents, (Table 1). The importance of this base-catalysed reaction (e.g. ROH/RONa at 25°C.) lies in the fact that it

TABLE 1

Compound:	CH_3CN	CH_2ClCN	C_6H_5CN	$p\text{-}O_2NC_6H_4CN$
Yields:	very poor	excellent	poor	excellent

complements the Pinner acid catalysed synthesis which yields amides with many such electronegatively substituted nitriles (see Section III, A). In addition, the presence of moisture is no great drawback in this reaction as the products are isolated in the form of the free imidates which are very much less sensitive to hydrolysis than are their salts (compare the Pinner

$$CCl_3CN \xrightarrow[\text{MeO}^-]{\text{MeOH}} CCl_3C \overset{\displaystyle \nearrow NH}{\underset{\displaystyle \searrow OMe}{}}$$

synthesis). However the reaction fails, like the Pinner method, for nitriles which are sterically hindered and in addition for nitriles with strongly acidic α-hydrogen atoms, e.g. alkyl cyanoacetates[68]. Although the simpler alcohols are effective, the reaction can employ higher primary or secondary alcohols[68] as well as in some instances tertiary alcohols (butyl or amyl)[69,70], thus differing in this last respect from the Pinner method.

In the aromatic series, a Hammett plot (log K vs σ) based on data for benzonitrile and its p-chloro and m-nitro derivatives is believed to predict adequately the reactivity of the commoner substituted benzonitriles[68]. Related studies[71] in the aliphatic series have been based on the inductive index, I, of the substituent *alpha* to the cyano group.

Much synthetic use has been made of this reaction over the last decade or so and it has come into prominence since the previous major review of this field[2]. Practical use has been made of the method to form perfluoroalkylthioimidates[72] (14) and aryl perfluoroalkylimidates[73] (15). In the

$$R_FC \overset{\displaystyle \nearrow NH}{\underset{\displaystyle \searrow SR}{}} \xleftarrow{\text{RSH}} R_FCN \xrightarrow{\text{PhOH}} R_FC \overset{\displaystyle \nearrow NH}{\underset{\displaystyle \searrow OPh}{}}$$

(14)   (16)   (15)

case of such nitriles (16) with very powerful electron withdrawing substituents, tertiary amines or alkali carbonates may be used as the base catalysts in place of the alkoxides[72,73]. Perfluoronitriles, however, fail to give imidates with tertiary alcohols[73], base catalysed trimerisation of the nitrile taking place in preference to imidate formation. Other synthetic

$$3R_FCN \longrightarrow \underset{\substack{R_F}}{\underset{|}{\overset{\displaystyle R_FC \overset{N}{\underset{N}{\diagdown}} CR_F}{\underset{N \diagdown C \diagup N}{\overset{|}{\underset{|}{}}}}}$$

applications of this method include the formation of fibrous cellulose 2,2,2-trichloro-acetimidates from trichloroacetonitrile and cotton cellulose pretreated with potassium hydroxide[74], and the formation of imidates from trichloroacetonitrile and partially fluorinated alcohols[75] among other examples[76-80]. A recent report draws attention to the selective formation of imidates from cyano groups in the α-position of a nitrogen heterocycle, groups in other positions failing to react—the catalyst in this case being a trace of sodium borohydride[81].

Extending the pioneering work of Nef[82] on the reaction of cyanogen with carbinols, Woodburn and his co-workers[30] showed that cyanogen and ethylene glycol monomethyl ether could give either a mono- or di-imidate depending on the catalyst employed.

$$(CN)_2 + CH_3OCH_2CH_2OH \xrightarrow[\text{Na}]{\text{KCN}} \begin{array}{l} NCC(=NH)OCH_2CH_2OCH_3 \\ \\ \text{---}[C(=NH)OCH_2CH_2OCH_3]_2 \end{array}$$

$$(CN)_2 + NH_2CH_2CH_2OH \xrightarrow[\text{KCN/EtOH}]{\text{KCN/H}_2\text{O}} \begin{array}{l} [NH_2CH_2CH_2OC(=NH)\text{---}]_2 \\ \\ [HOCH_2CH_2NHC(=NH\text{---}]_2 \end{array}$$

The reaction of cyanogen with ethanolamine[83] was also shown to be dependent on the catalyst employed.

A glycidic imidate has been shown to be one of the products of the reaction of a ketone with dichloroacetonitrile in the presence of iso-propoxide[84].

$$RR'C{=}O + Cl_2CHCN \xrightarrow[i\text{-PrO}^-]{i\text{-PrOH}} RR'C{-}CClC\overset{\nearrow NH}{\underset{\searrow OPr\text{-}i}{}}$$
$$\xrightarrow[\text{temp.}]{\text{lower}} RR'C(OH)CCl_2CN$$

Another interesting sequence of base-catalysed reactions leading to imidate formation is to be seen in the reaction of the aldehyde **17** with potassium cyanide[85]. In DMSO the product remains as the nitrile **18**, but in methanol it is converted into the imidate **19**.

In addition, in some instances, base catalysed secondary reactions can take place. One such example[86] is the base-catalysed cyclisation reaction which yields the cyclic imidate **20** from the open chain imidate intermediate **21**.

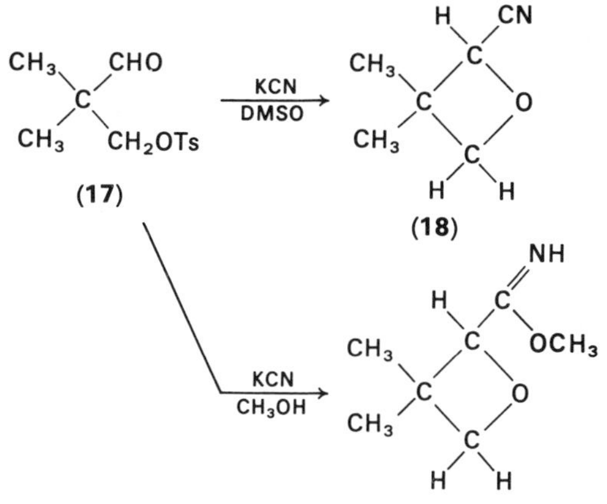

The reaction of nitriles with alcohols in neutral solution has not been studied to any extent and the reports that do exist are somewhat conflicting, e.g. Robinson[87] found that benzonitrile and ethylene glycol heated for several days in a sealed tube at an elevated temperature yielded the di-imidate **22**, whereas heating under reflux gave rise to ester believed

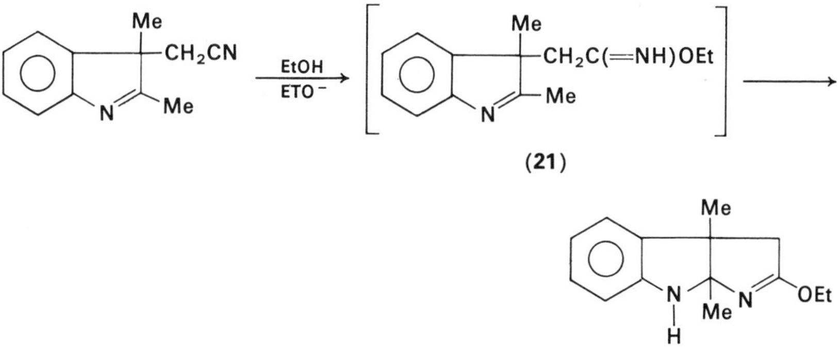

$$[PhC(=NH)OCH_2]_2$$
(22)

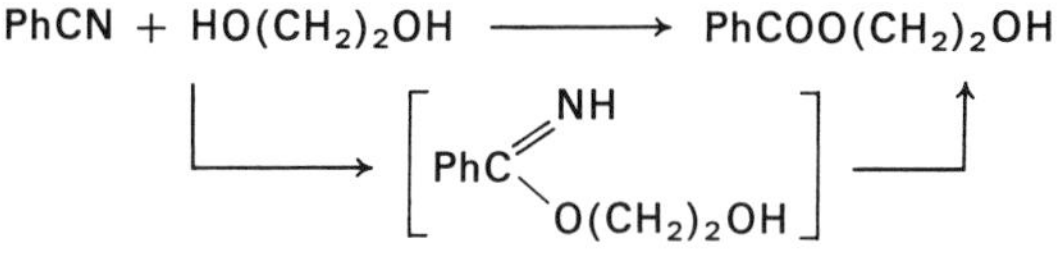

$$PhCN + HO(CH_2)_2OH \longrightarrow PhCOO(CH_2)_2OH$$

(23)

to be formed via the imidate intermediate **23**. On the other hand, Brown and his co-workers[72,73] have failed to observe any reaction between perfluoronitriles and alcohols or thiols in the absence of base catalysts.

C. Reaction of Imidoyl Halides with Alkoxides and Phenoxides

Alcohols and phenols, preferably as the alkoxides or phenoxides, (or their thio analogues) react readily with imidoyl halides, chlorides or bromides to give the corresponding imidates[2,88,89,90,91]. This synthetic

$$RC \overset{NR'}{\underset{X}{\diagdown}} \quad + \text{ NaOR''} \quad \longrightarrow \quad RC \overset{NR'}{\underset{OR''}{\diagdown}}$$

$$X = Cl \text{ or } Br$$

method has also been utilised to give N-amino and N-aryloxy imidates by the use of hydrazidoyl[89,92] and hydroxamoyl halides[89,93], respectively,

$$ArC \overset{NNHAr'}{\underset{Br}{\diagdown}} \quad \xrightarrow{\text{EtOH}} \quad ArC \overset{NNHAr'}{\underset{OEt}{\diagdown}}$$

$$ArC \overset{NOPh}{\underset{Cl}{\diagdown}} \quad \xrightarrow{\text{TlOR}} \quad ArC \overset{NOPh}{\underset{OR}{\diagdown}}$$

(in this latter case use being made of a thallium alkoxide). Other examples of compounds prepared by this procedure and in which the imidate nitrogen has unusual substituents can be seen in the phosphorus derivatives[94,95] **24,** and the boron compound[96] **25** among others[97].

$$ArC \overset{NPO(OR')_2}{\underset{XR'}{\diagdown}} \qquad\qquad CCl_3C \overset{NB(Me)Br}{\underset{SAlkyl}{\diagdown}}$$

$$X = S \text{ or } O$$

$$(24) \qquad\qquad\qquad (25)$$

Imidoyl fluorides have received scant attention but it has been observed that the action of alcoholic ammonia on the imidoyl fluoride **26** led to a mixture of imidate and amidine[98].

$$PhC \overset{NC_6H_{11}}{\underset{F}{\diagdown}} \quad \xrightarrow[\text{EtOH}]{\text{NH}_3} \quad PhC \overset{NC_6H_{11}}{\underset{OEt}{\diagdown}} \quad + \quad PhC \overset{NC_6H_{11}}{\underset{NH_2}{\diagdown}}$$

$$(26)$$

Cyclic imidates of the lactim ether type have also been extensively prepared by this method[3]; an interesting example being the cyclisation of the 4-aminocyclohexanecarboxylic acid **27** to the imidoyl chloride, which in turn can undergo alcoholysis to give the imidate[99] **28**.

(27)   (28)

Related to the above examples but in a special class of their own are the 1-chloroformimidates[89,100] which can undergo further nucleophilic attack with loss of halogen to give the acetals **29**.

(29)

Certain pyridine *N*-oxides have been reported to react with imidoyl halides to yield imidates but the reaction is not claimed to be general[101,102].

D. Conversion of Amides and Thioamides into Imidates

The chemistry of amides[103] and of thioamides[104,105] has recently been reviewed and it will suffice here to pick out the salient points regarding the imidate formation from these compounds, in particular giving references to the most recent work. Thioamides can be alkylated directly at the sulphur atom with alkyl halides—amides do not normally undergo similar attack at oxygen. The method appears to work well with primary, secondary, or tertiary thioamides[2,104,105], and also with N^2-substituted[106,107] as well as with N^1,N^2- and N^2,N^2-di-substituted thiohydrazides[108,109].

$$RC\overset{S}{\underset{NHNHPh}{\big<}} + EtI \xrightarrow{OEt^-} RC\overset{NNHPh}{\underset{SEt}{\big<}}$$

Dibromides, e.g. ethylene dibromide have also been employed but are reported to give different products (30–32) depending on the reaction conditions[110,111,112].

$$RC\overset{\overset{+}{N}H_2Br^-}{\underset{SCH_2CH_2Br}{\big<}} \qquad \left[RC\overset{NH}{\underset{SCH_2}{\big<}}\right]_2 \qquad RC\overset{N\text{---}}{\underset{S}{\big<}}$$

(30)   (31)   (32)

By contrast, amides can normally be alkylated successfully with alkyl halides only as their iminolate silver salts. Reactions of this type have been used to overcome some of the limitations of the Pinner synthesis caused by steric hindrance[2]. However, it must always be remembered that many imidate bases can rearrange to give N-substituted amides (see Chapman Rearrangement) particularly in the presence of alkyl halides, hence there has been, at times, some confusion as to whether direct N- or O-alkylation was taking place. Little use appears to have been made of this type of

$$RC\overset{NR'}{\underset{OAg}{\big<}} \xrightarrow{EtI} RC\overset{NR'}{\underset{OEt}{\big<}} \xrightarrow{\Delta} RC\overset{NR'Et}{\underset{O}{\big<}}$$

O-alkylation of late; however, some cyclic imidates of the type 33 have been prepared by the fusion of ω-bromo-amides—a process which involves direct O-alkylation[113].

$$Br(CH_2)_nCONHR \xrightarrow{\Delta} (CH_2)_n \quad C=\overset{+}{N}HR$$

(33)

Ethyl chloroformate[114] has been condensed with both O- and S-amides to yield imidates and it reacts in general with S-amides but fails to condense with aromatic O-amides[115].

$$MeC\overset{NHEt}{\underset{O}{\big<}} + ClCOOEt \longrightarrow MeC\overset{\overset{+}{N}HEt\ Cl^-}{\underset{OEt}{\big<}}$$

$$EtC\overset{NH_2}{\underset{S}{\big<}} + ClCOOEt \longrightarrow EtC\overset{\overset{+}{N}H_2\ Cl^-}{\underset{OEt}{\big<}}$$

Ohme and Schmitz[31, 116, 117] have recently developed a valuable method for obtaining formimidates using formamides and benzoyl chloride as starting materials. This route avoids the use of the unpleasant reagent, hydrogen cyanide, which is required by the Pinner process[1].

$$HC \underset{O}{\overset{NHR}{<}} + PhCOCl + R'OH \longrightarrow HC \underset{OR'}{\overset{\overset{+}{N}HR \; Cl^-}{<}}$$

(R = H or Me)   (R' = lower alkyl)

Both dimethyl sulphate and more recently trialkyloxonium fluoroborates have been used in the O- or S-alkylation of amides and thioamides[2, 3, 103]. Although the former reagent required the use of higher temperatures (up to 60°C.) the reaction appears to give exclusively O- or S-alkylation products[118–122] provided equivalent quantities of the reagents are employed, otherwise secondary products result[123, 124].

In addition to dialkyl sulphates, methyl fluorosulphate has also been employed[125]; however, triethyloxonium fluoroborate now appears to be superseding the dialkyl sulphates as the reagent of choice[3, 56] and it has been used with both amides[126–129] and thioamides[130]. Imidates unobtainable by the Pinner method have been synthesised by this route[56]. The advantages of triethyloxonium fluoroborate over dimethyl sulphate as a reagent for imidate formation[131] in some cyclic systems is illustrated by equation (6).

$$\tag{6}$$

A sophisticated synthetic application of this method of imidate formation is described by Hanessian[132] who selectively removed an *N*-acetyl group from an amino sugar containing both *N*- and *O*- acetyl groups.

Other reagents which have been utilised for *S*- or *O*-alkylation include diazomethane[133, 134, 135], used to obtain imidates from substituted amides, thioamides and *N*-chloroamides (it is probable that the reaction is not a general one) and alkane sultones[136, 137] which react with amides

or thioamides to yield imidates of the type 34.

$$(34)$$

E. Imidates from the Reaction of Amino Compounds and Ortho Esters

The chemistry of carboxylic ortho acid derivatives is the subject of a recent book by DeWolfe[138] and as this section will attempt to indicate only the broad principles involved in this method of synthesis, it is suggested that the reader should refer to DeWolfe's book for a more detailed description of some areas of this chemistry.

The reactions of ortho formates with aromatic primary amines are mechanistically quite complex although the products, N-substituted formimidates or N,N'-disubstituted formamidines can often be obtained in good yields and much synthetic use has been made of this reaction in recent

$$HC(OR)_3 + ArNH_2 \rightleftharpoons HC(=NAr)OR + 2\ ROH$$

(35)

$$\downarrow ArNH_2$$

$$HC(=NAr)NHAr + ROH$$

(36)

years[2, 138, 139]. Initially it was shown by Claisen[140] that the product most readily isolated in neutral solution was the amidine (36) but subsequent workers[141] were able to to get good yields of imidate (35) under conditions of acid catalysis. Hydrochloric, sulphuric, acetic and p-toluenesulphonic acids have all been employed successfully in this way as catalysts[2]. The mechanism of the reaction has been extensively studied by Roberts and his co-workers[2, 138, 142–144], who showed conclusively that the imidate (35) is in fact the initial product but that in neutral solution the imidate very rapidly reacts with further amine to give the amidine[144] (36). Support for this mechanism comes from the fact that hindered amines yield imidates even in the absence of acids[145–147], no amidine apparently being produced. In addition, under acidic conditions, it is known that N,N'-

diarylformamidines react with triethyl ortho formate/ethanol to give ethyl N-arylformimidates. DeWolfe[138] has summarised one set of equations for a mechanistic pathway leading to amidine and imidate products under differing reaction conditions but other interpretations have also been put forward[148]. The reaction of other ortho esters (37; R = Me,Et)

$$RC(OEt)_3 + R'NH_2 \longrightarrow RC(=NR')OEt$$

(37)

with aromatic primary amines has been shown to be sensitive to the presence of acid[138,149]. It is suggested that equimolar quantities of an amine and triethyl ortho acetate in the presence of traces of acid give rise to an imidate (from **37**; R = Me) but that two moles of amine to one of ortho acetate also in the presence of acid give an *N,N′*diarylacetamidine.

Aliphatic primary amines fail to yield imidates with either triethyl ortho formate or ortho acetate, amidines being formed under all experimental conditions investigated[138,149]. On the other hand, cyclic ortho

$$2\ C_6H_{11}NH_2 + HC(OEt)_3 \xrightarrow{\ H^+\ } HC(=NC_6H_{11})NHC_6H_{11}$$

esters of the type **38** react with both aliphatic and aromatic primary amines to give cyclic imidates (tetrahydrofuranimines, **39**)[150].

(38)   (39)

Higher ortho esters (**37**; R = Ph, Pr, Bu) have also been used particularly in conjunction with heterocyclic amines to give imidates[138,151].

A report by Wasfi[152] that amides (**40**; R = Ph, Me) react with ortho formates to yield *N*-acyl formimidates (**41**) must be looked upon with doubt in view of other more extensive work[123,153–156] which points to

(40)   (41)   (42)

structures of type **42** for the products of these reactions.

Sulphonamides, on the other hand, are readily converted into the corresponding imidates[2,138,157] (**43**), and in addition the imidates **44**

(43)

and **45** have been prepared from cyanamide[158] and phosphoramides[159] respectively.

(44)   (45)

Certain hydrazine derivatives can also react with ortho esters to yield imidates but the reaction is not general. For example[160], phenylhydrazine does not appear to react with triethyl ortho formate under acid-free conditions and yields diphenylformazan (**47**) when acetic acid is present,

$$PhNHNH_2 + HC(OEt)_3 \xrightarrow{H^+} \left[HC \begin{matrix} \diagup NNHPh \\ \diagdown NHNHPh \end{matrix} \right] \xrightarrow{[O]} HC \begin{matrix} \diagup NNHPh \\ \diagdown N=NPh \end{matrix}$$

(46)   (47)

this presumably being due to the irreversible oxidation of the dihydro derivative **46**. Other hydrazines, e.g. the 2- and 4-nitro- and 2,4-dinitro-phenylhydrazines are reported to yield imidates with ortho esters[161] and the hydrazine **48** was sucessfully converted into the corresponding imidate (**49**) with diethoxymethyl acetate[162].

(48)   (49)

Hydrazides and hydrazones similarly yield hydrazonate esters (**50 and 51**) on treatment with ortho esters[2,138,163,164]. The compounds **50** and **51** are particularly useful as synthetic intermediates.

$$HC \begin{matrix} \diagup NHNH_2 \\ \diagdown O \end{matrix} + HC(OEt)_3 \longrightarrow HC \begin{matrix} \diagup NHN \diagdown \\ \diagdown O \quad EtO \end{matrix} CH$$

(50)

$$RC \begin{matrix} \diagup NNH_2 \\ \diagdown H \end{matrix} + HC(OEt)_3 \longrightarrow RC \begin{matrix} \diagup NN \diagdown \\ \diagdown H \quad EtO \end{matrix} CH$$

(51)

F. Transesterification of Imidates

Imidates can be transesterified by heating with an alcohol of higher boiling point than that used in the original preparation[2,165–170]. The reaction is best carried out in the presence of some sodium alkoxide[165,166]. Thioimidates have been converted similarly into O-imidates[171].

$$HC \begin{matrix} \diagup NPh \\ \diagdown OEt \end{matrix} + HOCH_2CH_2Cl \longrightarrow HC \begin{matrix} \diagup NPh \\ \diagdown OCH_2CH_2Cl \end{matrix}$$

$$RC(=NH)SR' + R''OH \longrightarrow RC(=NH)\,OR'' + R'SH$$

The use of t-butanol gives rise to compounds not easily available by other routes[172].

G. Preparation of N-Substituted Imidates from Simple Imidates

The reaction of simple imidates with the esters of amino acids, first investigated by Schmidt[173], was later found to be a general reaction for the formation of N-substituted imidates provided that one equivalent of acid was present[2]. In the absence of acid the reaction produces imidazoles.

$$\underset{OR'}{\overset{NH}{RC}} + EtOOCCH_2\overset{+}{N}H_3\ Cl^- \longrightarrow \underset{OR'}{\overset{NCH_2COOEt}{RC}}$$

$$\underset{OR'}{\overset{NH}{RC}} + NCCH(NH_2)COOR'' \longrightarrow \underset{\underset{H}{N}}{\overset{N\text{---}CCOOR''}{RC}}CNH_2$$

Thioimidates, however, appear to give imidazoles[174,175] even in the presence of acid. The reactions of imidates with amino acid derivatives is discussed in more detail in Section IV.H.10.

Other reactions leading to N-substituted products include the formation of certain herbicides (52) by the action of halogen-substituted ketones on imidates[176];

$$\underset{OEt}{\overset{NH}{NCC}} + (CX_3)_2C{=}O \longrightarrow \underset{OEt}{\overset{NC(OH)(CX_3)_2}{NCC}}$$

$$(X = Halogen) \qquad (52)$$

the synthesis of imidatosilanes (54) from halosilanes[177] and the reaction

$$\underset{OEt}{\overset{NH}{MeC}} + MeSiCl_3 \xrightarrow{Et_3N} [EtOC(Me){=}N]_3SiMe$$
$$(54)$$

of benzimidates with chlorophosphines[178].

$$\underset{OEt}{\overset{NH}{PhC}} + Ph_2PCl \xrightarrow{Et_3N} \underset{OEt}{\overset{NPPh_2}{PhC}}$$

N-Hydroxy-imidates (55) undergo alkylation of the N-oxygen function with either halogen compounds[179,180] or olefins[180,181]. Activated aryl halides, e.g. 2,4-dinitrochlorobenzene have also been employed in this way[182].

$$RC \overset{NOH}{\underset{OR'}{\Bigg<}} \quad \overset{ClCH_2CHClCOOMe}{\longrightarrow} \quad RC \overset{NOCH_2CHClCOOMe}{\underset{OR'}{\Bigg<}}$$

$$\overset{CH_2=CHCHO}{\longrightarrow} \quad RC \overset{NOCH_2CH_2CHO}{\underset{OR'}{\Bigg<}}$$

(55)

H. Synthesis of Imidates from Unsaturated Systems

Several unsaturated systems can be used to synthesise imidates, but two main methods emerge. In the first of these, an aza-1,2-diene adds on an alcohol under basic conditions to yield an imidate[2, 183, 184]. In some

$$R_2C=C=NR' + MeOH \xrightarrow{MeO^-} R_2CHC \overset{NR'}{\underset{OMe}{\Bigg<}}$$

examples of this reaction, the diene has been formed *in situ*, e.g. by carbene addition to an isonitrile[185] or by base catalysed elimination of hydrogen

$$C_6H_{11}NC \xrightarrow{Cl_2C:} C_6H_{11}N=C=CCl_2 \xrightarrow{ROH} CHCl_2C \overset{NC_6H_{11}}{\underset{OR}{\Bigg<}}$$

cyanide from compounds of the type[186, 187] 56.

$$RCH=CHCH(CN)NHR' \xrightarrow[MeOH]{MeO^-} RCH_2CH_2C \overset{NR'}{\underset{OMe}{\Bigg<}}$$

(56)

In the second of these methods, use is made of an unsaturated ether such as ethoxyacetylene which adds on primary amines[2], or sulphonyl azides[188]

$$EtOC{\equiv}CH + RNH_2 \longrightarrow CH_3C \overset{NR}{\underset{OEt}{\Bigg<}}$$

to yield imidates. This latter reaction has been studied recently by Himbert and Regitz[189] who found that the products of it were the triazole 57 and the imidate 58 which co-exist in equilibrium in certain cases. Ethoxy-

$$R'C{\equiv}COEt + N_3SO_2R \longrightarrow \underset{\underset{SO_2R}{|}}{EtO} \overset{R'\text{---}N}{\underset{\underset{N}{\parallel}}{\boxed{}}}_{N} \rightleftharpoons \overset{N_2}{\underset{R'}{\diagdown}}C{-}C \overset{NSO_2R}{\underset{OEt}{\Bigg<}}$$

(57)   (58)

acetylene has also been used in a related reaction[190] to form the novel imidate 59.

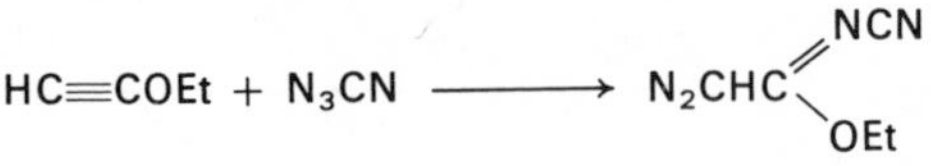

$$HC\equiv COEt + N_3CN \longrightarrow N_2CHC \underset{OEt}{\overset{NCN}{\diagup}}$$

(59)

An example of a different nature is found in the reaction of butyl vinyl ether with a primary amide to give imidates[191, 192] of type **60**.

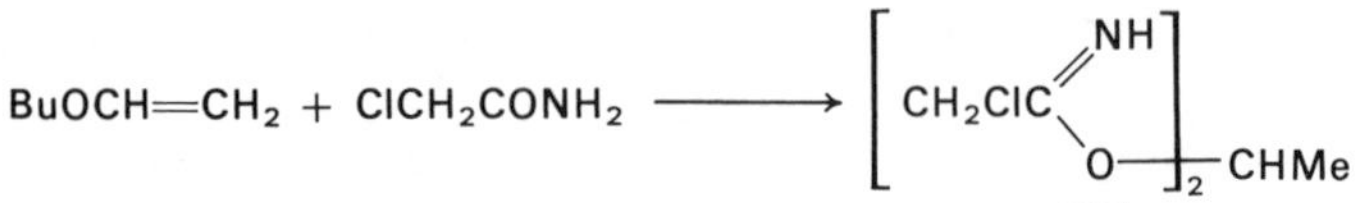

$$BuOCH\!=\!CH_2 + ClCH_2CONH_2 \longrightarrow \left[CH_2ClC \underset{O}{\overset{NH}{\diagup}} \right]_2 CHMe$$

(60)

Certain imines (**61**) which can be looked on as the imidoyl chlorides of chloroformic acid react with phenoxides to give imidates[193, 194] and these reactions are of general importance for the synthesis of formimidates.

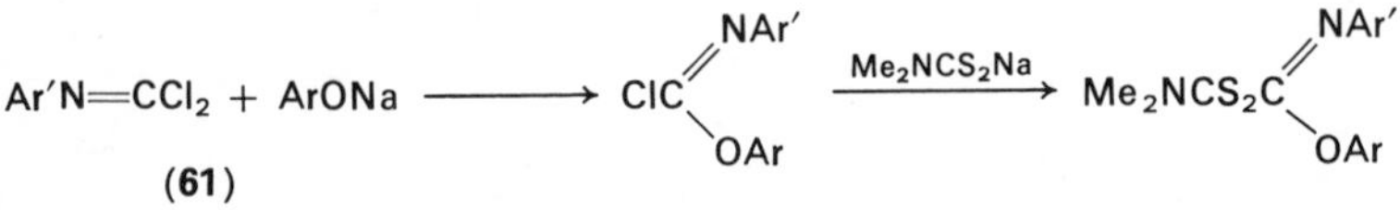

$$Ar'N\!=\!CCl_2 + ArONa \longrightarrow ClC \underset{OAr}{\overset{NAr'}{\diagup}} \xrightarrow{Me_2NCS_2Na} Me_2NCS_2C \underset{OAr}{\overset{NAr'}{\diagup}}$$

(61)

Other isolated examples of the use of unsaturated reagents in imidate formation include the addition reactions of isonitriles[195] (see the following section), the addition of alcohols to the fluoro-olefin[196] **62** and the oxidation[197, 198] by lead tetra-acetate (LTA) in methanol of aldehyde hydrazones.

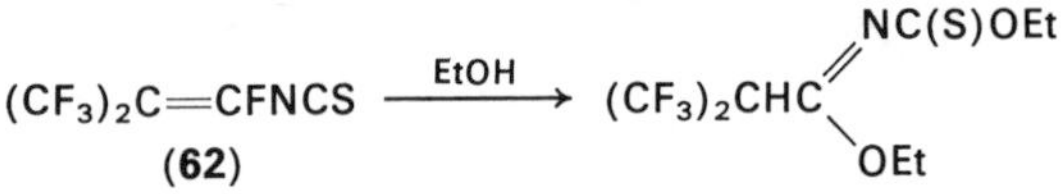

$$(CF_3)_2C\!=\!CFNCS \xrightarrow{EtOH} (CF_3)_2CHC \underset{OEt}{\overset{NC(S)OEt}{\diagup}}$$

(62)

$$RCH\!=\!NNHR' \xrightarrow[LTA]{MeOH} RC \underset{OMe}{\overset{NNHR}{\diagup}}$$

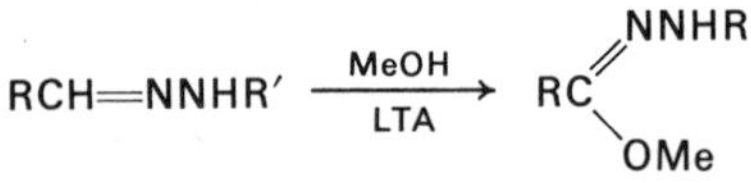

I. Imidates from Metal Complexes and Organometallic Compounds

Isonitriles have been successfully converted into N-substituted formimidates by their interaction with alcohols in the presence of metal catalysts[199–201]. Cuprous chloride is effective for β,γ-unsaturated alcohols

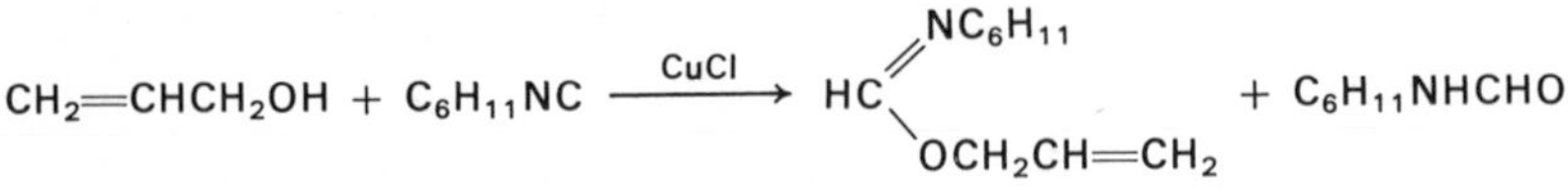

$$CH_2\!=\!CHCH_2OH + C_6H_{11}NC \xrightarrow{CuCl} HC \underset{OCH_2CH=CH_2}{\overset{NC_6H_{11}}{\diagup}} + C_6H_{11}NHCHO$$

but metallic copper or the copper oxides are more satisfactory for the saturated alcohols, high yields being claimed even when t-butanol was employed. This difference in behaviour has been rationalised in terms of the ability of the various alcohols to coordinate with the catalysts[202]. In the case of the reaction of thiols with isonitriles[203, 204], two separate, competing reactions take place. Of these, the reaction leading to thio-imidate formation (7a) is favoured if the thiol is primary and least favoured if the thiol is tertiary.

$$\text{RNC} + \text{R'SH} \quad \begin{cases} \text{(a)} \rightarrow \text{HC} \overset{\displaystyle NR}{\underset{\displaystyle SR'}{\diagdown}} \\ \text{(b)} \rightarrow \text{RN}{=}\text{C}{=}\text{S} + \text{R'H} \end{cases} \qquad (7)$$

In addition, imidate complexes, e.g. the palladium(II) complex **63**, have been reported from the action of methanol on palladium(III) isocyanide complexes[205].

$$\begin{array}{c} \text{Ph}_3\text{P} \diagdown \quad \diagup \text{Cl} \diagdown \quad \diagup \text{C(}{=}\text{NPh)OMe} \\ \text{Pd} \qquad \text{Pd} \\ \text{MeO(PhN}{=}\text{)C} \diagup \quad \diagdown \text{Cl} \diagup \quad \diagdown \text{PPh}_3 \end{array}$$

(**63**)

Nitrile complexes have also been successfully converted into imidates, e.g. cupric chloride in alcohol solution reacts with 2-cyanopyridine to give a complex identical with that obtained from the reaction of an alkyl pyridine-2-carboximidate and cupric chloride in hydrochloric acid[206]. Spectral studies suggest the structure **64** for complexes of this type which

(**64**)

unlike the parent imidates, appear quite stable to hydrolysis even in acid solution. Related pyridinecarboximidate complexes (**65**) have also been prepared[206].

M = Co, Ni or Fe

(**65**)

In addition, Clark and Manzer[207, 208] have shown that certain platinum and iridium complexes, e.g. compound **66**, react with perfluorobenzonitrile

$$\left[\begin{array}{c} Q \overset{Me}{\underset{OC}{\overset{|}{\diagdown}}} \overset{Cl}{\underset{\diagup}{\diagup}} \\ OC \diagup Ir \diagdown Q \\ Sol. \end{array}\right]^{+} (PF_6)^- + C_6F_5CN \xrightarrow{MeOH} \left[\begin{array}{c} Q \overset{Me}{\underset{OC}{\overset{|}{\diagdown}}} \overset{Cl}{\underset{\diagup}{\diagup}} \\ OC \diagup Ir \uparrow Q \\ NH \\ \| \\ C_6F_5COMe \end{array}\right]^{+} (PF_6)^-$$

(66)

$$Q = PMe_2Ph \text{ or } PMe(Ph)_2$$

in methanol to give imidate complexes. These reactions proceed via
π-bond intermediates in which the cyano group is activated towards
nucleophilic attack. Rhenium imidate complexes (67) have also been
isolated from the reaction of alcohols (but not phenols) with rhenium–nitrile
complexes[209] and the complex 67 (R = Et) was found to release ethyl

$$ReCl_4[MeC(=NH)OR]_2 \qquad R = Me \text{ or } Et$$
(67)

acetimidate on treatment with triphenylphosphine.

Of a somewhat different nature are the reactions of trialkyl tin derivatives[210, 211] or lead alkoxides[212] with electronegatively substituted nitriles
such as trichloroacetonitrile to yield imidates useful as biocides.

$$R_3SnOMe + CCl_3CN \longrightarrow CCl_3C\overset{NSnR_3}{\underset{OMe}{\diagdown}}$$

$$R_3PbOMe + CCl_3CN \longrightarrow CCl_3C\overset{NPbR_3}{\underset{OMe}{\diagdown}} + CCl_3C\overset{NPbR_3}{\underset{OPbR_3}{\diagdown}}$$
(68)

The use of bis-trialkyltin oxides, $(R_3Sn)_2O$, gives tin compounds[212]
analogous to the lead compound 68.

Other metal alkyls have also been utilised. Thus Tani, Yasuda and
Araki[213, 214] found that the interaction of trimethylaluminium with an
equimolar quantity of benzanilide gave the complex 69 which acted as a
highly stereospecific catalyst for the polymerisation of acetaldehyde. The

$$Me_3Al + PhCONHPh \longrightarrow [Me_2AlOC(=NPh)Ph]_2$$
(69)

imidate structure of compound 69 has since been confirmed by X-ray
analysis and the aldehyde adducts 70 and 71 have been similarly investigated[215, 216]. The species 69 and 70 were found to exist as dimers.

$$[Me_2AlOC(=NPh)Ph, MeCHO]_2 \qquad Me_2AlOC(=NPh)Ph, MeCHO, AlMe_3$$
(70)   (71)

1-Phenyl-1,2,3-triazole reacts with butyl-lithium via the intermediate **72** to give methyl N-phenylacetimidate[184].

$$(\bar{C}{\equiv}C{-}\bar{N}{-}Ph)(Li_2)^{2+} + MeOH \longrightarrow MeC\begin{smallmatrix}\nearrow NPh \\ \\ \searrow OMe\end{smallmatrix}$$

(72)

The chromium complex **73** reacts with anhydrous hydroxylamine in ether to produce (methyl acetimidato)pentacarbonylchromium(O), existing in the two forms **74a** and **b** which interconvert reversibly as the temperature alters[217, 218], and which split off methyl acetimidate on heating to 200°C.

$$(CO)_5Cr{=\!=\!=}C\begin{smallmatrix}\nearrow OMe \\ \\ \searrow Me\end{smallmatrix} + NH_2OH \longrightarrow \underset{MeO\;\;\;\;Me}{\overset{(CO)_5CrNH}{\underset{\parallel}{C}}} + \underset{Me\;\;\;\;OMe}{\overset{(CO)_5CrNH}{\underset{\parallel}{C}}}$$

(73)   **(74a)**   **(74b)**

J. Preparation of Boron–Imidate Derivatives

Several preparations of imidates involving boron intermediates or incorporating boron derivatives in the final imidate structure have been reported, e.g. boron thio-ethers react with amides[219] and nitriles[220] to form imidates.

$$2\,R_2BSR' + MeCN \longrightarrow \begin{smallmatrix} R'S\;\;\;\;Me \\ \diagdown C \diagup \\ \parallel \\ N \\ \diagup\;\;\;\diagdown \\ R_2B\;\;\;\;BR_2 \\ \diagdown\;\;\;\diagup \\ N \\ \parallel \\ C \\ \diagup\;\;\;\diagdown \\ R'S\;\;\;\;Me \end{smallmatrix}$$

Iminoboranes of the type **75** have been shown to react with alkane thiols to give thioimidates whereas thiophenols show preference for attack on the boron–halogen bond[221].

$$CCl_3C\begin{smallmatrix}\nearrow NBMeBr \\ \\ \searrow Br\end{smallmatrix} + RSH \longrightarrow CCl_3C\begin{smallmatrix}\nearrow NHBMeBr_2 \\ \\ \searrow SR\end{smallmatrix}$$

(75)

In addition, research into high energy propellant ingredients[222–224] has produced imidates of the type **76**.

$$MeC\begin{smallmatrix}\nearrow \overset{H}{N}{\rightarrow}BH_2N^+(Me)_3\;\;\;X^- \\ \\ \searrow OMe\end{smallmatrix} \qquad X = BF_4^- \text{ or } B_9H_{14}^-$$

(76)

K. Miscellaneous Preparations of Imidates

Thion esters react with amines to yield imidates[225], equation (8) effectively illustrating the difference in reactivity of the two ester groupings

$$CH_3OOCCH_2C(=S)OCH_3 \xrightarrow{RNH_2} CH_3OOCCH_2C\overset{NR}{\underset{OCH_3}{\diagup}} \qquad (8)$$

towards the amine. This reaction appears to be fairly general, but the reaction of hydrazine[226] or substituted hydrazines[227] has been shown to give rise to different products depending on the reaction conditions employed.

$$RC\overset{S}{\underset{OR'}{\diagup}} + NH_2NH_2 \longrightarrow RC\overset{NNH_2}{\underset{OR'}{\diagup}} + \left[RC\overset{N}{\underset{OR'}{\diagup}} \right]_2$$

$$HC\overset{S}{\underset{OEt}{\diagup}} + H_2NNRR' \longrightarrow HC\overset{NNRR'}{\underset{OEt}{\diagup}} \text{ or } HC\overset{S}{\underset{NHNRR'}{\diagup}} \text{ or } HC\overset{NNRR'}{\underset{NHNRR'}{\diagup}}$$

A modification of the foregoing procedure uses thio esters[228, 229] thus giving thioimidates as the end products.

Cyanic esters react with phenolates or with carbanions to yield imino

$$ArOCN + Ar'ONa \longrightarrow ArOC\overset{NH}{\underset{OAr'}{\diagup}}$$

$$(77)$$

$$ArOCN + \bar{C}HRR' \xrightarrow{Base} RR'CHC\overset{NH}{\underset{OAr}{\diagup}}$$

$$(78)$$

$$ArOH + HC(CN)RR' \longleftarrow RR'C=C(NH_2)OAr$$

$$(79)$$

R and R' = CN, COCH$_3$, etc.

carbonates (77) and imidates (78) respectively[230–232]. These latter compounds (78) can be represented by the tautomeric amino-ethylene structure 79; however in the basic conditions employed, decomposition often takes place resulting in the formation of phenol[230, 231, 233].

One or two photochemical reactions leading to imidates are also to be found in the literature. Among these[234] is a radical induced reaction of primary thiols with isonitriles[204] and the photoreduction of the tetra-

$$RNC + R'SH \longrightarrow HC{\overset{NR}{\underset{SR'}{}}}$$

methyl-dinitrobenzene **80** which gives amine and imidate as products[235].

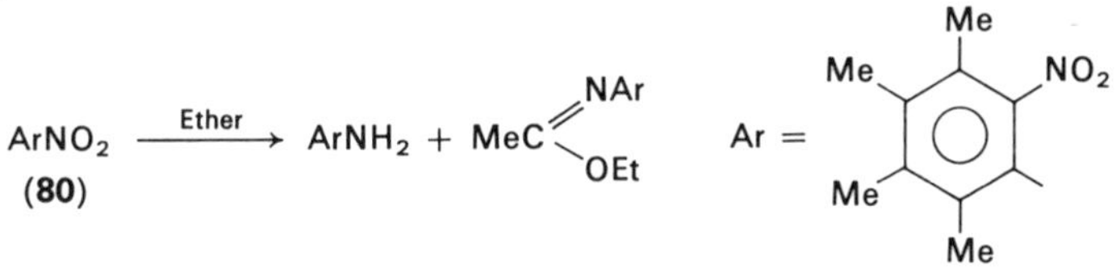

$$ArNO_2 \xrightarrow{\text{Ether}} ArNH_2 + MeC{\overset{NAr}{\underset{OEt}{}}} \qquad Ar =$$

(80)

In addition several photochemical reactions of cyclic imidates give rise to rearrangement products which are themselves imidates—these reactions are discussed in Section IV, O.

Thermal cycloaddition reactions of the diazo-imidate **81** have been investigated[190] and the products identified as the imidate derivatives, e.g. **82**.

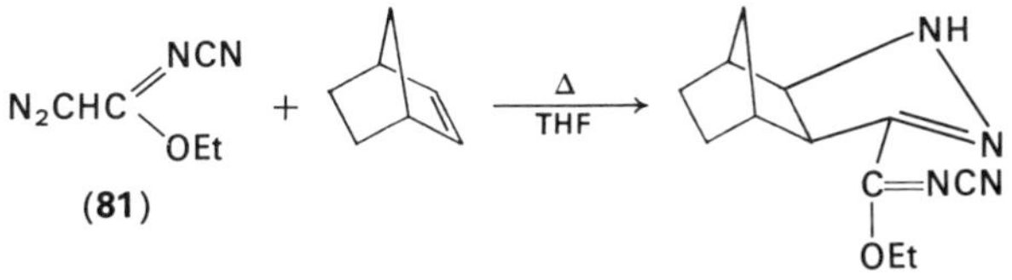

$$N_2CHC{\overset{NCN}{\underset{OEt}{}}} + \qquad \xrightarrow[\text{THF}]{\Delta} \qquad$$

(81)

(82)

Other preparations are to be found in the former review of imidates[2].

II. PROPERTIES OF IMIDATES

A. General Properties

The simpler imidate salts such as ethyl acetimidate hydrochloride tend to be hygroscopic in nature and thus hydrolyse fairly rapidly if left exposed to the atmosphere. Higher members, e.g. ethyl mandelimidate hydrochloride (**83**), appear much more stable in this respect. In addition, most imidate salts have decomposition points rather than true melting points.

$$PhCH(OH)C(\overset{+}{=}NH_2\overset{-}{Cl})OEt$$

(83)

Imidates are sometimes more readily handled as their bases which may be prepared by interaction of the salts with aqueous potassium carbonate or potassium or sodium hydroxide solutions[2], the imidate bases being extracted with ether. More recently, alcohol suspensions of the imidate salts when treated with ammonia at $-20°C$ have been shown to give good yields of imidate bases (see Section IV, H, 2). The lower aliphatic imidates can be distilled (e.g. ethyl acetimidate has b.p. 92–95°C) or are low melting solids (e.g. ethyl mandelimidate has m.p. 71–72°C) capable of being re-crystallised. However, aromatic imidates tend to decompose or rearrange on heating (see Section IV, D, 3).

Imidates are in the main weak bases although little study appears to have been made of their actual base strengths (see next section).

B. syn–anti Isomerism and Imidate Conformation

Although the evidence for the existence of geometric isomerism in compounds containing a carbon–nitrogen double bond has been re-viewed[236] recently (1970), little definite information was available regarding imidates at that time. More recently, n.m.r. studies have been carried out on

$$
\begin{array}{ccc}
\text{syn} & \text{anti} & \text{anti} \\
(84) & (85) & (86)
\end{array}
$$

cyclic imidates and thioimidates[237, 238]. For ring imidates of type **84** where $n = 2 \to 8$, the steric requirements are such that the compound must be in the *syn* form. However, spectral studies show that as the ring size increases, e.g. $n = 9 \to 13$ the imidates tend to exist in the *anti* form **85**. Similarities between the large ring compounds (**85**) and the open-chain analogues (**86**) point to these latter compounds also existing in the *anti* configuration. It is suggested that this is due to electron repulsion in the *syn* forms between oxygen non-bonding electrons and a lone pair localised in an sp^2 orbital on nitrogen. Equilibration studies carried out on the con-jugate acids pointed to high barriers for interconversion of the *syn* and *anti* forms of O-imidates[237]. In the case of the corresponding thioimidates

$$
\begin{array}{cc}
\text{anti} & \text{syn}
\end{array}
$$

the Coulombic repulsion curve for sulphur/nitrogen as against oxygen/nitrogen lone pair interaction is less steep and the activation energies for the interconversion of the cyclic compounds **87** and **88** are in the region of 19–22 kcal per mol[238]. The foregoing results confirm *anti* configurational

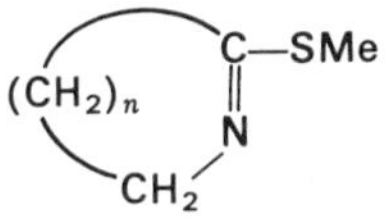

$$n = 8, 9, 10, 13$$

(87)   (88)

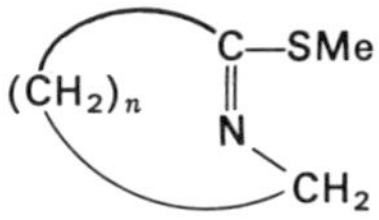

assignments given previously to imidates on the evidence of their dipole moments[239,240].

N-Halo-imidates have also been found to exist in *syn* and *anti* forms[2,241], e.g. compound **89** was a 9:1 mixture of *syn*:*anti* isomers.† In addition, it

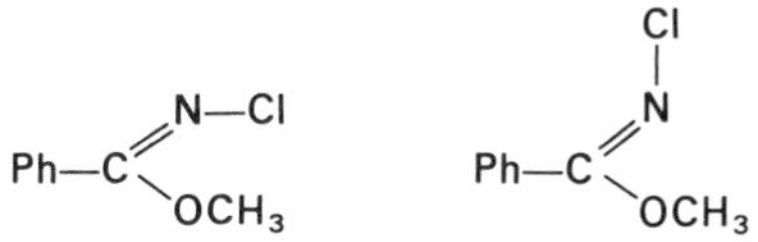

syn   *anti*

(89)

has been shown that hydroxamic acids can exist in *syn* and *anti* forms and the imidate derivatives have been studied[135,236,242]. Thus the reaction (equation 9) led to one single isomer which under irradiation gave both *syn* and *anti* forms separable on silica. On the basis of Beckmann type

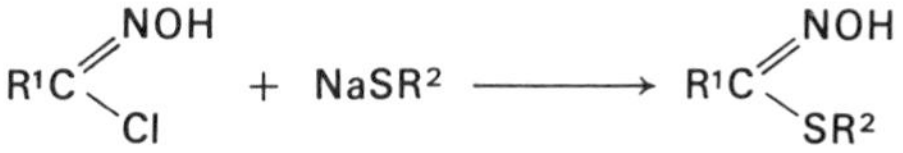

$$R^1 \text{ and } R^2 = \text{Me or Et}$$

reactions the initially formed product (which was the more stable one) was

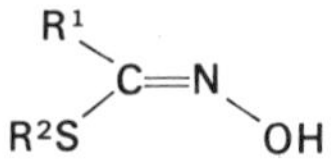

(90) 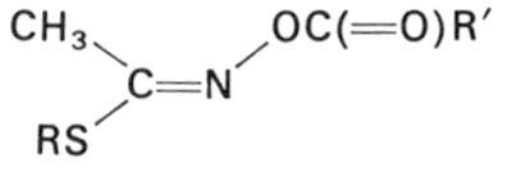   (91)

given the *syn* structure[243] (**90**). Related compounds (**91**) have been shown by X-ray analysis to have a similar *syn*-(alkylthio)-configuration[244]. Dipole studies[245] have also supported the *syn* structure for imidates of this N-alkoxy type and in addition suggest that for the molecule **92** the methoxy group is twisted by a dihedral angle of 45° and the *iso*-propyl group twisted in the opposite sense by about 30°.

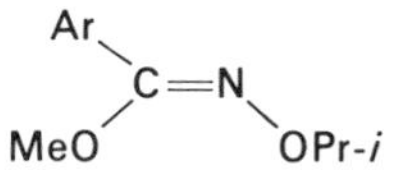

(92)

† When the imidate nitrogen carries a heteroatom substituent, the *syn–anti* nomenclature is reversed from the previous examples.

The mechanism of *syn–anti* isomerisation in imines is still obscure and it may be explained by (a) an inversion mechanism, (b) a torsion mechanism, or (c) a mechanism having both components. CNDO/2 calculations[246] made on compounds **93** and **94** show closely related inversion barriers but a much lower torsional barrier for the imidic acid (**94**).

(93)   (94)

C. Spectra of Imidates

The infrared spectra of methyl acetimidate (and its hydrochloride) and of methyl formimidate hydrochloride have been studied in detail[247,248], and comparisons made with the spectra of amide hydrochlorides[247]. Methyl acetimidate shows a doublet at 3372 and 3355 cm^{-1} in the vapour phase but a sharp singlet at 3343 cm^{-1} in dilute solution in carbon tetrachloride[248]; these bands being assigned to the NH group. The C=N stretching mode appears about 1661 cm^{-1} but shifts 12 cm^{-1} lower in frequency on protonation of the nitrogen. Both studies[247,248] point conclusively to structure **95** for imidate salts, there being no evidence

(95)   (96)

for protonation at oxygen. The frequencies of the stretching vibrations[249] for compounds of type **96** have been shown to depend on the nature of R and R'.

Measurement of the frequency and intensity of the NH bands points to the fact that the acidity of imidates increases in the series butyrimidate < benzimidate < phenylacetimidate[250,251]. In addition a Hammett plot[252] based on compounds of the type **97** shows an increasing acidity for X = *p*-Me < H < *m*-Cl < *p*-NO$_2$. Other general studies of infrared spectra of imidates have been carried out[2], e.g. on aliphatic imidates[253].

(97)

In a study of the electronic spectra of thioamides and their *N*- and *S*-derivatives, it has been found that for methyl thioacetimidate there is a

$\pi \rightarrow \pi^*$ transition at about 240 nm, but this is shifted to about 300 nm for methyl N-phenylbenzthioimidate[254].

D. Thermal Decomposition of Imidates

I. Unsubstituted imidate salts

Most unsubstituted imidate hydrohalides do not have true melting points but rather decompose on heating with evolution of alkyl halide leaving a residue of amide[2].

$$X = F, Cl, Br.$$

This thermal decomposition of imidate salts has been used as a method for the preparation of both alkyl halides[255,256] and of acid amides[36]. Alkyl thioimidate salts behave similarly, yielding thioamides. Less is known about imidate salts derived from phenols; however, it is reported that phenyl benzthioimidate hydrochloride on pyrolysis yields its primary components *viz.* benzonitrile, thiophenol and hydrogen chloride[28].

McElvain and Tate[257] have studied the thermal decomposition of several imidate salts in chloroform and in *t*-butanol. The rate of disappearance of halide ion was found to follow first order kinetics. These results could be explained either on the basis of an intramolecular attack by the halogen of an undissociated ion pair on the alkoxy group or, more probably by a bimolecular process exhibiting first order kinetics. Support

for this latter proposal came from the isolation of a *sec*-butyl chloride of high optical purity and of inverted configuration from the thermal decomposition of optically active *sec*-butyl acetimidate hydrochloride[22]. Subsequently it was shown using trichloroacetimidate salts (which permit the pyrolysis reaction to be carried out under very mild conditions) that the stereochemical course of the reaction depended mainly on the nature of the asymmetric centre of the alcohol. In fact the reaction[21] was found to proceed with either Walden inversion (octan-2-ol), racemisation (αphenylethanol), retention of configuration or Wagner–Meerwein re-arrangement (neopentanol). As well as hydrogen chloride, formic and acetic acids were used in these thermal decomposition studies[21].

$$RC\overset{\overset{+}{N}H_2\ Cl^-}{\underset{OCH_2C(CH_3)_3}{<}} \longrightarrow C_2H_5C(CH_3)_2Cl + (CH_3)_3CCH_2Cl \quad (10)$$

In addition to the rearranged products mentioned above[2, 21] arising from the neopentyl system (equation 10), imidate hydrochlorides derived from β-bromonitriles have been shown to give crossed products[258], (equation 11).

$$RCHBrCH_2C\overset{\overset{+}{N}H_2\ Cl^-}{\underset{OR'}{<}} \longrightarrow$$

$$R'Cl + R'Br + RCHClCH_2CONH_2 + RCHBrCH_2CONH_2 \quad (11)$$

It was pointed out in an earlier section (III, B) that imidates with strong electronegative substituents in the α-position normally tend to be un-stable as their salts and are better handled as the free bases. Shul'man[259]

$$CCl_3C\overset{\overset{+}{N}H_2\ X^-}{\underset{OCH_2R}{<}}$$

(98)

has shown that the compound **98** is stable when for example R = CCl$_3$ or CF$_3$ but not when R = alkyl. Such compounds (**98**) also appear to be more stable as the hydrobromides than as the hydrochlorides. The cor-responding imidate salt[260] derived from 2-nitro-ethanol (**98**, R = CH$_2$NO$_2$, X = Cl) pyrolised to give both CH$_2$ClCH$_2$NO$_2$ and CH$_2$=CHNO$_2$.

2. N-Substituted imidate salts

The thermal decomposition of N-substituted imidate salts appears complex and has evoked little study, e.g. phenyl N-phenylbenzimidate hydrochloride on pyrolysis gives the imidate base (main product), N,N'-diphenylbenzamidine hydrochloride, phenol, phenyl benzoate, benzanilide and hydrogen chloride[261].

N-Arylformimidates are reported to rearrange as their sulphates to give formyl derivatives of secondary amines[168].

$$HC\overset{NAr}{\underset{OEt}{<}} + H_2SO_4 \longrightarrow ArNEtCHO$$

3. Unsubstituted imidate bases

The simpler imidate bases are liquids (e.g. ethyl acetimidate, b.p. 90–93°C) which can normally be distilled unchanged. Imidates of higher molecular complexity may be solids (e.g. *t*-butyl trichloroacetimidate, m.p. 21°C). However, imidates derived from aryl systems decompose back to the parent nitriles and alcohols on heating[2].

$$ArC\overset{\displaystyle NH}{\underset{\displaystyle OR'}{\big\langle}} \longrightarrow ArCN + R'OH$$

4. *N*-Substituted imidate bases—the Chapman Rearrangement

Aryl *N*-aryl-arylimidates (**99**) rearrange thermally in an intramolecular manner involving a 1,3 shift of an aryl group from oxygen to nitrogen[2].

$$Ar^1C\overset{\displaystyle NAr^2}{\underset{\displaystyle OAr^3}{\big\langle}} \longrightarrow Ar^1C\overset{\displaystyle O}{\underset{\displaystyle NAr^2Ar^3}{\big\langle}} \qquad (12)$$

(99)

This reaction (equation 12) has become known as the Chapman rearrangement[262] (or sometimes the Chapman–Mumm rearrangement[263]) and is the subject of a comprehensive review[264]. Tetraglyme (b.p. 276°C) has recently been suggested as the solvent of choice in which to carry out the reaction under optimum conditions[265], and extensive use of the rearrangement has been made to obtain diarylamines[262, 264, 266].

The unimolecular nature of the reaction, proposed by Chapman[267] was later confirmed by Wiberg and Rowland who demonstrated the absence of crossed products when the imidates **100** and **101** were heated together[268]. More recently, tracer studies on the imidate **102** and its labelled counter-

$$PhC\overset{\displaystyle NPh}{\underset{\displaystyle OPh}{\big\langle}} \qquad PhC\overset{\displaystyle NC_6H_4Cl\text{-}p}{\underset{\displaystyle OC_6H_4Cl\text{-}p}{\big\langle}}$$

(100)   **(101)**

part (**103**) again demonstrated the intramolecular nature of this reaction[269].

$$PhC\overset{\displaystyle NC_6H_4CH_3\text{-}p}{\underset{\displaystyle OC_6H_4Br\text{-}p}{\big\langle}} \qquad PhC\overset{\displaystyle NPh}{\underset{\displaystyle O\overset{*}{C}_6H_4Br\text{-}p}{\big\langle}}$$

(102)   **(103)**

The rearrangement can be looked on as involving a four-membered transition state (**104**) in which there is nucleophilic attack by nitrogen on

$$\begin{array}{c} \overset{O\text{-----}Ar^3}{\underset{Ar^1C\text{====}NAr^2}{\text{ }}} \end{array}$$

(104)

the migrating aryl group[264, 268]. In general, the reaction is accelerated by electron attracting groups on the aryloxy ring (Ar^3) whereas similar electron attracting substituents on the arylimino ring (Ar^2) slow down the rearrangement. *ortho*-Substitution of the aryloxy ring (Ar^3) shows enhancement of reactivity over the corresponding *para*-substitution[268] and this has been explained in terms of an entropy effect—the restrictive nature of the *ortho*-substituent decreasing the entropy drop on going from reactant to intermediate. A recent study has shown that there is an increased rate of rearrangement for the imidates **105a** → **105b** → **105c** caused by steric acceleration due to hindered rotation in keeping with what was stated above regarding the entropy effect. However, for compound **105d**, a second factor—steric compression due to the bulky nature of the *t*-butyl group—becomes important and this causes a drop in rate[270].

Compound **99**, Ar^3 =

(105)

   (a) R = H, R' = H
   (b) R = H, R' = Me
   (c) R = Me, R' = Me
   (d) R = *t*-butyl, R' = *t*-butyl

Substituents on the *C*-aryl ring (Ar^1) have lesser effects but act in a similar sense to the arylimino ring (Ar^2) substituents.

In addition to arylamines and their aroyl derivatives[271] mentioned above, the reaction has been utilised to synthesise a variety of acridones[272] and benzacridones[273], ureas[274] **(106)**, and also a polymeric amide[275], derived

$$p\text{-}R_2N\underset{PhN}{\overset{NPh}{C}}OC_6H_4O\overset{\parallel}{C}NR_2 \longrightarrow p\text{-}R_2NC(=O)N\underset{Ph}{C_6H_4N}\underset{Ph}{C}(=O)NR_2$$

(106)

$$\left[C_6H_4C(Me_2)C_6H_4O\overset{Ph}{C}=NC_6H_4CH_2C_6H_4N=\overset{Ph}{C}O \right]_n$$

(107)

from the imidate **107**. In a few instances[264, 276], abnormal or secondary products are produced, e.g. equation 13.

$$(13)$$

The rearrangement of aryl N-arylthioimidates requires higher temperatures and the products tend to be more diverse[277] than in the case of the O-imidates and hence this reaction has aroused less interest.

In addition to the foregoing reactions, related rearrangements can occur with O-alkyl and O-allyl imidates[278] and these at times have been grouped with the Chapman rearrangement[2, 279], although not all workers would agree with this assignment as the reaction pathways are known to be different[264]. Thus, Mumm and Möller[280] showed that an O-allyl group migrated from oxygen to nitrogen with inversion of the allyl group—probably through a six-membered Claisen-like transition state (108). Hence in the rearrangement of the tritium labelled compound 109, all the

$$(108)$$

activity can be accounted for by stepwise degradation of the product 110 to formaldehyde[269], thus confirming the above inversion mechanism.

The course of O-alkyl imidate rearrangements depends on the nature of the alkyl group and olefins are often produced (equation 14) via a unimolecular *cis*-process[281], (although Chapman-type products are also known[282]; equation 15). O-Methyl and O-benzyl imidates which by their

$$(14)$$

$$\underset{\underset{OCH_2CH_2NEt_2}{}}{\overset{\overset{NAr}{\|}}{ArC}} \longrightarrow ArC(=O)N(Ar)CH_2CH_2NEt_2 \qquad (15)$$

nature cannot form olefins have been shown to rearrange in the case of the formimidates, **111** and **112**, via an intermolecular process to Chapman type products—crossed products having been observed from the pyrolysis[283] of mixtures of compound **112** and the labelled compound **111**.

$$\underset{\underset{OMe^*}{}}{\overset{\overset{NC_6H_4CH_3\text{-}p}{\|}}{HC}} \qquad\qquad \underset{\underset{OMe}{}}{\overset{\overset{NC_6H_4C_2H_5\text{-}p}{\|}}{HC}}$$

$$\text{(111)} \qquad\qquad\qquad \text{(112)}$$

O-Alkyl imidates also readily rearrange in the presence of alkyl halides but at lower temperatures[2, 264, 284] than those required when the imidate is heated alone. This intermolecular reaction (equation 16) has been the subject of a study by Arbuzov and co-workers[285, 286] who also extended their researches to include thioimidates and O-aryl imidates.

$$\underset{\underset{OCH_3}{}}{\overset{\overset{NCH_3}{\|}}{RC}} \xrightarrow[\text{MeI}]{100°C} \underset{\underset{NMe_2}{}}{\overset{\overset{O}{\|}}{RC}} \qquad (16)$$

In the case of cyclic imidates (e.g. compound **113**), it has been shown that rearrangement occurs readily only in the presence of catalysts such as dialkyl sulphates[287, 288].

$$\text{(113)} \qquad\qquad\qquad \text{Me}$$

E. Hydrolysis of Imidates

The mechanism of the hydrolysis of imidates has recently been studied extensively as it offers one method for the investigation of unstable tetra-hedral addition intermediates which are believed to be important in many acyl transfer reactions. The products as well as the rates of hydrolysis

$$\underset{\underset{OR''}{}}{\overset{\overset{NR'}{\|}}{RC}} + H_2O \longrightarrow \underset{\underset{OR''}{|}}{\overset{\overset{OH}{|}}{RC-NHR'}} \longrightarrow products$$

have been shown to be sensitive to pH and in addition in many cases to general acid–base catalysis[289–294]. From an acyclic imidate, the typical products at low pH are an ester and amine (cyclic imidates yielding amino-esters) while an amide along with alcohol are the products of hydrolysis at high pH[289, 291, 294–299]. However, in very strong acid solution (up to 65% aqueous sulphuric acid) the products of hydrolysis of methyl benz-

$$RC\overset{NR'}{\underset{OR''}{\diagdown}} + H_2O \left[\begin{array}{l} \longrightarrow RCOOR' + R'NH_2 \text{ (low pH)} \\ \\ \longrightarrow RCONHR' + R''OH \text{ (high pH)} \end{array} \right.$$

imidates (114) have been shown to be the corresponding benzamides[300].

$$PhC\overset{NRR'}{\underset{OMe}{\diagdown}} \qquad R \text{ and } R' = H \text{ or } Me$$

(114)

Moreover, in the case of imidates derived from very weakly basic amines[291, 295], ($pK_a < -6$, e.g. compounds 115 and 116) increasing yields of amine were obtained with increasing pH, contrary to the more usual behaviour mentioned above.

$$p\text{-}CH_3C_6H_4SO_2N{=}C\overset{CH_3}{\underset{OCH_3}{\diagdown}} \qquad\qquad 2,4,6\text{-}(NO_2)_3C_6H_2N{=}C$$

(115)   (116)

pH rate profiles determined for a number of imidates and thioimidates have been found to follow sigmoid curve characteristics[289, 294], and it has been observed that the pH range at which the reaction products change can differ from that at which reaction rates change[294, 297–299, 301]. The product transition normally takes place around neutral pH,[289, 294, 296]; however, when the imidate is derived from a phenol[292, 297, 298], or is an acyclic thioimidate[301, 302], the main product transition occurs around pH 2–3. In addition, alkyl thioimidates decompose fairly rapidly at high pH to the parent nitrile and thiol[302].

$$RC\overset{NH}{\underset{SR'}{\diagdown}} \longrightarrow RCN + R'SH$$

The effect of buffers has also been studied. Amine type buffers (e.g. imidazole) tend to cause only slight increases in the yield of amine and ester[289, 294], but substances which can act as bifunctional catalysts, e.g. phosphate or bicarbonate, cause much more marked increases in amine yield[290, 294, 296, 299, 302]. Recently, however, the opposite effects have been noted for certain specific imidates derived from weakly basic amines[291], e.g. all buffers examined led to a decrease in the amount of 2,4-dinitro-

$$HC \overset{NC_6H_3(NO_2)_2\text{-}2,4}{\underset{OEt}{\diagdown}}$$

$$R-\underset{\underset{OH}{|}}{\overset{\overset{OR''}{|}}{C}}-NHR'$$

(117)   (118)

aniline formed from the imidate **117**. These results have been interpreted in terms of mechanisms involving interaction between general acid–base catalysts and the tetrahedral intermediates.

Less attention has been paid to the effect of structure on the behaviour of the intermediates (**118**). However, in going from the *O*-ethyl imidate

$$CH_3C \overset{NMe}{\underset{OC_2H_5}{\diagdown}}$$

$$CH_3C \overset{NMe}{\underset{OCH_2CF_3}{\diagdown}}$$

(119)   (120)

119 to its trifluoro derivative[297] **120**, the mid point of the product transition shifted from pH 9·8 to pH 6·5, i.e. a lowering of the pH of transition with increasing acidity of the alcohol was observed. Increasing amine basicity has been deduced as facilitating amine expulsion from the tetrahedral intermediate[296] **118**; however *N*-alkylacetimidates are supposedly less reactive than *N*-phenylacetimidates at low pH although the reverse is true[294] at high pH. Recent studies[295] have shown up a diversity of effects

$$OH^- + RC \overset{\overset{+}{N}HR'}{\underset{OR''}{\diagdown}}$$

$$RC \overset{\overset{+}{N}HR'}{\underset{OR''}{\diagdown}} + H_2O \ \rightleftharpoons \ R-\underset{\underset{OH}{|}}{\overset{\overset{OR''}{|}}{C}}-NHR' \ \longrightarrow \ RCOOR'' + R'NH_2$$

(118)

$$RC \overset{NR'}{\underset{OR''}{\diagdown}} + H^+$$

$$R-\underset{\underset{O^-}{|}}{\overset{\overset{OR''}{|}}{C}}-NHR' \ \longrightarrow \ RCONHR' + R''OH$$

in the case of various N-substituted imidates in which there was a wide variation in the base strength of the parent aromatic amines.

The hydrolysis of imidates clearly presents a complex pattern which may well not be adequately represented by any one general mechanistic scheme; however the above shows a possible mechanistic pathway[294,303] but this may require to be modified to cover cationic as well as neutral and anionic tetrahedral intermediates[294,295]. Among the systems the kinetics of which have been studied are alkyl N-arylformimidates and acetimidates[291,295,303], alkyl benzimidates[294,300,304,305] as well as aryl[292,296,298] and alkyl[297] N-alkylacetimidates. In addition, various alkyl N-substituted thioimidates[301,302] as well as cyclic systems[289-291,295] and polymers[306] have been examined.

Practical use[2] of the hydrolysis of imidate salts has been made in order to obtain esters[258,307-309] or amines[132,310] and this reaction offers a

$$CCl_3C \overset{\overset{+}{N}H_2Cl^-}{\underset{OCH_2NO_2}{\diagup}} \xrightarrow{H_3O^+} CCl_3COOCH_2NO_2$$

route for the selective hydrolysis of an N-acetyl group in the presence of O-acetyl groups[132] on sugars (see Section III. D.). In the case of thioimidate salts, thiol esters[2,311] are obtained although thioamides have also been reported as products under acid hydrolysis conditions[312].

$$ROOCCH_2C \overset{\overset{+}{N}H_2\ Cl^-}{\underset{SR}{\diagup}} \xrightarrow{H_3O^+} ROOCCH_2C \overset{O}{\underset{SR}{\diagup}}$$

$$RC \overset{\overset{+}{N}H_2\ Cl^-}{\underset{SCHAr_2}{\diagup}} \xrightarrow{H_3O^+} RC \overset{S}{\underset{NH_2}{\diagup}} + Ar_2CHOH$$

Less commonly, imidate bases have been observed to form amidine salts by reaction with water in neutral or basic conditions[73,313,314,315]. This reaction is believed to proceed by hydrolysis of the imidate base to an ammonium salt (R_FCOONH_4) which reacts with a further molecule of imidate to give the final product[315].

$$R_FC \overset{NH}{\underset{OR}{\diagup}} \longrightarrow R_FC \overset{\overset{+}{N}H_2}{\underset{NH_2}{\diagup}} \quad R_FCOO^-$$

F. Action of Imidates with Hydrogen Sulphide

Imidates, on treatment with hydrogen sulphide in basic media, give rise to thion esters (121) although thioamides can also be present[2,311,316-319] as the products of secondary reactions. In the case of

$$RC\overset{NH}{\underset{OR'}{\diagdown}} + H_2S \longrightarrow RC\overset{S}{\underset{OR'}{\diagdown}} + NH_3 \longrightarrow RC\overset{S}{\underset{NH_2}{\diagdown}}$$

(121)

benzoylacetimidate (122) attack was found to take place at the benzoyl group in preference to the imidate[320]. Dithioesters[26, 29, 311, 319, 321] are

$$PhCOCH_2C\overset{NH}{\underset{OR}{\diagdown}} \xrightarrow{H_2S} PhCSCH_2C\overset{NH}{\underset{OR}{\diagdown}}$$

(122)

readily available from the attack of hydrogen sulphide on a solution of an imidate salt in pyridine at 0°C (However thioamides have also been found among the products of this reaction, especially if the conditions varied markedly from the above[2]).

$$RC\overset{\overset{+}{N}H_2\ Cl^-}{\underset{SR'}{\diagdown}} \xrightarrow[C_5H_5N]{H_2S} RC\overset{S}{\underset{SR'}{\diagdown}}$$

The corresponding thiol esters can be prepared by the action of thio-imidate salts with water[2] (see previous section).

$$RC\overset{\overset{+}{N}H_2\ Cl^-}{\underset{SR'}{\diagdown}} + H_2O \longrightarrow RC\overset{O}{\underset{SR'}{\diagdown}} + NH_4Cl$$

G. Alcoholysis of Imidate Salts—The Preparation of Ortho Esters

Imidate salts react at room temperature with alcohols to give simple (123) or mixed ortho esters[322] (124) depending on the choice of alcohol[323-325]. The reaction is sometimes known as the Pinner synthesis

$$RC\overset{\overset{+}{N}H_2\ X^-}{\underset{OR'}{\diagdown}} + 2\ R'OH \longrightarrow HC(OR')_3 + NH_4X$$

(123)

$$RC\overset{\overset{+}{N}H_2\ X^-}{\underset{OR'}{\diagdown}} + 2\ R''OH \longrightarrow HC(OR'')_2OR' + NH_4X$$

R' = alkyl   (124)

of ortho esters. The process is slow and the yields are often poor. However, they can be improved by refluxing the imidate salt with an excess of alcohol (up to tenfold) in ether[326] or by stirring a suspension of the imidate hydrochloride in alcohol/petrol at room temperature[327-329]. Such conditions ensure reaction temperatures below those at which the thermal decomposition of the imidate salt to an amide would become predominant and in addition the low solvent polarity decreases the ionisation of the ion pair (125) and hence prevents S_N attack of the halide at the ether site[138,257] of compound 125.

$$ RC\overset{\overset{+}{N}H_2\ X^-}{\underset{OR'}{\diagdown}} \longrightarrow R'Cl + RC\overset{O}{\underset{NH_2}{\diagdown}} $$

(125)

Our understanding of the alcoholysis reaction of imidates is due, in the main to McElvain and his co-workers[257, 326-329] and much of this work has been reviewed previously by Roger and Neilson[2] and more recently by DeWolfe[138] in his monograph on ortho esters. However it is worth stating some broad guide-lines.

It is essential that the imidate salt be free of excess hydrogen halide and that moisture be excluded or the competing reactions (17) and (18) will take place.

$$ RC(OR')_3 + HX \longrightarrow RCOOR' + R'OH + R'X \qquad (17) $$
$$ RC(OR')_3 + H_3O^+ \longrightarrow RCOOR' + 2\,R'OH + H^+ \qquad (18) $$

Under the conditions stated above, imidates derived from unbranched nitriles or those possessing only a single α-substituent, e.g. alkyl, phenyl, halogen or alkoxy give in the main good yields of ortho esters[25] but when two substituents (α,α or α,β) are present the yields drop considerably and amide and ester formation become competitive reactions[25, 330]. Such ester formation is most troublesome in those cases where the imidate has a bulky α-substituent such as a phenyl group. This ester formation may be due to attack on the ortho ester by the imidate salt, i.e. acid catalysed decomposition, and it has been shown that the acidity of the imidate salt is an important factor in this respect. Equation (19) illustrates this general reaction of an imidate salt with an ortho ester; however a fuller mechanism is discussed by DeWolfe[138].

$$ R'C(OR)_3 + R''C\overset{\overset{+}{N}H_2\ X^-}{\underset{OR'''}{\diagdown}} \rightleftharpoons R''C\overset{NH}{\underset{OR'''}{\diagdown}} + R'C(OR)_2\overset{+}{\underset{H}{O}}R + X^- $$

$$ R'COOR + RX + ROH $$

(19)

Interesting side reactions have been noticed in the case of β-bromo-propionimidate hydrohalide, which, when subjected to alcoholysis, gives a mixture of the esters **126a–c** whereas the γ-bromobutyrimidate hydro-halide gives the ortho ester **127**. These results[258] have been explained in

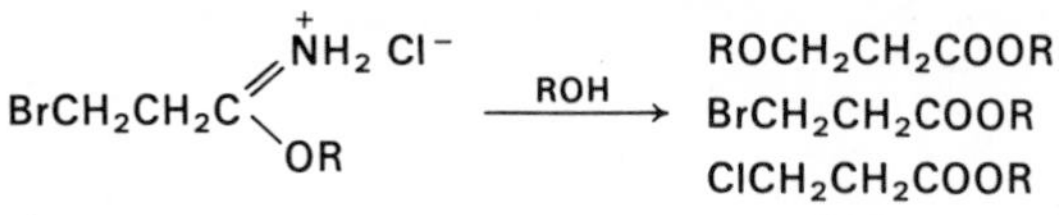

$$BrCH_2CH_2C\overset{+\!\!\!NH_2\ Cl^-}{\underset{OR}{}} \xrightarrow{ROH} \begin{array}{l} ROCH_2CH_2COOR \\ BrCH_2CH_2COOR \\ ClCH_2CH_2COOR \end{array}$$

(126 a–c)

$$BrCH_2CH_2CH_2C\overset{+\!\!\!NH_2\ Cl^-}{\underset{OR}{}} \longrightarrow BrCH_2CH_2CH_2C(OR)_3$$

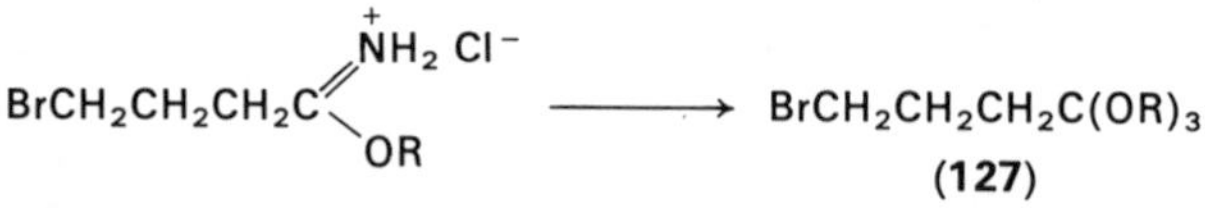

(127)

terms of the conjugated hybrid ion (**128**) which then reacts to form the addition products **126a–c**.

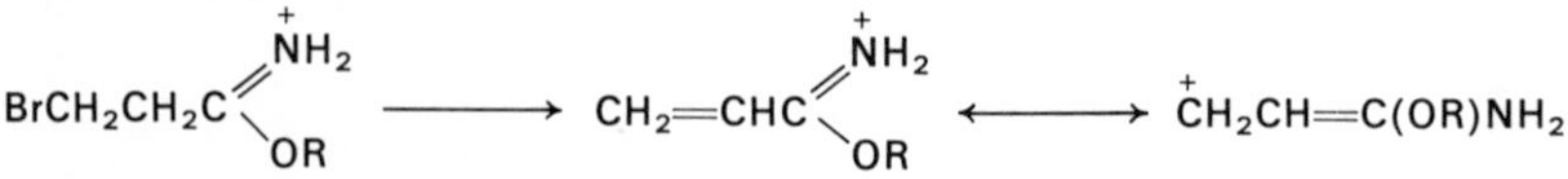

$$BrCH_2CH_2C\overset{+\!\!\!NH_2}{\underset{OR}{}} \longrightarrow CH_2{=}CHC\overset{+\!\!\!NH_2}{\underset{OR}{}} \longleftrightarrow \overset{+}{C}H_2CH{=}C(OR)NH_2$$

(128)

The alcoholysis of imidate salts using ethylene glycol[328,331] affords 2-alkoxy-1,3-dioxolanes (**129a** and **129b**). Another interesting example

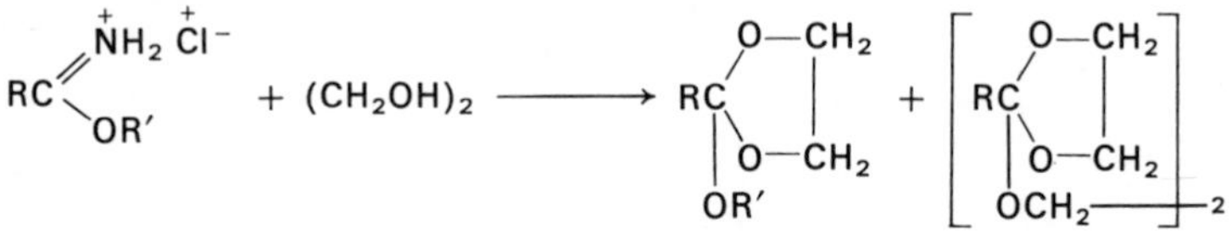

$$RC\overset{+\!\!\!NH_2\ \overset{+}{C}l^-}{\underset{OR'}{}} + (CH_2OH)_2 \longrightarrow RC\begin{array}{c} O{-}CH_2 \\ | \quad\quad | \\ O{-}CH_2 \\ OR' \end{array} + \left[RC\begin{array}{c} O{-}CH_2 \\ | \quad\quad | \\ O{-}CH_2 \\ OCH_2 \end{array} \right]_2$$

(129a)   **(129b)**

of the formation of a cyclic ortho ester[332] is to be seen in the synthesis of compound **130**.

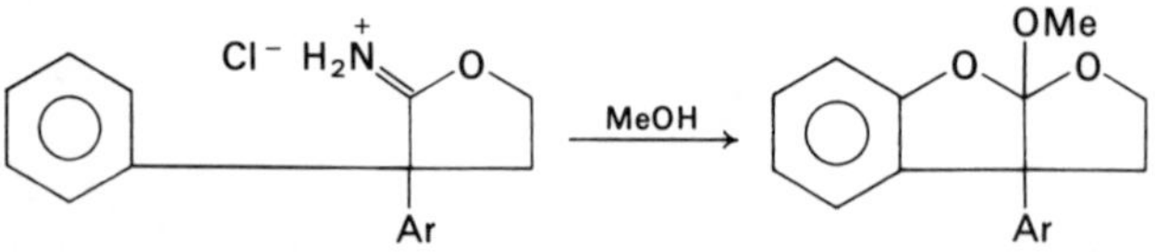

(130)

N-Substituted imidates have been converted also into ortho esters but little practical use has been made of this reaction[333].

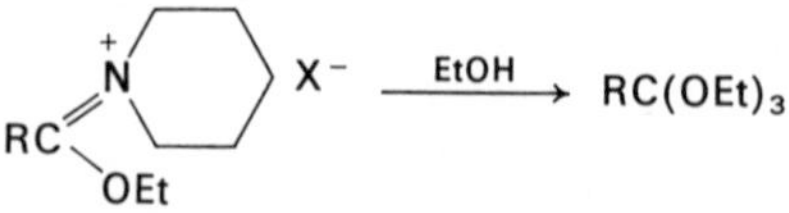

Industrial processes have also been described[12,334].

H. Reaction of Imidates with Ammonia and its Derivatives

1. General reactions

Imidates readily undergo nucleophilic attack with a wide range of amino compounds. The most usual pathway involves loss of alcohol from the imidate with formation of an amidine system **131** but the alter-

$$RC\!\!\begin{array}{c} \nearrow\overset{+}{N}H_2\ Cl^- \\ \searrow O\!-\!R' \end{array} + \underset{H}{\overset{R''}{N}}\!\!-\!R'' \longrightarrow RC\!\!\begin{array}{c} \nearrow\overset{+}{N}H_2\ Cl^- \\ \searrow NR''R'' \end{array} + R'OH$$

(131)

native, loss of ammonia and its replacement with the nucleophile can also take place in some cases.

$$RC\!\!\begin{array}{c} \nearrow\overset{+}{N}H_2\ Cl^- \\ \searrow OEt \end{array} + \underset{H}{\overset{R''}{N}}\!\!-\!R'' \longrightarrow RC\!\!\begin{array}{c} \nearrow\overset{+}{N}R''R''\ Cl^- \\ \searrow OEt \end{array} + NH_3$$

2. Reactions of imidates with ammonia

At low temperature (about $-20°C$) ammonia reacts with imidate salts to form the corresponding free bases. The reaction is conveniently carried out by suspending the imidate salt in ether, then treating it with dry ammonia and removing the ammonium chloride precipitate[316,335,336].

$$RC\!\!\begin{array}{c} \nearrow\overset{+}{N}H_2\ Cl^- \\ \searrow OEt \end{array} \xrightarrow[Et_2O/NH_3]{-20°C} NH_4Cl + RC\!\!\begin{array}{c} \nearrow NH \\ \searrow OEt \end{array}$$

At room temperatures the reaction follows a different course and amidine salts are formed[2,56,337,338]. In this case the imidate salt is normally dissolved or suspended in alcohol and treated with an excess of anhydrous ammonia. Alternatively, the imidate base can be treated with

$$RC\!\!\begin{array}{c} \nearrow\overset{+}{N}H_2\ X^- \\ \searrow OR' \end{array} \xrightarrow[EtOH]{NH_3} RC\!\!\begin{array}{c} \nearrow NH_2\ X^- \\ \searrow NH_2 \end{array} + R'OH$$

(X = halogen or BF$_4$)

an ammonium salt, (e.g. in aqueous alcohol, about 60°C), to yield the amidinium salt[2,337–339]. This procedure is especially useful when the imidate has a second functional group, e.g. ester, ketone or halogen

$$RC\!\!\begin{array}{c} \nearrow NH \\ \searrow OEt \end{array} + NH_4X \longrightarrow RC\!\!\begin{array}{c} \nearrow\overset{+}{N}H_2\ X^- \\ \searrow NH_2 \end{array} + EtOH$$

which could react with the free ammonia[2, 68, 240]. Much less work has been carried out using thioimidates but the thioimidate **132**, on treatment with a solution of 0·88 ammonia in alcohol, yields[340] the corresponding

(**132**)   (**133**)

amidinium iodide **133**, and the phosphorus derivatives **134** can also yield

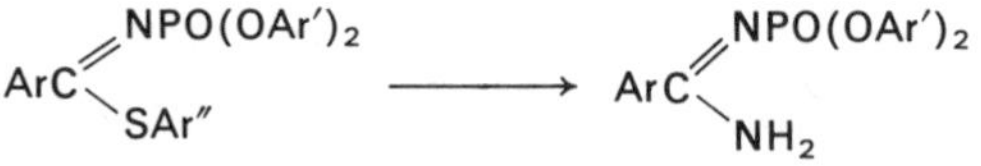

(**134**)

amidines[341], showing something of the utility of this reaction.

Lactim ethers react similarly to open chain compounds to form amidine salts[3, 342].

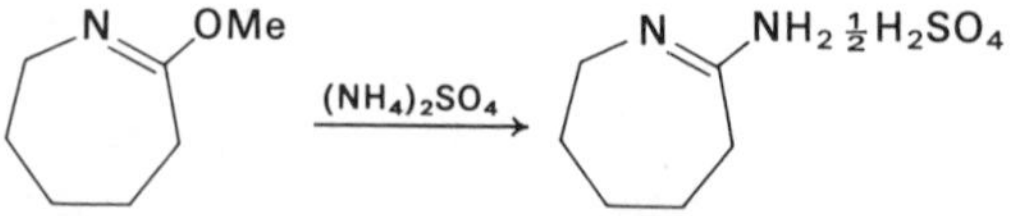

To a lesser extent the reaction has been used to obtain *N*-mono-substituted amidines by the use of *N*-substituted imidates[293].

The kinetics of the reaction of imidates with ammonia or amines (see following sections) have been studied by Hand and Jencks[293] and are best illustrated by the reaction of ammonia with ethyl *N*-methylbenzimidate. On the alkaline side (pH 10) the attack of amine is rate determining—the addition compound **125**, once formed, rapidly losing alcohol to form the monosubstituted amidine (route 20a). On the acid side of the pH rate

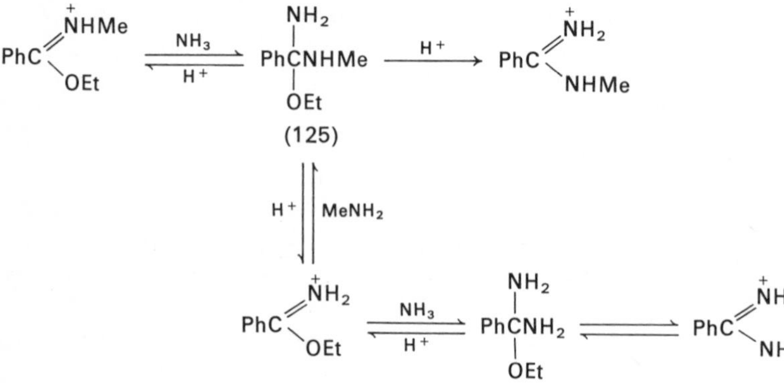

(20a)

(**125**)

(20b)

profile curve the product is the unsubstituted benzamidine, indicating equilibration of the intermediate with ammonia (route 20b).

3. Reaction of imidates with primary amines

Unsubstituted imidate or thioimidate salts react generally with primary amines to give either N-monosubstituted amidines (**136**), or N,N'-disubstituted amidines (**137**) if excess amine is present and prolonged heating applied[2, 56, 337, 343–346]; pH may also affect the reaction course[293] (see previous section).

$$ ArCH_2C \overset{\overset{+}{NH_2}\ Cl^-}{\underset{OMe}{\Big\langle}} \ +\ ArCH_2NH_2 \xrightarrow{\ MeOH\ } ArCH_2C \overset{\overset{+}{NHCH_2Ar}\ Cl^-}{\underset{NH_2}{\Big\langle}} $$

(**136**)

$$ ArC \overset{\overset{+}{NH_2}\ Cl^-}{\underset{OR}{\Big\langle}} \ +\ \triangleright\!-\!NH_2 \xrightarrow{\ ROH\ } ArC \overset{N\!-\!\triangleleft}{\underset{NH\!-\!\triangleleft}{\Big\langle}} $$

(**137**)

However, N-substituted imidates have also from time to time been isolated as products[347, 348] and Baiocchi and Palazzo[347] have suggested that weakly basic amines tend to give N-substituted imidates (**138**) whereas more basic amines give amidines (**139**) (see also Section IV, H, 2).

$$ ArC \overset{NH}{\underset{OEt}{\Big\langle}} \quad \begin{array}{c} \xrightarrow{\ PhNH_2\ } ArC \overset{NPh}{\underset{OEt}{\Big\langle}} \\[2mm] (\mathbf{138}) \\[4mm] \xrightarrow{\ cyclo\text{-}C_6H_{11}NH_2\ } ArC \overset{NH}{\underset{NHC_6H_{11}}{\Big\langle}} \end{array} $$

(**139**)

In addition to the N,N'-symmetrically disubstituted amidines mentioned above, the method can be extended by the use of suitable N-substituted imidates[349] to yield N,N'-disubstituted amidines (**140**) having different substituents on the nitrogen atoms. Trisubstituted amidines[321] can also

$$ HC \overset{NAr}{\underset{OEt}{\Big\langle}} \xrightarrow{\ Ar'NH_2\ } HC \overset{NAr}{\underset{NHAr'}{\Big\langle}} $$

(**140**)

be derived by use of thioimidates of type **141**.

$$PhC{\overset{\overset{+}{N}\diagdown O}{\diagup}}_{SMe} I^- \xrightarrow{\ PhNH_2\ } PhC{\overset{N\diagdown O}{\diagup}}_{\overset{+}{N}HPh} I^-$$

(141)

The mechanism of these foregoing reactions is discussed above (see the work of Hand and Jencks[293] in the previous section) and Roberts[2, 143] has shown that when care is taken to exclude all traces of acid the reactions (21a) and (21b) lead to the common product (142). However, in the

$$HC{\overset{NC_6H_5}{\diagup}}_{OEt} + p\text{-}NH_2C_6H_4CH_3 \xrightarrow{\ (a)\ }$$

$$HC{\overset{NC_6H_4CH_3\text{-}p}{\diagup}}_{OEt} + C_6H_5NH_2 \xrightarrow{\ (b)\ }$$

$$\longrightarrow HC{\overset{NC_6H_4CH_3\text{-}p}{\diagup}}_{NHC_6H_5} \qquad \text{(21)}$$

(142)

presence of acid disproportionation reactions take place and the product is a complex mixture of the three amidines **142**, **143**, and **144**. Sulphon-

$$HC{\overset{NC_6H_5}{\diagup}}_{NHC_6H_5}$$

(143)

$$HC{\overset{NC_6H_4CH_3\text{-}p}{\diagup}}_{NHC_6H_4CH_3\text{-}p}$$

(144)

amides react like primary amines with imidates to give amidines and appear to react preferentially to primary aryl amino groups when both are present on the same molecule[2, 350].

$$H_2NC_6H_4SO_2NH_2 + RC{\overset{NH}{\diagup}}_{OR'} \longrightarrow RC{\overset{NSO_2C_6H_4NH_2}{\diagup}}_{NH_2}$$

The reactions of imidates with amino acids and proteins are discussed in later sections.

4. Reaction of imidates with secondary amines

N,N'-Disubstituted amidines are formed from the reaction of secondary amines with imidates usually one or the other being present as its salt[77, 351].

$$ArOCH_2C{\overset{NH}{\diagup}}_{OCH_3} + R'R''\overset{+}{N}H_2\ Cl^- \longrightarrow ArOCH_2C{\overset{NH}{\diagup}}_{NR'R''}$$

Related tri-[352] and tetra-substituted[321] products have also been formed in this way by the reactions (22) and (23) respectively.

$$HC{\overset{NN=CHPh}{\diagup}}_{OEt} + NHR'R'' \longrightarrow HC{\overset{NN=CHPh}{\diagup}}_{NR'R''} \qquad \text{(22)}$$

$$\text{PhC}\!\!\underset{\text{SMe}}{\overset{\overset{+}{\text{N}}\diagup\diagdown\text{O}}{\diagup\diagdown}} \quad \xrightarrow{\qquad} \quad \text{PhC} \tag{23}$$

5. Reaction of imidates with tertiary amines

Tertiary amines do not react with imidates to give amidines[2,347] but can be used to form imidate bases from their salts[126].

$$\text{RC}\!\!\underset{\text{OR}'}{\overset{\overset{+}{\text{NH}_2}\ \text{BF}_4^-}{\diagdown}} \quad \xrightarrow{\text{NEt}_3} \quad \text{RC}\!\!\underset{\text{OR}'}{\overset{\text{NH}}{\diagdown}}$$

6. Reaction of imidates with hydrazine or its hydrates

The action of hydrazine on aryl imidates was first extensively studied by Pinner[2] and the corresponding reaction with alkylimidates was investigated by Oberhummer[353,354]. This chemistry has recently been

$$\underset{(145)}{\text{RC}\!\!\underset{\text{OR}'}{\overset{\overset{+}{\text{NH}_2}\ \text{Cl}^-}{\diagdown}}} \quad + \ \text{NH}_2\text{NH}_2 \quad \xrightarrow{\qquad} \quad \underset{(146)}{\text{RC}\!\!\underset{\text{NH}_2}{\overset{\text{NNH}_2}{\diagdown}}}$$

$$\text{RC}\!\!\underset{\text{OR}'}{\overset{\text{NH}}{\diagdown}} + \text{RC}\!\!\underset{\text{NH}_2}{\overset{\text{NNH}_2}{\diagdown}} \quad \xrightarrow{\qquad} \quad \underset{(147)}{\text{RC}\begin{smallmatrix}\text{N—N}\\ \diagup\ \ \diagdown \\ \text{H}_2\text{N}\ \ \text{NH}_2\end{smallmatrix}\text{CR}}$$

$$2\,\text{RC}\!\!\underset{\text{NH}_2}{\overset{\text{NNH}_2}{\diagdown}} \quad \xrightarrow{\text{EtOH}} \quad \underset{(148)}{\text{RC}\begin{smallmatrix}\text{N—NH}\\ \diagup\ \ \diagdown \\ \text{HN—N}\end{smallmatrix}\text{CR}}$$

(147)

(148)

Δ

[O]

$$\underset{(149)}{\text{RC}\begin{smallmatrix}\text{N——N}\\ \diagdown\ \ \diagup \\ \text{N}\\ |\\ \text{NH}_2\end{smallmatrix}\text{CR}} \qquad \underset{(150)}{\text{RC}\begin{smallmatrix}\text{N—N}\\ \diagup\ \ \diagdown \\ \text{N}=\text{N}\end{smallmatrix}\text{CR}}$$

reviewed[2,355] (see also Chapter 10 of this work). The reaction is not clean and hence the products obtained may be very diverse depending markedly on the ratio of reactants[353,354], temperature[353,354,356], pH[357], the solvent[358] and the nature of the R-group of the imidate[356,359] (145). The reaction would be expected to give an unsubstituted amidrazone (146) as an initial product but this can give rise to secondary products, e.g. a dihydrazidine (147) or dihydrotetrazine (148) either by further reaction with imidate and hydrazine[2] or by self condensation[360]. Moreover, since dihydrotetrazines can isomerise (e.g. on heating) to 4-amino-1,2,4-triazoles[361] (149) or oxidise readily on contact with air to the corresponding 1,2,4,5-tetrazines[361] (150), these compounds may also be found among the reaction products. In addition to the foregoing compounds 1,2,4-triazoles (151) and in the presence of excess hydrazine, dihydroformazans (152) have been isolated[2,355]. There appears to be no report, however, of the isolation of hydrazonate esters (153).

(151)   (152)   (153)

Favourable conditions for amidrazone formation exist in the treatment of an imidate base with anhydrous hydrazine (1:1) in anhydrous ethanol/ether[362]. Sometimes, however, the amidrazone is not isolated but used *in situ* for further synthesis[2,355,363].

Cyclic thioimidates have been reported to react similarly[364] and among the products isolated were compounds of the type 154.

(154)

Although the reaction of excess hydrazine[359] or its hydrate[357,365] with an imidate[365] or thioimidate[364] provides a synthetic route to the 1,2,4,5-tetrazines via their dihydro derivatives[361,366], the authors have shown that amidinium salts give cleaner products than the imidates[365,367].

X = OR or NH_2

Recently, successful syntheses of 3,6-unsymmetrically-disubstituted-1,2,4,5-tetrazines have been reported involving mixed imidate/amidine precursors[368].

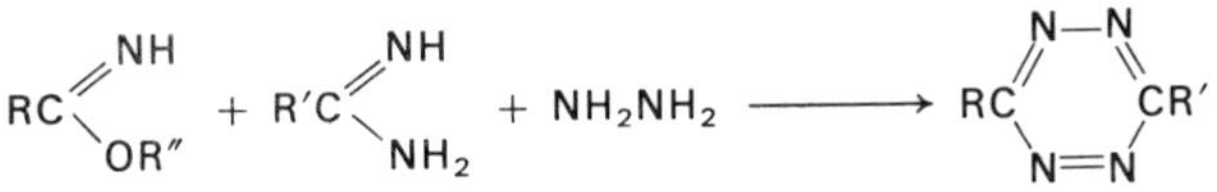

$$RC\overset{NH}{\underset{OR''}{\diagdown}} + R'C\overset{NH}{\underset{NH_2}{\diagdown}} + NH_2NH_2 \longrightarrow RC\overset{N-N}{\underset{N=N}{\diagdown}}CR'$$

Suitably N-substituted-imidates react with hydrazine to give rise to other nitrogen heterocycles[369], e.g. the imidazole[370] **155** or the 1,2,4-

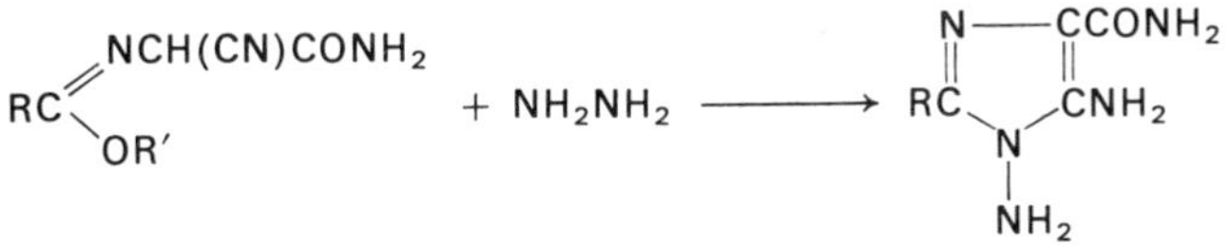

$$RC\overset{NCH(CN)CONH_2}{\underset{OR'}{\diagdown}} + NH_2NH_2 \longrightarrow$$

structure **(155)**

triazoles **157** and **159** from the chloroformimidate[371] **156** and the imidate[372] **158** respectively.

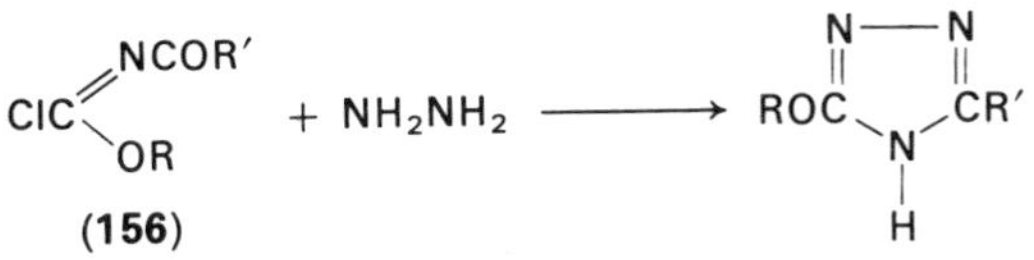

$$ClC\overset{NCOR'}{\underset{OR}{\diagdown}} + NH_2NH_2 \longrightarrow$$

(156)

structure **(157)**

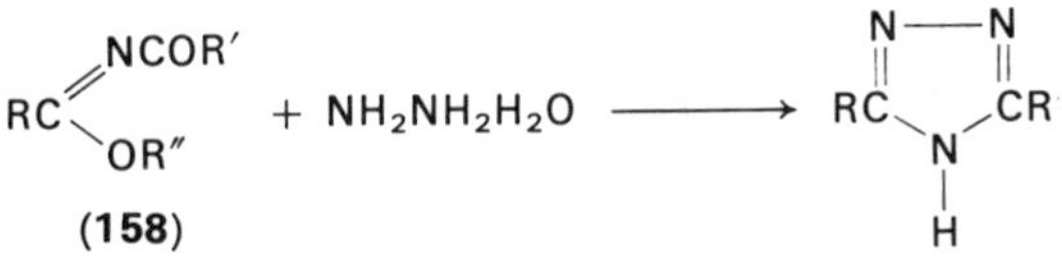

$$RC\overset{NCOR'}{\underset{OR''}{\diagdown}} + NH_2NH_2H_2O \longrightarrow$$

(158)

structure **(159)**

7. Reaction of imidates with monosubstituted hydrazines

Imidate or thioimidate salts react readily with monosubstituted hydrazines[2,355,373-377] to form N^1-substituted amidrazones (**161**) or with excess hydrazine to give formazans (**162**). In addition when for the

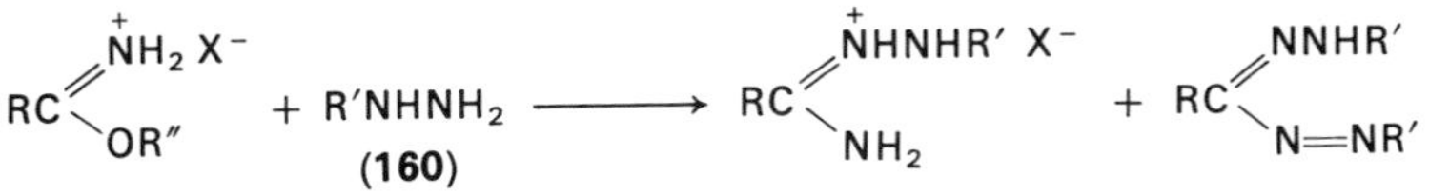

$$RC\overset{\overset{+}{N}H_2\ X^-}{\underset{OR''}{\diagdown}} + R'NHNH_2 \longrightarrow RC\overset{\overset{+}{N}HNHR'\ X^-}{\underset{NH_2}{\diagdown}} + RC\overset{NNHR'}{\underset{N=NR'}{\diagdown}}$$

(160)   **(161)**   **(162)**

hydrazine **160**, R' is an activating group, e.g. alkyl, the reaction can produce 1,4-dialkyl-1,2,4,5-tetrazines (**163**) among other products[378].

$$\underset{\text{HC}}{\overset{+}{\underset{\backslash OEt}{\overset{/NH_2\ Cl^-}{\diagup}}}} + \text{MeNHNH}_2 \longrightarrow$$

(structures for **163**)

$$\text{HC(O)NMeNH}_2$$

(163)

Hydrazonate esters (**164**) have also been recorded as products from the action of imidate salts and monosubstituted hydrazines but are best prepared by other routes[173, 379, 380].

$$\underset{\text{RC}}{\overset{+}{\underset{\backslash OR''}{\overset{/NH_2\ Cl^-}{\diagup}}}} + \text{R'NHNH}_2 \longrightarrow \underset{\text{RC}}{\overset{/NNHR'}{\underset{\backslash OR''}{\diagup}}}$$

(164)

Mono-acyl hydrazines react with imidate bases to give, under mild

$$\underset{\text{RC}}{\overset{/NH}{\underset{\backslash OR''}{\diagup}}} + \text{R'CONHNH}_2 \longrightarrow \underset{\text{RC}}{\overset{/NNHCOR'}{\underset{\backslash NH_2}{\diagup}}} \longrightarrow$$

(165)   **(166)**

conditions[374, 375, 381, 382, 383], N^1-acylamidrazones (**165**) which cyclise readily[2, 355, 374, 382] to the corresponding 3,5-disubstituted-1,2,4-triazoles (**166**). The reaction, however, appears to be sensitive to pH and under more acid conditions can give rise to 1,3,4-oxa-[380, 382] or 1,3,4-thia-diazoles[384–386] (**168**) in the case of the corresponding thio-compounds

$$\underset{\text{RC}}{\overset{+}{\underset{\backslash OR''}{\overset{/NH_2\ Cl^-}{\diagup}}}} + \text{R'C(X)NHNH}_2 \longrightarrow$$

(167)   **(168)**   (X = O or S)

$$\underset{\text{PhC}}{\overset{+}{\underset{\backslash OEt}{\overset{/NH_2\ Cl^-}{\diagup}}}} + \text{PhRP(O)NHNH}_2 \xrightarrow[\text{(R = Me)}]{}$$

(169)

(R = Ph)

$$\underset{\text{PhC}}{\overset{/NNHP(O)Ph_2}{\underset{\backslash NH_2}{\diagup}}}$$

(**167**; X = S). These latter compounds (**168**) probably arise via hydrazonate ester intermediates[380].

Triazoles have also been found among the products of the interaction of the hydrazine (**169**) with benzimidate but the reaction is not general[387]. *N*-Substituted-imidates give rise either to the corresponding N^1,N^3-

$$HC\underset{OR'}{\overset{NAr}{\diagdown}} + Ar'CONHNH_2 \longrightarrow HC\underset{NHNHCOAr'}{\overset{NAr}{\diagdown}}$$

(**170**)

disubstituted-amidrazones[388] (**170**) or, if the imidate carries a reactive group[372,389,390], e.g. CN or COOR, to 1,2,4-triazole derivatives (**171**).

$$HC\underset{OEt}{\overset{NCN}{\diagdown}} + PhNHNH_2 \xrightarrow{C_6H_6} \text{1,2,4-triazole}$$

(**171**)

However, from imidates with more extensive functional groups (**172**), 1,2-dihydro-1,2,4-triazines (**173**) have been reported[370].

$$MeC\underset{OEt}{\overset{NCH(CN)CONH_2}{\diagdown}} + MeNHNH_2 \longrightarrow \text{triazine}$$

(**179**)   (**173**)

8. Reaction of imidates with disubstituted hydrazines

Imidates or their thio-analogues react with 1,1-disubstituted hydrazines

$$RC\underset{SMe}{\overset{\overset{+}{N}H_2\ I^-}{\diagdown}} + Me_2NNH_2 \longrightarrow RC\underset{NH_2}{\overset{NNMe_2}{\diagdown}}$$

(**174**)

to give N^1,N^1-disubstituted-amidrazones[370,376,391,392] (**174**) or under more vigorous conditions and particularly in the presence of ammonium

$$RC\underset{OEt}{\overset{NH}{\diagdown}} + Me_2NNH_2 \longrightarrow RC\underset{NHNMe_2}{\overset{NNMe_2}{\diagdown}}$$

(**175**)

salts[392], dihydroformazans (**175**). Dihydroformazans have also been noted as the products of reaction of hydrazonate esters (**176**) and N^1,N^1-disubstituted hydrazines[108].

$$RC\overset{\overset{+}{N}R^1NHR^2\ I^+}{\underset{SMe}{\diagdown}} \quad + R^3R^4NNH_2 \quad \longrightarrow \quad RC\overset{NNR^3R^4}{\underset{NR^1NHR^2}{\diagdown}}$$

(176)

α-Hydrazino-acids[393, 394] or their derivatives[394] **(177)** undergo condensation reactions with imidates on heating to yield 1,2,4-triazines.

$$RC\overset{NH}{\underset{OEt}{\diagdown}} \quad + (EtOOC)_2C{=}NNH_2 \quad \longrightarrow$$

(177)

1,2-Disubstituted-hydrazines on treatment with imidates[393] yield N^1,N^2-disubstituted-amidrazones **(178)**.

$$PhC\overset{NH}{\underset{OR'}{\diagdown}} \quad + PhNHNHCH_2COOH \quad \longrightarrow \quad PhC\overset{NH}{\underset{N(CH_2COOH)NHPh}{\diagdown}}$$

(178)

The chemistry of amidrazones[335] is discussed further in Chapter 10 of this volume.

9. Reaction of imidates with hydroxylamine

The imino group of an imidate[395] or thioimidate[396] can be replaced by the oximino group when an aqueous solution of hydroxylamine is

$$MeC\overset{NH}{\underset{SMe}{\diagdown}} \quad + NH_2OH \quad \longrightarrow \quad MeC\overset{NOH}{\underset{SMe}{\diagdown}}$$

shaken with an ethereal solution of the imidate. The resultant N-hydroxy-imidates readily undergo O-alkylation of the oximino group[395, 397].

$$RC\overset{NOH}{\underset{OEt}{\diagdown}} \quad + R'X \quad \xrightarrow{MeO^-} \quad RC\overset{NOR'}{\underset{OEt}{\diagdown}}$$

Although amidoximes **(179)** have been reported to have been prepared both by the action of hydroxylamine on an imidate, (equation 24a), or by the action of ammonia on an N-hydroxy-imidate (equation 24b), these compounds **(179)** are more usually synthesised from nitriles or thioamides and hydroxylamine[398].

More recently, Aurich[399], using N-substituted hydroxylamines, obtained as products the nitrones **(180)**, unambiguously available for identification by an alternate route.

$$\text{ArC} \begin{array}{c} \nearrow \text{NH} \\ \searrow \text{OR}' \end{array} + \text{NH}_2\text{OH} \longrightarrow \text{ArC} \begin{array}{c} \nearrow \text{NOH} \\ \searrow \text{NH}_2 \end{array} \qquad (24a)$$
$$(179)$$

$$\text{ArC} \begin{array}{c} \nearrow \text{NOH} \\ \searrow \text{OR}' \end{array} + \text{NH}_3 \longrightarrow \qquad\qquad (24b)$$

$$\text{HC} \begin{array}{c} \nearrow \text{NR}' \\ \searrow \text{OEt} \end{array} + \text{R}''\text{NHOH} \longrightarrow \text{HC} \begin{array}{c} \nearrow \overset{+}{\text{N}} \diagdown \text{R}'' \\ \\ \searrow \text{NHR}' \end{array}$$
$$(180)$$

10. Reaction of imidates with amino acids and their simple derivatives

The interaction of imidates with amino acids and their simple derivatives can follow one of two main courses depending on the conditions of the experiment.

a. *Free amino acids.* The direct interaction of an imidate base with an α-amino acid gives rise to either an amidine which possesses an imino-peptide structure[400] (181) or to an imidazolone (182). The nature of the group R probably influences the pathway. The investigation has been

$$\text{RC} \begin{array}{c} \nearrow \text{NH} \\ \searrow \text{OEt} \end{array} + \text{H}_2\text{NCH}_2\text{COOH} \longrightarrow \text{RC} \begin{array}{c} \nearrow \overset{+}{\text{N}}\text{H}_2 \\ \searrow \text{NHCH}_2\text{COO}^- \end{array} \quad \text{or} \quad \text{RC} \begin{array}{c} \text{N}\text{---}\text{C}=\text{O} \\ \| \qquad | \\ \searrow \text{CH}_2 \\ \text{NH} \end{array}$$
$$(181) \qquad\qquad (182)$$

extended to β-, γ-, δ-, and ε-amino acids and the corresponding amidines were isolated in each case although β-amino acids may give rise to cyclic

$$\text{ArC} \begin{array}{c} \nearrow \text{NH} \\ \searrow \text{OEt} \end{array} + \text{NH}_2\text{CHRCHR}'\text{COOH} \longrightarrow \text{ArC} \begin{array}{c} \text{N}\text{---}\text{C}=\text{O} \\ \diagdown \qquad \diagup \text{H} \\ \qquad \text{C} \diagdown \text{R}' \\ \text{HN}\text{---}\text{C} \\ | \quad \diagdown \\ \text{H} \quad \text{R} \end{array}$$
$$(183)$$

products[401, 402] (183). Cyclisation reactions also take place in the cases of anthranilic acid[403] and of *o*-aminophenylacetic acid[404] (see next page).

b. *Amino acid derivatives in non-acidic conditions.* Ethyl glycinate (184; R = H) reacts under mild, non-acidic conditions with ethyl phenyl-acetimidate to yield 2-benzyl-4(5)-imidazolone[405] (185, R = H). The use of, for example, diethyl α-aminomalonate[2] in place of glycine gives rise

to 5(4)-carbethoxy-4(5)-imidazolones (185; R = COOEt). When the condensation is carried out in a ketonic solvent such as acetone the corresponding 5(4)-ylidene derivative is obtained[406] and this condensation reaction has been suggested as a synthetic route from glycine[407] to more highly substituted α-amino acids (equation 25).

(25)

Thioimidates may successfully replace their oxygen analogues in these reactions[2].

c. *Amino acid derivatives in the presence of acid.* Based on the initial observation by Schmidt[173], it has since been shown, with few exceptions[408], that amino acid derivatives such as esters or amides react with imidates in the presence of one equivalent of acid to yield, with loss of ammonia, *N*-substituted-imidates[2, 409, 410].

$$\text{MeC}\underset{\text{OEt}}{\overset{\text{NH}}{\diagdown}} + \text{EtOOCCH}_2\overset{+}{\text{N}}\text{H}_3 \ \text{Cl}^- \longrightarrow \text{MeC}\underset{\text{OEt}}{\overset{\text{NCH}_2\text{COOEt}}{\diagdown}}$$

$$\text{MeC}\underset{\text{OR}'}{\overset{\overset{+}{\text{N}}\text{H}_2 \ \text{Cl}^-}{\diagdown}} + \text{NH}_2\text{CN} \longrightarrow \text{MeC}\underset{\text{OR}'}{\overset{\text{NCN}}{\diagdown}}$$

II. Reactions of imidates with proteins and related smaller molecules

Since the previous review on imidates[2], this field of work has come into prominence. Credit for this must go in part to Hunter and Ludwig[411] who demonstrated the greater reactivity in this field of imidates over the closely related O-alkyl iso-ureas. These workers also found that, of all the reactive groups present in proteins, only the amino groups reacted with imidates in aqueous solution. The rate of reaction was found to be strongly dependent on pH (see also reference 293) giving a maximum rate on the pH scale related to the nature of the amine and the imidate. This

$$\text{RC}\underset{\text{OMe}}{\overset{\text{NH}}{\diagdown}} + \text{NH}_2 \text{ protein} \longrightarrow \text{RC}\underset{\text{NH protein}}{\overset{\text{NH}}{\diagdown}}$$

finding permitted the rate of imidate attack at an α-amino group as against an ε-amino group to be varied by the appropriate choice of pH. However it is also possible to cause amidination to take place at both these centres, e.g. insulin/imidate reactions carried out in the pH range 7–10 resulted in the complete blocking of α- and ε-amino groups. Although amidination alters the protein it does not *per se* alter its charge and hence has minimal effect on the conformation of the protein[412, 413]. In addition, Ludwig and Byrne[414] were able to show that the amidination process could be reversed by treating the modified protein with ammonia/acetic acid at about pH 11·3, i.e. under conditions which will not normally break peptide bonds. Moreover, as amidination of an ε-amino group prevents tryptic digestion of the adjacent peptide bond[411, 412], this process has potential for the stepwise degradation of proteins.

Further, it was suggested that di-imidates might be used as reagents for the modification of proteins by cross-linking[411, 415], and this process was later shown to be applicable selectively[416]. The methodology of this cross-linking of proteins has been discussed[417, 418].

Ethyl acetimidate[412] is often the reagent of choice for the amidination because of its convenient molecular size and α-lactalbumin[419], bovine pancreatic ribonuclease A[413] and γ-G-globulin[420] have been treated with that reagent. However, other imidates have also been used[419, 421] and it has been suggested that the fluorescent[127] imidate (186) or the coloured[422] imidate (187) might be useful in this respect.

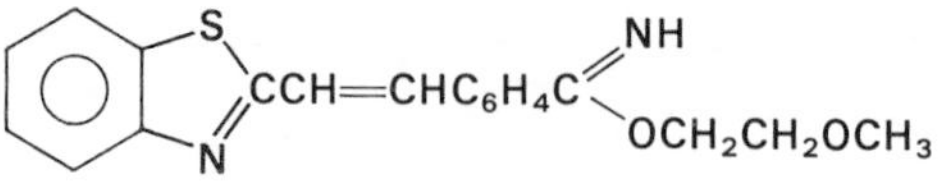

(186)

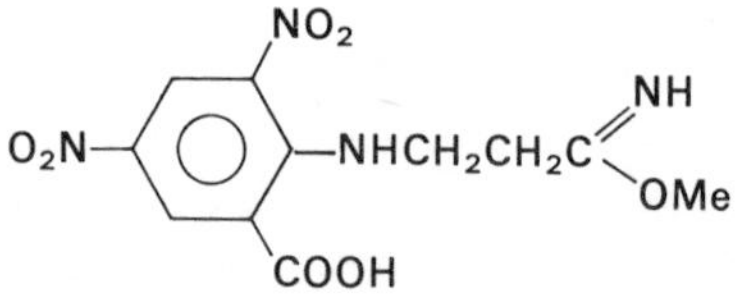

(187)

For the purpose of cross-linking proteins, dialkyl malondi-imidate[415], suberdi-imidates[416, 423–425] and adipdi-imidates[426] have all been used. Alteration of the di-imidate chain length may well prove useful in the formation of modified proteins with different properties or give some idea of the availability or relative position of amino groups within a protein.

12. Formation of imidazoles, imidazolones and imidazolines

In addition to certain reactions already described in Section IV, H, 10 and which give rise directly to 4(5)-imidazolones, these compounds (188) can be derived by the action of ammonia[427] on N-substituted-imidates of

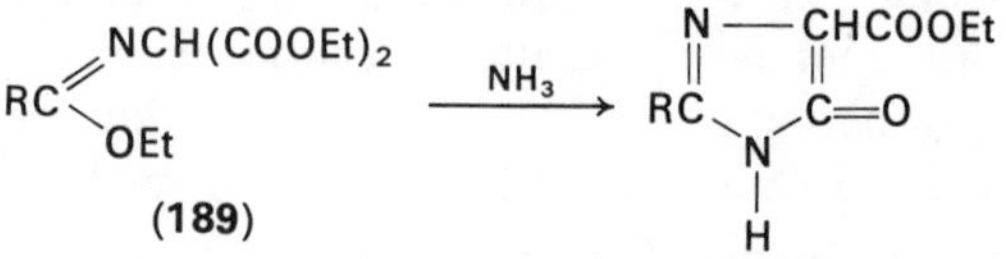

(189)

(188)

type 189. Other related imidate derivatives (190) which do not possess carbethoxy groups react with amines[428] or hydrazines[369] (191; R″ = Me_2N) to afford directly either 1,2,5-trisubstituted (192; R = H) or 1,2,4,5-tetrasubstituted imidazoles[2].

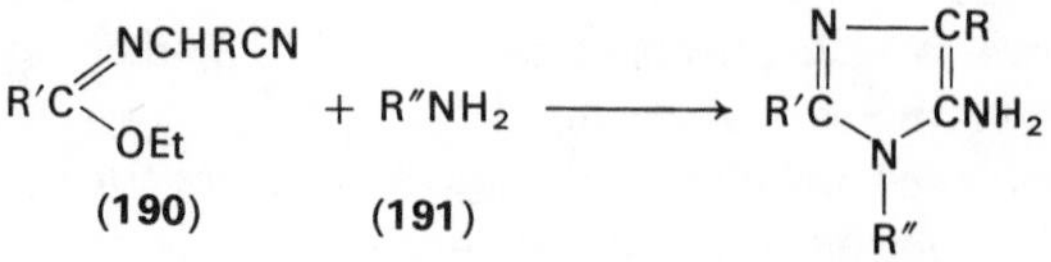

(190)   (191)

(192)

α-Amino-aldehydes or their acetals react in neutral solution with a wide range of imidates[429, 430] to give first of all, amidines (193) and then by cyclisation in acid, the imidazoles 194, but α-amino-ketones[431] tend to

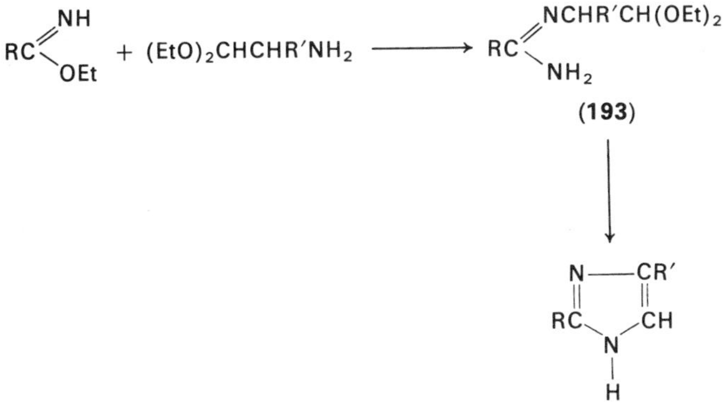

$$RC\underset{OEt}{\overset{NH}{\diagup}} + (EtO)_2CHCHR'NH_2 \longrightarrow RC\underset{NH_2}{\overset{NCHR'CH(OEt)_2}{\diagup}}$$

(193)

(194)

give a mixture of two products, a neutral one identified as an oxazole (195) and a basic product—an imidazole (196).

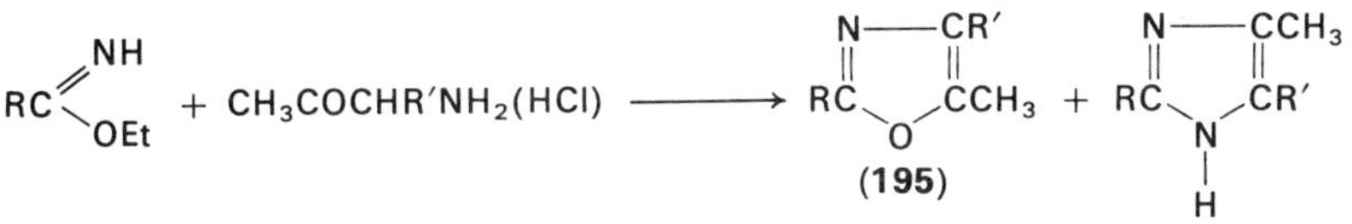

$$RC\underset{OEt}{\overset{NH}{\diagup}} + CH_3COCHR'NH_2(HCl) \longrightarrow (195) + (196)$$

(195)

(196)

An alternate route leading to imidazoles is illustrated by the following reaction sequence involving formylation of an N-substituted imidate[2, 432, 433].

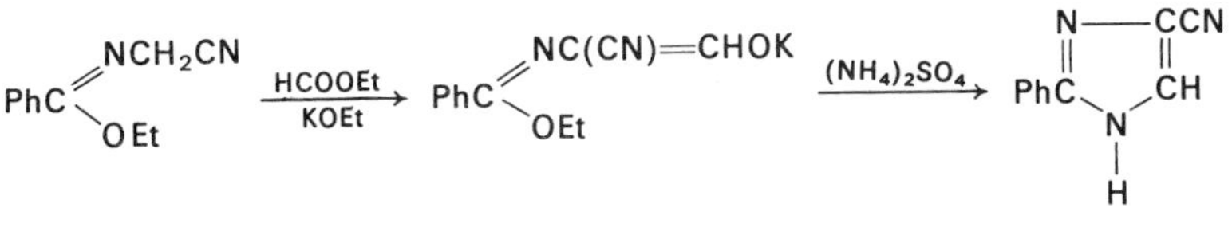

$$PhC\underset{OEt}{\overset{NCH_2CN}{\diagup}} \xrightarrow[KOEt]{HCOOEt} PhC\underset{OEt}{\overset{NC(CN)=CHOK}{\diagup}} \xrightarrow{(NH_4)_2SO_4}$$

Imidazolines, a biologically important series of compounds, are readily accessible from the reaction of an imidate salt and a 1,2-alkyldiamine under mild conditions[2, 434–436]. The reaction has been shown by Bristow[434] to proceed via an amidinium intermediate (197) which in the particular example illustrated was isolated (equation 26). Compound 198 has also been prepared in an optically active form by a similar reaction sequence (equation 26) using (−)-mandelonitrile (from amygdalin) as starting material[35].

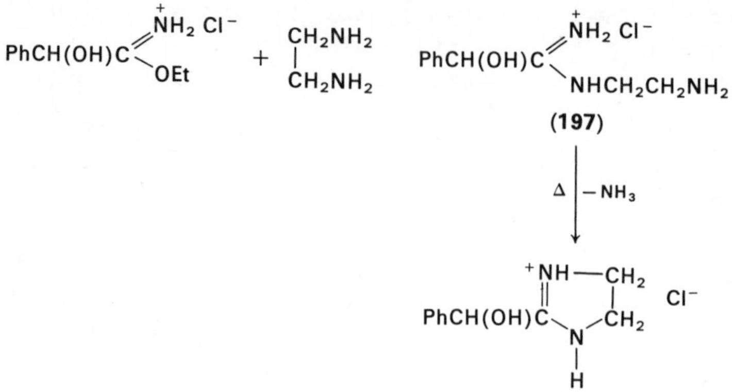

(197)

$$\Delta \;\Big|\; -NH_3 \qquad\qquad (26)$$

(198)

A variation of this general procedure is to be found in the use of 1,2-diaminocyclohexane[437].

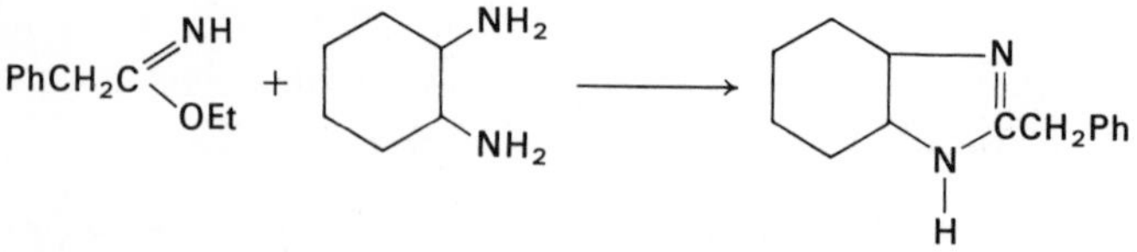

13. Formation of benzimidazoles

o-Phenylenediamines condense with imidates or thioimidates in the presence of one or two equivalents of acid to yield 2-substituted-benz-

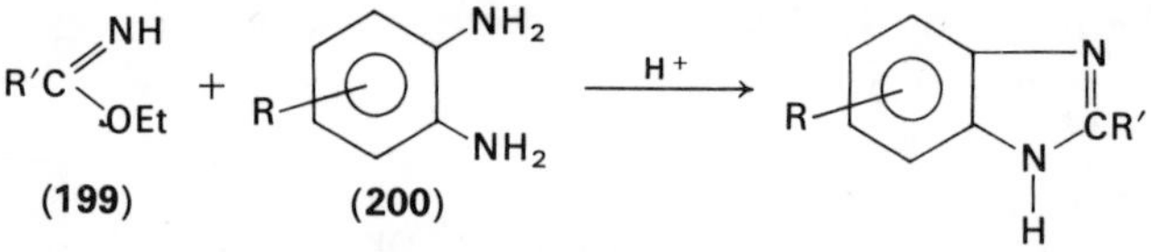

(199)   **(200)**

imidazoles[2, 438]. Although the method is least successful when, for the compound **199**, R' = alkyl or the diamine **200** carries an electronegative substituent in the 4-position[439], the synthesis has found fairly general application[440-442], and has been adapted to the preparation of poly-benzimidazoles[443, 444] of type **201**.

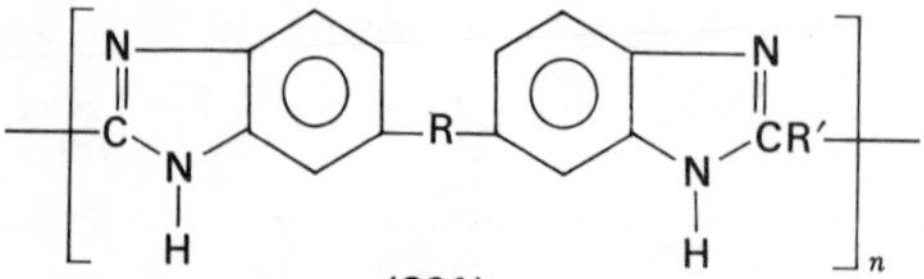

$(R = CH_2 \text{ or } CMe_2, \; R' = C_6H_4)$

(201)

N-Substituted-benzimidazoles (**202**) are similarly available from *N*-monosubstituted-*o*-phenylenediamines[438, 445–447] but at times it has been noted[438] that this reaction can lead to amidines (**203**).

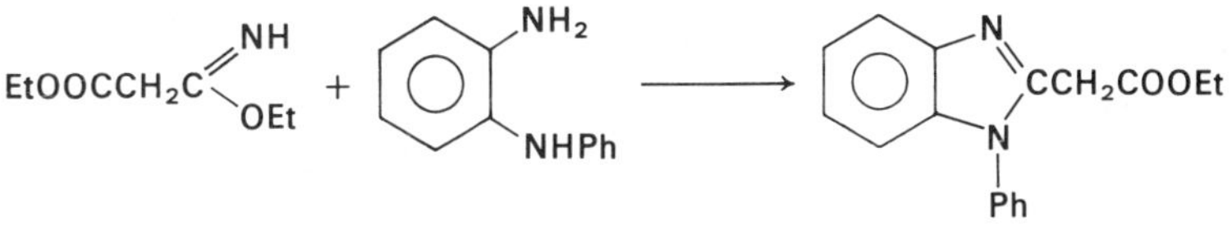

(**202**)

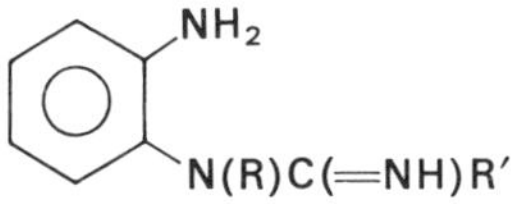

(**203**)

14. Formation of oxazoles and oxazolines

Imidates condense with α-amino acids in the presence of an equivalent of acid to yield *N*-substituted imidates[2] (**204**; see Section IV, H, 10). Cornforth and Cornforth[433] have utilised this reaction by formylating the intermediate **204** and have thus obtained the parent member of this

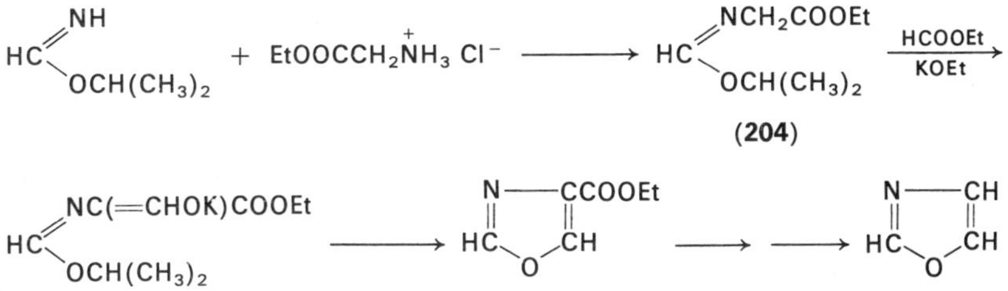

(**204**)

(**205**)

series—oxazole (**205**). The method is of fairly general application[2] and, for example, can be adapted to give 4-cyano-oxazoles by the utilisation of aminoacetonitrile in place of the amino ester[448].

The dithio-diimidate **206** condenses with aryl aldehydes to give exclusively 5-amino-oxazoles (**207**) or their benzylidene derivatives[449] and not the corresponding imidazoles **208**.

There is also an isolated report which indicates that the imidate **209** forms on basification, 2-phenyl-4-methyloxazole[19].

2-Oxazolines, which are in fact cyclic imidates, can be prepared either by the ring closure of imidates prepared from β-halo-alcohols or by the reaction of imidates with alkanolamines[2, 450]. In the first of these reactions

$$ArCHO + \left[C(=NH)SR\right]_2 \xrightarrow{\Delta}$$

(206)

(207)

$$\begin{array}{c} N\text{----}CSR \\ \parallel \qquad \parallel \\ ArC \qquad CNH_2 \\ \diagdown O \diagup \end{array}$$

(208)

$$\begin{array}{c} N\text{----}CSR \\ \parallel \qquad \parallel \\ ArC \qquad CSR \\ \diagdown N \diagup \\ \mid \\ H \end{array}$$

$$\underset{(209)}{PhC\diagdown \underset{OCH_2C\equiv CH}{\overset{\overset{+}{N}H_2 \ Cl^-}{\diagup}}} \xrightarrow{base} \begin{array}{c} N\text{----}CMe \\ \parallel \qquad \parallel \\ PhC \qquad CH \\ \diagdown O \diagup \end{array}$$

it has been shown by Wislicenus and Körber[451] that the oxazoline (210) is in fact an intermediate in the formation[313, 452] of the corresponding

$$\underset{OCHR'CHR''X}{RC\diagdown \overset{NH}{\diagup}} \longrightarrow \begin{array}{c} \overset{+}{HN}\text{----}CHR'' \\ \parallel \qquad \mid \\ RC \qquad CHR' \\ \diagdown O \diagup \end{array} \longrightarrow \underset{NHCHR''CHR'X}{RC\diagdown \overset{O}{\diagup}}$$

(210)   (211)

X = Cl or Br but not F (see reference[75])

amide 211. Brown and Wetzel[313] have shown that for perfluoro-alkyl systems the reaction can best be carried out as a one-step base catalysed process from the nitrile.

$$R_FCN + HOCH_2CH_2Cl \xrightarrow{Me_3N} \begin{array}{c} N\text{----}CH_2 \\ \parallel \qquad \mid \\ R_FC \qquad CH_2 \\ \diagdown O \diagup \end{array}$$

The alternate route to 2-oxazolines involves the treatment of an imidate or thioimidate (or its salt) with an ethanolamine[2, 450].

$$\underset{OR'}{RC\diagdown \overset{NH}{\diagup}} + \underset{CH_2NH_2}{\overset{R''CHOH}{\mid}} \longrightarrow \begin{array}{c} N\text{----}CH_2 \\ \parallel \qquad \mid \\ RC \qquad CHR'' \\ \diagdown O \diagup \end{array}$$

The scope of this reaction has been extended by the use of (±)-nor-ephedrine[453], ethyl 11-amino-10-hydroxy-undecanoate[454] and the cis-

and *trans*-2-aminocyclohexanols[455]. In this last case[455], the *cis*-compound gave exclusively the corresponding 2-phenyl-oxazoline from ethyl benzimidate whereas the *trans*-compound gave a mixture of the amidine

(212)   **(213)**

(212) and the oxazoline **(213)**. In addition, *trans*-2-aminocyclopentanol fails to cyclise whereas the *cis*-compound readily forms an oxazoline[455]. This work[455] along with that on substituted serine esters[456] has shown clearly that the reaction normally proceeds with retention of configuration at both asymmetric centres and application of this has been made in the elucidation of the stereochemistry of elaiomycin[457]. Oxazoline chemistry, based on imidate intermediates, has also been widely utilised in the synthesis of the important antibiotic, chloromycetin **(214)**, and this work, much of it in the form of patents, is well documented[450].

$$Ar = p\text{-}O_2NC_6H_4$$

(214)

15. Formation of benzoxazoles

The condensation of *o*-aminophenols with imidates gives rise to 2-substituted benzoxazoles[2]. Recent applications[458] of this synthesis

have led to 2-carbethoxymethyl-[459] and 2-styryl-benzoxazoles[43]; dibenzoxazoles[460] **(215)** and polybenzoxazoles derived from terephthalimidate[444] and perfluoroalkyl-di-imidates[461] (e.g. compound **216**).

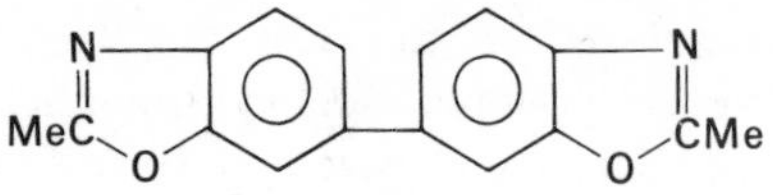

(215)

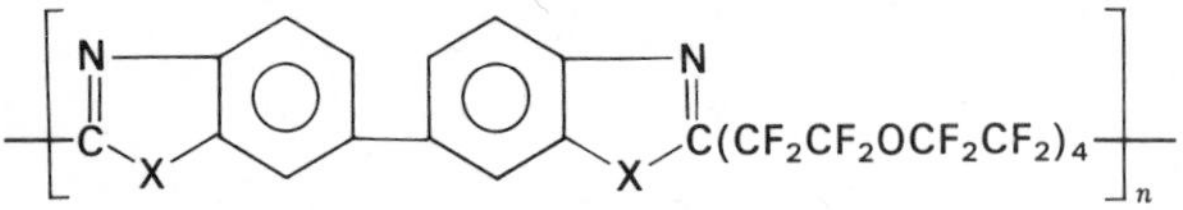

(216)   (X = O or S)

16. Formation of thiazoles, benzothiazoles, isothiazoles and their reduced derivatives

Imidate salts react with α,β-mercapto-amine salts to yield thiazolines[2]— the bases giving only unresolvable oils[462]. However, the reaction is not entirely general and thioimidates in particular may fail to react[463]. The reaction was used extensively in the early studies in the penicillin field[464] but has not been applied recently to any great extent.

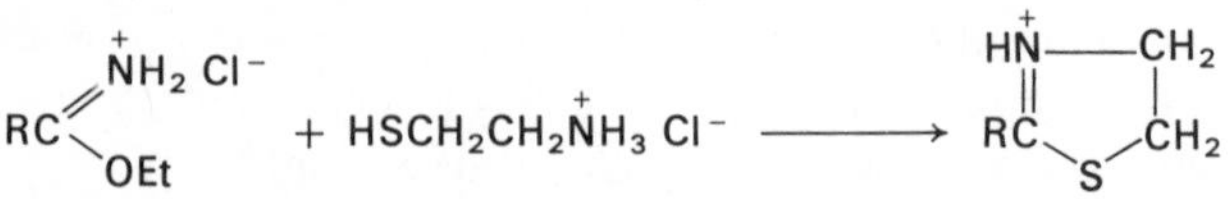

2-Substituted thiazolin-4-ones arise from the cyclisation of imidates of the type **217** on treatment with base[465] or by heating in an inert solvent[466]. This reaction is most successful with aryl imidates. The closely related ketonic compounds (**218**, R = alkyl or aryl) cyclise similarly but yield 2,4-disubstituted thiazoles[467].

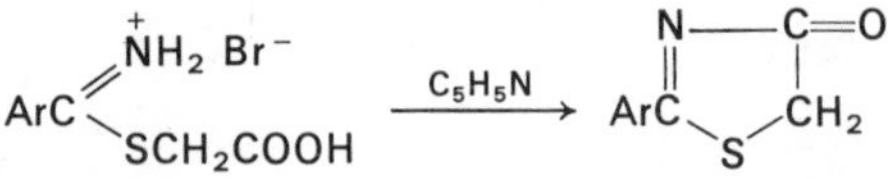

(217)

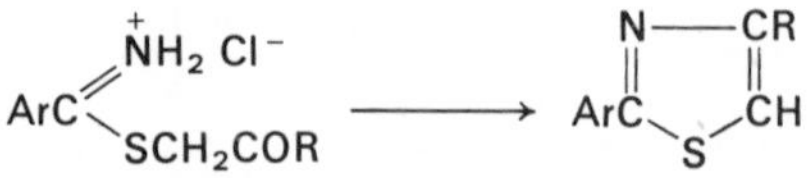

(218)

Benzothiazoles[468] can be prepared by the action of 2-aminothiophenols with imidates and the reaction has been extended to give polybenzothiazoles[461] (compound **216**, X = S). The formation of another fused thiazole system[469] is exemplified by reaction (27).

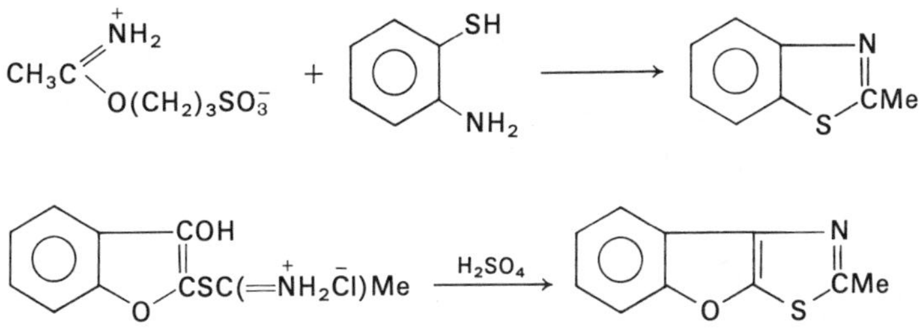

$$CH_3C \overset{\overset{+}{N}H_2}{\underset{O(CH_2)_3SO_3^-}{\diagdown}} \quad + \quad \left[\text{benzene ring with SH and NH}_2\right] \quad \longrightarrow \quad \left[\text{benzothiazole ring, N, S}\right]\text{CMe}$$

$$\left[\text{benzene ring with COH and } CSC(\overset{+}{=}NH_2\bar{Cl})Me \text{ on O}\right] \quad \xrightarrow{H_2SO_4} \quad \left[\text{benzofuran-thiazole ring}\right]\text{CMe} \qquad (27)$$

Although less extensively studied, isothiazoles have been derived from the action of a halogen on an imidate possessing a suitable sulphur substituent (e.g. —S—S— or C=S) in the β-position[320, 470].

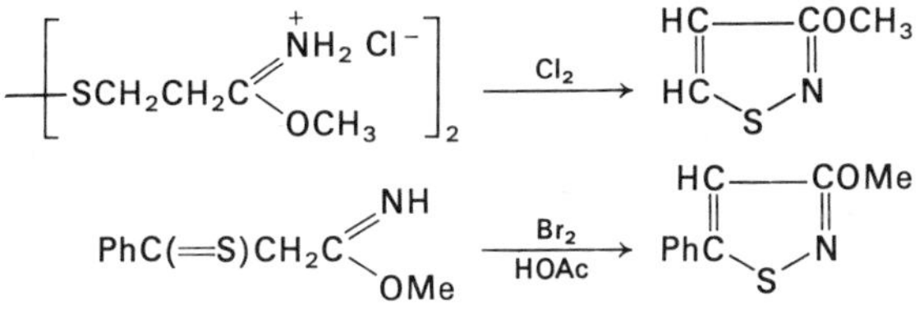

$$\left[\begin{array}{c} -\text{SCH}_2\text{CH}_2\text{C} \overset{\overset{+}{N}H_2 \ Cl^-}{\underset{OCH_3}{\diagdown}} \end{array} \right]_2 \quad \xrightarrow{Cl_2} \quad \begin{array}{c} HC\text{——}COCH_3 \\ \| \quad\quad \| \\ HC \diagdown_S\diagup N \end{array}$$

$$PhC(=S)CH_2C \overset{NH}{\underset{OMe}{\diagdown}} \quad \xrightarrow[HOAc]{Br_2} \quad \begin{array}{c} HC\text{——}COMe \\ \| \quad\quad \| \\ PhC \diagdown_S\diagup N \end{array}$$

17. Formation of oxadiazoles

Both 1,2,4- and 1,3,4-oxadiazoles have been prepared from imidates although the latter have been more extensively studied.

1,2,4-Oxadiazoles have been formed from the reaction of imidates with either aryl nitrile oxides[471] or with hydroxamoyl halides[472].

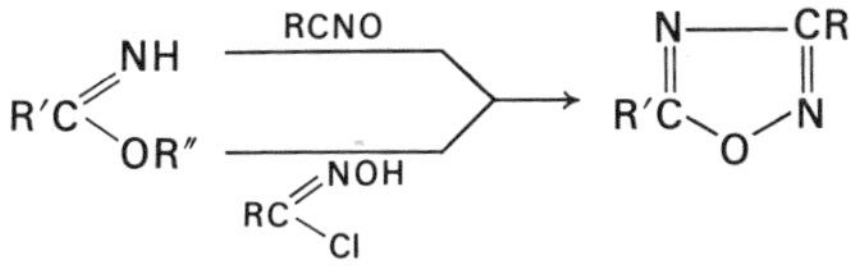

$$R'C \overset{NH}{\underset{OR''}{\diagdown}} \quad \overset{RCNO}{\underset{RC \diagdown_{Cl}^{NOH}}{\Big\rangle}} \quad \longrightarrow \quad \begin{array}{c} N\text{——}CR \\ \| \quad\quad \| \\ R'C \diagdown_O\diagup N \end{array}$$

On the other hand, the following reaction sequence (28) affords a convenient route to either monosubstituted or 2,5-unsymmetrically disubstituted-1,3,4-oxadiazoles[163, 473, 474].

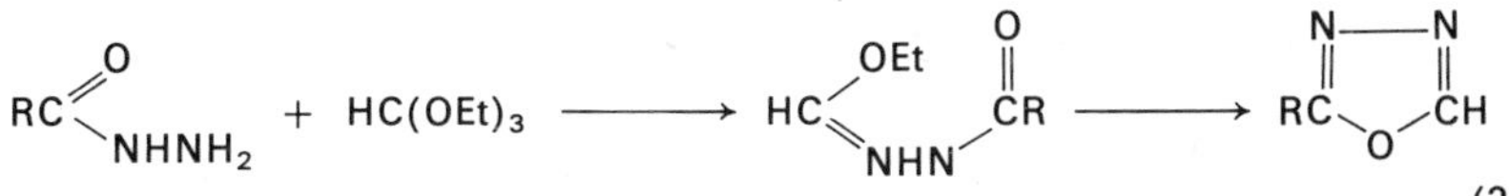

$$RC \overset{O}{\underset{NHNH_2}{\diagdown}} \quad + \quad HC(OEt)_3 \quad \longrightarrow \quad HC \overset{OEt}{\underset{NHN}{\diagdown}} \overset{O}{\underset{}{\overset{\|}{C}}}R \quad \longrightarrow \quad \begin{array}{c} N\text{——}N \\ \| \quad\quad \| \\ RC \diagdown_O\diagup CH \end{array}$$

$$(28)$$

Alternatively, the acid hydrazide can be treated with an imidate salt[472] above 100°C when cyclisation occurs; this latter route has been used to prepare poly-1,3,4-oxadiazoles[382, 475]. However the condensation appears

$(CH_2)_n(CONHNH_2)_2 + p\text{-}C_6H_4[C(\!=\!NH)OEt]_2 2HCl \xrightarrow[C_5H_5N]{\Delta}$

to be sensitive to conditions such as pH, for condensation of an imidate base and a hydrazide gives rise to a poly-1,2,4-triazole[382].

18. Formation of thiadiazoles

Imidate salts react with compounds of the general type NH_2NHCSR ($R = SH$[384] or NH_2[385]) to yield 1,3,4-thiadiazoles (**219**).

(**219**)

In addition, certain specific imidates, e.g. oxaldiimidates or cyano-formimidates when treated with sulphur dichloride have been shown to furnish the isomeric 1,2,5-thiadiazoles[476].

19. Formation of 1,2,4-triazoles

1,2,4-Triazoles can be synthesised from imidates usually via amidrazone intermediates which need not be isolated[355] (see Chapter 10).

20. Formation of tetrazoles

Tetrazoles may be prepared from imidates either directly[2, 24, 477, 478] through the action of hydrazoic acid or more usually through the inter-action of an amidrazone intermediate with nitrous acid[355, 479]. An imide azide (**220**) is recognised as the intermediate in these reactions[479].

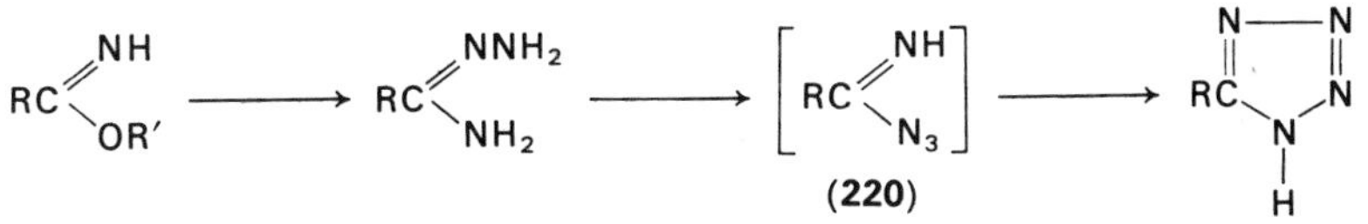

(**220**)

21. Formation of azines

In the past few years there has been a marked increase in the study of this area of imidate chemistry. This has led to the production of, in par-ticular, diazines, triazines and tetrazines, some of which exemplify new or unusual substitution patterns for these compounds.

a. *Pyridines.* There are occasional reports of the synthesis of pyridine derivatives from imidates[2, 480, 481], e.g. the quinoline **221** results from the treatment of the imidate **222** with diethyl malonate[482]. However, no systematic study appears to have been made of pyridine synthesis from imidates.

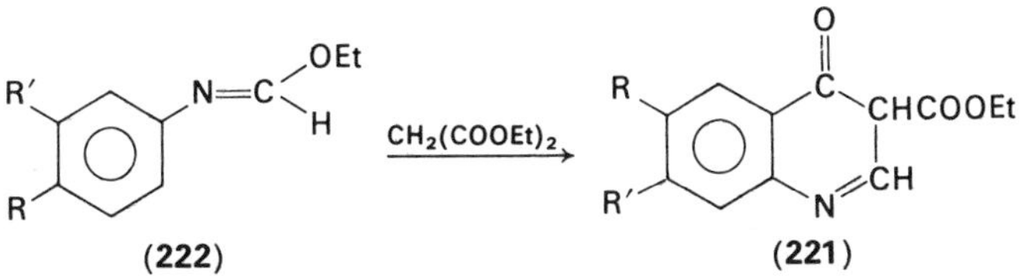

(**222**)   (**221**)

b. *Pyrimidines.* Three main routes to pyrimidines are available from imidates. In the classical procedure[337, 483] the imidate is converted into an amidine which, in turn, is reacted with e.g. a β-keto-ester, β-diketone or malonic acid derivative to give the pyrimidine.

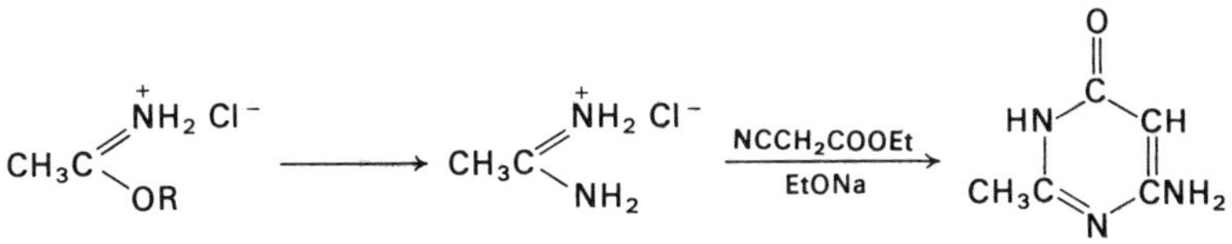

The second procedure involves the condensation of a β-amino acid or ester with an imidate and is illustrated by equation (29). This method

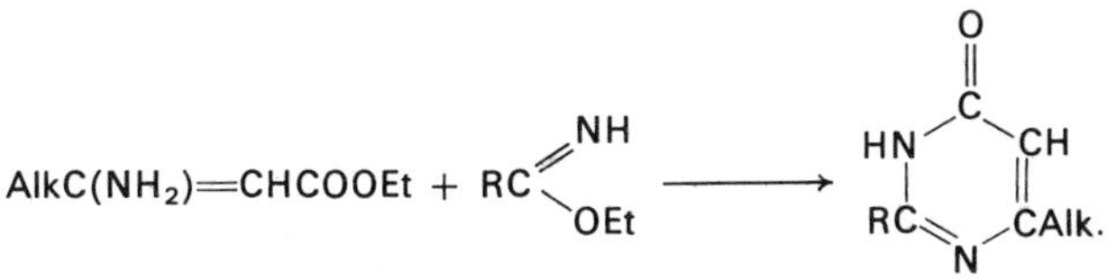

$$\text{AlkC(NH}_2\text{)=CHCOOEt} + \text{RC} \overset{\text{NH}}{\underset{\text{OEt}}{}} \longrightarrow \text{(29)}$$

has been used extensively by Ried and his co-workers[484] to form pyrimidones fused to other ring systems, e.g. the compounds **223a** → **223c** from the appropriate aminopyridine carboxylate esters[485] among others[486-489].

(223a)   (223b)   (223c)

Modifications of this procedure are illustrated[490] by equation (30) and also include the use of β-aminopropionitrile to give 6-amino-4,5-dihydropyrimidines[491] and of saturated β-amino acids to give tetrahydro-

$$(30)$$

pyrimidine derivatives. In this latter case, reaction at lower temperatures yields amidines in place of the pyrimidines[401, 401, 402].

The third main route to pyrimidines requires the use of a compound having an amino group adjacent to either a cyano or an amide group—these often being substituents of a ring system. The imidate is first formed from the amine by treatment with an ortho ester and the cyclisation is then promoted either by heating or with ammonia[492-496]. Two examples are chosen to illustrate this procedure[497, 498]; in the second of these the ortho ester is also used to build the imidazole nucleus of the purine product.

H$_2$NOCC(NH$_2$)=C(NH$_2$)$_2$ + PhC(OMe)$_3$ ⟶

Pyrimidines have also been synthesised by the interaction of an acyl halide and diethyl malondiimidate (or its thio analogue)[499] and from the reaction of s-triazine with imidates possessing an active methylene group[500, 501].

R = CN or COOEt

The synthesis of various fused pyrimidines from the condensation of lactim ethers with amidines or guanidines is discussed by Glushkov and Granik in their recent review[3].

c. *Pyridazines.* Fused pyridazine derivatives (**224**) have been obtained from the action of hydrazine hydrate on lactim ethers but no simple pyridazines derived from imidates have been noted[502].

d. *1,3,5-Triazines.* Direct trimerisation of imidates gives rise to 2,4,6-symmetrically trisubstituted-1,3,5-triazines. The reaction appears to

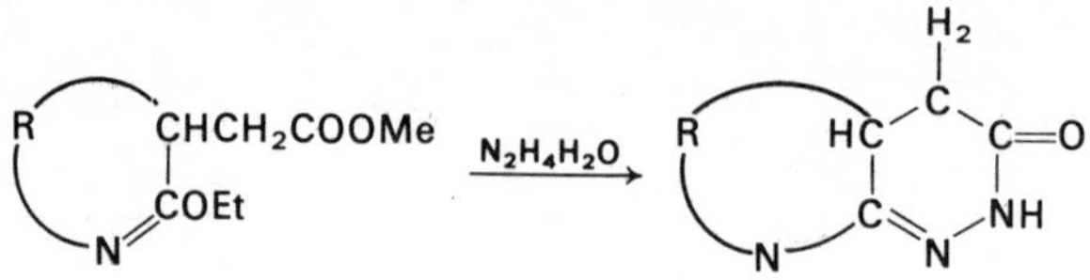

(224)

go readily with an imidate base in the presence of some acid[100,503-506] such as acetic or trifluoroacetic acid.

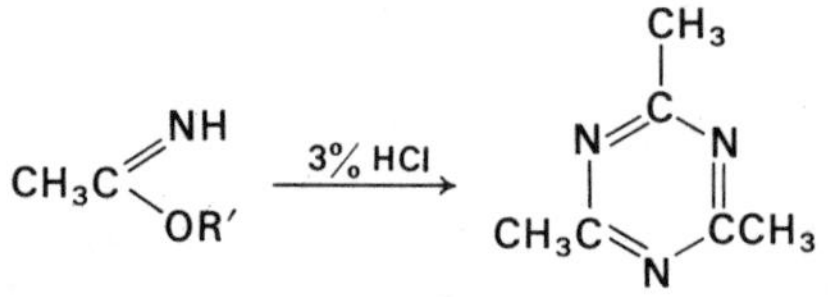

In the case of some cyclic imidates (225) attempts to form the free bases resulted in trimerisation reactions and hence in the isolation of the corresponding triazines[62] (226).

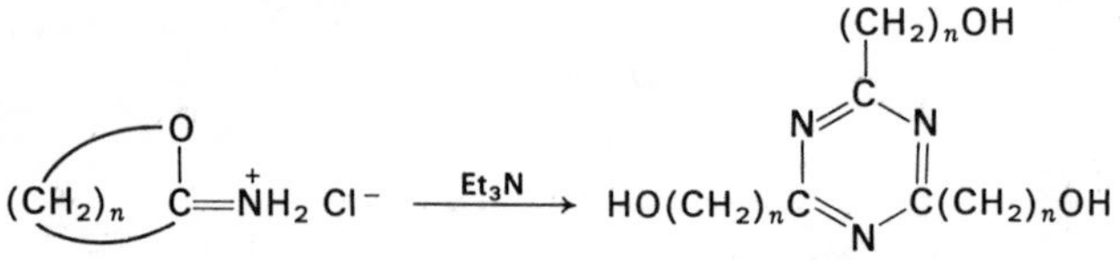

(225)   **(226)**

$(n = 3-5)$

2,4,6-Unsymmetrically trisubstituted-1,3,5-triazines of the type 227 have been formed from the co-trimerisation of two imidates[507], or better, from the interaction of an amidine salt with a lower aliphatic imidate[508]. Schaefer found in this latter case[508] that the predominant product had one substituent derived from the amidine and two from the imidate. In

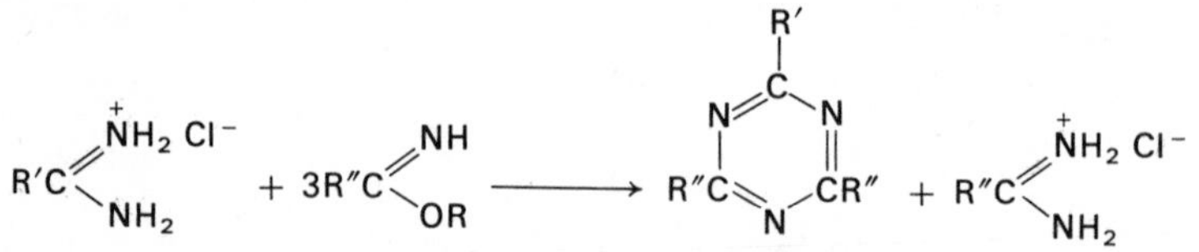

(227)

much the same way, the condensation of N-cyano-acetimidate with either an amidine or amidoxime led to the isolation of a 2-amino-1,3,5-triazine or its 1-N-oxide, respectively. A further modification of these procedures is to be found in the use of an acyl derivative of an imidate; this permits the synthesis of 1,3,5-triazines with either two[509] or three[510,511] different substituents. The reaction however is not clean as transacylation reactions (imidate–amidine) take place.

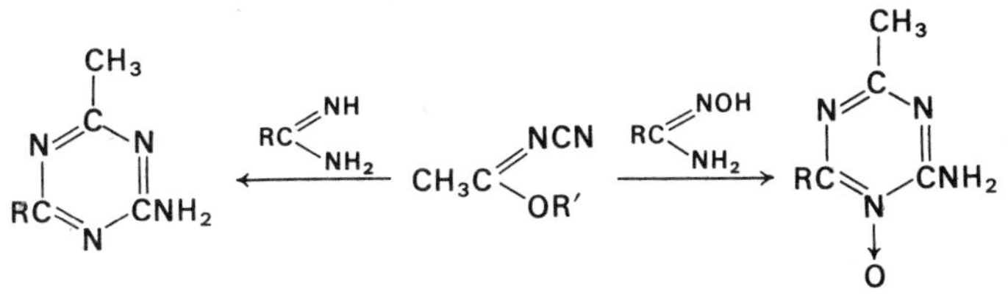

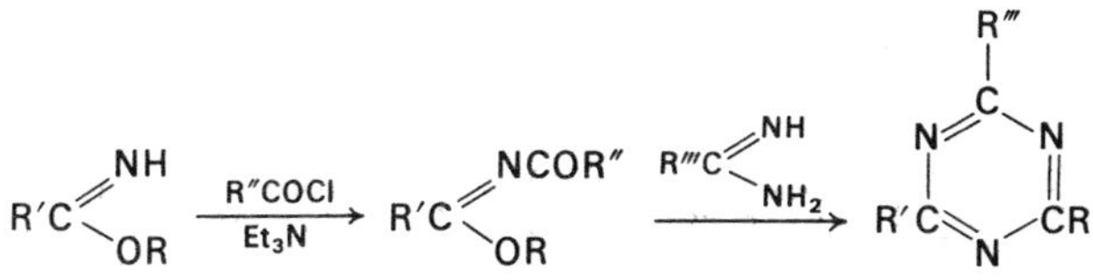

A route to monosubstituted-1,3,5-triazines has been devised through the interaction of 1,3,5-triazine with an imidate. Compound (228) is proposed as an intermediate of this reaction[512].

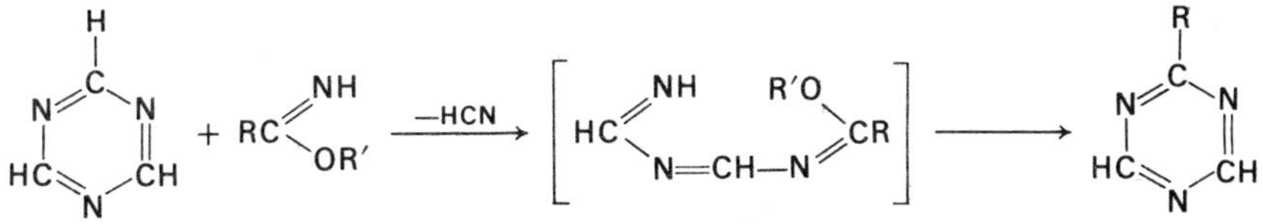

(228)

e. *1,2,4-Triazines.* Several direct but little used routes to 1,2,4-triazines have been reported. For example, s-tetrazines undergo Diels–Alder addition reactions with a variety of compounds possessing double bonds; the addition is then followed by loss of nitrogen in a retro Diels–Alder reaction. Imidates have been used in this reaction to form 1,2,4-triazines[513].

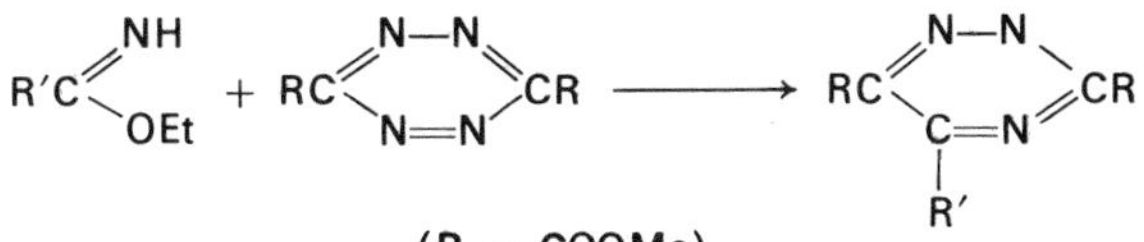

(R = COOMe)

A 4-N-oxide derivative of a 1,2,4-triazine has been formed from the interaction of the diketone derivative (229) with an imidate salt[514] and

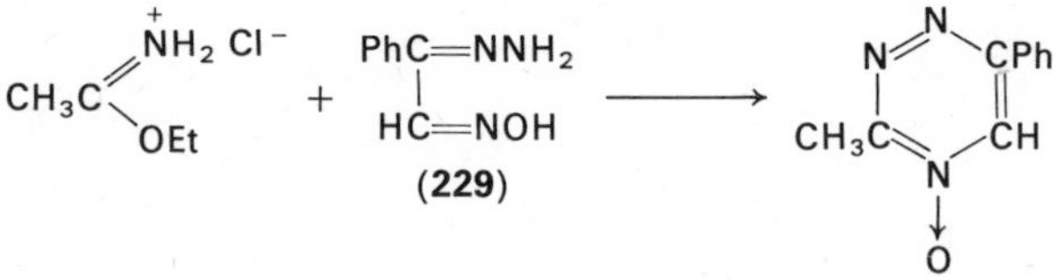

(229)

nucleophilic attack[515] by hydrazine on the compounds of the type (230) gives rise to pyrimido[5,4e]-1,2,4-triazines (231).

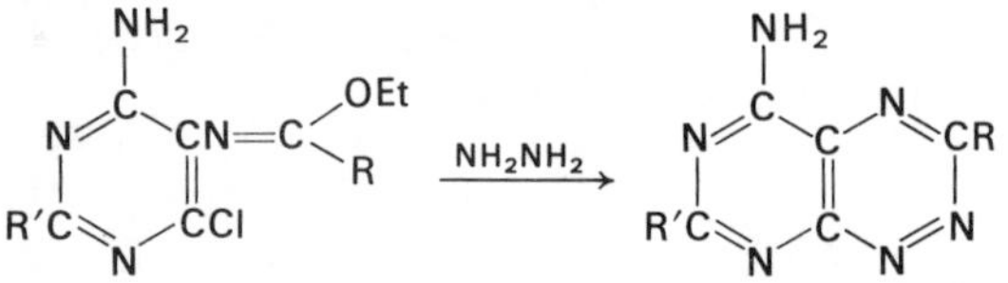

(230)   (231)

The most general route to 1,2,4-triazines is, however, via cyclisation reactions of amidrazones which already have the basic nitrogen structure of a 1,2,4-triazine. These reactions, typified by the following equation, are discussed in detail in Chapter 10 and in previous reviews[355, 394].

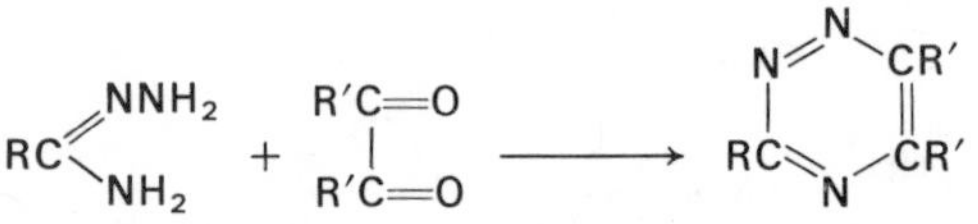

f. s-*Tetrazines.* Excess hydrazine (usually as the hydrate) reacts with imidate salts under mild conditions to yield 1,4-dihydro-s-tetrazines (232) which are readily oxidised, e.g. by nitrous acid to the parent tetrazine[2, 359, 366, 367, 516]. It has been suggested, however, that amidines give

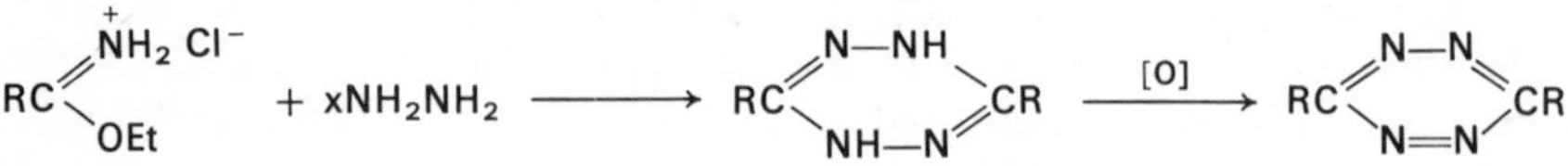

(232)

cleaner products[367, 517]. Polymeric 1,4-dihydro-s-tetrazines (233), useful as precursors of polymeric 4-aminotriazoles (234), have been synthesised in this way from diimidates[518].

Mono-alkyl hydrazines such as methylhydrazine react similarly with ethyl formimidate hydrochloride to yield e.g. 1,4-dimethyl-1,4-dihydro-s-tetrazine[378] among other products but arylhydrazines yield *N'*-aryl-

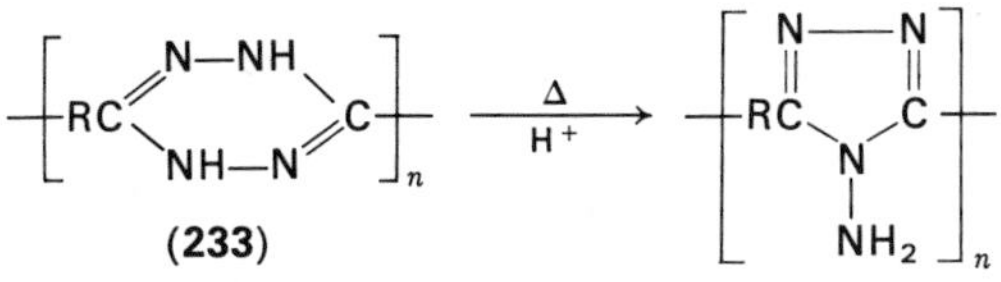

(233) → (234)

amidrazones due to the decreased activity of the aryl nitrogen of the hydrazine.

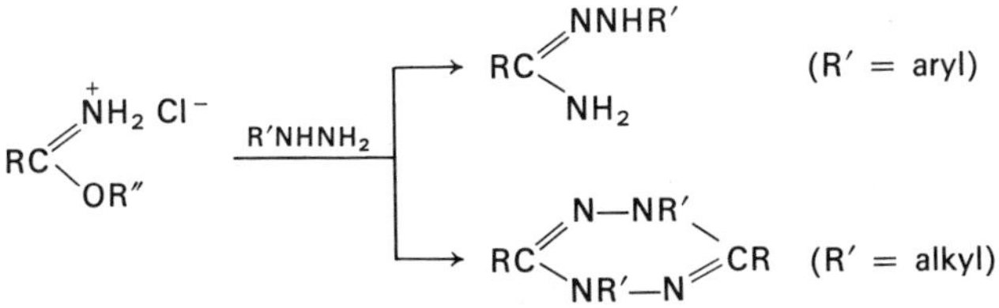

Unsymmetrically 3,6-disubstituted-*s*-tetrazines are as yet quite novel compounds; however 3-phenyl-*s*-tetrazine (235) has been reported as the product of the interaction of formamidine acetate with benzimidate in the presence of hydrazine[368].

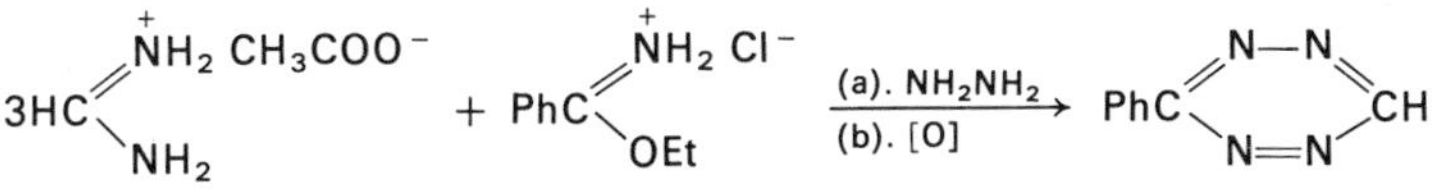

(235)

As tetrazines[519] are intermediates in the synthesis of many other heterocyclic systems, the above reactions provide routes from imidates to 1,2,4-triazoles and their 4-amino derivatives, 1,3,4-oxadiazoles, and pyridazines among others.

22. Formation of oxazines

The substituted propanolamine 236 and ethyl benzimidate[520] condense to yield the dihydro-1,3-oxazine 237.

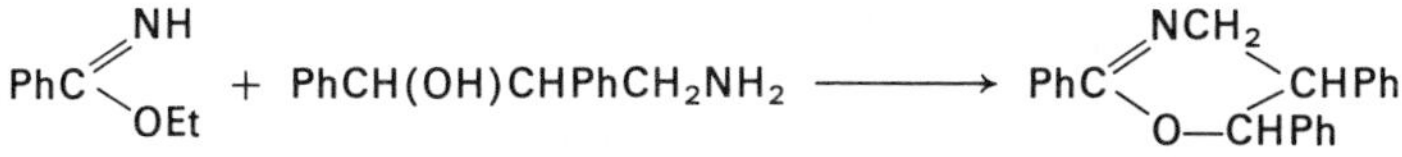

(236) (237)

23. Formation of azepines

Tetrahydro-1,3-diazepines have been prepared by the condensation of 1,4-diaminobutanes with appropriate reagents including imidates[521] and this reaction has been extended to give fused azepine derivatives[468,522].

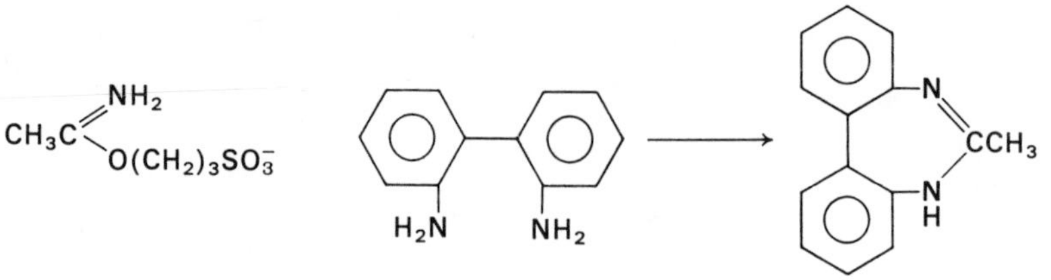

o-Aminophenylacetic acid has also been used in related condensation reactions to give diazepines (e.g. **238**) useful as hypnotics[404].

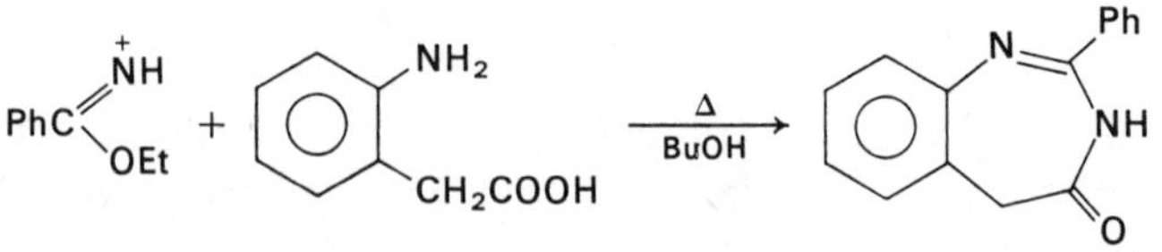

(238)

I. Reaction of Imidates with Grignard Reagents and Metal Alkyls

Imidates undergo attack by Grignard reagents[2]; in the main displacement of the alkoxy group of the imidate takes place with formation of the corresponding imine which can be hydrolysed to the parent aldehyde or ketone. Indeed, the interaction of aryl Grignard reagents with ethyl N-phenylformimidate has been shown to be a convenient route to aryl

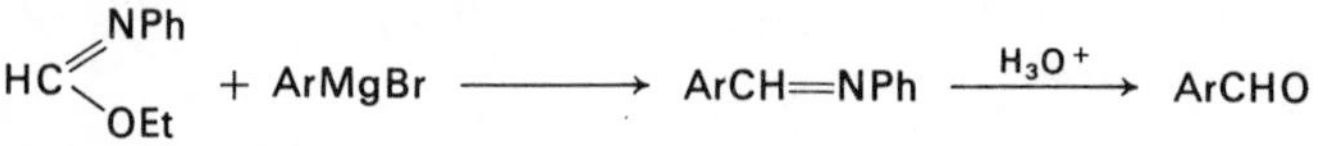

aldehydes[523]. However, in some cases these reactions can proceed beyond the aldimine stage[524,525] and thus give secondary amines as products, equation (31). Similar results have been obtained by the use of $CH_2{=}CHCH_2Li$ or $CH_2{=}CHCH_2ZnBr$ in place of the corresponding magnesium compound[524].

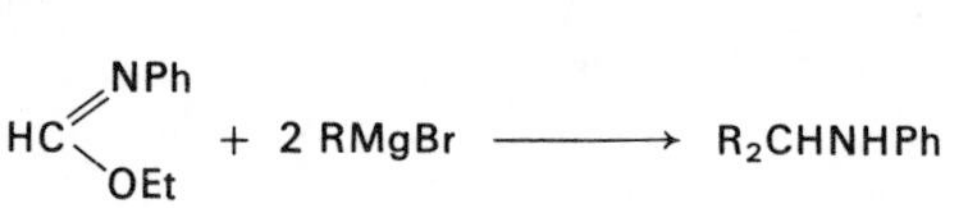

(31)

The reaction of Grignard reagents with imidates is fairly general in scope and has been used to give ketols[526], e.g. (−)-benzoin (**239**) and hydrazine derivatives[478] of the type **240**. In addition, imidates (**241**)

$$(-)\text{-PhCH(OH)C}\underset{\text{OEt}}{\overset{\text{NH}}{\diagup}} + \text{PhMgBr} \longrightarrow (-)\text{-PhCH(OH)C}(=\text{O})\text{Ph}$$

$$(\textbf{239})$$

$$\text{HC}\underset{\text{OR}}{\overset{\text{NN}=\text{CHPh}}{\diagup}} + (\text{Ph})_3\text{CMgBr} \longrightarrow \text{PhC}\underset{\text{H}}{\overset{\text{N}-\text{N}}{\diagup}}\underset{\text{H}}{\diagdown}\text{CCPh}_3$$

$$(\textbf{240})$$

derived from sulphonamides have been used as a source of primary amines[527] as outlined in equation (32).

$$\text{PhSO}_2\text{NH}_2 + \text{HC(OEt)}_3 \longrightarrow \text{HC}\underset{\text{OEt}}{\overset{\text{NSO}_2\text{Ph}}{\diagup}} \xrightarrow{2\ \text{RMgBr}}$$

$$(\textbf{241})$$

$$\text{R}_2\text{CHNHSO}_2\text{Ph} \longrightarrow \text{R}_2\text{CHNH}_2 \quad (32)$$

Although the reaction of ethyl N-phenylformimidate with alkyl lithiums[528] gave, in the main products **242** and **243** rather than the desired iso-nitrile, the Grignard reagent (CH$_3$CH$_2$)$_2$NMgBr afforded the desired iso-nitrile in high yield[525].

$$\text{HC}\underset{\text{OEt}}{\overset{\text{NPh}}{\diagup}}\ \begin{cases} \xrightarrow{n\text{-BuLi}} & n\text{-Bu}_2\text{CHNHPh} + \text{PhNC} \\ & (\textbf{242}) \\ \xrightarrow{\text{MeLi}} & \text{HC}\underset{\text{NHPh}}{\overset{\text{NPh}}{\diagup}} + \text{PhNC} \\ & (\textbf{243}) \end{cases}$$

J. Oxidation

Little is reported in the literature on the oxidation of imidates. However, N-substituted thioimidates (**244**) which appear in two independent studies involving selenium dioxide[529] and benzoyl peroxide[530] yield, on oxidation, the corresponding N-substituted amides **245**, and peracid

$$\text{ArC}\underset{\text{SR}'}{\overset{\text{NR}}{\diagup}} \longrightarrow \text{ArC}\underset{\text{NHR}}{\overset{\text{O}}{\diagup}}$$

$$(\textbf{244}) \qquad\qquad (\textbf{245})$$

oxidation with *m*-chloroperbenzoic acid of *N*-substituted and cyclic imidates yields oxaziranes which are useful as synthetic intermediates[531].

K. Reduction

The reduction of imidates has been much more widely studied than their oxidation. Adapting earlier work of Henle[532] on simple imidates, de Ruggieri and his co-workers[533, 534] found that *N*-substituted imidates could be readily reduced to primary amines with zinc or sodium amalgams

(R = alkyl or cycloalkyl, including steroidal residues)

in acid solution. More recently, work by Borch has led to the conversion in high yields of nitriles[67] or *N*-mono- or *N,N*-disubstituted amides[125, 128, 535] into amines via imidate intermediates (equations 33 and 34). However, when for compound **246**, R′ = R″ = H, the boro-

$$RCN \xrightarrow[\text{(b). EtOH}]{\text{(a). Et}_3O^+ \text{ BF}_4^-} RC\diagdown_{OEt}^{NEt} \xrightarrow{NaBH_4} RCH_2NHEt \qquad (33)$$

$$RC\diagdown_{NR'R''}^{O} \xrightarrow[\text{CH}_2\text{Cl}_2]{\text{Et}_3O^+BF_4^-} RC\diagdown_{OEt}^{NR'R''} \xrightarrow{NaBH_4} RCH_2NHR'R'' \qquad (34)$$

(246)   R′ = H or alkyl, R″ = alkyl

hydride causes smooth dehydration of the amide to give the corresponding nitrile[128].

In the case of cyclic imidates, borohydride or deuteride reduction of 2-alkyl-dihydro-1,3-oxazines (e.g. compound **247**) affords a useful synthetic route [536-540] to substituted aldehydes or their $C_{(1)}$ deuterated derivatives as illustrated in equation (35).

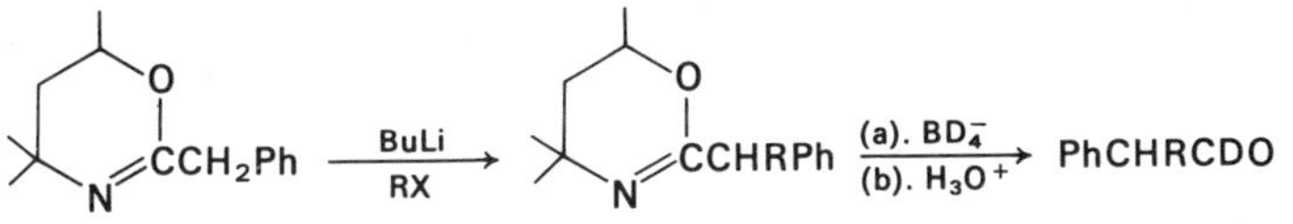

$$\text{(35)}$$

On the other hand, reduction of cyclic imidates of the tetrahydrofuranimine type (248) with lithium aluminium hydride gives rise to alkanolamines [541-543].

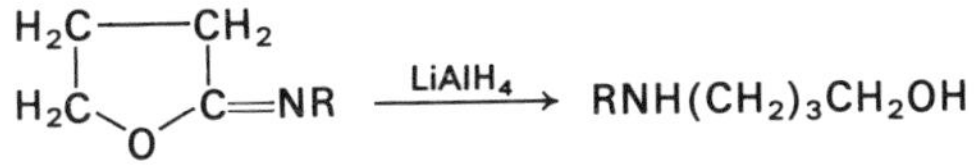

$$H_2C \xrightarrow{} CH_2, \quad H_2C \diagdown O \diagup C{=}NR \xrightarrow{LiAlH_4} RNH(CH_2)_3CH_2OH$$

(248)

The third type of cyclic imidates i.e. those based on lactams (249) can be reduced by borohydride [128] to give cyclic amines.

(249)

L. Preparation and properties of acyl and sulphonyl derivatives of imidates

Simple imidates react with acid halides including phosgene to give the corresponding acyl derivative of the imidate [2,510,544]. These compounds (250) are very susceptible to hydrolysis and hence diacylamines are often the final products of these reactions [2]. N-Substituted imidates react

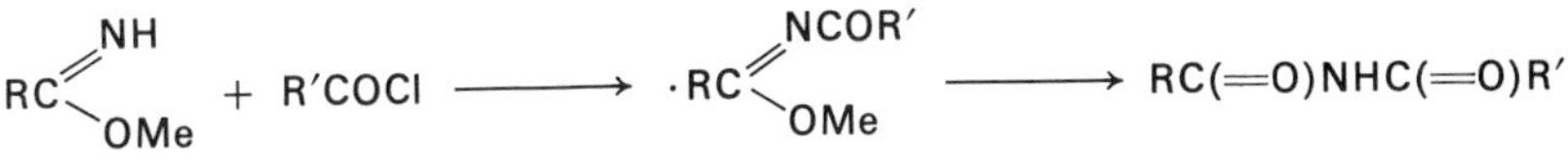

$$RC{\diagup}^{NH}_{\diagdown OMe} + R'COCl \longrightarrow \cdot RC{\diagup}^{NCOR'}_{\diagdown OMe} \longrightarrow RC({=}O)NHC({=}O)R'$$

(250)

similarly with acid chlorides and compounds such as chloroformate esters [545] to give diacylamines [147], and related compounds (251). Synthetic

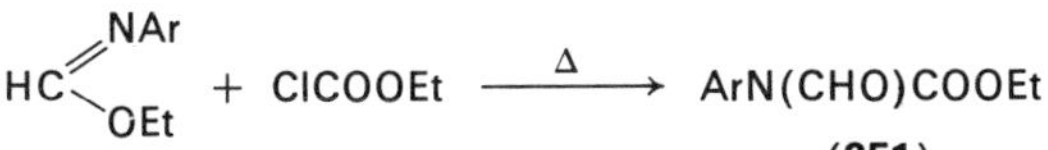

$$HC{\diagup}^{NAr}_{\diagdown OEt} + ClCOOEt \xrightarrow{\Delta} ArN(CHO)COOEt$$

(251)

use of these reactions has been made to give acyl isocyanates[546] (**252**) as illustrated in equation (36) and triazines[510] (see Section IV, H, 21, e).

$$RC{\overset{NH}{\underset{OEt}{\diagup}}} \xrightarrow{(COCl)_2} RC{\overset{NCOCOCl}{\underset{OEt}{\diagup}}} \xrightarrow{\Delta} RCONCO \qquad (36)$$

(**252**)

Although acetic anhydride has been reported to react with imidates to give diacylamines[1,2], little work has been done in this area of imidate chemistry.

$$RC{\overset{NH}{\underset{OR'}{\diagup}}} + (CH_3CO)_2O \longrightarrow RCONHCOCH_3$$

Sulphonyl halides condense with imidates to yield a mixture of products[547] which includes the sulphonyl derivative of the imidate (equation 37). However better routes to these sulphonyl derivatives of imidates are known, e.g. treatment of a sulphonamide with an ortho ester[527].

$$RC{\overset{NH}{\underset{OEt}{\diagup}}} + R'SO_2Cl \longrightarrow$$

$$RC{\overset{NSO_2R'}{\underset{OEt}{\diagup}}} + RC{\overset{O}{\underset{NHSO_2R'}{\diagup}}} + RC{\overset{\overset{+}{N}H_2\ Cl^-}{\underset{OEt}{\diagup}}} + C_2H_5Cl \quad (37)$$

$$PhSO_2NH_2 + HC(OEt)_3 \longrightarrow HC{\overset{NSO_2Ph}{\underset{OEt}{\diagup}}}$$

M. Properties and Reactions of N-Halo-imidates

Sodium hypochlorite[547] and hypobromite[547, 548], *t*-butyl hypochlorite[549], bromine and iodine[550] have all been used to convert imidates into their *N*-halo derivatives. *N*-Fluoro-imidates have also been reported[9, 551] but are prepared by other methods, e.g. equation (**38**). In

$$RC{\overset{NH}{\underset{OR'}{\diagup}}} + HOX \longrightarrow RC{\overset{NX}{\underset{OR'}{\diagup}}} + H_2O$$

(X = Cl, Br)

$$RC\diagdown\kern-0.7em^{NF}_{F} + R'ONa \longrightarrow RC\diagdown\kern-0.7em^{NF}_{OR'} \qquad (38)$$

$$(R = CN \text{ or } R_F) \qquad (252)$$

particular, the perfluoro compounds (252; $R = R_F$) are of interest as on hydrolysis they are reported[9] to give the N-fluoro-imidates 253 rather than compounds of the tautomeric fluoro-amide structure, 254.

$$R_FC\diagdown\kern-0.7em^{NF}_{OH} \qquad R_FC\diagdown\kern-0.7em^{NHF}_{O}$$

$$(253) \qquad\qquad (254)$$

N-Bromo-imidates have also been identified as intermediates in the bromination of olefins with N-bromacetamide (NBA) which unlike N-bromosuccinimide fails to give allylic bromination[552]. Thus cyclohexane and NBA yield the imidate intermediate (255) which decomposes to give the bromo-compounds 256 and 257.

Although known from the early days of imidate chemistry, it is only recently that N-halo-imidates have become of synthetic importance. For example, compounds of this type possessing α-hydrogen atoms undergo Neber type rearrangements to form α-amino-esters[549] and a recent modification[553] of this reaction has given ortho esters of α-amino acids, (equation 39).

Papa[241] has shown that N-chlorobenzimidates react with dialkyl sulphides to give either sulphilimines (259; $R' = Me$) or sulphenamides

$$RCH_2C\overset{NCl}{\underset{OR'}{\Big\langle}} \quad \xrightarrow{R'O^-} \quad \left[RCH\!-\!\underset{N}{\overset{}{\text{COR}'}} \right]$$

$$\overset{RCHCOOR'}{\underset{\overset{|}{NH_3}\ Cl^-}{\underset{+}{}}} \quad \xleftarrow{H_3O^+} \quad R\overset{H}{\underset{NH}{\overset{|}{C}\!-\!C(OR')_2}} \tag{39}$$

$$\xrightarrow{\ H_3O^+\ } \quad RCH(NH_2)C(OR')_3$$

(**260**; R' = *n*-Pr). He also demonstrated[241] by n.m.r. techniques that the *N*-halo-imidates (**258**) existed as a mixture of *syn* and *anti* isomers.

$$ArC\overset{NCl}{\underset{OMe}{\Big\langle}} \quad \xrightarrow{R_2'S} \quad ArCON\!=\!SR_2' \text{ or } ArCONHSR'$$

$$(\textbf{258}) \qquad\qquad\qquad (\textbf{259}) \qquad\qquad (\textbf{260})$$

Much of the recent work in the field of *N*-halo-imidates is of Russian origin and is concerned with the reaction of *N*-halo-imidates with various phosphorus compounds. For example, trialkyl phosphites react with *N*-chloro-imidates to give products of the type **261** via an Arbuzov type

$$RC\overset{NCl}{\underset{OR'}{\Big\langle}} + (R''O)_3P \quad \longrightarrow \quad R''Cl + RC\overset{NPO(OR'')_2}{\underset{OR'}{\Big\langle}}$$

$$(\textbf{261})$$

rearrangement[554-557]. Related reactions have been carried out with *N,N'*-dichloro-oxaldiimidate[558]; however, triaryl phosphites[559] cannot undergo Arbuzov type rearrangements and hence yield, under similar conditions, compounds of type **262**.

$$RC\overset{NCl}{\underset{OR'}{\Big\langle}} + P(OAr)_3 \quad \longrightarrow \quad R'Cl + RCON\!=\!P(OAr)_3$$

$$(\textbf{262})$$

Phosphorus trichloride reacts with *N*-chloro-arylimidates to give compounds described as *N*-(tetrachlorophosphoramyl)carboximidates (**263**);

$$\underset{\underset{OR}{\big\backslash}}{\overset{\overset{NCl}{\big\diagup}}{ArC}} + PCl_5 \longrightarrow Cl_2 + \underset{\underset{OR}{\big\backslash}}{\overset{\overset{\overset{+}{N}PCl_3\ Cl^-}{\big\diagup}}{ArC}}$$

$$(263)$$

phosphorus pentachloride reacts similarly but with evolution of chlorine[560]. Chlorophenylphosphines and related compounds[560-563] have also been studied and shown to react as in equation (40).

$$\underset{\underset{OR'}{\big\backslash}}{\overset{\overset{NCl}{\big\diagup}}{RC}} + P(Ph)_{3\to1}Cl_{0\to2} \longrightarrow R'Cl + RCON{=}PPh_{3\to1}Cl_{0\to2} \quad (40)$$

N-Halo-imidates of the type **264** have been prepared for studies on syn–anti isomerisation[564].

$$\underset{\underset{OSiMe_3}{\big\backslash}}{\overset{\overset{NX}{\big\diagup}}{RC}}$$

$$(264)$$

N. Phosphorus and Antimony Derivatives of Imidates

Several methods applicable to the formation of simple imidates have also been applied to give imidates incorporating nitrogen–phosphorus bonds. For example, imidoyl halides having N–P bonds have been treated with phenolates[565] or thiols[95, 341] in basic conditions to yield imidates.

$$\underset{\underset{Cl}{\big\backslash}}{\overset{\overset{NPO(Me)Cl}{\big\diagup}}{RC}} \xrightarrow{NaOAr} \underset{\underset{OAr}{\big\backslash}}{\overset{\overset{NPO(Me)OAr}{\big\diagup}}{RC}}$$

$$\underset{\underset{Cl}{\big\backslash}}{\overset{\overset{NPO(OPh)_2}{\big\diagup}}{ArC}} \xrightarrow[Et_3N]{PhCH_2SH} \underset{\underset{SCH_2Ph}{\big\backslash}}{\overset{\overset{NPO(OPh)_2}{\big\diagup}}{ArC}}$$

$$(265)$$

Thioimidates of the type **265** may also be made by the direct alkylation of the corresponding thioamide[566] (**266**) and related compounds (**267**) have been prepared by the action of vinyl ethers with phosphorazidates[567].

$$\underset{\underset{NHPO(OPh)_2}{\big\backslash}}{\overset{\overset{S}{\big\diagup}}{ArC}} \xrightarrow[Et_3N]{RX} \underset{\underset{SR}{\big\backslash}}{\overset{\overset{NPO(OPh)_2}{\big\diagup}}{ArC}}$$

$$(266)$$

$$R_2PON_3 + CH_2{=}CH(OR') \longrightarrow HC{\overset{\displaystyle NP(O)R_2}{\underset{\displaystyle OR'}{\big<}}} + CH_2N_2$$

(267)

Another synthetic route to imidates with phosphorus substituents involves the interaction of alkyl or aryl thiocyanates with phosphorus pentachloride[568,569] to give the chloroformic thioimidates **268** and **269**.

$$RSCN + PCl_5 \longrightarrow ClC{\overset{\displaystyle NPCl_4}{\underset{\displaystyle SR}{\big<}}}$$

(R = alkyl)   (268)

$$ArSCN + PCl_5 \longrightarrow ClC{\overset{\displaystyle \overset{+}{N}PCl_3\ PCl_6^-}{\underset{\displaystyle SAr}{\big<}}}$$

(269)

Other phosphorus derivatives have been prepared by the direct interaction of imidate bases, e.g. benzimidates with phosphorus pentachloride[570, 571] or by the interaction of that reagent with N-halo-imidates[560]. However, not all compounds of the type **270** are stable and N-acyl-

$$RC{\overset{\displaystyle NH}{\underset{\displaystyle OEt}{\big<}}} + PCl_5 \longrightarrow RC{\overset{\displaystyle NPCl_4}{\underset{\displaystyle OEt}{\big<}}}$$

(270)

phosphorimidic chlorides (**271**) are obtained in some instances; equation (41) illustrates a route to related compounds[572, 573].

Phosphorus trichloride also reacts with imidates[559] as does chloro-

$$RC{\overset{\displaystyle NH}{\underset{\displaystyle OR'}{\big<}}} + PCl_5 \longrightarrow HCl + R'Cl + RCON{=}PCl_3$$

(271)

$$RC{\overset{\displaystyle NH}{\underset{\displaystyle OR'}{\big<}}} + Ph_{1\to2}PCl_{4\to3} \longrightarrow HCl + R'Cl + RCON{=}P(Ph)_{1\to2}Cl_{2\to1} \quad (41)$$

$$ArC{\overset{\displaystyle NH}{\underset{\displaystyle OEt}{\big<}}} + Ph_2PCl \longrightarrow ArC{\overset{\displaystyle NPPh_2}{\underset{\displaystyle OEt}{\big<}}} \quad (42)$$

diphenylphosphine[178] (equation 42). Other related reactions of imidates with various chlorophosphines[574–576] and with triaminophosphines[577] are illustrated by the following equations (43–45).

$$
\begin{array}{ccc}
\underset{RO}{\overset{HN}{\diagdown}}C-C\underset{OR}{\overset{NH}{\diagup}} & \xrightarrow[Et_3N]{R'PCl_2} & \underset{N}{\overset{ROC-COR}{\underset{P}{\diagup}}}N \\
& & \qquad R'
\end{array} \qquad (43)
$$

$$
RC\underset{OMe}{\overset{NH}{\diagup}} + (EtO)_2PCl \longrightarrow RC\underset{OMe}{\overset{NP(OEt)_2}{\diagup}} \qquad (44)
$$

$$
RC\underset{OR'}{\overset{NH}{\diagup}} + P(NR_2)_3 \longrightarrow RC\underset{OR'}{\overset{NP(NR_2)_2}{\diagup}} + RC\underset{NR_2}{\overset{NP(OR')NR_2}{\diagup}} \qquad (45)
$$

Some related work involving antimony derivatives, e.g. antimony pentachloride[578] has been published and a novel synthesis of formimidate involves the use of that reagent along with hydrogen cyanide and alcohol[579].

$$
HCN + SbCl_5 + EtOH \longrightarrow HC\underset{OEt}{\overset{NH_2^+ \ SbCl_6^-}{\diagup}}
$$

O. Photochemistry of Imidates

Little study has been made of the photo-reactions of simple imidates although it has been shown that the imidate **271** on irradiation yields the corresponding N-substituted amide[580].

$$
\underset{HC}{\overset{HC-CH}{\underset{O}{\diagdown}}}\underset{CC}{\overset{}{\diagup}}\underset{OR}{\overset{NH}{\diagup}} \xrightarrow{h\nu} \underset{HC}{\overset{HC-CH}{\underset{O}{\diagdown}}}CCONHR
$$

(271)

The photochemistry of several cyclic imidates[581–584] has been studied and, for example, the azabullvalene derivatives **272** and **273** have been found to be the products of irradiation of the imidate **274**.

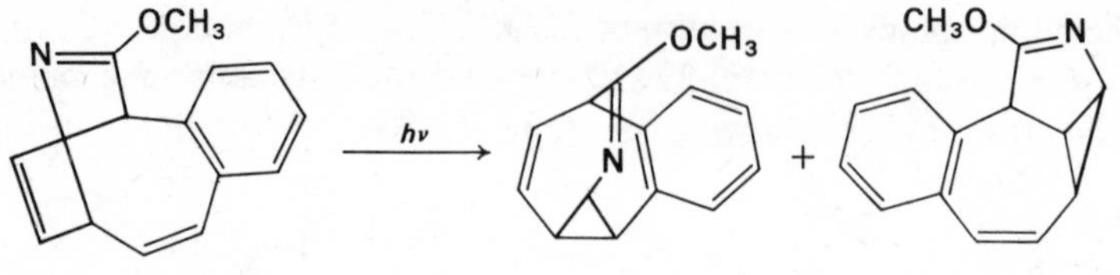

(274)   (272)   (273)

Photoisomerisations of 2-ethoxy-3*H*-azepines (**275**) to 2-azabicyclo-[3,2,0]hepta-2,6-dienes have also been reported[585]. On the other hand,

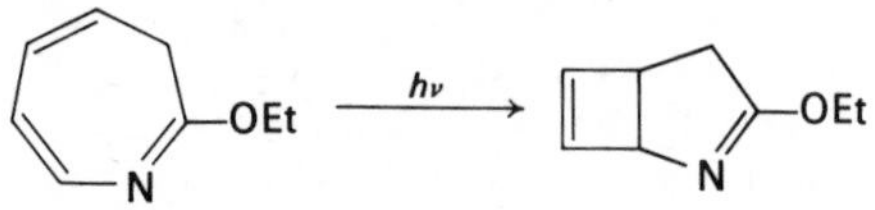

(275)

the 4,5-dihydro-3*H*-azepine **276** was found to be photostable but could be converted into azabicyclo[3,2,0]hept-6-ene (**277**) in the presence of triplet sensitisers[586]. The related compound **278** on photolysis in the

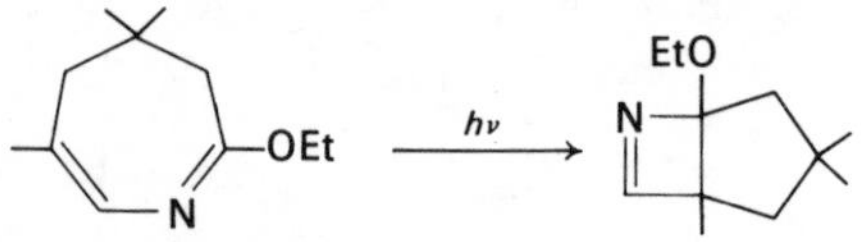

(276)   (277)

presence of sensitisers gave products the nature of which depended on the solvent used in the reaction[587].

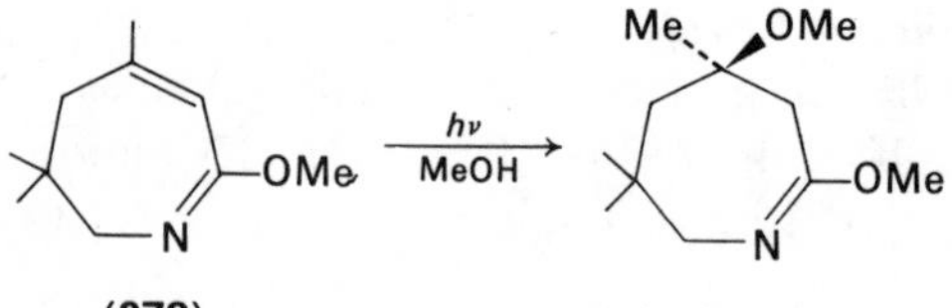

(278)

P. Iminoanhydrides

Iminoanhydrides (isoimides) have been observed only infrequently in organic synthesis. Indeed, only when the iminoanhydride grouping, **279**,

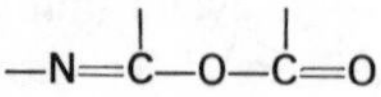

$$-N{=}\overset{|}{C}-O-\overset{|}{C}{=}O$$

(279)

is stabilised by certain structural features can these compounds be isolated[588-590], otherwise they rearrange readily to diacylamines, and it has
been suggested that it is the *syn* form of these compounds[589] which is

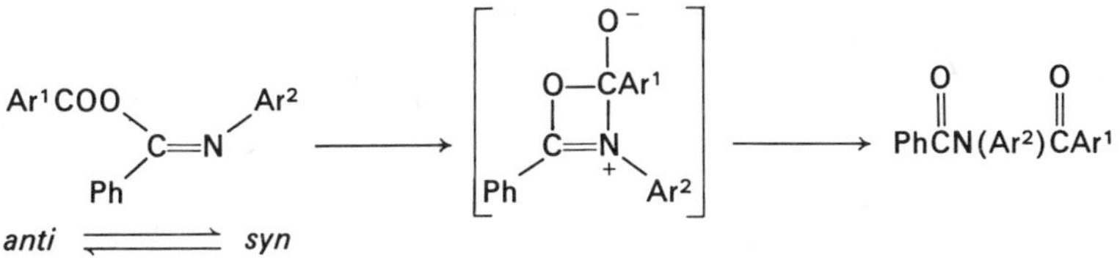

anti $\rightleftharpoons$ syn   (46)

(280)

responsible for the rearrangement, (equation 46). Factors which help to
stabilise the anhydride system include the presence of electron withdrawng
substituents on the aryl group (Ar^2 of **280**) attached to the nitrogen[589-591]
or the enclosing of the anhydride group in a cyclic system (**281**) in which
attack by the imino nitrogen on the carbonyl group is inhibited sterically.

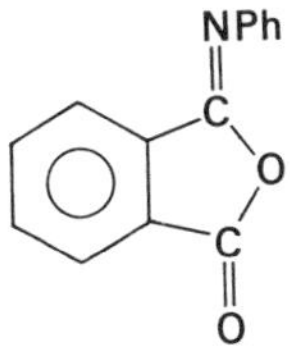

(281)

Studies on the rearrangement of these anhydrides[590, 592] [and of the related dithio compounds[593] (**282**)] have shown that the process follows
first order kinetics.

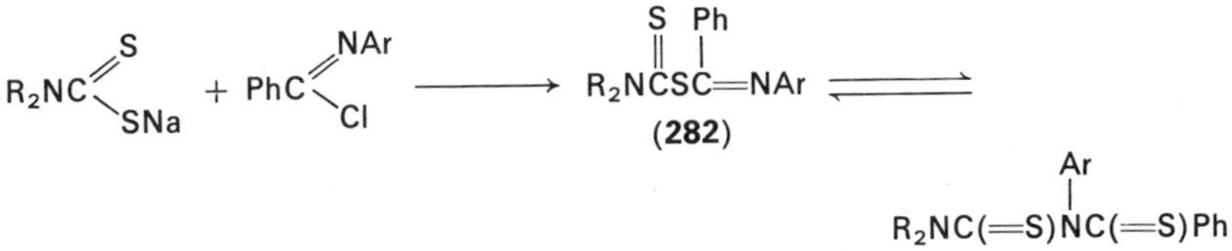

$$R_2NC(=S)NC(=S)Ph$$
with Ar substituent on the N

Q. Miscellaneous Reactions of Imidates

Malononitrile condenses with imidates[594] and thioimidates[595] to give
dicyanoethylene derivatives (**283**). Other reactive groups have been utilised
to synthesise pyridines (see Section IV, H, 21, a) and isooxazolones[596].

$$CH_2(CN)_2 + CH_3C\underset{SR}{\overset{NCH_2Ph}{\diagdown}} \longrightarrow \underset{NC}{\overset{NC}{\diagdown}}C=C\underset{NHCH_2Ph}{\overset{CH_3}{\diagup}}$$

(283)

$$RC\underset{OR'}{\overset{NH}{\diagup}} + \underset{CH_3C=NOH}{\overset{H_2CCOOR''}{\diagup}} \longrightarrow CH_3C\!\!-\!\!-\!\!CHC(=NH)R$$

Amidrazones have been found to react with imidates. Thus the amidrazone **284** and the related imidate **285** condense to give an imidine[597] (**286**). Unsubstituted amidrazones also react with imidates, and the

$$HC\underset{NH_2}{\overset{NNMePh}{\diagup}} + HC\underset{OEt}{\overset{NNMePh}{\diagup}} \longrightarrow \begin{array}{c} HC\overset{NNMePh}{\diagup}\\ \diagdown NH\\ HC\underset{NNMePh}{\diagdown} \end{array}$$

(284)   (285)   (286)

products have been identified as dihydrazidines, e.g. compound **287**, which can be readily cyclised in acid to the corresponding triazoles[598].

$$\left[-C\underset{OEt}{\overset{NH}{\diagup}}\right]_2 + RC\underset{NH_2}{\overset{NNH_2}{\diagup}} \longrightarrow RC\underset{NH_2 \; H_2N}{\overset{N-\!\!-\!\!N}{\diagup}}C\!\!-\!\!C\underset{NH_2 \; H_2N}{\overset{N-\!\!-\!\!N}{\diagdown}}CR$$

(287)

Cyclo addition reactions of imidates with nitrilimines have also been reported[599].

$$\underset{PhC}{\overset{N-NPh}{\diagup}} + CH_3C\underset{OCH_3}{\overset{NH}{\diagup}} \longrightarrow PhC\underset{N}{\overset{N-\!\!-\!\!NPh}{\diagup}}CCH_3$$

Other imidate reactions have been reviewed[2].

R. Uses of Imidates

In addition, to their wide utility in synthetic work, imidates have from time to time found use in industrial processes although their general instability tends to preclude their wide usage in this way. However, recently, imidates of the type **288** and related compounds have proved useful as catalysts for the formation of stereoregular methacrylonitrile

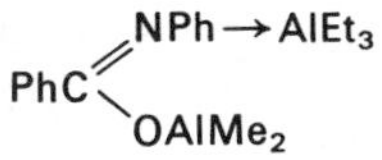

$$PhC \overset{NPh \rightarrow AlEt_3}{\underset{OAlMe_2}{\diagdown}}$$

(288)

polymers[600, 601]. In addition ethyl N-methylbenzimidate has been used as a catalyst in the polymerisation of ε-caprolactam[602], the silicon derivative **289** has been utilised as a vulcanising agent[603] and diimidates of the type **290** have been found to be useful as adhesive additives[604]. A thermo-

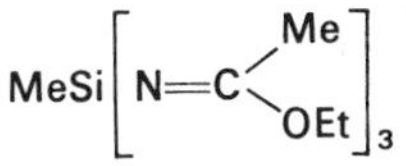

$$MeSi \left[N{=}C \overset{Me}{\underset{OEt}{\diagdown}} \right]_3$$

(289)

$$\overset{HN}{\underset{R^1O}{\diagdown}} C{-}R{-}C \overset{NH}{\underset{OR^2}{\diagdown}}$$

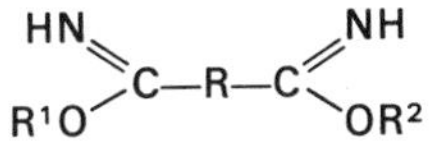

(R, R^1, R^2 = C$_{1-25}$ alkyl, etc.)

(290)

setting imidate polymer derived from hexafluoropentanediol and perfluorosuberonitrile has been utilised as an intermediate in the formation of polytriazines[70].

Another area in which imidates have found use in recent times is in the realm of agricultural products such as herbicides[591,605,606], bactericides[607] and pesticides[608, 609], e.g. compounds of the type **291**.

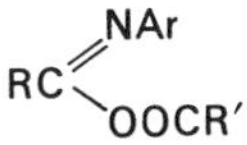

$$RC \overset{NAr}{\underset{OOCR'}{\diagdown}}$$

(291)

Other miscellaneous applications include the use of imidates in acid plating solutions for copper[610], in impregnating varnishes for electrical insulation[611] and as fluorescent materials[127].

V. ACKNOWLEDGEMENT

The author gratefully acknowledges the assistance of Dr K. M. Watson and Mr J. Fraser in preparing this chapter.

VI. REFERENCES

1. A. Pinner, *Die Imidoäther und ihre Derivate*, Oppenheim, Berlin (1892).
2. R. Roger and D. G. Neilson, *Chem. Revs.*, **61**, 179 (1961).
3. R. G. Glushkov and V. G. Granik, *Adv. Heterocyclic Chem.*, **12**, 185 (1970).

4. W. Seelinger, E. Aufderhaar, W. Diepers, R. Feinauer, R. Nehring, W. Thier and H. Hellmann, *Angew. Chem. Int. Ed.*, **5**, 875 (1966).
5. S. R. Sandler and W. Karo, *Organic Functional Group Preparation*, Vol. 3, Academic Press, New York, 1972, Chap. 8.
6. M. B. Robin, F. A. Bovey and H. Basch, *The Chemistry of Amides* (Ed. J. Zabicky), Interscience Publishers, New York, 1970, pp. 1–72.
7. D. B. Brown, R. D. Burbank and M. B. Robin, *J. Am. Chem. Soc.*, **91**, 2895 (1969).
8. E. Allenstein and A. Schmidt, *Z. Anorg. Allgem. Chem.*, **344**, 113 (1966).
9. B. L. Dyatkin, K. N. Makarov and I. L. Knunyants, *Tetrahedron*, **27**, 51 (1971).
10. R. Roger and D. G. Neilson, *J. Chem. Soc.*, 688 (1959).
11. R. Roger and D. G. Neilson, *J. Chem. Soc.*, 3181 (1961).
12. A. C. Flisik and R. W. Handy, *U.S. Pat.* 3,121,751 (1964); *Chem. Abstr.*, **60**, 11902d (1964).
13. H. J. Hagemeyer, Jr. and W. J. Gammans, *U.S. Pat.* 3,538,139 (1970); *Chem. Abstr.*, **74**, 53090p (1971).
14. R. P. Mull, *U.S. Pat.* 3,189,601 (1965); *Chem. Abstr.*, **63**, 11522d (1965).
15. F. C. Schaefer, *The Chemistry of the Cyano Group* (Ed. Z. Rappoport) Interscience Publishers, New York, 1970, pp. 239–305.
16. E. J. Tillmanns and J. J. Ritter, *J. Org. Chem.*, **22**, 839 (1957).
17. E. N. Zil'berman and A. M. Sladkov, *Zh. Obshch. Khim.*, **31**, 245 (1961).
18. I. A. Pearl and D. L. Beyer, *J. Am. Chem. Soc.*, **74**, 3188 (1952).
19. Y. Yaru, *Chem. Pharm. Bull.* (*Tokyo*), **10**, 1094 (1962).
20. V. V. Dovlatyan and T. O. Chakryan, *Izv. Akad. Nauk Arm. SSR Khim. Nauki.*, **17**, 81 (1964); *Chem. Abstr.*, **61**, 3014g (1964).
21. F. Cramer and H. J. Baldauf, *Chem. Ber.*, **92**, 370 (1959).
22. C. L. Stevens, D. Morrow and J. Lawson, *J. Am. Chem. Soc.*, **77**, 2341 (1955).
23. G. A. Shvekhgeimer and M. L. Shul'man, *Zh. Org. Khim.*, **4**, 1734 (1968).
24. M. L. Shul'man, G. A. Shvekhgeimer and R. A. Miftakhova, *J. Org. Chem. USSR.*, **3**, 840 (1967).
25. S. M. McElvain and B. Fajardo-Pinzon, *J. Am. Chem. Soc.*, **67**, 690 (1945).
26. R. Mayer, S. Scheithauer, and D. Kunz, *Chem. Ber.*, **99**, 1393 (1966).
27. H. Behringer and D. Weber, *Ann. Chem.*, **682**, 196 (1965).
28. R. H. Hartigan and J. B. Cloke, *J. Am. Chem. Soc.*, **67**, 709 (1945).
29. C. S. Marvel, P. De Radzitzky, and J. J. Brader, *J. Am. Chem. Soc.*, **77**, 5997 (1955).
30. H. M. Woodburn, A. B. Whitehouse and B. G. Pautler, *J. Org. Chem.*, **24**, 210 (1959).
31. R. Ohme and E. Schmitz, *Ann. Chem.*, **716**, 207 (1968).
32. N. S. Drozdov and A. F. Bekhli, *J. Gen. Chem.*, **14**, 480 (1944).
33. Farbenindustrie A.-G., I. G., *French Pat.* 781,001 (1935); *Chem. Abstr.*, **29**, 5951 (1935).
34. D. G. Neilson and D. A. V. Peters, *J. Chem. Soc.*, 4455 (1963).
35. D. G. Neilson, D. A. V. Peters, and L. H. Roach, *J. Chem. Soc.*, 2272 (1962).
36. H. E. Johnson and D. G. Crosby, *J. Org. Chem.*, **27**, 798 (1962).

37. M. Mengelberg, *Chem. Ber.*, **89**, 1185 (1956).
38. D. F. Ewing and D. G. Neilson, *J. Chem. Soc. (C)*, 390 (1966).
39. D. W. Woolley, J. W. B. Hershey and H. A. Jodlowski, *J. Org. Chem.*, **28**, 2012 (1963).
40. A. K. Bose, F. Greer, J. S. Gots and C. C. Price, *J. Org. Chem.*, **24**, 1309 (1959).
41. J. Poupaert, A. Bruylants and P. Croody, *Synthesis*, 622 (1972).
42. W. O. Foye and J. M. Kauffman, *J. Pharm. Soc.*, **57**, 1614 (1968).
43. G. A. Shvekhgeimer, E. I. Gorbatov, and V. D. Tyurin, *Zh. Org. Khim.*, **7**, 815 (1971).
44. J. R. Merchant and A. S. U. Choughuley, *Chem. Ber.*, **95**, 1792 (1962).
45. I. K. Fel'dman, N. G. Volikova, and L. A. Aleksandrova, *Khim. Geterotsikl. Soedin.*, **2**, 125 (1970); *Chem. Abstr.*, **77**, 19464r (1972).
46. W. O. Kenyon and C. C. Unruh, *U.S. Pat.* 2,499,392 (1950); *Chem. Abstr.*, **44**, 6877 (1950).
47. W. Steinkopf and W. Malinowski, *Chem. Ber.*, **44**, 2898 (1911).
48. D. G. Neilson and K. M. Watson, unpublished results.
49. K. R. Huffman and F. C. Schaefer, *J. Org. Chem.*, **28**, 1813 (1963).
50. A. Pinner and F. Klein, *Chem. Ber.*, **11**, 1475 (1878).
51. A. Pinner, *Chem. Ber.*, **23**, 2917 (1890).
52. A. Pinner, *Chem. Ber.*, **23**, 2942 (1890).
53. G. D. Lander and F. T. Jewson, *J. Chem. Soc.*, **83**, 766 (1903).
54. A. P. T. Easson and F. L. Pyman, *J. Chem. Soc.*, 2991 (1931).
55. J. Guy and J. Paris, *Bull. Soc. Chim. France*, 406 (1947).
56. L. Weintraub, S. R. Oles, and N. Kalish, *J. Org. Chem.*, **33**, 1679 (1968).
57. F. Walls and X. Arevalo, *Bol. Inst. Quim. Univ. Nacl. Auton. Mex.*, **15**, 3 (1963); *Chem. Abstr.*, **61**, 8260c (1964).
58. M. Nakanishi, C. Tanaka and A. Naraki, *Japan. Pat.*, 6782 (1963); *Chem. Abstr.*, **59**, 11518 (1963).
59. M. Nakanishi and C. Tashiro, *Japan Pat.*, 3690 (1964); *Chem. Abstr.*, **61**, 3121g (1964).
60. M. Renson and R. Collienne, *Bull. Soc. Chim. Belges.*, **73**, 419 (1964).
61. D. V. Claridge, *British Pat.* 1,272,733 (1972); *Chem. Abstr.*, **77**, 62491j (1972).
62. H. Nohira, Y. Nishikawa, Y. Furuya and T. Mukaiyama, *Bull. Chem. Soc. Japan*, **38**, 897 (1965).
63. S. P. McManus and J. T. Carroll, *Org. Preparations and Procedures*, **2**, 71 (1970).
64. H. Chosho, K. Ichimura and M. Ohta, *Bull. Chem. Soc. Japan*, **37**, 1670 (1964).
65. H. J. Barber and R. Slack, *J. Chem. Soc.*, 82 (1947).
66. P. E. Spoerri and A. S. DuBois, *Organic Reactions*, **5**, 387 (1949).
67. R. F. Borch, *J. Org. Chem.*, **34**, 627 (1969).
68. F. C. Schaefer and G. A. Peters, *J. Org. Chem.*, **26**, 412 (1961).
69. F. Cramer, K. Pawelzik and H. J. Baldauf, *Chem. Ber.*, **91**, 1049 (1958).
70. E. Dorfman and C. T. Bean, *U.S. Pat.* 3,523,132 (1970); *Chem. Abstr.*, **73**, 110,334b (1970).
71. M.-C. Chiang and T.-C. Tai, *Hua Hsueh Hsueh Pao*, **30**, 312 (1964); *Chem. Abstr.*, **61**, 13169d (1964).
72. H. C. Brown and R. Pater, *J. Org. Chem.*, **27**, 2858 (1962).

73. H. C. Brown and C. R. Wetzel, *J. Org. Chem.*, **30**, 3724 (1965).
74. T. Vigo and C. M. Welch, *Carbohyd. Res.*, **17**, 145 (1971).
75. A. Baklouti, *Tetrahedron Letters*, 241 (1973).
76. G. Trolo and G. Gambaretto, *Ann. Chim. (Rome)*, **58**, 25 (1968).
77. W. J. Haggerty and W. J. Rost, *J. Pharm. Soc.*, **58**, 50 (1969).
78. T. O. Stevens, *J. Org. Chem.*, **33**, 2660 (1968).
79. R. B. LaCount and C. E. Griffin, *J. Chem. Soc. (C)*, 2071 (1966).
80. H. Cordes, J. Kranz, D. Neubauer and H. Weidinger, *Ger. Pat.* 1,443,432; *Chem. Abstr.*, **74**, 87386 (1971).
81. H. Watanabe, Y. Kikugawa and S. Yamada, *Chem. Pharm. Bull.*, **21**, 465 (1973).
82. J. U. Nef, *Ann. Chem.*, **287**, 274 (1895).
83. H. M. Woodburn and E. L. Graminski, *J. Org. Chem.*, **23**, 819 (1958).
84. P. Coutrot, J. C. Gombret and J. Villieras, *C.R. Acad. Sci. Paris*, **270**, 1674 (1970).
85. F. Nerdel, P. Weyerstahl and K. Lucas, *Tetrahedron Letters*, 5751 (1968).
86. M. Nakazaki, *Bull. Chem. Soc. Japan*, **32**, 588 (1959).
87. B. Robinson, *J. Chem. Soc.*, 2417 (1963).
88. R. Bonnet, *The Chemistry of the Carbon–Nitrogen Double Bond* (Ed. S. Patai), Interscience Publishers, New York, 1970, pp. 597–662.
89. H. Ulrich, *The Chemistry of the Imidoyl Halides*, Plenum Press, New York, 1968.
90. H. Paul, A. Weise and R. Dettmer, *Chem. Ber.*, **98**, 1450 (1965).
91. A. P. T. Easson, *J. Chem. Soc.*, 1029 (1961).
92. J. B. Aylward and F. L. Scott, *J. Chem. Soc. (C)*, 968 (1970).
93. E. C. Taylor and F. Kienzle, *J. Org. Chem.*, **36**, 233 (1971).
94. G. I. Derkach, A. M. Lepesa and A. V. Kirsanov, *Zh. Obshch. Khim.*, **31**, 3424 (1961).
95. V. A. Shokol, G. I. Derkach and E. S. Gubnitskaya, *J. Gen. Chem.*, **33**, 2984 (1963).
96. A. Meller and W. Maringgele, *Monatsh. Chem.*, **102**, 121 (1971).
97. H. W. Roesky and H. H. Giere, *Chem. Ber.*, **102**, 3707 (1969).
98. J. R. Norell, *J. Org. Chem.*, **35**, 1619 (1970).
99. C. F. Koelsch, *J. Org. Chem.*, **25**, 164 (1960).
100. E. Grigat and R. Puetter, *Chem. Ber.*, **97**, 3012 (1964).
101. W. E. Parham and K. B. Sloan, *Tetrahedron Letters*, 1947 (1971).
102. R. A. Abramovitch and R. B. Rogers, *Tetrahedron Letters*, 1951 (1971).
103. B. C. Challis and J. A. Challis, *The Chemistry of Amides* (Ed. J. Zabicky), Interscience Publishers, New York, 1970, pp. 731–857.
104. W. Walter and J. Voss, *The Chemistry of Amides* (Ed. J. Zabicky), Interscience Publishers, New York, 1970, pp. 383–475.
105. K. A. Petrov and L. N. Andreev, *Russian Chem. Revs.*, **40**, 505 (1971).
106. W. Walter and K. J. Reubke, *The Chemistry of Amides* (Ed. J. Zabicky), Interscience Publishers, New York, 1970, pp. 477–514.
107. T. Sato and M. Ohta, *Bull. Chem. Soc. Japan*, **27**, 624 (1954).
108. R. Grashey, M. Baumann and H. Bauer, *Chem. Ztg.*, **96**, 224 (1972).
109. Shell International, *Neth. Appl.* 66 08261 (1966); *Chem. Abstr.*, **69**, 35785m (1968).
110. S. Gabriel and P. Neymann., *Chem. Ber.*, **24**, 788 (1891).

111. G. Pinkus, *Chem. Ber.*, **26**, 1077 (1893).
112. Shell Research Ltd., *Belg. Pat.* 612,252 (1962); *Chem. Abstr.*, **58**, 3362 (1963).
113. C. J. M. Stirling, *J. Chem. Soc.*, 255 (1960).
114. W. Hechelhammer, *Ger. Pat.* 948,973 (1956); *Chem. Abstr.*, **53**, 6088d (1959).
115. F. H. Suydam, W. E. Greth and N. R. Langerman, *J. Org. Chem.*, **34**, 292 (1969).
116. R. Ohme and E. Schmitz, *Angew. Chem. Int. Ed.*, **6**, 566 (1967).
117. R. Ohme and E. Schmitz, *Ger.* (*East*) *Pat.* 57,119 (1967); *Chem. Abstr.*, **68**, 114092r (1968).
118. H. Bredereck, F. Effenberger and E. Henseleit, *Chem. Ber.*, **98**, 2754 (1965).
119. H. Bredereck, F. Effenberger and E. Henseleit, *Angew. Chem.*, **75**, 790 (1963).
120. J. Yates and E. Haddock, *Brit. Pat.* 1,039,459 (1966); *Chem. Abstr.*, **65**, 14353 (1966).
121. P. Reynaud, R. C. Moreau and N. H. Thu, *Compt. Rend.*, **253**, 1968 (1961).
122. H. Bredereck, G. Simchen and W. Kantlehner, *Chem. Ber.*, **104**, 924 (1971).
123. H. Bredereck, R. Gompper, H. Rempfer, K. Klemm and H. Keck, *Chem. Ber.*, **92**, 329 (1959).
124. H. Bredereck, F. Effenberger and G. Simchen, *Chem. Ber.*, **96**, 1350 (1963).
125. S. Julia and R. J. Ryan, *Compt. Rend.*, **274**, 1207 (1972).
126. W. Ried and E. Schmidt, *Ann. Chem.*, **695**, 217 (1966).
127. W. Ried, D. Piechaczek and E. Vollberg, *Ann. Chem.*, **734**, 13 (1970).
128. R. F. Borch, *Tetrahedron Letters*, 61 (1968).
129. R. R. Fraser and R. B. Swingle, *Canad. J. Chem.*, **48**, 2065 (1970).
130. S. Petersen and E. Tietze, *Ann. Chem.*, **623**, 166 (1959).
131. R. G. Glushkov and O. Yu. Magidson, *Khim. Geterotsikl. Soedin.*, **2**, 240 (1965).
132. S. Hanessian, *Tetrahedron Letters*, 1549 (1967).
133. N. Stojanac and V. Hahn, *Bull. Sci. Conseil. Acad. RSF Yugoslavie*, **11**, 98 (1966); *Chem. Abstr.*, **65**, 20084 (1966).
134. O. O. Orazi, R. A. Corral and H. Schuttenberg, *Tetrahedron Letters*, 2639 (1969).
135. R. Blaser, P. Imfeld and O. Schindler, *Helv. Chim. Acta.*, **52**, 569 (1969).
136. N. V. Phillip's Gloeilampenfabrieken, *Neth. Appl.* 6,514,351 (1967); *Chem. Abstr.*, **68**, 68752f (1968).
137. T. Sakazume and K. Mine, *Japan. Pat.* 72 09100; *Chem. Abstr.*, **77**, 50165t (1972).
138. R. H. DeWolfe, *Carboxylic Ortho Acid Derivatives*, Academic Press, New York, 1970.
139. R. M. Roberts, *Organic Synthesis, Collected*, Vol. IV, 464 (1963).
140. L. Claisen, *Ann. Chem.*, **287**, 360 (1895).
141. F. M. Hamer, R. J. Rathbone, and B. S. Winton, *J. Chem. Soc.*, 954 (1947).

142. R. M. Roberts, *J. Am. Chem. Soc.*, **71**, 3848 (1949).
143. R. M. Roberts, *J. Am. Chem. Soc.*, **72**, 3603 (1950).
144. R. M. Roberts and R. H. DeWolfe, *J. Am. Chem. Soc.*, **76**, 2411 (1954).
145. R. H. DeWolfe, *J. Org. Chem.*, **27**, 490 (1962).
146. J. F. Olin, *Belg. Pat.* 660,485 (1965); *Chem. Abstr.*, **64**, 2024g (1966).
147. J. P. Chupp, J. F. Olin and H. K. Landwehr, *J. Org. Chem.*, **34**, 1192 (1969).
148. E. H. Cordes, *The Chemistry of Carboxylic Acids and Esters* (Ed. S. Patai), Interscience Publishers, New York, 1969, pp. 656–658.
149. E. C. Taylor and W. A. Ehrhart, *J. Org. Chem.*, **28**, 1108 (1963).
150. T. Mukaiyama and K. Sato, *Bull. Chem. Soc. Japan*, **36**, 99 (1963).
151. M. Kanaoka, *Pharm. Bull.* (*Tokyo*), **5**, 385 (1957).
152. A. S. Wasfi, *J. Indian Chem. Soc.*, **45**, 750 (1968).
153. H. Bredereck, R. Gompper, F. Effenberger, H. Keck, and H. Heise, *Chem. Ber.*, **93**, 1398 (1960).
154. H. Bredereck, F. Effenberger and H. J. Treiber, *Chem. Ber.*, **96**, 1505 (1963).
155. S. S. Novikov, L. I. Khmel'nitskii, S. N. Shvedova, E. V. Shepelev and O. V. Lebedev, *Izv. Akad. Nauk. SSSR, Ser. Khim.*, 888 (1971); *Chem. Abstr.*, **75**, 76118p (1971).
156. D. G. Neilson, K. M. Watson and M. Butt, unpublished results.
157. P. H. L. Wei, S. C. Bell and S. J. Childress, *U.S. Pat.* 3,574,739 (1971); *Chem. Abstr.*, **75**, 63414g (1971).
158. F. C. Schaefer and K. P. Huffman, *U.S. Pat.* 3,225,077 (1965); *Chem. Abstr.*, **64**, 9602 (1966).
159. V. V. Moskva, A. I. Maikova and A. I. Razumov, *Zh. Obshch. Khim.*, **38**, 2586 (1968).
160. D. G. Neilson and D. A. V. Peters, unpublished results.
161. C. Runti and C. Nisi, *J. Med. Chem.*, **7**, 814 (1964).
162. A. Kovačič, B. Stanovnik and M. Tišler, *J. Heterocyclic Chem.*, **5**, 351 (1968).
163. C. Ainsworth, *J. Am. Chem. Soc.*, **87**, 5800 (1965).
164. I. Hagedorn and H.-D. Winkelmann, *Chem. Ber.*, **99**, 850 (1967).
165. P. Benko and L. Pallos, *J. Prakt. Chem.*, **313**, 179 (1971).
166. P. Benko and L. Pallos, *J. Prakt. Chem.*, **314**, 636 (1972).
167. W. I. Awad, F. A. Hussein and M. T. A. Bashi, *J. Chem. U.A.R.*, **10**, 153 (1967).
168. F. A. Hussein and K. S. Al-Dulaimi, *J. Chem. U.A.R.*, **9**, 287 (1966).
169. P. Benko, L. Pallos, F. Ordogh and B. Rosdy, *Hung. Pat.* 155,431 (1968); *Chem. Abstr.*, **71**, 3284x (1969).
170. P. Benko and L. Pallos, *Magy. Kem. Foly.*, **78**, 111 (1972); *Chem. Abstr.*, **77**, 19498e (1972).
171. R. J. Kaufmann and R. Adams, *J. Am. Chem. Soc.*, **45**, 1744 (1923).
172. R. M. Roberts, T. D. Higgins, Jr. and P. R. Noyes, *J. Am. Chem. Soc.*, **77**, 3801 (1955).
173. E. Schmidt, *Chem. Ber.*, **47**, 2545 (1914).
174. H. Bader and J. D. Downer, *J. Chem. Soc.*, 1636 (1953).
175. A. H. Cook, A. C. Davis, I. Heilbron, and G. H. Thomas, *J. Chem. Soc.*, 1071 (1949).

176. F. W. Stacey, *U.S. Pat.* 3,115,513 (1963); *Chem. Abstr.*, **60**, 6754f (1964).
177. M. L. Evans, *Ger. Offen.* 2,011,937 (1970); *Chem. Abstr.*, **74**, 22997k (1971).
178. A. Schmidpeter and W. Zeiss, *Chem. Ber.*, **104**, 1199 (1971).
179. G. Zinner, G. Nebel and M. Hitze, *Arch. Pharm.* (*Weinheim*), **303**, 317 (1970); *Chem. Abstr.*, **73**, 14048r (1970).
180. R. M. Khomutov, *Zh. Obshch. Khim.*, **31**, 1992 (1961).
181. M. Y. Karpeiskii, R. M. Khomutov and E. S. Severin, *Zh. Obshch. Khim.*, **32**, 1357 (1962).
182. A. O. Ilvespää and A. Marxer, *Helv. Chim. Acta*, **46**, 2009 (1963).
183. R. J. W. Cremlyn, *J. Chem. Soc.*, 1805 (1961).
184. R. Raap, *Canad. J. Chem.*, **49**, 1792 (1971).
185. A. Halleux, *Angew. Chem.*, **76**, 889 (1964).
186. J. S. Walia, P. S. Walia, L. Heindl and H. Lader, *Chem. Commun.*, 1290 (1967).
187. J. S. Walia, D. H. Rao, M. Singh and G. R. Nath, *Chem. and Ind.*, 583 (1967).
188. P. Grünanger, P. V. Finzi and C. Scotti, *Chem. Ber.*, **98**, 623 (1965).
189. G. Himbert and M. Regitz, *Chem. Ber.*, **105**, 2975 (1972).
190. M. E. Hermes and F. D. Marsh, *J. Am. Chem. Soc.*, **89**, 4760 (1967).
191. S. V. Adomaitene, A. M. Sladkov and V. P. Shishkov, *Zh. Obshch. Khim.*, **34**, 432 (1964).
192. S. V. Adomaitene, A. M. Sladkov and V. P. Shishkov, *Zh. Obshch. Khim.*, **34**, 2958 (1964).
193. Farbenfabriken Bayer A-G, *Belg. Pat.* 610,175 (1962); *Chem. Abstr.*, **57**, 13694 (1962).
194. A. Kaji and K. Miyazaki, *Nippon Kagaku Zasshi*, **87**, 727 (1966); *Chem. Abstr.*, **65**, 15255h (1966).
195. T. Saegusa, Y. Ito, S. Tomita, H. Kinoshita and N. Taka-ishi, *Tetrahedron*, **27**, 27 (1971).
196. S. R. Sterlin, B. L. Dyatkin, L. G. Zhuravkova and I. L. Knunyants, *Izv. Akad. Nauk SSSR Ser. Khim.*, **5**, 1176 (1969); *Chem. Abstr.*, **71**, 38233g (1969).
197. W. A. F. Gladstone, J. B. Aylward and R. O. C. Norman, *J. Chem. Soc.* (*C*), 2587 (1969).
198. R. N. Butler, F. L. Scott and T. A. F. O'Mahony, *Chem. Revs.*, **73**, 93 (1973).
199. T. Saegusa, Y. Ito, S. Kobayashi and K. Hirota, *Tetrahedron Letters*, 521 (1967).
200. T. Saegusa, Y. Ito, S. Kobayashi, N. Takeda and K. Hirota, *Tetrahedron Letters*, 1273 (1967).
201. T. Saegusa and Y. Ito, *Japan. Pat.* 70 21,685; *Chem. Abstr.*, **73**, 87482n (1970).
202. T. Saegusa, Y. Ito, S. Kobayashi, K. Hirota and N. Takeda, *Canad. J. Chem.*, **47**, 1217 (1969).
203. T. Saegusa, S. Kobayashi, K. Hirota, Y. Okumura and Y. Ito, *Bull. Chem. Soc., Japan*, **41**, 1638 (1968).
204. T. Saegusa, S. Kobayashi and Y. Ito, *J. Org. Chem.*, **35**, 2118 (1970).
205. B. Crociani and T. Boschi, *J. Organometal. Chem.*, **24**, Cl (1970).

206. P. F. B. Barnard, *J. Chem. Soc. (A)*, 2140 (1969).
207. H. C. Clark and L. E. Manzer, *Inorg. Chem.*, **10**, 2699 (1971).
208. H. C. Clark and L. E. Manzer, *J. Organometal. Chem.*, **47**, C17 (1973).
209. G. Rouschias and G. Wilkinson, *J. Chem. Soc. (A)*, 489 (1968).
210. A. J. Bloodworth, A. G. Davies and S. C. Vasishtha, *J. Chem. Soc. (C)*, 1309 (1967).
211. A. G. Davies, *U.S. Pat.* 3,347,890 (1967); *Chem. Abstr.*, **68**, 49775z (1968).
212. A. G. Davies and R. J. Puddephatt, *J. Organometal. Chem.*, **5**, 590 (1966).
213. H. Tani, H. Araki and H. Yasuda, *J. Polym. Sci., Part B*, **6**, 389 (1968).
214. H. Tani and H. Yasuda, *J. Polym. Sci., Part B*, **7**, 17 (1969).
215. Y. Kai, N. Yasuoka, N. Kasai and M. Kakudo, *J. Organometal. Chem.*, **32**, 165 (1971).
216. Y. Kai, N. Yasuoka, N. Kasai and M. Kakudo, *Bull. Chem. Soc. Japan*, **45**, 3403, 3397, 3388 (1972).
217. E. O. Fischer and R. Aumann, *Angew. Chem. Int. Ed.*, **6**, 181 (1967).
218. E. O. Fischer and R. Aumann, *Chem. Ber.*, **101**, 963 (1968).
219. B. M. Mikhailov and V. A. Dorokhov, *Izv. Akad. Nauk SSSR Ser. Khim.*, 1446 (1970); *Chem. Abstr.*, **74**, 53881x (1971).
220. A. Meller and A. Ossko, *Monatsh. Chem.*, **102**, 131 (1970).
221. A. Meller and W. Maringgele, *Monatsh. Chem.*, **102**, 121 (1971).
222. S. F. Stafiej and E. A. Takacs, *U.S. Pat.* 3,341,582 (1967); *Chem. Abstr.*, **68**, 68486x (1968).
223. S. F. Stafiej and E. A. Takacs, *U.S. Pat.* 3,383,399 (1968); *Chem. Abstr.*, **69**, 76615z (1968).
224. L. E. Benzamin, D. A. Carvalho, S. F. Stafiej and E. A. Takacs, *Inorg. Chem.*, **9**, 1844 (1970).
225. R. Raap, *Canad. J. Chem.*, **46**, 2255 (1968).
226. G. Barnikow and G. Strickmann, *Chem. Ber.*, **100**, 1661 (1967).
227. W. Walter, H. Weiss and K. J. Reubke, *Ann. Chem.*, **736**, 166 (1970).
228. K. Hartke and B. Seib, *Tetrahedron Letters*, 5523 (1968).
229. K. Hartke and B. Seib, *Arch. Pharm.* **303**, 625 (1970); *Chem. Abstr.*, **73**, 98859s (1970).
230. E. Grigat, R. Pütter and E. Mühlbauer, *Chem. Ber.*, **98**, 3777 (1965).
231. E. Grigat and R. Pütter, *Angew. Chem. Int. Ed.*, **6**, 206 (1967).
232. E. Grigat, *Angew. Chem. Int. Ed.*, **11**, 949 (1972).
233. D. Martin, K.-H. Schwarz, J. Rackow, P. Reich, and E. Gründemann, *Chem. Ber.*, **99**, 2302 (1966).
234. A. Reiker, R. Beutler, B. Narr, and E. Mueller, *Ann. Chem.*, **761**, 1 (1972).
235. H. Hart and J. W. Link, *J. Org. Chem.*, **34**, 758 (1969).
236. C. G. McCarty, *The Chemistry of the Carbon–Nitrogen Double Bond* (Ed. S. Patai), John Wiley and Sons, New York, 1970, pp. 363–463.
237. R. Moriarty, C.-L. Yeh, K. C. Ramey, and P. W. Whitehurst, *J. Am. Chem. Soc.*, **92**, 6360 (1970).
238. E.-L. Yeh, R. M. Moriarty, C.-L. Yeh and K. C. Ramey, *Tetrahedron Letters*, 2655 (1972).
239. H. Lumbroso, D. Bretin, and P. Reynaud, *Compt. Rend.*, **261**, 399 (1965).
240. H. Lumbroso and D. Bretin, *Bull. Soc. Chim. France*, 1728 (1970).
241. A. J. Papa, *J. Org. Chem.*, **35**, 2837 (1970).

242. D. Šnobl and O. Exner, *Collect. Czech. Chem. Comm.*, **34**, 3325 (1969).
243. J. H. Davies, R. H. Davis and P. Kirby, *J. Chem. Soc. (C)*, 431 (1968).
244. M. G. Waite and G. A. Sim, *J. Chem. Soc. (B)*, 752 (1971).
245. O. Exner and O. Schindler, *Helv. Chim. Acta.*, **52**, 577 (1969).
246. M. Raban, *Chem. Communications*, 1415 (1970).
247. E. Spinner, *Aust. J. Chem.*, **19**, 2153 (1966).
248. W. H. Prichard and W. J. Orville-Thomas, *J. Chem. Soc. (A)*, 1102 (1967).
249. B. Baccar, R. Mathis, A. Secches, J. Barrans and F. Mathis, *J. Mol. Structure*, **7**, 369 (1971).
250. R. Mathis, B. Baccar, J. Barrans and F. Mathis, *J. Mol. Structure*, **7**, 355 (1971).
251. R. Pujol and F. Mathis, *C.R. Acad. Sci. Paris (B)*, **269**, 975 (1969).
252. R. Pujol and F. Mathis, *C.R. Acad. Sci. Paris (B)*, **269**, 1049 (1969).
253. P. Reynaud and R. C. Moreau, *Bull. Soc. Chim. France.*, 2997 (1964).
254. N. Stojanac and N. Trinajstić, *Monatsh. Chem.*, **98**, 2263 (1967).
255. F. Cramer, K. Pawelzik and F. W. Lichtenthaler, *Chem. Ber.*, **91**, 1555 (1958).
256. H. Felkin, *C.R. Acad. Sci. Paris*, **240**, 2322 (1955).
257. S. M. McElvain and B. E. Tate, *J. Am. Chem. Soc.*, **73**, 2233 (1951).
258. J. P. Schroeder, D. C. Schroeder, J. Hardin and J. K. Marshall, *J. Org. Chem.*, **34**, 3332 (1969).
259. M. L. Shul'man and G. A. Schvekhgeimer, *Dokl. Akad. Nauk SSSR*, **173**, 378 (1967).
260. M. L. Shul'man, G. A. Shvekhgeimer and R. A. Miftakhova, *Zh. Org. Khim.*, **3**, 874 (1967).
261. A. W. Chapman, *J. Chem. Soc.*, **123**, 1150 (1923).
262. A. W. Chapman, *J. Chem. Soc.*, 569 (1929).
263. O. Mumm, H. Hesse and H. Volquartz, *Chem. Ber.*, **48**, 379 (1915).
264. J. W. Schulenberg and S. Archer, *Org. Reactions*, **14**, 1 (1965).
265. O. H. Wheeler, F. Roman, M. V. Santiago and F. Quiles, *Canad. J. Chem.*, **47**, 503 (1969).
266. A. L. J. Beckwith, *The Chemistry of Amides* (Ed. J. Zabicky), Interscience Publishers, New York, 1970, pp. 73–185.
267. A. W. Chapman, *J. Chem. Soc.*, **127**, 1992 (1925).
268. K. B. Wiberg and B. I. Rowland, *J. Am. Chem. Soc.*, **77**, 2205 (1955).
269. O. H. Wheeler, F. Roman and O. Rosado, *J. Org. Chem.*, **34**, 966 (1969).
270. H. M. Relles, *J. Org. Chem.*, **33**, 2245 (1968).
271. G. Bock and W. Deuschel, *Belg. Pat.* 628,295 (1963); *Chem. Abstr.*, **60**, 16031h (1964).
272. M. M. Jamison and E. E. Turner, *J. Chem. Soc.*, 1954 (1937).
273. J. Cymerman-Craig and J. W. Loder, *J. Chem. Soc.*, 4309 (1955).
274. G. Bock, *Chem. Ber.*, **100**, 2870 (1967).
275. R. Barclay, Jr., *Canad. J. Chem.*, **43**, 2125 (1965).
276. H. M. Blatter, H. Lukaszewski and G. deStevens, *J. Org. Chem.*, **30**, 1020 (1965).
277. A. W. Chapman, *J. Chem. Soc.*, 2296 (1926).
278. C. S. Benton, *Dissertation Abstr.*, **18**, 1251 (1958).
279. C. G. McCarty, *The Chemistry of the Carbon–Nitrogen Double Bond* (Ed. S. Patai), Interscience Publishers, New York, 1970, pp. 439–447.

280. O. Mumm and F. Möller, *Chem. Ber.*, **70**, 2214 (1937).
281. N. P. Marullo, C. D. Smith and J. F. Terapane, *Tetrahedron Letters*, 6279 (1966).
282. T. Taguchi, Y. Kawazoe, K. Yoshihira, H. Kanayama, M. Mori, K. Tabata and K. Harano, *Tetrahedron Letters*, 2717 (1965).
283. K. B. Wiberg, T. M. Shryne and R. R. Kintner, *J. Am. Chem. Soc.*, **79**, 3160 (1957).
284. G. D. Lander, *J. Chem. Soc.*, **83**, 406 (1903).
285. A. E. Arbuzov and V. E. Shishkin, *Dokl. Acad. Nauk SSSR.*, **141**, 81, 349, 611 (1961).
286. A. E. Arbuzov, V. E. Shishkin and S. S. Tyulenev, *Zh. Org. Khim.*, **1**, 1442 (1965).
287. R. E. Benson and T. L. Cairns, *J. Am. Chem. Soc.*, **70**, 2115 (1948).
288. J. W. Ralls and C. A. Eliger, *Chem. Ind. (London)*, 20 (1961).
289. G. L. Schmir and B. A. Cunningham, *J. Am. Chem. Soc.*, **87**, 5692 (1965).
290. B. A. Cunningham and G. L. Schmir, *J. Am. Chem. Soc.*, **88**, 551 (1966).
291. T. Okuyama, D. J. Sahn and G. L. Schmir, *J. Am. Chem. Soc.*, **95**, 2345 (1973).
292. M. Kandel and E. H. Cordes, *J. Org. Chem.*, **32**, 3061 (1967).
293. E. S. Hand and W. P. Jencks, *J. Am. Chem. Soc.*, **84**, 3505 (1962).
294. R. K. Chaturvedi and G. L. Schmir, *J. Am. Chem. Soc.*, **90**, 4413 (1968).
295. T. Okuyama, T. C. Pletcher, D. J. Sahn and G. L. Schmir, *J. Am. Chem. Soc.*, **95**, 1253 (1973).
296. G. M. Blackburn and W. P. Jencks, *J. Am. Chem. Soc.*, **90**, 2638 (1968).
297. T. C. Pletcher, S. Koehler and E. H. Cordes, *J. Am. Chem. Soc.*, **90**, 7072 (1968).
298. W. P. Jencks and M. Gilchrist, *J. Am. Chem. Soc.*, **90**, 2622 (1968).
299. T. Okuyama and G. L. Schmir, *J. Am. Chem. Soc.*, **94**, 8805 (1972).
300. C. R. Smith and K. Yates, *J. Am. Chem. Soc.*, **94**, 8811 (1972).
301. R. K. Chaturvedi and G. L. Schmir, *J. Am. Chem. Soc.*, **91**, 737 (1969).
302. R. K. Chaturvedi, A. E. MacMahon and G. L. Schmir, *J. Am. Chem. Soc.*, **89**, 6984 (1967).
303. R. H. DeWolfe, *J. Org. Chem.*, **36**, 162 (1971).
304. J. T. Edwards and S. C. R. Meacock, *J. Chem. Soc.*, 2009 (1957).
305. R. H. DeWolfe and F. B. Augustine, *J. Org. Chem.*, **30**, 699 (1965).
306. T. L. Vigo, D. J. Daigle and C. H. Mack, *J. Appl. Polym. Sci.*, **15**, 2051 (1971).
307. V. Ružička and A. Marhoul, *Collect. Czech. Chem. Commun.*, **33**, 622 (1968).
308. G. A. Shvekhgeimer and M. L. Shul'man, *Zh. Org. Khim.*, **3**, 600 (1967).
309. A. I. Shreibert, V. E. Shishkin and N. V. Kryukov, *Zh. Org. Khim.*, **7**, 2439 (1971).
310. R. A. Chittenden and G. H. Cooper, *J. Chem. Soc. (C)*, 49 (1970).
311. S. Scheithauer and R. Mayer, *Chem. Ber.*, **100**, 1413 (1967).
312. W. Walter and J. Krohn, *Chem. Ber.*, **102**, 3786 (1969).
313. H. C. Brown and C. R. Wetzel, *J. Org. Chem.*, **30**, 3724, 3729 (1965).
314. J. E. MacKenzie, *J. Chem. Soc.*, **113**, 1 (1918).
315. H. G. Rule, *J. Chem. Soc.*, **113**, 3 (1918).
316. P. Reynaud, *Brit. Pat.* 1,080,879 (1967); *Chem. Abstr.*, **68**, 95566e (1968).

317. S. Mizukami and K. Nagata, *Chem. Pharm. Bull.*, **14**, 1249 (1966).
318. G. Barnikow and G. Strickmann, *Chem. Ber.*, **100**, 1428 (1967).
319. R. N. Hurd and G. DeLaMater, *Chem. Revs.*, **61**, 45 (1961).
320. J. Goerdeler and W. Mittler, *Chem. Ber.*, **96**, 944 (1963).
321. D. A. Peak and F. Stansfield, *J. Chem. Soc.*, 4067 (1952).
322. H. W. Post, *The Chemistry of Aliphatic Ortho Esters*, Reinhold Publishing Co., New York, 1963.
323. A. Pinner, *Chem. Ber.*, **16**, 1643 (1883).
324. P. P. T. Sah, *J. Am. Chem. Soc.*, **50**, 516 (1928).
325. P. P. T. Sah, S. Y. Ma and C. H. Kao, *J. Chem. Soc.*, 305 (1931).
326. S. M. McElvain and J. W. Nelson, *J. Am. Chem. Soc.*, **64**, 1825 (1942).
327. S. M. McElvain and C. L. Aldridge, *J. Am. Chem. Soc.*, **75**, 3987 (1953).
328. S. M. McElvain and R. E. Starn, Jr., *J. Am. Chem. Soc.*, **77**, 4571 (1955).
329. S. M. McElvain and D. H. Clemens, *J. Am. Chem. Soc.*, **80**, 3915 (1958).
330. S. M. McElvain and C. L. Stevens, *J. Am. Chem. Soc.*, **69**, 2663 (1947).
331. S. M. McElvain and C. L. Aldridge, *J. Am. Chem. Soc.*, **75**, 3993 (1953).
332. H. E. Zaugg, V. Papendick and R. J. Michaels, *J. Am. Chem. Soc.*, **86**, 1399 (1964).
333. H. Eilingsfield, M. Seefelder and H. Weidinger, *Angew. Chem.*, **72**, 836 (1960).
334. F. Moulin, *Swiss Pat.* 420,100 (1967); *Chem. Abstr.*, 67, 53707a (1967).
335. M. Berçot-Vatteroni, R. C. Moreau and P. Reynaud, *Bull. Soc. Chim. France.*, 1820 (1961).
336. P. Reynaud, R. C. Moreau and J.-P. Samama, *Bull. Soc. Chim. France.*, 3628 (1965).
337. R. L. Shriner and F. W. Neumann, *Chem. Revs.*, **35**, 351 (1944).
338. S. R. Sandler and W. Karo, *Organic Functional Group Preparation*, Vol. III, Academic Press, New York, 1972, pp. 205–237.
339. E. J. Poziomek, *J. Org. Chem.*, **28**, 590 (1963).
340. Welcome Foundation Ltd., *France*, 1,572,961 (1969); Chem. Abstr., 72, 111091m (1970).
341. A. V. Kirsanov and V. A. Shokol, *Zh. Obshch. Khim.*, **31**, 582 (1961).
342. R. G. Glushkov and O. Yu. Magidson, *Zh. Obshch. Khim.*, **31**, 189 (1961).
343. G. Szilagyi, E. Kasztreiner, L. Vargha and J. Borsy, *Acta Chim.*, (Budapest) **65**, 325 (1970); *Chem. Abstr.*, **74**, 22523c (1971).
344. P. Reynaud, R. C. Moreau and N. H. Thu, *C.R. Acad. Sci. Paris*, **253**, 2540 (1961).
345. E. C. Taylor and P. K. Loeffler, *J. Am. Chem. Soc.*, **82**, 3147 (1960).
346. D. L. Garmaise, R. W. Kay, R. Gaudry, H. A. Baker and A. F. McKay, *Canad. J. Chem.*, **39**, 1493 (1961).
347. L. Baiocchi and G. Palazzo, *Ann. Chim.* (*Rome*), **58**, 608 (1968); *Chem. Abstr.*, **69**, 76859g (1968).
348. W. Lwowski, *Synthesis*, 263 (1971).
349. M. Ruccia, N. Vivona and G. Cusmano, *J. Hetero. Chem.*, **8**, 137 (1971).
350. M. Miyazaki, *Japan Pat.* 4477 (1952); *Chem. Abstr.*, **48**, 8259 (1954).
351. S. E. Callander and J. Yates, *Brit. Pat.* 1,018,308 (1966); *Chem. Abstr.*, **64**, 17499f (1966).
352. H. Neunhoeffer and H. Hennig, *Chem. Ber.*, **101**, 3947 (1968).
353. W. Oberhummer, *Monatsh. Chem.*, **57**, 106 (1931).

354. W. Oberhummer, *Monatsh. Chem.*, **63**, 285 (1933).
355. D. G. Neilson, R. Roger, J. W. M. Heatlie and L. R. Newlands, *Chem. Revs.*, **70**, 151 (1970).
356. S. Mahmood, D. G. Neilson and K. M. Watson, unpublished results.
357. D. D. Libman and R. Slack, *J. Chem. Soc.*, 2253 (1956).
358. R. H. Wiley, C. H. Jarboe, Jr., and F. N. Hayes, *J. Org. Chem.*, **22**, 835 (1957).
359. P. Westermann, *Chem. Ber.*, **97**, 523 (1964).
360. S. Kubota, O. Kirino, Y. Koida and K. Miyake, *J. Pharm. Soc. Japan*, **92**, 275 (1972).
361. V. P. Wystrach, *Heterocyclic Compounds*, Vol. 8 (Ed. R. C. Elderfield), John Wiley and Sons, New York, 1967, pp. 105–161.
362. E. C. Taylor and S. F. Martin, *J. Org. Chem.*, **37**, 3958 (1972).
363. M. Brugger, H. Wamhoff and F. Korte, *Ann. Chem.*, **755**, 101 (1972).
364. P. Mukaiyama and S. Ono, *Tetrahedron Letters*, **32**, 3569 (1968).
365. J. L. Fahey, P. A. Foster, D. G. Neilson, K. M. Watson, J. L. Brokenshire and D. A. V. Peters, *J. Chem. Soc.* (*C*), 719 (1970).
366. P. Yates, O. Meresz and H. Morrison, *Tetrahedron Letters*, **1**, 77 (1967).
367. D. G. Neilson, S. Mahmood and K. M. Watson, *J. Chem. Soc., Perkin. Trans.* **1**, 335 (1973).
368. O. Meresz and P. A. Foster-Verner, *Chem. Comm.*, 950 (1972).
369. C. L. Leese and G. M. Timmis, *J. Chem. Soc.*, 3816 (1961).
370. R. N. Naylor, G. Shaw, D. V. Wilson and D. N. Butler, *J. Chem. Soc.*, 4845 (1961).
371. E. Grigat and R. Puetter, *Ger. Offen.* 1,940,366 (1971); *Chem. Abstr.*, **74**, 100,065 (1971).
372. B. G. Baccar and R. Mathis, *C.R. Acad. Sci. Paris*, **261**, 174 (1965).
373. P. Westermann, H. Paul and G. Hilgetag, *Chem. Ber.*, **97**, 528 (1964).
374. H. Paul, G. Hilgetag and G. Jähnchen, *Chem. Ber.*, **101**, 2033 (1968).
375. H. G. O. Becker, G. Goermar and H. J. Timpe, *J. Prakt. Chem.*, **312**, 610 (1970).
376. P. Reynaud, R. C. Moreau and T. Gousson, *Compt. Rend.*, **259**, 4067 (1964).
377. Yu. A. Rybakova and N. P. Bednyagina, *Khim. Geterotsikl. Soedin. Akad. Nauk Latv. SSR.*, 287 (1965); *Chem. Abstr.*, 63, 6992a (1965).
378. H. Kohn and R. A. Olofson, *J. Org. Chem.*, **37**, 3504 (1972).
379. M. Pesson, S. Dupin and M. Antoine, *C.R. Acad. Sci. Paris*, **253**, 285 (1961).
380. H. Weidinger and J. Kranz, *Chem. Ber.*, **96**, 1049 (1963).
381. C. S. Davis, *U.S. Pat.* 3,458,500 (1969); *Chem. Abstr.*, **71**, 81223n (1969).
382. G. Caraculaca, L. Stoicescu-Crivetz and I. Zugravescu, *Rev. Roum. Chim.*, **12**, 1021 (1967); *Chem. Abstr.*, **69**, 77827g (1968).
383. J. J. Baldwin and F. C. Novello, *Ger. Offen.* 2,147,794 (1972); *Chem. Abstr.*, **77**, 5480f (1972).
384. K. Karigome, H. Ueda and T. Furuhashi, *Japan Pat.* 72 32,071 (1972); *Chem. Abstr.*, **78**, 16190h (1973).
385. H. Weidinger and J. Kranz, *Chem. Ber.*, **96**, 1059 (1963).
386. E. J. Browne and J. B. Polya, *J. Heterocyclic Chem.*, **3**, 523 (1966).
387. E. Steininger, *Monatsh. Chem.*, **97**, 1195 (1966).

388. F. A. Hussein and A. A. Kadir, *J. Indian Chem. Soc.*, **45**, 729 (1968).
389. R. Gompper, H. E. Noppel and H. Schaefer, *Angew. Chem.*, **75**, 918 (1963).
390. B.-G. Baccar and F. Mathis, *C.R. Acad. Sci. Paris*, **258**, 6470 (1964).
391. G. S. Gol'din, V. G. Poddubnyi, A. A. Simonova and A. B. Kamenskii, *U.S.S.R.* 253,064 and 254,513 (1969); *Chem. Abstr.*, **72**, 121,687z and 132,945h (1970).
392. G. S. Shor and E. A. Rybakova, *Zh. Org. Khim.*, **5**, 1404 (1969).
393. W. Ried and A. Czack, *Ann. Chem.*, **676**, 121 (1964).
394. R. L. Jones and J. R. Kershaw, *Rev. Pure and Applied Chem.*, **21**, 23 (1971).
395. R. M. Khomutov, E. S. Severin, N. V. Gnuchev and T. Ya. Derevyanko, *Izv. Akad. Nauk SSSR. Ser. Khim.*, 1820 (1967); *Chem. Abstr.*, **68**, 39053y (1968).
396. J. B. Buchanan, *U.S. Pat.* 3,374,260 (1968); *Chem. Abstr.*, **69**, 43,482a (1968).
397. Yu. V. Markova, N. G. Ostroumova and M. N. Shchukina, *Zh. Org. Khim.*, **3**, 1207 (1967).
398. F. Eloy and R. Lenaers, *Chem. Revs.*, **62**, 155 (1962).
399. H. G. Aurich, *Chem. Ber.*, **101**, 1761 (1968).
400. W. Ried, W. Stephan and W. von der Emden, *Chem. Ber.*, **95**, 728 (1962).
401. W. Ried and D. Piechaczek, *Ann. Chem.*, **696**, 97 (1966).
402. I. K. Fel'dman and E. I. Boksiner, *Chem. Abstr.*, **65**, 3953h (1966).
403. W. Ried and W. Stephan, *Chem. Ber.*, **95**, 3042 (1962).
404. W. Taub and A. Loeffler, *Ger. Offen.*, 1,947,062 (1970); *Chem. Abstr.*, **72**, 111,521b (1970).
405. A. Kjaer, *Acta. Chem. Scand.*, **7**, 1017 (1953).
406. H. Lehr, S. Karlan and M. W. Goldberg, *J. Am. Chem. Soc.*, **75**, 3640 (1953).
407. A. R. Kidwai and G. M. Devasia, *J. Org. Chem.*, **27**, 4527 (1962).
408. C. S. Miller, S. Gurin and D. W. Wilson, *J. Am. Chem. Soc.*, **74**, 2892 (1952).
409. K. R. Huffman and F. C. Schaefer, *J. Org. Chem.*, **28**, 1816 (1963).
410. G. Shaw and R. N. Warrener, *Proc. Chem. Soc.*, 193 (1958).
411. M. J. Hunter and M. L. Ludwig, *J. Am. Chem. Soc.*, **84**, 3491 (1962).
412. L. Wofsy and S. J. Singer, *Biochemistry*, **2**, 104 (1963).
413. J. H. Reynolds, *Biochemistry*, **7**, 3131 (1968).
414. M. L. Ludwig and R. Byrne, *J. Am. Chem. Soc.*, **84**, 4160 (1962).
415. A. Dutton, M. Adams and S. J. Singer, *Biochem. and Biophys. Res. Commun.*, **23**, 730 (1966).
416. G. Davis and G. R. Stark, *Proc. Nat. Acad. Sci.*, *U.S.A.*, **66**, 651 (1970).
417. F. Wold, *Methods in Enzymology*, **11**, 617 (1967).
418. M. L. Ludwig and M. J. Hunter, *Methods in Enzymology*, **11**, 595 (1967).
419. F. M. Robbins, M. J. Kronman and R. E. Andreotti, *Biochim. Biophys. Acta.*, **109**, 223 (1965).
420. M. Ljaljevic, J. Ljaljevic and C. W. Parker, *J. Immunol.*, **100**, 1041 (1968).

421. S. P. Agarwal, C. J. Martin, T. T. Blair and M. A. Marini, *Biochem. Biophys. Res. Commun.*, **43**, 510 (1971).
422. H. J. Schramm, *Hoppe-Seyler's Z. Physiol. Chem.*, **348**, 289 (1967); *Chem. Abstr.*, **67**, 108348s (1967).
423. L. I. Slobin, *Proc. Nat. Acad. Sci.*, *U.S.A.*, **69**, 3769 (1972).
424. L. I. Slobin, *J. Mol. Biol.*, **64**, 297 (1972).
425. F. H. Carpenter and K. T. Harrington, *J. Biol. Chem.*, **247**, 5580 (1972).
426. W. G. Niehaus, Jr., and F. Wold, *Biochem. Biophys. Acta*, **196**, 170 (1970).
427. J. W. Cornforth, *The Chemistry of Penicillin*, Princeton University Press, Princeton, N. J., 1949, p. 727.
428. G. Shaw and D. N. Butler, *J. Chem. Soc.*, 4040 (1959).
429. A. Lawson, *J. Chem. Soc.*, 4225 (1957).
430. M. R. Grimmett, *Advances in Heterocyclic Chem.*, **12**, 119 (1970).
431. J. W. Cornforth and H. T. Huang, *J. Chem. Soc.*, 1960 (1948).
432. H. W. van Meeteren and H. C. van den Plas, *Rec. Trav. Chim.*, *Pays-Bas*, **87**, 1089 (1968).
433. J. W. Cornforth and R. H. Cornforth, *J. Chem. Soc.*, 96 (1947).
434. N. W. Bristow, *J. Chem. Soc.*, 513 (1957).
435. E. R. Freiter, L. E. Begin and A. H. Abdallah, *J. Heterocyclic Chem.*, **10**, 391 (1973).
436. A. B. Sen and K. Shanker, *J. Prakt. Chem.*, **29**, 309 (1965).
437. Societe pour L'Industrie Chimique à Bale, *Swiss Pat.* 204,730 (1939); *Chem. Abstr.*, **35**, 2680 (1941).
438. F. E. King and R. M. Acheson, *J. Chem. Soc.*, 1396 (1949).
439. R. P. Thomas and G. J. Tyler, *J. Chem. Soc.*, 2197 (1957).
440. G. A. Shvekhgeimer, E. I. Gorbatov and V. D. Tyurin, *Zh. Org. Khim.*, **7**, 815 (1971).
441. Z. F. Solomko and G. A. Polinovskii, *Khim. Geterotsikl. Soedin.*, 874 (1969).
442. G. Holan and E. L. Samuel, *U.S. Pat.* 3,560,195 (1971); *Chem. Abstr.*, **74**, 141802c (1971).
443. V. V. Korshak, S. V. Vinogradova and V. A. Pandratov, *Vysokomol Soedin Ser. B.*, **10**, 481 (1968); *Chem. Abstr.*, **69**, 77824d (1968).
444. V. V. Korshak, S. V. Vinogradova and V. A. Pandratov, *Vysokomol. Soedin Ser. B.*, **13**, 550 (1971); *Chem. Abstr.*, **75**, 141, 204e (1971).
445. A. Hunger, J. Kebrle, A. Rossi and K. Hoffmann, *Helv. Chim. Acta.*, **43**, 1727 (1960).
446. C. R. Ganellin, H. F. Ridley and R. G. W. Spickett, *J. Heterocyclic Chem.*, **3**, 278 (1966).
447. G. Paglietti and F. Sparatore, *Farmaco. Ed. Sci.*, **27**, 333 (1972); *Chem. Abstr.*, **77**, 48338h (1972).
448. J. W. Cornforth, E. Fawaz, L. J. Goldsworthy and R. Robinson, *J. Chem. Soc.*, 1549 (1949).
449. A. R. Martin and R. Ketcham, *J. Org. Chem.*, **31**, 3612 (1966).
450. J. A. Frump, *Chem. Revs.*, **71**, 483 (1971).
451. W. Wislicenus and H. Körber, *Chem. Ber.*, **35**, 164 (1902).
452. S. Inaba, K. Ishizumi, T. Okamoto and H. Yamamoto, *Chem. Pharm. Bull.* (*Japan*), 20, 1628 (1972).

453. T. Taguchi and M. Kojima, *Pharm. Bull., Japan*, **3**, 4 (1955).
454. K. Bruns, *J. Prakt. Chem.*, **24**, 74 (1964).
455. G. E. McCasland and E. C. Horswill, *J. Am. Chem. Soc.*, **73**, 3744 (1951).
456. D. F. Elliot, *J. Chem. Soc.*, 589 (1949).
457. C. L. Stevens, B. T. Gillis and T. H. Haskell, *J. Am. Chem. Soc.*, **81**, 1435 (1959).
458. C. I. Braz, G. V. Myasnikova and A. Y. Yakubovich, *Khim. Geterosikl. Soedin*, 147 (1965); *Chem. Abstr.*, **63**, 5622g (1965).
459. E. I. Gorbatov, *Neft. Gaz. Ikh. Prod.*, 179 (1971); *Chem. Abstr.*, **78**, 3644j (1973).
460. R. N. Gitina, G. I. Braz, V. P. Bazov and A. Y. Yakubovich, *Vysokomol. Soedin*, **8**, 1535 (1966); *Chem. Abstr.*, **6**, 10869b (1967).
461. C. D. Burton and N. L. Madison, *U.S. Pat.* 3,560,438 (1971); *Chem. Abstr.*, **74**, 100, 447x (1971).
462. G. Sosnovsky and P. Schneider, *Tetrahedron*, **19**, 1313 (1963).
463. A. H. Cook, J. A. Elvidge, A. R. Graham and G. Harris, *J. Chem. Soc.* 3220 (1949).
464. A. H. Cook, *Quarterly Revs.*, **2**, 234 (1948).
465. K. A. Jensen and I. Crossland, *Acta Chem. Scand.*, **17**, 144 (1963).
466. F. N. Stepanov and Z. Z. Moiseeva, *Zh. Obshch. Khim.*, **25**, 1170 (1955).
467. I. Simiti and M. Farkas, *Chem. Ber.*, **98**, 3446 (1965).
468. W. Ried and E. Schmidt, *Ann. Chem.*, **676**, 114 (1964).
469. L. T. Bogolyubskaya and M. A. Alperovich, *Zh. Obshch. Khim.*, **34**, 3119 (1964).
470. G. A. Miller and M. Hausman, *J. Heterocyclic Chem.*, **8**, 657 (1971).
471. P. Rajagopalan, *Tetrahedron Letters*, 311 (1969).
472. F. Eloy and R. Lenaers, *Bull. Soc. Chim. Belg.*, **72**, 719 (1967).
473. A. Hetzheim and K. Möckel, *Adv. in Heterocyclic Chem.*, **7**, 183 (1966).
474. R. H. DeWolfe, *Carboxylic Ortho Acid Derivatives*, Academic Press, New York, 1970, p. 201.
475. J. Kram, H. Pohlemann, F. Schauder and H. Weidinger, *Ger. Pat.* 1,154,626 (1963); *Chem. Abstr.*, **59**, 15,443 (1963).
476. L. M. Weinstock, P. Davis, B. Handelsman and R. Tull, *J. Org. Chem.*, **32**, 2823 (1967).
477. C. Ainsworth, *J. Am. Chem. Soc.*, **75**, 5728 (1953).
478. I. Hagedorn, K. E. Lichtel and H. D. Winkelmann, *Angew. Chem. Int. Ed.*, **4**, 702 (1965).
479. F. R. Benson, *Heterocyclic Compounds*, Vol. 8 (Ed. R. C. Elderfield), John Wiley and Sons, New York, 1967, pp. 1–104.
480. S. Rajappa, B. G. Advani and R. Sreenivasan, *Indian J. Chem.*, **10**, 323 (1972).
481. J. Egri, J. Halmos and J. Rakoczi, *Acta. Chim. (Budapest)*, **74**, 351 (1973).
482. J. Egri, J. Halmos, A. Jeszenszky, B. Majerko and J. Rakoczi, *Ger. Pat.* 2,058,002 (1971); *Chem. Abstr.*, **75**, 63636f (1971).
483. G. W. Kenner and A. Todd, *Heterocyclic Chemistry*, Vol. 6, John Wiley and Sons, New York, 1957, p. 234.
484. W. Ried and P. Stock, *Ann. Chem.*, **700**, 87 (1966).
485. W. Ried and J. Valentin, *Ann. Chem.*, **707**, 250 (1967).
486. W. Ried and W. Stephan, *Chem. Ber.*, **95**, 3042 (1962).

487. W. Ried and R. Giesse, *Ann. Chem.*, **713**, 143 (1968).
488. W. Ried and R. Giesse, *Ann. Chem.*, **713**, 149 (1968).
489. H. Wamhoff, *Chem. Ber.*, **101**, 3377 (1968).
490. S. Rajappa and B. G. Advani, *Tetrahedron*, **29**, 1299 (1973).
491. Y. Okamoto, T. Tsuji and T. Ueda, *Chem. Pharm. Bull., Japan*, **17**, 2273 (1969).
492. N. J. Leonard, K. L. Carraway and J. P. Helgeson, *J. Heterocyclic Chem.*, **2**, 291 (1965).
493. E. C. Taylor and R. W. Hendess, *J. Am. Chem. Soc.*, **87**, 1995 (1965).
494. E. C. Taylor, A. McKillop and S. Vromen, *Tetrahedron*, **23**, 885 (1967).
495. E. C. Taylor, A. McKillop, Y. Shvo and G. H. Hawks, *Tetrehedron*, **23**, 2081 (1967).
496. K. Hartke and L. Peshkar, *Arch. Pharm.*, **301**, 611 (1968).
497. E. Richter, J. E. Loeffler and E. C. Taylor, *J. Am. Chem. Soc.*, **82**, 3144 (1960).
498. Y. Ohtsuka, *Bull. Chem. Soc. Japan*, **43**, 187 (1970).
499. H. Eilingsfeld, M. Patsch and H. Scheuermann, *Chem. Ber.*, **101**, 2426 (1968).
500. F. C. Schaefer, K. R. Huffman and G. A. Peters, *J. Org. Chem.*, **27**, 548 (1962).
501. F. C. Schaefer, K. R. Huffman and G. A. Peters, *J. Org. Chem.*, **27**, 551 (1962).
502. V. G. Granik and R. G. Glushkov, *Khim. Pharm. Zh.*, **2**, 16 (1968).
503. E. L. Zaitseva, G. I. Braz, A. Ya. Yakubovich, V. P. Bazov, R. M. Gitina, L. G. Petrova and I. M. Filatova, *Zh. Vses. Khim. Obshch. im D.I. Mendeleeva*, 8, 353 (1962); *Chem. Abstr.*, **59**, 8748h (1963).
504. G. A. Shvekhgeimer and A. P. Kryuchkova, *Tr. Mosk. Inst. Neftekhim. Gazov. Pron.*, **72**, 87 (1967); *Chem. Abstr.*, **68**, 78,345 (1968).
505. A. Ya. Yakubovich, E. L. Zaitseva, G. I. Braz and V. P. Bazov, *J. Gen. Chem.*, **32**, 3345 (1962).
506. F. C. Schaefer and G. A. Peters, *J. Org. Chem.*, **26**, 2778 (1961).
507. F. C. Schaefer, *J. Org. Chem.*, **27**, 3362 (1962).
508. F. C. Schaefer, *J. Org. Chem.*, **27**, 3608 (1962).
509. H. Bader, *U.S. Pat.* 3,444,137 (1967); *Chem. Abstr.*, **68**, 95,861x (1968).
510. H. Bader, *J. Org. Chem.*, **30**, 707 (1965).
511. P.-G. Baccar, *C.R. Acad. Sci. Paris.*, **264**, 352 (1967).
512. F. C. Schaefer and G. A. Peters, *J. Org. Chem.*, **26**, 2784 (1961).
513. P. Roffey and J. P. Verge, *J. Heterocyclic Chem.*, **6**, 497 (1969).
514. H. Neunhoffer, F. Weischedel and V. Böhnisch, *Ann. Chem.*, **750**, 12 (1971).
515. C. Temple, Jr., and J. A. Montgomery, *J. Org. Chem.*, **28**, 3038 (1963).
516. J. Allegretti, J. Hancock and R. Knutson, *J. Org. Chem.*, **27**, 1463 (1962).
517. S. Mahmood, *Ph.D. Thesis*, University of Dundee, 1973.
518. L. Stoicescu-Crivet, E. Mantaluta, G. Neamtu and I. Zugravescu, *J. Polymer. Sci. (C)*, **22**, 761 (1967).
519. V. P. Wystrach, *Heterocyclic Compounds*, Vol. 8 (Ed. R. C. Elderfield), John Wiley and Sons, New York, 1957, pp. 105–161.
520. G. Drefahl and H.-H. Hörhold, *Chem. Ber.*, **94**, 1641 (1961).
521. J. A. Faust, A. Mori and M. Sahyun, *J. Am. Chem. Soc.*, **81**, 2214 (1959).

522. F. D. Popp and A. Catala Noble, *Adv. in Heterocyclic Chem.*, **8**, 21 (1967).
523. L. I. Smith and J. Nichols, *J. Org. Chem.*, **6**, 489 (1941).
524. J. Pornet and L. Miginiac, *C.R. Acad. Sci. Paris*, **271**, 381 (1970).
525. J. Pornet and L. Miginiac, *Tetrahedron Letters.*, 967 (1971).
526. A. S. Martin, *Ph.D. Thesis*, University of St. Andrews, 1952.
527. H. Stetter and D. Theisen, *Chem. Ber.*, **102**, 1641 (1969).
528. N. Koga, G. Koga and J. P. Anselme, *Tetrahedron Letters*, 3309 (1970).
529. R. Boudet, *Bull. Soc. Chim. France*, **18**, 377 (1951).
530. W. Walter, J. Voss and J. Curts, *Ann. Chem.*, **695**, 77 (1966).
531. D. Thomas and D. H. Aue, *Tetrahedron Letters*, 1807 (1973).
532. F. Henle, *Chem. Ber.*, **38**, 1362 (1905).
533. P. de Ruggieri, C. Gandolfi and D. Chiaramonti, *Gazz. Chim. Ital.*, **91**, 665 (1961).
534. P. de Ruggieri, C. Gandolfi and D. Chiaramonti, *U.S. Pat.* 3,137,710 (1964); *Chem. Abstr.*, **61** 7075c (1964).
535. R. F. Borch and H. D. Durst, *J. Am. Chem. Soc.*, **91**, 3996 (1969).
536. A. I. Meyers and A. C. Kovelesky, *Tetrahedron Letters*, **22**, 1783 (1969).
537. A. I. Meyers, A. Nabeya, H. W. Adickes and I. R. Politzer, *J. Am. Chem. Soc.*, **91**, 763 (1969).
538. A. I. Meyers, A. Nabeya, I. R. Politzer, H. W. Adickes, J. M. Fitzpatrick and G. R. Malone, *J. Am. Chem. Soc.*, **91**, 764 (1969).
539. A. I. Meyers, H. W. Adickes, I. R. Politzer and W. N. Beverung, *J. Am. Chem. Soc.* **91**, 765 (1969).
540. H. W. Adickes, I. R. Politzer and A. I. Meyers, *J. Am. Chem. Soc.*, **91**, 2155 (1969).
541. C. J. M. Stirling, *J. Chem. Soc.*, 255 (1960).
542. H. E. Zaug and R. J. Michaels, *J. Org. Chem.*, **31**, 1332 (1968).
543. N. R. Easton and V. B. Fish, *J. Am. Chem. Soc.*, **77**, 1776 (1955).
544. L. I. Samarai, V. P. Belaya, O. V. Vishnevskii and G. I. Derkach, *Zh. Org. Khim.*, **4**, 720 (1968).
545. J. F. Olin, *Brit. Pat.* 1,115,270 (1968); *Chem. Abstr.*, **69**, 96,177m (1968).
546. L. I. Samarai, V. P. Belaya, G. F. Galenko and G. I. Derkach, *Zh. Org. Khim.*, **6**, 85 (1970).
547. H. J. Barber, *J. Chem. Soc.*, 101 (1943).
548. J. Houben and E. Schmidt, *Chem. Ber.*, **46**, 3616 (1913).
549. H. E. Baumgarten, J. E. Dirks, J. M. Petersen and R. L. Zey, *J. Org. Chem.*, **31**, 3708 (1966).
550. H. S. Wheeler and P. T. Waldon, *Am. Chem. J.*, **19**, 129 (1897).
551. D. L. Ross, C. L. Coon and M. E. Hill, *J. Org. Chem.*, **35**, 3093 (1970).
552. S. Wolfe and D. V. C. Awang, *Canad. J. Chem.*, **49**, 1384 (1971).
553. W. H. Graham, *Tetrahedron Letters*, **27**, 2223 (1969).
554. K. A. Petrov, A. A. Neimysheva, M. G. Fomenko, L. M. Chernushevich and A. D. Kuntsevich, *Zh. Obshch. Khim.*, **31**, 516 (1961).
555. G. I. Derkach, A. M. Lepesa and A. V. Kirsanov, *J. Gen. Chem.*, **32**, 167 (1962).
556. G. I. Derkach, E. S. Gubnitskaya, L. J. Samarai and V. A. Shokol, *J. Gen. Chem.*, **33**, 551 (1963).
557. G. I. Derkach and E. S. Gubnitskaya, *J. Gen. Chem.*, **35**, 1014 (1965).

558. G. I. Derkach, E. S. Gubnitskaya, V. A. Shokol and A. V. Kirsanov, *J. Gen. Chem.*, **32**, 1176 (1962).
559. G. I. Derkach, L. I. Samarai and V. A. Shokol, *J. Gen. Chem.*, **32**, 2039 (1962).
560. G. I. Derkach and L. I. Samarai, *J. Gen. Chem.*, **34**, 1152 (1964).
561. G. I. Derkach, E. S. Gubnitskaya, V. A. Shokol and A. V. Kirsanov, *J. Gen. Chem.*, **32**, 1853 (1962).
562. G. I. Derkach, L. I. Samarai, A. S. Shtepanek and A. V. Kirsanov, *J. Gen. Chem.*, **32**, 3686 (1962).
563. G. I. Derkach, G. K. Fedorova and E. S. Gubnitskaya, *J. Gen. Chem.*, **33**, 1006 (1963).
564. L. Birkofer and H. Dickopp, *Chem. Ber.*, **101**, 2585 (1968).
565. V. A. Shokol, V. F. Gamaleya and G. I. Derkach, *J. Gen. Chem.*, **38**, 1815 (1968).
566. V. A. Shokol, G. I. Derkach and E. S. Gubnitskaya, *Zh. Obshch. Khim.*, **33**, 3058 (1963).
567. K. D. Berlin and M. A. R. Khayat, *Tetrahedron*, **22**, 975 (1964).
568. V. I. Shevchenko, N. K. Kulibaba and A. V. Kirsanov, *Zh. Obshch. Khim.*, **38**, 850 (1968).
569. V. I. Shevchenko, N. K. Kulibaba and A. V. Kirsanov, *Zh. Obshch. Khim.*, **38**, 326 (1968).
570. G. I. Derkach, V. A. Shokol, L. I. Samarai and A. V. Kirsanov, *J. Gen. Chem.*, **32**, 155 (1962).
571. G. I. Derkach, L. I. Samarai and A. V. Kirsanov, *J. Gen. Chem.*, **32**, 3689 (1962).
572. V. A. Shokol, G. I. Derkach and A. V. Kirsanov, *J. Gen. Chem.*, **32**, 162 (1962).
573. G. I. Derkach, V. A. Shokol and E. S. Gubnitskaya, *J. Gen. Chem.*, **33**, 547 (1963).
574. Y. Charbonnel, J. Barrans and R. Burgada, *Bull. Soc. Chim. France.*, 1363 (1970).
575. G. F. Dregval and G. I. Derkach, *J. Gen. Chem.*, **33**, 2880 (1963).
576. G. I. Derkach and Yu. V. Piven, *J. Gen. Chem.*, **36**, 1101 (1966).
577. Y. Charbonnel, R. Burgada and J. Barrans, *C.R. Acad. Sci. Paris*, **266**, 1241 (1968).
578. G. I. Derkach and L. I. Samarai, *J. Gen. Chem.*, **32**, 2038 (1962).
579. E. Allenstein and A. Schmidt, *Chem. Ber.*, **97**, 1863 (1964).
580. H. Hiraoka, *Tetrahedron*, **29**, 2955 (1973).
581. L. A. Paquette and T. J. Barton, *J. Am. Chem. Soc.*, **89**, 5480 (1967).
582. L. A. Paquette and G. R. Krow, *J. Am. Chem. Soc.*, **90**, 7149 (1968).
583. L. A. Paquette and J. R. Malpass, *J. Am. Chem. Soc.*, **90**, 7151 (1968).
584. L. A. Paquette, J. R. Malpass and G. R. Krow, *J. Am. Chem. Soc.*, **92**, 1980 (1970).
585. R. A. Odum and B. Schmall, *Chem. Comm.*, 1299 (1969).
586. T. H. Koch and D. A. Brown, *J. Org. Chem.*, **36**, 1934 (1971).
587. T. H. Koch, M. A. Geigel and C. Tsai, *J. Org. Chem.*, **38**, 1090 (1973).
588. R. J. Cotter, C. K. Sauers and J. M. Whelan, *J. Org. Chem.*, **26**, 10 (1961).
589. D. Y. Curtin and L. L. Miller, *Tetrahedron Letters*, **23**, 1869 (1965).

590. D. Y. Curtin and L. L. Miller, *J. Am. Chem., Soc.* **89**, 637 (1967).
591. A. Galat. *U.S. Pat.* 3,694,503 (1972); *Chem. Abstr.*, **77**, 164,252z (1972).
592. D. J. Hoy and E. J. Poziomek, *J. Org. Chem.*, **33**, 4050 (1968).
593. K. Miyazaki, *Bull. Chem. Soc. Japan.*, **41**, 1001 (1968).
594. R. K. Howe, *J. Org. Chem.*, **34**, 230 (1969).
595. K. Hartke, *Angew. Chem.*, **76**, 781 (1964).
596. W. Ried and A. Czack, *Ann. Chem.*, **676**, 130 (1964).
597. W. Walter, H. Weiss and K.-J. Reubke, *Ann. Chem.*, **736**, 167 (1970).
598. W. Ried and P. Schomann, *Ann. Chem.*, **714**, 122 (1968).
599. R. Huisgen, R. Grashey, E. Aufderhaar and R. Kunz, *Chem. Ber.*, **98**, 642 (1965).
600. Y. Kai, N. Yasuoka, M. Kakudo, N. Kasai, H. Yasuda, and H. Tani, *J. Chem. Soc. (D)*, 1243 (1970).
601. K. Nakatsuka, F. Ite, Y. Jo and Y. Furutake, *Japan Pat.*, 70,28,583; *Chem. Abstr.*, **74**, 42,858u (1971).
602. Stamicarbon N. V., *Belg. Pat.* 634,391 (1964); *Chem. Abstr.*, **61**, 739 (1964).
603. M. L. Evans, *Ger. Offen.*, 2,011,704 (1970); *Chem. Abstr.*, **74**, 14,015y (1971).
604. J. F. Cordes and D. Neubauer, *Ger. Offen.*, 1,158,258 (1963); *Chem. Abstr.*, **60**, 6999e (1964).
605. J. J. Van Daalen, J. Daams and J. Wijma, *S. African Pat.*, 68,00404; *Chem. Abstr.*, **72**, 90,095y (1970).
606. G. A. Miller, *Fr. Pat.* 1,505,972 (1967); *Chem. Abstr.*, **70**, 19,131k (1969).
607. Y. Miyazaki, K. Hashimoto and T. Noguchi, Japan, 69 26098; *Chem. Abstr.*, **72**, 55,048p (1970).
608. S. Janiak and V. Dittrich, *S. African Pat.*, 68,01,921; *Chem. Abstr.*, **70**, 96380t (1969).
609. Shell International, *Neth. Appl.*, 66 15,725 (1967); *Chem. Abstr.*, **69**, 35430s (1968).
610. W. Strauss and W. D. Willmund, *Ger. Offen.*, 1,084,098 (1957); *Chem. Abstr.*, 56, 3280 (1962).
611. Beck and Co., G.m.b.H., *Brit. Pat.*, 1,010,050 (1965); *Chem. Abstr.*, **64**, 8488 (1966).

CHAPTER **10**

The chemistry of amidrazones

K. M. WATSON AND D. G. NEILSON

Chemistry Department, The University, Dundee,
Scotland

$$\overset{X}{\underset{Z}{\diagdown}}C{=}Y$$

I. INTRODUCTION

Amidrazones (1) may be regarded as the hydrazides of the hypothetical imidic acids (2) and hence are related to the esters (imidates, 3), amides (amidines, 4) and acid halides (imidoyl halides, 5; X = halogen). For

$$RC\overset{NNR^1R^2}{\underset{NR^3R^4}{\diagup}} \quad RC\overset{NH}{\underset{OH}{\diagup}} \quad RC\overset{NH}{\underset{OR^1}{\diagup}} \quad RC\overset{NH}{\underset{NH_2}{\diagup}} \quad RC\overset{NR^1}{\underset{X}{\diagup}}$$

(1)   (2)   (3)   (4)   (5)

compound 1, the substituents R, R^1, R^2, R^3 and R^4 can be hydrogen or any of a very wide range of atoms or groups. Thus by the introduction of the appropriate functional groups into structure 1, aminoguanidine (6), isosemicarbazides (7; X = O) and isothiosemicarbazides (7; X = S) may all be looked on as amidrazones. However, these will not be discussed in detail here. In addition 1,2,4-triazoles (8) and 1,2,4-triazines (9)

$$H_2NC \underset{NH_2}{\overset{NNH_2}{{<}}} \qquad RXC \underset{NR^3R^4}{\overset{NNR^1R^2}{{<}}}$$

$$(6) \qquad\qquad (7)$$

can also be regarded as cyclic amidrazones and while their chemistry will not be reviewed, it will be shown that amidrazones are frequent precursors in the synthesis of these heterocycles.

$$(8) \qquad\qquad (9)$$

A comprehensive review of amidrazone chemistry[1] appeared in 1970 and therefore this article will tend to deal with the general aspects of the preparations and properties of amidrazones and the reader should refer to the earlier review[1] for further references.

II. NOMENCLATURE

The nomenclature used to describe compounds of structure 1 has been and indeed still is somewhat confusing. For example, amidrazones are found under the term 'hydrazidines' in *Chemical Abstracts* but this name has also been applied to compounds of the structure 10 although these (10) are also sometimes called dihydroformazans or hydrazide hydrazones. Amidrazones may be of types 11 or 12 which have been

$$RC \underset{NHNH_2}{\overset{NNH_2}{{<}}} \qquad RC \underset{NR^1_2}{\overset{NNR^2_2}{{<}}} \qquad RC \underset{NR^1}{\overset{NR^1NR^2_2}{{<}}}$$

$$(10) \qquad\qquad (11) \qquad\qquad (12)$$

termed amide hydrazones and hydrazide imides respectively. When, however, for compounds 11 and 12, $R^1 = H$, tautomerism is possible so that strictly these terms cannot be applied and the name amidrazone is preferable.

An amidrazone is named after the acid theoretically obtained from it on hydrolysis[1,2] and the nitrogen atoms are numbered as in formula

$$\underset{\underset{3}{\overset{\overset{2\ 1}{NNHPh}}{\parallel}}{\underset{NHMe}{\overset{}{HC}}}$$

$$\underset{H_2N}{\overset{H_2NN}{\underset{}{C-C}}}\overset{NNH_2}{\underset{NH_2}{}}$$

$$\text{(13)} \qquad\qquad \text{(14)}$$

13, which represents N^3-methyl-N^1-phenyl-formamidrazone. Whilst oxaldiamidrazone (**14**) is thus a true diamidrazone, the compounds represented by formulae **15** and **16** are not. These compounds (**15** and **16**) are dihydrazidines or, in the case of **15**, amide azines[3].

$$\underset{NR^1R^2 \quad R^2R^1N}{\overset{N\!\!-\!\!-\!\!-\!\!-\!\!-\!\!-\!\!N}{RC\qquad\qquad CR}}$$

$$\underset{NR^2 \quad R^2N}{\overset{NR^1\!\!-\!\!R^1N}{RC\qquad\qquad CR}}$$

$$\text{(15)} \qquad\qquad\qquad \text{(16)}$$

III. FORMATION OF AMIDRAZONES

A. Introduction

The chief methods for preparing amidrazones fall into two main categories: addition reactions and substitution reactions.

In the addition reactions a hydrazine can be added, for example, to a cyano group or to a carbon multiply linked as in a ketimine or carbodiimide.

In the second type of reaction, a suitable leaving group, attached to an imino-carbon, may be substituted by a hydrazine residue (equation 1)

$$\underset{X}{\overset{NR^1}{RC}} + R^2NHNH_2 \longrightarrow \underset{NHR^1}{\overset{NNHR^2}{RC}} \qquad (1)$$

or conversely, ammonia or an amine may act as the nucleophile and attack a hydrazone possessing a suitable leaving group (equation 2).

$$\underset{X}{\overset{NNR^1_2}{RC}} + NH_3 \longrightarrow \underset{NH_2}{\overset{NNR^1_2}{RC}} \qquad (2)$$

Apart from these two main methods of formation, amidrazones may be obtained by the ring opening of certain heterocyclic systems, usually after quaternization, or by the use of compounds in which the carbon and nitrogen atoms are already suitably positioned in the reactant molecule, as for example in the case of the reduction of a formazan to an amidrazone (equation 3).

$$RC \overset{\text{NNHR}^1}{\underset{\text{N=NR}^1}{\diagup}} \quad \xrightarrow{\text{[H]}} \quad RC \overset{\text{NNHR}^1}{\underset{\text{NH}_2}{\diagup}} \tag{3}$$

B. Preparation of Amidrazones by Addition of Hydrazines

I. Addition of hydrazines to nitriles

The primary product of the nucleophilic attack of hydrazine on a nitrile is an amidrazone and these compounds are obtained where an electron withdrawing group, such as a heterocyclic[4-7] or perfluoro-alkyl[8]

$$RCN + NH_2NH_2 \longrightarrow RC \overset{\text{NNH}_2}{\underset{\text{NH}_2}{\diagup}}$$

residue, is attached to the cyano moiety. Both anhydrous hydrazine[5,9] and hydrazine hydrate[7,10] have been used, normally in alcohol solution at room temperature. Secondary reactions can take place however; e.g. the amidrazone can attack unchanged nitrile giving dihydrazidine (17), which itself can undergo ring closure by loss of ammonia to form the

$$RCN + RC \overset{\text{NNH}_2}{\underset{\text{NH}_2}{\diagup}} \longrightarrow RC \overset{\text{N---N}}{\underset{\text{NH}_2 \ H_2N}{\diagup}} CR \longrightarrow RC \overset{\text{N---NH}}{\underset{\text{N}}{\diagup}} CR$$
$$\qquad\qquad\qquad\qquad\qquad (\textbf{17}) \qquad\qquad\qquad (\textbf{18})$$

triazole (18). Moreover it has been noted that, in the absence of hydrazine, amidrazones can undergo self-condensations to give dihydrazidines (17) and dihydrotetrazines (19). For example, Libman and Slack[11] isolated

$$RC \overset{\text{NNH}_2}{\underset{\text{NH}_2}{\diagup}} \longrightarrow (\textbf{17}) + RC \overset{\text{N---N}}{\underset{\text{NH---NH}}{\diagup}} CR$$
$$\qquad\qquad\qquad\qquad\qquad\qquad (\textbf{19})$$

the triazole 18 and the dihydrotetrazine 19 (R = 4-pyridyl), by heating pyridine-4-carboxamidrazone hydrochloride in a sealed tube and Geldard and Lions[12] obtained compound **19** (R = 2-pyridyl), in almost quantitative yield, by heating picolinamidrazone in aqueous or ethanolic solution. Recently, a more detailed study[13] has shown that picolinamidrazone, under reflux in an atmosphere of nitrogen, gives rise to the dihydro-*s*-tetrazine (**19**; R = 2-pyridyl) and the dihydrazidine (**17**; R = 2-pyridyl), but introduction of oxygen during the reaction produces two additional products, namely the *s*-tetrazine **20** and the compound **21** (equation 4).

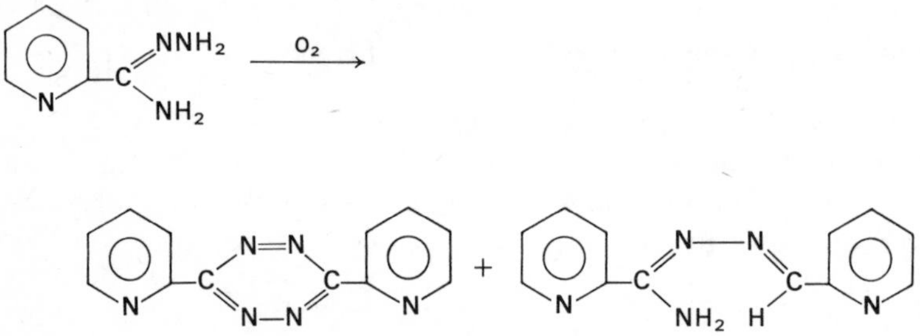

(4)

(20)　　　　　　　　　　　　　　　(21)

+ compounds **17** and **19** (R = 2-pyridyl)

The formation of the tetrazine (**20**) is by no means unexpected, as dihydrotetrazines are themselves precursors of tetrazines (by oxidation) and also of 4-amino-1,2,4-triazoles (**22**) (by the action of acid or of heat)[1, 14] and hence these compounds are recognized from time to time as byproducts. In particular, prolonged treatment, higher temperatures and excess hydrazine tend to give dihydrotetrazines or aminotriazoles along with, or in place of, amidrazones. The final product, however, is dependent not only on the conditions employed but also on the nature of the group R, since when R is an electron withdrawing group the amidrazone appears to be, in general, more stable, or the dihydrotetrazine, if formed, less prone to isomerise to the aminotriazole (**22**).

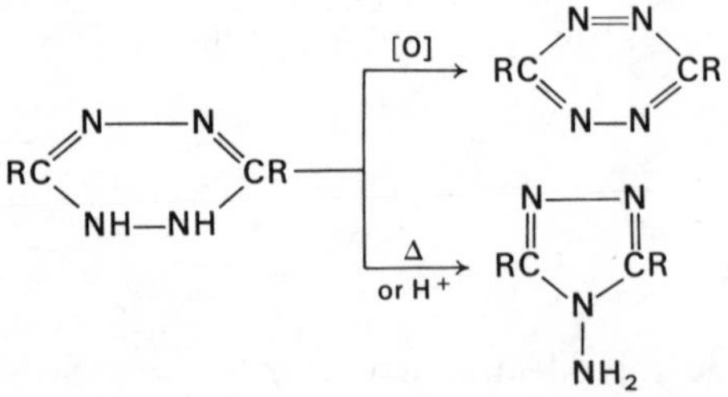

(22)

Cyanogen, an industrially important dinitrile, reacts similarly with hydrazine as the cyano group itself can act as an electron acceptor. Two moles of hydrazine[1, 15] lead to the formation of oxaldiamidrazone (23) but with one mole of hydrazine[16, 17], cyanoformamidrazone (24) can be obtained, particularly when reaction conditions permit of its precipitation and removal from reaction.

$$(CN)_2 \quad \begin{array}{c} \xrightarrow{2\ NH_2NH_2} \\ \\ \xrightarrow{NH_2NH_2} \end{array}$$

$$\begin{array}{cc} H_2NN & NNH_2 \\ \diagdown C\!-\!C \diagup \\ H_2N \diagup & \diagdown NH_2 \end{array}$$
(23)

$$\begin{array}{c} NNH_2 \\ NCC \diagup \\ \diagdown NH_2 \end{array}$$
(24)

2. Addition of substituted hydrazines to nitriles

Substituted hydrazines react similarly with nitriles to give substituted amidrazones. Monosubstituted hydrazines (25) possess two possible sites for attack and hence it might be expected that such hydrazines would give rise to products of the types 26 and 27, depending on the nature of the

$$RCN + R^1NHNH_2 \longrightarrow RC \begin{array}{c} NNHR^1 \\ \diagup \\ \diagdown NH_2 \end{array} + RC \begin{array}{c} NR^1NH_2 \\ \diagup \\ \diagdown NH \end{array}$$

(25)   (26)   (27)

group R^1 (25). However, it would appear that in the majority of cases, N^1-substituted amidrazones (26) are the predominant products of these reactions irrespective of whether R^1 is electron withdrawing or repelling. This appears to be so, also, when amidrazones are formed by substitution reactions (section C). Thus 2-cyanopyridine was found to react with methylhydrazine to give N^1-methyl-picolinamidrazone (28). No proof of the structure was given other than that the product 28 was reported to cyclise to compound 29 on treatment with carbon disulphide[5].

A series of compounds for use as herbicides has also been prepared by this route[18] (equation 5) and perfluoropolytriazoles have been obtained via amidrazone intermediates by the action of perfluorohydrazides of dicarboxylic acids on perfluorodinitriles[19].

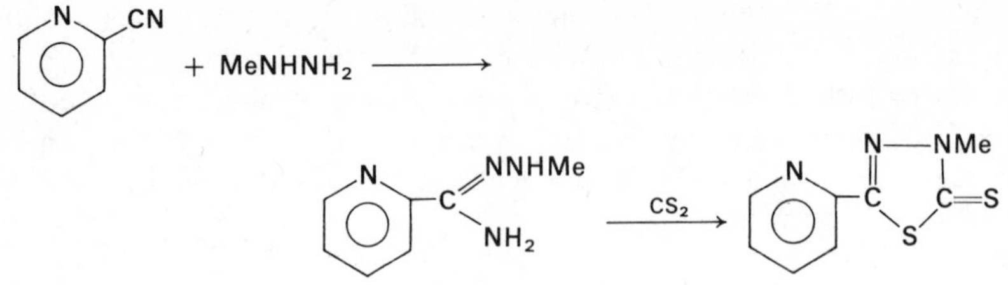

(28)   (29)

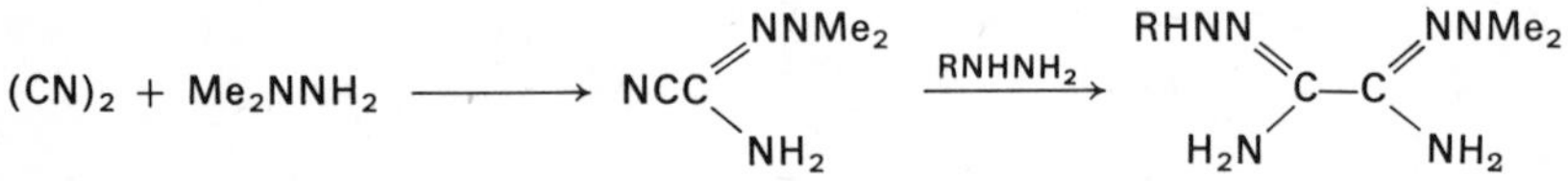

(5)

3. Addition of hydrazines to nitrile complexes

The bonding of a nitrile function to the electrophile $B_{10}H_{12}$ activates the nitrile, thus facilitating nucleophilic attack at the carbon atom, e.g. acetonitrile when complexed with decaborane reacts as readily with hydrazine as does a perfluoroalkyl nitrile[20,21] (equation 6).

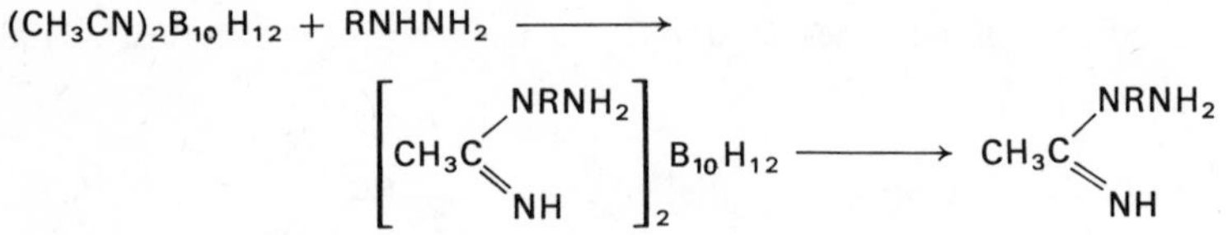

(R = H or Me)

A cyano group can also be activated by a Lewis acid such as aluminium chloride[22] and rapid addition of 1-phenyl-1-methylhydrazine then occurs to give an amidrazone even when an aliphatic nitrile is employed (equation 7).

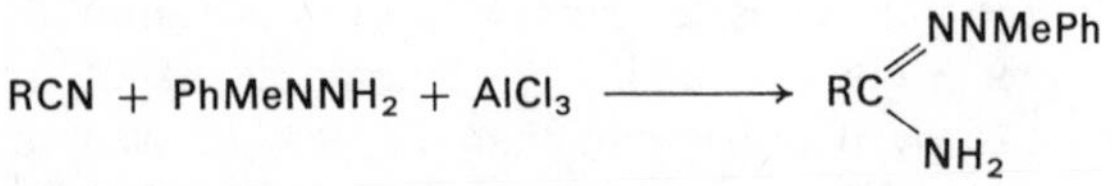

(7)

Recently the ferric chloride complex of a nitrile[23] has been treated, first with an alkyl halide and then with an amine, to give a disubstituted amidine (equation 8) and it may well prove possible to adapt this synthesis to give substituted amidrazones.

$$RCN \cdot FeCl_3 \xrightarrow{\ R^1Cl\ } RC{\equiv}\overset{\oplus}{N}R^1 \ FeCl_4^{\ominus} \xrightarrow{\ R^2R^3NH\ }$$

$$RC\overset{NHR^1}{\underset{\overset{|}{N}R^2R^3}{{\Big\langle}}{}^{\oplus}} \quad FeCl_4^{\ominus} \longrightarrow RC\overset{NR^1}{\underset{NR^2R^3}{\diagup\!\!\diagdown}} \qquad (8)$$

4. The addition of alkali hydrazides to nitriles

The use of sodium hydrazides in place of hydrazines has extended the range of nitriles which yield amidrazones, since the attacking nucleophile is, in this case, the more reactive hydrazide anion. Kauffmann and his co-workers[24] used this reaction to obtain amidrazone-like products, e.g. 2-hydrazinopyridines from the action of sodium hydrazides in the presence of the corresponding hydrazine on pyridine (equation 9). The suggested mechanism of the reaction is of the Tschitschibabin type and the nature of the product **30** suggests that the charge on the attacking hydrazide ion resides on the unsubstituted nitrogen (see Section III, B, 2).

(30)

(9)

The method has been extended to give good yields of unsubstituted amidrazones from both alkyl and aryl nitriles[25] by the use of sodium hydrazide and hydrazine in inert solvents under nitrogen. The method appears to be fairly generally applicable except where the nitrile can yield a stabilised anion as for example, that from malononitrile[26] or 2-phenyl-2-(2-pyridyl)-acetonitrile[27]. Another instance[27] in which the expected product is not obtained, is provided by the 3-aminopropionitriles (**31**) where in some cases, e.g. compound **31** R^1 = Ph, R^2 = Et, a retro-Michael reaction causes cleavage of the aminonitrile (equation 10).

Lithium[28] and barium[29] hydrazides have also been used instead of the sodium compound to react with nitriles; in these cases slight alterations in the conditions led to different products, e.g. diphenylhydrazidine was obtained from benzonitrile, but phenylacetamidrazone from phenylacetonitrile.

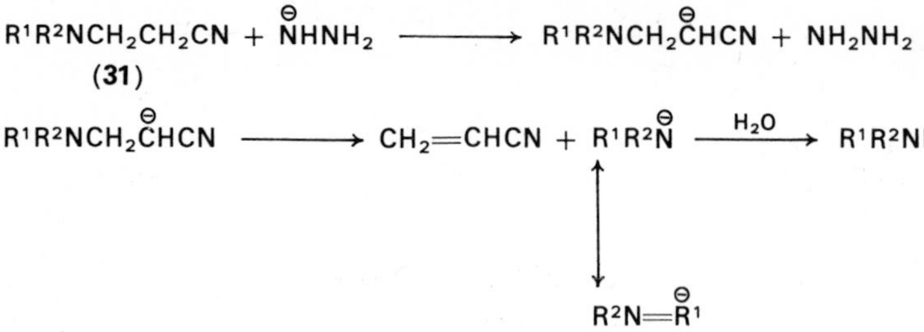

$$R^1R^2NCH_2CH_2CN + \overset{\ominus}{N}HNH_2 \longrightarrow R^1R^2NCH_2\overset{\ominus}{C}HCN + NH_2NH_2$$

(31)

$$R^1R^2NCH_2\overset{\ominus}{C}HCN \longrightarrow CH_2{=}CHCN + R^1R^2\overset{\ominus}{N} \xrightarrow{H_2O} R^1R^2NH \quad (10)$$

$$R^2N{=}\overset{\ominus}{R^1}$$

5. Addition of hydrazines to ketimines, carbodi-imides and s-triazine

Ketimines and carbodi-imides can both yield amidrazones by addition of hydrazine; for example, N^3-p-tolyl-diphenylacetamidrazone is obtained from the corresponding ketimine[30] (equation 11) while either one or two molecules of a carbodi-imide[31,32] can react with a hydrazine to give products of the types **32** and **33**.

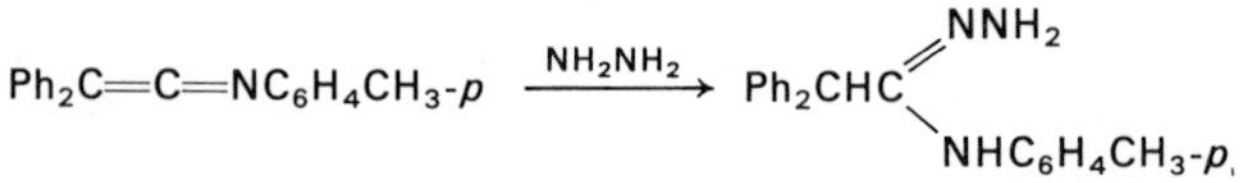

$$Ph_2C{=}C{=}NC_6H_4CH_3\text{-}p \xrightarrow{NH_2NH_2} Ph_2CHC\overset{\displaystyle {=}NNH_2}{\underset{\displaystyle NHC_6H_4CH_3\text{-}p}{\Big\backslash}} \quad (11)$$

$$RN{=}C{=}NR \longrightarrow$$

$$\xrightarrow{NH_2NHCOOEt} RNHC\overset{\displaystyle {=}NNHCOOEt}{\underset{\displaystyle NHR}{\Big\backslash}}$$

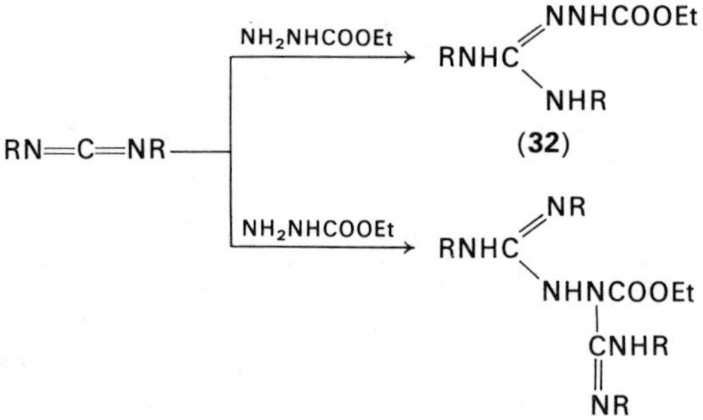

(32)

$$\xrightarrow{NH_2NHCOOEt} RNHC\overset{\displaystyle {=}NR}{\underset{\displaystyle NHNCOOEt}{\Big\backslash}}$$

$$\underset{\displaystyle NR}{\overset{\displaystyle CNHR}{\|}}$$

(33)

s-Triazine can react with hydrazines to yield a variety of products depending on the hydrazine employed. Amidrazones are formed when 1,1-disubstituted hydrazines are used[33].

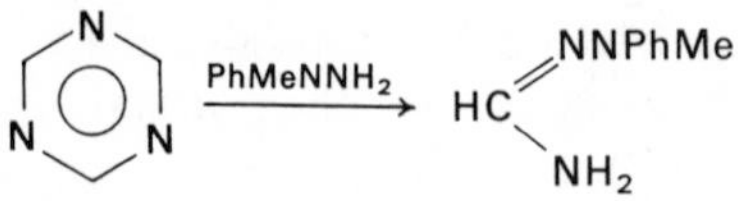

$$\xrightarrow{PhMeNNH_2} HC\overset{\displaystyle {=}NNPhMe}{\underset{\displaystyle NH_2}{\Big\backslash}}$$

C. Preparation of Amidrazones by Substitution Reactions of Hydrazines

I. Reaction of hydrazine on imidates

Imidates are more susceptible to nucleophilic attack than are nitriles and hence, as expected, a wider range of imidates can react with hydrazine[1,34] to yield amidrazones by replacement of the alkoxy group.

$$RC\overset{NH}{\underset{OR^1}{\big\langle}} + NH_2NH_2 \longrightarrow RC\overset{NNH_2}{\underset{NH_2}{\big\langle}} + R^1OH$$

(34)

As indicated in Section B (see also Chapter 9), however, the reaction may not stop at the amidrazone but, depending on the experimental conditions and on the nature of the group R in compound **34**, a dihydrazidine (**17**) or occasionally a triazole (**18**) may be obtained. Dihydro-s-tetrazines (**19**) are also frequently produced and these may oxidise to the corresponding s-tetrazines or rearrange to 4-amino-1,2,4-triazoles (**22**). Moreover dihydroformazans (**35**) can be formed by the action of two moles of

$$RC\overset{NH}{\underset{OR^1}{\big\langle}} + 2\,NH_2NH_2 \longrightarrow RC\overset{NNH_2}{\underset{NHNH_2}{\big\langle}}$$

(35)

hydrazine on the imidate. Thus the reaction is not always clean and the amidrazone may not be the major product, but nevertheless good yields can be obtained in many instances. Optimum conditions for amidrazone formation appear to be the interaction, at or below 0°C in ethanol or in an ethanol/ether mixture, of equimolar amounts of anhydrous hydrazine with imidate base; however sometimes the imidate salt has been used. There is a greater tendency for side products to be produced when hydrazine hydrate acts on an imidate at room temperature. This general method has been employed in the preparation of unsubstituted amidrazones, or their hydrochlorides, from aliphatic, aromatic and heterocyclic imidates[34,35,36] (equations 12–14). However, the products are not

$$CH_3C\overset{\overset{\oplus}{N}H_2\ Cl^{\ominus}}{\underset{OEt}{\big\langle}} \xrightarrow[\text{below 0°C}]{NH_2NH_2} CH_3C\overset{\overset{\oplus}{N}HNH_2\ Cl^{\ominus}}{\underset{NH_2}{\big\langle}} \qquad (12)$$

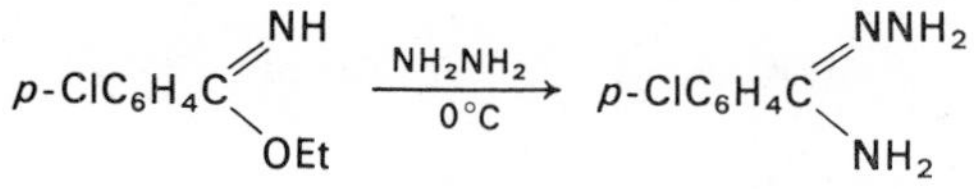

$$p\text{-}ClC_6H_4C\overset{NH}{\underset{OEt}{\Big\langle}} \xrightarrow[0°C]{NH_2NH_2} p\text{-}ClC_6H_4C\overset{NNH_2}{\underset{NH_2}{\Big\langle}} \qquad (13)$$

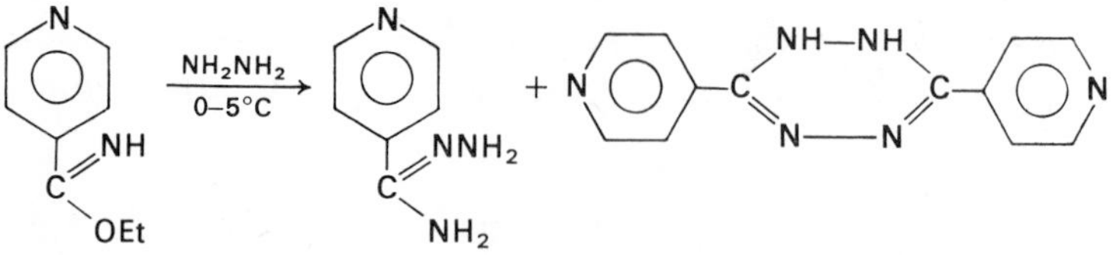

(14)

always isolated but may be used *in situ* for further synthesis[37] (equation 15).

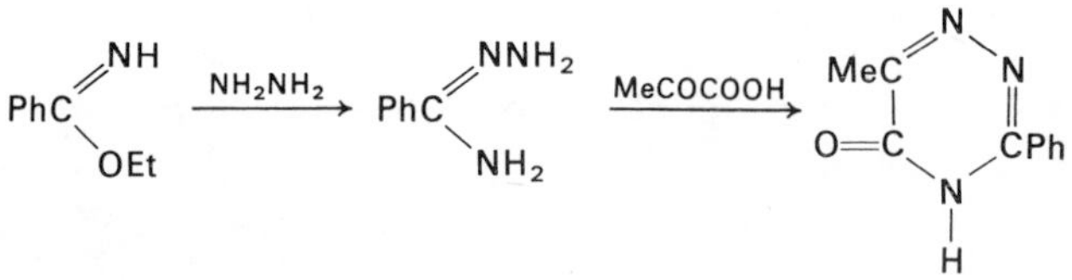

(15)

When *N*-substituted imidates are used as starting materials, reaction with hydrazine gives the corresponding N^3-substituted amidrazones. This synthesis has been applied to give triazinones[38] (**36**) by reaction with α-keto-esters of the amidrazone intermediates which were not isolated.

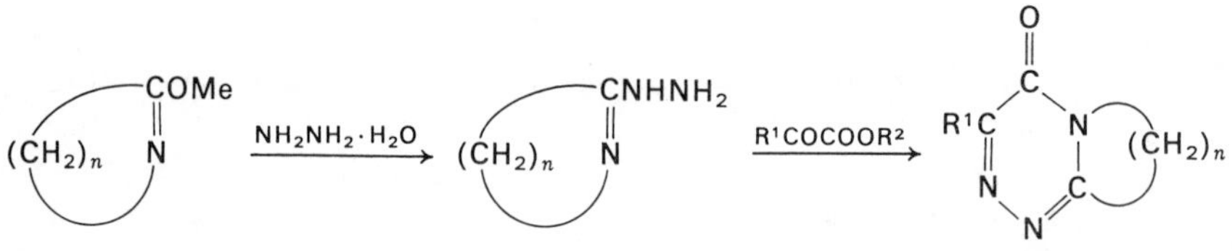

(**36**)

2. Reaction of substituted hydrazines on imidates

Monosubstituted hydrazines react smoothly at room temperature in alcohol with equimolar quantities of imidates or their salts[1, 39] (equation 16). The number of possible by-products is reduced but, since formazans

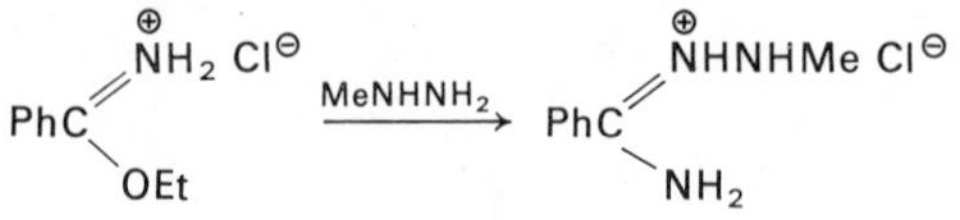

(16)

can be formed by the reaction of an imidate with two moles of the hydrazine (equation 17), they, or occasionally dihydroformazans, may be produced along with the amidrazone[1].

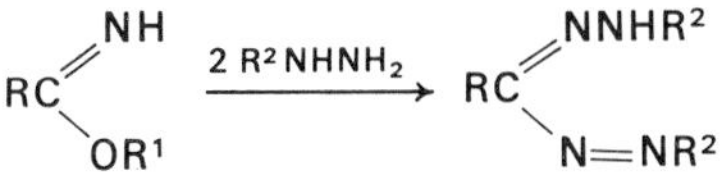

$$RC\overset{NH}{\underset{OR^1}{\diagup}} \xrightarrow{2\ R^2NHNH_2} RC\overset{NNHR^2}{\underset{N=NR^2}{\diagup}} \tag{17}$$

Acyl hydrazines can similarly give N^1-acyl-amidrazones[40] which usually cyclise to 1,2,4-triazoles under alkaline conditions[1]. By this route, Becker and his co-workers[41] obtained N^1-acyl-N^3-hydroxyamidrazones and from them, 4-hydroxy-1,2,4-triazoles (equation 18). When

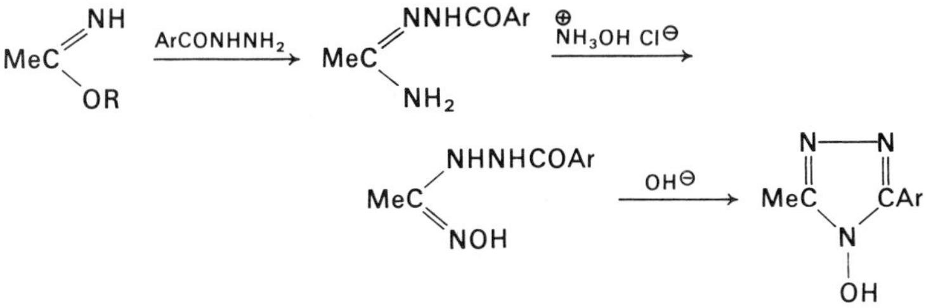

$$MeC\overset{NH}{\underset{OR}{\diagup}} \xrightarrow{ArCONHNH_2} MeC\overset{NNHCOAr}{\underset{NH_2}{\diagup}} \xrightarrow{\overset{\oplus}{NH_3OH}\ Cl^\ominus}$$

$$MeC\overset{NHNHCOAr}{\underset{NOH}{\diagup}} \xrightarrow{OH^\ominus} \tag{18}$$

formylhydrazine is used as the reagent, the acyl–amidrazone can be converted into the corresponding unsubstituted amidrazone by acid hydrolysis of the formyl group[27, 42].

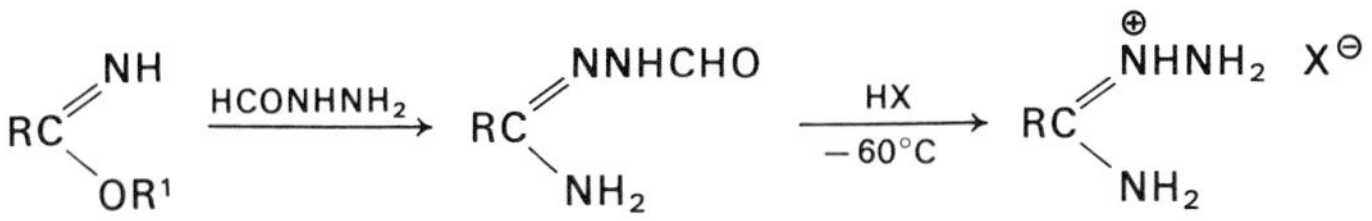

$$RC\overset{NH}{\underset{OR^1}{\diagup}} \xrightarrow{HCONHNH_2} RC\overset{NNHCHO}{\underset{NH_2}{\diagup}} \xrightarrow[-60^\circ C]{HX} RC\overset{\overset{\oplus}{N}HNH_2}{\underset{NH_2}{\diagup}}\ X^\ominus$$

A novel application of this reaction[43] is seen in the preparation of the cyclic amidrazone **38** from the action of phenylhydrazine on the imidate **37**.

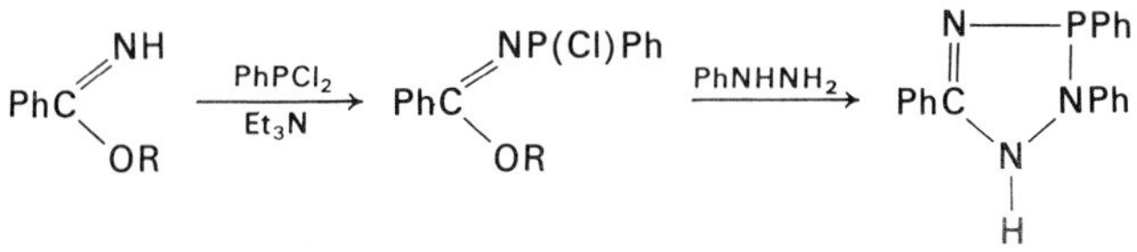

$$PhC\overset{NH}{\underset{OR}{\diagup}} \xrightarrow[Et_3N]{PhPCl_2} PhC\overset{NP(Cl)Ph}{\underset{OR}{\diagup}} \xrightarrow{PhNHNH_2}$$

(37)

(38)

At room temperature imidates react with 1,1-disubstituted hydrazines to give N^1,N^1-disubstituted-amidrazones[1, 44] while at higher temperatures and in the presence of ammonium salts, dihydroformazans are formed

$$
\underset{\text{OEt}}{\overset{\text{NH}}{RC}} \quad
\begin{array}{c}
\xrightarrow{\text{Me}_2\text{NNH}_2} \quad \underset{\text{NH}_2}{\overset{\text{NNMe}_2}{RC}} \hspace{2em} (19) \\[2em]
\xrightarrow[]{2\ \text{Me}_2\text{NNH}_2} \quad \underset{\text{NHNMe}_2}{\overset{\text{NNMe}_2}{RC}} \hspace{2em} (20)
\end{array}
$$

(equations 19 and 20). This reaction has been adapted[45] to give a series of silyl amidrazones (39). A further interesting example of this type of

$$
\underset{\text{OR}^1}{\overset{\text{NH}}{R_3\text{Si(CH}_2)_n\text{C}}} \xrightarrow{R_2^2\text{NNH}_2} \underset{\text{NH}_2}{\overset{\text{NNR}_2^2}{R_3\text{Si(CH}_2)_n\text{C}}}
$$

(39)

R and R^2 = alkyl ; n = 2 or 3

reaction is provided by Gol'din and his co-workers[46], who treated dialkylamino-alkylimidates with 1,1-dimethyl hydrazine and obtained, on heating, the acrylamidrazone (40) which reacted with excess of the hydrazine to from the dihydroformazan (41). This reaction is reminiscent of a similar cleavage of a related nitrile reported in Section III, B, 4.

$$
\underset{\text{OR}^1}{\overset{\text{NH}}{R_2\text{N(CH}_2)_2\text{C}}} \xrightarrow{\text{Me}_2\text{NNH}_2} \underset{\text{NH}_2}{\overset{\text{NNMe}_2}{R_2\text{N(CH}_2)_2\text{C}}} \xrightarrow{-\ R_2\text{NH}}
$$

$$
\underset{\text{NH}_2}{\overset{\text{NNMe}_2}{\text{CH}_2=\text{CHC}}} \xrightarrow[\text{excess}]{\text{Me}_2\text{NNH}_2} \underset{\text{NHNMe}_2}{\overset{\text{NNMe}_2}{\text{Me}_2\text{NNH(CH}_2)_2\text{C}}}
$$

(40)   (41)

When substituted hydrazines act on N-substituted imidates, this provides a route to N^1,N^3-disubstituted[47] or N^1,N^1,N^3-trisubstituted amidrazones[22] (equation 21).

$$
\underset{\text{OR}}{\overset{\text{NPh}}{HC}} + \text{Me}_2\text{NNH}_2 \longrightarrow \underset{\text{NHPh}}{\overset{\text{NNMe}_2}{HC}} \hspace{2em} (21)
$$

3. Reaction of hydrazines on thioimidates

Less use has been made of thioimidates than of imidates as starting materials for the synthesis of amidrazones, indeed thioamides have been more frequently used[1]. With hydrazine, thioimidates readily give dihydrotetrazines, for example Mukaiyama and Ono[48] found that N-substituted thioimidate salts, e.g. compound **42** yielded dihydrotetrazines via dihydrazidine intermediates, as shown in equation 22. They also used this route to obtain related polymeric compounds[48].

$$\tag{22}$$

On the other hand, substituted hydrazines were found to react satisfactorily with cyclic thioimidates to give amidrazones[48] (equation 23).

$$\tag{23}$$

A novel application[49] of the use of a thioimidate precursor for the formation of an amidrazone is illustrated by the reaction sequence (24).

$$\tag{24}$$

4. Reaction of hydrazines on imidoyl halides

Hydrazine itself can react with imidoyl halides to give N^3-substituted amidrazones; this type of reaction is most frequently illustrated in the literature by the formation of amidrazone-like derivatives of heterocycles[50], e.g. 1-hydrazino-isoquinoline[51] (**43**).

(43)

In the case of monosubstituted hydrazines the position is more complex. Thus, when phenylhydrazine reacts with N-phenyl-benzimidoyl chloride, two products (44 and 45) can be isolated[1]—not only the expected N^1-substituted amidrazone (44) but also the N^2-substituted isomer (45). It may be that, due to the extreme reactivity of the imidoyl halide, the hydrazine reacts less selectively with this type of compound than with other substrates (see Section III, B, 4).

Smith and his co-workers[52, 53], for their study of the alkylation and tautomerism of amidrazones, and Walter and Weiss[22] for their spectroscopic studies, have used imidoyl halides to obtain amidrazones of unambiguous structure (compounds 46 and 47). However, it is of interest to note that while reaction of the halide with a large excess of 1,1-dimethylhydrazine gave the amidrazone 47, with 2 mol of the hydrazine[52] the product 48 was obtained along with 47.

5. Reaction of hydrazines on amidines

Amidines are less readily attacked by nucleophiles than are imidates or imidoyl halides and hence have been used less frequently in the preparation of amidrazones[1, 8], but illustrative examples of this type are seen in equations 25–27.

$$C_3F_7C\overset{NH}{\underset{NH_2}{{}}} + NH_2NH_2 \longrightarrow C_3H_7C\overset{NNH_2}{\underset{NH_2}{{}}} \tag{25}$$

$$CH_3CH(OH)C\overset{\overset{\oplus}{NH_2}\ Cl^{\ominus}}{\underset{NH_2}{{}}} + PhMeNNH_2 \longrightarrow CH_3CH(OH)C\overset{NNPhMe}{\underset{NH_2}{{}}} \tag{26}$$

$$HC\overset{NPh}{\underset{NHPh}{{}}} + PhNHNH_2 \longrightarrow HC\overset{NNHPh}{\underset{NHPh}{{}}} \tag{27}$$

Formamidine salts have been used to give amidrazone intermediates[54] for the synthesis of quinazoline derivatives (**49**). Predictably the time required for the completion of the reaction was reduced when R (compound **50**) was an electron donor and considerably increased when R was an acceptor group.

(**50**)   (**49**)

6. Reaction of hydrazines on amides

The formation of amidrazones from amides may be thought of as the substitution of a potential hydroxyl group ($RCONH_2 \rightleftharpoons RC(OH){=}NH$). This can be achieved indirectly via the imidoyl halide (see Section III, C, 4) or more directly by the reaction of hydrazines with, for example, disubstituted amides in the presence of phosphorus oxychloride (equation

28) or p-toluenesulphonyl chloride (equation 29) to provide intermediates with better leaving groups[52, 55].

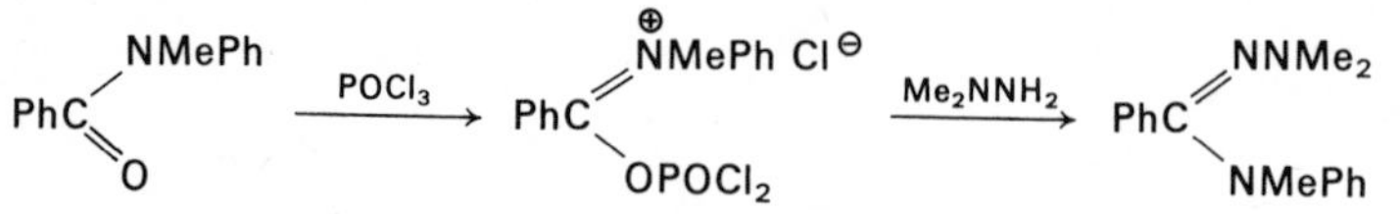

(28)

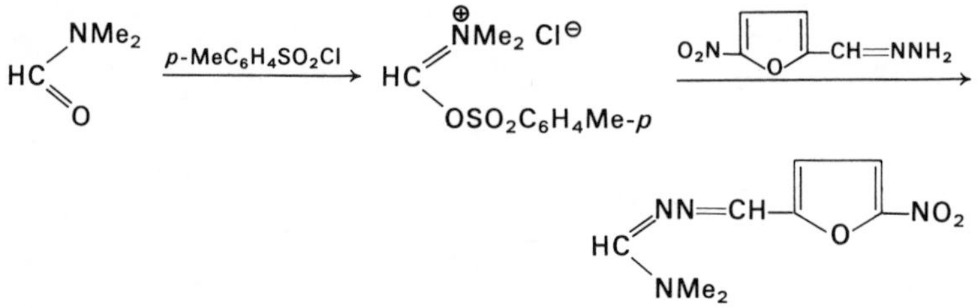

(29)

A somewhat different approach has led from N,N-dimethylformamide via the imidate fluoroborate (51) to the amide acetal (52), which with hydrazine yields the amidrazone[56].

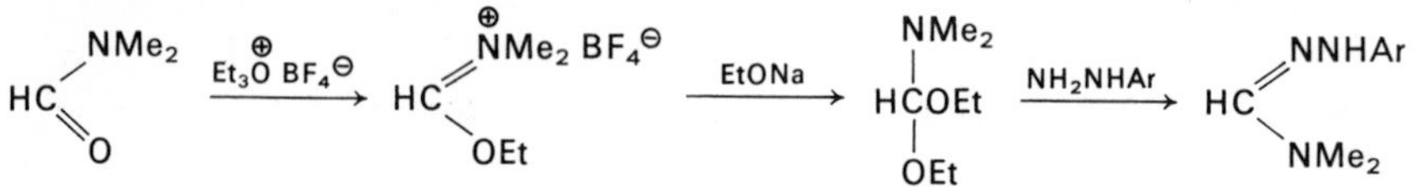

(51)   (52)

7. Reaction of hydrazines on thioamides

Since thioamides react much more readily with hydrazine than do amides, they have been used in the preparation of unsubstituted and substituted amidrazones. For example, isonicotinic acid thioamide gives the corresponding amidrazone on brief heating with hydrazine hydrate. However, prolonged treatment at room temperature[57] results in the formation of the thiadiazole 53.

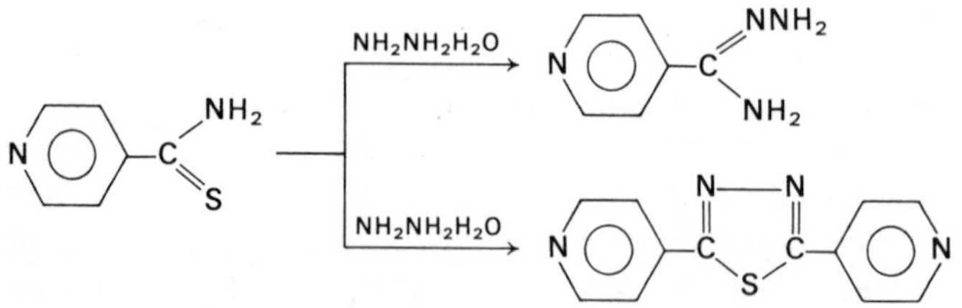

(53)

The reaction of hydrazine hydrate on arylthiocarboxanilides yields amidrazones in the cold, but on heating dihydrotetrazines are produced[58, 59].

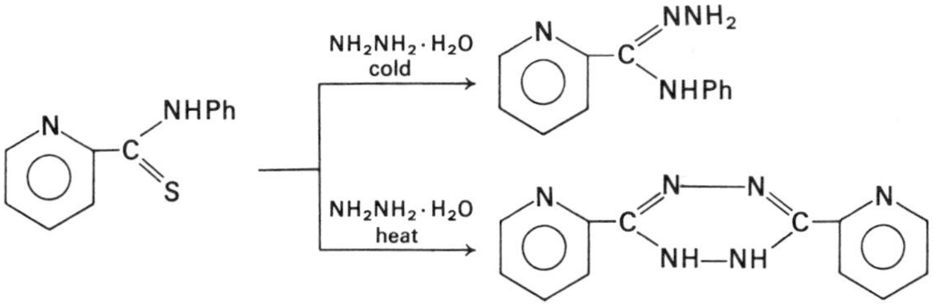

D. Preparation of Amidrazones by Substitution Reactions of Ammonia and Amines

I. Reaction of ammonia and amines on hydrazonate esters and thioesters

Parallelling the substitution by a hydrazino residue of the group X in —CX=NR is the substitution by an amino group of X in a hydrazone derivative of the type —CX=NNR$_2$. Thus hydrazonate esters can react with ammonia or amines to give amidrazones[60, 61] (equation 30) but these reagents may also cause cleavage of the C=N as well as of the C—O bond and hence produce amidines[60, 62] (equation 31). Neunhoeffer[60] has

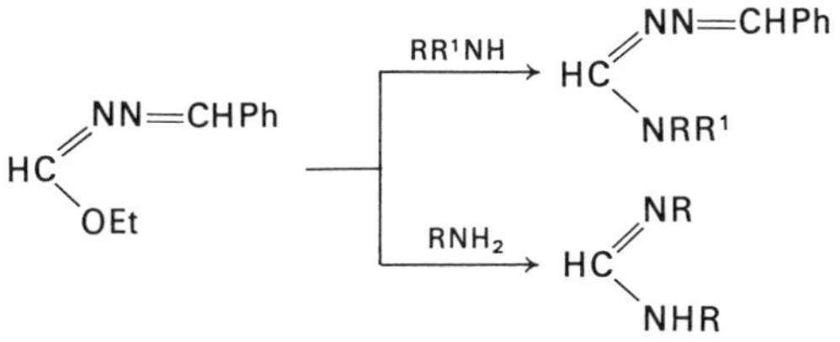

(30)

(31)

adapted this method in order to obtain formamidrazone hydrochloride as seen in the following reaction sequence (equation 32).

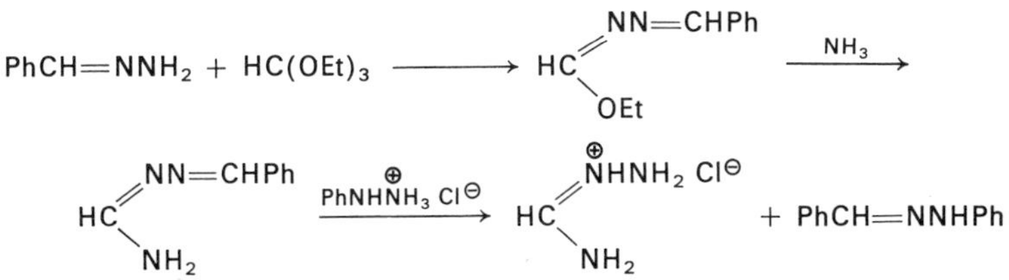

(32)

Ethyl thioformate reacts, in ethylamine at 0°C, with 1-methyl-1-phenylhydrazine to give the thiohydrazide (54) while in ethanol at lower temperatures, the hydrazonate ester (55) is formed. At higher temperatures, either in ethanol solution or in the absence of solvent, these compounds (54 and 55) are produced along with the amidrazones 56 and 57. It was shown that the amidrazone 57 could arise from the reaction on the hy-

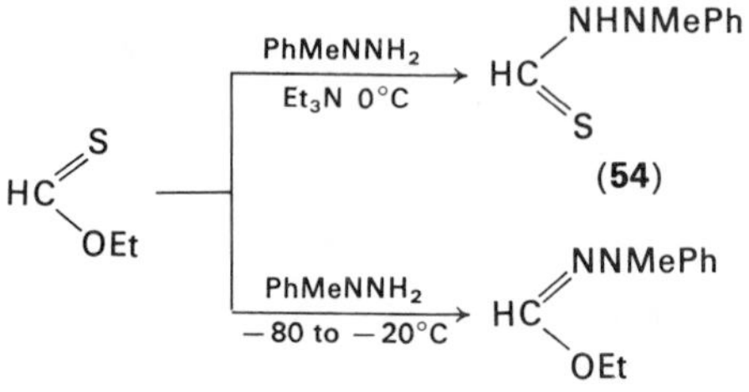

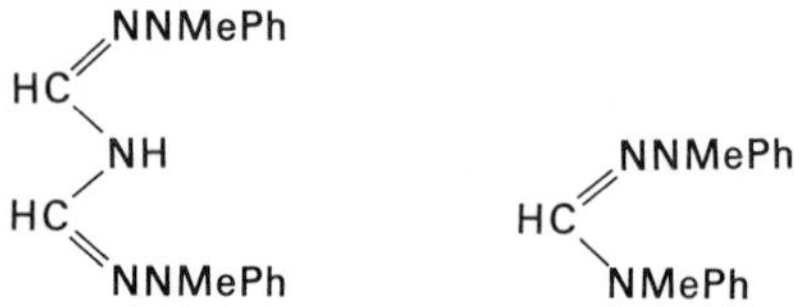

drazonate ester of methylphenylamine formed through reductive cleavage of methylphenylhydrazine by hydrogen sulphide released in the overall reaction[61]. Grashey and his co-workers[63] have also found evidence of amidrazone formation due to similar reductive cleavage of the N—N bond of a hydrazine.

2. Reaction of ammonia and amines on other hydrazonyl derivatives

The very reactive halogen of hydrazonoyl halides can be readily replaced on treatment with ammonia or amines[22,64] as illustrated in equations 33 and 34.

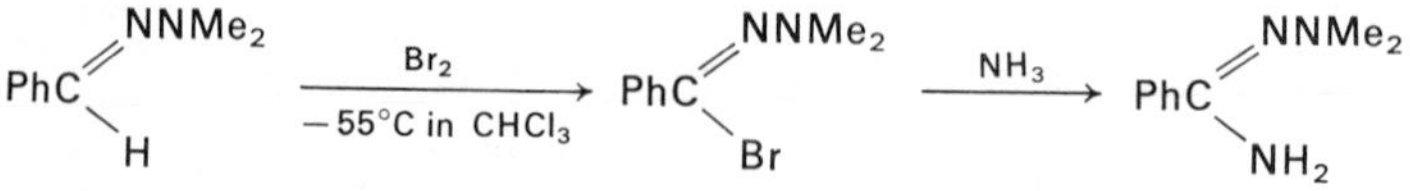

(33)

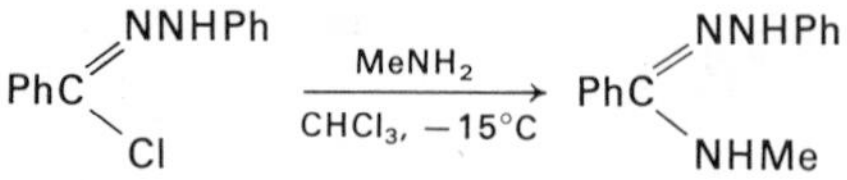

(34)

Use[65] has been made of this reaction to prepare a series of fungicides, of the type **58**.

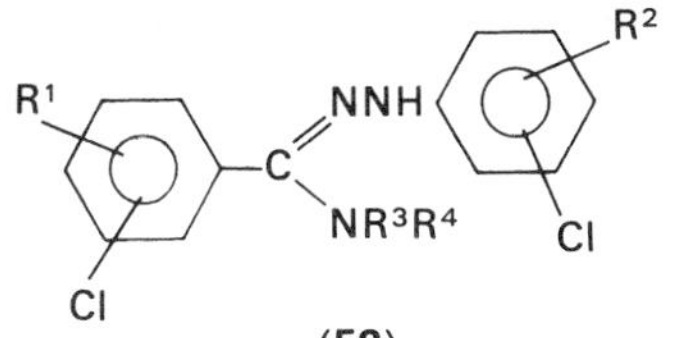

(58)

The nitro groups of 1-nitroaldehyde hydrazones also undergo aminolysis with resultant amidrazone formation[66].

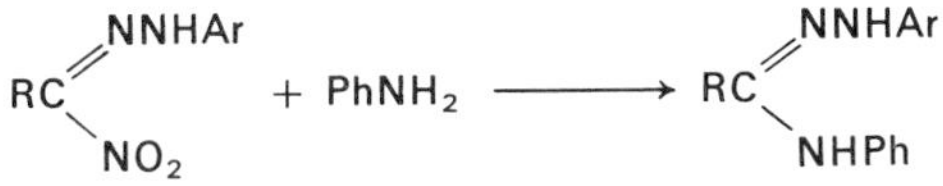

E. Formation of Amidrazones from Heterocyclic Compounds

Amidrazones may be obtained from a number of heterocyclic systems by ring opening reactions due to nucleophilic attack involving hydrolysis (e.g. of triazoles) or aminolysis (e.g. of oxazoles). Usually the heterocycle requires to be quaternized to facilitate this nucleophilic reaction.

I. Formation of amidrazones from oxadiazoles or oxadiazolium salts

1,3,4-Oxadiazoles bearing strong electron withdrawing substituents can undergo ring opening by nucleophilic attack[67]. Thus 3,5-perfluoro-dialkyl-1,3,4-oxadiazoles when treated with ammonia give the amidrazone **59** but with primary amines give the dihydrazidine **60**. Usually, however,

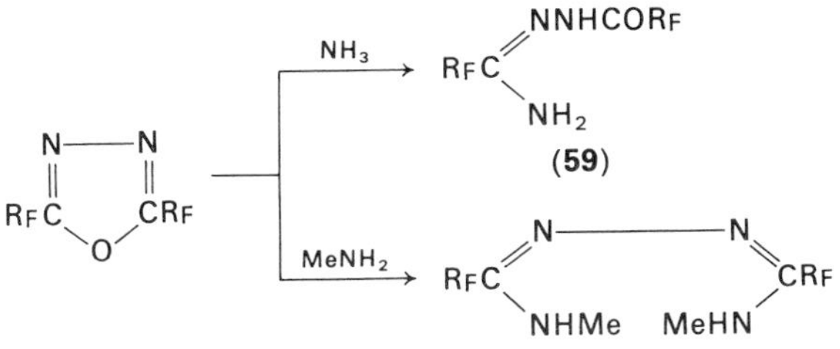

(59)

(60)

it is necessary to increase the electrophilicity of the heterocycle, by first converting it to a quaternary salt, before nucleophilic attack can effect ring opening. The product normally obtained is a recyclisation product, namely a triazole or triazolium salt, but amidrazone intermediates have been isolated. In a reaction reminiscent of the formation of pyridinium salts from their pyrylium counterparts, oxadiazolium perchlorates react with ammonia and with primary aromatic amines to give triazolium salts[68] (equation 35). Boyd and Summers[68] have isolated the amidrazone (**61**,

$$
\begin{array}{c}
\text{N}\!-\!\!-\!\text{NPh} \\
\text{PhC}\overset{\oplus}{\underset{\text{O}}{\diamond}}\text{CPh} \quad \text{ClO}_4^{\ominus} \;+\; \text{RNH}_2 \longrightarrow \quad
\text{PhC}\overset{\text{N}\!-\!\!-\!\text{NPh}}{\underset{\underset{\text{R}}{\overset{\oplus}{\text{N}}}}{\diamond}}\text{CPh} \quad \text{ClO}_4^{\ominus} \qquad (35)
\end{array}
$$

$$
\begin{array}{c}
\text{NH}\!-\!\!-\!\text{NPh} \\
\text{PhC}\diagdown\underset{\text{O} \quad \text{RHN}}{\diamond}\overset{+}{}\text{CPh} \quad \text{ClO}_4^{-}
\end{array}
$$

(**61**)

R = Ph) from the reaction of aniline on 1,3,5-triphenyl-1,2,4-oxadiazolium perchlorate, indicating that nucleophilic attack occurs at $C_{(2)}$ of the oxadiazolium salt. When for compound **61**, R = alkyl, however, the amidrazonium salt reverts to the oxadiazolium perchlorate and does not recyclise to a triazolium salt.

Shvaika and Fomenko[69] have studied related recyclisation reactions brought about by the action of hydrazines on the alkyl tosylates of various heterocycles, e.g. 2,5-diaryl-1,3-oxazoles, thiazoles and 1,3,4-oxadiazoles, but were unable to isolate the amidrazone intermediates in most cases.

2. Formation of amidrazones from triazolium salts

Triazolium salts, likewise, can undergo ring opening reactions to form amidrazones[64]. However, since amidrazones may be the starting materials for the synthesis of triazolium salts, this is not a very useful synthetic route. Tetraphenyltriazolium perchlorate (**62**) with alkali forms the amidrazone **63**, but as the reaction is reversible, strong acid converts the amidrazone back into the triazolium salt (**63** → **62**).

$$
\begin{array}{c}
\text{N}\!-\!\!-\!\text{NPh} \\
\text{PhC}\overset{\oplus}{\underset{\underset{\text{Ph}}{\text{N}}}{\diamond}}\text{CPh} \quad \text{ClO}_4^{\ominus} \quad \underset{\text{H}^+}{\overset{\text{OH}^-}{\rightleftarrows}} \quad
\text{PhC}\overset{\text{NNHPh}}{\underset{\text{N(Ph)COPh}}{\diagup}}
\end{array}
$$

(**62**) \qquad\qquad\qquad (**63**)

On the other hand, a triazolium salt with a carbonyl substituent on $C_{(3)}$ is attacked irreversibly at the carbonyl and the ring is opened by cleavage of the N—N bond to give an amidine in place of an amidrazone (equation 36).

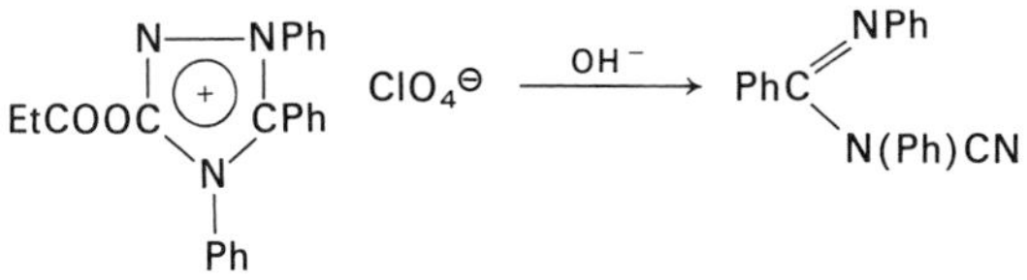

$$\underset{\substack{| \\ Ph}}{\overset{\substack{N\text{---}NPh \\ | \quad (+) \quad |}}{EtCOOC \overset{\displaystyle\bigcirc}{\underset{N}{}} CPh}} \quad ClO_4^{\ominus} \quad \xrightarrow{OH^-} \quad PhC\overset{NPh}{\underset{N(Ph)CN}{\diagup}} \qquad (36)$$

Fused ring compounds such as s-triazolo[3,4-f]1,2,4-triazine can also undergo related ring opening reactions to yield amidrazone-like products[70].

3. Formation of amidrazones from s-tetrazines

Kohn and Olofson[71] demonstrated that the ring opening with alkali of a quaternary salt of 1,4-dimethyl-1,4-dihydro-s-tetrazine (**64**) occurred by cleavage of an N—N bond thus leading to an amidrazone. Brief treatment with alkali was sufficient to give the amidrazone (**65**) whereas secondary products (**66** and **67**) were formed with more prolonged reaction times.

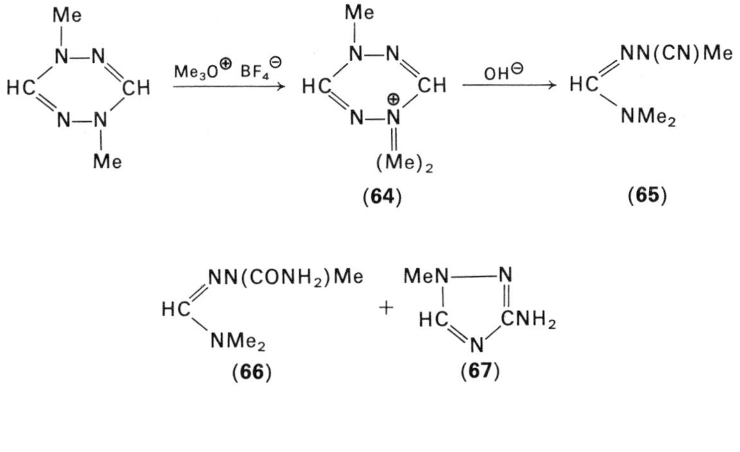

(**64**)   (**65**)

(**66**)   (**67**)

F. Miscellaneous Methods of Preparation of Amidrazones

Amidrazones have been obtained by the reduction of a number of types of compounds where the carbon and nitrogen atoms are already suitably positioned. Thus amidrazones have been prepared by the reduction of nitrazones[72] (equation 37). Similarly formazans are reduced by a variety

$$PhC\underset{NO_2}{\overset{NNHC_6H_4Me\text{-}p}{=}} \quad \xrightarrow{\text{Raney Nickel}} \quad PhC\underset{NH_2}{\overset{NNHC_6H_4Me\text{-}p}{=}} \qquad (37)$$

of reagents such as mercaptals[73] and phenyl hydrazine[74]. The stepwise reduction of tetrazolium salts by catalytic hydrogenation or by sodium dithionite has been studied by Jerchel and his co-workers[72, 75, 76] (equation 38).

$$\qquad \longrightarrow RC\underset{N=NR}{\overset{NNHR}{=}} \longrightarrow RC\underset{NHNHR}{\overset{NNHR}{=}} \longrightarrow RC\underset{NH_2}{\overset{NNHR}{=}} \qquad (38)$$

Amidrazones of the structure **68** have been prepared by the action of sodamide in liquid ammonia on 4-arylmethylene-1,2,4-triazoles; this reaction possibly involves a mechanism similar to that of the Tschitschibabin reaction[70].

$$\xrightarrow{NaNH_2/NH_3}$$

(68)

Aryl diazonium salts can couple with acylaminomalonic acid monoesters and similar compounds to form amidrazones as illustrated[72,77] in equation 39.

$$EtOOC\underset{NHCOMe}{\overset{COOH}{C}}H \quad \xrightarrow[Na_2CO_3]{p\text{-}O_2NC_6H_4N_2^{\oplus}} \quad EtOOCC\underset{NHCOMe}{\overset{NNC_6H_4NO_2\text{-}p}{=}} \qquad (39)$$

IV. PROPERTIES OF AMIDRAZONES

A. Tautomerism

Amidrazones may be classified as those capable of existing in tautomeric forms, e.g. **69** ⇌ **70**, and those which cannot so tautomerise and therefore must have either an amide hydrazone structure (**71**; R^3, $R^4 \neq H$) or a hydrazide imide structure (**72**; R^3, $R^4 \neq H$). Even where tautomerism is

$$RC\overset{NHNR^1R^2}{\underset{NR^3}{\diagup}}\rightleftharpoons RC\overset{NNR^1R^2}{\underset{NHR^3}{\diagup}}$$

$$\text{(69)} \qquad\qquad \text{(70)}$$

$$RC\overset{NNR^1R^2}{\underset{NR^3R^4}{\diagup}} \qquad RC\overset{NR^3NR^1R^2}{\underset{NR^4}{\diagup}}$$

$$\text{(71)} \qquad\qquad\qquad \text{(72)}$$

possible, however, it appears, from spectroscopic studies, that only one tautomer is normally obtained, there being no evidence for the presence of tautomeric mixtures.

Amidrazones which are unsubstituted, or substituted only on N^1 have been shown by infrared and n.m.r. studies[8, 44, 55, 78, 79] to exist in the amidrazone form (**73**). The structure of some substituted amidrazones,

$$RC\overset{NNRR}{\underset{NH_2}{\diagup}} \quad PhC\overset{NHNMe_2}{\underset{NPh}{\diagup}} \quad PhC\overset{NNMe_2}{\underset{NHMe}{\diagup}} \quad R^1C\overset{NNR^2R^3}{\underset{N}{\diagup}}\overset{H}{\underset{R^4}{\diagdown}}$$

$$\text{(73)} \qquad\quad \text{(74)} \qquad\quad \text{(75)} \qquad\quad \text{(76)}$$

however, is less clear cut. Thus Smith and his co-workers[52, 53] prepared the amidrazone **74** from the reaction of dimethyl hydrazine on N-phenyl-benzimidoyl chloride and concluded from comparison of its ultraviolet spectrum with those of model compounds, that the compound had the hydrazide imide structure **74** whereas the corresponding N^3-methyl derivative had the amidrazone structure **75**. Compound **74**, however, was also included in a series of di- and tri-substituted amidrazones prepared by Walter and Weiss[22]. These workers made a study of the infrared spectra of these compounds and their 'half deuterated' derivatives (e.g. **76**, $R^4 = D$) and found their results to be consistent with these compounds having the amidrazone structure **76**, with hydrogen bonding between N^1 and the hydrogen atom on N^3.

B. General Properties

Amidrazones are either colourless liquids or solids. Unsubstituted amidrazones, although fairly stable when pure, tend to be unstable in solution and may turn red due to conversion into s-tetrazines. They retain the reducing properties of the parent hydrazines. Substitution on the nitrogen atoms normally enhances the stability of amidrazones.

Amidrazones in general are mono-acid bases which form salts with inorganic acids; the hydrochlorides are by far the most frequently prepared but nitrates, sulphates, bromides, iodides, picrates, tosylates, perchlorates and fluoroborates have all been reported[1, 42, 52, 67]. The basicities of aliphatic amidrazones do not differ greatly from those of aliphatic amines, for example N^1,N^1-dimethylacetamidrazone has a pK_a of 10·0 in water and 9·6 in methanol[80]. The corresponding perfluoro-compound[8] has pK_a 10·2. The pK_a values of a series of cyclic amidrazones and vinylogous amidrazones, used as photographic dye developers, have been determined[81], and these values, the light stabilities and coupling activities were found to be related to the electron densities of the nitrogen atoms.

Amidrazone protonation can give rise to the amidinium-like cation **77** where spreading of the charge would be expected to lead to enhanced stability and there is n.m.r. evidence to support this[52, 53]. However,

$$
RC\begin{matrix} \nearrow NHNR^1R^2 \\ + \\ \searrow NR^3R^4 \end{matrix}
\qquad
PhC\begin{matrix} \diagup NHNPhMe \\ + \\ \diagdown NH_2 \end{matrix}
$$

(77)   **(78)**

infrared studies[79] suggest that some amidrazones, for example N^1-methyl-N^1-phenylbenzamidrazone, give salts of the structure **78**.

C. Spectral Properties

In their infrared spectra, amidrazones which are not fully substituted show strong bands assigned to NH_2 and NH stretching vibrations lying between 3500 and 3100 cm^{-1}, and these bands in particular have been used in structure determinations[22, 44, 78, 79]. Bands assigned to C=N stretching vibrations vary from 1690 to 1590 cm^{-1} and a low wave number has been cited as indicative of the presence of hydrazone ($\diagdown$C=NN—) rather than imine ($\diagdown$C=N—) bonding[79]. Bands at around 1660 and 1590 cm^{-1} have been assigned to NH_2 and NH deformations respectively but do not appear to have found much application in structure determination.

N.m.r. studies[22, 52, 53] have also found use in structure determinations (see Sections IV, A and B).

D. Alkylation and Silylation

Amidrazones can be methylated with either methyl iodide or methyl tosylate to give the corresponding salts; these salts, on basification, yield the methylated bases (equation 40). Smith and his co-workers[52, 53] have

$$PhC \begin{matrix} NNMe_2 \\ \\ NHMe \end{matrix} \xrightarrow{MeI} PhC \begin{matrix} NMeNMe_2 \\ \oplus \\ N-H \\ | \\ Me \end{matrix} I^\ominus \xrightarrow{OH^\ominus} PhC \begin{matrix} NMeNMe_2 \\ \\ NMe \end{matrix} \qquad (40)$$

made a study of the methylation of substituted amidrazones and hydrazide imides and have shown that, in most cases, methylation occurs either at N^2 or N^3 to give amidinium type ions which are stabilised by charge delocalisation (equations 40 and 41). On the other hand, they[52, 53]

$$PhC \begin{matrix} NMeNMe_2 \\ \\ NPh \end{matrix} \xrightarrow{MeI} PhC \begin{matrix} NMeNMe_2 \\ \oplus \\ NMePh \end{matrix} I^\ominus \qquad (41)$$

found that compounds of the type **79** undergo methylation on N^1 and they attribute this difference in behaviour to steric crowding. However, it is probable that a number of different, but finely balanced, factors must influence the position of alkylation.

$$RC \begin{matrix} NNMe_2 \\ \\ NR^1Me \end{matrix} \xrightarrow{MeI} RC \begin{matrix} \overset{\oplus}{N}NMe_3 \\ \\ NR^1Me \end{matrix} I^\ominus$$

(79)

R = R^1 = Ph; R = Ph, R^1 = Me; R = Et, R^1 = Me

Silyl-substituted amidrazones[40, 82] can be prepared by treating an amidrazone with chlorotrimethylsilane or with the corresponding dimethylamino compound **80**.

$$R_2N(CH_2)_2C \begin{matrix} NNMe_2 \\ \\ NH_2 \end{matrix} \xrightarrow[Et_3N]{Me_3SiCl} R_2N(CH_2)_2C \begin{matrix} NNMe_2 \\ \\ NHSiMe_3 \end{matrix}$$

$$CH_2{=}CHC \begin{matrix} NNMe_2 \\ \\ NH_2 \end{matrix} + Et_2NSiMe_3 \longrightarrow CH_2{=}CHC \begin{matrix} NNMe_2 \\ \\ NHSiMe_3 \end{matrix}$$

(80)

E. Nucleophilic Substitution Reactions

Nucleophilic substitution, usually of the N^3-amino residue, is possible in amidrazones which thus can undergo hydrolysis, aminolysis and hydrazinolysis.

I. Hydrolysis

Amidrazones and their salts can undergo hydrolysis by the action of both acids[52,53,83] and bases[52,53]. Prolonged treatment is necessary for the completion of the reaction with fully substituted amidrazones (equations 42 and 43). Hydrolysis of amidrazones in acetic acid has also been

$$PhC \overset{\overset{\oplus}{N}NMe_3}{\underset{NMePh}{\big<}} X^{\ominus} \xrightarrow[\Delta\ 2\ days]{6\ M—HCl} PhCOOH + PhNHMe + Me_3\overset{\oplus}{N}NH_2\ X^{\ominus} \quad (42)$$

$$EtC \overset{NNMePh}{\underset{NMe_2}{\big<}} \xrightarrow[\Delta\ 8\ h]{6\ M—OH^{\ominus}} EtC \overset{NHNMePh}{\underset{O}{\big<}} + Me_2NH \quad (43)$$

reported to give acid hydrazides[84]. By a similar reaction hydrogen sulphide has been shown to yield thiohydrazides[84a].

2. Aminolysis and related reactions

The amidrazone **81** is reported to give product **82** on treatment with ammonia[60]. Amidrazones also react with primary or secondary amines,

$$HC \overset{NN=CHPh}{\underset{NH_2}{\big<}} \xrightarrow{NH_3} \begin{array}{c} HC \overset{NN=CHPh}{\underset{NH}{\big<}} \\ HC \underset{NN=CHPh}{\big<} \end{array}$$

$$(81) \qquad\qquad\qquad (82)$$

on heating in the presence of ammonium salts as catalysts, but in these cases the products are N^3-substituted amidrazones[85].

$$MeC \overset{NNMe_2}{\underset{NH_2}{\big<}} \xrightarrow{RR^1NH} MeC \overset{NNMe_2}{\underset{NRR^1}{\big<}}$$

Hydroxylamine reacts smoothly with amidrazones to give N^3-hydroxy-amidrazones[41,78,86] (**83**). These compounds have been used to prepare 4-hydroxy-1,2,4-triazoles[41] and 1,2,4-triazine-oxides[86].

$$PhC \underset{NH_2}{\overset{NNPhMe}{\big<}} \ + \ NH_2OH \longrightarrow PhC \underset{NHOH}{\overset{NNPhMe}{\big<}}$$

(83)

Amidrazones also react with hydrazines. Thus during the preparation of an amidrazone, excess hydrazine may attack the newly formed product to give further reaction. If hydrazine itself is the nucleophile used, then a dihydrotetrazine (**19**) can be formed and from it secondary products may also arise (see Section III, B, 1). When the nucleophile is a substituted hydrazine, further reaction leads to the production of a formazan or its dihydro derivative (see equation 17).

These reactions exemplify replacement of the amino group of the amidrazone by the nucleophile. However, in other cases either the hydrazino group or both the amino and the hydrazino functions[44, 87] may be displaced, (equations 44 and 45). In the latter reaction (equation 45),

$$2 \ RC \underset{NH_2}{\overset{NNH_2}{\big<}} \longrightarrow RC \underset{NH_2 \ H_2N}{\overset{N \longrightarrow N}{\big<}} CR \qquad (44)$$

$$MeC \underset{NH_2}{\overset{NNMe_2}{\big<}} \xrightarrow[NH_4^{\oplus}]{H_2NNEt_2} MeC \underset{NHNEt_2}{\overset{NNMe_2}{\big<}} \xrightarrow{H_2NNEt_2} MeC \underset{NHNEt_2}{\overset{NNEt_2}{\big<}} \qquad (45)$$

(84)   **(85)**

the resultant dihydroformazan **85** is unrelated to the starting material **84**, in its nitrogen substitution pattern.

F. Reaction of Amidrazones with Acid Derivatives and Carbonyl Compounds

I. General

The most nucleophilic position, at least in unsubstituted amidrazones, is N^1 and hence it is this nitrogen which usually attacks electrophilic centres such as the carbon atom of carbonyl groups. The resultant open chain primary products frequently cyclise under the reaction conditions employed to give 1,2,4-triazole derivatives. Thus acylamidrazones, for example, are important precursors for triazoles and other five-membered heterocyclic systems[1]. Moreover the reactions of 1,2-dicarbonyl or α,β-unsaturated compounds with amidrazones can give access to 1,2,4-

triazines[1]. The chemistry of both the 1,2,4-triazoles[88] and of the 1,2,4-triazines[89, 90] has been the subject of comprehensive reviews.

2. Monofunctional acid derivatives containing a carbonyl group

Acid chlorides, anhydrides, esters and even acids themselves all react with unsubstituted amidrazones. When acid chlorides are used, acyl-amidrazones are the usual products[49, 51, 64, 91–95]. These can lose water on heating or on treatment with base to form the corresponding 1,2,4-triazoles. Indeed, when esters or acids are used, the triazole and not the intermediate acylamidrazone is the product normally isolated[51, 92, 96, 97].

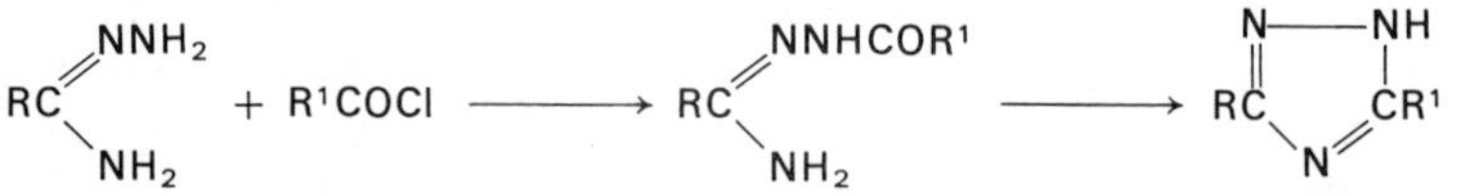

(86)

Both alkyl and aryl acid chlorides have been used and this approach to 3,5-disubstituted-1,2,4-triazoles (86) and to fused ring systems of the type 87 is mentioned in the patent literature[92]. N^1-Substituted[88] and N^3-

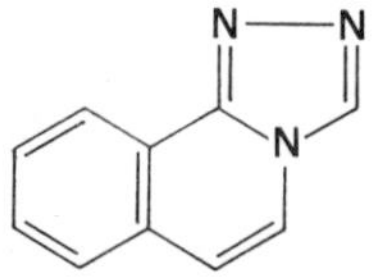

(87)

substituted[98] amidrazones can react similarly and give rise to 1,2,4-triazoles, but N^1,N^1-disubstituted amidrazones[78] merely acylate on N^3, (equations 46–48 respectively).

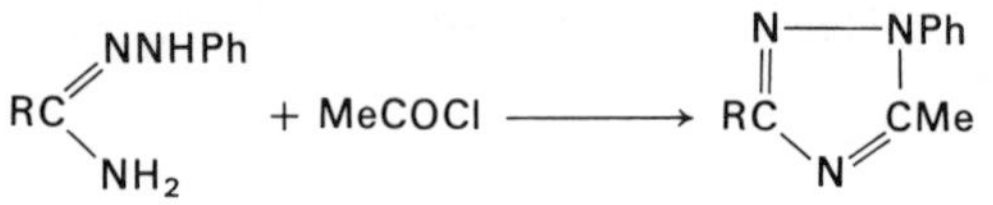

(46)

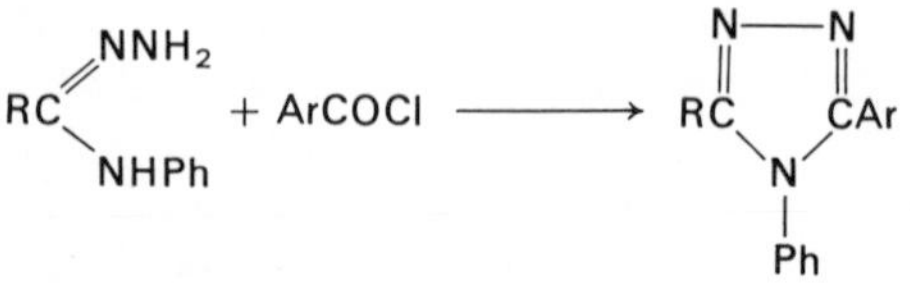

(47)

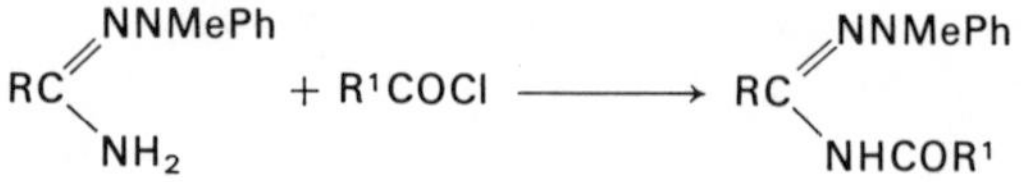

(48)

When N^1,N^3-diphenylbenzamidrazone is heated with an acid chloride it yields, on acidification with perchloric acid, a triazolium salt[64] (**88**).

$$\text{PhC}\begin{array}{c}\nearrow \text{NNHPh} \\ \searrow \text{NHPh}\end{array} \xrightarrow[\text{(b). HClO}_4]{\text{(a). RCOCl}} \quad \text{PhC}\underset{\underset{\text{Ph}}{\mid}}{\overset{\text{N---NPh}}{\bigoplus}}\text{CR} \quad \text{ClO}_4^{\ominus}$$

(88)

Acid anhydrides react very similarly to acid chlorides and here again acylamidrazones may be isolated. Thus 5-nitrofuran-carboxamidrazone[99] reacts with acetic or propionic acid anhydrides in the cold, to give the corresponding N^1-acyl amidrazones, but heating with excess anhydride yields the 1-acyl-1,2,4-triazoles (**89**) which hydrolyse, on heating in water, to the parent compounds **90**.

$$\text{RC}\begin{array}{c}\nearrow \text{NNH}_2 \\ \searrow \text{NH}_2\end{array} \xrightarrow[\text{excess}]{\text{(R}^1\text{CO)}_2\text{O}} \quad \text{RC}\overset{\text{N---NCOR}^1}{\underset{\text{N}}{\diagup\diagdown}}\text{CR}^1 \xrightarrow{\text{H}_2\text{O}} \quad \text{RC}\overset{\text{N---NH}}{\underset{\text{N}}{\diagup\diagdown}}\text{CR}^1$$

(89)   **(90)**

$$\text{R} = \text{O}_2\text{NC}\overset{\text{HC---CH}}{\underset{\text{O}}{\diagup\diagdown}}\text{C} \quad ; \text{R}^1 = \text{CH}_3 \text{ or } \text{CH}_3\text{CH}_2$$

Formic acid and ethyl formate have been used to give 1,2,4-triazoles unsubstituted in the 5-position, but this method has sometimes proved less satisfactory than that employing ethyl orthoformate[100, 101]. Higher acids, e.g. acetic or propionic acids, also react with amidrazones; however with acetic acid both the 1,2,4-triazole and the 1,3,4-oxadiazole are obtained[99] (equation 49).

$$\text{RC}\begin{array}{c}\nearrow \text{NNH}_2 \\ \searrow \text{NH}_2\end{array} \xrightarrow{\text{MeCOOH, HNO}_3} \quad \text{RC}\overset{\text{N---NH}}{\underset{\text{N}}{\diagup\diagdown}}\text{CMe} + \text{RC}\overset{\text{N---N}}{\underset{\text{O}}{\diagup\diagdown}}\text{CMe} \quad (49)$$

(R as for compound **89**)

Dithio-esters can also react with amidrazones to form 1,2,4-triazoles[51].

As suggested in equation 49 the reaction is not always straight-forward and, depending on the conditions, oxadiazoles may be formed[94, 102]: for example, thermal cyclisation of the compound **91** can give, not only the 1,2,4-triazole (**92**), but also some 1,3,4-oxadiazole (**93**). Indeed when for compound **91**, R = H, a 2-amino-1,3,4-oxadiazole is obtained[41], presumably via dehydration of compound **91** before cyclisation.

$$
\underset{(91)}{\overset{NHNHCOR^1}{\underset{NOH}{RC}}} \longrightarrow \underset{(92)}{\overset{N\text{---}N}{\underset{N}{\underset{OH}{RC}}}CR^1} + \underset{(93)}{\overset{N\text{---}N}{\underset{O}{RC}}CR^1}
$$

3. Bifunctional acid derivatives containing carbonyl groups

Derivatives of dicarboxylic acids, such as diacid chlorides, or diesters, on reaction with amidrazones give access to compounds containing two triazole rings[103] (equation 50) whereas the half acid chloride, half ester or anhydride provide a route to a triazolosubstituted acid[104] (**94**).

$$
\underset{NHAr}{\overset{NNH_2}{PhC}} \xoverset{ClOC(CH_2)_nCOCl}{\longrightarrow} \underset{\underset{Ar}{N}}{\overset{N\text{---}N}{PhC}}C\text{---}(CH_2)_n\text{---}C\underset{\underset{Ar}{N}}{\overset{N\text{---}N}{}}CPh \quad (50)
$$

$$
\underset{\underset{Ar}{N}}{\overset{N\text{---}N}{PhC}}C(CH_2)_nCOOH
$$
$$
(94)
$$

By varying the conditions of cyclisation, either 1,2,4-triazole or 1,3,4-oxadiazole rings can be incorporated in the product[102, 105] and when diamidrazones are used, polymeric products can be obtained[105].

$$
\underset{NH_2}{\overset{NNH_2}{RC}} + m\text{-}ClOCC_6H_4COCl \longrightarrow RC\underset{NH_2}{\overset{N\text{---}HN}{}}CC_6H_4C\underset{H_2N}{\overset{NH\text{---}N}{}}CR
$$

R = 2-pyridyl   (95)

4. Other acid derivatives and related compounds

Certain acid derivatives which do not possess carbonyl groups, such as orthoesters, imidates and hydrazonate esters, are very susceptible to attack by nucleophiles and hence it is not unexpected that they should react with amidrazones. Extensive use has been made of orthoesters, particularly of ethyl orthoformate, in the preparation of 1,2,4-triazoles from amidrazones [42, 51, 70, 93, 97, 106, 107]. Although the 1,2,4-triazole is

the product normally obtained, the intermediate triazoline **96** may sometimes be isolated [106]. Where the nucleophilicity of N^3 is reduced, as in 2-chloro-6-hydrazino-pyridine, the imidate **97** can be isolated [51]. It has

also been reported that diethoxymethyl acetate can effect ring closure when the reaction of the amidrazone with ethyl orthoformate does not proceed beyond the imidate stage [108]. The reactions of orthoesters with amino-compounds have been reviewed recently [109, 110].

Amidrazones can also react with imidates[51,111,112] to form triazoles, for example oxaldi-imidate[111] reacts to give the intermediate **98** which, on treatment with acid, loses ammonia to form the ditriazolyl **99**.

$$RC \underset{NH_2}{\overset{NNH_2}{<}} + \underset{HN}{\overset{EtO}{>}}C-C\underset{NH}{\overset{OEt}{<}} \longrightarrow$$

$$RC\overset{N---N}{\underset{NH_2 \quad H_2N}{<}}C-C\overset{N---N}{\underset{NH_2 \quad H_2N}{>}}CR \longrightarrow RC\overset{N---NH \quad N---NH}{\underset{\underset{H}{N} \quad \underset{H}{N}}{<}}C-C\overset{}{>}CR$$

(98) (99)

N^1-Methyl-N^1-phenyl formamidrazone[61] acts with a large excess of the hydrazonate ester **100** to give the compound **101**.

$$HC\underset{NH_2}{\overset{NNMePh}{<}} + HC\underset{OEt}{\overset{NNMePh}{<}} \longrightarrow \begin{array}{c} HC\overset{NNMePh}{\underset{NH}{<}} \\ HC\underset{NNMePh}{\overset{}{<}} \end{array}$$

(100) (101)

Alkyl and aryl sulphinyl chlorides (RSOCl) may be thought of as the counter parts of acid chlorides so it is not surprising that these compounds react smoothly with amidrazones to form N^1-sulphinylamidrazones; attempts to cyclise these products with base, however, gave no cyclic products but only Schiff's bases[113].

$$PhC\underset{NHPh}{\overset{NNH_2}{<}} + ClSOR \longrightarrow PhC\underset{NHPh}{\overset{NNHSOR}{<}} \overset{OH^-}{\longrightarrow} PhCH{=}NPh$$

5. Aldehydes and ketones

Unsubstituted amidrazones react with aldehydes to form compounds which have been described as having the structure of either Schiff's bases (**102**) or of 1,2,4-triazolines (**103**), and it would appear that both types of compounds may be formed depending on the particular reagents and conditions used[1,83,101,114,115]. However, irrespective of their structure, these compounds (**102** and **103**) can be oxidised to give 1,2,4-triazoles[1,88,101] (**104**). Ketones similarly yield Schiff's bases or 1,2,4-

triazolines[36,83,115]. Usually only 1 mol of the carbonyl compound acts with the amidrazone[101] but formaldehyde[1] can react with both amino functions to give compounds of the structure (105). The reaction has also been extended to the preparation of polytriazolines[115] (equation 51).

$$RC \overset{NN=CHR^1}{\underset{NH_2}{<}}$$

(102)

$$\begin{array}{c} N\text{---}NH \\ \| \quad \| \\ RC \quad CHR^1 \\ \diagdown N \diagup \\ | \\ H \end{array}$$

(103)

$$\begin{array}{c} N\text{---}N \\ \| \quad \| \\ RC \quad CR^1 \\ \diagdown N \diagup \\ | \\ H \end{array}$$

(104)

$$RC \overset{NN=CH_2}{\underset{N=CH_2}{<}}$$

(105)

$$\begin{array}{c} H_2NN \diagdown \qquad \diagup NNH_2 \\ \quad C \quad N \quad C \\ H_2N \diagup \bigcirc \diagdown NH_2 \end{array} + OHCC_6H_4CHO \longrightarrow$$

$$\left[\begin{array}{c} HN\text{---}N \qquad N\text{---}NH \\ | \quad \| \qquad \| \quad | \\ HC \quad C \quad N \quad C \quad CHC_6H_4 \\ \diagdown N \diagup \bigcirc \diagdown N \diagup \\ \quad | \qquad\qquad | \\ \quad H \qquad\qquad H \end{array} \right]_n$$

(51)

N^1-Monosubstituted amidrazones can also react with carbonyl compounds to give products which have been reported both as 1,2,4-triazolines and as Schiff's bases and which on oxidation yield the corresponding 1,2,4-triazoles. In some instances, however, the 1,2,4-triazoles appear to be formed directly, e.g. by the reaction of an aldehyde with N^1-phenyl-mandelamidrazones[1].

$$PhCH(OH)C \overset{NNHPh}{\underset{NH_2}{<}} + RCHO \longrightarrow PhCH(OH)C \begin{array}{c} N\text{---}NPh \\ \| \quad | \\ \quad CR \\ \diagup N \diagup \end{array}$$

Since N^3-substituted amidrazones possess a free amino group at N^1, they also can react with aldehydes to give coloured Schiff's bases which can be oxidised by mercuric oxide[98] or ferric chloride[116] to the corresponding

$$ArC \overset{NNH_2}{\underset{NHAr^1}{<}} + Ar^2CHO \longrightarrow ArC \overset{NN=CHAr^2}{\underset{NHAr^1}{<}} \longrightarrow ArC \begin{array}{c} N\text{---}N \\ \| \quad \| \\ \quad CAr^2 \\ \diagdown N \diagup \\ | \\ Ar^1 \end{array}$$

1,2,4-triazoles. Likewise with aldehydes and ketones, cyclic amidra-zones[51, 116–118] which have N^3 within a heteroaromatic ring, give hy-drazones (106) but these bear a hydrogen atom on N^2 and hence are not conjugated throughout. On heating or on oxidation, compounds of the type 106 cyclise to give triazolo-derivatives by elimination of R^1, the substituent with the greater steric requirement[117].

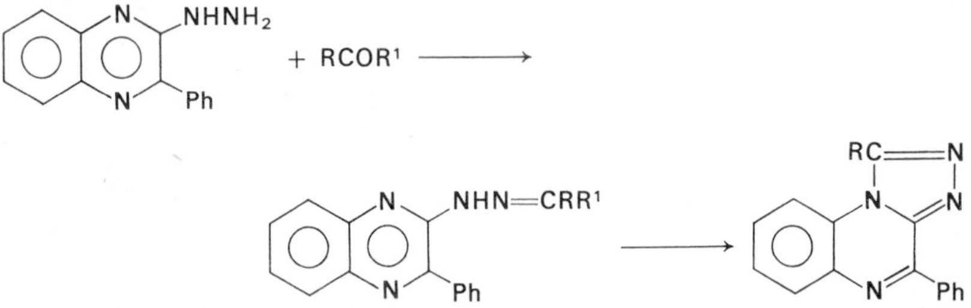

NHNH₂ ... + RCOR¹ ⟶

NHN=CRR¹

(106)

Fusco[64] heated N^1,N^3-diphenyl-benzamidrazone with aldehyde acetals and obtained triazolines which, on oxidation, gave triazolium salts (equation 52).

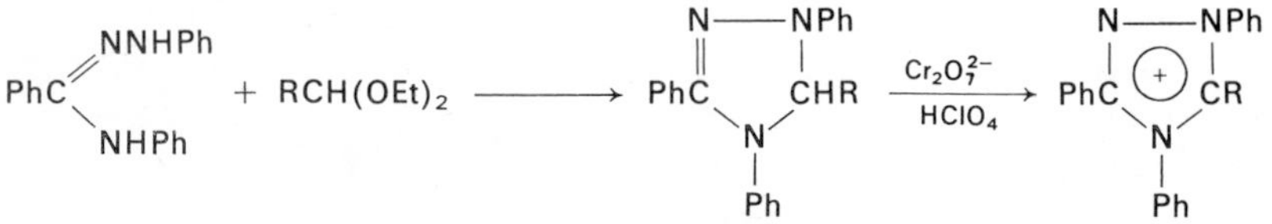

PhC(NNHPh)(NHPh) + RCH(OEt)₂ ⟶ [N——NPh ∥ PhC CHR N Ph] →Cr₂O₇²⁻/HClO₄→ [N——NPh PhC(+)CR N Ph]

(52)

The compound 107 which is an N^1,N^3-disubstituted amidrazone, gives Schiff's bases with aldehydes and these on treatment with acid undergo rearrangement to from 1,2,4-triazoles[118] (equation 53).

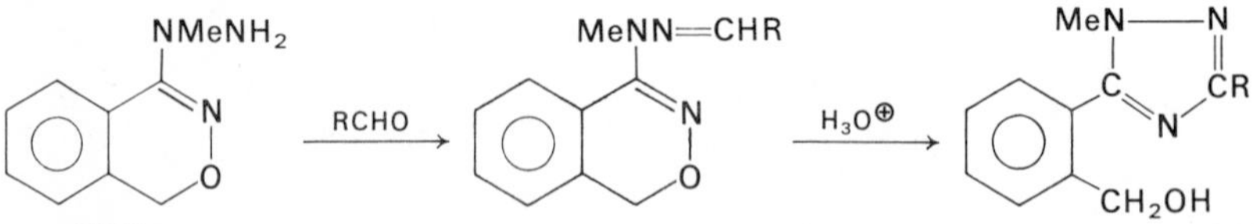

NMeNH₂ →RCHO→ MeNN=CHR →H₃O⊕→ [MeN——N C CR N] CH₂OH

(107)   (53)

Reactions involving α-haloketones have also been investigated. Thus α-bromocarbonyl derivatives have been reacted with N^1-acylamidrazones[40] and cyclic products obtained, e.g. when the reactant was phenacyl bromide, imidazoles (108) were formed but when α-bromopropiophenone

was used, a considerably higher temperature was required and the product
was a 1,2,4-triazole (**109**).

A series of fused ring compounds of the imidazole type has also been
prepared by the action of α-bromo-aldehydes or ketones on 3-amino-
6-methyl-pyridazine which has an amidrazone-type structure[119].

6. 1,2-Dicarbonyl and related compounds

The reaction between unsubstituted amidrazones and dicarbonyl
compounds has been used extensively in the synthesis of 1,2,4-
triazines[1, 89, 90, 106] which are the normal products of this reaction.
Glyoxal can at times give open chain compounds[1], but has been used
successfully in the preparation of 3-substituted-1,2,4-triazines[35] and the
parent member of this series, 1,2,4-triazine, itself has been obtained by
the condensation of glyoxal and formamidrazone[120], (equation 54,
R, R^1 and $R^2 = H$). Where unsymmetrical diketo compounds or keto-

$$(54)$$

aldehydes are used, the two possible isomeric compounds **110** and **111** may
sometimes be obtained, but frequently only one isomer is isolated[84, 120–122].

The reaction is normally carried out under basic conditions but when the condensation takes place in the presence of acid, the open chain osazone (**112**) is obtained. The osazone can readily be converted into the 1,2,4-triazine by heating in an inert solvent[123].

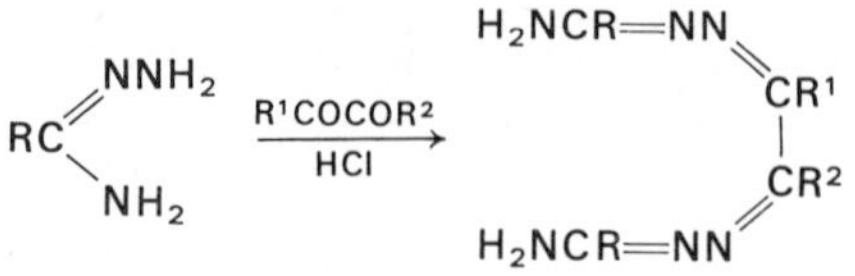

(**112**)

The method has been extended to give a series of 1,2,4-triazines substituted by hetero-aromatic residues which have been investigated as possible complexing agents[4, 83, 101, 114], and triazine polymers have also been formed[10, 124].

Where α-keto-acids or esters are used the corresponding triazinones

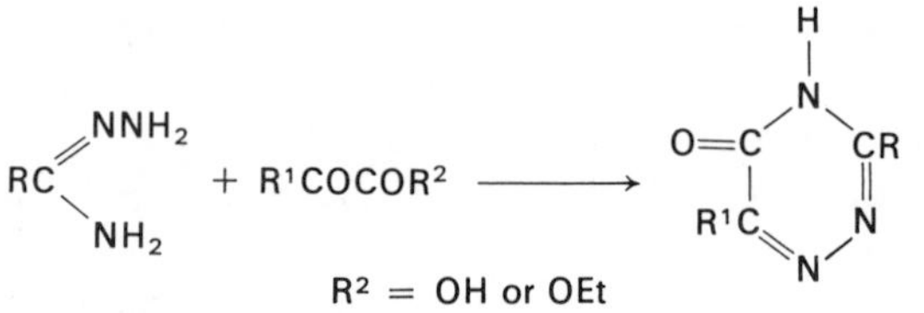

$$R^2 = OH \text{ or } OEt \tag{55}$$

are obtained[10,37,38] (equation 55) and the monoxime of a diketone provides a route to the N^4-oxides of 3,6-disubstituted-1,2,4-triazines[86]

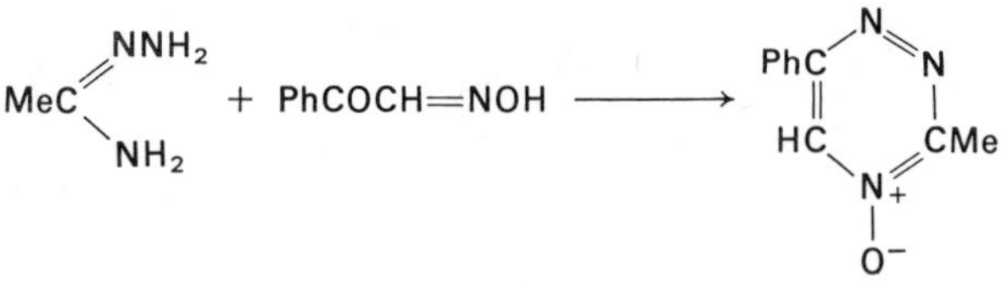

(56)

(equation 56). When an amidrazone reacts with the nitrile of an α-keto acid[125] the resulting Schiff's base can be converted into the 4-amino-1,2,4-triazine **114**. If, however, the Schiff's base **113** is first hydrolysed to the acid **115**, cyclisation[37] yields the triazinone **116**.

Unsubstituted amidrazones react in the same way with 1,2,3-tricarbonyl as with 1,2-dicarbonyl compounds to form 1,2,4-triazines[36, 126–128] which possess a carbonyl substituent in position 5. However, when a pyridyl carboxamidrazone (**117**, R = pyridyl) is used, intermediate amidrazone adducts can be isolated[126]. The triazine formed

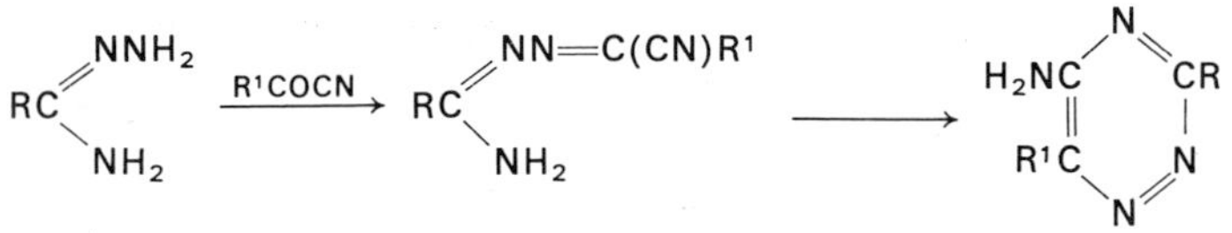

(113)　(114)

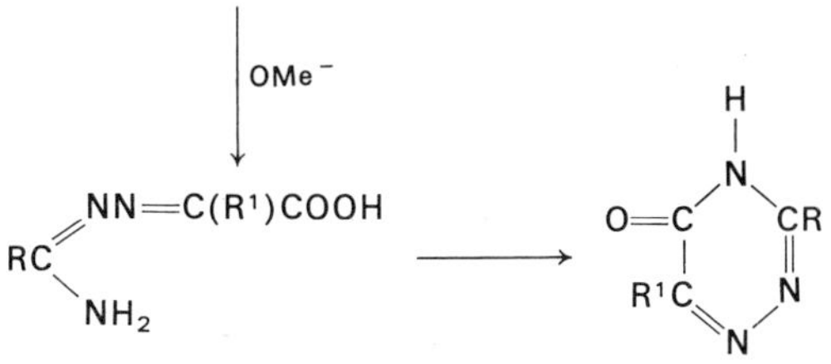

(115)　(116)

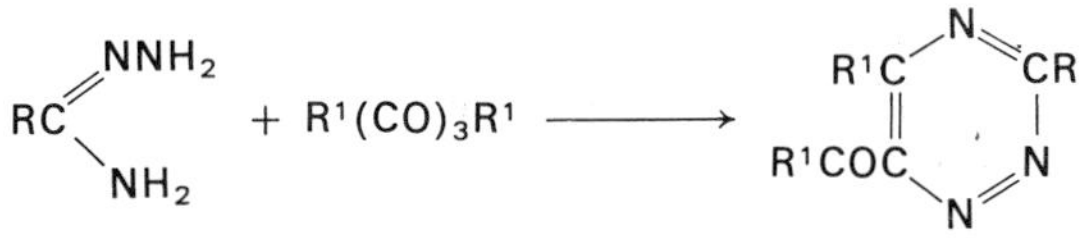

(117)　(118)

when the tricarbonyl derivative is diethyl mesoxalate (**118**, R^1 = OEt) provides a convenient route to 6-azapteridines[128].

7. 1,3- and 1,4-dicarbonyl compounds

The reactions of 1,3-dicarbonyl compounds with cyclic amidrazones such as 2-hydrazinoquinoxalines[117] and 1-hydrazinoisoquinolines[51] have been studied. The primary products of these reactions can be cyclised, either to pyrazoles (**119**), or to fused ring triazole systems (**120**).

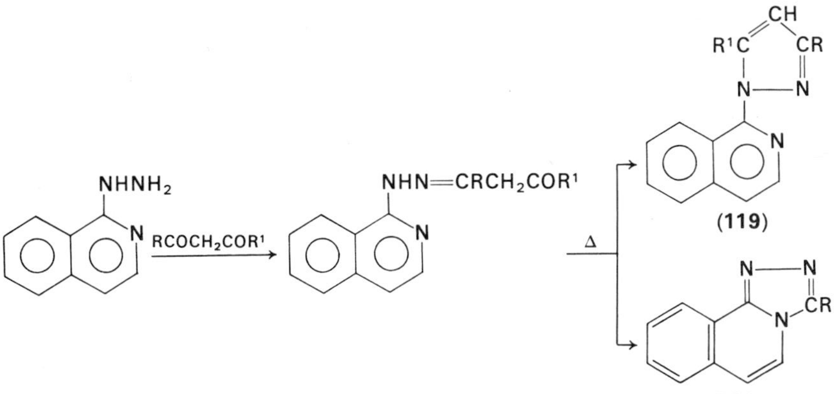

(119)

(120)

It has been suggested that the course of the reaction depends, not only on the ease of elimination of the anionic moiety, but that it can also be influenced by steric factors[51, 117].

The same types of heterocyclic systems are produced where open chain amidrazones are the starting materials. N^3-Phenyl-amidrazones of the type 121 give products (122) which cyclise on heating to 1,2,4-triazoles[129].

$$
\underset{\textbf{(121)}}{\text{ArC}\langle\overset{\text{NNH}_2}{\underset{\text{NHPh}}{}}} \xrightarrow{\text{RCOCHR}^1\text{COR}^2} \underset{\textbf{(122)}}{\text{ArC}\langle\overset{\text{NN=CRCHR}^1\text{COR}^2}{\underset{\text{NHPh}}{}}} \xrightarrow{\Delta} \underset{\textbf{(Ph)}}{\text{ArC}}\begin{array}{c}\text{N---N}\\ \| \quad \| \\ \text{CR} \\ \text{N} \\ | \\ \text{Ph}\end{array}
$$

Ar = aryl or pyridyl

$+ \text{R}^1\text{CH}_2\text{COR}^2$

On the other hand, acetamidrazone reacted with benzoylacetophenone, in the presence of acid under anhydrous conditions, to give the open chain compound 123, which on treatment with alkali gave the pyrazole 124; in this case the amidrazone molecule itself has also undergone hydrolytic cleavage[130]. Triazepine derivatives do not appear to be formed by the action of 1,3-dicarbonyl compounds on amidrazones[111, 130].

$$
\text{CH}_3\text{C}\langle\overset{\text{NNH}_2}{\underset{\text{NH}_2}{}} \xrightarrow[\text{HCl}]{\text{PhCOCH}_2\text{COPh}} \underset{\textbf{(123)}}{\text{CH}_3\text{C}\langle\overset{\text{NN=CPhCH}_2\text{COPh}}{\underset{\text{NH}_2}{}}} \xrightarrow{\text{NaOH}} \underset{\textbf{(124)}}{\text{PhC}\begin{array}{c}\text{HC----CPh}\\ \| \quad \| \\ \text{N} \\ \text{N} \\ | \\ \text{H}\end{array}}
$$

Cyclic amidrazones of the type of 1-hydrazino-isoquinoline (125) react with 1,4-dicarbonyl compounds to give pyridazine derivatives[51, 117] (equation 57). Maleic anhydride also yields a pyridazine with compound 125, showing the anhydride reacts here, not as an unsaturated carbonyl compound, but as a 1,4-dicarbonyl derivative[51] (see next section).

(125) + MeCO(CH$_2$)$_2$COOH $\longrightarrow$

NHN=CMe(CH$_2$)$_2$COOH

$\longrightarrow$ (57)

8. α,β-Unsaturated carbonyl compounds

Unsubstituted amidrazones add readily to activated double bonds; α,β-unsaturated ketones (126, R^1 = Ph) thus undergo Michael additions to give adducts of the type 127. On the other hand, when for compound 126 R^1 is an alkyl group, the double bond is less activated and the N^1 of the amidrazone attacks the carbonyl group to give Schiff's bases[131]

of the type 128. The very reactive phenylcyclobutenedione 129 behaves as an unsaturated ketone rather than as a diketone and reacts with amidrazones to give fused triazine derivatives[132], e.g. compound 130.

N^1-Phenyl-benzamidrazone has also been found to react with α,β-unsaturated ketones, but in this case the substituted amidrazone intermediate decomposed and the pyrazoline 131 was isolated[131].

Amidrazones can also react with α,β-unsaturated ketones which have an acetylenic rather than an ethylenic bond. p-Toluamidrazone[130] thus acts with the carbonyl group of propargylaldehyde, in the cold, to give the Schiff's base 132. When esters of acetylenic carboxylic acids react

$$p\text{-MeC}_6\text{H}_4\text{C}\overset{\displaystyle =\text{NNH}_2}{\underset{\displaystyle \text{NH}_2}{}} + \text{HC}\equiv\text{CCHO} \longrightarrow p\text{-MeC}_6\text{H}_4\text{C}\overset{\displaystyle =\text{NN}=\text{CHC}\equiv\text{CH}}{\underset{\displaystyle \text{NH}_2}{}}$$

$$\textbf{(132)}$$

with cyclic amidrazones, however, either triazine or pyrazole derivatives
are obtained. A series of fused ring triazinones has been prepared from
cyclic amidrazones of the structure **133** and dimethyl acetylenedicarboxy-
late. Here it is suggested that N^1 attack on the triple bond is followed by
nucleophilic attack of the ring nitrogen on the neighbouring carbonyl
group[133].

On the other hand, 1-hydrazino-isoquinoline reacts with ethyl propiolate
to form a pyrazolone[51] (equation 58).

$$(58)$$

9. Compounds of the general type $\overset{\displaystyle X}{\underset{\displaystyle Z}{\diagdown}}\text{C}=\text{Y}$

Compounds of the general formula XZC=Y where X and Z represent
good leaving groups, react with amidrazones to give open chain products
which cyclise, on heating, to 1,2,4-triazole derivatives. Phosgene and
thiophosgene are obvious examples of this class of compounds, but X
and Z need not be identical and ethyl chloroformate can be included in
this classification. Furthermore, the concept can be extended so that the
central atom of XZC=Y is not carbon but a hetero atom such as sulphur.

Thus amidrazones can react with ethyl chloroformate[51,117,134], ethyl orthocarbonate[70] or phosgene[51,91,135] to give N^1-substituted amidrazones (134) which can cyclise to 1,2,4-triazolin-5-ones (135, Y = O) and with thiophosgene[51,91,100,135] and trithiocarbonate[100] to give the correspond-

(134)   (135)

ing thiones (135, Y = S). Related syntheses include 1,2,4-triazoles (135, R = PhCONH) from amidrazones and N-dichloro-methylene-benzamide (PhCON=CCl$_2$)[51,91,135].

Compounds which have a central atom other than carbon can undergo similar reactions thus providing a synthetic route to some interesting five-membered heterocycles. Thus thionyl chloride[51,91] reacts with N^1,N^3-diphenyl-benzamidrazone[136] to give the thiatriazole 136; with

(136)

acyl-1-hydrazino-isoquinoline, however, compound 137 is formed but rearranges on heating to the 1,2,4-triazolo-derivative 138, with loss of sulphur dioxide[51].

(137)   (138)

10. Compounds of the general type X=C=Y

Compounds of the general structure X=C=Y such as carbon di-sulphide, ketenes, isocyanates etcetera react with amidrazones in a

similar way to those of structure XZC=Y and have been used to synthesise 1,2,4-triazoline derivatives (139). Where for compound 139
R^1 = H, the 1,2,4-triazole (140) may be formed[51,91,137]. For example,

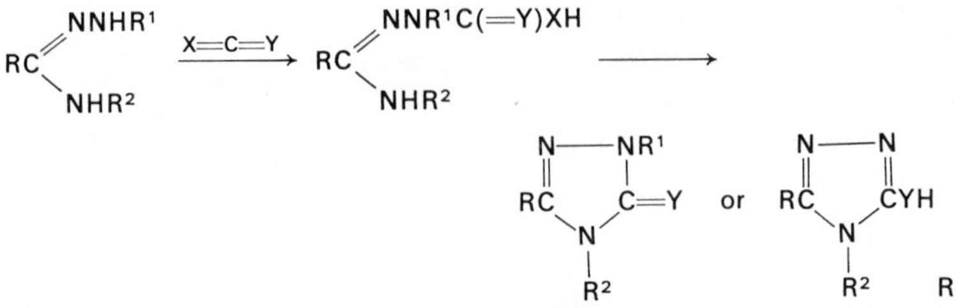

$$\text{(59)}$$

(139)   (140)

the same 1,2,4-triazole (140, R = R^2 = Ph, Y = CPh_2) can be obtained
from N^3-phenyl-benzamidrazone by reaction with diphenylketene, di-
phenylthioketene, N-(p-tolyl)-diphenylketimine. Dicyclohexylcarbodi-
imide reacts similarly[91,137].

Carbon disulphide reacts with cyclic amidrazones such as 1-hydrazino-
isoquinoline on heating to give fused ring 1,2,4-triazolinthiones[51,100]
(141, Y = S) but at room temperatures unsubstituted amidrazones react
to form 1,3,4-thiadiazolin-5-thiones[5] (equation 60).

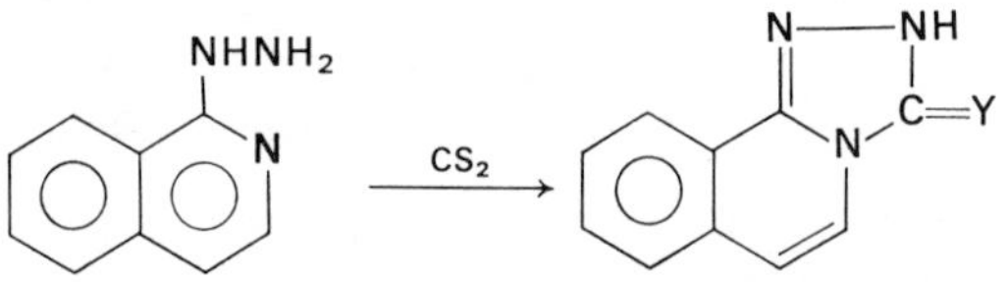

(141)

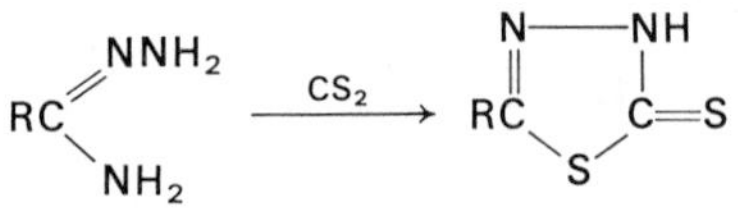

$$\text{(60)}$$

Reimlinger and his co-workers[51,91] have suggested that isocyanates
and isothiocyanates also behave as X=C=Y type compounds and react
with amidrazones to give triazolin-ones and thiones respectively according to equation 59. Thus N^3-phenylbenzamidrazone[91] reacts with
phenyl isothiocyanate to give 1,3,4-triphenyl-1,2,4-triazolin-5-thione
(139, R = R^1 = R^2 = Ph, Y = S) and phenylisocyanate[51], on heating
with 1-hydrazino-isoquinoline, gives the expected s-triazolo[3,4-a]iso-
quinoline (141, Y = O). However, other workers[138] have found that
adducts of the type 142 can be isolated when amidrazones or their hydro-

chlorides are treated at room temperature with isocyanates or isothiocyanates. These adducts (142) cyclise to the triazoline derivatives 143, which must be formed by nucleophilic attack on the carbon of the amidrazone group, by the nitrogen originating from the isocyanate or isothiocyanate moiety[138–140].

$$
\begin{array}{ccc}
RC\!\!\begin{array}{c}{}^{\diagup NNHR^1}\\{}_{\diagdown NH_2}\end{array} + R^2NCX & \longrightarrow & RC\!\!\begin{array}{c}{}^{\diagup NNR^1CXNHR^2}\\{}_{\diagdown NH_2}\end{array} \\
& & X = O \text{ or } S \\
& & (142)
\end{array}
\quad \xrightarrow{\Delta} \quad
\begin{array}{c}
\text{(143)}
\end{array}
$$

Practical use has been made of this type of reaction to prepare polymeric products from diisocyanates[141].

N^1,N^1-Disubstituted amidrazones[131, 138, 142] cannot react in the same way as the foregoing compounds but undergo condensation reactions at N^3.

$$
RC\!\!\begin{array}{c}{}^{\diagup NNMe_2}\\{}_{\diagdown NH_2}\end{array} + R^1NCO \longrightarrow RC\!\!\begin{array}{c}{}^{\diagup NNMe_2}\\{}_{\diagdown NHCONHR^1}\end{array}
$$

Akin to compounds of the type X=C=Y are the ynamines (e.g. compound 144), which react readily with amidrazones to form 3-alkyl-1,2,4-triazoles[51, 91, 143].

$$
PhC\!\!\begin{array}{c}{}^{\diagup NNH_2}\\{}_{\diagdown NHPh}\end{array} + Et_2NC\!\!\equiv\!\!CMe \longrightarrow
$$

(144)

$$
PhC\!\!\begin{array}{c}{}^{\diagup NNHC(=CHMe)NEt_2}\\{}_{\diagdown NHPh}\end{array} \longrightarrow
$$

G. Synthesis of Miscellaneous Ring Systems from Amidrazones

In addition to their wide application in the synthesis of 1,2,4-triazoles and 1,2,4-triazines mentioned in the previous section, amidrazones have been used from time to time in the formation of other heterocyclic systems[1]. Among these reactions is the formation of tetrazoles by the action of nitrous acid or ethyl nitrite on unsubstituted, N^1-substituted and cyclic amidrazones[1, 106, 117, 144] (equations 61 and 62).

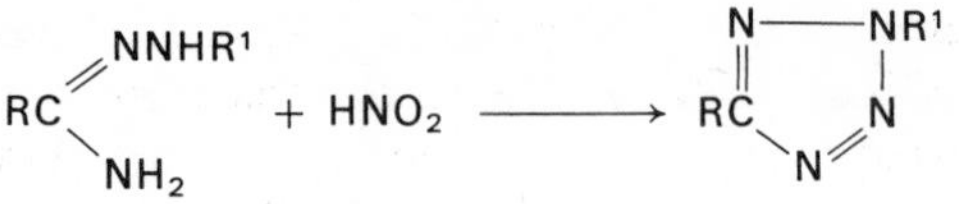

$$RC\underset{NH_2}{\overset{NNHR^1}{<}} + HNO_2 \longrightarrow RC\underset{\underset{N}{\diagdown}}{\overset{N-NR^1}{<}}\underset{N}{\diagup} \qquad (61)$$

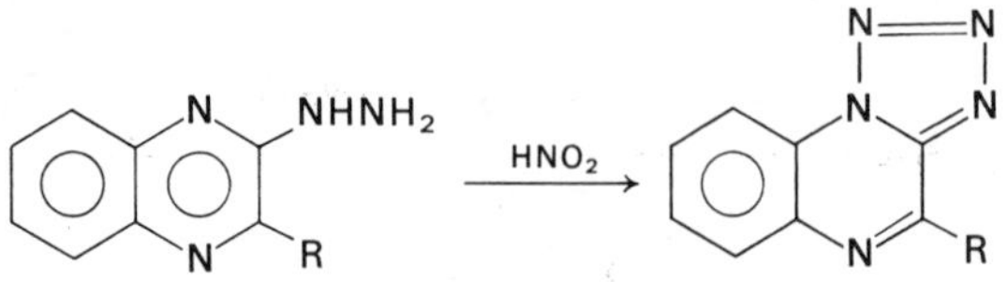

(62)

Phosphorus has been incorporated into five-membered nitrogen heterocycles by the use of amidrazones[43] and some interesting spiro compounds[145] (145) have been prepared.

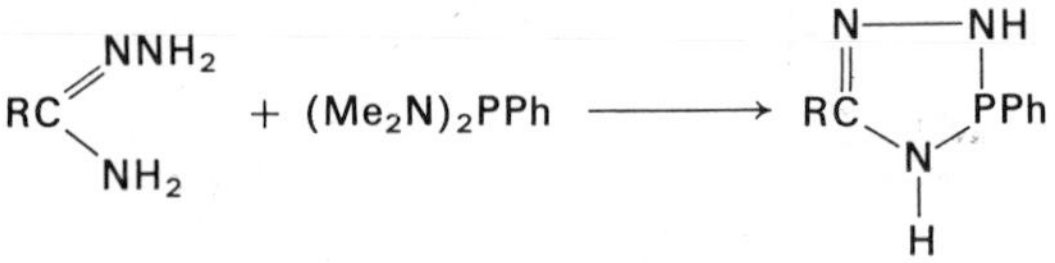

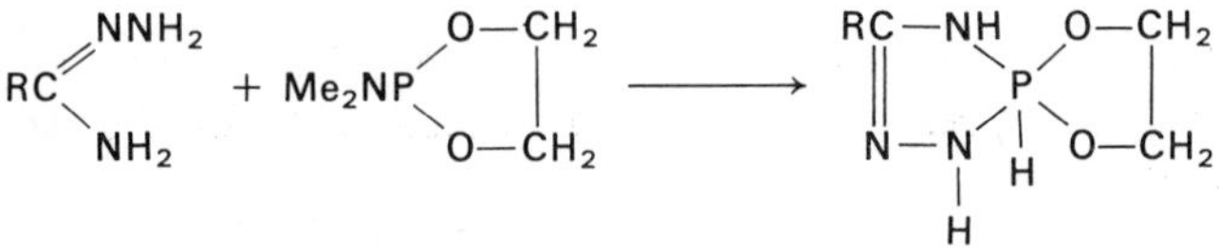

$$RC\underset{NH_2}{\overset{NNH_2}{<}} + (Me_2N)_2PPh \longrightarrow RC\underset{\underset{H}{N}}{\overset{N-NH}{<}}PPh$$

(145)

The aromatic 5,1,3,4-boratriazoles 147 have been formed by heating an amidrazone in benzene with a boronic acid derivative[146] (146, X = Cl, OH, OMe, NMe₂).

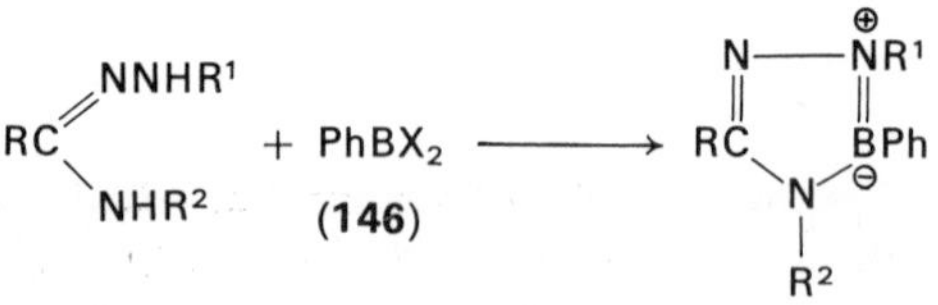

$$RC\underset{NHR^2}{\overset{NNHR^1}{<}} + PhBX_2 \longrightarrow RC\underset{\underset{R^2}{N}}{\overset{N-\overset{\oplus}{N}R^1}{<}}\overset{\ominus}{B}Ph$$

(146)   (147)

Fused pyrimidine systems have been synthesised from amidrazones of type 148 by cyclisation between suitably situated carbonyl groups and N² of the amidrazone[54].

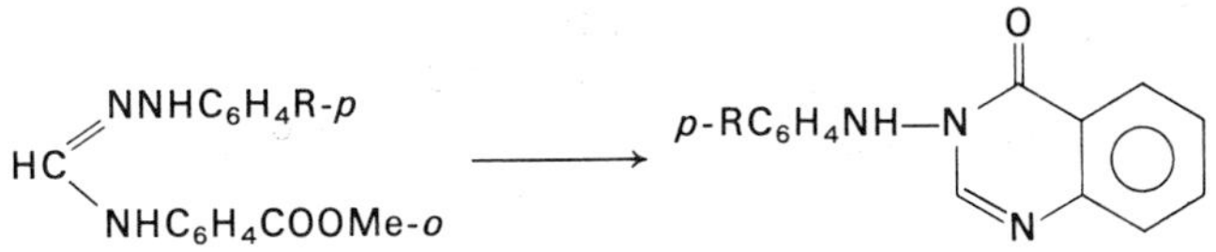

(148)

H. Oxidation

Mercuric oxide has been reported to oxidise amidrazones[147] of the type **149**.

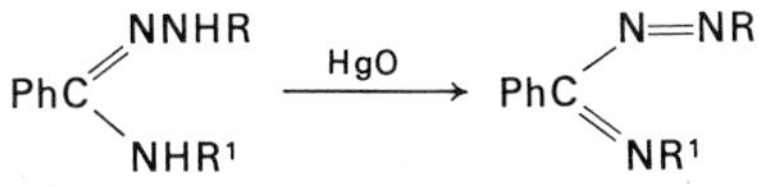

(149)

Perfluoroamidrazones are converted by hydrogen peroxide into the corresponding perfluoroacids, although hydrolysis does not occur in the absence of the oxidant[8], while pyridine-2-carboxamidrazone[13] with the same reagent gives the amide **151** and the substituted amidrazone **152**.

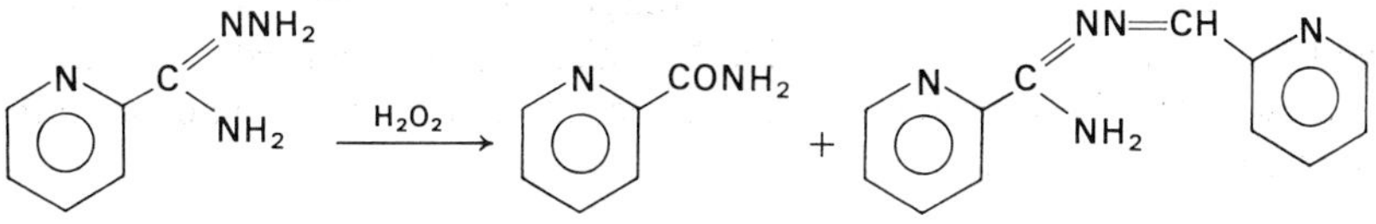

(150)   (151)   (152)

When the amidrazone **150** is refluxed in ethanol in the presence of oxygen[13] the corresponding dihydrotetrazine, tetrazine, and dihydrazidine are formed along with compound **152** (see equation 4). Since 1,2,4-triazoles may arise from dihydrazidines, this can account for the presence of these compounds among the products formed by the prolonged heating of amidrazones[91, 111].

Cyclic amidrazones undergo oxidative coupling with aromatic amines, phenols and reactive methylene compounds (equation 63). This work has been the subject of a recent review[148].

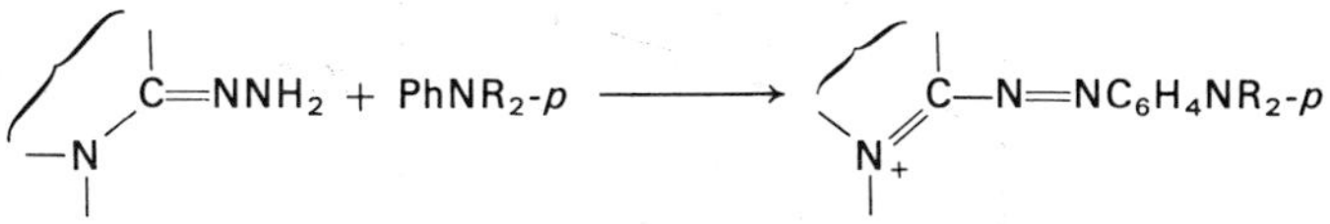

The stable free radical **154** has been obtained by the nucleophilic attack of N^1 of the cyclic amidrazone **153** on 1,2,3-trinitrobenzene under oxidative conditions[149].

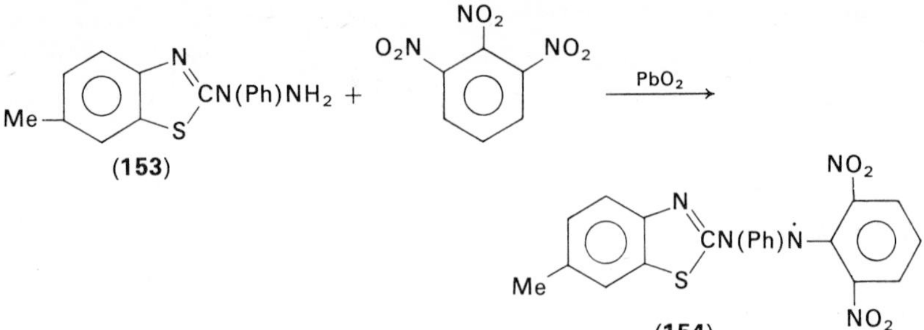

(153)

(154)

I. Metal Complexes

Many amidrazones form deeply coloured complexes with transition metal ions[1], e.g. Atkinson and Polya[39] tested for the presence of amidrazones by treating a chloroform solution of the base with an aqueous solution of cobalt chloride and observing a purple colour develop in the organic phase. However, the nature of the amidrazone appears to be important for the compound **155** formed well defined complexes[150] with Cd^{2+}, Co^{2+} and Fe^{3+} whereas the compound **156** formed a complex[151] with Cu^+ but not with Co^{2+} or Fe^{2+}. It also appears that redox reactions can take place between the ligand and the cation, but these do not seem to have been seriously investigated[150].

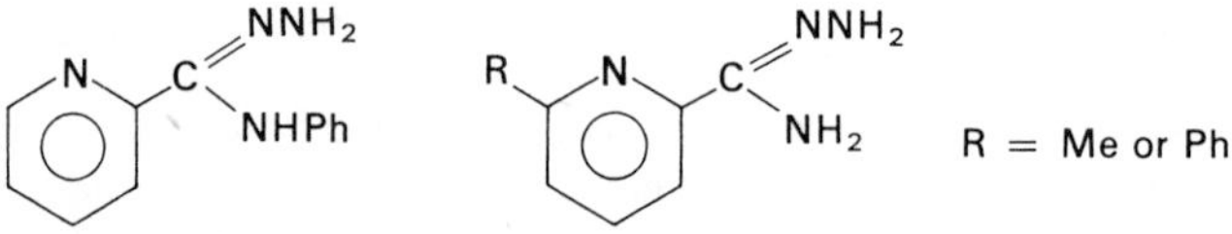

(155)   (156)   R = Me or Ph

Tridentate ligands[152] which have the structure of cyclic amidrazones have been used to prepare zinc and iron complexes of type **157**.

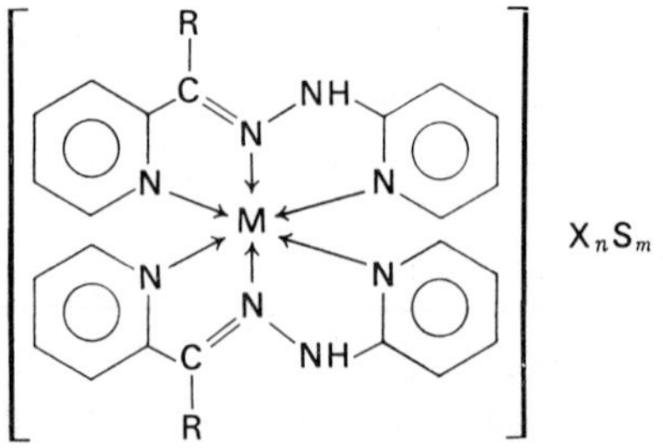

(157)

X = anion, S = solvent ; n and m = 0, 1, 2, etc.

When excess methyl isocyanide and hydrazine are added to an aqueous solution of potassium tetrachloroplatinate(II), bright orange crystals are obtained (Chugaev's salt). This salt is believed to have the structure **158** in which a dihydrazidine acts as one of the ligands[153].

$$\begin{array}{c} \text{N}\text{------}\text{N} \\ \text{MeNH}\text{==}\text{C} \qquad \qquad \text{C}\text{==}\text{NHMe} \\ \text{Pt} \\ \text{MeNC} \qquad \text{CNMe} \end{array}$$

(158)

In addition some aluminium[1,154] and boron[20,21] derivatives of amidrazones have been reported, e.g. compound **159**.

J. Uses

In addition to their value in synthesis discussed above, amidrazones, because of their wide variety of substitution patterns, have found both industrial and medicinal applications[1].

Various amidrazones, among them the compounds **159** and **160**, have been tested as potential rocket fuels or fuel additives[20,21,155]. Other

$$\left[\text{MeC} \begin{array}{c} \text{NMeNH}_2 \\ \text{NH} \end{array} \right]_2 \quad \text{B}_{10}\text{H}_{12}$$

(159)

$$\begin{array}{c} \text{NHNH}_2 \\ \text{N} \quad \text{N} \\ \text{N} \quad \text{N} \quad \text{N} \\ \text{H}_2\text{NHN} \quad \text{NHNH}_2 \\ \text{N} \quad \text{N} \end{array}$$

(160)

heavy industrial applications which have been investigated include the synthesis of polymers[1,105,115,124,141,156,157], e.g. from oxaldiamidrazone and the di-acid chloride of terephthalic acid. Amidrazones have also found use in photographic processing[1,81]. Other uses include the application of compounds of the type **58** as fungicides[65] and of some substituted oxaldiamidrazones and cyanoformamidrazones as herbicides[18].

Medicinal investigations include the testing of amidrazones, particularly β-substituted-propionamidrazones, as nasal decongestants and hypotensive agents[27,158]. In addition some amidrazones appear to have potential as anti-irradiating agents[159, 160], and some cyclic amidrazone (**161**) have been tested as anti-cancer agents[161].

(**161**)

V. ACKNOWLEDGEMENTS

We wish to thank Mr John Fraser and Mrs D. G. Neilson for help in preparing this manuscript.

VI. REFERENCES

1. D. G. Neilson, R. Roger, W. M. Heatlie and L. R. Newlands, *Chem. Revs.*, **70**, 151 (1970).
2. H. Rapoport and R. M. Bonner, *J. Am. Chem. Soc.*, **72**, 2783 (1950).
3. Nomenclature Section, *Chem. Abstracts*, **56**, 51N (1962).
4. F. H. Case, *J. Org. Chem.*, **30**, 931 (1965).
5. S. Kubota, Y. Koida, T. Kosaka and O. Kirino, *J. Pharm. Bull.* (*Japan*), **18**, 1696 (1970).
6. F. H. Case, *J. Heterocyclic Chem.*, **5**, 223 (1968).
7. Y. Kato and I. Hirao, *Bull. Kyushu Inst. Tech.*, **15**, 57 (1965), referred to in reference 106.
8. H. C. Brown and D. Pilipovich, *J. Am. Chem. Soc.*, **82**, 4700 (1960).
9. D. Genthe, *Deut. Luft-Raumfahrt, Forschungber.*, 1970 DLR FB 70 (52); *Chem. Abstr.*, 75, 65760r (1971).
10. P. M. Hergenrother, *J. Polym. Sci.*, *Part A1*, **7**, 945 (1969).
11. D. D. Libman and R. Slack, *J. Chem. Soc.*, 2253 (1956).
12. J. F. Geldard and F. Lions, *J. Org. Chem.*, **30**, 318 (1965).
13. S. Kubota, O. Kirino, Y. Koida and K. Miyake, *J. Pharm. Soc. Japan*, **92**, 275 (1972).
14. V. P. Wystrach, *Heterocyclic Compounds*, Vol. 8 (Ed. R. C. Elderfield), John Wiley and Sons, New York, 1967, p. 113.
15. T. K. Brotherton and J. W. Lynn, *Chem. Revs.*, **59**, 841 (1959).
16. K. Matsuda and L. T. Morin, *J. Org. Chem.*, **26**, 3783 (1961).
17. L. T. Morin and K. Matsuda, *U.S. Patent* 3,033,893 (1962); *Chem. Abstr.*, **57**, 14948d (1962).
18. R. G. Haldeman, L. T. Morin and K. Matsuda, *U.S. Patent* 3,075,013 (1963); *Chem. Abstr.*, **58**, 11276a (1963).
19. G. W. McNeely, *Diss. Abstr.*, *Int. B.*, **31**, 2570 (1971).
20. M. M. Fein, J. Bobinski, J. E. Paustian, D. Grafstein and M. S. Cohen, *Inorg. Chem.*, **4**, 422 (1965).
21. J. Bobinski, M. M. Fein and D. Grafstein, *U.S. Patent* 3,288,802 (1966); *Chem. Abstr.*, **66**, 39498f (1967).

22. W. Walter and H. Weiss, *Ann. Chem.*, **758**, 162 (1972).
23. R. Fuks, *Tetrahedron*, **29**, 2147 (1973).
24. T. Kauffmann, J. Hansen, C. Kosel and W. Schoeneck, *Ann. Chem.*, **656**, 103 (1962).
25. T. Kauffmann, S. Spaude and D. Wolf, *Chem. Ber.*, **97**, 3436 (1964).
26. T. Kauffmann and L. Ban, *Chem. Ber.*, **99**, 2600 (1966).
27. G. C. Wright, R. P. Halliday and C. S. Davis, *J. Pharm. Sci.*, **59**, 105 (1970).
28. K. H. Linke, R. Taubert, K. Bister, W. Bornatsch and B. J. Liem, *Z. Naturforsch. B.*, **26**, 296 (1971).
29. V. S. Garkusha-Bozhko, S. M. Baranov and O. P. Shvaika, *Ukr. Khim. Zh.*, **38**, 171 (1972); *Chem. Abstr.*, **77**, 34426y (1972).
30. C. L. Stevens, R. C. Freeman and K. Noll, *J. Org. Chem.*, **30**, 3718 (1965).
31. F. Kurzer and D. R. Hanks, *J. Chem. Soc. (C)*, 1375 (1968).
32. F. Kurzer and K. Douraghi-Zadeh, *Chem. Revs.*, **67**, 107 (1967).
33. C. Grundmann, *Angew. Chem. (Int. Ed.)*, **2**, 309 (1963).
34. R. Roger and D. G. Neilson, *Chem. Revs.*, **61**, 179 (1961).
35. H. Neunhoeffer and H.-W. Frühauf, *Ann. Chem.*, **760**, 102 (1972).
36. E. C. Taylor and S. F. Martin, *J. Org. Chem.*, **37**, 3958 (1972).
37. V. Uchytilova, P. Fiedler, M. Prystaš and J. Gut, *Coll. Czech. Chem. Commun.*, **36**, 1955 (1971).
38. M. Brugger, H. Wamhoff and K. Korte, *Ann. Chem.*, **755**, 101 (1972).
39. M. R. Atkinson and J. B. Polya, *J. Chem. Soc.*, 3319 (1954).
40. E. E. Glover, K. T. Rowbottom and D. C. Bishop, *J. Chem. Soc., Perkin I*, 2927 (1972).
41. H. G. O. Becker, G. Goermar and H.-J. Timpe, *J. Prakt. Chem.*, **312**, 601 (1970).
42. R. Kraft, H. Paul and G. Hilgetag, *Chem. Ber.*, **101**, 2028 (1968).
43. Y. Charbonnel and J. Barrans, *C.R. Acad. Sci. Paris, C*, **272**, 1675 (1971).
44. G. S. Gol'din, V. G. Poddubnyi, A. A. Simonova, G. S. Shor and E. A. Rybakov, *Zh. Org. Khim.*, **5**, 1404 (1969).
45. G. S. Gol'din, V. G. Poddubnyi, A. A. Simonova, A. B. Kamenskii and G. S. Shor, *J. Gen. Chem.*, **40**, 1281 (1970).
46. G. S. Gol'din, M. V. Maksakova and A. N. Kol'tsova, *J. Gen. Chem.*, **43**, 317 (1973).
47. F. A. Hussein and A. A. Kadir, *J. Indian Chem. Soc.*, **45**, 729 (1968).
48. T. Mukaiyama and S. Ono, *Tetrahedron Letters*, 3569 (1968).
49. E. P. Nesynov, M. M. Besprozvannaya and P. S. Pel'kis, *Zh. Org. Khim.*, **6**, 805 (1970).
50. E. Bisagni, J. P. Marquet and J. André-Louisfert, *Bull. Soc. Chim. France*, 1483 (1972).
51. H. Reimlinger, J. M. Vandewalle and W. R. F. Lingier, *Chem. Ber.*, **103**, 1960 (1970).
52. R. F. Smith, D. S. Johnson, C. L. Hyde, T. C. Rosenthal and A. C. Bates, *J. Org. Chem.*, **36**, 1155 (1971).
53. R. F. Smith, D. S. Johnson, R. A. Abgott and M. J. Madden, *J. Org. Chem.*, **38**, 1344 (1973).

54. Z. Csuros, R. Soos, I. Bitter and J. Palinkas, *Acta Chim. Acad. Sci. Hung.*, **69**, 361 (1971); *Chem. Abstr.*, **75**, 129,751y (1971).
55. W. Hoyle, *J. Chem. Soc.* (*C*), 690 (1967).
56. H. Meerwein, W. Florian, N. Schön and G. Stopp, *Ann. Chem.*, **641**, 1 (1961).
57. W. J. van der Burg, *Rec. Trav. Chem.*, **74**, 257 (1955).
58. A. Spassov and E. Golovinsky, *J. Gen. Chem.*, **32**, 3330 (1962).
59. A. Spassov, E. Golovinsky and G. Demirov, *Chem. Ber.*, **99**, 3734 (1966).
60. H. Neunhoeffer and H. Hennig, *Chem. Ber.*, **101**, 3947 (1968).
61. W. Walter, H. Weiss and K.-J. Reubke, *Ann. Chem.*, **736**, 166 (1970).
62. I. Hagedorn and H.-D. Winkelmann, *Chem. Ber.*, **99**, 850 (1966).
63. V. R. Grashey, M. Baumann and H. Bauer, *Chem. Ztg.*, **96**, 224 (1972).
64. R. Fusco and P. Dalla Croce, *Gazz. Chim. Ital.*, **99**, 69 (1969).
65. J. Perronet and P. Girault, *German Patent* 2,146,192, (1972); *Chem. Abstr.*, **77**, 34128c (1972).
66. G. Ponzio, *Gazz. Chim. Ital.*, **40**, 77 (1910).
67. H. C. Brown and M. T. Cheng, *J. Org. Chem.*, **27**, 3240 (1962).
68. G. V. Boyd and A. J. H. Summers, *J. Chem. Soc.* (*C*), 409 (1971).
69. O. P. Shvaika and V. I. Fomenko, *Dokl. Acad. Nauk SSSR* (*Chem.*), **200**, 134 (1971); *Chem. Abstr.*, **76**, 59542a (1972).
70. H. G. O. Becker, D. Beyer, G. Israel, R. Müller, W Riediger and H.-J. Timpe, *J. Prakt. Chem.*, **312**, 669 (1970).
71. H. Kohn and R. A. Olofson, *J. Org. Chem.*, **37**, 3504 (1972).
72. D. Jerchel and H. Fischer, *Ann. Chem.*, **574**, 85 (1951).
73. F. Moczar and L. Mester, *Bull. Soc. Chim. France*, 186 (1962).
74. M. Regitz and B. Eistert, *Chem. Ber.*, **96**, 3120 (1963).
75. D. Jerchel and R. Kuhn, *Ann. Chem.*, **568**, 185 (1950).
76. D. Jerchel and W. Wotichy, *Ann. Chem.*, **605**, 191 (1957).
77. H. Hellmann and W. Schwiersch, *Chem. Ber.*, **94**, 1868 (1961).
78. B.-G. Baccar and J. Barrans, *C.R. Acad. Sci. Paris*, *C*, **263**, 743 (1966).
79. B.-G. Baccar, R. Mathis, A. Secches, J. Barrans and F. Mathis, *J. Mol. Struct.*, **7**, 369 (1971).
80. R. R. Tarasyants, G. M. Petrova, G. S. Gol'din, V. G. Poddubnyi, S. G. Fedorov, and A. A. Simonova, *J. Gen. Chem.*, **43**, 495 (1973).
81. H.-J. Hofmann and M. Scholz, *J. Prakt. Chem.*, **313**, 349 (1971).
82. G. S. Gol'din, M. V. Maksakova and A. N. Kol'tsova, *J. Gen. Chem.*, **42**, 1351 (1972).
83. F. H. Case, *J. Heterocyclic Chem.*, **10**, 353 (1973).
84. A. S. Shawali, A. Osman and H. H. Hassaneen, *Indian J. Chem.*, **10**, 965 (1972).
84a. H. C. Brown and R. Pater, *J. Org. Chem.*, **30**, 3739 (1965).
85. G. S. Gol'din, V. G. Poddubnyi, A. A. Simonova, E. V. Orlova and G. S. Shor, *Zh. Org. Khim.*, **5**, 1411 (1969).
86. H. Neunhoeffer, F. Weischedel and V. Böhnisch, *Ann. Chem.*, **750**, 12 (1971).
87. G. S. Gold'din, V. G. Poddubnyi, E. V. Orlova and A. V. Kisin, *Zh. Org. Khim.*, **9**, 517 (1973).
88. J. H. Boyer, *Heterocyclic Compounds*, Vol. 7 (Ed. R. C. Elderfield), John Wiley and Sons, New York, 1961, pp. 384–461.

89. J. P. Horwitz, *Heterocyclic Compounds*, Vol. 7 (Ed. R. C. Elderfield), John Wiley and Sons, New York, 1961, pp. 720–796.
90. R. L. Jones and J. R. Kershaw, *Rev. Pure Appl. Chem.*, **21**, 23 (1971).
91. H. Reimlinger, W. R. F. Lingier and J. J. M. Vandewalle, *Chem. Ber.*, **104**, 639 (1971).
92. J. E. Francis, *U.S. Patent* 3,354,164 (1967); *Chem. Abstr.*, **69**, 36142m (1968).
93. H. Moeller, *German Patent* 2,113,731 (1972); *Chem. Abstr.*, **77**, 164709d (1972).
94. M. Saga and T. Shono, *J. Polymer. Sci.*, **B4**, 869 (1966).
95. M. L. Hoefle and A. Holmes, *French Patent* 2,100,863 (1972); *Chem. Abstr.*, **77**, 164710x (1972).
96. M. Yanai, T. Kinoshita, S. Takeda, M. Nishimura and T. Kuraishi, *Chem. Pharm. Bull.*, *Japan*, **20**, 1617 (1972).
97. K. T. Potts and C. R. Surapaneni, *J. Heterocyclic Chem.*, **7**, 1019 (1970).
98. A. Spassov, E. Golovinsky and G. Russev, *Chem. Ber.*, **96**, 2996 (1963).
99. I. Hirao, Y. Kato and H. Tateishi, *Bull. Chem. Soc. Japan*, **45**, 208 (1972).
100. K. T. Potts and S. W. Schneller, *J. Heterocyclic Chem.*, **5**, 485 (1968).
101. F. H. Case, *J. Heterocyclic Chem.*, **7**, 1001 (1970).
102. P. M. Hergenrother, *J. Heterocyclic Chem.*, **6**, 965 (1969).
103. A. Spassov and G. Demirov, *Chem. Ber.*, **101**, 4238 (1968).
104. A. Spassov and G. Demirov, *Chem. Ber.*, **102**, 2530 (1969).
105. P. M. Hergenrother, *Macromolecules*, **3**, 10 (1970).
106. I. Hirao, Y. Kato, T. Hayakawa and H. Tateishi, *Bull. Chem. Soc. Japan*, **44**, 780 (1971).
107. P. Westermann, H. Paul and G. Hilgetag. *Chem. Ber.*, **97**, 528 (1964).
108. B. Stanovnik and M. Tišler, *Tetrahedron Letters*, 2403 (1966).
109. V. V. Mezheritskii, E. P. Olekhnovich and G. N. Dorofeenko, *Russian Chem. Revs.*, **42**, 392 (1973).
110. R. H. DeWolfe, *Carboxylic Ortho Acid Derivatives*, Academic Press, New York, 1970.
111. W. Ried and P. Schomann, *Ann. Chem.*, **714**, 122 (1968).
112. H. Weidinger and J. Krantz, *Chem. Ber.*, **96**, 1064 (1963).
113. A. Spassov and G. Demirov, *Chem. Ber.*, **103**, 3867 (1970).
114. F. H. Case, *J. Heterocyclic Chem.*, **8**, 173 (1971).
115. P. M. Hergenrother and L. A. Carlson, *J. Polymer Sci.*, **A-18**, 1003 (1970).
116. S. Naqui and V. R. Srinivasan, *Indian J. Chem.*, **3**, 162 (1965).
117. D.-I. Shiho and S. Tagami, *J. Am. Chem. Soc.*, **82**, 4044 (1960).
118. G. Pifferi and P. Consonni, *J. Heterocyclic Chem.*, **9**, 581 (1972).
119. A. Pollack, B. Stanovnik and M. Tišler, *Tetrahedron*, **24**, 2623 (1968).
120. H. Neunhoeffer and H. Hennig, *Chem. Ber.*, **101**, 3952 (1968).
121. H. Neunhoeffer, H. Hennig, N.-W. Frühauf and M. Mutterer, *Tetrahedron Letters*, 3147 (1969).
122. H. Paul, S. Chatterjee and G. Hilgetag, *Chem. Ber.*, **101**, 3696 (1968).
123. H. Neunhoeffer, L. Motitschke, H. Hennig and K. Ostheimer, *Ann. Chem.*, **760**, 88 (1972).

124. P. M. Hergenrother and D. E. Kiyohara, *J. Macromol. Sci. Chem.*, **5**, 365 (1971).
125. V. Uchytilova, P. Fiedler and J. Gut, *Coll. Czech. Chem. Comm.*, **37**, 2221 (1972).
126. W. Ried and P. Schomann, *Ann. Chem.*, **714**, 128 (1968)
127. F. H. Case, *J. Heterocyclic Chem.*, **9**, 457 (1972).
128. M. Brugger, H. Wamhoff and F. Korte, *Ann. Chem.*, **758**, 173 (1972).
129. A. Spassov, E. Golovinsky and G. Russev, *Chem. Ber.*, **99**, 3728 (1966).
130. H. Neunhoeffer and F. Weischedel, *Ann. Chem.*, **761**, 34 (1972).
131. W. Ried and P. Schomann, *Ann. Chem.*, **701**, 92 (1967).
132. W. Ried and P. Schomann, *Ann. Chem.*, **714**, 140 (1968).
133. M. Brugger, H. Wamhoff and F. Korte, *Ann. Chem.*, **757**, 100 (1972).
134. A. Spassov, E. Golovinsky and G. Demirov, *Chem. Ber.*, **98**, 932 (1965).
135. H. Reimlinger, W. R. F. Lingier, J. J. M. Vandewalle and E. Goes, *Synthesis*, 433, (1970).
136. R. Huisgen, R. Grashey, M. Seidel, H. Knupfer and R. Schmidt, *Ann. Chem.*, **658**, 169 (1962).
137. H. Reimlinger, F. Billiau and W. R. F. Lingier, *Synthesis*, 260 (1970).
138. W. Walter and H. Weiss, *Ann. Chem.*, 1294 (1973).
139. T. Bany, *Rocz. Chem.*, **42**, 247 (1968); *Chem. Abstr.*, **69**, 67340d (1968).
140. T. Bany and H. Dobosz, *Rocz. Chem.*, **46**, 1123 (1972); *Chem. Abstr.*, **77**, 164587n (1972).
141. D. Frank, W. Brodowski and P. Hentschel, *German Patent* 2,061,895 (1972); *Chem. Abstr.*, **77**, 102430x (1972).
142. G. S. Gol'din, V. G. Poddubnyi, E. S. Smirnova and V. P. Kozyukov, *J. Gen. Chem.*, **43**, 342 (1973).
143. H. Reimlinger, *Chem. and Ind.*, 1082 (1970).
144. T. Kuraishi and R. N. Castle, *J. Heterocyclic Chem.*, **1**, 42 (1964).
145. Y. Charbonnel and J. Barrans, *C.R. Acad. Sci. Paris*, *C.*, **274**, 2209 (1972).
146. M. J. S. Dewar, R. Golden and P. A. Spanninger, *J. Am. Chem. Soc.*, **93**, 3298 (1971).
147. M. Busch and R. Ruppenthal, *Ber. dt. chem. Ges.*, **43**, 3001 (1910).
148. S. Hünig, W. Brenninger, H. Geiger, G. Kaupp, W. Kniese, W. Lampe, H. Quast, R. D. Rauschenbach and A. Schütz, *Angew. Chem.* (*Int. Ed*), **7**, 335 (1968).
149. N. I. Abramova, R. O. Matevosyan, Yu. A. Abramov, V. N. Yakovleva, A. K. Chirkov, L. A. Perelyaeva, V. A. Gubanov and V. I. Koryakov, *Khim. Geterotsikl.*, *Soedin*, **7**, 1484 (1971); *Chem. Abstr.*, **77**, 34409v (1972).
150. M. Genchev, R. Pelova and G. Demirev, *Natura* (*Plovdiv*) **2**, 41 (1968); *Chem. Abstr.*, **72**, 96145m (1970).
151. A. A. Schilt, W. E. Dunbar, B. W. Gandrud and S. E. Warren, *Talanta*, **17**, 649 (1970).
152. J. F. Geldard and F. Lions, *Inorg. Chem.*, **2**, 270 (1963).
153. G. Rouschias and B. L. Shaw, *J. Chem. Soc.* (*A*), 2097 (1971).
154. T. Kauffmann, L. Bán and D. Kuhlmann, *Angew. Chem.* (*Int. Ed.*), **6**, 256 (1967).
155. C. D. Wright, *U.S. Patent* 3,202,659 (1965); *Chem. Abstr.*, **63**, 18123f (1965).

156. J. Kunde and R. Gehm, *German Patent* 1,186,618 (1965); *Chem. Abstr.*, **62,** 10649e (1965).
157. A. Schoepf and G. Meyer, *U.S. Patent* 3,560,453 (1971); *Chem. Abstr.*, **74,** 88572z (1971).
158. C. S. Davis, *U.S. Patent* 3,458,500 (1969); *Chem. Abstr.*, **71,** 81223r (1969).
159. T. L. Fridinger and J. E. Robertson, *J. Med. Chem.*, **12,** 1114 (1969).
160. T. Pantev, N. Bokova, I. Baev and I. Nikolov, *Rentgenol. Radiol.*, **10,** 11 (1971); *Chem. Abstr.*, **75,** 95153n (1971).
161. B. Prescott and G. Caldes, *J. Pharm. Sci.*, **59,** 101 (1970).

CHAPTER **11**

Estimation of the thermochemistry of imidic acid derivatives

Robert Shaw

Physical Sciences Division, Stanford Research Institute, Menlo Park, California, 94025, U.S.A

> During the past decade there have been significant developments of empirical methods for estimating the thermodynamic properties of organic compounds . . . group additivity methods have been the most successful schemes developed . . . Benson has been the principal architect.
> —W. C. Herndon, *Chem. Rev.*, **72**, 157 (1972)

I. INTRODUCTION

It is almost embarrassing for a thermochemist to report the lack of experimental thermochemical data on imidic acid derivatives, but a search of the best and most recent reviews of published thermochemical work[1–3], and of work in progress[4], revealed no experiments on the thermochemistry of imidic acid derivatives.

The absence of good data hurts twice, because present methods of estimation are empirical and rely heavily on a selected data base. Measurement of the heat of formation, or even heat of hydrogenation, of a few

547

key compounds would greatly reduce the uncertainty in the estimates described below.

The thermochemical quantity under consideration is the heat of formation, ΔH_f^0, for the ideal gas state at 25°C. Entropies and heat capacities for the ideal gas state can be calculated with reasonable accuracy by atom or bond additivity[5]. Heats of formation of liquids are calculated from the ideal gas values and heats of vaporization estimated from the formula[6]

$$\Delta H_{\mathrm{vap}25°C} = S_{\mathrm{T}}[(1\cdot76 \times 10^{-3})t_{\mathrm{B}} + 0\cdot253]$$

where $\Delta H_{\mathrm{vap},25°C}$ is the heat of vaporization in kcal/mol, S_{T} is the Trouton constant (22 cal mol^{-1} deg^{-1} for most substances), and t_{B} is the boiling point in degrees centigrade.

Heats of fusion are typically small, usually less than 5 kcal/mol and can be estimated by comparison with a hydrocarbon of similar molecular weight.

II. ESTIMATION BY GROUP ADDITIVITY

Group Additivity postulates that the chemical thermodynamic properties of molecules consist of contributions from the individual groups that make up the molecule. Group Additivity is therefore an extension of the series: atom additivity, bond additivity, . . ., and turns out to be an excellent compromise between simplicity and accuracy. For a detailed treatment of the additivity principle as applied to thermochemistry, see an early paper by Benson and Buss[5], and a more recent Chemical Review article[6]. The latter contains all the group values that could be derived from gas-phase thermochemical data up to 1969. In addition, the present author has discussed Group Additivity while reviewing recent advances in the thermochemistry of organic halogen compounds[7], thiols[8] and azo compounds[9]. Given an adequate experimental data base, group values usually permit ΔH_f^0 to be estimated to within ± 1 kcal/mol for most compounds. However, in the case of the imidic acid derivatives, the absence of good experimental data reduces the accuracy to at least ± 5 kcal/mol, and in some cases, ± 10 kcal/mol. This is getting close to the accuracy of bond additivity. Bond additivity works well for non-polar materials, such as hydrocarbons, and at first I was very tempted to stick to bond additivity and leave it at that. However, for polar compounds, bond additivity does not work very well. For example, bond additivity predicts that the heats of hydrogenation* of reactions (1) to (5) should be the same.

---

* Heats of formation were taken from Cox and Pilcher[1]. If not listed there, they were taken from Stull, Westrum and Sinke[2], or from the Chemical Reviews paper[6].

$$CH_3COOH + H_2 \longrightarrow CH_3CH(OH)_2 \quad -0.2 \text{ kcal/mol} \qquad (1)$$
$$-103.3 \qquad\qquad -103.1$$

$$HCOOH + H_2 \longrightarrow H_2C(OH)_2 \quad +2.9 \text{ kcal/mol} \qquad (2)$$
$$-90.6 \qquad\qquad -93.5$$

$$CH_3COCH_3 + H_2 \longrightarrow CH_3CHOHCH_3 \quad +13.2 \text{ kcal/mol} \qquad (3)$$
$$-51.9 \qquad\qquad -65.1$$

$$CH_3CHO + H_2 \longrightarrow CH_3CH_2OH \quad +16.5 \text{ kcal/mol} \qquad (4)$$
$$-39.7 \qquad\qquad -56.2$$

$$CH_2O + H_2 \longrightarrow CH_3OH \quad +22.1 \text{ kcal/mol} \qquad (5)$$
$$-26.0 \qquad\qquad -48.1$$

The heats of hydrogenation should be the same if bond additivity is obeyed, because in each case the reaction is $\diagdown\!C{=}O \rightarrow \diagdown\!CHOH$. However, the range of heats is more than 22 kcal/mol, so clearly, bond additivity is likely to be a poor approximation. In addition, it seemed best to use the Group Additivity approach even though many of the groups have to be estimated, because it will make the analysis easier when experimental data finally becomes available.

Let us now consider how Group Additivity operates. The process has two distinct phases. In the first phase, the groups are derived from known data, and in the second phase, the groups are used to predict new data. Take a simple example, the heats of formation of alkanes. The n-hexane molecule is made up of two methyls bonded to carbon atoms, indicated by $2[C—(C)(H)_3]$, and four methylenes bonded to two carbon atoms, indicated by $4[C—(C)_2(H)_2]$. Similarly, n-heptane is composed of $2[C—(C)(H)_3] + 5[C—(C)_2(H)_2]$; i.e. for units of kcal/mol,

$$2[C—(C)(H)_3] + 4[C—(C)_2(H)_2] = -39.9$$
$$2[C—(C)(H)_3] + 5[C—(C)_2(H)_2] = -44.9$$

Solving, $[C—(C)_2(H)_2] = -5.0$ and $[C—(C)(H)_3] = -10.0$. The foregoing was the first phase derivation of the groups. We can now use these groups to predict the heat of formation of any linear alkane. For example:

$$\Delta H_f^0 \text{ (n-octane)} = 2[C—(C)(H)_3] + 6[C—(C)_2(H)_2]$$
$$= \quad -20 \qquad\qquad -30$$
$$= -50 \text{ kcal/mol} \quad (cf. -49.9 \text{ observed})$$

In practice, when there is a large amount of measured data, the groups are derived using a least squares regression analysis. There is obviously a lot of data on the alkanes, and the analysis gives the group values

$$[C—(C)_2(H)_2] = -4·95 \quad \text{and} \quad [C—(C)(H)_3] = -10·08$$

Repeating the example of n-octane, $\Delta H_f^0 = (2 \times -10·08) + (6 \times -4·95)$ $= -20·16 - 29·70 = -49·86$ (cf. $-49·9$ observed!). To take an example at the other end of the scale:

$$\Delta H_f^0 \; (n\text{-dotriacontane, } C_{32}H_{66}) = (2 \times -10·08) + (30 \times -4·95)$$
$$= -20·16 - 148·5$$
$$= -168·66 \qquad (cf. \; -166·7 \text{ observed})$$

For highly branched compounds there are some additional small next-nearest-neighbour corrections, but the lack of experimental data for the imidic acid derivatives makes consideration of these corrections unnecessary at present.

III. ASSUMPTIONS, DERIVATION OF GROUP VALUES AND SOME EXAMPLES OF THEIR USE

A. Imines and Imidic Acids

Imidic acids have the structure

There are two groups that are not known and have to be estimated. They are the $[N_I—(C_d)(H)]$ and $[C_d—(N_I)(O)(R)]$ where R is H, or C*. Consider the $N_I—(C_d)(H)$ group first. This group is in the structure

The heats of formation of these compounds are unknown, but they are closely related to the imines:

* The notation closely follows the Chemical Reviews paper[6]. For completeness the group attached by the double bond to the carbon atom and nitrogen atom is included specifically, whereas in the Chemical Reviews paper, it was implied.

whose heats of formation have been discussed by Benson and Walsh[6]. They estimated that the heat of hydrogenation of the imines was 21·5 kcal/mol. If the assumption is made that the heat of hydrogenation of RCH=NH is the same as the heat of hydrogenation of RCH=NR', then we can estimate ΔH_f^0(RCH=NH) from ΔH_f^0(RCH$_2$NH$_2$). To take a specific example, ΔH_f^0(ethylamine) is $-11\cdot0$ kcal/mol; therefore, ΔH_f^0 (ethylimine) is $10\cdot5$ kcal/mol.

$$
\underset{10\cdot5}{CH_3C\overset{NH}{\underset{H}{\diagdown}}} + H_2 \longrightarrow \underset{-11\cdot0}{CH_3CH\overset{NH_2}{\underset{H}{\diagdown}}} + 21\cdot5\ kcal/mol
$$

From ΔH_f^0(ethylimine) we obtain the $[N_I—(C_d)(H)]$ as follows:

$$
\begin{aligned}
[C—(C_d)(H)_3] &= -10\cdot1 \\
[C_d—(N_I)(C)(H)] &= 8\cdot6 \text{ assumed}^6 = [C_d—(C_d)(C)(H)] \\
[N_I—(C_d)(H)] &= \underline{X} \\
&10\cdot5
\end{aligned}
$$

i.e. $-10\cdot1 + 8\cdot6 + X = 10\cdot5$
i.e. $\quad X = 12\cdot0 = [N_I—(C_d)(H)]$

This is a reasonable value by comparison with $[N_I—(C_d)(C)]$ which is known[6]. Thus

$$
\begin{aligned}
[N_I—(C_d)(C)] &= 21\cdot3 \\
[N_I—(C_d)(H)] &= \underline{12\cdot0} \\
\text{difference} &= 9\cdot3
\end{aligned}
$$

compare

$$
\begin{aligned}
[N—(C)_2(H)] &= 15\cdot5 \quad \text{and} \quad [N—(C)_3] &= 24\cdot4 \\
[N—(C)(H)_2] &= \underline{4\cdot8} \qquad\qquad [N—(C)_2(H)] &= \underline{15\cdot4} \\
\text{difference} &= 10\cdot6 \qquad\qquad\quad \text{difference} &= 9\cdot0
\end{aligned}
$$

Clearly the effect of substituting a hydrogen atom for a carbon atom is comparable in all three cases.

The next step is to estimate the $[C_d—(N_I)(O)(R)]$ groups where R is H, C, C$_B$ etc. For example if R is C, then the group is present in acetimidic acid,

$$
CH_3—C\overset{NH}{\underset{OH}{\diagdown}}
$$

and we need to estimate its heat of formation. There are two ways to estimate it.

In one way, consider the insertion of an oxygen atom into the C—H bond of acetaldehyde:

$$CH_3C\!\!\begin{array}{c}O\\ \diagup\\ \diagdown\\ H\end{array} + \tfrac{1}{2}O_2 \longrightarrow CH_3C\!\!\begin{array}{c}O\\ \diagup\\ \diagdown\\ OH\end{array} + 63\cdot6 \text{ kcal/mol}$$

$$-39\cdot7 \qquad\qquad\qquad\qquad -103\cdot3$$

Assume that the insertion of an oxygen atom into ethylimine causes the same increase in stability:

$$CH_3C\!\!\begin{array}{c}NH\\ \diagup\\ \diagdown\\ H\end{array} + \tfrac{1}{2}O_2 \longrightarrow CH_3\!\!-\!\!C\!\!\begin{array}{c}NH\\ \diagup\\ \diagdown\\ OH\end{array} + 63\cdot6 \text{ kcal/mol}$$

$$10\cdot5 \qquad\qquad\qquad\qquad X$$

i.e. $10\cdot5 = X + 63\cdot6$

i.e. $X = \Delta H_f^0 \, (CH_3C(NH)OH) = -53\cdot1$

In the other way, consider acetamide, which has a heat of formation of $-57\cdot8$ kcal/mol. Acetimidic acid is the isomer of acetamide, and we can go from the one to the other by successive hydrogenation and dehydrogenation.

$$CH_3C\!\!\begin{array}{c}O\\ \diagup\\ \diagdown\\ NH_2\end{array} \xrightarrow{(+H_2)} CH_3CH\!\!\begin{array}{c}OH\\ \diagup\\ \diagdown\\ NH_2\end{array} \xrightarrow{(-H_2)} CH_3C\!\!\begin{array}{c}OH\\ \diagup\\ \diagdown\\ NH\end{array}$$

The heat of formation of the intermediate, the hydroxy amine, is unknown, so we assume that the heat of hydrogenation is the same as that for acetone:

$$CH_3C\!\!\begin{array}{c}O\\ \diagup\\ \diagdown\\ CH_3\end{array} + H_2 \longrightarrow CH_3CH\!\!\begin{array}{c}OH\\ \diagup\\ \diagdown\\ CH_3\end{array} + 13\cdot2 \text{ kcal/mol}$$

$$-51\cdot9 \qquad\qquad\qquad\qquad -65\cdot1$$

i.e. $CH_3C\!\!\begin{array}{c}O\\ \diagup\\ \diagdown\\ NH_2\end{array} + H_2 \longrightarrow CH_3CH\!\!\begin{array}{c}OH\\ \diagup\\ \diagdown\\ NH_2\end{array} + 13\cdot2 \text{ kcal/mol}$

$$-57\cdot8 \qquad\qquad\qquad\qquad X$$

i.e. $-57\cdot8 = X + 13\cdot2$

i.e. $X = \Delta H_f^0(CH_3CHOHNH_2) = -57\cdot8 - 13\cdot2 = -71\cdot0$ kcal/mol

then, if the heat of hydrogenation of acetimidic acid is $21\cdot5$ kcal/mol,

$$CH_3C\overset{\displaystyle OH}{\underset{\displaystyle NH}{\Big\backslash}} + H_2 \longrightarrow CH_3CH\overset{\displaystyle OH}{\underset{\displaystyle NH_2}{\Big\backslash}} + 21\cdot5$$

$$\underset{X}{} \qquad\qquad \underset{-71\cdot0}{}$$

$$X = \Delta H_f(CH_3C(NH)OH) = -71\cdot0 + 21\cdot5 = -49\cdot5 \text{ kcal/mol.}$$

We therefore have two values for the heat of formation of acetimidic acid, $-53\cdot1$ and $-49\cdot5$, and select -52 kcal/mol as the best value. This is a reasonable value, as it means that acetimidic acid is about 6 kcal/mol less stable than acetamide, and that is consistent with the lack of detection of the acid form.

We can now derive the value of the $[C_d\text{—}(N_I)(C)(O)]$ group:

$$CH_3\text{—}C\overset{\displaystyle OH}{\underset{\displaystyle NH}{\Big\backslash}}$$

Groups:

$$
\begin{aligned}
[C\text{—}(C)(H)_3] &= -10\cdot1\\
[N_I\text{—}(C_d)(H)] &= 12\cdot0\\
[O\text{—}(C_d)(H)] &= -37\cdot9 \quad (\text{assigned} \equiv [O\text{—}(C)(H)])\\
[C_d\text{—}(N_I)(C)(O)] &= \underline{X}\\
&-52
\end{aligned}
$$

i.e. $-48\cdot0 + 12 + X = -52$
i.e. $X = [C_d\text{—}(N_I)(C)(O)] = -52 + 36 = -16$ kcal/mol.

Estimation of the heat of formation of formidic acid,

$$H\text{—}C\overset{\displaystyle OH}{\underset{\displaystyle NH}{\Big\backslash}}$$

will give the value for the group $[C_d\text{—}(N_I)(O)(H)]$. We estimate that heat of formation of formimidic acid by comparing it with acetimidic acid and estimating the effect of the CH_2 group:

$$CH_3C\overset{\displaystyle O}{\underset{\displaystyle OH}{\Big\backslash}} \xrightarrow{(-CH_2)} HC\overset{\displaystyle O}{\underset{\displaystyle OH}{\Big\backslash}} \qquad -12\cdot7 \text{ kcal/mol}$$

$$\underset{-103\cdot3}{} \qquad\qquad \underset{-90\cdot6}{}$$

$$CH_3C\overset{\displaystyle O}{\underset{\displaystyle NH_2}{\Big\backslash}} \xrightarrow{(-CH_2)} HC\overset{\displaystyle O}{\underset{\displaystyle NH_2}{\Big\backslash}} \qquad -13\cdot3 \text{ kcal/mol}$$

$$\underset{-57\cdot8}{} \qquad\qquad \underset{-44\cdot5}{}$$

Taking an average value of $-13 \cdot 0$ kcal/mol,

$$\underset{\substack{\displaystyle \\ -52 \cdot 0}}{CH_3C{\overset{\displaystyle NH}{\underset{\displaystyle OH}{\Big<}}}} \qquad \underset{\substack{\displaystyle \\ X}}{HC{\overset{\displaystyle NH}{\underset{\displaystyle OH}{\Big<}}}} \qquad -13 \cdot 0 \text{ kcal/mol}$$

$$X - 13 \cdot 0 = -52 \cdot 0$$
$$X = \Delta H_f^0(HC(NH)OH) = -52 \cdot 0 + 13 \cdot 0 = -39 \cdot 0 \text{ kcal/mol}.$$

We then obtain the group value $[C_d\!-\!(N_I)(O)(H)]$.

$$\underset{\substack{\displaystyle \\ -39 \cdot 0}}{H\!-\!C{\overset{\displaystyle NH}{\underset{\displaystyle OH}{\Big<}}}}$$

$$[N_I\!-\!(C_d)(H)] = 12 \cdot 0$$
$$[O\!-\!(C_d)(H)] = -37 \cdot 9$$
$$[C_d\!-\!(N_I)(O)(H)] = \underline{X}$$
$$= -39 \cdot 0$$

i.e. $-25 \cdot 9 + X = -39 \cdot 0$

i.e. $X = [C_d\!-\!(N_I)(O)(H)] = -13 \cdot 1$ kcal/mol

As a last example of imidic acid estimation, consider benzimidic acid:

$$Ph\!-\!C{\overset{\displaystyle NH}{\underset{\displaystyle OH}{\Big<}}}$$

It may be assumed that the difference between benzimidic acid and acetimidic acid is the same as that between benzoic acid and acetic acid.

$$\underset{\substack{\displaystyle \\ -70 \cdot 1}}{Ph\!-\!C{\overset{\displaystyle O}{\underset{\displaystyle OH}{\Big<}}}} \quad \longrightarrow \quad \underset{\substack{\displaystyle \\ -103 \cdot 3}}{CH_3C{\overset{\displaystyle O}{\underset{\displaystyle OH}{\Big<}}}} \qquad +33 \cdot 2 \text{ kcal/mol}$$

$$\underset{\substack{\displaystyle \\ -18 \cdot 8}}{Ph\!-\!C{\overset{\displaystyle NH}{\underset{\displaystyle OH}{\Big<}}}} \quad \longrightarrow \quad \underset{\substack{\displaystyle \\ -52}}{CH_3C{\overset{\displaystyle NH}{\underset{\displaystyle OH}{\Big<}}}} \qquad +33 \cdot 2 \text{ kcal/mol}$$

For benzimidic acid the groups are:

$$
\begin{aligned}
5[C_B\!-\!(H)] &= 16\cdot5 \\
1[C_B\!-\!(C_d)] &= 5\cdot7 \\
1[O\!-\!(C_d)(H)] &= -37\cdot9 \\
1[C_d\!-\!(N_I)(O)(C_B)] &= X \\
\hline
&-18\cdot8
\end{aligned}
$$

i.e. $X - 15\cdot7 = -18\cdot8$

i.e. $X = [C_d - (N_I)(O)(C_B)] = -3\cdot1$ kcal/mol

Summarizing, we have derived the following groups:

$$
\begin{aligned}
{[N_I\!-\!(C_d)(H)]} &= 12\cdot0 \\
{[C_d\!-\!(N_I)(O)(C)]} &= -16\cdot0 \\
{[C_d\!-\!(N_I)(O)(H)]} &= -13\cdot1 \\
{[C_d\!-\!(N_I)(O)(C_B)]} &= -3\cdot1
\end{aligned}
$$

and assumed

$$[C_d\!-\!(N_I)(X)(Y)] = [C_d\!-\!(C_d)(X)(Y)], \text{ where } X,Y = C, H$$

(a) To estimate propionimidic acid:

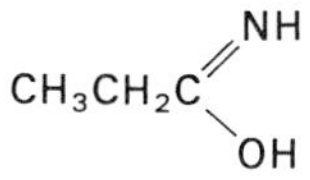

$$
CH_3CH_2C\!\!\underset{OH}{\overset{NH}{<}}
$$

the groups are:

$$
\begin{aligned}
{[C\!-\!(C)(H)_3]} &= -10\cdot1 \\
{[C\!-\!(C)(C_d)(H)_2]} &= -4\cdot8 \\
{[C_d\!-\!(N_I)(O)(C)]} &= -16\cdot0 \\
{[N_I\!-\!(C_d)(H)]} &= 12\cdot0 \\
{(O\!-\!(C_d)(H))} &= -37\cdot9 \\
\hline
&= -68\cdot8 + 12\cdot0 = -56\cdot8
\end{aligned}
$$

The heat of formation of propionimidic acid is $-56\cdot8$ kcal/mol.

(b) To estimate p-chlorobenzimidic acid:

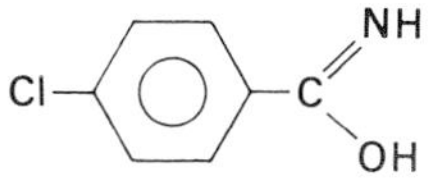

the groups are: $[C_B-(Cl)] = -3.8$
$4[C_B-(H)] = 13.2$
$[C_B-(C_d)] = 5.7$
$(C_d-(C_B)(N_I)(O)] = -3.1$
$[N_I-(C_d)(H)] = 12.0$
$[O-(C_d)(H)] = -37.9$

$$-44.8 + 30.9 = -13.9$$

The heat of formation of p-chlorobenzimidic acid is -13.9 kcal/mol.

B. Amidines

Amidines are the amides of the corresponding imidic acids. For example, acetamidine has the structure:

$$CH_3-C\overset{\displaystyle NH}{\underset{\displaystyle NH_2}{<}}$$

There are two new groups to be estimated; namely, $[C_d-(N_I)(C)(N)]$ and $N-(C_d)(H)_2$. First we need to estimate the heat of formation of the amidine. The effect of replacing a carbonyl hydrogen by an amino group is shown by

$$CH_3-C\overset{\displaystyle O}{\underset{\displaystyle H}{<}} \longrightarrow CH_3-C\overset{\displaystyle O}{\underset{\displaystyle NH_2}{<}} \qquad +18.1 \text{ kcal/mol}$$

$$\qquad -39.7 \qquad\qquad\qquad -57.8$$

Assuming that replacing a hydrogen atom by an NH_2 in ethylimine is also worth 18·1 kcal/mol:

$$CH_3-C\overset{\displaystyle NH}{\underset{\displaystyle H}{<}} \longrightarrow CH_3-C\overset{\displaystyle NH}{\underset{\displaystyle NH_2}{<}} \qquad +18.1 \text{ kcal/mol}$$

$$\qquad 10.5 \qquad\qquad\qquad X$$

$X + 19.1 = 10.5$
$X = \Delta H_f^0(CH_3C(NH)NH_2) = 10.5 - 18.1 = -7.6 \text{ kcal/mol}$

In order to separate the values for the two unknown groups $[C_d-(N_I)(C)(N)]$ and $[N-(C_d)(H)_2]$, we will follow current practice[6], and assign a value to one of them. In this case, let $[N-(C_d)(H)_2] = [N-(C)(H)_2] = 4.8$ kcal/mol.

Then for:

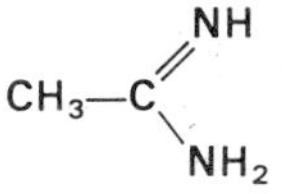

$$CH_3-C \overset{NH}{\underset{NH_2}{<}}$$

the groups are:

$$
\begin{aligned}
[C-(C_d)(H)_3] &= -10.1 \\
[N_I-(C_d)(H)] &= 12.0 \\
[N-(C_d)(H)_2] &= 4.8 \\
[C_d-(N_I)(C)(N)] &= X \\
\hline
&\quad -7.6
\end{aligned}
$$

i.e. $16.8 - 10.1 + X = -7.6$

i.e. $X = [C_d-(N_I)(C)(N)] = -7.6 - 6.7 = -14.3 \text{ kcal/mol}$

The amide of formimidic acid, formamidine, will give the group $[C_d-(N_I)(H)(N)]$. Assuming the difference between the amides of formimidic and acetimidic acid are the same as in the parent imidic acids,

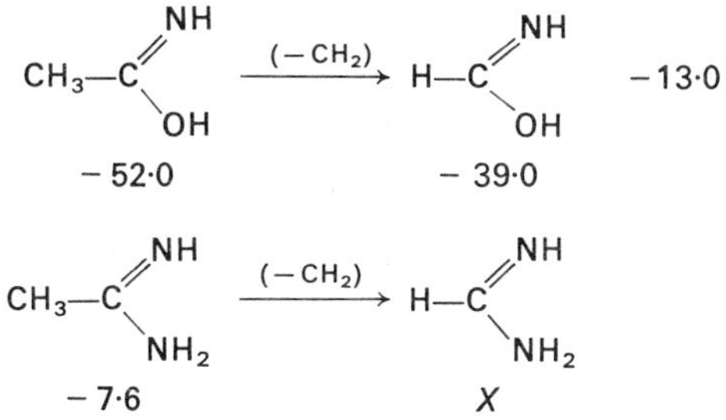

$$CH_3-C \overset{NH}{\underset{OH}{<}} \quad \xrightarrow{(-CH_2)} \quad H-C \overset{NH}{\underset{OH}{<}} \quad -13.0$$
$$-52.0 \qquad\qquad\qquad -39.0$$

$$CH_3-C \overset{NH}{\underset{NH_2}{<}} \quad \xrightarrow{(-CH_2)} \quad H-C \overset{NH}{\underset{NH_2}{<}}$$
$$-7.6 \qquad\qquad\qquad X$$

i.e. $-7.6 = X - 13.0$

i.e. $X = \Delta H_f^0(HC(NH)NH_2) = -7.6 + 13.0 = 5.4 \text{ kcal/mol.}$

Then for,

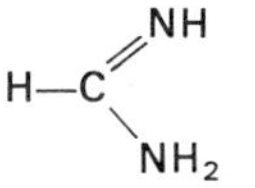

$$H-C \overset{NH}{\underset{NH_2}{<}}$$

the groups are:

$$
\begin{aligned}
[N_I-(C_d)(H)] &= 12.0 \\
[N-(C_d)(H)_2] &= 4.8 \\
[C_d-(N_I)(H)(N)] &= X \\
\hline
&\quad 5.4
\end{aligned}
$$

i.e. $16.8\ 16.8 + X = 5.4$

i.e. $X = [C_d-(N_I)(H)(N)] = 5.4 - 16.8 = -11.4$

The difference between the values of the groups $[C_d—(N^I)(H)(N)]$ and $[C_d—(N_I)(C)(N)]$ is $-11 \cdot 4 - (-14 \cdot 3) = 2 \cdot 9$ kcal/mol. That is, the effect of substituting a carbon for a hydrogen in the group $[C_d—(N_I)(X)(N)]$ is $2 \cdot 9$ kcal/mol. This is the same difference as that between the groups, $[C_d—(N_I)(H)(O)]$ and $[C_d—(N_I)(C)(O)] = -13 \cdot 1 - (-16 \cdot 0) = 2 \cdot 9$, because the assumptions were the same. Therefore, the effect of substituting a phenyl carbon will also be the same.

$$\text{i.e. } [C_d—(N_I)(H)(N)] - [C_d—(N_I)(C_B)(N)]$$
$$= [C_d—(N_I)(H)(O)] - [C_d—(N_I)(C_B)(O)]$$
$$= \qquad -13 \cdot 1 \qquad - \qquad (-3 \cdot 1)$$
$$= \qquad -10 \cdot 0$$
$$\therefore [C_d—(N_I)(C_B)(N)] = [C_d—(N_I)(H)(N)] + 10 \cdot 0$$
$$= -11 \cdot 4 + 10 \cdot 0$$
$$= -1 \cdot 4$$

Summarizing the group values for estimating heats of formation of the amidines:

$$[C_d—(N_I)(N)(H)] = -11 \cdot 4$$
$$[C_d—(N_I)(N)(C)] = -14 \cdot 3$$
$$[C_d—(N_I)(N)(C_B)] = -1 \cdot 4$$

and $[N—(C_d)(X)(Y)] \equiv [N—(C)(X)(Y)]$ for $X, Y = C$ or H.

For an example of calculating the heat of formation of an amidine consider[10]

The groups are:
$$10[C_B—(H)] = \qquad 33 \cdot 0$$
$$[C_B—(C_d)] = \qquad 5 \cdot 7$$
$$[C_B—(N_I)] = -\ 0 \cdot 5$$
$$[N_I—(C_B)(C_d)] = \qquad 14 \cdot 1$$
$$[C_d—(N_I)(N)(C_B)] = -\ 1 \cdot 4$$
$$[N—(C_d)(C)_2] = \qquad 24 \cdot 4$$
$$4[C—(C)(H)_3] = -40 \cdot 3$$
$$2[C—(C)_2(N)(H)] = -10 \cdot 4$$

$$= \qquad 77 \cdot 2 - 52 \cdot 6 = 24 \cdot 6 \text{ kcal/mol.}$$

That is, neglecting steric interactions,

$$\Delta H_f^0(PhCN(Ph)N[CHMe_2]_2) = 24 \cdot 6 \text{ kcal/mol.}$$

C. Amidrazones

The simplest amidrazone is:

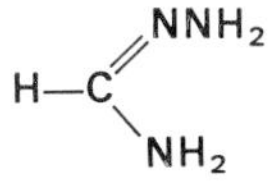

$$H-C\overset{\displaystyle NNH_2}{\underset{\displaystyle NH_2}{\diagdown}}$$

This compound has the groups $[C_d-(N_I)(H)(N)]$ and $[N-(C_d)(H)_2]$ which are known, and the groups $[N_I-(C_d)(N)]$ and $[N-(N_I)(H)_2]$, which are not. Following previous practice, one of the groups is assigned, leaving the other to be determined. The $[N-(N_I)(H)_2]$ group may be assigned equal to $[N-(N)(H)_2]$ which is 11·4 kcal/mol. The $[N_I-(C_d)(N)]$ group then may be obtained from the heat of formation of $CH_2=NNH_2$, which in turn is obtained by estimating its heat of hydrogenation. Benson[11] has estimated its heat of hydrogenation to be 20 kcal/mol, i.e.

$$CH_2=NNH_2 \xrightarrow{\text{H}_2} CH_3NHNH_2 \quad + 20 \text{ kcal/mol}$$
$$X \phantom{\xrightarrow{\text{H}_2}} 22·6$$

i.e. $X = \Delta H_f(CH_2NNH_2) = 22·6 + 20 = 42·6$ kcal/mol.
For the structure,

$$CH_2=NNH_2$$

the groups are:
$$[N-N_I(H)_2] = 11·4$$
$$[C_d-(N_I)(H)_2] = 6·3$$
$$[N_I-(C_d)(N)] = X$$
$$\overline{42·6}$$

i.e. $17·7 + X = 42·6$
i.e. $X = [N_I-(C_d)(N)] = 42·6 - 17·7 = 24·9$ kcal/mol.

The heat of formation of the amidrazone can now be calculated:

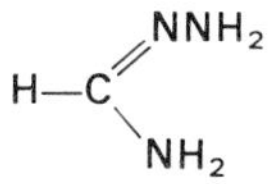

$$H-C\overset{\displaystyle NNH_2}{\underset{\displaystyle NH_2}{\diagdown}}$$

The groups are:
$$[C_d-(N_I)(H)(N)] = -11·4$$
$$[N-(C_d)(H)_2] = 6·3$$
$$[N_I-(C_d)(N)] = 24·9$$
$$[N-(N_I)(H)_2] = 11·4$$
$$\overline{41·2}$$

i.e. $\Delta H_f^0(HC(NH_2)NNH_2) = 41·2$ kcal/mol.

The amidrazone has an isomer obtained by shifting the double bond and a hydrogen atom.

$$\begin{array}{ccc}
\text{H—C}\begin{array}{c}\nearrow \text{NNH}_2\\ \searrow\\ \text{NH}_2\end{array} & \rightleftharpoons & \text{H—C}\begin{array}{c}\nearrow \text{NHNH}_2\\ \searrow\\ \text{NH}\end{array}\\
\mathbf{A} & & \mathbf{B}
\end{array}$$

The heat of formation of **B** can be estimated by considering the effect of inserting an NH group into a C—N bond:

$$CH_3NH_2 \xrightarrow{(+NH)} CH_3NHNH_2 \qquad -28{\cdot}1 \text{ kcal/ mol}$$
$${-5{\cdot}5} \qquad\qquad 22{\cdot}6$$

Then assume the same for formamidine:

$$\text{H—C}\begin{array}{c}\nearrow \text{NH}_2\\ \searrow\\ \text{NH}\end{array} \xrightarrow{(+NH)} \text{H—C}\begin{array}{c}\nearrow \text{NHNH}_2\\ \searrow\\ \text{NH}\end{array} \qquad -28{\cdot}1$$
$$5{\cdot}4 \qquad\qquad\qquad\qquad X$$

That is, the **B** form is some 8 kcal/mol more stable than the **A** form. The **B** form has three groups that are known and one that is unknown:

$$\text{H—C}\begin{array}{c}\nearrow \text{NHNH}_2\\ \searrow\\ \text{NH}\end{array}$$

The groups are:
$$[C_d\text{—}(N_I)(H)N] = -11{\cdot}4$$
$$[(N_I\text{—}(C_d)(H))] = 12{\cdot}0$$
$$[(N\text{—}(N)(H)_2)] = 11{\cdot}4$$
$$[N\text{—}(C_d)(N)(H)] = \phantom{-11{\cdot}}X$$
$$\overline{\phantom{[N\text{—}(C_d)(N)(H)] = -1}33{\cdot}5}$$

i.e. $12{\cdot}0 + X = 33{\cdot}5$

i.e. $X = [N\text{—}(C_d)(N)(H)] = 33{\cdot}5 - 12{\cdot}0 = 21{\cdot}5 \text{ kcal/mol}$

This group value may be compared to that already[6] derived for $[N\text{—}(C)(N)(H)]$ which is 20·9 kcal/mol. The difference of only 0·6 kcal/mol is so small that it seems to be a good approximation to take $[N\text{—}(C_d)(N)(C)] = [N\text{—}(N)(C)_2] + 0{\cdot}6 = 29{\cdot}2 + 0{\cdot}6 = 29{\cdot}8 \text{ kcal/mol}$.

Summarizing for the amidrazones, the following groups were derived:

$$[N_I\!\!-\!\!(C_d)(N)] = 24\!\cdot\!9$$
$$[N\!\!-\!\!(C_d)(N)(H)] = 21\!\cdot\!5$$
$$[N\!\!-\!\!(C_d)(N)(C)] = 29\!\cdot\!8$$

and assigned, $[N\!\!-\!\!(N_I)(X)(Y)] = [N\!\!-\!\!(N)(X)(Y)]$, where $X, Y = C$ or H. For an example of calculating the heat of formation of an amidrazone, consider the structure[12]:

$$PhC \overset{\displaystyle NNH_2}{\underset{\displaystyle NH_2}{\big<}}$$

The groups are:

$$
\begin{aligned}
5[C_B\!\!-\!\!(H)] &= && 16\!\cdot\!5 \\
[C_B\!\!-\!\!(C_d)] &= && 5\!\cdot\!7 \\
[C_d\!\!-\!\!(N_I)(C_B)(N)] &= && -\ 1\!\cdot\!4 \\
[N\!\!-\!\!(C_d)(H)_2] &= && 4\!\cdot\!8\ (= [N\!\!-\!\!(C)(H)_2]) \\
[N_I\!\!-\!\!(C_d)(N)] &= && 24\!\cdot\!9 \\
[N\!\!-\!\!(N_I)(H)_2] &= && 11\!\cdot\!4\ (= [N\!\!-\!\!(N)(H)_2]) \\
\hline
&= && 63\!\cdot\!3\ -\ 1\!\cdot\!4 = 61\!\cdot\!9\ \text{kcal/mol}
\end{aligned}
$$

i.e. $\Delta H_f^0(PhC(NH_2)NNH_2) = 61\!\cdot\!9$ kcal/mol.

D. Amidoximes

Amidoximes have a general structure:

$$R\!\!-\!\!C \overset{\displaystyle NOH}{\underset{\displaystyle NH_2}{\big<}}$$

The key group is $[N_I\!\!-\!\!(C_d)(OH)]$, and its value has already been derived by Benson and Walsh[6] to be $-5\!\cdot\!0$ kcal/mol. Moving straight to an example of an amidoxime[13]:

$$Ph\!\!-\!\!C \overset{\displaystyle NOH}{\underset{\displaystyle NH_2}{\big<}}$$

The groups are:

$$
\begin{aligned}
5[C_B\!\!-\!\!(H)] &= && 16\!\cdot\!5 \\
[C_B\!\!-\!\!(C_d)] &= && 5\!\cdot\!7 \\
[C_d\!\!-\!\!(N_I)(C_B)(N)] &= && -\ 1\!\cdot\!4 \\
[N_I\!\!-\!\!(C_d)(OH)] &= && -\ 5\!\cdot\!0 \\
[N\!\!-\!\!(C_d)(H)_2] &= && 4\!\cdot\!8 \quad (= [N\!\!-\!\!(C)(H)_2]) \\
\hline
&= && 27\!\cdot\!0\ -\ 6\!\cdot\!4 = 20\!\cdot\!6\ \text{kcal/mol}
\end{aligned}
$$

i.e. $\Delta H_f^0(PhC(NOH)NH_2) = 20\!\cdot\!6$ kcal/mol.

E. Imidoyl Halides

This section is concerned with imidoyl halides that have the structure:

$$R-C{\overset{\displaystyle NH}{\underset{\displaystyle Hal}{}}}$$

The basic assumption is that the difference in the heats of formation of a carboxylic acid and acyl halide,

$$R-C{\overset{\displaystyle O}{\underset{\displaystyle OH}{}}} \longrightarrow R-C{\overset{\displaystyle O}{\underset{\displaystyle Hal}{}}}$$

is the same as the difference between an imidic acid and the imidoyl halide:

$$R-C{\overset{\displaystyle NH}{\underset{\displaystyle OH}{}}} \longrightarrow R-C{\overset{\displaystyle NH}{\underset{\displaystyle Hal}{}}}$$

The differences in measured heats of formation of carboxylic acids and the corresponding acyl halides are:

Acid	(Acyl fluoride)	(Acyl chloride)	(Acyl bromide)	(Acyl iodide)
Acetic	3·1	−44·9	−57·7	−72·6
Benzoic		−44·0	−58·5	−73·2
Average	3·1	−44·5	−58·0	−73·0

From the assumption mentioned above, it follows that:

imidic acid minus imidoyl halide =

$$[C_d—(N_I)(O)(X)] + [O—(C_d)(H)] − [C_d—(N_I)(Hal)(X)]$$

where Hal is F, Cl, Br, or I and X is H, C, or C_B.

$$= 3·1 \text{ for Hal} = F$$
$$= −44·5 \text{ for Hal} = Cl$$
$$= −58·0 \text{ for Hal} = Br$$
$$= −73·0 \text{ for Hal} = I$$

For Hal = F,

i.e. $[C_d—(N_I)(F)(X)] = [C_d—(N_I)(O)(X)] + [O—(C_d)(H)] - 3\cdot1$
$= [C_d—(N_I)(O)(X)] - 37\cdot9* - 3\cdot1$
$= [C_d—(N_I)(O)(X)] - 41\cdot0$

Group values for $[C_d—(N_I)(O)(X)]$ were derived earlier in this section. That is,

$$[C_d—(N_I)(F)(H)] = -13\cdot1 - 41\cdot0 = -54\cdot1$$
$$[C_d—(N_I)(F)(C)] = -16\cdot0 - 41\cdot0 = -57\cdot0$$
$$[C_d—(N_I)(F)(C_B)] = -3\cdot1 - 41\cdot0 = -44\cdot1$$

Similar calculations for Hal = Cl give:

$$Cd—(N_I)(Cl)(X) = [C_d—(N_I)(O)(X)] - 37\cdot9 + 44\cdot5$$
$$= [C_d—(N_I)(O)(X)] + 6\cdot6$$

That is, $$[C_d—(N_I)(Cl)(H)] = -13\cdot1 + 6\cdot6 = -6\cdot5$$
$$[C_d—(N_I)(Cl)(C)] = -16\cdot0 + 6\cdot6 = -9\cdot4$$
$$[C_d—(N_I)(Cl)(C_B)] = -3\cdot1 + 6\cdot6 = +3\cdot5$$

Similarly, for Hal = Br:

$$[C_d—(N_I)(Br)(X)] = [C_d—(N_I)(O)(X)] - 37\cdot9 + 58\cdot0$$
$$= [C_d—(N_I)(O)(X)] + 10\cdot1$$

That is, $$[C_d—(N_I)(Br)(H)] = -13\cdot1 + 10\cdot1 = -3\cdot0$$
$$[C_d—(N_I)(Br)(C)] = -16\cdot0 + 10\cdot1 = -5\cdot9$$
$$[C_d—(N_I)(Br)(C_B)] = -3\cdot1 + 10\cdot1 = 7\cdot0$$

and finally, for Hal = I,

$$[C_d—(N_I)(I)(X)] = [C_d—(N_I)(O)(X)] - 37\cdot9 + 73\cdot0$$
$$= [C_d—(N_I)(O)(X)] + 35\cdot1$$

That is, $$[C_d—(N_I)(I)(H)] = -13\cdot1 + 35\cdot1 = 22\cdot0$$
$$[C_d—(N_I)(I)(C)] = -16\cdot0 + 35\cdot1 = 19\cdot1$$
$$[C_d—(N_I)(I)(C_B)] = -3\cdot1 + 35\cdot1 = 32\cdot0$$

As an example of a calculated heat of formation, consider the imidoyl halide:

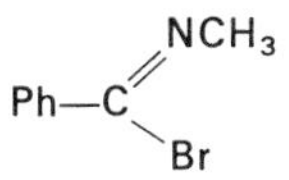

* $[O—(C_d)H] \equiv [O—(C)(H)]$, ref. 14.

The groups are:
$$5[C_B-(C)] = 16\cdot5$$
$$[C_B-(C_d)] = 5\cdot7$$
$$[C_d-(N_I)(Br)(C_B)] = 7\cdot0$$
$$[N_I-(C_d)(C)] = 21\cdot3 \quad (\text{ref. 6})$$
$$[C-(N_I)(H_3)] = -10\cdot1$$
$$\overline{50\cdot5 - 10\cdot1 = 40\cdot4}$$

That is, $\Delta H_f^0(PhC(Br)NCH_3) = 40\cdot4$ kcal/mol.

IV. ACKNOWLEDGEMENTS

It gives me pleasure to acknowledge the many valuable discussions with Sidney W. Benson and David M. Golden, and to express my thanks to Elaine Adkins for typing the manuscript.

V. REFERENCES

1. J. D. Cox and G. Pilcher, *Thermochemistry of Organic and Organometallic Compounds*, Academic Press, New York, 1970.
2. D. R. Stull, E. F. Westrum and G. C. Sinke, *The Chemical Thermodynamics of Organic Compounds*, John Wiley and Sons, Inc., New York, 1969.
3. JANAF Thermochemical Data, Dow Chemical Co., Midland, Michigan.
4. Bulletin of Thermodynamics and Thermochemistry published annually by INPAC.
5. S. W. Benson and J. H. Buss, *J. Chem. Phys.*, **29**, 546 (1958).
6. S. W. Benson, F. R. Cruickshank, D. M. Golden, G. R. Haugen, H. E. O'Neal, A. S. Rodgers, R. Shaw and R. Walsh, *Chem. Rev.*, **69**, 279 (1969).
7. R. Shaw, Thermochemistry of Organic Halides, *Chemistry of the Carbon–Halogen Bond* (Ed. S. Patai), John Wiley and Sons, Chichester, 1973. 1973.
8. R. Shaw, Thermochemistry of Thiols, *Chemistry of the Thiol Group* (Ed. S. Patai), John Wiley and Sons, Chichester, 1974.
9. R. Shaw, Thermochemistry of Hydrazo, Azo, and Azoxy Groups, *Chemistry of Hydrazo, Azo, and Azoxy Compounds* (Ed. S. Patai), John Wiley and Sons, Chichester, 1975.
10. Z. Rappoport and R. Ta-Shma, *Tetrahedron Lett.*, **52**, 5281 (1972).
11. S. W. Benson, private communication, 1974.
12. M. Brugger, H. Wamhoff and F. Korte, *Justus Liebigs Ann. Chem.* **757**, 100 (1972).
13. R. Takaoka, S. Okade, T. Nakahama, H. Sakada and A. Hongo, *Japanese Patent* 71 42, 562 (1971).
14. H. K. Eigenmann, D. M. Golden and S. W. Benson, *J. Phys. Chem.*, **77**, 1687 (1973).

CHAPTER **12**

Complex formation, H-bonding and basicity of imidic acid derivatives

J. Ševčík and F. Grambal

Department of Physical Chemistry, Palacký University, Olomouc, ČSSR

I. INTRODUCTION

The available data concerning complex-forming abilities of amidines as well as of some other imidic acid derivatives are comparatively recent. Lately there is an enhanced interest in this group of substances, and especially amidines are finding increasing application in various branches of chemistry, pharmacology and clinical therapeutics.

Shriner and Neumann[1] in their survey make no reference to the complex-forming abilities of amidines. At that time only salts of amidines and iminoethers[2,3] were known and described, (1, X = halogen). Other salts

$$R-C\begin{smallmatrix} \diagup NR^1 \\ \diagdown NR^1R^2 \cdot HX \end{smallmatrix}$$

(1)

of amidines with the structure 2 were described in detail in the work of Holy[4]. These data were of qualitative character only, pertaining first of all to information obtained by syntheses.

$$\left[\begin{smallmatrix} R^3 \\ \diagdown \\ R^4 \diagup \end{smallmatrix} N-\underset{\underset{R}{|}}{C}=N \begin{smallmatrix} \diagup R^1 \\ \diagdown R^2 \end{smallmatrix} \right]^+ \quad X^-$$

(2)

Amidoxime complexes, on the other hand, were studied much earlier, chiefly because of the interesting colour reactions of amidoximes with ions of various metals which have found broad application in analytical chemistry[5-22]. The structure of the resulting chelates, however, has not been satisfactorily confirmed. Red-coloured complexes with Ni ions, obtained in oxidative media have remained an unsolved problem till the present day.

A similar situation prevailed also in information concerning the hydrogen bond. Some references to hydrogen bonding with amidines are found in the literature[1,23]. Dimer formation may be anticipated with N,N'-disubstituted amidines, as it is in the case of carboxylic acids. This problem has been studied in detail in recent years only.

Hydrogen bonding with amidoximes was studied by Hall and Llewellyn[24] and by Mollin[25] as well as by some other authors, especially with respect to the need of determining isomer conformations.

Older data concerning the basicity of this group of substances are exclusively qualitative. They were obtained mainly in synthetic experiments, resulting from analogy with the respective oxygen derivatives, and are limited, at best, to a very rough comparison of basicity of the investigated derivatives, e.g. with ammonia. A systematic quantitative study of basicity was started in about 1960.

II. COMPLEX FORMATION

A. *Complexes of Amidines*

As already mentioned, complexes of amidines have been studied intensively during the last 20 years only. Even so, the works dealing with these problems are not numerous.

Among the first papers on amidine complexes the studies of Bradley and co-workers[26, 27] should be mentioned. In these the authors describe metallic *N,N'*-diarylamidine derivatives and some of their chemical reactions. For the Cu-complex of *N,N'*-di-2-anthraquinonylformamidine derivative the structure 4 has been proposed. This derivative has been prepared by means of various processes and its identity verified by using spectral data. The Cu-derivative 4 has been prepared from *N,N'*-di-2-anthraquinonylformamidine (3) using cuprous chloride or cupric acetate

(3)

or Cu-bronze. Thus, e.g. when heating one mol of cupric acetate with two mol of 3 in nitrobenzene, 4 was isolated and half the amount of 3 remained unchanged[26]. The resulting complex exhibits very good solubility in organic solvents; it hydrolyses only slowly in acids and it does not react with aniline or other organic bases.

The high chemical stability points to a high degree of covalence between the copper and nitrogen atoms. Therefore, the authors[26] have proposed the structure 4 for this compound.

It is worth mentioning that in an experiment to obtain the Cu derivative from 5 the monomethyl derivative of 3 through an analogous reaction, demethylation took place and again 4 was formed.

A similar reaction occurred also with *N,N'*-di-2-anthraquinonyl-benzamidine with formation of the corresponding Cu derivative. The

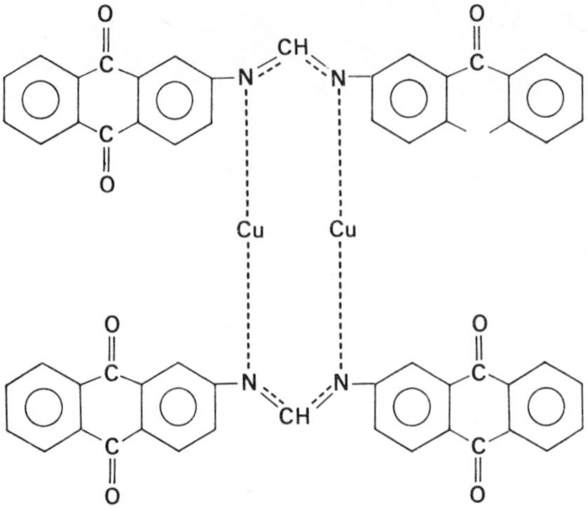

(4)

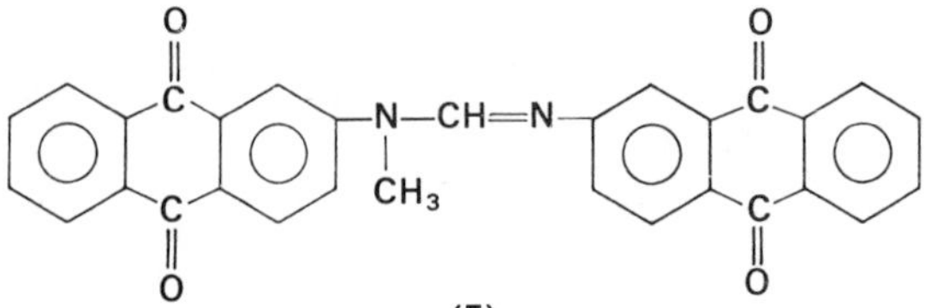

(5)

resulting crystalline salt is dark-green under reflected light and red in transmitted light. It is less stable than **4**, and on heating with pyridine or with acetic acid, it decomposes.

The isomeric *N,N'*-1-anthraquinonylformamidines do not yield the corresponding Cu derivatives.

Some other metallic derivatives of e.g. *N,N'*-diarylacetamidine and *N,N'*-diarylformamidine have been investigated[27]. Thus, cuprous derivatives as well as silver and mercury derivatives have been prepared. All these derivatives are less stable than **4**. Furthermore, cupric *N,N'*-diarylformamidine derivatives have been synthesized and studied. They are green-coloured and unstable. On the basis of the study of their chemical reactions they have been given the structure (**6**).

An attempt to substitute two hydrogen atoms in two molecules of *N,N'*-di-2-anthraquinonylformamidine with one Cu-atom was unsuccessful. However the tetrapyridine adduct of Cu and Ni derivatives from *N,N'*-di-*p*-nitrophenylformamidine[27] was isolated. This is most probably a coordination compound which may be formulated as **7**.

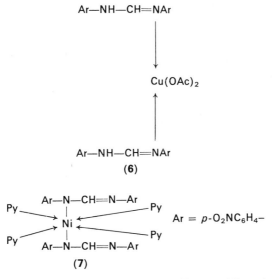

(6)

(7)

$Ar = p\text{-}O_2NC_6H_4\text{-}$

For the mercury derivative of $N,N'\text{-}p$-tolylformamidine the following wing structure (8) was proposed:

(8)

All formamidines which have been investigated reacted with cuprous as well as with silver salts under formation of stable inner-complex salts[26], the latter well resistant to water as well as to aqueous ammonia. The determination of molecular weights in these complexes is not decisive, since it is presumed that they form cyclic dimers as well as linear polymers[23, 28]. The same is true of the determination of the molecular weights of N,N'-diarylacetamidines as well as N,N'-diarylbenzamidines[23, 29]. Ebullioscopically determined molecular weights for all three derivatives correspond to tetramers[27]. On the basis of similar chemical properties and by analogy with Cu-1,3-diphenyltriazine, but in the absence of X-ray crystallographic structure studies the structure (9) has been proposed. In the case of formamidines, the anion-polarizability[30] contributes to the stability of 9.

$$\left[\begin{array}{c} \text{Ar—N—CH=N—Ar} \\ | \\ \text{Cu} \\ \uparrow \\ \text{Ar—N=CH—N—Ar} \\ | \\ \text{Cu} \end{array} \right]_2$$

(9)

The facts given above, while interesting with respect to the complex-forming ability of amidines, are predominantly of descriptive character. Their limitation may be seen, in the first place, in the fact that hypothetical structures could not be confirmed through detailed physico-chemical study of these complexes, e.g. by X-ray diffraction, etc.

I. Complexes of α-hydroxyamidines

In 1960 the first paper in a series of studies concerning α-hydroxy-amidine complexes[31-36] was published.

Alpha-hydroxyamidines form stable complexes with ions of transition metals[31] in strongly alkaline solutions. For such a complex formation it is necessary that the OH group be located in the α-position. This has been verified, e.g. through unsuccessful experiments with phenylacetamidine, as well as with β-hydroxy substituted amidine complexes[37] and through the finding that the hydrogen in the α-OH group is slightly acid, and that in this case the anion shows some complex forming ability[38]. Therefore α-hydroxyamidines may participate in simple chelate formation, through the oxygen of the hydroxyl group and amino-nitrogen of the amidine group[33].

Between the complexes of transition-metal ions with α-hydroxyamidines on the one hand and with amino acids on the other, there appears to be a close analogy (the O and NH functions being interchanged).

Cu- and Ni-complexes with α-hydroxy-α-phenylamidinium ions with

$$\text{HO—}\underset{\underset{R}{|}}{\overset{\overset{Ph}{|}}{C}}\text{—C}\underset{\diagdown NH_2}{\overset{\diagup \overset{+}{N}H_2}{}}$$

R = H (mandelamidinium ion, abbreviated to mdH_2^+), with R = CH_3 (atrolactamidinium ion; alH_2^+) as well as with R = C_2H_5 (α-hydroxy-α-phenylbutyramidinium ion; hbH_2^+) have been investigated by using Job's method of continuous variations and pH titrations[33, 34]. It was found by means of conductometric measurements that the complexes show electroneutral character. The complexes are easily soluble in alcohol, formamide as well as in piperidine, but only very slightly in pyridine. The complexes are completely insoluble in non-polar solvents, e.g. in ether, dioxane, benzene, acetone, etc. The complexes could be extracted from aqueous solutions by using water-immiscible alcohols. By means of the titration method, stability constants of the complexes (Figure 1, Table 1) have been determined.

On the basis of the experiments performed as well as on the basis of elementary analyses the authors have come to the conclusion[33, 34] that the complexes formed in aqueous solutions between Cu^{II} or Ni^{II} and α-hydroxyamidines exhibit great stability in this medium. The two types of complexes are entirely analogous: the chelate comprises four hydroxyl and two amidine ions for each metal ion. Neither the formation of higher complexes with great excess of amidine nor the formation of 1 : 1 complexes in absence of hydroxide surplus have been observed. From the fact that with phenylacetamidine the formation of similar complexes does not take

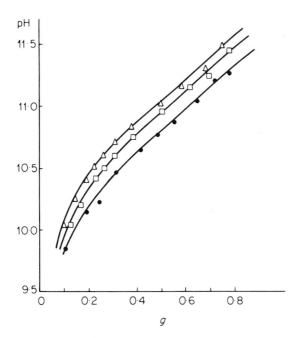

FIGURE 1. pH titrations of α-hydroxyamidinium chlorides. Theoretical curves, calculated for pK values in Tables 1 and 14 and experimental points: (●) mdH₂Cl; (□) alH₂Cl; (△) hbH₂Cl. [Reproduced by permission from R. O. Gould and R. F. Jameson, *J. Chem. Soc.*, 296 (1962).]

TABLE 1. Stability constants of copper and nickel complexes with α-hydroxyamidines at 25°C and ionic strength of 0·1[33,34,a]

Complex	log K_1	log K_2
Cumd$_2$	12·50	11·30
Cual$_2$	12·73	11·57
Cuhb$_2$	12·86	11·70
Nimd$_2$	7·38	7·02
Nial$_2$	7·87	7·53
Nihb$_2$	8·06	7·74

[a] Reproduced by permission from R. O. Gould and R. F. Jameson, *J. Chem. Soc.*, 15 (1963) and 5215 (1963).

place[31], chelate formation involving the hydroxyl groups, (10) and (11), was presumed.

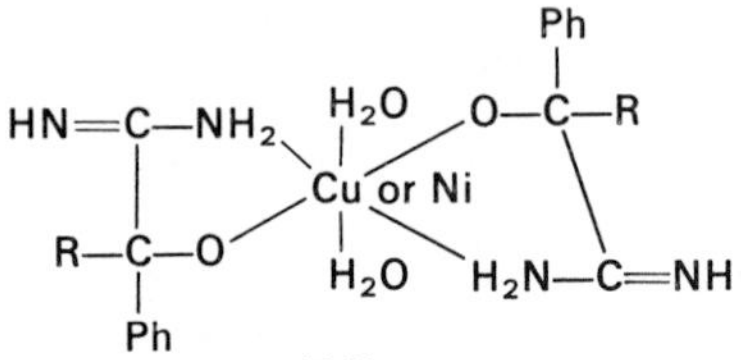

(10)

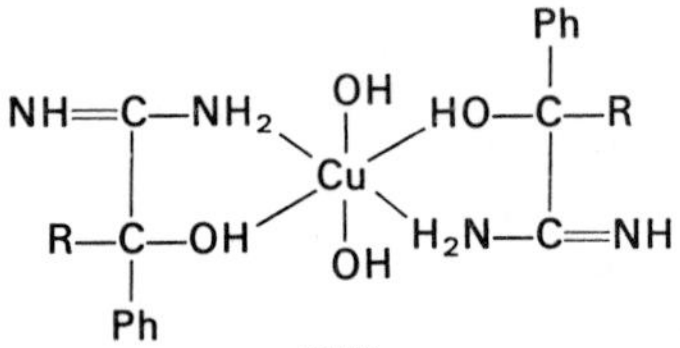

(11)

Considering the great stability of these complexes, the alternative coordination through the imino nitrogen atoms appears to be rather improbable.

The structure 10 is favoured by the results concerning the role of aliphatic hydroxyl groups in chelate systems[33], and by the insolubility of the complex in all except polar solvents, as well as by the ability of the hydrated complexes to lose water, which could not be explained with the structure 11. The possibility of a strong hydrogen bond existing between the water molecules of the hydrated complexes and an adjacent imino group, or an adjacent oxygen giving a five-membered chelate, should also be considered. The existence of such a hydrogen bond may be supported by the fact that the dehydration of the hydroxy phenylbutyramidine complex is rather difficult.

Later on, these presumptions[33] were fully corroborated and the structure of the complex **10** was proved. Iball and Morgan[41] studied the structure of the Cu-α-hydroxy-α-phenylbutyramidine complex using three-dimensional X-ray analysis. It is interesting to note that contrary to expectation this complex did not contain one (+) molecule and one (−) molecule of amidine having the centre of symmetry at the Cu atom, though this complex had been synthesized from a racemic mixture of the amidines. The complex always contained either two (+) or two (−) molecules of amidines. This is a single case found in the literature: in complex syntheses with analogues of the ligands in a series of cases[42] it has been found that complexes obtained from racemic mixtures always have a centre of symmetry on the metal atom and containing in every case one (+) as well as one (−) molecule of the ligand.

For the given complex the formula $C_{20}H_{26}CuN_4O_2 \cdot 2H_2O$ was proved. From the fact that two water molecules are associated with one molecule of the complex, octahedral coordination around the Cu atom was presumed, this fact having been entirely proved also by the X-ray study. The existence of an H-bond between an amino group and the water molecule has been confirmed, (the distances of the oxygens in the water molecules from the amino groups in the molecule are 2·83 Å and 2·80 Å). The distances of the oxygen atoms of the five-membered rings from the oxygen atoms of the water molecules are 2·72 Å and 2·73 Å, giving evidence also of the existence of a strong hydrogen-bonded hydrogen atom of water[41].

The given case is an unusual one of an H-bond existing between fully coordinately-bonded atoms. However, similar cases have already been described in the literature and the structure of such compounds has been determined[43-47].

From comparing stability constants for Cu^{II} complexes (evaluated on the bases of structure **10**) with those of 1:1 complexes of simple amino acids, it can be seen that stability constants of amidine complexes are much higher[39, 40]. Evidently, a much stronger bond exists between alkoxy oxygens and metal ions, than is in the case between carbonyl oxygens and metal ions.

With Ni^{II} complexes this difference is not so evident as it is in the case of Cu^{II} complexes. From the magnitude of the stability constants given for Ni^{II} complexes of amidines it may be concluded that in this case no interaction exists between 2:1 complexes and neutral amidine molecules in solutions with higher pH values.

The drop of stability appearing between Ni and Cu complexes means that complexes of first order transitional bivalent ions will be far less stable.

Recently Ag^I, Cd^{II} as well as Hg^{II} complexes with hydroxyamidines[35, 36] have been described, and studied, using titration methods. It has been found that complexes of the type AgL^+, AgL_2^+, AgL_2OH, CdL_2^+, $CdLOH^-$, HgClL, HgL_2 and HgLOH (where L stands for the amidine base) are formed in these cases.

Complexes with mercury(II) are formed not only with hydroxyamidines (see Table 2). It is well known that mercury(II) forms complexes with organic bases quite easily, even in an acid medium. With amidine, in the presence of mercury chloride titrations in chloride or in nitrate medium showed some differences, yielding evidence of the coordination of chloride ions with mercury, the chloride ions being substituted in turn by the ligand[55]:

$$L + HgCl_2 \xrightleftharpoons{} HgClL + Cl^- \qquad (1)$$

$$L + HgClL \xrightleftharpoons{} HgL_2 + Cl^- \qquad (2)$$

Since no formation of higher complexes (HgL_3) has been observed, it appears to be necessary to assume that simultaneously with reaction (2) hydrolysis also occurs:

$$HgClL + OH^- \xrightleftharpoons{} HgLOH + Cl^- \qquad (3)$$

For the titrations taking place in excess chloride (where the Cl^- concentration appears to be constant) the following constants are given:

$$K_1' = \frac{[HgClL]}{[HgCl_2][L]}, \qquad K_2' = \frac{[HgL_2]}{[HgClL][L]}, \qquad K_h' = \frac{[HgLOH]}{[HgClL][OH^-]}.$$

These results show that hydroxyamidine bases are able—as in the case with the ions of imidates, to displace Cl^- ions from mercury chloride so that the second substitution competes with the hydrolysis. Experimental results[35, 36] given for amidine complexes (Table 2) corresponded very well with those of the work[56], where it has also been found that the absence of the hydroxy group in the alpha position caused no anomalous behaviour. Hence it is reasonable to assume that the OH group does not participate considerably in mercury coordination.

TABLE 2. Stability constants and hydrolytic constants for complexes of silver(I), cadmium(II) and mercury(II) with hydroxy-amidines in 0.1 M-KNO₃ or in 0.1 M-KCl[a]

Ligand	Metal	$\log K_1$	$\log \beta_2$[b]	$\log K_h$
Phenylglycolamidine	Ag	4·40 (8)	9·32 (1)	5·02 (4)
Phenyllactamidine	Ag	4·16 (7)	8·86 (1)	3·90 (6)
α-Hydroxyisobutyramidine	Ag	4·05 (7)	8·61 (2)	3·45 (10)
α-Hydroxy-α-methylbutyramidine	Ag	4·23 (15)	9·14 (3)	3·56 (6)
Phenylglycolamidine	Cd	2·71 (9)		4·98 (9) $\beta_{2h} = 10\cdot80(20)$[c]
Phenyllactamidine	Cd	2·90 (6)		4·88 (6)
Phenylglycolamidine	Hg	4·86 (3)	4·96 (3)	5·42 (1)
Phenyllactamidine	Hg	5·00 (3)	4·93 (5)	5·65 (7)
α-Hydroxyisobutyramidine	Hg	5·02 (3)	4·57 (9)	5·86 (3)
α-Hydroxy-α-methylbutyramidine	Hg	5·27 (3)	4·95 (6)	5·76 (3)
Acetamidine	Hg	6·06 (3)	5·98 (15)	5·97 (6)

[a] Reproduced by permission from R. O. Gould and H. M. Sutton, J. Chem. Soc. (A), 1184 (1970)[35] and J. Chem. Soc. (A), 1439 (1970)[36].

[b] $\beta_2 = K_2$

[c] $\beta_{2h} = \dfrac{[\text{Cd L(OH)}_2]}{[\text{Cd}^{2+}][\text{OH}^-]^2[\text{L}]}$

In the titration curves given for the silver-amidine system[36] the experimental data have been interpreted under the following conditions:

$$K_1 = \frac{[\text{AgL}^+]}{[\text{Ag}^+][\text{L}]}; \quad K_2 = \frac{[\text{AgL}_2^+]}{[\text{AgL}^+][\text{L}]}; \quad K_h = \frac{[\text{AgL}_2\text{OH}]}{[\text{AgL}_2^+][\text{OH}^-]}.$$

In this case the hydrolysis is expressed only formally as an addition reaction. There is no evidence of further hydrolysis in the range of the solutions studied.

For Cd^{II} complexes the data have been compiled on the basis of the defined equilibrium constants:

$$K_1 = \frac{[\text{CdL}_2^+]}{[\text{CdL}^+][\text{L}]}; \quad K_h = \frac{[\text{CdLOH}^+]}{[\text{CdL}^{2+}][\text{OH}^-]},$$

where complex formation between Cd and Cl^- has been neglected.

2. Complexes of other amidines

The complex of acetamidine with mercury(II) has been already mentioned (Table 2).

In the reactions of lithium benzamidines with halogen derivatives of metal organic compounds of the groups IV and V, di- as well as monosubstituted benzamidines resulted, some of them exhibiting in their n.m.r. spectrum intramolecular ligand exchange reactions[57]. According to equation 4, N-trimethylsilyl-N,N'-dimethylbenzamidine (12) as well as N-trimethylstannyl-N,N'-dimethylbenzamidine are obtained, where the structure of complexes having intrinsic symmetry has been corroborated through n.m.r.[57]

(M = Sn or Si)

Boylan, Nelson and Deeney[58] have prepared Fe^{II} complexes with N-(2-pyridylmethyl)picolinamidine (abbreviated to ppa) and its two methyl derivatives (Meppa and Me$_2$ppa) and studied in detail their spectroscopic and magnetic properties. The possible tautomerism of the ligands has been solved by means of i.r. spectroscopy as well as electron spectroscopy of the Fe^{II} complexes.

R^1—⬡—N—C(=N—CH$_2$)—⬡—N—R, with NH$_2$

(13)

R^1—⬡—N—C(=NH)—NH—CH$_2$—⬡—N—R

(14)

On the basis of the obtained experimental data as well as by comparison with earlier works discussing similar problems [59,60,61], proof was furnished that the ligand exists in the complex in the tautomeric form **13**[58]. The two ligands in the complex stand in planar conformation at an angle of 90° to each other. In the complexes of the type [M(ligand)$_2$]X$_{2n}$H$_2$O, where M denotes Ni^{II} or Fe^{II}, the ligand is ppa, Meppa or Me$_2$ppa, and X = Cl$^-$, Br$^-$, NCS$^-$, ClO$_4^-$, BF$_4^-$, PF$_6^-$; n varies from 1 to 4, depending upon the complex structure. These complexes are soluble in water as well as in polar organic solvents. They have been studied also by means of conductometry and it was found that the best proton acceptor among all the counterions mentioned is the Br-ion[58].

The electronic spectra are practically identical in the solid state, in water, or in nitrobenzene for all the complexes[58] under investigation and independent of the character of the counterion.

Recently a new fast condensation of coordinated ligands with tridentate amidines[62] has been published. The reaction of

$$cis\text{-}[Co(en)_2(NH_2CH_2CN)Cl]^{2+}$$

(where en = ethylendiamine) at pH 7·31 to 8·94 was carried out (at ionic strength $\mu = 1\cdot0$ NaClO$_4$) and a purple complex with the structure

$$\left[\begin{array}{c} Co(en)NH_2CH_2C-NH_2 \\ \diagdown\diagdown \\ NCH_2CH_2NH_2Cl \end{array} \right]^{2+}$$

(abbreviated to I-Cl) was isolated. This complex was investigated by means of the three-dimensional X-ray analysis as well as by spectrophotometry. The kinetics of the complex formation were also followed. In an analogous manner, the complex I-Br was also obtained.

The mechanism of formation of the tridentate amidine complexes consists in deprotonation of the NH$_2$ group, which is in a *trans* position to a Cl$^-$ or Br$^-$ bond. This deprotonation is associated with a nucleophilic attack of the coordinated amine on the C-atom of the nitrile. Through

J. Ševčík

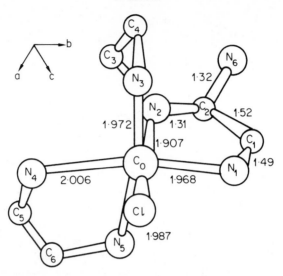

Figure 2. The molecular structure of the tridentate amidine complex. [Reproduced by permission from D. A. Buckingham, B. M. Foxman, A. M. Sargeson and A. Zanella, *J. Am. Chem. Soc.*, **94**, 1007 (1972).]

consecutive shifting of protons, there takes place the formation of an exo NH_2 group, which has been verified through n.m.r. studies[62]. Since the proton exchange is fast compared to the rate of cyclization, the latter is rate-limiting. Since three reactive sites may be taken into account, but only one isomer results, it may be possible in this case to talk about a certain stereospecificity of the reaction[62]. In Figure 2 the model of the tridentate amidine complex as well as the data concerning the lengths of the ligand bonds are represented.

In studying intermediates occurring in the biosynthesis of *de novo* purine derivatives it was found that acid catalysed decarboxylation occurring in aqueous solutions, may be fully inhibited through the presence of transition-metal ions[63, 64]. The hydrolysis of a cyclohexyl derivative (15) taking place in borate buffer medium (pH = 8) was studied in the presence of Cu^{II}, Ni^{II}, Co^{II} as well as Mn^{II} ions at various concen-

$$C_6H_{11}-NH-\underset{\underset{NH}{\|}}{C}-CH_2NH_2$$

(15)

trations. Amidine **15** in the presence of transition metal ions, yields complexes of the structure $ML_2(NO_3)_2$, where M stands for Cu, Ni, Co and L stands for **15** (Table 3).

According to e.s.r. studies, the Cu as well as the Ni complexes show square planar configuration[63].

3. Complexes of amidinourea derivatives

Since amidinourea derivatives will be discussed separately, here we will give only some fundamental information concerning the problems of complex formation with amidinourea derivatives.

a. *Complexes of O-alkyl-1-amidinourea.* Dutta and Ray[66] synthesized some complexes of transition elements which they regarded as guanylurea derivatives. Their results were criticized, and a re-investigation[70] showed that the compounds in question were not guanylurea derivatives (**16**), but *O*-alkyl-1-amidinourea derivatives (**17**) and that complex formation occurred with two nitrogen ligands and six-membered ring formation.

TABLE 3. Absorption spectra of metal complexes of α-amino-N-cyclohexylacetamidine and related compounds[a]

Complex	Colour	λ_{max}/nm	ε_{max}
$Cu[II]_2(NO_3)_2$[b]	Purple	555	57
$Cu[II]_2(NO_3)_2$[c]	Purple	545	54
$Cu[II]_2(NO_3)_2$[d]	Purple	559	50·4
$Ni[II]_2(NO_3)_2$[b]	Orange	447	58
$Ni[II]_2(NO_3)_2$[d]	Orange	455	70
$Co[II]_2(NO_3)_2$[d]	Orange	498	
$Co[III]_2(NO_3)_2$[e]	Purple		
$Ni[III]_2(NO_3)_2$[e]	Light blue		
$Cu[III]_2(NO_3)_2$[b]	Dark blue	655	52
$Cu[IV]_3(NO_3)_2$[d]	Dark blue	666	30·7
$Ni[IV]_3(NO_3)_2$[d]	Light green	407	22·1
$Co[IV]_3(NO_3)_2$[d]	Purple	524	28·8

[a] Reproduced by permission from L. A. Mulligan, G. Shaw and P. J. Staples, *J. Chem. Soc. (C)*, 1585 (1971).
[b] In water.
[c] In acetone.
[d] In ethanol
[e] Not sufficiently soluble for spectral determinations. II denotes C_6H_{11}—NH—C—CH_2NH_2; III, C_6H_{11}—NH—CO—CH_2NH_2;

           ||
          NH
IV, C_6H_{11}—NH—CO—CH_2NHCHO.

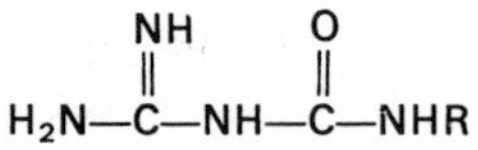

$$\underset{(16)}{\text{H}_2\text{N}-\overset{\overset{\displaystyle\text{NH}}{\|}}{\text{C}}-\text{NH}-\overset{\overset{\displaystyle\text{O}}{\|}}{\text{C}}-\text{NHR}} \qquad \underset{(17)}{\text{H}_2\text{N}-\overset{\overset{\displaystyle\text{NH}}{\|}}{\text{C}}-\text{NH}-\overset{\overset{\displaystyle\text{NH}}{\|}}{\text{C}}-\text{OR}}$$

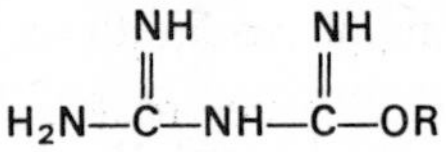

In other work[71, 72] Ni^{II} as well as Cu^{II} complexes with O-alkyl-l-amidino-urea having the general formula $Ni(R\text{-}au)_2X_2$ were investigated. In these complexes R-au represents O-alkyl-l-amidinourea and X denotes a monovalent anion. From the study of the magnetic properties of these complexes it may be concluded that R-au forms diamagnetic complexes having planar square form around the central metal atom with very likely axial anion coordination[71, 72]. The results of i.r. spectroscopic studies fully confirmed the structure of these complexes as O-alkyl-l-amidinourea derivatives. The complexes were also studied by X-ray methods and the structures **18** and **19** were proposed for the cation and for the neutral Ni^{II} complex.

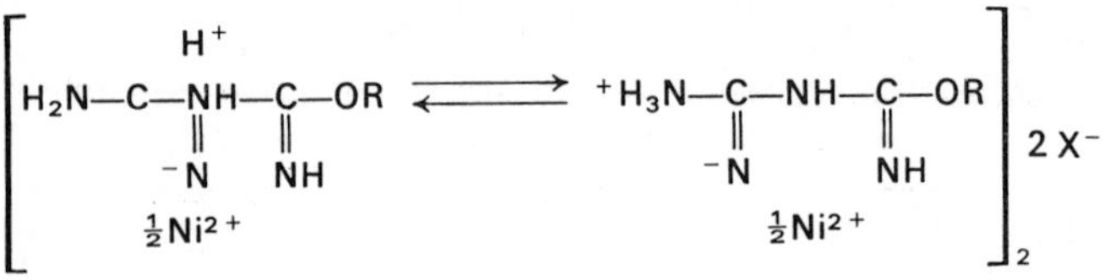

(18)

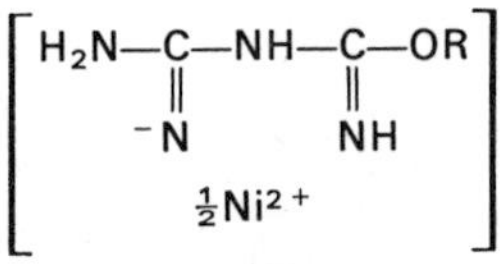

(19)

b. *Complexes of 1-amidinourea.* Ni^{II} as well as Cu^{II} complexes with amidinourea were synthesized already in the last century[73]. Nearly a hundred years later, further Co^{II} and Co^{III} as well as Pd^{II} complexes[74] were described. For these complexes the structures **20** and **21**[74, 75] were proposed, from which the structure **21** should be preferred partly by

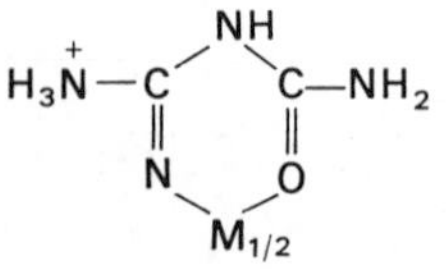

(20) 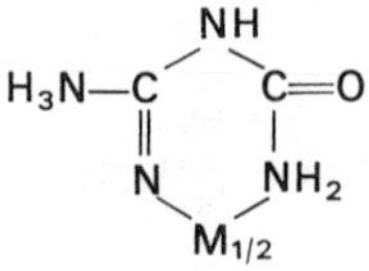   (21)

analogy with biguanide complexes[74], partly on the basis of u.r. spectroscopical data which showed that the donor atoms appearing in the amidinourea complexes were the N atoms[76]. The complexes having the form $[M(H\text{-}au)_2]X_2$ where M stands for Co^{II}, Ni^{II} and Pd^{II} and where X denotes Cl, OH, $\frac{1}{2}SO_4$ showed diamagnetic character. On the basis of detailed spectroscopic and magnetic studies they were assigned a square planar configuration, in full agreement with the existence of a strong ligand field around the central metal ion[76] (Table 4).

c. *Complexes of 1-amidino-2-thiourea.* Several studies[77-80], deal with the metallic complexes of 1-amidino-2-thiourea (ATU, **22**), which may be present as a bidentate ligand in the forms **23** and **24** or in a unidentate ligand form, either sulphur- or nitrogen-coordinated.

$$\begin{array}{ccc} \text{(22)} & \text{(23)} & \text{(24)} \end{array}$$

For the very stable Ni^{II} complex with ATU, which does not change even under boiling in alkaline solution, the structure **25** was proposed[77] on the basis of chemical reactions. To verify this structure further ATU complexes with Cu^{II}, Mn^{II} and Pd^{II} were also studied[80].

(25)

On the basis of i.r. spectroscopic data it has been concluded that in these complexes a metal–sulphur bond is present, as in **26**. In support of this structure, the Ni^{II} chelate with bis(dithiobiuret) (**27**) has been investigated[81] and it was found that the i.r. spectra of both the ligands[27] and the chelates **26** showed very similar characteristics. This led to the conclusion

$$\begin{array}{cc} \text{(26)} & \text{(27)} \end{array}$$

TABLE 4. Infrared absorption spectra of amidinourea (au) complexes of some bivalent ions[a].

Amidinourea	Cu(Hau)$_2$Cl$_2$·H$_2$O	Ni(Hau)$_2$(OH)$_2$	Pd(Hau)$_2$Cl$_2$·$\frac{1}{2}$H$_2$O	Possible assignments
735 w	760 m	765 m	790 m	δ(CO) and π(NH)
—	860 m	800 m	825 m	ring def.
970 w	970 m	990 w	970 m	ν(CN)
1130 m	1100 w	1115 m	1050 m	δ(NH) and ν(CN)
1360 v.s.	1265 v.s.	1280 v.s.	1270 v.s.	δ(NH) and ν(CN)
1460 v.s.	1400 s	1420 s	1400 s	ν(CN)
1585 s	1540 s	1525 s	1525 s	ν(NH)
1630 w	—	—	—	ν(C=NH) and amide 2
1680 v.s.	1650 v.s.	1650 v.s.	1650 v.s.	ν(CO)
1730 v.s.	—	—	—	amide 1
3180 s	3200 s	3200 s	3200 s	NH group
3330 s	3320 s	3300 v.s.	3300 s	ν(NH$_2$)
	3400 sh	3400 sh	3400 sh	ν(OH)

s = strong, v.s. = very strong, m = medium, w = weak, sh = shoulder; all bands in cm^{-1}.
[a] Reproduced by permission from A. Syamal, *Z. Naturforsch. B*, 1514 (1969).

that the only structure justified for the Ni^{II} Cu^{II} Mn^{II} as well as Pd^{II} chelates appears to be structure **26**[80]. Cd^{II} complexes with ATU appear to involve monodentate ligand–sulphur coordination[79].

4. Boron complexes of amidines

Jefferson and co-workers[48] followed in detail the boronation process of di-*p*-tolylcarbodiimide occurring under various conditions. It follows from their work that complex formation between boron and amidines may also be expected. This presumption was later fully justified by the synthesis of the inner complex (**28**) was obtained either from butylmercaptodipropylboron and acetonitrile[49] or through the reaction taking place between acetaminodipropylboron and acetonitrile[50]:

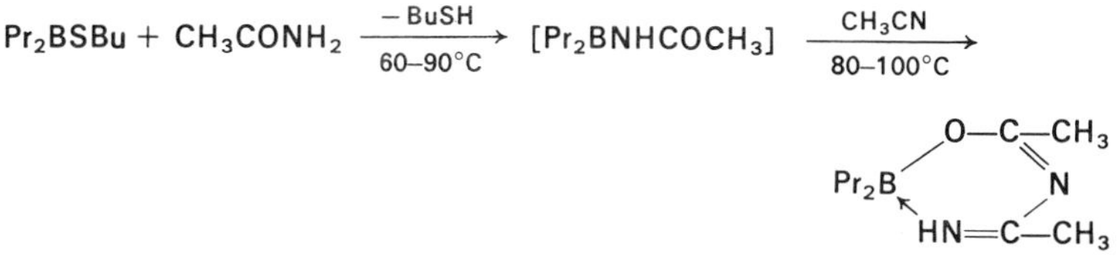

$$Pr_2BSBu + CH_3CONH_2 \xrightarrow[60-90°C]{-BuSH} [Pr_2BNHCOCH_3] \xrightarrow[80-100°C]{CH_3CN}$$

(5)

(28)

In this case the chelate-forming entity appears to be the *N*-acetylacetamidine. Structure **28** was confirmed through n.m.r. and i.r. studies. Similar reactions were observed also with other alkylmercaptoborons as well as primary amides and nitriles[50, 54].

On heating mixtures containing benzamidine, benzonitrile and a trialkylboron, transitional complexes with trialkylborate (**29**) were obtained at first. These were transformed on further heating to **30** and finally, on reaction with benzonitrile to dialkylboron benzimidoyl-benzimidinate (**31**)[51]:

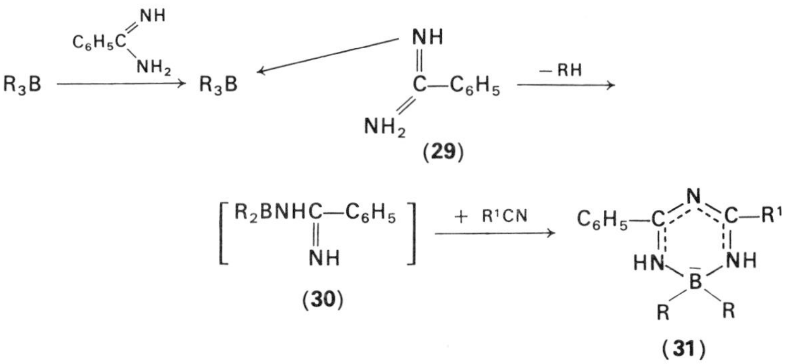

(6)

(29)

(30)

(31)

The structure of these complexes can be represented (31) with more precision by assuming half-coordinated N—B bonds[52].

For syntheses of complexes obtained from low-boiling nitriles a modified form of reaction (6) is used[51]. In such cases the starting material is an alkylmercaptodialkylboron, which eliminates a mercaptan on reaction with benzamidine to yield 30, which in turn, on reaction with a nitrile gives the complex 31.

Aliphatic amidines are not very suitable for the preparation of complexes of type 31, because of their low stability as well as difficulties in the isolation of the free bases[53]. Nevertheless, the acetamidine complex with tri-n-propylboron was isolated (32), from the reaction of acetamidine hydrochloride with the sodium methylate and tri-n-propylboron in methanolic medium[51].

$$CH_3-C\begin{matrix} \nearrow NH \\ \searrow NH_2 \end{matrix} \longrightarrow B(n\text{-}C_3H_7)_3$$

(32)

The corresponding imidoylamidinates (31) can be obtained from the complex 32 by heating with nitriles to 130–150°C (Table 5).

The complexes 31a–e are crystalline substances easily soluble in ether, alcohol and benzene, slightly in hexane and iso-pentane, and insoluble in water. They are all stable in the atmosphere, not undergoing hydrolysis under boiling with water and bases. They give salts of the type 31. HX, from

TABLE 5. Some complexes of the type 31, given in reference 51.

$R-C\overset{N}{\underset{HN}{\diagdown}}\overset{}{\underset{\diagdown B \diagup}{C}}-R^1$ $R^2\quad R^2$ (31)	R	R^1	R^2
a	C_6H_5	C_6H_5	$n\text{-}C_3H_7$
b	C_6H_5	C_6H_5	$i\text{-}C_3H_7$
c	C_6H_5	C_6H_5	$n\text{-}C_4H_9$
d	C_6H_5	CH_3	$n\text{-}C_4H_9$
e	CH_3	CH_3	$n\text{-}C_3H_7$
f	C_6H_5	CH_3	$n\text{-}C_3H_7$

which the base may be recovered. Their structure has been corroborated by i.r. and ^{1}H and ^{11}B n.m.r. studies.

Spectrophotometric studies show that with these complexes no association caused by H-bonds takes place.

It is worth mentioning that the tri-*n*-propylboron complex with acetamidine (32) represents the first known case of a boron complex having an unsubstituted amidine as the ligand[51].

B. Complexes of Amidoximes

The practical application of amidoximes is wide-spread, especially from the analytical point of view. Their reactions with a great number of metallic ions have been investigated, and lead to variously coloured products. Nevertheless the structure of the chelates formed has not been always satisfactorily solved.

Many products resulting from the reactions of amidoximes with metal ions, formerly formulated as basic salts, may be regarded as complex compounds. Table 6 surveys some salts as well as amidoxime complexes.

Even though many formulations concerning salts or complexes given in Table 6 appears to be disputable at present, the data are nevertheless very illustrative, depicting very well the development of views concerning complex formation of amidoximes.

In recent years benzamidoxime complexes were studied by Manolov and his co-workers[93-102]. Their work represents a fundamental contribution to this problem.

Manolov studied the very stable blue or green coloured benzamidoxime complexes (abbreviated to Bz) with copper(II) and nickel(II)[95] in alkaline media. Since the depolarization of NH_2 group was found occurring in pH range 4–6[96], complex formation could be expected even in neutral solutions. In order to obtain the complexes in a crystalline form, it was necessary to use $NH_4Ag(SCN)_2$ solution[93] as additional reagent. The complexes with $[Co(Bz)_2Ag(SCN)_2)_2$, $[Ni(Bz)_2Ag(SCN)_2]_2$ and $[Cd(Bz)_2Ag(SCN)_2]_2$ were investigated using i.r. spectroscopy. The i.r. spectra supported the view that coordination between the metal ions and N atoms was taking place, not involving the O atom. On the basis of i.r. analysis (absorption bands near 1590 and 1650 cm^{-1}) it was assumed that the nitrogen atom of the NH_2 group took part in the complex formation[93].

On the basis of a roentgenographic study the Co^{II} as well as the Ni^{II} complexes were assigned the structure 33, Cd^{II} complexes the structure 34 (Table 6). According to these structures, no changes in the configuration of benzamidoxime molecules take place. The position of these

TABLE 6. Some salts and complexes of amidoximes

Amidoxime	Salt or complex	Other metal ions which also react with the amidoxime	Ref.
Formamidoxime		Hg^I, Hg^{II}, Pb^{II}	7
Homoveratric acid amidoxime		Hg^{II}, Ni^{II}, Fe^{III}	8
α-Hydroxyiso-butyric acid amidoxime		Hg^I, Cu^{II}	10
Phenylacetic acid amidoxime		Fe^{II}, Fe^{III}, Hg^{II}	11, 82
Malonic acid hydroxamide amidoxime		UO_2^{II}, Cu^{II}, Hg^I, Hg^{II}, Co^{II}, Fe^{II}, Fe^{III},	13

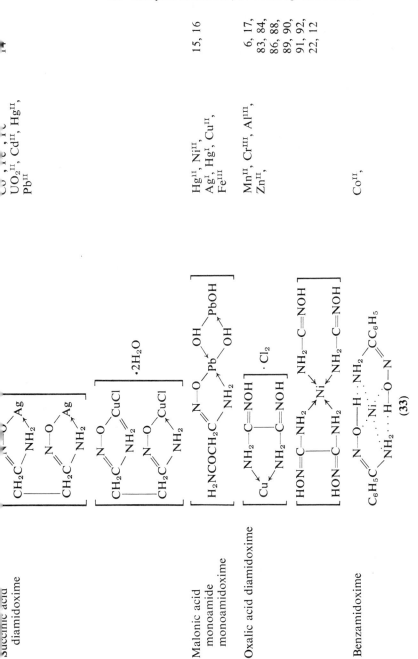

Ligand	Metal ions	References
Succinic acid diamidoxime	UO_2^{II}, Cd^{II}, Hg^{II}, Pb^{II}	
Malonic acid monoamide monoamidoxime	Hg^{II}, Ni^{II}, Ag^I, Hg^I, Cu^{II}, Fe^{III}	15, 16
Oxalic acid diamidoxime	Mn^{II}, Cr^{III}, Al^{III}, Zn^{II},	6, 17, 83, 84, 86, 88, 89, 90, 91, 92, 22, 12
Benzamidoxime	Co^{II},	

(33)

(continued)

TABLE 6 *cont.*

Amidoxime	Salt or complex	Other metal ions which also react with the amidoxime	Ref.
Benzamidoxime	$C_6H_5C=N-O-H\cdots NH_2$ ... Cd ... $NH_2-CC_6H_5$ $H\cdots NH_2$ ... $H-O-N$ $N-O$ ... NH_2 ... H C_6H_5C **(34)**	Ba^{II}, Ca^{II}, Ag^I, Fe^{III}, Cu^I, Co^{II}, Ni^{II}, Cd^{II}, $UO_2{}^{II}$	5, 89, 93, 103, 104
o-, *p-*Toluic acid amidoxime	$\left[CH_3C_6H_4C\begin{smallmatrix}NH_2\\NO\end{smallmatrix} \right]_2 Ni^{IV}=O$		85, 87
*m-*Nitrobenz-amidoxime	$NO_2C_6H_4C\begin{smallmatrix}NO\\NH_2\end{smallmatrix}CuOH\cdot H_2O$	Ag^I, Hg^{II}, Au^{III}, Ni^{II}, Fe^{III}, $UO_2{}^{II}$	19
Aminoacetic acid amidoxime	$\left[NO_2C_6H_4C\begin{smallmatrix}N-O\cdots H\\NH_2\end{smallmatrix} \right] PdCl_2$		86
Cinnamic acid amidoxime	$HCl\left[H_2NCH_2C\begin{smallmatrix}NO\\NH_2\end{smallmatrix} \right] Ni\cdot 2 H_2O$	Ag^I, Hg^{II}, Cu^{II}, Cd^{II}, Fe^{III}, Co^{II}, Ni^{II}, $UO_2{}^{II}$, Pd^{II}	20

hydrogen atoms of the NOH groups renders it possible to form additional hydrogen bonds with the nitrogen atoms of the second benzamidoxime molecule.

When studying complex formation between Co^{II} and benzamidoxime in alkaline medium, Manolov[94] worked out a new spectrophotometric method aimed at the determination of ligand numbers and complex stability constants. In alkaline medium very often some precipitation of metal ions M in the form of hydroxide takes place. Equation (7) applies for the formation of a mononuclear complex:

$$M + nA \rightleftharpoons MA_n \qquad (7)$$

where A stands for a ligand. Assuming that both the stability constant of the complex as well as the ligand concentration are small, the value

$$K = \frac{[MA_n]}{[M][A]^n}$$

of the solubility product of the hydroxide will be exceeded and some precipitation will take place. Solubility products for various hydroxides were tabulated[178, 179]. The number of ligands may be determined photometrically in case the complex is coloured. If the absorption of a given solution is expressed as a function of ligand concentration, frequently the stability constant of the complex may be evaluated. The first complex formation constant was evaluated according to the expression

$$\beta_1 = \frac{1}{[Co^{2+}]\dfrac{a-x}{x}} = 7 \cdot 15 \times 10^4 \qquad (8)$$

For the constant $K = \beta_1 \cdot \beta_2$ the values of $K = 6 \cdot 00 \pm 0 \cdot 02$ and $n = 2$ for $\beta_2 = K/\beta_1 = 14$ were found[94].

The above new method enables us to follow the complex forming processes in media which have not been accessible to direct investigation. This method was applied[105, 106] to the complex of p-methylbenzamidoxime (abbreviated to pMBz) with Co^{II} ions. A clear, intensively blue-coloured solution is formed in strongly alkaline medium (pH = 10), in the presence of a great surplus of pMBz. At low pMBz concentration, partial precipitation of $Co(OH)_2$ takes place, while part of the Co^{II} remains bound in the form of a complex[98]

The complex formation constant for this reactions was determined[94]

$$pMBz + Co^{2+} \rightleftharpoons [Co-pMBz] \qquad (9)$$

$$\beta_1 = (2 \cdot 95 \pm 0 \cdot 06) \times 10^5,$$

$$\beta_2 = \frac{K}{\beta_1} = 21 \cdot 8 \pm 0 \cdot 3 \text{ at } 25°C \text{ and } \mu = 0 \cdot 2.$$

With o-methylbenzamidoxime (abbreviated to oMBz) a yellow complex is formed with MoO_4^{2-} ions[100], according to equation (10)[105,106].

$$MoO_4^{2-} + 3\ oMBz \ \rightleftharpoons \ [MoO_4^{2-}(oMBz)_3] \qquad (10)$$

From the experimental data[94] it follows that the complex formation 1:3 takes place. For the complex given in equation (10) the stability constant values have been found[100]:

$$\log K = 5\cdot04\ (\text{for } MoO_4^{2-}\ \text{concentration } 10^{-2}\ \text{M})$$
$$\log K = 5\cdot18\ (\text{for } MoO_4^{2-}\ \text{concentration } 5\cdot10^{-3}\ \text{M})$$
$$\log K = 5\cdot11 \pm 0\cdot009.$$

In alkaline medium, Co^{II} also forms complexes with a great surplus of oMBz[101]. From spectrophotometric data, by using Manolov's method[94] the reaction according to equation (11) was postulated and the constants

$$Co^{2+} + oMBz \ \rightleftharpoons \ [CooMBz]^{2+} \qquad (11)$$

were determined[101]:

$$\beta_1 = (4\cdot1 \pm 0\cdot6)10^4,\ \log K = 4\cdot23 \pm 0\cdot03$$
$$\beta_2 = 0\cdot41\ \text{at } 25°C \text{ and } \mu = 1.$$

Finally Ni^{II} complexes with pMBz were studied in neutral as well as in alkaline media[102]. In neutral medium the complex is very unstable. In alkaline medium mononuclear [Ni(pMBz)] complex formation takes place for which the constants[102] $\beta_1 = 4\cdot10^5$, $\log K = 4\cdot82$, $\beta_2 = 0\cdot2$ have been determined.

III. HYDROGEN BONDING

A. Hydrogen Bonds Involving Amidines

I. Intermolecular bonds

Amidines may form dimers, analogous to those of carboxylic acids. These dimers may be formulated[23] as follows:

$$(12)$$

It is possible that dimerization is assisted by resonance stabilization in the corresponding cation and anion after salt-formation[1]:

$$(13)$$

The reaction of aromatic amines with the ethyl orthoformate leads to N,N'di-arylformamidines. Although these products were earlier formulated differently[108,109], later studies[28,110] determined the correct structures. The neutralization equivalent of N,N'-diarylformamidine hydrochlorides as well as cryoscopic determination of the molecular weight in benzene and in naphthalene showed[28] that N,N'-diphenylformamidine as well as N,N'-di-p-chlorophenylformamidine were associated in both solvents (far more in naphthalene). Owing to steric hindrance[111] the association of N,N'-di-o-chlorphenylformamidine occurs to a much lesser extent.

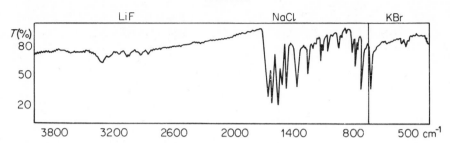

FIGURE 3. Infrared spectrum of N,N'-diphenylbenzamidine. Prism material (LiF, NaCl, KBr) indicated on top of the figure. [Reproduced by permission from P. Sohár, *Acta. chim. Acad. Sci. Hung.*, **54**, 91 (1967).]

The first study confirming the existence of hydrogen bonds with amidines was the work of Sohár[29], using infrared spectroscopic data. In continuation of earlier i.r. spectroscopic studies of N-monosubstituted amides[112–114] N,N'-disubstituted amidines **35a** and **35b** were investigated:

$$R^2—C{\overset{N—R^3}{\underset{NHR^1}{}}}$$

(a) $R^1 = R^2 = R^3 = C_6H_5$

(b) $R^1 = R^3 = C_6H_5$, $R^2 = CH_3$

(35)

From the occurrence of the band of N—H stretching vibrations found between 3400 and 3200 cm^{-1} for **35a** (Figure 3) Sohár[29] deduced that in this case a simple intermolecular association takes place and not a cyclic dimerization. This statement is supported by the fact that a weakening of electron donation occurs in the C=N group owing to the effect of the substituents. The spectrum of **35b** (Figure 4) shows clearly the significant intensity increase as well as the shifting of the (N—H) band. From the

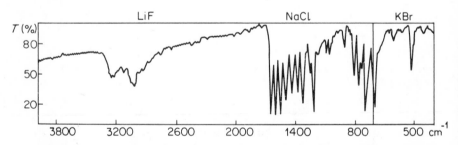

FIGURE 4. Infrared spectrum of N,N'-diphenylacetamidine. Prism material (LiF, NaCl, KBr) indicated on top of the figure. [Reproduced by permission from P. Sohár, *Acta chim. Acad. Sci. Hung.*, **54**, 91 (1967).]

character of this band (3350–2500 cm^{-1}) it was determined[29] that most of the molecules were present as cyclic dimers 36, while the lesser part existed in simple intermolecular association.

$$
\begin{array}{cc}
C_6H_5 & C_6H_5 \\
| & | \\
N\text{---}H\text{---}N \\
\end{array}
$$

CH$_3$—C$\diagup$ $\diagdown$C—CH$_3$

$$
\begin{array}{cc}
N\text{—}H\text{---}N \\
| & | \\
C_6H_5 & C_6H_5 \\
\end{array}
$$

(36)

The electron-donating methyl group increases the density of electrons in the C=N group, enabling it to form strong hydrogen bonds and the broadening or distribution of N—H—C=N bonds is ascribed to the two types of association. The γ(N—H) band found near 510 cm^{-1} is ascribed to the molecules in the form of cyclic dimers in which inner rotation is impossible. The corresponding band in molecules having simple inter-molecular hydrogen bonds appears to be diffusive owing to internal rotation, shifting to the region of 900 to 600 cm^{-1}.

The hindered rotation about the C—N bond in N,N-dimethylbenz-amidine derivatives 37a–d was studied by n.m.r. techniques.

C$_6$H$_5$—C$\diagup$N(CH$_3$)$_2$
      $\diagdown$NR

(37)

(a) R = H
(b) R = COC$_6$H$_5$
(c) R = SO$_2$C$_6$H$_5$
(d) R = PO(OC$_6$H$_5$)$_2$

The magnetic non-equivalence of the methyl protons found in the n.m.r. spectrum made it clear that these protons hindered the rotation, so that the C—N bond showed partial double bond character[115]. The activation energy E_a was evaluated from the temperature dependence of the signals as well as from their position[116]. It was found that 37a was able to form hydrogen bonds but the energy of this hydrogen bond did not exceed 3 kcal/mol[115].

2. Intramolecular bonds

Using a cryoscopic method Hunter and Marriott[23] examined the existence of N—H—N bonds with glyoxalines (38) and benzimidazoles (39) which are cyclic amidines, and also with some non-cyclic amidines.

CH——NH
‖ |
CH   C—R
  `N´

(38)

NH
  \
   C—R
  /
N

(39)

In accordance with the generally accepted statement that the hydrogen bond is the more evident the more acid is the respective hydrogen atom[107], they found that the tendency to create hydrogen bonds was far greater with **38** and **39** than with non-cyclic amidines.

Substitution of the imino hydrogen by alkyl or aryl groups decreases the association considerably. Trisubstituted amidines are not associated at all.

Studying the influence of substituents on the association of benzimidazoles **39** it was found[23] that the benzoyl derivative gave five-membered chelates of the type **40**:

H---O
         ‖
N   C—C—C$_6$H$_5$
  \   /
   C
  /
N

(40)

In the case of phenylhydrazone **41** the formation of a six-membered chelate was postulated:

NH
   \
    C—C—C$_6$H$_5$
   /   \\
  N   N
          |
H—N—C$_6$H$_5$

(41)

Neither the cryoscopic method of molecular weight determination, nor the study of solubility are able to give detailed information concerning the character of associates. This is especially true in the case of amidines, which are basic and give only weak hydrogen bonds so that no unambiguous results may be obtained without detailed spectroscopic investigations.

Hill and Rabinowitz[117] studied the reactions of various N,N'-disubstituted amidines with isocyanates, in which urea derivatives were formed. For the product obtained from N,N'-dimethylbenzamidine with phenyl or methyl isocyanate, structure **42** was proposed on the basis of i.r. spectroscopic data. The presence of a strong intramolecular hydrogen bond was deduced from the width as well as from the position of ν(N—H) vibrations (3000 cm^{-1}) and from the disappearance of this band on deuteration (ν(ND) near 2210 cm^{-1})[118,119]. The value of the frequency

$$
\begin{array}{ccc}
\underset{\underset{C_6H_5}{|}}{\overset{CH_3\,\diagdown\,N\,\diagup\,H}{\underset{\underset{CH_3}{|}}{C}}}=N & + \; RNCO \; \rightleftharpoons & \text{(42)}
\end{array}
\tag{14}
$$

$$R = C_6H_5, CH_3$$

of the amide band I excluded the possibility that the carbonyl group may be the acceptor of the hydrogen bond.

Constantly increasing interest in pharmacologically active sulphonyl-amidines has resulted in much attention to this group of substances. Determination of the structure of sulphonylamidines is complicated by uncertainties concerning the existence of tautomeric equilibria and of geometric isomers. On the basis of some studies[120, 121] the imino form of sulphonylamidines (44) has been preferred.

(43)   (44)

On the other hand, in cyclic amidines of this type especially in alkaline media, some preponderance of form 43 was observed[121–124]. This problem was the subject of several almost simultaneous studies[125–127], the results of which were fully identical, and generally contradict the results obtained by Barber[120], who preferred the form 44. The tautomeric equilibrium with amidines is known to be very rapid[128] and in the older literature many unsuccessful experiments concerning the isolation of single forms are described. Separate isomers cannot be isolated but there have been attempts to identify the predominant tautomer in an equilibrium mixture in mono and di-substituted derivatives[-29,130].

Taking into account the possibility of geometrical isomerism in sulphonylamidines, the two structures in the equilibrium (15) may also occur in the other possible isomeric form, i.e. 43a and 44a.

(43a)   (44a)

From Table 7 it is clear that the region of N—H vibrations given for substituted sulphonylamidines shows one sharp maximum and one diffusion band, the intensity of which does not change practically on dilution. Hence it may be deduced that a strong intramolecular hydrogen bond is formed and that the sulphonylamidines are present largely in the form **43** which alone is able to form such bonds. A weak absorption maximum has been ascribed to the presence of a low equilibrium concentration of the geometrical isomeric from **43a**. On the basis of the spectrophotometric data as well as by analogy with the o-alkanesulphonyl-anilines[131], the structure **43b** has been proposed[125-127]:

$$R^3\underset{\underset{O}{\overset{\parallel}{S}}\overset{}{\diagdown}O\cdots H}{\overset{\diagup N=C\overset{\diagup R^4}{\diagdown}}{}}N-R^2$$

(43b)

An exception in this series of sulphonylamidines is found in the N-t-butyl-N'-alkylsulphonylamidines[125], in which the character of N—H absorption band testifies to some intermolecular hydrogen bonding. In this type of compounds the *trans* form **45** has been proved as a preponderant structure by use of spectroscopy.

$$R-SO_2-N=C\overset{\overset{\displaystyle NHBu\text{-}t}{\diagup}}{\diagdown C_6H_5}$$

(45)

From the intensity decrease of the absorption band on dilution, with three sulphonylamidines having an unsubstituted amino group (N-methylsulphonylacetamidine, N-phenyl- and N-p-tolyl-sulphonylbenz-amidine) an additional intermolecular hydrogen bond has been deduced[126].

If one of the hydrogen atoms of the amino group is substituted, in the *syn–anti* isomer pair the *anti* form is the more stable one[126, 132].

syn **(46)**   *anti* **(47)**

TABLE 7. The i.r. spectra of substituted sulphonylamidines
$R^1SO_2N{=}CPhNHR^{2}$ [a]

R^1	R^2	νN—H	ν_{as}	νSO$_2$	ν_s
CH$_2$I	s-Bu	c 3428 3308br	1298br	1119	1110
		d 3421 3296br			
CH$_2$I	C$_6$H$_{11}$	c 3426 3301br	1294br	1117	
				1112sh	
		d 3429 3304br	1300	1116	1110
CH$_2$I	Ph$_2$CH	c 3436 3311br	1304br	1119	
		d 3432 3298br	1304br	1120	
CH$_2$I	t-Bu	c 3435 3350	1201br	1117	
		d 3431			
		m 3298 3106	1279	1111	
Me	t-Bu†	c 3441 3338	1295br	1128	
		d 3445	1295br	1128	
Me	t-Bu	c 3346 3350	1290br	1126	
		d 3346			
CH$_2$I	Me	c 3451 3333	1298	1117	1110
			1294		
		d 3451 3331	1299br	1118	1108
CH$_2$I	PhCH$_2$	c 3433 3319	1300br	1118 + sh	
		d 3435 3315	1302br	1119	1111
Me	Ph	c 3424 3289br			
		d 3422 3287br	1280br	1107	
		m 3264 3123	1276	1118	1108
CH$_2$I	p-MeOC$_6$H$_4$	d 3411 3291	1295br	1107	
CH$_2$I	Et	c	1299br	1111	
		m	1291br	1127	1119

c = concentrated solution in CHCl$_3$, d = dilute solution in CHCl$_3$, m = Nujol-mull: 0·1 mm cells (NaCl); † o–ClC$_6$H$_4$ derivative.
[a] Reproduced by permission from R. B. Tinkler, *J. Chem. Soc.* (B), 1053 (1970).

B. Hydrogen Bonds of Imidates

With aliphatic imidates the existence of two forms, the imino from **48** and the enamino form **49**, may be presumed:

$$RCH_2-C\overset{\displaystyle NH}{\underset{\displaystyle OR^1}{}} \qquad RCH{=}C\overset{\displaystyle NH_2}{\underset{\displaystyle OR^1}{}}$$

(48)   (49)

From energy considerations and on the basis of dipole moment data it was proved that the only correct structure is the imino form **48**[133].

Structure **48** shows *syn–anti* isomerism. The conversion is very rapid in this case and the activation energy does not exceed 20 kcal/mol[134, 135]. Each of the two isomers **50** and **51** can be present in principle in two planar conformations, namely *s-cis*, *s-trans*, *a-cis* and *a-trans*.

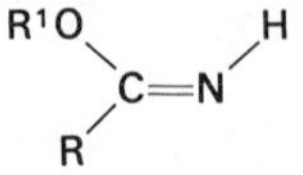

syn (**50**) 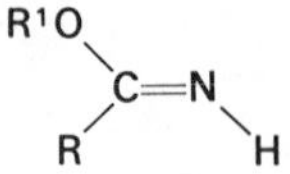   *anti* (**51**)

These problems were throughly discussed by Lumbroso and Bertin[133]. On the basis of a detailed discussion of calculated as well as measured dipole moments two hydrogen bonded forms were proposed, e.g. in triethylamine solution:

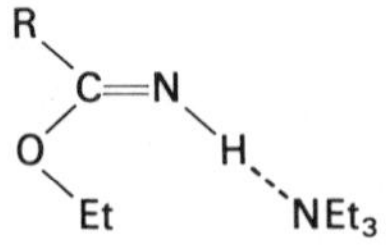

(**52**)   (**53**)

(*syn-cis* ··· *n*)   (*anti-cis* ··· *n*)

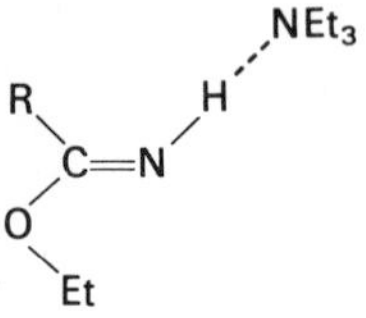

For steric reasons structure **53** is given preference.

From the character of N—H vibrations the ability of *O*-ethylbutyrimidate, *O*-ethylbenzimidates as well as *O*-ethylphenylacetimidates to form hydrogen bonds[136] was determined. From the existence of a double absorption band ν(N—H) given for the phenylacetimidate and on the basis of previously obtained data[137] the structures **54** and **55** were proposed for this compound:

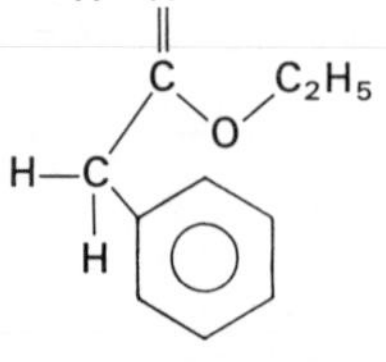

(**54**) 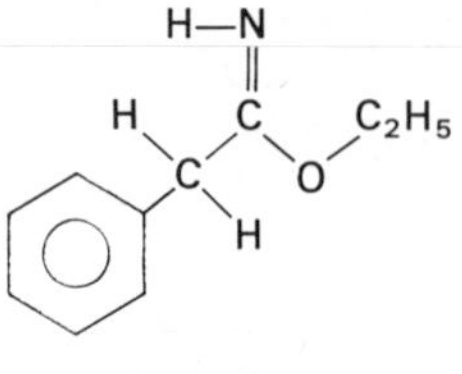   (**55**)

The band lying near 3320 cm^{-1} was ascribed to N—H... π association in the form **55**, the second band showing variable frequency (depending on the solvent—3340 cm^{-1} for hexane and 3267 for pyridine) was attributed to the form **54**, where hydrogen bonding of the NH group with the solvent[136] took place. For the butyrimidate and the phenylacetimidates the ν(NH) bands lying near 3337 cm^{-1} were ascribed to the free NH groups, the bands near 3274 cm^{-1} to the —N—H...N groups. The acidity of these imidates[136] was shown to be in the sequence butyrimidate < benzimidate < phenylacetimidate.

Analogously, substituted *O*-ethylbenzimidates **56** were studied in 12 solvents[138]:

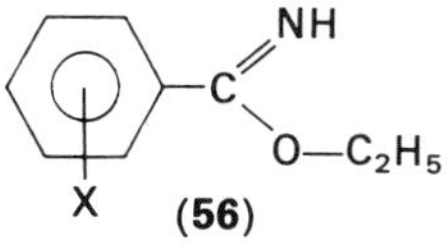

(a) X = H
(b) X = *p*-NO$_2$
(c) X = *p*-CH$_3$
(d) X = *m*-Cl
(e) X = *m*-CH$_3$

The following sequence was found according to the increasing acidity of the NH group:

56e < 56c < 56a < 56d < 56b

C. Intramolecular Hydrogen Bonds of Amidoximes

The tautomerism of amidoximes was not studied quantitatively until recently. Some previously isolated amidoximes were obtained in the modifications showing two different melting points, which was also taken for proof of the existence of tautomerism. Hall and Llewelyn[24] tried to solve this problem by structural analysis. They studied both crystallographic modifications of formamidoxime (m.p. 105°C and 114°C). The values of the interatomic distances C—N^1 and C—N^2 point to resonance, i.e. neither of these bonds can be taken exclusively as simple or as double. From the values of the distances (O—H^3 = 0·40 Å, the angle H^3—O—H^4 = 142°; and O—H^4 = 0·49 Å) a strong intramolecular bond was proved,

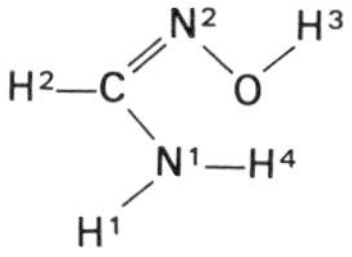

(57)

the existence of which was also verified by using the spectroscopic data for benzamidoxime[93].

Summarizing the results of various studies[24, 93, 139–143] it may be stated that the amidoximes are present in the *syn*-hydroxyimino form, which is stabilized through a strong intramolecular hydrogen bond.

The intramolecular hydrogen bond has been found also in 3-amino-amidoximes[144] (58) and (59) as well as in their acylated derivatives[145, 146].

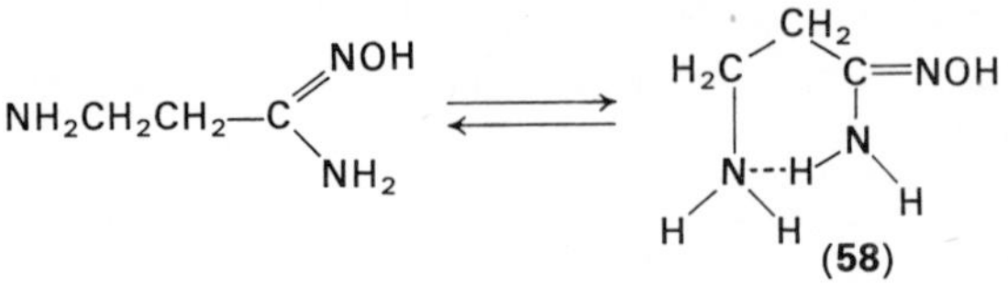

(16)

Through the analysis of i.r. absorption bands obtained in a dilute solution it has been proved that they exist in the form of chelates, illustrated by the structures 58–61.

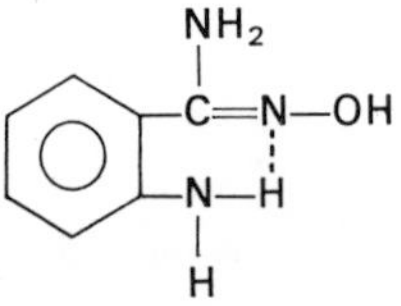

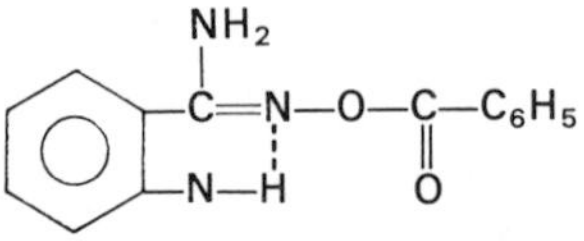

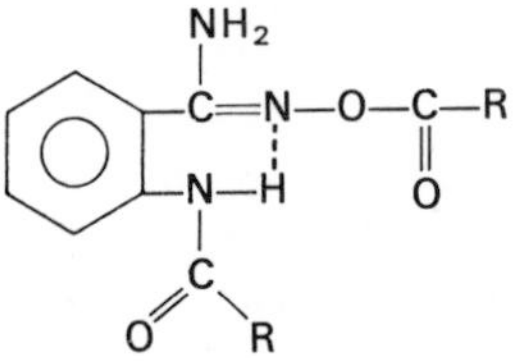

(61)

IV. BASICITY

Amidines and amidoximes each contain two nitrogen atoms in their functional groups. The fact that each of these N atoms has a free pair of electrons confers basic properties to these substances.

Moreover, in the case of amidoximes some dissociation of the proton from the oxygen atom may take place, so that this group may show both acidic and basic properties.

The basicity of iminoethers has not been studied quantitatively probably owing to their very low stability. The available data appear to show that the basicity of iminoethers fluctuates very much depending on the presence of various substituents[148,149].

A. Basicity of Amidines

Unsubstituted amidines are strong bases[149]. The same is true for the asymmetrical N,N-diphenylbenzamidine but the symmetric N,N'-diphenylbenzamidine is neutral to litmus, and is thus a weaker base than ammonia[150]. Both acetamidine as well as benzamidine give hydrochlorides even in solutions containing excess of ammonia, i.e., they are stronger bases than the latter[149]. On the basis of scattered data, largely obtained in synthetic experiments the sequence of increasing basicity of amidines is as follows: N,N'-trisubstituted amidines; N,N'-disubstituted amidines; N-mono-; N,N-di- and unsubstituted amidines. However, this classification is not valid if strongly electron-donating or attracting substituents are present (see Tables 8–10).

TABLE 8. pK_a values for N-monosubstituted benzamidines in 50% water-ethanol solvent ($c = 5 \cdot 10^{-5}$ M)[a]

R^1	λ(nm)	pK_a
$p\text{-}NO_2C_6H_4$	226, 228, 324	$6 \cdot 84 \pm 0 \cdot 05$
$m\text{-}NO_2C_6H_4$	224, 260	$7 \cdot 05 \pm 0 \cdot 05$
$p\text{-}ClC_6H_4$	234, 256	$7 \cdot 35 \pm 0 \cdot 09$
$p\text{-}C_2H_5O_2CC_6H_4$	244, 246, 282, 284	$7 \cdot 43 \pm 0 \cdot 08$
$p\text{-}BrC_6H_4$	234, 240, 244, 248	$7 \cdot 49 \pm 0 \cdot 07$
$m\text{-}IC_6H_4$	230, 232, 262, 266	$7 \cdot 50 \pm 0 \cdot 07$
$p\text{-}IC_6H_4$	220, 236, 238, 240	$7 \cdot 52 \pm 0 \cdot 06$
C_6H_5	222, 236, 238, 240	$7 \cdot 71 \pm 0 \cdot 09$
$p\text{-}CH_3C_6H_4$	240, 242, 266, 268, 270	$8 \cdot 06 \pm 0 \cdot 08$

[a] Reproduced by permission from J. Ševčík, *Chem. Zvesti*, **26**, 49 (1972).

I. Benzamidine derivatives

In studying the influence of substituents on pK values, the pK-value of a series of N-mono-[151], N,N'-di- as well as N,N'-trisubstituted benzamidines[152] has been determined using the photometric method. In Tables 8–10 the pK values obtained as well as the wavelengths used, are given.

An interesting problem, closely connected with the basicity of all types of amidines, is represented by the question of the site of attachment of the added proton. According to Sidgwick[153] the charge is carried on the

imino nitrogen, and the tautomerism of amidinium ions may be represented by (17) for unsubstituted amides, by (18) for the N-mono-
substituted ones and by (19) for the N,N'-disubstituted ones:

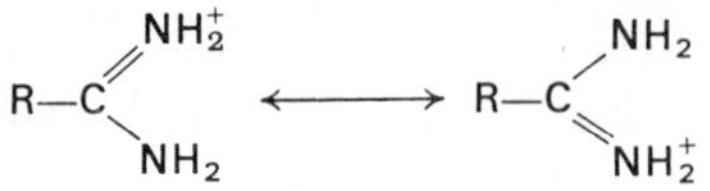

$$
R-C\overset{NH_2^+}{\underset{NH_2}{\big\langle}} \quad\longleftrightarrow\quad R-C\overset{NH_2}{\underset{NH_2^+}{\big\langle}} \tag{17}
$$

$$
\left[R-C\overset{N-R^1}{\underset{NH_2}{\big\langle}} + H^+ \right] \;\rightleftharpoons\; \left[R-C\overset{\overset{H}{|}{N-R}}{\underset{NH}{\big\langle}} H^+ \right] \quad \begin{array}{l} R = C_6H_5 \\ R^1 = XC_6H_4 \end{array} \tag{18}
$$

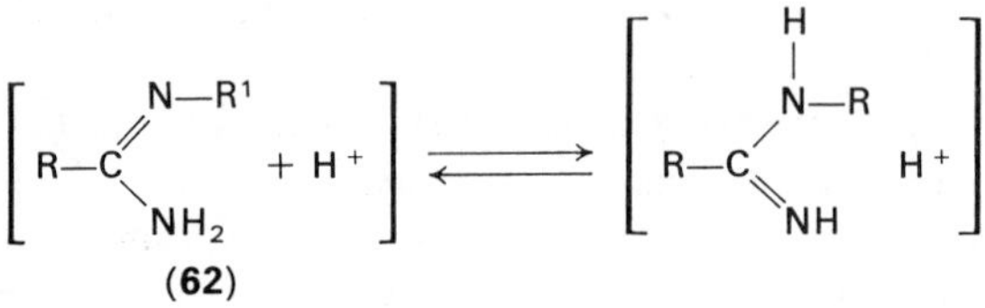

(62)

$$
R-C\overset{N-R^1}{\underset{N-R^2}{\big\langle}} + H^+ \;\rightleftharpoons\; \left[R-C\overset{NH-R^1}{\underset{NH-R^2}{\big\langle}} \right]^+
$$

(63)

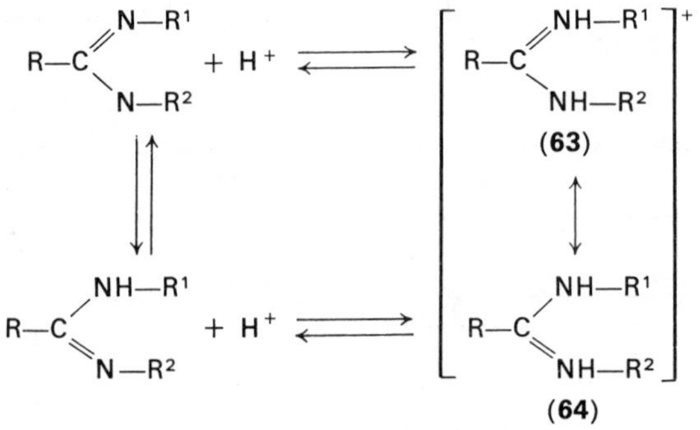

$$
\updownarrow \qquad\qquad \updownarrow \tag{19}
$$

$$
R-C\overset{NH-R^1}{\underset{N-R^2}{\big\langle}} + H^+ \;\rightleftharpoons\; \left[R-C\overset{NH-R^1}{\underset{NH-R^2}{\big\langle}} \right]
$$

(64)

In order to gain further information, the correlation of pK-values of
N-mono-, N,N'-di- as well as N,N'-trisubstituted benzamidines with
Hammett's ρ constants was studied. It was found that the pK-values of the
investigated amides obeyed Hammett's relation (Figure 5). By the least
squares method, the following slopes have been obtained:

for N-monosubstituted benzamidines[151] $\rho = 1\cdot070$
for N,N'-disubstituted benzamidines[152] $\rho = 0\cdot802$
for N,N'-trisubstituted benzamidines[152] $\rho = 2\cdot480$.

The values obtained for the dissociation constants of N-mono- as well
as of N,N'-disubstituted benzamidines characterize only a general dissociation equilibrium (20), and it is impossible to decide which of the two

$$
B + H^+ \;\rightleftharpoons\; BH^+ \tag{20}
$$

nitrogen atoms is being protonated, even though some sources state that
the protonation occurs at the sp_2 nitrogen[172]. Charge distribution be-

TABLE 9. pK_a values for N,N'-disubstituted benzamidines in 50% water–ethanol solvent (R^1 is C_6H_5)[180]

R^2	λ(nm)	pK_a
p-$C_2H_5O_2CC_6H_4$	254, 258, 260, 284, 286	6·46 ± 0·02[a]
m-$NO_2C_6H_4$	260, 264, 268	6·53 ± 0·09
o-$CH_3C_6H_4$	220, 232, 240, 242	6·54 ± 0·05
p-IC_6H_4	228, 246, 244	6·59 ± 0·05
CH_3	224, 232, 236, 240	6·66 ± 0·04
C_6H_5	228, 234, 238	6·92 ± 0·09
m-$CH_3C_6H_4$	228, 232, 234	7·04 ± 0·09
p-$CH_3C_6H_4$	234, 238, 242	7·18 ± 0·03
p-$CH_3OC_6H_4$	232, 236, 248	7·18 ± 0·02

$c = 5 \times 10^{-5}$ M,
[a] $c = 7·5 \times 10^{-5}$ M.

tween the two atoms nitrogen may lead a resonance stabilized symmetric cation (*cf.* the equations 18 and 19).

Regarding the existence of tautomeric equilibria, e.g. in equation (18) the predominant form appears to be **62**[60]. With N,N'-disubstituted benzamidines there are two canonical forms of the cation **63** and **64** and the real ion corresponds to their resonance hybrid. From the study of amidine tautomerism the conclusion has been drawn[60] that the resonance

TABLE 10. pK_a values for N,N'-trisubstituted benzamidines in 50% water–ethanol solvent. NR^2R^3 is piperidyl)[180]

R^1	λ(nm)	pK_a
p-$NO_2C_6H_4$	284, 290, 300, 370, 375, 395	5·37 ± 0·02
m-$NO_2C_6H_4$	254, 266, 268, 270	6·13 ± 0·09
p-$C_2H_5OCC_6H_4$	286, 288, 290	6·39 ± 0·09
p-ClC_6H_4	226, 228, 232,	6·44 ± 0·09
m-IC_6H_4	222, 224, 238	6·70 ± 0·07
p-IC_6H_4	228, 230, 234	7·07 ± 0·09
p-BrC_6H_4	234, 236, 238	7·10 ± 0·05
C_6H_5	220, 258, 260, 262	7·56 ± 0·05
p-$CH_3C_6H_4$	220, 222, 226	7·86 ± 0·04
C_6H_5[a]	220, 226, 232	6·94 ± 0·09

$c = 5 \times 10^{-5}$ M;
[a] R^2 is —C_6H_5 and R^3 is —$COCH_3$

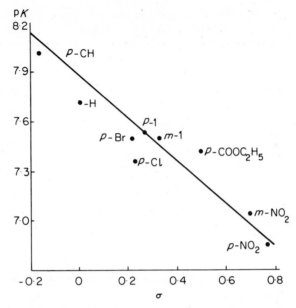

FIGURE 5. The correlation of pK_a values of N-monosubstituted benzamidines with Hammett's σ-constants. [Reproduced by permission from J. Ševčík, *Chem. Zvesti*, **26**, 49 (1972).]

of symmetrically disubstituted amidinium ion resembles that of the un-substituted amidinium cations. Partial stabilization of one resonance form takes place only in the case when the two atoms of nitrogen are substituted with extremely different substituents[153].

The ρ-value given for the N,N'-trisubstituted amidines shows that in this case protonation of the imino nitrogen occurs

$$R-C\begin{matrix} \nearrow N-R^1 \\ \searrow N-R^2 \end{matrix} + H^+ \rightleftharpoons R-C\begin{matrix} \nearrow \overset{+}{N}-R^1 \\ \searrow N-R^2 \end{matrix}$$

(21)

$$R^1 = C_6H_4X, \quad R^2 = CH_2\begin{matrix} \nearrow CH_2-CH_2- \\ \searrow CH_2-CH_2- \end{matrix}$$

2. S-amidine analogues

Tinkler used[125], pK-values for differentiating the tautomeric forms **43** and **44** of sulphonylamidines. He attributed the pK-value of 12·5 to

N-methylsuphonylamidine in the form **43**, and the pK-value of 6–9 to the form **44**. The preparation of the so-called sulphinamidines is described in the literature[154].

$$\text{Ar—S} \overset{\displaystyle \text{N—SO}_2\text{Ar}}{\underset{\displaystyle \underset{\displaystyle \text{R}^1}{\mid}}{\text{N—R}^2}}$$

(a) R^1 = R^2 = H
(b) R^1 = H
(c) R^1 = SO$_2$Ar

(65)

Compounds of the type **65** showed, contrary to expectation, some acidic properties, forming water-soluble salts with bases. In the presence of acids they hydrolyse very easily. No quantitative data concerning the acidity of sulphinamidines have been presented up to now.

3. P-Amidine analogues

The phosphor analogues of amidines—the phosphamidines also show tautomerism. They are basic, giving crystalline salts[155–158]. The position of the tautomeric equilibrium is again very strongly dependent on the influence of the substituents R and R^1.

$$\underset{\text{(66a)}}{\text{C}_2\text{H}_5 \overset{\displaystyle \text{N—R}}{\underset{\displaystyle \text{NH—R}^1}{\diagup\!\!\!\diagdown}} \text{P}} \quad \rightleftharpoons \quad \underset{\text{(66b)}}{\text{C}_2\text{H}_5 \overset{\displaystyle \text{NH—R}}{\underset{\displaystyle \text{N—R}^1}{\diagup\!\!\!\diagdown}} \text{P}}$$

Thus, for R = COCH$_3$ and R^1 = C$_6$H$_5$ the equilibrium is shifted to **66a**, while with R = COCH$_3$ and R^1 = PO(C$_2$H$_5$)$_2$ to **66b**. With R = COCH$_3$ being constant, the basicity is changed in dependence on R^1: e.g. for R^1 = C$_6$H$_5$ the form **66a** is the less basic one, while with R^1 = PO(C$_2$H$_5$)$_2$ the form **66b** shows weaker basicity[158]. The influence of substituents on the tautomeric equilibrium was quantitatively studied with model compound of the type **66**, in which R = C$_6$H$_4$X and R^1 = p-C$_6$H$_4$Y. Since the phosphamidines under investigation showed strongly basic character, the titration method was used for the study, and the phosph-amidinium cation was formulated[158] as **67**:

$$\left[\text{C}_2\text{H}_5 \overset{\displaystyle \text{NH—C}_6\text{H}_4\text{X}}{\underset{\displaystyle \text{NH—C}_6\text{H}_4\text{Y}}{\diagup\!\!\!\diagdown}} \text{P}\right]^+$$

(67)

The data show that all the phosphamidines investigated are strongly basic in nitromethane medium. N,N'-Diphenylphosphamidine ($X = Y = H$) is more basic than diphenylguanidine, and nearly equal in basicity to triethylamine[158, 159].

The evaluation of the tautomeric equilibrium constant was carried out by using three methods[158]. All these were in very good agreement and the results showed that donor substituents shifted the equilibrium in the direction of that form in which the proton was located nearer to the donor substituent. In the case of acceptor substituents the opposite was true[158]. In addition, it was found that the tautomeric forms **66** are less basic than the corresponding methyl derivatives[158].

4. Diacidic benzamidines

N-(3-Dialkylaminopropyl) benzamidines show strong antihistaminic effects[160]. These amidines contain an additional basic tertiary amino group:

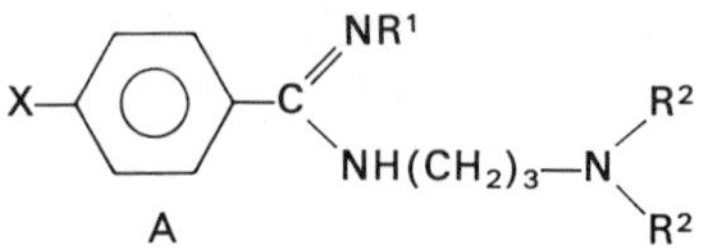

(68)

In a series of these substances the pK_1 values as well as pK_2 values were titrimetrically determined in 50% aqueous ethanolic medium[161] (Tables 11 and 12).

It may be assumed that in structure **68**, changes in the substituent X will influence the pK-values of amidino group, while this change should hardly influence the pK-values of tertiary amino group. The values given in Table 11 show stronger dependence on the substituent X in the pK_2 values than in the pK_1 values.

In addition to that, it is evident from Table 11 that the pK_1 values given for the series of dimethyl derivatives **a–c** and **k–o** are generally lower than with the series of diethyl derivatives, which is consistent with the weaker basic character of N,N-dimethylamines compared with their N,N-diethyl homologues[162]. The pK_2 values of these homologues are, however, nearly identical. This again supports the assignment of the pK_1 values to the tertiary amino group, and the pK_2 value to the amidino group.

In Table 12 on the other hand, the change in X provokes greater changes with the pK_1 values than with the pK_2 values, and pK_2 of the dimethyl homologues is lower than that of the corresponding diethyl

TABLE 11. pK_a values of N-(3-dialkylaminopropyl)benzamidines (**68**), with $R^1 = H$ or alkyl[a]

68	X	R^1	R^2	Salt	pK_{a_1}	pK_{a_2}
a	H	H	Me	di-HCl	7·8	11·4
b	Cl	H	Me	di-HCl	7·6	10·9
c	Br	H	Me	di-HCl	7·7	10·9
d	Me	H	Me	di-HCl	7·9	11·6
e	MeO	H	Me	di-HCl	7·9	11·6
f	H	H	Et	di-HCl	8·2	11·2
g	Cl	H	Et	di-HCl	8·1	10·8
h	Br	H	Et	di-HCl	7·9	10·7
i	Me	H	Et	di-HCl	8·3	11·6
j	MeO	H	Et	di-HCl	8·3	11·6
k	H	Et	Me	di-HCl	7·8	11·0
l	Cl	Et	Me	di-HCl	7·6	10·7
m	Br	Et	Me	di-HCl	7·7	10·6
n	Me	Et	Me	di-HCl	7·8	11·1
o	MeO	Et	Me	di-HCl	8·0	11·5
p	H	Et	Et	di-HBr	8·0	11·0
r	Cl	Et	Et	di-HBr	7·8	10·6
s	Br	Et	Et	di-HBr	7·9	10·6
t	Me	Et	Et	di-HBr	8·0	11·1
u	MeO	Et	Et	di-HBr	8·0	11·2

[a] Reproduced by permission from J. A. Smith and H. Taylor, *J. Chem. Soc.* (*B*), 64 (1969).

TABLE 12. pK_a values of N-(3-dialkylaminopropyl)benzamidines (**68**), with $R^1 = $ aryl[a]

68	X	R^1	R^2	Salt	pK_{a_1}	pK_{a_2}
a'	H	PhCH$_2$	Me	di-HCl	6·9	9·2
b'	Cl	Ph	Me	di-HCl	6·7	9·1
c'	Br	Ph	Me	di-HCl	6·5	8·9
d'	Me	Ph	Me	di-HCl	7·2	9·3
e'	MeO	Ph	Me	di-HCl	7·1	9·3
f'	H	Ph	Et	di-HBr	6·9	9·7
g'	Cl	Ph	Et	di-HBr	6·6	9·5
h'	Br	Ph	Et	di-HBr	6·6	9·4
i'	Me	Ph	Et	di-HBr	7·2	9·7
j'	MeO	Ph	Et	di-HBr	7·4	9·9

[a] Reproduced by permission from J. A. Smith and H. Taylor, *J. Chem. Soc.* (*B*), 64 (1969).

J. Ševčík

TABLE 13. pK_a values of N- and N'-aryl substituted benzamidines **69**[a]

69	R^1	R^2	R^3	pK_{a_1}	pK_{a_2}
a	Ph	Me	$(CH_2)_3NEt_2$	6·5	9·4
b	Ph	Me	Me	7·8	
c	$(CH_2)_3NEt_2$	Me	Ph	7·7	10·1
d	2,6-xylyl	Me	$(CH_2)_3NEt_2$	7·0	9·4
e	2,6-xylyl	H	$(CH_2)_3NEt_2$	7·2	9·6
f	2,6-xylyl	Et	Et	7·7	
g	Ph	H	H	8·2	
h	H	n-Bu	Ph	10·4[164]	
i	H	H	H	11·2[165]	
j	Ph	H	$(CH_2)_3NEt_2$	6·9[161]	

[a] Reproduced by permission from J. A. Smith and H. Taylor, *J. Chem. Soc. (B),* 66 (1969).

homologues while the pK_1 values remain the same. Therefore with compounds **68a'–j'** the assignment of pK_1 and pK_2 is the reverse than it was in the preceding case[161].

Similarly, the basicity of benzamidines of type **69** was also studied[163]. The results are given in the Table 13. On the basis of analysis of the data

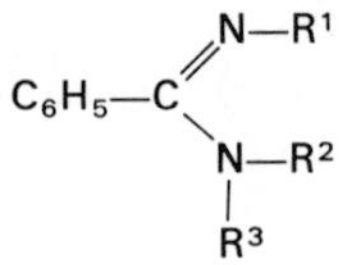

(69)

given in the Table 13 as well as on the basis of u.v. spectrophotometrical data, the pK_1 values were ascribed to the amidino group.

5. Variously substituted amidines

In a study of the complexes of α-hydroxyamidines pK_1 as well as pK_2 values for the amidino and for the α-hydroxy group have been determined using a titration method[32]. Substances of the type **70** have been studied, and the results are given in Table 14.

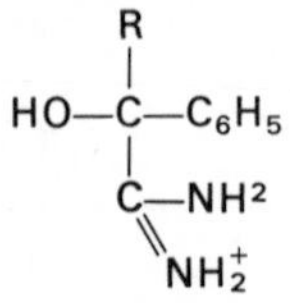

(70)

From the data given in the Table 14 it appears that increasing the length of the side chain R, both the values of pK_1 and pK_2 are increased[32].

In a study of the tautomerism and isomerism of N-halogenoamidines it is stated[167] that this group of substances shows a low basicity (pK about 4·5). More precise data are, however, not given.

TABLE 14. Acid dissociation constants for some α-hydroxy amidines (**70**), measured at 25°C and at an ionic strength of $0·1^a$

R	pK_1	pK_2
H	10·82 ± 0·01	12·52 ± 0·05
CH_3	10·96 ± 0·01	12·72 ± 0·05
C_2H_5	11·06 ± 0·01	12·96 ± 0·05

^{a} Reproduced by permission from R. O. Gould and R. F. Jameson, *J. Chem. Soc.*, 296 (1962).

B. Basicity of Amidoximes

I. Unsubstituted and substituted amidoximes

The data concerning basicity of unsubstituted as well as substituted amidoximes are given in Table 15.

While it is well known that amidoximes add a proton in acid media, the site of protonation is a matter for controversy[24, 157, 168, 171, 173]. According to the latest opinion[144] it is presumed, on the basis of the comparison of intensities as well as from the position of absorption bands, that the nitrogen atom N^1 is very nearly sp^2 hybridized and conjugated with the π orbit of the C=N double bond. In the case of sp^3 hybridization of the N^1 atom, some shifting of the valence vibrations of the NH_2 group should take place to lower wave number values (3520 and 3410 cm^{-1} → 3380 and 3310 cm^{-1})[144]. On the basis of the analogy found between the amides and amidoximes the following scheme was proposed:

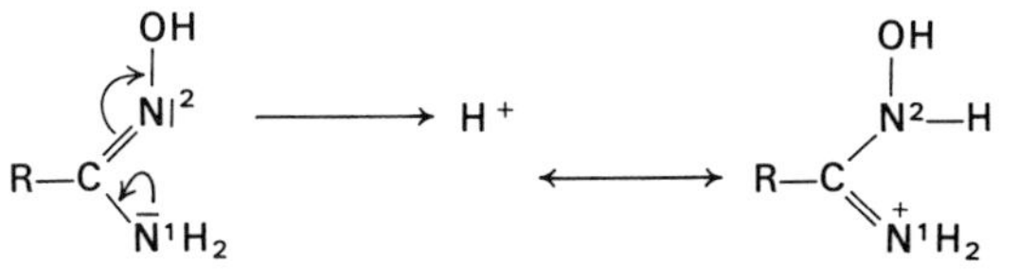

(22)

TABLE 15. pK_a values for unsubstituted and substituted amidoximes

Amidoxime	pK_a		Ref.
	Photo-metrical	Potentio-metrical	
Oxalic diamidoxime		3·02	166
	2·96[a]	2·95[a]	167
	11·31[a]	11·37[a]	167
		10·62	168
Malonic acid amidoxime		4·77	166
Phenylacetic acid amidoxime		5·24	166
Benzamidoxime		4·99	166
		5·03[b]	169
Salicyclic amidoxime		4·99[b]	169
o-Toluic amidoxime		4·03	166
p-Toluic amidoxime		5·14	166
		5·03[b]	171
Cinnamic amidoxime		4·98[b]	170
N-methylbenzamidoxime	5·38	5·36	25
N-ethylbenzamidoxime		5·42	25
N-diethylbenzamidoxime		5·62	25
N-Oximinobenzyl piperidine[c]	5·16	5·10	25
N-Oximinobenzyl morpholine[d]	4·11	4·07	25
N-phenylbenzamidoxime	4·11	4·35	171
N-β-naphthylbenzamidoxime	4·03		25
N,N'-diphenylbenzamidoxime	4·14	4·29	25
N,N-diethylbenzamidoxime		5·25	17
N-phenyl-N'-m-tolylbenzamidoxime	4·29		25
N-phenyl-N'-p-tolylbenzamidoxime	4·40		25
N-Phenyl-N'-m-chlorphenylbenzamidoxime	4·03		25
N-Phenyl-N'-p-chlorphenylbenzamidoxime	4·05		25
N-Phenyl-N'-m-bromphenylbenzamidoxime	4·04		25
N-Phenyl-N'-p-bromphenylbenzamidoxime	4·09		25

[a] Water solution
[b] 25% Ethanol–water solution

$$
\begin{array}{c}
\text{CH}_2\text{---CH}_2 \\
\end{array}
$$

c PhC with structure: N–(CH₂—CH₂)₂–CH₂ ring (piperidine), =NOH

d PhC with structure: N–(CH₂—CH₂)₂–O ring (morpholine), =NOH

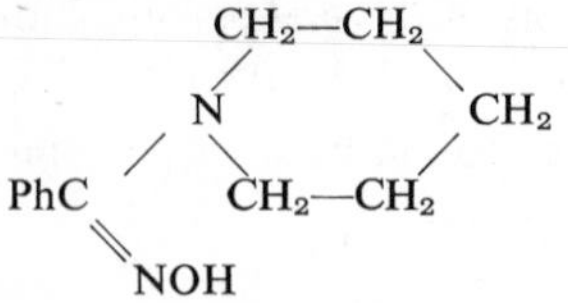

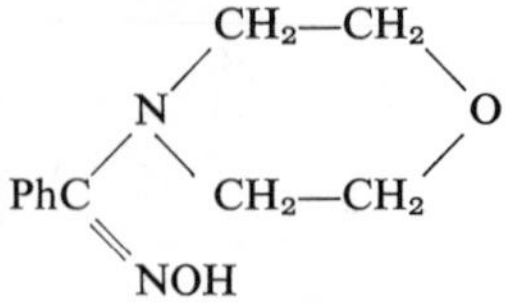

TABLE 16. pK_a values for some malonic amidoxime hydrazide derivatives (from ref. 174)[a]

Amidoxime-hydrazide	pK_a
HON$\diagdown$CCH$_2$CONHN=C$\diagup^{\text{H}}$, H$_2$N$\diagup$, CH=CHC$_6$H$_4$N(CH$_3$)$_2$-p	4.58
HON$\diagdown$CCH$_2$CONHN=C$\diagup^{\text{H}}$, H$_2$N$\diagup$, CH=CHC$_6$H$_5$	4·30
HON$\diagdown$CCH$_2$CONHN=C$\diagup^{\text{H}}$, H$_2$N$\diagup$, C$_6$H$_4$OH-o	3·91
HON$\diagdown$CCH$_2$CONHN=C$\diagup^{\text{CH}_3}$, H$_2$N$\diagup$, CH$_3$	4·03
HON$\diagdown$CCH$_2$CONHN=C$\diagup^{\text{H}}$, H$_2$N$\diagup$, C$_6$H$_5$	4·11

[a] Reproduced with permission from J. Mollin, J. Ševčík, J. Rubín and E. Ružička, *Monatsh. Chem.*, **92**, 1201 (1961).

2. Amidoximehydrazides of malonic acid

This group of nitrogen-rich substances also undergoes protonation in acid media[174]. With the use of a titration method the pK-values were determined and are given in Table 16. On the basis of these data, equation 23 was proposed for the protonation:

$$\text{HO}-\overset{+}{\text{N}}\diagup\underset{\text{H}}{}\overset{}{\text{C}}-\text{CH}_2-\overset{\text{O}}{\text{C}}-\text{NH}-\text{N}=\text{CH}\diagup^{\text{R}}_{\text{R}^1} \underset{\overset{}{}}{\overset{}{\rightleftharpoons}}$$

$$\text{HO}-\text{N}\diagdown\overset{}{\text{C}}-\text{CH}_2-\overset{\text{O}}{\text{C}}-\text{NH}-\text{N}=\text{CH}\diagup^{\text{R}}_{\text{R}^1} + \text{H}^+ \quad (23)$$

3. 3-Aminoamidoximes

Several papers[144, 175–177] deal with the study of the structure as well as the basicity of 3-aminoamidoximes. For the dissociation equilibria with this group of substances equations 24 and 25 have been proposed:

$$^+NH_3-R-C{\overset{\displaystyle ^+NH_2}{\underset{\displaystyle N-OH \atop H}{}}} \rightleftharpoons\ ^+NH_3-R-C{\overset{\displaystyle NH_2}{\underset{\displaystyle N-OH}{}}} + H^+ \qquad (24)$$

$$^+NH_3-R-C{\overset{\displaystyle NOH}{\underset{\displaystyle NH_2}{}}} \rightleftharpoons\ NH_2-R-C{\overset{\displaystyle NOH}{\underset{\displaystyle NH_2}{}}} + H^+ \qquad (25)$$

and the respective pK_1 and pK_2-values referring to the above equations were determined. On the basis of comparisons with aminonitriles and with amines, the pK_1 value was ascribed to the dissociation of the amidoxime group (equation 24), and the pK_2 value was ascribed to the dissociation of the amino group (equation 25). Comparing the pK values of the corresponding aminonitriles with the pK_2 value of amino amidoximes it becomes clear that the basicity[144], is lower in the nitrile, owing to the stronger electrophilic effect of the nitrile group compared with the amidoxime group. On the other hand, the pK_1 and pK_2 values given of aminoamidoximes are lower than the pK-values of amines and amidoximes.[144]

Evidently, with aminoamidoximes basicity decrease of the amino function takes place through the influence of the electrophilic effect of the amidoxime group. The electrophilic inductive effect of the positively charged group again lowers the basicity of the amidoxime group[144] (equation 25).

V. REFERENCES

1. R. L. Shriner and F. W. Neumann, *Chem. Revs.*, **35**, 351 (1944).
2. A. Pinner and F. Klein, *Ber.*, **10**, 1889 (1877).
3. A. Gautier, *Ann. Chim. Phys.*, **17** (4), 103 (1869).
4. A. Holý, *Chem. Listy*, **58**, 261 (1964).
5. M. Kuraš, *Chem. Listy*, **38**, 38 (1944).
6. M. Kuraš, *Coll. Czech. Chem. Commun.*, **12**, 198 (1947).
7. M. Kuraš and E. Ružička, *Chem. Listy*, **44**, 41 (1950).
8. M. Kuraš and E. Ružička, *Chem. Listy*, **44**, 90 (1950).
9. M. Kuraš and E. Ružička, *Coll. Czech. Chem. Commun.*, **15**, 147 (1950).

10. M. Kuraš and E. Ružička, *Chem. Listy*, **45**, 37 (1951).
11. M. Kuraš and E. Ružička, *Chem. Listy*, **46**, 482 (1952).
12. M. Kuraš and E. Ružička, *Chem. Listy*, **46**, 91 (1952).
13. M. Kuraš and E. Ružička, *Chem. Listy*, **47**, 1870 (1953).
14. M. Kuraš and E. Ružička, *Chem. Listy*, **48**, 1257 (1954).
15. M. Kuraš and E. Ružička, *Chem. Listy*, **49**, 1897 (1955).
16. M. Kuraš and E. Ružička, *Coll. Czech. Chem. Commun.*, **21**, 1075 (1956).
17. M. Kuraš and J. Mollin, *Chem. Listy*, **52**, 344 (1958).
18. M. Kuraš and J. Mollin, *Coll. Czech. Chem. Commun.*, **24**, 290 (1959).
19. M. Kuraš and J. Bartoň, *Coll. Czech. Chem. Commun.*, **24**, 1720 (1959).
20. M. Kuraš, V. Stužka, F. Kašpárek, J. Mollin and J. Slouka, *Coll. Czech. Chem. Commun.*, **26**, 315 (1961).
21. H. Grisollet and M. Servigne, *Ann. chim.*, **12**, 321 (1930).
22. R. Pallaud, *Chim. anal.*, **33**, 239 (1951).
23. L. Hunter and J. A. Marriott, *J. Chem. Soc.*, 777 (1941).
24. D. Hall and F. J. Llewellyn, *Acta Cryst.*, **9**, 108 (1956).
25. J. Mollin, *Dissertation*, Charles' University, Praha, 1963.
26. W. Bradley and E. Leete, *J. Chem. Soc.*, 2147 (1956).
27. W. Bradley and I. Wright, *J. Chem. Soc.*, 640 (1956).
28. R. M. Roberts, *J. Am. Chem. Soc.*, **72**, 3608 (1950).
29. P. Sohár, *Acta Chim. Acad. Sci. Hung.*, **54**, 91 (1967).
30. J. E. B. Randles, *J. Chem. Soc.*, 802 (1941).
31. R. O. Gould, R. F. Jameson and D. G. Neilson, *Proc. Chem. Soc.*, 314 (1960).
32. R. O. Gould and R. F. Jameson, *J. Chem. Soc.*, 296 (1962).
33. R. O. Gould and R. F. Jameson, *J. Chem. Soc.*, 15 (1963).
34. R. O. Gould and R. F. Jameson, *J. Chem. Soc.*, 5211 (1963).
35. R. O. Gould and H. M. Sutton, *J. Chem. Soc. (A)*, 1184 (1970).
36. R. O. Gould and H. M. Sutton, *J. Chem. Soc. (A)*, 1439 (1970).
37. C. C. Price and J. Zomlefer, *J. Org. Chem.*, **14**, 210 (1949).
38. J. L. Hall and W. E. Dean, *J. Am. Chem. Soc.*, **80**, 4123 (1958).
39. J. Bjerrun, G. Schwarzenbach and P. Silén, Stability Constants, *Chem. Soc. Special Publ. No. 7*, Part I, pp. 8, 15, 25.
40. R. O. Gould and W. C. Vosburgh, *J. Am. Chem. Soc.*, **64**, 1630 (1942).
41. J. Iball and C. H. Morgan, *J. Chem. Soc. (A)*, 52 (1967).
42. J. Iball and C. H. Morgan, *Nature*, **202**, 689 (1964).
43. M. Spencer, W. Fuller, M. H. F. Wilkins and G. L. Brown, *Nature*, **194**, 1014 (1962).
44. R. Langridge and P. J. Gomatos, *Science*, **141**, 694 (1963).
45. N. Comerman and J. Trotter, *Acta Cryst.*, **18**, 203 (1965).
46. T. R. R. Donald and C. A. Beevers, *Acta Cryst.*, **5**, 654 (1952).
47. H. Mc D. McGeachin and C. A. Beevers, *Acta Cryst.*, **10**, 227 (1957).
48. R. Jefferson, M. F. Lappert, B. Prokai and B. P. Tilley, *J. Chem. Soc. (A)*, 1584 (1966).
49. V. A. Dorochov, B. M. Lavrinovic and B. M. Michajlov, *Dokl. Akad. Nauk SSSR*, **195**, 1100 (1970).
50. B. M. Michajlov and V. A. Dorochov, *Izv. Akad. Nauk SSSR, ser. chim.*, 1446 (1970).

51. B. M. Michajlov, V. A. Dorochov and V. I. Seredenko, *Dokl., Akad. Nauk SSSR*, **199**, 1328 (1971).
52. B. M. Michajlov, *Chimia borovodorodov, Izd. Nauka, Moskva*, 1967, p. 81.
53. M. Davies and A. E. Parsons, *Chem. and Ind.*, 628 (1958).
54. B. M. Michajlov and V. A. Dorochov, *Izv. Akad. Nauk SSSR, ser. chim.*, 201 (1971).
55. J. A. Partridge, R. M. Izatt and J. J. Christensen, *J. Chem. Soc.*, 4231 (1965).
56. H. Irving and H. S. Rossoti, *Acta Chem. Scand.*, **10**, 72 (1956).
57. O. J. Scherrer and P. Hornig, *Chem. Ber.*, **101**, 2533 (1968).
58. M. J. Boylan, S. M. Nelson and F. A. Deeney, *J. Chem. Soc. (A)*, 976 (1971).
59. D. C. Prevorsek, *Bull. Soc. Chim. France*, **788** (1958).
60. D. C. Prevorsek, *J. Phys. Chem.*, **66**, 769 (1962).
61. S. M. Nelson and J. Rodgers, *J. Chem. Soc. (A)*, 272 (1968).
62. D. A. Buckingham, B. M. Foxman, A. M. Sorgeson and A. Zanella, *J. Am. Chem. Soc.*, **94**, 1007 (1972).
63. L. A. Mulligan, G. Shaw and P. J. Staples, *J. Chem. Soc. (C)*, 1585 (1971).
64. M. Franks, C. P. Green, G. Shaw and G. J. Litchfield, *J. Chem. Soc. (C)*, 2270 (1966).
65. D. A. Buckingham, P. A. Marzilli and A. M. Sargeson, *Inorg. Chem.*, **8**, 1595 (1969).
66. R. L. Dutta and P. Ray, *J. Indian. Chem. Soc.*, **36**, 499 (1959).
67. R. L. Dutta and P. Ray, *J. Indian. Chem. Soc.*, **36**, 367 (1959).
68. R. L. Dutta and P. Ray, *J. Indian. Chem. Soc.*, **36**, 576 (1959).
69. P. Ray, *Chem. Revs.*, **61**, 313 (1961).
70. G. D. Diana, E. Zalay and R. A. Cutler, *J. Org. Chem.*, **30**, 298 (1965).
71. V. Rasmussen and W. A. Baker, Jr., *J. Chem. Soc. (A)*, 580 (1967).
72. V. Rasmussen and W. A. Baker, Jr., *J. Chem. Soc. (A)*, 1712 (1967).
73. J. Haag, *Ann. Chem.*, **122**, 31 (1862).
74. P. Ray and G. Bandopadhya, *J. Indian. Chem. Soc.*, **29**, 865 (1952).
75. L. Tschugajeff, *Ber.*, **40**, 1975 (1907).
76. A. Syamal, *Z. Naturforsch. (B)*, **24**, 1514 (1969).
77. P. Ray and A. K. Choudbury, *J. Indian. Chem. Soc.*, **27**, 673 (1950).
78. P. Ray and S. N. Podar, *J. Indian. Chem. Soc.*, **29**, 279 (1959).
79. A. Syamal, *J. Indian. Chem. Soc.*, **44**, 1084 (1967); *Current Sci.*, **36**, 666 (1967).
80. A. Syamal, *Z. Naturforsch. (B)*, **24**, 1192 (1969).
81. H. Luth, E. A. Hall, W. A. Spofford and E. L. Amma, *Chem. Commun.*, 520 (1969).
82. P. Kundsen, *Ber.*, **18**, 1068 (1886).
83. M. Kuraš, *Chem. Obzor*, **19**, 9 (1944).
84. R. Chaterjee, *J. Indian. Chem. Soc.*, **15**, 608 (1938).
85. J. V. Dubský and M. Kuraš, *Chem. Listy*, **24**, 464 (1930).
86. J. Trtílek, *Coll. Czech. Chem. Comm.*, **7**, 6 (1935).
87. M. Kuraš, *Chem. Obzor*, **18**, 1 (1943).
88. L. Tschugaeff and J. Surenjanz, *Ber.*, **40**, 182 (1907).

89. J. V. Dubský, J. Trtílek and M. Kuraš, *Coll. trav. Chim. Tch.*, **7**, 1 (1935).
90. M. Kuraš, *Chem. Obzor*, **19**, 183 (1944).
91. L. Tschugaeff, *Z. anorg. Chem.*, **46**, 168 (1905).
92. J. V. Dubský and Z. Okáč, *Coll. Czech. Chem. Comm.*, **4**, 388 (1932).
93. K. R. Manolov, *Monatsh.*, **99**, 2416 (1968).
94. K. R. Manolov, *Monatsh.*, **99**, 1774 (1968).
95. K. R. Manolov, *Nautschni Tr. Vissch. Ped. Inst.*, *Plovdiv*, **6**, 81 (1968).
96. K. R. Manolov, *Nautschni Tr. V. J. H. V. Plovdiv*, **14**, 247 (1967).
97. K. R. Manolov, *C.R. Acad. Sci. Bulg.*, **17**, 833 (1964).
98. K. R. Manolov and D. W. Kovatschev, *Monatsh.*, **100**, 304 (1969).
99. K. R. Manolov, *Nautschni Tr. Vissch. Ped. Inst.*, *Plovdiv*, **5**, 75 (1967).
100. K. R. Manolov and P. Wassileva, *Monatsh.*, **100**, 1184 (1969).
101. K. R. Manolov and D. W. Kovatschev, *Monatsh.*, **100**, 1233 (1969).
102. K. R. Manolov and A. T. Kozhukharova, *Monatsh.*, **100**, 2033 (1969).
103. P. Krüger, *Ber.*, **18**, 1053 (1885).
104. F. Tiemann, *Ber.*, **17**, 128 (1884).
105. W. D. Kingory and D. N. Hume, *J. Am. Chem. Soc.*, **71**, 3186 (1949).
106. L. Newman and D. N. Hume, *J. Am. Chem. Soc.*, **79**, 4571 (1957).
107. L. Pauling, *The Nature of Chemical Bond*, Corneíl, 1940, pp. 287, 307.
108. H. W. Post, *The Chemistry of the Aliphatic Orthoesters*, Reinhold Publishing Corp., New York, 1943, p. 92.
109. C. D. Lewis, R. C. H. Krupp, H. Tieckelmann and H. W. Post, *J. Org. Chem.*, **12**, 303 (1947).
110. R. M. Roberts, *J. Am. Chem. Soc.*, **72**, 3603 (1950).
111. R. M. Roberts, *J. Org. Chem.*, **14**, 277 (1949).
112. P. Sohár, *Acta Chim. Acad. Sci. Hung.*, **40**, 317 (1964).
113. P. Sohár and L. Farkas, *Acta Chim. Acad. Sci. Hung.*, **54**, 79 (1967).
114. P. Sohár, *Magyar Kém. Foly.*, **71**, 415 (1965).
115. G. Schwenker and H. Rosswag, *Tetrahedron Lett.*, 2691 (1969).
116. H. S. Gutowsky and C. H. Holm, *J. Chem. Phys.*, **25**, 1228 (1956).
117. A. J. Hill and J. Rabinowitz, *J. Am. Chem. Soc.*, **48**, 732 (1926).
118. G. Schwenker and R. Kolb, *Tetrahedron*, **25**, 5549 (1969).
119. G. Schwenker, *Mitt. Pharm. Ges.*, **41**, 305 (1971).
120. H. J. Barber, *J. Chem. Soc.*, 101 (1943).
121. S. J. Angyal and W. K. Warburton, *Austral. J. Sci. Res.*, **4A**, 93 (1951).
122. R. A. Jones and A. R. Katritzky, *J. Chem. Soc.*, 378 (1961).
123. R. Shephard, A. Bratton and K. Blanchard, *J. Am. Chem. Soc.*, **64**, 2532 (1942).
124. A. Lawson and R. B. Tinkler, *J. Chem. Soc. (C)*, 652 (1969).
125. R. B. Tinkler, *J. Chem. Soc. (C)*, 1052 (1970).
126. G. Schwenker and K. Bösl, *Arch. Pharm.*, **303**, 980 (1970).
127. J. C. Danilewicz, M. J. Sewell and J. C. Thurman, *J. Chem. Soc. (C)*, 1704 (1971).
128. C. K. Ingold, *Structure and Mechanism in Organic Chemistry*, Cornell, New York, 1953, ch. 10.
129. J. C. Grivas and A. Taurins, *Canad. J. Chem.*, **37**, 795 (1959).
130. J. A. Smith and H. Taylor, *J. Chem. Soc.*, 66 (1969).
131. A. N. Hambly and B. V. O' Grady, *Austral. J. Chem.*, **17**, 860 (1964).

132. G. Schwenker and K. Bösl, *Pharmazie*, **24**, 653 (1969).
133. H. Lumbroso and D. M. Bertin, *Bull. Soc. Chim. France*, 1729 (1970).
134. N. P. Marullo, C. D. Smith and F. Terapave, *Tetrahedron Lett.* 6279 (1966).
135. O. Exner and V. Jehlička, *Coll. Czech. Chem. Comm.*, **30**, 639 (1965).
136. R. Pujol and F. Mathis, *C. R. Acad. Sci. Paris, Ser. B.*, **269**, 975 (1969).
137. H. Lumbroso, D. M. Bertin and P. Reynaud, *C. R. Acad. Sci. Paris, ser. B*, **261**, 399 (1965).
138. R. Pujol and F. Mathis, *C. R. Acad. Sci. Paris, ser. B*, **269**, 1049 (1969).
139. J. Barrans, R. Mathis-Noel and F. Mathis, *C. R. Acad. Sci. Paris, ser. B*, **245**, 419 (1957).
140. W. J. Orville-Thomas and A. E. Parsons, *Trans. Faraday Soc.*, **54**, 460 (1958).
141. D. Prevorsek, *C. R. Acad. Sci. Paris, ser. B.*, **247**, 1333 (1958).
142. O. Exner, *Coll. Czech. Chem. Commun.*, **30**, 652 (1965).
143. M. H. Millen and W. A. Waters, *J. Chem. Soc. (B)*, 408 (1968).
144. H. Goncalves and A. Secches, *Bull. Soc. Chim. France*, 2589 (1970).
145. H. Goncalves, F. Mathis and C. Foulcher, *Bull. Soc. Chim. France*, 2599 (1970).
146. J. D. Aubort and R. F. Hudson, *J. Chem. Soc. (D)*, 1342 (1969).
147. G. A. Schwechgejmer and M. L. Schulman, *Dokl. Akad. Nauk SSSR, ser. chim.*, **173**, 378 (1967).
148. M. L. Schulman and G. A. Schwechgejmer, *Zhr. Org. Chim.*, **5**, 229 (1969).
149. *Organic Syntheses, Coll.*, Vol. I, 2 Ed., Wiley, New York, 1941, p. 5.
150. A. Bernthsen, *Ann. Chem.* **192**, 1 (1878).
151. J. Ševčík, *Chem. Zvesti*, **26**, 49 (1972).
152. J. Ševčík, *Acta Univ. Palackanae*, RN **39** (in press).
153. N. V. Sidgwick, *The Organic Chemistry of Nitrogen* (T. W. J. Taylor, W. Baker, Eds.), The Claredon Press, Oxford, 1937, p. 155.
154. E. S. Levchenko and L. V. Seleznenko, *Zhr. Org. Chim.*, **4**, 153 (1968).
155. V. A. Gilyarov and M. I. Kabachnik, *Zhr. Obs. Khim.*, **36**, 282 (1966).
156. M. I. Kabachnik, V. A. Gilyarov and R. V. Kudrjacev, *Zhr. Obs. Khim.*, **36**, 57 (1966).
157. M. I. Kabachnik, V. A. Gilyarov and R. V. Kudrjacev, *Zhr. Obs. Khim.*, **35**, 1476 (1965).
158. M. I. Kabachnik, V. A. Gilyarov, B. A. Korolev and T. A. Rajevskaja, *Izv. Akad. Nauk SSSR, ser. chim.*, 772 (1970).
159. J. Ševčík, *Monatsh.*, **100**, 1307 (1969).
160. J. A. Smith and H. Taylor, *J. Pharm. Pharmacol.*, **15**, 548 (1963).
161. J. A. Smith and H. Taylor, *J. Chem. Soc. (B)*, 64 (1969).
162. N. A. Lange, *Handbook of Chemistry*, McGraw Hill, London, 1961, 10th edn., p. 1203.
163. J. A. Smith and H. Taylor, *J. Chem. Soc. (B)*, **66**, (1969).
164. E. Lorz and R. Baltzly, *J. Am. Chem. Soc.*, **71**, 3992 (1949).
165. J. N. Baxter and J. Cymerman-Craig, *J. Chem. Soc.*, 1490 (1959).
166. G. A. Pearse and R. T. Pflaum, *J. Am. Chem. Soc.*, **81**, 6505 (1959).
167. P. E. Wenger, D. Monnier and I. Kapétanidis, *Helv. Chim. Acta*, **40**, 1456 (1957).

168. H. E. Ungnade, L. W. Kissinger, A. Narath, and D. C. Barham, *J. Org. Chem.*, **28**, 134 (1963).
169. J. Mollin and F. Kašpárek, *Coll. Czech. Chem. Commun.*, **25**, 451 (1960).
170. J. Mollin and F. Kašpárek, *Coll. Czech. Chem. Commun.*, **26**, 2438 (1961).
171. J. Mollin and F. Kašpárek, *Acta Univ. Palackianae*, RN **7**, 125 (1961).
172. R. Mecke and W. Kutzelnigg, *Spectrochim. Acta*, **16**, 1216 (1960).
173. H. Lund, *Acta Chem. Scand.*, **13**, 249 (1959).
174. J. Mollin, J. Ševčík, J. Rubín and E. Ružička, *Monatsh.*, **92**, 1201 (1961).
175. H. Goncalves and F. Mathis, *C. R. Acad. Sci. Paris*, **258**, 3056 (1964).
176. H. Goncalves, *C. R. Acad. Sci. Paris*, **264** 1206 (1967).
177. H. Goncalves and C. Foulcher, *C. R. Acad. Sci. Paris*, **264**, 1320 (1967).
178. *Handbook of Chemistry and Physics*, 52nd Edition 1971–1972, p. B232, The Chemical Rubber Co., Cleveland, Ohio.
179. J. Číhalík, *Potenciometrie*, NČSAV, Praha, 1961, p. 712.
180. J. Ševčík, unpublished results.

Author Index

This index is designed to enable the reader to locate an author's name and work with the aid of the reference numbers appearing in the text. The page numbers are printed in normal type in ascending numerical order, followed by the reference numbers in parentheses. The numbers in *italics* refer to the pages on which the references are actually listed.

Subject Index

657